Lecture Notes in Computer Science 16296

The series Lecture Notes in Computer Science (LNCS), including its subseries Lecture Notes in Artificial Intelligence (LNAI) and Lecture Notes in Bioinformatics (LNBI), has established itself as a medium for the publication of new developments in computer science and information technology research, teaching, and education.

LNCS enjoys close cooperation with the computer science R & D community, the series counts many renowned academics among its volume editors and paper authors, and collaborates with prestigious societies. Its mission is to serve this international community by providing an invaluable service, mainly focused on the publication of conference and workshop proceedings and postproceedings. LNCS commenced publication in 1973.

Michael H. F. Wilkinson · Jiří Kosinka
Editors

Discrete Geometry and Mathematical Morphology

4th International Joint Conference, DGMM 2025
Groningen, The Netherlands, November 3–6, 2025
Proceedings

Springer

Editors
Michael H. F. Wilkinson
University of Groningen
Groningen, The Netherlands

Jiří Kosinka
University of Groningen
Groningen, The Netherlands

ISSN 0302-9743 ISSN 1611-3349 (electronic)
Lecture Notes in Computer Science
ISBN 978-3-032-09543-5 ISBN 978-3-032-09544-2 (eBook)
https://doi.org/10.1007/978-3-032-09544-2

This Springer imprint is published by the registered company Springer Nature Switzerland AG
The registered company address is: Gewerbestrasse 11, 6330 Cham, Switzerland

Preface

The 4th IAPR International Conference on Discrete Geometry and Mathematical Morphology (DGMM 2025) was held in Groningen, the Netherlands, 3–6 November 2025. The DGMM series is the main conference of the IAPR TC 18 on Discrete Geometry and Mathematical Morphology, and is a continuation of two previous separate series of conferences, i.e., the International Conference on Discrete Geometry for Computer Imagery (DGCI), and the International Symposium on Mathematical Morphology (ISMM). The current volume covers the key areas of research within the joint community. This includes theoretical work, models, and new algorithms, but also many applications. Unsurprisingly, there is an increase in papers relating the field to machine learning.

The conference consisted of one day of tutorials and "state of the art reports" (STARs), mainly aimed at graduate students. STARs are a new addition, complementing the tutorials, that focus on the foundations of mathematical morphology and discrete geometry, by providing a good overview of current research directions. After this, during the three main conference days there was a keynote speech each day, followed by oral and poster sessions.

DGMM 2025 received 52 regular submission from authors in 17 countries: Algeria, Australia, Austria, Brazil, China, France, Germany, Hungary, India, Italy, the Netherlands, Norway, Russia, Serbia, Spain, Sweden, and the USA. All papers underwent a single-blind peer review process involving at least 3 reviewers each, followed by a rebuttal phase. In all cases when program committee members were included in the author list, they declared a conflict of interest within the EasyChair submission system, which then ensured the identity of reviewers was properly hidden from these PC members. The review process led to a total of 37 papers being accepted, of which 26 were oral presentations, and 11 posters.

Apart from the regular papers, three internationally renowned keynote speakers presented on key topics in the field:

- Joost Batenburg, of the Leiden Institute of Advanced Computer Science (LIACS), Leiden University.
- Beatriz Marcotegui Iturmendi, of MINES ParisTech, Centre de Morphologie Mathématique (CMM)
- Georgios Ouzounis, of Atlas AI, Palo Alto, California

As is traditional, both in the preceding ISMM and DGCI conferences, and in the DGMM series, the proceedings of the DGMM 2025 conference appear in Springer's LNCS series. A special issue of the Journal of Mathematical Imaging and Vision will be published with a selection of extended versions of papers from these proceedings.

We would like to thank all members of the Steering Committee for accepting our bid to organize the conference, and the Program Committee and the many reviewers who invested their time to ensure the scientific quality of the proceedings by providing detailed feedback. For the local organization, Charlotte Hardy and Sarah van Wouwe did sterling work in handling many aspects of the organization. David Coeurjolly, Bernard Burgeth, Yukiko Kenmochi, and Samy Bluseau were instrumental in organizing the STARs and tutorials. Further thanks go out to the keynote speakers Joost Batenburg, Beatriz Marcotegui Iturmendi, and Georgios Ouzounis.

We are grateful to the International Association for Pattern Recognition for their sponsorship and support, and to Springer for their continued help in publishing the DGMM proceedings. A special thanks to Atlas AI, Palo Alto, California, for their sponsorship of the conference. The EasyChair conference management system was instrumental in providing a smooth submission, review, and rebuttal process. Finally, the authors of the papers deserve special thanks, as without their combined efforts there would be no conference.

November 2025

Michael H. F. Wilkinson
Jiří Kosinka

Organization

Program Chairs

Michael Wilkinson	University of Groningen, The Netherlands
Jiří Kosinka	University of Groningen, The Netherlands

Steering Committee Chairs

Yukiko Kenmochi Greyc	CNRS, France
Michael Wilkinson	University of Groningen, The Netherlands

Steering Committee

Jesús Angulo Mines	ParisTech, France
Isabelle Bloch	Sorbonne Université, France
Gunilla Borgefors	Uppsala University, Sweden
Srecko Brlek	University of Quebec at Montreal, Canada
Bernhard Burgeth	Saarland University, Germany
David Coeurjolly	LIRIS, CNRS, France
Isabelle Debled-Rennesson	Université de Lorraine, France
Andrea Frosini	Università di Firenze, Italy
Marie José Jiménez	University of Seville, Spain
Walter Kropatsch	TU Wien, Austria
Jacques-Olivier Lachaud	University of Savoie Mont Blanc, France
Cris Luengo Hendriks	Deepcell, Inc., USA
Petros Maragos	National Technical University of Athens, Greece
Laurent Najman	Université Gustave Eiffel, France
Sara Brunetti	University of Siena, Italy
Dan Schonfeld	University of Illinois Chicago, USA
Hugues Talbot	Université Paris-Saclay, France

Program Committee

Jesús Angulo	Mines ParisTech, France
Michela Ascolese	Università degli Studi di Firenze, Italy

Samy Blusseau	Mines ParisTech, France
Nicolas Boutry	EPITA Research Laboratory, France
Michael Breuss	Brandenburg Technical University, Germany
Jean Cousty	Université Gustave Eiffel, France
Paolo Dulio	Politecnico di Milano, Italy
Andrea Frosini	Università di Firenze, Italy
Yan Gerard	Université Clermont Auvergne, France
Maria-Jose Jimenez	University of Seville, Spain
Yukiko Kenmochi	GREYC, CNRS, France
Bertrand Kerautret	Université de Lyon 2, France
Jiří Kosinka	University of Groningen, The Netherlands
Jacques-Olivier Lachaud	University of Savoie Mont Blanc, France
Laurent Najman	Université Gustave Eiffel, France
Phuc Ngo	Université de Lorraine, France
Akihiro Sugimoto	National Institute of Informatics, Japan
Michael Wilkinson	University of Groningen, The Netherlands

Additional Reviewers

Andreas Alpers
Eric Andres
Arya Babai
Peter Balazs
Felipe Belém
Jin Ben
Arindam Biswas
Isabelle Bloch
Samy Blusseau
Antoine Bottenmuller
Stéphane Breuils
Sara Brunetti
Edwin Carlinet
Aditya Challa
Giovanni Chierchia
Guillaume Damiand
Sravan Danda
Pascal Desbarats
Niccolò Di Marco
El Hadji S. Diop
Eva Dokladalova
Eric Domenjoud
Mousumi Dutt
Alexandre Falcão
Thomas Fernique
Fabien Feschet
Konstantinos Fotopoulos
Laurent Fuchs
Aysylu Gabdulkhakova
Rocio Gonzalez-Diaz
Thierry Grandpierre
Ronaldo Fumio Hashimoto
Nina S. T. Hirata
Damien Jamet
Marvin Kahra
Antti Laaksonen
Bastien Laboureix
Raphaël Lapertot
Josselin Lefèvre
Etienne Le Quentrec
Olivier Lezoray
Kelly Maggs
Paola Magillo
Petros Maragos
Diego Marcondes
Beatriz Marcotegui

Arnold Meijster
Julien Mendes Forte
Paulo Miranda
Daiki Miyahara
Pascal Monasse
David Mosquera Lois
Benoît Naegel
Benedek Nagy
Keiller Nogueira
Georgios Ouzounis
Silvia Pagani
Eduardo Paluzo-Hidalgo
Nicolas Passat
Valentin Penaud-Polge
Benjamin Perret
Christophe Picouleau
Organization IX
Kacper Pluta
Elodie Puybareau
Jos Roerdink
Gonzalo Romero-García
Pascal Romon
Tristan Roussillon
Christos Sakaridis
Deise Santana Maia
Shima Shabani
Isabelle Sivignon
Robin Strand
Imants Svalbe
Hugues Talbot
Lama Tarsissi
Paul Teeninga
Marcos Eduardo Valle
Santiago Velasco-Forero
Martin Welk
Petra Wiederhold
Rita Zrour

Contents

Discrete Geometry – Models, Transforms, and Visualization

Computational Aspects of Discrete Structures and Tilings

Discrete and Combinatorial Topology

Discrete Geometry – Models, Transforms, and Visualization

An Analytical Definition of Discrete Circles in the Triangular Grid

Lidija Čomić[1(✉)], Rita Zrour[2], Eric Andres[2], and Gaëlle Largeteau-Skapin[2]

[1] University of Novi Sad, Faculty of Technical Sciences, Novi Sad, Serbia
comic@uns.ac.rs

[2] University of Poitiers, Laboratory XLIM, ASALI, UMR CNRS 7252, BP 30179, 86962 Futuroscope Chasseneuil, France
{rita.zrour,eric.andres,gaelle.largeteau.skapin}@univ-poitiers.fr

Abstract. We introduce and investigate the concept of a discrete circle in the triangular grid, taking into account the presence of two distinct types of triangles, depending on their orientation. We propose an analytical definition of a discrete circle in this grid, incorporating a non-constant thickness that depends on both the type of the considered triangle and the position of the circle's center. This varying thickness ensures the edge connectivity of the resulting circle. Building on this analytical definition, we also propose an incremental generation algorithm that is linear in the number of triangles on the circle.

Keywords: Triangular grid · Arithmetic discrete circles · Flake model · Incremental algorithm

1 Introduction

Among the regular 2D grids, the square and hexagonal grids received a fair share of research attention related to discrete geometry and image processing problems. The triangular grid has been relatively overlooked, probably due to the fact that its basic cell, the equilateral triangle, appears in two different orientations, which makes the reasoning about this grid more technically demanding.

Digital line and circle, as basic geometric primitives, have been important topics for many years in the computer graphics, computer vision, pattern recognition, image processing and digital geometry communities, used in generation, characterization and recognition tasks in the square and hexagonal grids. Here, we are interested in definition, characterization and generation of digitized circles in the triangular grid.

Reveillès initiated the study of discrete arithmetic geometry in the square grid, where geometric primitives are defined as sets of discrete points between two continuous (Euclidean) curves, with the thickness defined through the distance between the two curves, and introduced the arithmetic discrete lines [12].

M. H. F. Wilkinson and J. Kosinka (Eds.): DGMM 2025, LNCS 16296, pp. 3–15, 2026.
https://doi.org/10.1007/978-3-032-09544-2_1

Andres [1] proposed an analytical characterization of discrete circles with constant thickness, while Fiorio et al. [6] used a non-constant thickness. Both these approaches allowed extensions to higher dimensions [2,14]. Recently, Zrour et al. [15] introduced and studied discrete circles with non-constant thickness in the hexagonal grid.

In the square grid, Bresenham [4] proposed an iterative algorithm for circle generation. In the hexagonal grid, such algorithms have been proposed by Liu [9], Manoharan et al. [10] and Zrour et al. [15]. They differ in the way they choose the next point on the circle starting from the current one. Only the last approach proposes an analytical characterization of the discrete circle.

Here, we propose analytical characterizations of discrete circles of real radius in the triangular grid, with a non-constant thickness, based on the approach proposed in [6] for circles and in [5,6] for parabolas in the square grid and the one in [15] for circles in the hexagonal grid. The circle may be centered at any point in the Cartesian plane and may have an arbitrary radius greater than 1.

We compute the thicknesses using the flake model, that is a morphological type digitization scheme with adjacency flake as structuring element [7,13,14]. One of the main advantages of our proposed method is that it gives analytical characterizations of discrete circles. This allows to decide in constant time the membership of a point to a circle independently of the algorithm, by checking the two inequalities, and enables to design simple incremental algorithms for circle generation.

In Sect. 2, we present some preliminaries on the triangular grid. In Sects. 3 and 4, we present the equations of discrete vertex flake circles and we study their topological properties. In Sect. 5, we present the generation algorithms. We give a brief discussion and present perspectives in the last section.

2 The Triangular Grid

The triangular grid is a tessellation of the plane into regular (equilateral) triangles of side length 1. The triangles are of two types, mutually congruent with respect to translation. We consider them as oriented upward or downward, and call them up- and down-triangles, respectively, as shown in Fig. 1.

Each triangle shares an edge with three triangles of opposite type (1-neighbors or edge-neighbors), and it shares exactly one vertex with nine triangles (vertex-neighbors), three across each vertex, two of the same type (2-neighbors) and one of the opposite type (3-neighbor). For example, the three 1-neighbors of the up-triangle labelled $(0, 0)$ in Fig. 1(a) are down-triangles $(1, 0)$, $(-1, 0)$ and $(0, -1)$, its two 2-neighbors across its upper vertex are up-triangles $(-1, 1)$ and $(1, 1)$, and its 3-neighbor across the same vertex is down-triangle $(0, 1)$.

A set S of triangles is edge connected (vertex connected) if for any two triangles in S there is a sequence of triangles in S connecting them, such that any two consecutive triangles in the sequence are edge neighbors (vertex neighbors).

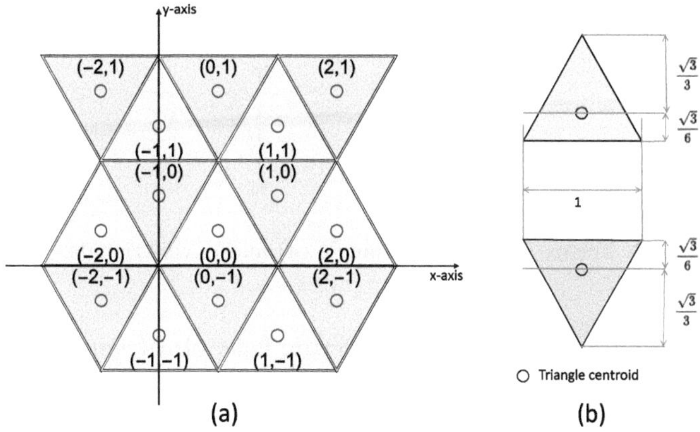

Fig. 1. The triangular grid: coordinate system (a) and triangle size (b).

We place the origin of Cartesian coordinate system on one grid vertex. In the implementation of the circle generation algorithm, to navigate the triangular grid, we will also use an integer valued addressing scheme, as shown in Fig. 1, which we will call triangular coordinates. Our motivation is to use simple binary matrix representation of objects in the triangular grid.

The sum of triangular coordinates is even for up-triangles and odd for down-triangles. The transition from triangular (x_t, y_t) to Cartesian (x, y) coordinates is given by

$x = \frac{1}{2}(x_t + 1)$, $y = \sqrt{3}\left(\frac{1}{2}y_t + \frac{1}{6}(1 + (x_t + y_t)(\text{mod } 2))\right)$.

Here, $(x_t + y_t)(\text{mod } 2)$ is the remainder when dividing $(x_t + y_t)$ with 2: it is equal to 0 for up-triangles and it is equal to 1 for down-triangles.

The inverse conversion, from Cartesian to triangular coordinates, is

$x_t = 2x - 1$, $y_t = \lfloor \frac{2\sqrt{3}}{3} y \rfloor$,

where $\lfloor a \rfloor$ is the largest integer not greater than a.

3 Vertex Flake Circles Definition

We adopt the flake digitization model [3,13] to define the discrete arithmetic circles. The structuring elements we will consider are called adjacency flakes [13], which are the unions of a finite number of straight line segments. The vertex flake of a triangle consists of three segments connecting the triangle centroid to the three triangle vertices (see Fig. 2).

Definition 1. *A flake model discrete arithmetic primitive P^* in the triangular grid is the set of triangles such that their vertex flake has a non-empty intersection with the corresponding Euclidean primitive P.*

The primitives we consider here are circles $(x - x_c)^2 + (y - y_c)^2 - r^2 = 0$ of radius $r \in \mathbb{R}^+$, centered at an arbitrary (Euclidean) point (x_c, y_c). To avoid

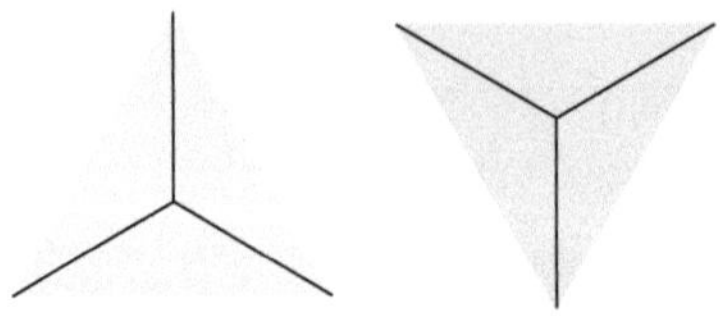

Fig. 2. Vertex flake for up-triangle (a) and down-triangle (b).

situations where a circle intersects the vertex flake of a triangle, but all three triangle vertices are outside of the circle, we assume $r > 1$, which are interesting values in applications. This definition of flake-based digitized circles leads to two Diophantine inequalities. To avoid unnecessary triangles in degenerate cases, we will put "$<$" on one side of the double inequality, and "$\leq$" on the other side in the circles definition.

3.1 Vertex Flake Circle with Arbitrary Center

Let $(x_c, y_c) \in \mathbb{R}^2$ and $r > 1$ be the center and diameter of the Euclidean circle $C : (x - x_c)^2 + (y - y_c)^2 - r^2 = 0$, (x_0, y_0) the centroid of (up or down) triangle T and $X_0 = x_0 - x_c$, $Y_0 = y_0 - y_c$.

Proposition 1 (Vertex flake circle with arbitrary center)
A triangle T belongs to the vertex flake circle C^* iff

$$-w_\triangle^+ - \frac{1}{3} \leq X_0^2 + Y_0^2 - r^2 < -w_\triangle^- - \frac{1}{3},$$

where $w_\triangle^- = \min A$, $w_\triangle^+ = \max A$,
$A = \{X_0 - \frac{\sqrt{3}}{3}Y_0, -X_0 - \frac{\sqrt{3}}{3}Y_0, \frac{2\sqrt{3}}{3}Y_0\}$ *if T is an up-triangle and*
$w_\nabla^- = \min B$, $w_\nabla^+ = \max B$,
$B = -A = \{-X_0 + \frac{\sqrt{3}}{3}Y_0, X_0 + \frac{\sqrt{3}}{3}Y_0, -\frac{2\sqrt{3}}{3}Y_0\}$ *if T is a down-triangle.*

Proof. For $r > 1$, a triangle T belongs to the vertex flake circle C^* if its vertices are on the different sides of the Euclidean circle C, i.e., some are inside and some are outside C. The signed distance between a point $A = (a_x, a_y)$ and the circle $(x - x_c)^2 + (y - y_c)^2 - r^2 = 0$ is equal to $\sqrt{(a_x - x_c)^2 + (a_y - y_c)^2} - r$. It is positive for points outside the circle and negative for points inside.

Let us consider an up-triangle $T = (x_0, y_0)$. The three vertices of T are $(x_0 \pm \frac{1}{2}, y_0 - \frac{\sqrt{3}}{6})$ and $(x_0, y_0 + \frac{\sqrt{3}}{3})$, and the three signed distances are given by:

$$
\begin{aligned}
Dist &= \{\sqrt{(X_0+\frac{1}{2})^2+(Y_0-\frac{\sqrt{3}}{6})^2}-r, \sqrt{(X_0-\frac{1}{2})^2+(Y_0-\frac{\sqrt{3}}{6})^2}-r,\\
&\quad \sqrt{X_0^2+(Y_0+\frac{\sqrt{3}}{3})^2}-r\}\\
&= \{\sqrt{d_0+X_0-\frac{\sqrt{3}}{3}Y_0}-r, \sqrt{d_0-X_0-\frac{\sqrt{3}}{3}Y_0}-r, \sqrt{d_0+\frac{2\sqrt{3}}{3}Y_0}-r\}
\end{aligned}
$$

where $d_0 = X_0^2+Y_0^2+\frac{1}{3}$. If the vertices are at the opposite sides of the circle, then

$$\min_{1\le i\le 3}\{d_i\} < 0 \le \max_{1\le i\le 3}\{d_i\}. \tag{1}$$

We have that $\min_{1\le i\le 3}\{d_i\} = \sqrt{d_0+w_\triangle^-}-r$; $\max_{1\le i\le 3}\{d_i\} = \sqrt{d_0+w_\triangle^+}-r$ and

$$\sqrt{d_0+w_\triangle^-}-r < 0 \le \sqrt{d_0+w_\triangle^+}-r,$$

$$d_0+w_\triangle^- < r^2 \le d_0+w_\triangle^+,$$

$$-w_\triangle^+ \le d_0 - r^2 < -w_\triangle^-,$$

$$-w_\triangle^+ \le X_0^2+Y_0^2+\frac{1}{3}-r^2 < -w_\triangle^-.$$

This last inequality gives the necessary and sufficient condition that the up-triangle (x_0, y_0) lies on the circle.

For a down-triangle (x_0, y_0), the vertices are $(x_0 \pm \frac{1}{2}, y_0+\frac{\sqrt{3}}{6})$ and $(x_0, y_0 - \frac{\sqrt{3}}{3})$. The proof is analogous to the proof for up-triangles and is omitted.

The inequalities in the previous proposition can be obtained in an alternative form, by taking $d_0 = x_0^2+y_0^2+\frac{1}{3}$ in the proof.

Corollary 1 (Vertex flake circle with arbitrary center - alternative) *A triangle $T=(x_0,y_0)$ belongs to the vertex flake circle C^* iff*

$$-w_\triangle^+ - x_c^2 - y_c^2 - \frac{1}{3} \le x_0^2+y_0^2-r^2 < \quad w_\triangle^- - x_c^2 - y_c^2 - \frac{1}{3},$$

where $w_\triangle^+ = \max C$, $w_\triangle^- = \min C$,
$C = \{x_0(1-2x_c) - y_0(\frac{\sqrt{3}}{3}-2y_c) - x_c - \frac{\sqrt{3}}{3}y_c, -x_0(1+2x_c) - y_0(\frac{\sqrt{3}}{3}-2y_c) + x_c - \frac{\sqrt{3}}{3}y_c, -2x_0x_c + 2y_0(\frac{\sqrt{3}}{3}-y_c) - \frac{2\sqrt{3}}{3}y_c\}$ *if T is an up-triangle and*
$w_\triangledown^+ = \max D$, $w_\triangledown^- = \min D$,
$D = \{x_0(1-2x_c) + y_0(\frac{\sqrt{3}}{3}-2y_c) - x_c - \frac{\sqrt{3}}{3}y_c, -x_0(1+2x_c) + y_0(\frac{\sqrt{3}}{3}-2y_c) + x_c - \frac{\sqrt{3}}{3}y_c, -2x_0x_c - 2y_0(\frac{\sqrt{3}}{3}+y_c) + \frac{2\sqrt{3}}{3}y_c\}$ *if T is a down-triangle.*

In Fig. 3, we show two examples, a circle centered at $(0.6, 0.2)$ of radius 8 and a circle centered at $(1.6, 0.4)$ of radius 5.5.

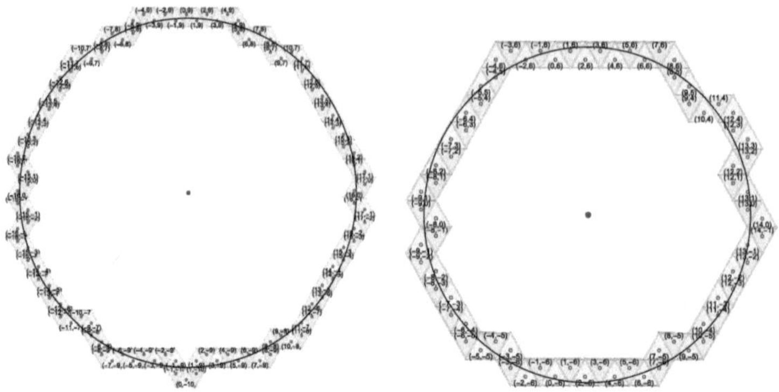

(a) (b)

Fig. 3. Vertex flake circle with radius 8 centered at (0.6, 0.2) (a) and of radius 5.5 centered at (1.6, 0.4) (b). The triangles are labelled by their integer matrix coordinates.

3.2 Vertex Flake Circle Centered at Triangle Centroid

As a special case, we will state the formulas for circles centered at centroid of both up-triangle $(\frac{1}{2}, \frac{\sqrt{3}}{6})$ and down-triangle $(0, \frac{\sqrt{3}}{3})$. Circles centered at other triangle centroids can be obtained by translation.

Corollary 2 (Vertex flake circle centered at up-triangle)
If $(x_c, y_c) = (\frac{1}{2}, \frac{\sqrt{3}}{6})$, *then*
$A = \{x_0 - \frac{\sqrt{3}}{3}y_0 - \frac{1}{3}, -x_0 - \frac{\sqrt{3}}{3}y_0 + \frac{2}{3}, \frac{2\sqrt{3}}{3}y_0 - \frac{1}{3}\}$
$B = \{-x_0 + \frac{\sqrt{3}}{3}y_0 + \frac{1}{3}, x_0 + \frac{\sqrt{3}}{3}y_0 - \frac{2}{3}, -\frac{2\sqrt{3}}{3}y_0 + \frac{1}{3}\}$
$C = \{-\frac{2\sqrt{3}}{3}y_0, -2x_0 - \frac{2\sqrt{3}}{3}y_0 + 1, -x_0 + \frac{\sqrt{3}}{3}y_0\}$
$D = \{-1, -2x_0, -x_0 - \sqrt{3}y_0\}$.

Corollary 3 (Vertex flake circle centered at down-triangle)
If $(x_c, y_c) = (0, \frac{\sqrt{3}}{3})$, *then*
$A = \{x_0 - \frac{\sqrt{3}}{3}y_0 + \frac{1}{3}, -x_0 - \frac{\sqrt{3}}{3}y_0 + \frac{1}{3}, \frac{2\sqrt{3}}{3}y_0 - \frac{2}{3}\}$
$B = \{-x_0 + \frac{\sqrt{3}}{3}y_0 - \frac{1}{3}, x_0 + \frac{\sqrt{3}}{3}y_0 - \frac{1}{3}, -\frac{2\sqrt{3}}{3}y_0 + \frac{2}{3}\}$
$C = \{x_0 - \sqrt{3}y_0, -x_0 - \sqrt{3}y_0, -1\}$
$D = \{x_0 - \frac{\sqrt{3}}{3}y_0, -x_0 - \frac{\sqrt{3}}{3}y_0, -\frac{4\sqrt{3}}{3}y_0 + 1\}$.

4 Vertex Flake Circles Properties

Lemma 1. *If a triangle T belongs to the flake circle C^*, then at least two of the edge neighbors of T are also in C^*.*

Proof. If the triangle $T = pqr$ is on the flake circle C^*, one vertex p is on one side, other two vertices q and r on the other side of the Euclidean circle C.

Then two edge neighbors of pqr, triangles prt $(t \neq q)$ and pqs $(s \neq r)$ have their vertices p and q, and p and r, respectively, on different sides of the circle, so they are also on the circle C^*.

Let us recall that supercover (upper Jordan) digitization S^+ of a Euclidean set S consists of all (closed) triangles having nonempty intersection with the set.

Lemma 2. *Supercover circle C^+ of the Euclidean circle C with $r > 1$ is edge-connected.*

Proof. As C is connected, C^+ is (vertex-)connected. It is also edge-connected. Suppose that, on the contrary, two triangles T_1 and T_2 in C^+ are vertex-adjacent through a vertex v but are not edge-adjacent. Then either C passes through the shared vertex v (and thus all six triangles incident to v belong to C^+ and T_1 and T_2 are edge-connected), or C passes through one of the two edge connected fans of triangles incident to v, so again T_1 and T_2 are edge-connected in C^+.

Proposition 2. *The vertex flake circle C^* centered at (x_c, y_c) with radius $r > 1$ is edge-connected.*

Proof. The flake circle C^* is contained in the super cover digitization C^+, which is edge-connected.

We can obtain C^* by deleting from C^+ triangles not belonging to C^*. We will show that deleting such triangles does not affect edge-connectivity of the set of remaining triangles, so C^* is edge-connected.

Let T be a triangle in C^+ and not in C^*. Then T is intersected by C and its vertex flake is not intersected by C. Thus, C intersects an edge e of T. Let T' be the triangle edge adjacent to T across e. As $r > 1$, C intersects the vertex flake of T', so T' is in C^*. By Lemma 1, the two triangles edge-adjacent to T' different from T are in C^* as well. The two triangles edge-adjacent to T different from T' cannot belong to C^+ as they are on the opposite side of the line determined by the edge e from the circle C (save the part of C inside T). Thus, removing T from C^+ does not affect edge-connectivity of $C^+ \setminus T$, so C^* is edge-connected.

5 Incremental Circle Generation

To design an incremental algorithm for circle generation, we need to

- determine a starting point, i.e., a triangle which belongs to the circle, and
- determine the rule for progressing to the next triangle from the current one.

5.1 Starting Point

Let (x_c, y_c) be the center and r the radius of the Euclidean circle. The starting point of the generation algorithm is the triangle which contains the point $(x_c, y_c + r)$. Thus, we need to determine the centroid (x_0, y_0) of the triangle containing a given point $(X, Y) \in \mathbb{R}^2$.

Let $k = \lfloor \frac{2\sqrt{3}}{3} Y \rfloor$ be the index of the horizontal strip containing (X, Y). We further subdivide the Cartesian plane into vertical strips of width $\frac{1}{2}$ through the points $x = l$ and $x = l + \frac{1}{2}$, $l \in \mathbb{Z}$. The rectangle at the intersection of the horizontal and vertical strips containing (X, Y) is further subdivided by a slanted triangular grid line into two right triangles. Which of them contains (X, Y) is determined by the slope of the grid line ($\sqrt{3}$ or $-\sqrt{3}$) and the position of (X, Y) with respect to this line (below or above).

If k is even and $0 \le X - \lfloor X \rfloor < \frac{1}{2}$ then if
$Y \le \sqrt{3}X + k\frac{\sqrt{3}}{2} - \sqrt{3}\lfloor X \rfloor = \sqrt{3}X + \sqrt{3}(\frac{k}{2} - \lfloor X \rfloor)$, (X, Y) belongs to up-triangle with centroid $(\lfloor X \rfloor + \frac{1}{2}, k\frac{\sqrt{3}}{2} + \frac{\sqrt{3}}{6})$;
otherwise if $Y > \sqrt{3}X + \sqrt{3}(\frac{k}{2} - \lfloor X \rfloor)$, (X, Y) belongs to down-triangle with centroid $(\lfloor X \rfloor, k\frac{\sqrt{3}}{2} + \frac{\sqrt{3}}{3})$.

If k is even and $\frac{1}{2} \le X - \lfloor X \rfloor < 1$ then if
$Y \le -\sqrt{3}X + \sqrt{3}(\frac{k}{2} + \lfloor X \rfloor + 1)$, (X, Y) belongs to up-triangle with centroid $(\lfloor X \rfloor + \frac{1}{2}, k\frac{\sqrt{3}}{2} + \frac{\sqrt{3}}{6})$;
otherwise if
$Y > -\sqrt{3}X + \sqrt{3}(\frac{k}{2} + \lfloor X \rfloor + 1)$, (X, Y) belongs to down-triangle with centroid $(\lfloor X \rfloor + 1, k\frac{\sqrt{3}}{2} + \frac{\sqrt{3}}{3})$.

If k is odd and $\frac{1}{2} \le X - \lfloor X \rfloor < 1$ then if
$Y \le \sqrt{3}X + \sqrt{3}(\frac{k-1}{2} - \lfloor X \rfloor)$, (X, Y) belongs to up-triangle with centroid $(\lfloor X \rfloor + 1, k\frac{\sqrt{3}}{2} + \frac{\sqrt{3}}{6})$;
otherwise if
$Y > \sqrt{3}X + \sqrt{3}(\frac{k-1}{2} - \lfloor X \rfloor)$, (X, Y) belongs to down-triangle with centroid $(\lfloor X \rfloor + \frac{1}{2}, k\frac{\sqrt{3}}{2} + \frac{\sqrt{3}}{3})$.

If k is odd and $0 \le X - \lfloor X \rfloor < \frac{1}{2}$ then if
$Y \le -\sqrt{3}X + \sqrt{3}(\frac{k+1}{2} + \lfloor X \rfloor)$, (X, Y) belongs to up-triangle with centroid $(\lfloor X \rfloor, k\frac{\sqrt{3}}{2} + \frac{\sqrt{3}}{6})$;
otherwise if
$Y > -\sqrt{3}X + \sqrt{3}(\frac{k+1}{2} + \lfloor X \rfloor)$, (X, Y) belongs to down-triangle with centroid $(\lfloor X \rfloor + \frac{1}{2}, k\frac{\sqrt{3}}{2} + \frac{\sqrt{3}}{3})$.

5.2 Incremental Algorithm

To determine if the triangle (x_0, y_0) belongs to the flake circle, we need to compute maximum and minimum of the sets A and B. To this aim, we consider the elements of A and B separately for the two types of triangles.

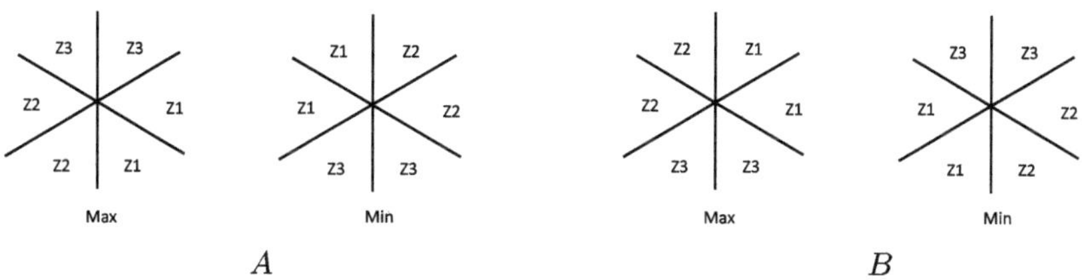

Fig. 4. Maxima and minima of A and B.

For up-triangles, we consider the elements of set A, $z_1 = X - \frac{\sqrt{3}}{3}Y$, $z_2 = -X - \frac{\sqrt{3}}{3}Y$ and $z_3 = \frac{2\sqrt{3}}{3}Y$, $X = x - x_c$, $Y = y - y_c$. We solve the equations $z_1 = z_2$, $z_2 = z_3$ and $z_1 = z_3$ to get $X = 0$, $Y = -\frac{\sqrt{3}}{3}X$, and $Y = \frac{\sqrt{3}}{3}X$, i.e., $x = x_c$, $y = -\frac{\sqrt{3}}{3}(x - x_c) + y_c$ and $y = \frac{\sqrt{3}}{3}(x - x_c) + y_c$, respectively. These three lines intersect at point (x_c, y_c) and they divide the plane into six regions (sextants). Then, $zmin = \min A$ and $zmax = \max A$ inside each region are easily obtained, see Fig. 4. We call first sextant the sextant to the right of the positive y-axis, and we number them clockwise.

For down-triangles, as $B = -A$, we get the same three lines as before.

Thus, to check if a triangle T centered at (x_0, y_0) belongs to the flake circle, it is enough to find the sextant (centered at (x_c, y_c)) containing (x_0, y_0) and to check if the two Diophantine inequalities are satisfied for appropriate $zmin$ and $zmax$, without computing all three elements and the extrema of each of A and B. For example, to check if an up-triangle (x_0, y_0) in the first sextant belongs to the circle, it suffices to verify if $-z_3 - \frac{1}{3} \leq X_0^2 + Y_0^2 - r^2 < -z_2 - \frac{1}{3}$, with $X_0 = x_0 - x_c$, $Y_0 = y_0 - y_c$.

For each current triangle T, only two of its edge neighbors T_1 and T_2 are candidates for the next triangle of the flake circle. We check e.g. T_1. If it satisfies the inequalities, then T_1 is the next triangle. Otherwise, the next triangle is T_2.

Another optimization is also possible. Inside each sextant, the next neighbor for one type of triangles is predetermined, and no computation is needed. If the tracing of the circle is clockwise then the next neighbor of

- down triangle is its right neighbor in the first sextant,
- up-triangle is its downwards neighbor in the second sextant
- down-triangle is its left neighbor in the third sextant
- up-triangle is its left neighbor in the fourth sextant
- down-triangle is its upwards neighbor in the fifth sextant
- up-triangle is its right neighbor in the sixth sextant and

We will prove e.g. the third statement. The remaining proofs are analogous.

Let $T = (x_0, y_0)$ be a down-triangle on the flake circle in the third sextant, i.e.,
$\frac{2\sqrt{3}}{3}Y_0 - \frac{1}{3} = -z_3 \leq X_0^2 + Y_0^2 - r^2 < -z_2 = X_0 - \frac{\sqrt{3}}{3}Y_0 - \frac{1}{3}$.
and let $T_1 = (x_0 - \frac{1}{2}, y_0 - \frac{\sqrt{3}}{6})$ be its left neighbor. Then, T_1 also belongs to the

circle because the inequalities for T_1
$-z_1 - \frac{1}{3} \leq (X_0 - \frac{1}{2})^2 + (Y_0 - \frac{\sqrt{3}}{6})^2 - r^2 < -z_3 - \frac{1}{3}$, i.e.,
$-(X_0 - \frac{1}{2}) + \frac{\sqrt{3}}{3}(Y_0 - \frac{\sqrt{3}}{6}) - \frac{1}{3} \leq (X_0 - \frac{1}{2})^2 + (Y_0 - \frac{\sqrt{3}}{6})^2 - r^2 < -\frac{2\sqrt{3}}{3}(Y_0 - \frac{\sqrt{3}}{6}) - \frac{1}{3}$,
$-X_0 + \frac{1}{2} + \frac{\sqrt{3}}{3}Y_0 - \frac{1}{6} - \frac{1}{3} \leq X_0^2 - X + \frac{1}{4} + Y_0^2 - \frac{\sqrt{3}}{3}Y_0 + \frac{1}{12} - r^2 < -\frac{2\sqrt{3}}{3}Y_0 + \frac{1}{3} - \frac{1}{3}$,
$\frac{2\sqrt{3}}{3}Y_0 - \frac{1}{3} \leq X_0^2 + Y_0^2 - r^2 < X_0 - \frac{\sqrt{3}}{3}Y_0 - \frac{1}{3}$
are the same as the inequalities for T.

We can also show that not both of the remaining two triangles T_2 and T_3 edge-adjacent to T can belong to C^*.

For $T_2 = (x_0, y_0 + \frac{\sqrt{3}}{3})$, $z_1 = X_0 - \frac{\sqrt{3}}{3}Y_0 - \frac{1}{3}$, $z_3 = \frac{2\sqrt{3}}{3}Y_0 + \frac{2}{3}$, and the inequalities for T_2 are $-z_1 - \frac{1}{3} \leq X_0^2 + (Y_0 + \frac{\sqrt{3}}{3})^2 - r^2 < -z_3 - \frac{1}{3}$, i.e.,
$-X_0 + \frac{\sqrt{3}}{3}(Y_0 + \frac{\sqrt{3}}{3}) + \frac{1}{3} - \frac{1}{3} \leq X_0^2 + (Y_0 + \frac{\sqrt{3}}{3})^2 - r^2 < -\frac{2\sqrt{3}}{3}(Y_0 + \frac{\sqrt{3}}{3}) - \frac{2}{3} - \frac{1}{3}$
$-X_0 + \frac{\sqrt{3}}{3}Y_0 - \frac{1}{6} \leq X_0^2 + Y_0^2 + \frac{2\sqrt{3}}{3}Y_0 + \frac{1}{3} - r^2 < -\frac{2\sqrt{3}}{3}Y_0 - 1$
$-X_0 - \frac{\sqrt{3}}{3}Y_0 - \frac{1}{3} \leq X_0^2 + Y_0^2 - r^2 < -\frac{4\sqrt{3}}{3}Y_0 - \frac{4}{3}$

For $T_3 = (x_0 + \frac{1}{2}, y_0 - \frac{\sqrt{3}}{6})$, $z_1 = X_0 + \frac{1}{2} - \frac{\sqrt{3}}{3}Y_0 + \frac{1}{6}$, $z_3 = \frac{2\sqrt{3}}{3}Y_0 - \frac{1}{3}$, and the inequalities for T_3 are
$-z_1 - \frac{1}{3} \leq (X_0 + \frac{1}{2})^2 + (Y_0 - \frac{\sqrt{3}}{6})^2 - r^2 < -z_3 - \frac{1}{3}$, i.e.,
$-(X_0 + \frac{1}{2}) + \frac{\sqrt{3}}{3}(Y_0 - \frac{\sqrt{3}}{6}) - \frac{1}{3} \leq (X_0 - \frac{1}{2})^2 + (Y_0 - \frac{\sqrt{3}}{6})^2 - r^2 < -\frac{2\sqrt{3}}{3}(Y_0 - \frac{\sqrt{3}}{6}) - \frac{1}{3}$
$-X_0 - \frac{1}{2} + \frac{\sqrt{3}}{3}Y_0 - \frac{1}{6} + \frac{1}{6} - \frac{1}{3} \leq X_0^2 + X_0 + \frac{1}{4} + Y_0^2 - \frac{\sqrt{3}}{3}Y_0 + \frac{1}{12} - r^2 < -\frac{2\sqrt{3}}{3}Y_0 + \frac{1}{3} - \frac{1}{3}$
$-2X_0 + \frac{2\sqrt{3}}{3}Y_0 - \frac{2}{3} \leq X_0^2 + Y_0^2 - r^2 < -X_0 - \frac{\sqrt{3}}{3}Y_0 - \frac{1}{3}$

The left side of inequalities for T_2 is the same as the right side of the inequalities for T_3, thus both inequalities cannot be satisfied simultaneously and only one of T_2, T_3 belongs to C^*.

Analogously, it can be shown that for an up-triangle T on C^* in the third sextant, its right neighbor belongs to C^*, and only one of its downwards and left neighbors belongs to C^*.

The proof is analogous for other sextants. Thus, vertex flake circle C^* contains no triangles of side 2, i.e., a triangle and all three of its edge neighbors.

Another optimization is possible, which we describe for the third sextant. For each down triangle $T = (x_0, y_0)$ on C^*, its left neighbor $T_1 = (x_0 - \frac{1}{2}, y_0 + \frac{\sqrt{3}}{6})$ is added to C^*. Which one of the two neighbor candidates of T_1 is in C^* can be determined by comparing $X_0^2 + Y_0^2 - r^2$ with the discriminant $D = 2X_0 + \frac{2\sqrt{3}}{3}Y_0 - \frac{4}{3}$ (obtained in a similar manner as above). If $X_0^2 + Y_0^2 - r^2 \geq D$, then the left neighbor $(x_0 - 1, y_0)$ of T_1 is in C^*. Otherwise, if $X_0^2 + Y_0^2 - r^2 < D$, the downwards neighbor $(x_0 - \frac{1}{2}, y_0 - \frac{\sqrt{3}}{2})$ is in C^*. Thus, only one comparison suffices to determine two triangles in C^*.

Our algorithm is obviously linear in the number of generated points.

6 Discussion

We have proposed analytical characterization of discrete vertex flake circles in the triangular grid and an incremental circle generation algorithm based on this definition.

An obvious generalization of analytical circle definition from the square [1, 12] to the triangular grid would be to use constant thickness $w = 1$. We show some examples in Fig. 5(a). Although such circles with integer radius tile the plane, they are less visually pleasing than the flake circles, and have weaker topological properties. They are vertex-connected but not necessarily edge-connected, and they may contain simple points, which are not necessary to maintain vertex-connectivity.

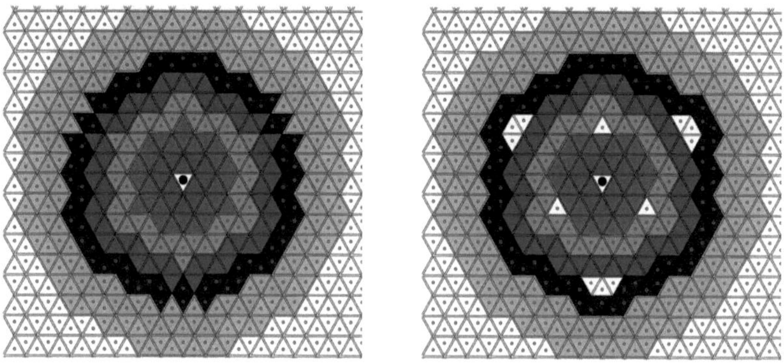

(a) (b)

Fig. 5. Circles of radius 1 to 7 centered at a down-triangle, with constant thickness 1 (a) and generated according to our vertex flake algorithm (b).

Another approach, based on neighborhood sequences, has been proposed by Nagy [11]. The idea is to propagate the triangles starting from the central one, where at each step a predetermined type of neighbors (1-, 2- or 3-neighbors) of current triangles are selected. Thus, the circles defined by any infinite neighborhood sequence tile the triangular grid. In Fig. 6(a), we show an example of circles defined by a non unique neighborhood sequence 1,2,1,2 (or 3),1, 2 (or 3), 2 (or 3). The approach is completely discrete, the center is considered to be a triangle, and not an arbitrary Cartesian point as in our approach. If the radius is considered to be equal to the length of the neighborhood sequence, then it is constrained to be integer, while our flake circles may have arbitrary real radius. We can notice also that such integer radii do not correspond in an obvious way to Euclidean distances between the centroids of triangles on circles and the centroid of the central triangle. The algorithm for circle generation is also not straightforward: to construct a circle of radius $n \in \mathbb{N}$, the circles of all radii less than n have to be constructed. The obtained circles are vertex-connected, but not necessarily edge-connected. They also may contain simple points, whose removal keeps the circle connected and does not cause the merging of circle interior and exterior. In Fig. 6, we show some circles proposed in [11], defined by (non-unique) sequence 1,2,1,2 (or 3),1, 2 (or 3), 2 (or 3), and our vertex-flake circles of the same radius.

A related notion of chamfer discs has been defined and investigated as a solution to a linear integer programming problem [8]. Chamfer or weighted dis-

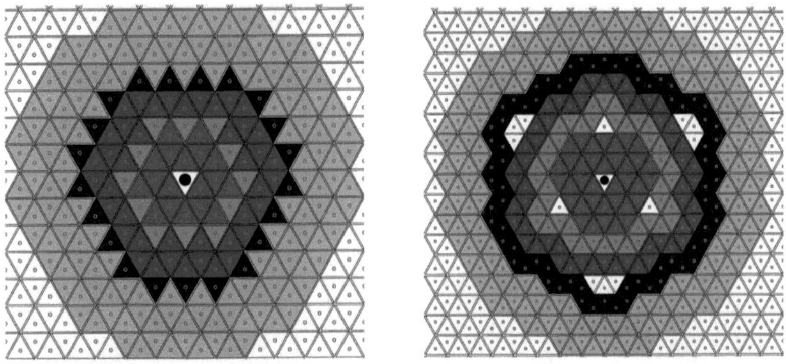

Fig. 6. Circles of radius 1 to 7 centered at a down-triangle, generated according to: (a) algorithm by Nagy [11] with neighborhood sequence 1,2,1,2 (or 3),1, 2 (or 3), 2 (or 3) and (b) our vertex flake algorithm.

tance is a path-based distance where the steps from a triangle to its 1-, 2- and 3-neighbors are assigned different weights.

In the future we would like to extend this work to other arithmetic discrete primitives such as parabolas, ellipses and hyperbolas.

Acknowledgement. This research has been partially supported by the Science Fund of the Republic of Serbia through PRIZMA program, GRANT No 7632, project "Mathematical Methods in Image Processing under Uncertainty" - MaMIPU.

References

1. Andres, E.: Discrete circles, rings and spheres. Comput. Graph. **18**(5), 695–706 (1994)
2. Andres, E., Jacob, M.-A.: The discrete analytical hyperspheres. IEEE Trans. Visual Comput. Graphics **3**, 75–86 (1997)
3. Andres, E., Largeteau-Skapin, G.: Digital surfaces of revolution made simple. In: Normand, N., Guédon, J., Autrusseau, F. (eds.) DGCI 2016. LNCS, vol. 9647, pp. 244–255. Springer, Cham (2016). https://doi.org/10.1007/978-3-319-32360-2_19
4. Bresenham, J.: A linear algorithm for incremental digital display of circular arcs. Commun. ACM **20**(2), 100–106 (1977)
5. Debled-Rennesson, I., Domenjoud, E., Jamet, D.: Arithmetic discrete parabolas. In: Bebis, G., et al. (eds.) ISVC 2006. LNCS, vol. 4292, pp. 480–489. Springer, Heidelberg (2006). https://doi.org/10.1007/11919629_49
6. Fiorio, C., Jamet, D., Toutant, J.-L.: Discrete circles: an arithmetical approach with non-constant thickness. In: Vision Geometry XIV, vol. 6066, pp. 87–98. International Society for Optics and Photonics, SPIE (2006)
7. Fiorio, C., Toutant, J.-L.: Arithmetic discrete hyperspheres and separatingness. In: Kuba, A., Nyúl, L.G., Palágyi, K. (eds.) DGCI 2006. LNCS, vol. 4245, pp. 425–436. Springer, Heidelberg (2006). https://doi.org/10.1007/11907350_36

8. Kovács, G., Nagy, B., Vizvári, B.: On disks of the triangular grid: an application of optimization theory in discrete geometry. Discret. Appl. Math. **282**, 136–151 (2020)
9. Liu, Y.-K.: The generation of circular arcs on hexagonal grids. Comput. Graph. Forum **12**(1), 21–26 (1993)
10. Manoharan, P., Ray, B.: An improved circle drawing algorithm on a hexagonal grid. J. Graph. Tools **17**, 5–15 (2013)
11. Nagy, B.: Characterization of digital circles in triangular grid. Pattern Recognit. Lett. **25**(11), 1231–1242 (2004)
12. Reveillès, J.P.: Calcul en Nombres Entiers et Algorithmique. Ph.D. thesis, Université Louis Pasteur, Strasbourg, France (1991)
13. Toutant, J.-L., Andres, E., Largeteau-Skapin, G., Zrour, R.: Implicit digital surfaces in arbitrary dimensions. In: Barcucci, E., Frosini, A., Rinaldi, S. (eds.) DGCI 2014. LNCS, vol. 8668, pp. 332–343. Springer, Cham (2014). https://doi.org/10.1007/978-3-319-09955-2_28
14. Toutant, J.-L., Andres, E., Roussillon, T.: Digital circles, spheres and hyperspheres: from morphological models to analytical characterizations and topological properties. Discret. Appl. Math. **161**(16–17), 2662–2677 (2013)
15. Zrour, R., Čomić, L., Andres, E., Largeteau-Skapin, G.: Discrete circles: analytical definition and generation in the hexagonal grid. J. Comb. Optim. **49**(1), 14 (2025)

Fig. 1. 3D illustration of notations and concepts, from left to right: input convex shape X, its digitization $X_h := X \cap h\mathbb{Z}^d$ seen as a collection of cubes (visually equivalent to the h-boundary $\partial_h X$), its digitized convex hull $Y_h := \mathrm{CvxH}(X_h)$, the h-boundary $\partial_h X$ with the normals of the closest facet on Y_h. The three left images are rendered with material "normal", i.e. the displayed color corresponds to the normal vector direction.

paring its practical accuracy to state-of-the-art digital normal estimators is an essential next step, but for space reasons could not be developed here.

2 Basic Notions

We are only concerned with convex sets of the Euclidean space that are compact. Hence compact or bounded is implicit in all statements. The topological boundary of a compact set S is denoted ∂S, its interior $\mathring{S} := S \setminus \partial S$. The d-dimensional volume of S is written $\mathrm{Vol}^d(S)$ (here taking the Lebesgue or the Hausdorff measure is equivalent in our context). The scalar product of two vectors $\mathbf{u}, \mathbf{v}$ of $\mathbb{R}^n$ is denoted by $\mathbf{u} \cdot \mathbf{v}$ and the Euclidean norm of $\mathbf{u}$ is $\|\mathbf{u}\| := \sqrt{\mathbf{u} \cdot \mathbf{u}}$. The Euclidean distance between a point $\mathbf{x}$ and the set S is $\mathrm{d}_{\mathbf{E}}(\mathbf{x}, S) := \min_{\mathbf{s} \in S} \|\mathbf{s} - \mathbf{x}\|$.

Support Function, Normal Cone, Normal Vector. Given a convex set S, we denote by ϕ_S its *support function*, such that $\phi_S : \mathbf{w} \in \mathbb{R}^d \mapsto \max_{\mathbf{x} \in S} \mathbf{w} \cdot \mathbf{x} \in \mathbb{R}$. We recall the main property of the support function [27, §1.7.1]:

Theorem 1. *For $S \subset T$ two convex sets of $\mathbb{R}^d$, $\forall \mathbf{w} \in \mathbb{R}^d, \phi_S(\mathbf{w}) \leqslant \phi_T(\mathbf{w})$.*

For a given unit vector $\mathbf{w} \in \mathbb{R}^d$, assume the point $\mathbf{p} \in S$ satisfies $\mathbf{p} \cdot \mathbf{w} = \phi_S(\mathbf{w})$, then $\mathbf{p} \in \partial S$ and we say that $\mathbf{w}$ is a *normal vector to S at* $\mathbf{p}$. For any $\mathbf{p} \in \partial S$, the *normal cone to S at* $\mathbf{p}$ is the set of normal vectors to S at $\mathbf{p}$, and is denoted by $\mathrm{N}_S(\mathbf{p})$. Note that the normal cone is reduced to one vector when ∂S is twice differentiable at $\mathbf{p}$ and we speak of *the normal vector to S at* $\mathbf{p}$.

Gauss Digitization, Digitized Boundary, Digitized Convex Hull. The *gridstep* is a real positive number denoted by h. The lattice $h\mathbb{Z}^d$ is denoted by $\mathcal{L}_h$. The set X always designates a non-empty compact convex subset of $\mathbb{R}^d$. Its *Gauss digitization at step h* is the finite set $\mathrm{D}_h(X) := \mathcal{L}_h \cap X$. Following standard notations, the number of lattice points hitting X is written $\mathcal{L}_h(X) := \#(\mathrm{D}_h(X))$.

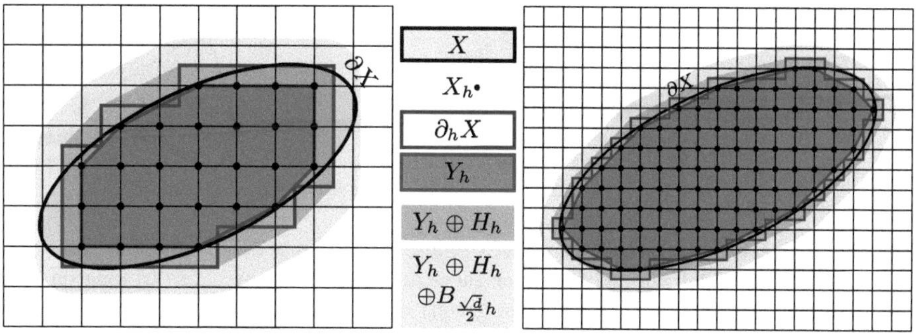

Fig. 2. 2D illustration of the main notations: the convex shape X, its Gauss digitization X_h, the convex hull of the digitization Y_h and convex sets used in Theorem 4, Sect. 4 ($Y_h \oplus H_h$, and $Y_h \oplus H_h \oplus B_{\frac{\sqrt{d}}{2}h}$).

Let H_h be the hypercube $h[-\frac{1}{2}, \frac{1}{2}]^d$. For any *digital set* $Z \subset h\mathbb{Z}^d$, the *voxel representation of* Z is $\mathcal{Q}_h Z := Z \oplus H_h$ (and is a union of axis-aligned cubes with edge length h centered on each digital point of Z). The *(digitized) h-boundary* of a subset $S \subset \mathbb{R}^d$ is $\partial_h S := \partial \mathcal{Q}_h \mathrm{D}_h(S)$ (e.g. see [14]). Let $(\mathbf{e}_i)_{i=1,\dots,d}$ be the canonical orthonormal basis of $\mathbb{Z}^d$. For any $z \in Z$, and if $z' := z \pm h\mathbf{e}_i$ is not in Z, we call the pair (z, z') a *surfel* of Z. Clearly, the segment $[z, z']$ crosses $\partial_h Z$ at exactly a point, which is the center of the face of $z \oplus H_h$ common with $z' \oplus H_h$.

Finally, the digital set $\mathrm{D}_h(X)$ is written X_h to shorten notations. The *digitized convex hull* Y_h *of* X is $Y_h := \mathrm{CvxH}(X_h)$ (where $\mathrm{CvxH}(\cdot)$ stands for the convex hull). *The objective of the paper is to show how we can infer the geometry of* X *from the geometry of* Y_h. Refer to Figs. 2 and 1 for 2D and 3D illustrations.

3 Properties for Digitized General Convex Shapes

This section focuses on geometric properties of digitized convex shape, without specific assumptions on the input convex shape. Let X be a compact convex shape of $\mathbb{R}^d$ (smooth or not). The next lemma states that if the shape boundary ∂X is close to a facet of an included polyhedron, then the normal of this facet and the normal of a nearby point $\mathbf{x}$ of ∂X are close to each other, and, the closer the points, the closer the normals. Besides, the further the projection of $\mathbf{x}$ is from the boundary of the facet, the closer are the normals.

Lemma 1. *Let* $Y \subset X$ *be a convex polyhedron (i.e.* $Y = \mathrm{CvxH}(V)$*, where* V *is a finite subset of* X*). Let* $\mathbf{x}$ *be an arbitrary point of* ∂X*, and* $\mathbf{n} \in \mathrm{N}_X(\mathbf{x})$*. Let* $\mathbf{y}$ *be the closest point of* $\mathbf{x}$ *on* ∂Y *(which is unique). Assume* $\mathbf{y}$ *is contained in only one facet (say* σ*) of* Y*, with unit normal vector* $\mathbf{n}_\sigma$*. Then the normal vector* $\mathbf{n}_\sigma$ *to* Y *and the normal vector* $\mathbf{n}$ *to* X *are related as:*

$$\begin{array}{c} \mathbf{n} \cdot \mathbf{n}_\sigma \geqslant 0, \\ \sin^2 \angle(\mathbf{n}, \mathbf{n}_\sigma) \leqslant \frac{\epsilon^2}{\epsilon^2 + r^2}, \end{array} \quad \textit{with} \quad \begin{cases} \epsilon := \|\mathbf{x} - \mathbf{y}\|, \\ r := \mathrm{d}_{\mathbf{E}}(\mathbf{y}, \partial\sigma). \end{cases}$$

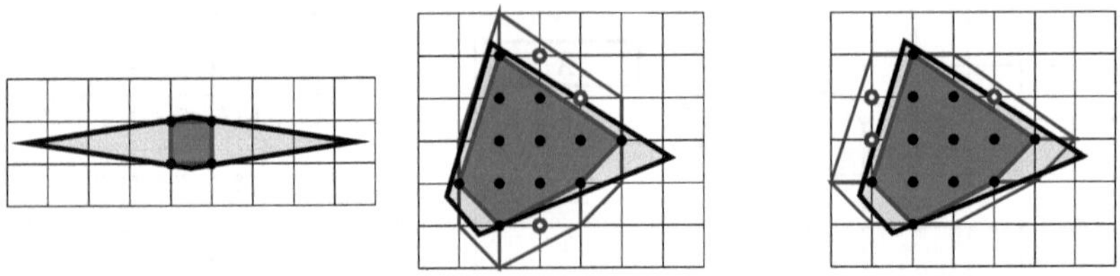

Fig. 3. Digitizations of arbitrary convex shapes: ∂Y_h (in red) can be very far away from ∂X (in black, left figure); however there are exterior points close to long enough edges, depending on the x- (middle, in green) or y-component (right, in blue) of the edge vector, so ∂X is at a distance less than h near these points. (Color figure online)

Proof. Since $\mathbf{y} \in Y \subset X$ and using the support function of X, we have immediately (see Theorem 1): $\mathbf{y} \cdot \mathbf{n} \leqslant \phi_Y(\mathbf{n}) \leqslant \phi_X(\mathbf{n}) = \mathbf{x} \cdot \mathbf{n}$. But $\mathbf{x} = \mathbf{y} + \epsilon \mathbf{n}_\sigma$ by construction. Substituting above gives

$$\mathbf{y} \cdot \mathbf{n} \leqslant (\mathbf{y} + \epsilon \mathbf{n}_\sigma) \cdot \mathbf{n} \quad \Rightarrow \quad 0 \leqslant \epsilon \mathbf{n}_\sigma \cdot \mathbf{n}.$$

Either $\epsilon > 0$ and we conclude for the relation $\mathbf{n} \cdot \mathbf{n}_\sigma \geqslant 0$, or $\epsilon = 0$ and $\mathbf{x} = \mathbf{y}$ which implies $\sigma \subset \partial X$ and $\mathrm{N}_X(\mathbf{x}) = \mathrm{N}_Y(\mathbf{x}) = \{\mathbf{n}_\sigma\}$, hence $\mathbf{n}_\sigma = \mathbf{n}$ and we also conclude.

For the second relation, let us project $\mathbf{n}$ onto the plane containing σ. This gives the vector $\mathbf{n}' := \mathbf{n} - (\mathbf{n} \cdot \mathbf{n}_\sigma)\mathbf{n}_\sigma$. If $\mathbf{n}' = \mathbf{0}$ then $\mathbf{n} = \mathbf{n}_\sigma$ and the relation is obvious. Otherwise, by the convexity and compactness of σ, the ray from point $\mathbf{y}$ in direction $\mathbf{n}'$ hits the boundary of σ at one point $\mathbf{a}$. We can write $\mathbf{a} = \mathbf{y} + s\frac{\mathbf{n}'}{\|\mathbf{n}'\|}$ with s the (positive) distance from $\mathbf{y}$ to $\mathbf{a} \in \partial\sigma$. Note that $s \geqslant r$ by hypothesis.

Since $\mathbf{a} \in \sigma \subset Y \subset X$, we have $\mathbf{a} \in X$ and using the support function of X, we get: $\mathbf{a} \cdot \mathbf{n} \leqslant \sup_{\mathbf{p} \in X} \mathbf{p} \cdot \mathbf{n} = \phi_X(\mathbf{n}) = \mathbf{x} \cdot \mathbf{n}$. It implies $(\mathbf{x} - \mathbf{a}) \cdot \mathbf{n} \geqslant 0$. We can decompose $\mathbf{x} - \mathbf{a} = \mathbf{x} - \mathbf{y} + \mathbf{y} - \mathbf{a} = \epsilon \mathbf{n}_\sigma - s\frac{\mathbf{n}'}{\|\mathbf{n}'\|}$. It follows that

$$-s\frac{\mathbf{n}' \cdot \mathbf{n}}{\|\mathbf{n}'\|} + \epsilon \mathbf{n}_\sigma \cdot \mathbf{n} \geqslant (\mathbf{x} - \mathbf{a}) \cdot \mathbf{n} \geqslant 0 \quad \Rightarrow \quad -s\sqrt{1 - (\mathbf{n}_\sigma \cdot \mathbf{n})^2} + \epsilon \mathbf{n}_\sigma \cdot \mathbf{n} \geqslant 0,$$

since simple calculations give $\mathbf{n}' \cdot \mathbf{n} = \mathbf{n}' \cdot \mathbf{n}' = 1 - (\mathbf{n}_\sigma \cdot \mathbf{n})^2$ (both $\mathbf{n}$ and $\mathbf{n}_\sigma$ are unit vectors). Posing $c = \mathbf{n}_\sigma \cdot \mathbf{n}$ and recalling that the first relation implies that $c \geqslant 0$, we derive

$$\epsilon c \geqslant s\sqrt{1 - c^2} \quad \Rightarrow \quad c^2 \geqslant \frac{s^2}{s^2 + \epsilon^2} \quad \Rightarrow \quad 1 - c^2 \leqslant \frac{\epsilon^2}{s^2 + \epsilon^2}.$$

Now $\sin \angle(\mathbf{n}_\sigma, \mathbf{n}) = \frac{\|\mathbf{n}'\|}{\|\mathbf{n}\|} = \sqrt{1 - c^2}$. The result follows since $r \leqslant s$. □

In the following corollary (proof is in Appendix A), we consider as convex polyhedron Y the convex hull Y_h of $X_h := \mathrm{D}_h(X)$.

Corollary 1. *Let* $(\mathbf{z}, \mathbf{z}')$ *be a surfel of* X_h. *Let* $\mathbf{y}$ *be the nearest point on* ∂Y_h *to* $\mathbf{z}'$ *and let* σ *be a facet of* Y_h *containing* $\mathbf{y}$, *with normal vector* $\mathbf{n}_\sigma$. *We have:*

- *there exists* $\mathbf{x} \in \partial X \cap [\mathbf{y}, \mathbf{z}'[$, *that is at a distance less than* h *from* $\mathbf{y}$,
- *for any* $\mathbf{n} \in \mathrm{N}_X(\mathbf{x})$, $\sin^2 \angle(\mathbf{n}, \mathbf{n}_\sigma) \leqslant \frac{1}{1+(r/h)^2}$, *if* $r := \mathrm{d}_\mathbf{E}(\mathbf{y}, \partial\sigma)$.

The previous corollary looks like a proof of convergence of normal vectors of Y_h towards normal vectors of X. It means that the point where the best estimation will be obtained is around the center of the inscribed circle/sphere of the facet. However it leaves unclear where the exterior lattice points of boundary surfels are projected onto their nearest facet: perhaps no such point exists above this center, especially if the facet is small or elongated. The following theorem shows that there are indeed exterior lattice points just above the interior of facets of Y_h, for long enough edges in 2D or wide enough triangles in 3D.

We say that a d-simplex σ of vertices $(p_1, \ldots, p_d)$ in $h\mathbb{Z}^d$ is *primitive* if $\mathrm{CvxH}(\sigma) \cap h\mathbb{Z}^d = \{p_1, \ldots, p_d\}$. The $d-1$-dimensional measure of the projection of σ onto a plane orthogonal to the axis $i \in \{1, \ldots, d\}$ is denoted by $A_i(\sigma)$ (informally the length or the area of the projected simplex). See Fig. 3 for a 2D illustration of where the lattice points are not in X but close to Y_h.

Theorem 2. *For* $d \in \{2, 3\}$, *let* σ *be a primitive edge (*$d = 2$*) or a primitive triangle (*$d = 3$*) of* ∂Y_h. *Then there are (at least)* k *lattice points of* $h\mathbb{Z}^d$ *not in* X *at a distance less than* h *from* σ, *where* k *follows:*

- $d = 2$*: then* $k = \max_{i \in \{1,2\}}(A_i(\sigma)/h) - 1$,
- $d = 3$*: then* $k \geqslant \frac{-7+\sqrt{25+48 \max_{i \in \{1,2,3\}} A_i(\sigma)/h^2}}{2}$.

Otherwise said, a segment of ∂Y_h *with one component greater or equal to* $2h$, *and a triangle with one projected area greater or equal to* $\frac{7}{6}h^2$, *have at least one close exterior lattice point not in* X *that projects in* $\mathring{\sigma}$ *along some* $\pm\mathbf{e}_i$.

Proof. Let us start with the case $d = 2$. Let $\sigma = (p, q)$ be a primitive edge of ∂Y_h. Let $\mathbf{t} = \pm h\mathbf{e}_i$, $i \in \{1, 2\}$ be an axis lattice vector such that $p' := p + \mathbf{t} \notin X$ and $q' := q + \mathbf{t} \notin X$ (at least one such vector exists, otherwise p and q are not vertices of Y_h). Let P be the parallelogram $\mathrm{CvxH}(\{p, q, p', q'\})$ and we denote by k the number of lattice points in the interior of P. We use Pick's theorem to bound k from below:

$$\begin{aligned}
\mathrm{Vol}^2\left(\frac{1}{h}P\right) &= \mathcal{L}_h(\mathring{P}) + \frac{1}{2}\mathcal{L}_h(\partial P) - 1 && \text{(Pick's theorem)} \\
&= k + 1 && \text{(since } \partial P \text{ hits 4 points exactly)}
\end{aligned}$$

These k interior points are necessarily outside X, otherwise they would be in Y_h and (p, q) would not be an edge of ∂Y_h. Noticing that $\mathrm{Vol}^2\left(\frac{1}{h}P\right) = \frac{1}{h^2}|\det(q - p, \mathbf{t})| = \frac{1}{h}|(q-p)_i| = A_i(\sigma)/h$ and using both directions $i \in \{1, 2\}$, we have found k lattice points not in X but strictly inside a parallelogram touching Y_h with width at most h. This concludes.

The case $d = 3$ is harder, especially because there are lattice polyhedra with infinite volume that hits a few lattice points.[1] Let $\sigma = (p, q, r)$ be a primitive

[1] For instance the tetrahedron $\{(0,0,0), (1,0,0), (0,1,0), (1,1,j)\}$ with $j \in \mathbb{Z}$.

triangle of ∂Y_h. Let $\mathbf{t} = \pm h\mathbf{e}_i$, $i \in \{1,2,3\}$ be an axis lattice vector such that $p' := p + \mathbf{t} \notin X$, $q' := q + \mathbf{t} \notin X$ and $r' := r + \mathbf{t} \notin X$ (at least one exists for the reason mentioned above). Let P be the triangular prism $\mathrm{CvxH}(\{p,q,r,p',q',r'\})$ and we denote by k the number of lattice points in the interior of P. We exploit here Reeve's result [26, Theorem I], valid for any integer $n \geqslant 1$ and full-dimensional lattice convex polyhedron P:

$$2(n^3 - n)\mathrm{Vol}^3\left(\frac{1}{h}P\right) = 2\left(\mathcal{L}_{\frac{h}{n}}(P) - n\mathcal{L}_h(P)\right) - \left(\mathcal{L}_{\frac{h}{n}}(\partial P) - n\mathcal{L}_h(\partial P)\right), \quad (1)$$

$$\text{and} \quad 2(1 - n^2) = \mathcal{L}_{\frac{h}{n}}(\partial P) - n^2\mathcal{L}_h(\partial P). \quad (2)$$

Note that we have $\mathcal{L}_{\frac{h}{n}}(P) = \mathcal{L}_h(nP)$ (where nP stands for homothety of P centered at the origin by the factor n). We use the relations above with the dilation factor $n = 2$.

$\mathcal{L}_h(P) = 6 + k$ and $\mathcal{L}_h(\partial P) = 6$ are obvious from the fact that the triangular face is primitive and k is the number of interior lattice points. Dilating by two the lattice prism creates 5 lattice points per quadrangular face, and 3 per triangular face, but most are shared between two faces (except one per quadrangle). We have 3 quadrangular faces, so $3(\frac{4}{2} + 1)$ new lattice points. We have 2 triangular faces, so $2(\frac{3}{2})$ new lattice points. Summing with the initial 6 lattice points gives $\mathcal{L}_{\frac{h}{2}}(\partial P) = \mathcal{L}_h(\partial 2P) = 6 + 9 + 3 = 18$. We can already check (2).

Finally, each of the k interior lattice points of P can be combined with any lattice point of ∂P and their middle point gives an interior lattice point of $\mathcal{L}_h(2P)$, hence at most $6k$ points. The middle point of any pair of interior lattice points of P is also a lattice point of $\mathcal{L}_h(2P)$ and there are at most $\binom{k}{2}$ of them that are distinct (some of them could be the same). We get $\mathcal{L}_{\frac{h}{2}}(P) = \mathcal{L}_h(2P) \leqslant 18 + 6k + \binom{k}{2}$. Inserting the found values in (1):

$$12\mathrm{Vol}^3\left(\frac{1}{h}P\right) \leqslant 2\left(18 + 6k + \frac{k(k-1)}{2} - 2(6+k)\right) - (18 - 12) = 6 + 7k + k^2.$$

Otherwise said, k satisfies $k^2 + 7k + 6 - 12\mathrm{Vol}^3\left(\frac{1}{h}P\right) \geqslant 0$. A short computation gives $k \geqslant (-7 + \sqrt{25 + 48\mathrm{Vol}^3\left(\frac{1}{h}P\right)})/2$.

We notice again that $\mathrm{Vol}^3\left(\frac{1}{h}P\right) = \frac{1}{2h^3}|\det(q-p, r-p, \mathbf{t})| = \frac{1}{h^2}A_i(\sigma)$ to conclude the argument. □

The results in this section show that there are points on Y_h where the normal vector of the underlying shape can be pretty well estimated, and we know where the estimation will be good: above large enough facets and near their inscribed sphere center. However we have no control on the radius of the inscribed sphere. The purpose of the next section is to get better estimates by considering digitization of convex shapes with smoothness properties.

4 Properties for Digitized Smooth Convex Shapes

If we assume smoothness of the convex shape X, we can use the previous result to show the convergence of normal vectors in 2D around most edges of Y_h.

Theorem 3. *If X is a convex shape of $\mathbb{R}^2$ with C^3-smooth boundary and positive curvature, then any edge σ of ∂Y_h having at least the average length of edges of ∂Y_h has a normal $\mathbf{n}_\sigma$ close to the normal $\mathbf{n}$ of the closest point of ∂X to the edge center, and more precisely $\angle(\mathbf{n}, \mathbf{n}_\sigma) \leqslant \Theta(h^{\frac{1}{3}})$.*

Proof. [1, Theorem 2] tells that the number of vertices $N(Y_h)$ of Y_h is some $\Theta(h^{-\frac{2}{3}})$. So the average length $L(Y_h)$ of the edges of Y_h is the perimeter of X divided by $N(Y_h)$, so $L(Y_h) = \Theta(h^{\frac{2}{3}})$. According to Theorem 2 in the case $d = 2$, there are almost as many exterior lattice points as the discrete length of the nearby edge. Except for a constant number of edges (maximum 8, the ones having directions $(1, 0)$ and $(1, 1)$ and their 4 rotations), all the other edges have an exterior lattice point close to the center of the edge. For an edge σ with approximately the average length $L(Y_h)$, the estimation error between n_σ and the closest point on ∂X is (Corollary 1): $\sin^2 \angle(\mathbf{n}, \mathbf{n}_\sigma) \leqslant \frac{1}{1+(L(Y_h)/2h)^2} = \frac{1}{1+\Theta(h^{-2/3})}$. Taylor expansion of the previous relation gives $\angle(\mathbf{n}, \mathbf{n}_\sigma) \leqslant \Theta(h^{\frac{1}{3}})$. □

Unfortunately, this result cannot be easily extended to higher dimensions, since there exist small or elongated facets starting from 3D. In fact, by using a completely different approach, we achieve below a much stronger result on normal convergence.

Before that, let us try to better understand how the geometry of Y_h and the geometry of X exhibit more relations in the case where X has a smooth boundary. The *reach* of $S \subset \mathbb{R}^d$, denoted $\mathrm{reach}(S)$, is the infimum of the distance of the set S to its medial axis [7] (the medial axis gathers points that have at least two nearest neighbors on S). From now on, we change the smoothness condition on X and we assume that $\mathrm{reach}(\partial X)$ is greater than some $\rho > 0$. In our context, it means that ∂X has its principal curvatures between 0 (included) and $1/\rho$ (excluded).

We start by sandwiching X between Y_h and $Y_h \oplus H_h \oplus B_{\frac{\sqrt{d}}{2}h}$, where B_r is the ball centered on $\mathbf{0}$ and of radius r. Note that it is not true in general that $X \subset Y_h \oplus H_h$, as illustrated by Fig. 2.

Theorem 4. *Assume $X \subset \mathbb{R}^d$, compact, convex with* $\mathrm{reach}(\partial X) > \rho$. *For all gridsteps h, $0 < h < \frac{2\rho}{\sqrt{d}}$, we have $Y_h \subset X \subset Y_h \oplus H_h \oplus B_{\frac{\sqrt{d}}{2}h}$. It follows that the convex boundary ∂X lies in the strip $Y_h \oplus H_h \oplus B_{\frac{\sqrt{d}}{2}h} \setminus \mathrm{Int}(Y_h)$. Furthermore* $\mathrm{d}_H(X, Y_h) = \mathrm{d}_H(\partial Y_h, \partial X) \leqslant \sqrt{d}h$.

Proof. We have $X_h \subset X$ so $Y_h = \mathrm{CvxH}\,(X_h) \subset \mathrm{CvxH}\,(X) = X$ and the first inclusion follows. By definition, the h-boundary $\partial_h X$ of X is the topological boundary of $X_h \oplus H_h$, i.e. $\partial_h X = \partial(X_h \oplus H_h)$. It implies

$$\mathrm{CvxH}\,(\partial_h X) = \mathrm{CvxH}\,(\partial(X_h \oplus H_h)) = \mathrm{CvxH}\,(X_h \oplus H_h) = Y_h \oplus H_h. \quad (3)$$

According to [14, Theorem 1], $\mathrm{d}_H(\partial_h X, \partial X) \leqslant \frac{\sqrt{d}}{2}h$ for $h < \frac{2\rho}{\sqrt{d}}$. Hence

$$\partial X \subset \partial_h X \oplus B_{\frac{\sqrt{d}}{2}h}. \quad (4)$$

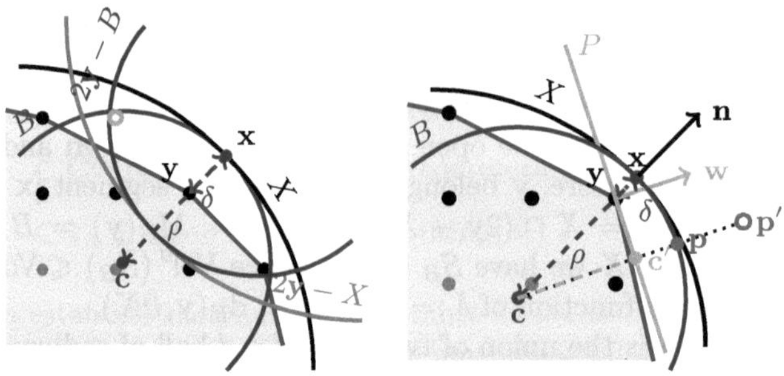

Fig. 4. Illustrations for the proofs of Theorem 5 (left) and Theorem 6 (right).

Theorem 6 below is our main result: it shows that the normal vectors to the facets of the digitized convex hull converge towards the normal vectors of the smooth convex shape, for small enough gridstep h, and that the speed of convergence is proportional to $\sqrt{h/\rho}$.

Theorem 6. *Assume the convex set $X \subset \mathbb{R}^d$ has* $\mathrm{reach}(\partial X) > \rho$. *Let $\mathbf{y}$ be any point on the boundary ∂Y_h. The point $\mathbf{x} := \pi_{\partial X}(\mathbf{y})$ is its closest point on ∂X, and it is well known that the outer normal $\mathbf{n}$ to X at $\mathbf{x}$ is aligned with $\mathbf{x} - \mathbf{y}$. Let $\mathbf{w} \in \mathrm{N}_{Y_h}(\mathbf{y})$ be any normal vector to Y_h at $\mathbf{y}$. Let $\delta := \|\mathbf{x} - \mathbf{y}\|$. Then for gridsteps h, $0 < h < \frac{\rho}{\sqrt{d}}$, it holds that $0 \leqslant \delta < \sqrt{d}h$ and:*

$$\mathbf{n} \cdot \mathbf{w} \geqslant \frac{1 - \sqrt{d}\frac{h}{\rho}}{1 - \frac{\delta}{\rho}} \geqslant 1 - \sqrt{d}\frac{h}{\rho} > 0 \qquad \textit{i.e.} \qquad \angle(\mathbf{n}, \mathbf{w}) \leqslant O\left(\sqrt{\frac{h}{\rho}}\right).$$

Proof. See Fig. 4 (right) for an illustration. Let B be the ball centered on $\mathbf{c} := \mathbf{x} - \rho\mathbf{n}$ and of radius ρ. Since ∂X has positive reach ρ, B lies inside X. Let P be the plane containing $\mathbf{y}$ and orthogonal to $\mathbf{w}$ and let $\mathbf{c}' := \pi_P(\mathbf{c}) = \mathbf{c} - ((\mathbf{c} - \mathbf{y}) \cdot \mathbf{w})\mathbf{w}$, i.e. the projection of $\mathbf{c}$ onto P. Letting $\delta := \|\mathbf{x} - \mathbf{y}\|$, we have $\|\mathbf{c} - \mathbf{y}\| = \rho - \delta$ since $\mathbf{c}, \mathbf{y}, \mathbf{x}$ are aligned.

$$\begin{aligned}(\mathbf{c}' - \mathbf{c}) \cdot \mathbf{w} &= (\mathbf{c} - ((\mathbf{c} - \mathbf{y}) \cdot \mathbf{w})\mathbf{w} - \mathbf{c}) \cdot \mathbf{w} \\ &= (\mathbf{y} - \mathbf{c}) \cdot \mathbf{w} = (\rho - \delta)\mathbf{n} \cdot \mathbf{w}. \end{aligned} \tag{5}$$

On one side, we know that $\mathbf{p} := \mathbf{c} + \rho\mathbf{w}$ belongs to B hence also to X. On the other side, Corollary 2 entails that $\mathbf{p}' := \mathbf{c}' + \sqrt{d}h\mathbf{w}$ does not belong to X and more precisely that $\phi_X(\mathbf{w}) < \mathbf{p}' \cdot \mathbf{w}$. We have

$$\begin{aligned}\mathbf{p} \cdot \mathbf{w} = (\mathbf{c} + \rho\mathbf{w}) \cdot \mathbf{w} \leqslant \phi_X(\mathbf{w}) < \mathbf{p}' \cdot \mathbf{w} = (\mathbf{c}' + \sqrt{d}h\mathbf{w}) \cdot \mathbf{w} \\ \Rightarrow \quad \rho - \sqrt{d}h < (\mathbf{c}' - \mathbf{c}) \cdot \mathbf{w} = (\rho - \delta)\mathbf{n} \cdot \mathbf{w}. \quad (\text{using}(5)) \end{aligned} \tag{6}$$

Corollary 2 also implies that $\delta < \sqrt{d}h$ (otherwise $\mathbf{x}$ would be outside X). Since $h < \rho/\sqrt{d}$ by hypothesis, we have $\delta < \rho$ and $\sqrt{d}h < \rho$, so the left hand side of

(5) is positive, and so is its right hand. This entails $\mathbf{n} \cdot \mathbf{w} > 0$. Moreover

$$\mathbf{n} \cdot \mathbf{w} > \frac{\rho - \sqrt{d}h}{\rho - \delta} = \frac{1 - \sqrt{d}h/\rho}{1 - \delta/\rho} \geqslant 1 - \sqrt{d}\frac{h}{\rho}. \tag{7}$$

Let $\alpha := \angle(\mathbf{n}, \mathbf{w})$. So $\cos \alpha > 1 - \sqrt{d}\frac{h}{\rho}$. From the above expression, it can be seen that α tends to zero as h tends to zero. Taylor expansion of $\arccos(1 - \sqrt{d}\frac{h}{\rho})$ around $h = 0$ gives $\alpha < \sqrt{\frac{2\sqrt{d}}{\rho}h} + O(h^{\frac{3}{2}})$, which concludes. □

The result is tight in convergence order, and we can exhibit a simple example where the constant is almost reached (gap is less than $2^{1/4} \approx 20\%$). Just take a disk X of radius $\rho = h(k + 1 - \epsilon)$, k a positive integer, ϵ an arbitrary small positive real number. The convex hull Y_h has a vertical edge symmetric about the x-axis going through the lattice point $(hk, 0)$. Due to Pythagoras theorem, its two vertices are close to $\pm(hk, h\sqrt{2k})$ (up to negligible terms). Let $\mathbf{x}$ be the closest point on ∂X to the upper vertex, its normal is $\mathbf{n} \approx \frac{(k,\sqrt{2k})}{\sqrt{k^2+2k}}$. The normal to edge is $\mathbf{w} = (1, 0)$. We get $\mathbf{n} \cdot \mathbf{w} = \frac{1}{\sqrt{1+2/k}}$, thus $\angle(\mathbf{n}, \mathbf{w}) \approx \sqrt{\frac{2}{k}} \approx \sqrt{\frac{2h}{\rho}}$, to compare with $\sqrt{\frac{2\sqrt{2}h}{\rho}}$ of the theorem.

5 Conclusion

In this paper, we have explored links between geometrical quantities estimated at the boundary of a convex set $X \subset \mathbb{R}^d$ and similar quantities on the boundary of the convex hull of the digitization of X. We have shown the proximity of the digitized convex hull to X in terms of Hausdorff distance and highlighted that convex hull vertices are much closer to ∂X than their incident faces, explaining the visual quality of convex hulls. Our main result is that the normal vector to each facet of the digital convex hull converges towards the normal vectors of the convex shape, for small enough gridstep h, with explicit convergence speed (proportional to $\sqrt{h/\rho}$ for smooth convex shapes). This result indicates that the normal vector to facets of the digitized convex hull could be used as a discrete normal estimator on digitized convex shapes, and more generally in convex or concave parts of digitized shapes.

This result is related to the properties of the projection function π_K onto the nearest point on a compact set K, where the difference between π_K and $\pi_{K'}$ is shown proportional to the square root of their Hausdorff distance [4,22]. Even if we have shown that vertices are closer than $O(h)$, this is not the case elsewhere on the shape. Our achieved result is then tight, with a constant much better than the one of Voronoi Covariance Measure [21,22], which also requires kernel integration.

The next step is to evaluate the practical accuracy of this normal estimator with respect to state-of-the art approaches [5,9,15]. Furthermore, normal vectors to facets could be combined, convolved or interpolated to design more accurate normal estimators. We intend to pursue this line of work in an extension or a subsequent paper.

Acknowledgments. This work is partially supported by the French National Research Agency within the StableProxies project (ANR-22-CE46-0006).

A Proofs of Some Properties

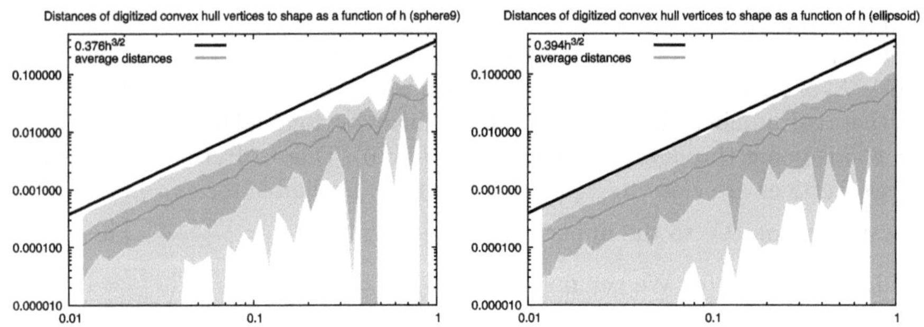

Fig. 5. Distance of digitized convex hull vertices to input `sphere9` shape (left) and `ellipsoid` shape (right), as a function of the gridstep h: convex hull vertices are much closer to the smooth input shape boundary than most digitized points, i.e. $O(h^{\frac{3}{2}})$. The expected constant is $2/\sqrt{\pi\rho}$, that is $2/\sqrt{9\pi} \approx 0.376$ for `sphere9` and $2/\sqrt{45\pi/\sqrt{30}} \approx 0.394$ for `ellipsoid`. Lightest green zone indicates min/max values, medium green zone indicates the standard deviation around the median value.

Corollary 1. *Let* $(\mathbf{z}, \mathbf{z}')$ *be a surfel of* X_h*. Let* $\mathbf{y}$ *be the nearest point on* ∂Y_h *to* $\mathbf{z}'$ *and let* σ *be a facet of* Y_h *containing* $\mathbf{y}$*, with normal vector* $\mathbf{n}_\sigma$*. We have:*

- *there exists* $\mathbf{x} \in \partial X \cap [\mathbf{y}, \mathbf{z}'[$*, that is at distance less than* h *from* $\mathbf{y}$*,*
- *for any* $\mathbf{n} \in \mathrm{N}_X(\mathbf{x})$*,* $\sin^2 \angle(\mathbf{n}, \mathbf{n}_\sigma) \leqslant \frac{1}{1+(r/h)^2}$*, if* $r := \mathrm{d}_{\mathbf{E}}(\mathbf{y}, \partial\sigma)$*.*

Proof. We have $\|\mathbf{y} - \mathbf{z}'\| \leqslant \|\mathbf{z} - \mathbf{z}'\| = h$ since $\mathbf{y}$ is the closest point of Y_h to $\mathbf{z}'$ and $\mathbf{z}$ belongs also to Y_h. Since $\mathbf{y} \in Y_h = \mathrm{CvxH}\,(X_h) \subset \mathrm{CvxH}\,(X) = X$ and $\mathbf{z}' \notin X_h$ so $\mathbf{z}' \notin X$, there must be a point $\mathbf{x}$ on the boundary of X on the straight segment joining $\mathbf{y}$ to $\mathbf{z}'$. It follows that $\|\mathbf{y}-\mathbf{x}\| < \|\mathbf{y}-\mathbf{z}'\| \leqslant h$, the strict relation coming from $\mathbf{x} \neq \mathbf{z}'$. The relation between normal vectors follows from Lemma 1, replacing ϵ with the distance h, if $\mathbf{y}$ lies in the interior of the facet σ. However if $\mathbf{y} \in \partial\sigma$, then $r := \mathrm{d}_{\mathbf{E}}(\mathbf{y}, \partial\sigma) = 0$, and the angle relation reduces to $\sin^2 \angle(\mathbf{n}, \mathbf{n}_\sigma) \leqslant 1$, which is always true. □

Corollary 2. *For all gridsteps* h*,* $0 < h < \frac{2\rho}{\sqrt{d}}$*, for* $\mathbf{y} \in \partial Y_h$ *and any normal vector* $\mathbf{w} \in \mathrm{N}_{Y_h}(\mathbf{y})$*, define* P *as the plane orthogonal to* $\mathbf{w}$ *and containing* $\mathbf{y}$*. then for any point* $\mathbf{y}' \in P$*, we have that* $\mathbf{y}' + t\mathbf{w}$ *is outside* X *for* $t \geqslant \sqrt{d}h$*.*

Proof. Indeed $\mathbf{w} \in \mathrm{N}_{Y_h}(\mathbf{y})$ implies by definition that $\phi_{Y_h}(\mathbf{w}) = \mathbf{y} \cdot \mathbf{w}$. The Hausdorff distance between Y_h and X is less than $\sqrt{d}h$ (Theorem 4), so

$$\phi_X(\mathbf{w}) < \phi_{Y_h}(\mathbf{w}) + \sqrt{d}h. \tag{8}$$

Let $\mathbf{y}'' := \mathbf{y}' + t\mathbf{w}$, for $t \geqslant \sqrt{d}h$. Observe that $\mathbf{y}' \cdot \mathbf{w} = \mathbf{y} \cdot \mathbf{w}$ since $\mathbf{y}'$ belongs to P. It follows that $\mathbf{y}'' \cdot \mathbf{w} = \mathbf{y}' \cdot \mathbf{w} + t\mathbf{w} \cdot \mathbf{w} = \mathbf{y} \cdot \mathbf{w} + t \geqslant \phi_{Y_h}(\mathbf{w}) + \sqrt{d}h$. Using (8), we get $\mathbf{y}'' \cdot \mathbf{w} > \phi_X(\mathbf{w})$ and $\mathbf{y}' \notin X$ by definition of the support function. □

References

1. Balog, A., Bárány, I.: On the convex hull of the integer points in a disc. In: Proceedings of the 7th Annual Symposium on Computational Geometry, pp. 162–165 (1991)
2. Bárány, I.: Random points, convex bodies, lattices. arXiv preprint math/0304462 (2003), iCM 2002
3. Brimkov, V., Coeurjolly, D., Klette, R.: Digital planarity a review. Discret. Appl. Math. **155**(4), 468–495 (2007)
4. Chazal, F., Cohen-Steiner, D., Mérigot, Q.: Boundary measures for geometric inference. Found. Comput. Math. **10**, 221–240 (2010)
5. Cuel, L., Lachaud, J.-O., Thibert, B.: Voronoi-based geometry estimator for 3D digital surfaces. In: Barcucci, E., Frosini, A., Rinaldi, S. (eds.) DGCI 2014. LNCS, vol. 8668, pp. 134–149. Springer, Cham (2014). https://doi.org/10.1007/978-3-319-09955-2_12
6. Esbelin, H.-A., Malgouyres, R.: Convergence of binomial-based derivative estimation for C^2 noisy discretized curves. In: Brlek, S., Reutenauer, C., Provençal, X. (eds.) DGCI 2009. LNCS, vol. 5810, pp. 57–66. Springer, Heidelberg (2009). https://doi.org/10.1007/978-3-642-04397-0_6
7. Federer, H.: Curvature measures. Trans. Am. Math. Soc. **93**(3), 418–491 (1959)
8. Fernique, T.: Generation and recognition of digital planes using multi-dimensional continued fractions. Pattern Recogn. **42**(10), 2229–2238 (2009)
9. Fourey, S., Malgouyres, R.: Normals estimation for digital surfaces based on convolutions. Comput. Graph. **33**(1), 2–10 (2009)
10. Huxley, M.N.: Area, lattice points, and exponential sums, vol. 13. Clarendon Press (1996)
11. Kim, C.E.: Digital convexity, straightness, and convex polygons. IEEE Trans. Pattern Anal. Mach. Intell. **4**(6), 618–626 (1982)
12. Klette, R., Rosenfeld, A.: Digital straightness a review. Discret. Appl. Math. **139**(1–3), 197–230 (2004)
13. Klette, R., Žunić, J.: Multigrid convergence of calculated features in image analysis. J. Math. Imaging Vision **13**, 173–191 (2000)
14. Lachaud, J.-O., Thibert, B.: Properties of gauss digitized shapes and digital surface integration. J. Math. Imaging Vision **54**(2), 162–180 (2015). https://doi.org/10.1007/s10851-015-0595-7
15. Lachaud, J.-O., Coeurjolly, D., Levallois, J.: Robust and convergent curvature and normal estimators with digital integral invariants. In: Najman, L., Romon, P. (eds.) Modern Approaches to Discrete Curvature. LNM, vol. 2184, pp. 293–348. Springer, Cham (2017). https://doi.org/10.1007/978-3-319-58002-9_9

16. Lachaud, J.O., Provençal, X., Roussillon, T.: Two plane-probing algorithms for the computation of the normal vector to a digital plane. J. Math. Imaging Vision **59**(1), 23–39 (2017)
17. Lachaud, J.-O., Vialard, A.: Geometric measures on arbitrary dimensional digital surfaces. In: Nyström, I., Sanniti di Baja, G., Svensson, S. (eds.) DGCI 2003. LNCS, vol. 2886, pp. 434–443. Springer, Heidelberg (2003). https://doi.org/10.1007/978-3-540-39966-7_41
18. Lachaud, J.O., Vialard, A., de Vieilleville, F.: Fast, accurate and convergent tangent estimation on digital contours. Image Vis. Comput. **25**(10), 1572–1587 (2007)
19. Lu, J.T., Roussillon, T., Lachaud, J.O., Coeurjolly, D.: Delaunay property and proximity results of the l-algorithm for digital plane probing. Theoret. Comput. Sci. **1011**, 114719 (2024)
20. Macbeath, A.M.: A theorem on non-homogeneous lattices. Ann. Math. **56**(2), 269–293 (1952)
21. Mérigot, Q., Ovsjanikov, M., Guibas, L.: Robust voronoi-based curvature and feature estimation. In: 2009 SIAM/ACM Joint Conference on Geometric and Physical Modeling, pp. 1–12 (2009)
22. Mérigot, Q., Ovsjanikov, M., Guibas, L.J.: Voronoi-based curvature and feature estimation from point clouds. IEEE Trans. Visual Comput. Graphics **17**(6), 743–756 (2010)
23. Minkowski, H.: Geometrie der zahlen. Teubner, Leipzig (1896),§30
24. Pottmann, H., Wallner, J., Yang, Y.L., Lai, Y.K., Hu, S.M.: Principal curvatures from the integral invariant viewpoint. Comput. Aided Geom. Des. **24**(8–9), 428–442 (2007)
25. Provot, L., Gérard, Y.: Estimation of the derivatives of a digital function with a convergent bounded error. In: Debled-Rennesson, I., Domenjoud, E., Kerautret, B., Even, P. (eds.) DGCI 2011. LNCS, vol. 6607, pp. 284–295. Springer, Heidelberg (2011). https://doi.org/10.1007/978-3-642-19867-0_24
26. Reeve, J.E.: On the volume of lattice polyhedra. Proc. Lond. Math. Soc. **3**(1), 378–395 (1957)
27. Schneider, R.: Convex bodies: the Brunn–Minkowski theory, vol. 151. Cambridge University Press (2013)
28. Stelldinger, P., Köthe, U.: Towards a general sampling theory for shape preservation. Image Vis. Comput. **23**(2), 237–248 (2005)
29. Stelldinger, P., Latecki, L.J., Siqueira, M.: Topological equivalence between a 3D object and the reconstruction of its digital image. IEEE Trans. Pattern Anal. Mach. Intell. **29**(1), 126–140 (2006)
30. Tellier, P., Debled-Rennesson, I.: 3D discrete normal vectors. In: Bertrand, G., Couprie, M., Perroton, L. (eds.) DGCI 1999. LNCS, vol. 1568, pp. 447–458. Springer, Heidelberg (1999). https://doi.org/10.1007/3-540-49126-0_35
31. Wills, M.D.: Hausdorff distance and convex sets. J. Convex Anal. **14**(1), 109–117 (2007)

Fast and Exact Visibility on Digitized Shapes and Application to Saliency-Aware Normal Estimation

Romain Negro(✉) and Jacques-Olivier Lachaud

University Savoie Mont Blanc, CNRS, LAMA, 73000 Chambéry, France
{romain.negro,jacques-olivier.lachaud}@univ-smb.fr

Abstract. Computing visibility on a geometric object requires heavy computations since it requires to identify pairs of points that are visible to each other, i.e. there is a straight segment joining them that stays in the close vicinity of the object boundary. We propose to exploit a specific representation of digital sets based on lists of integral intervals in order to compute efficiently the complete visibility graph between lattice points of the digital shape. As a quite direct application, we show then how we can use visibility to estimate the normal vector field of a digital shape in an accurate and convergent manner while staying aware of the salient and sharp features of the shape.

Keywords: Visibility · Geometric inference · Digital normal estimation · Digital geometry

1 Introduction

Visibility is a fundamental concept in computational geometry and digital topology, with applications ranging from computer vision (occlusions), geometric modeling (feature detection, geodesics) to computer graphics (path tracing, shadowmap). Visibility within polygons in the plane has been extensively studied in the litterature [9] and is still studied in higher dimensional spaces [20]. The general problem has a high complexity. For instance, the 3d visibility complex between n spheres and occluding spheres has a complexity $O(n^4)$ [6]. Exact 3d visibility between polygons and occluding polygons is often cast into 5d Plücker space [19], which nicely represents 3d lines and their respective positions. The computational cost remains high even with decomposition into coarse cells for speed up (several days of computation for 1M triangles at the time).

To increase efficiency, several authors have cast the visibility problem into a digital space, generally $\mathbb{Z}^3$ or its decomposition into cubical cells. Visibility is simplified as whether there is a connected digital straight line joining a pair of cells without occluding cell(s) (e.g. Soille [21] uses Bresenham segments). Coeurjolly *et al.* [4] use a non-symmetric visibility definition and preimage computations to speed up these algorithms. Chica [2] uses a half-cell erosion of the complementary space and Bresenham lines to decide visibility.

M. H. F. Wilkinson and J. Kosinka (Eds.): DGMM 2025, LNCS 16296, pp. 31–43, 2026.
https://doi.org/10.1007/978-3-032-09544-2_3

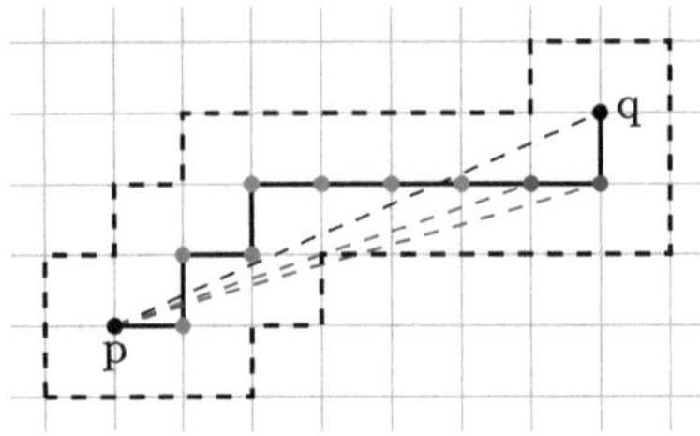

Fig. 2. The set of visible points in X from source p is not 1-connected.

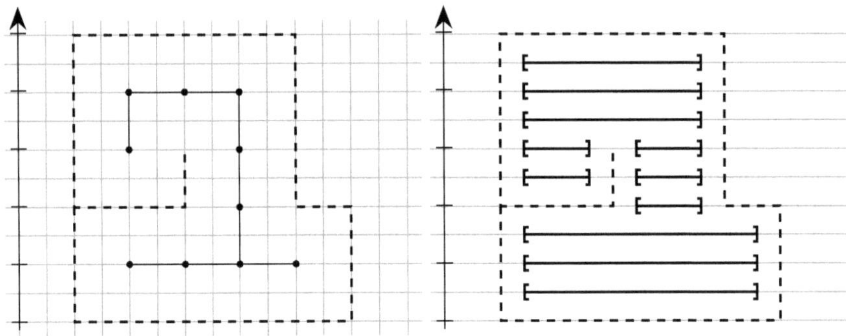

Fig. 3. Representation of a 2d cell complex (here the star of a given curve) as a lattice map. From 63 cells (with two coordinates) on the left, we get 11 intervals on the right.

3 Fast Computation Using Integer Intervals Intersections

As said in the introduction, any cell of $\mathcal{C}$ can be characterized as a point with integer coordinates, aka *lattice point*. For instance, a standard way is to double the coordinates of the centroid of the cell (often called its Khalimsky code). Now we have to represent an arbitrary set K of lattice points.

Let $(\mathbf{e}_j)_{j=1,\ldots,d}$ be the canonical basis of $\mathbb{Z}^d$. We may choose an arbitrary projection *axis* $j \in \{1, \ldots, d\}$. Letting π_j be the projector along this axis, we can collect all the cells of K that projects onto the same points. Let $p \in K$. Then $\pi_j^{-1}(p) \cap \mathrm{Star}(K)$ is a set of cells having all the same coordinates except along coordinate j. This set can be ordered increasingly and stored as a list of integer intervals (see Fig. 3 for an illustration).

A *sequence of intervals* is an ordered list of intervals $L = ([a_i, b_i])_{i=1,\ldots,n}$ such that $a_i \in \mathbb{Z}, b_i \in \mathbb{Z}, b_i + 1 < a_{i+1}$, i.e., disjoint intervals with at least a missing integer. We denote by $\mathbb{L}$ the set of sequences of intervals. An integer k belongs to L iff there exists i such that $a_i \leqslant k \leqslant b_i$. For a finite set of integers, there is a unique sequence of intervals representing it.

A *lattice map along axis* j is a set of pairs (q, L_q), with $q \in \mathbb{Z}^{d-1}$ and $L_q \in \mathbb{L}$. The lattice map M_K *represents* K when, for any point $p \in K$, the pair $(\pi_j(p), L_{\pi_j(p)})$ exists in M_K and $p_j \in L_{\pi_j(p)}$, with p_j the j-th coordinate of p, and reciprocally, $\forall (q, L_q) \in M_K, \forall i \in L_q, q + i\mathbf{e}_j \in K$. A lattice map is thus made of pairs (q, L_q), where q is the *shift* while L_q is the *intervals*. For a point q which is a shift of M_K, we write $M_K[q]$ for the corresponding intervals L_q.

The global algorithm for computing visibility is given in Algorithm 1. Given a digital input surface C (a subcomplex), we compute the lattice map Ω of Star (C) once, generally along the axis where C is the most elongated. In order to compute visibility up to a given distance r, we consider all non-null primitive vectors D with infinity norm no greater than r. For each vector $\mathbf{v}$, we compute the lattice map $M_{\mathbf{v}}$ of its intersected cells (as a vector from $\mathbf{0}$ to $\mathbf{v}$ in $\mathbb{R}^d$). The visibility along the direction $\mathbf{v}$ reduces to computing for each pair of $M_{\mathbf{v}}$ the possible translations of intervals in Ω, and then intersecting all the translations.

Algorithm 1 Given a subcomplex C and an integer r, returns the visibility from every point of C up to distance r. The main axis is supposed to be z, while x, y are the auxiliary axes.

```
function VISIBILITY(C: Subcomplex, r: Integer)
    Ω ← M_STAR(C)                     ▷ Lattice map of the star of input subcomplex
    Directions ← GETALLPRIMALDIRECTIONS(r)
    V : vector of boolean ← [0, ..., 0]     ▷ length Size(Directions) × #C.pointels
    low, high ← BOUNDINGBOXZ(Ω)
    for all v in Directions do
        for all shift S in Ω do
            R ← [low, high]
            for all pair P in M_STAR([0,v]) do
                R ← R ∩ TRANSLATIONS(P.intervals, Ω[S + P.shift])
            UPDATEVISIBILITY(V, R)
    return V
```

Figure 4 shows an example of execution of this algorithm in an elementary case. It reduces to determining all the possible inclusions by translation of sequences of intervals in another sequence of intervals (note that for pairwise visibility, it is enough to consider all the possible inclusions by translation of one interval in another sequence of intervals), and then intersecting progressively the results. Of course, the process is stopped as soon as the intersection is empty. Some results of visible points are displayed on Fig. 5 on 3d shapes.

To efficiently compute the intersection of two sequences of intervals K and L (Algorithm 2), we go through the first 2 unvisited elements of the 2 lists. If they do not overlap (i.e. the start and end of the 1st interval are both smaller than the start of the other), then we can discard the smaller interval. Else, we construct the intersection of these 2 intervals as one of the resulting intervals. We then can skip the interval with the smallest end. If both intervals have the same end, then both intervals can be skipped. Doing this computation until one of the lists is empty returns the intersection of both lists of intervals.

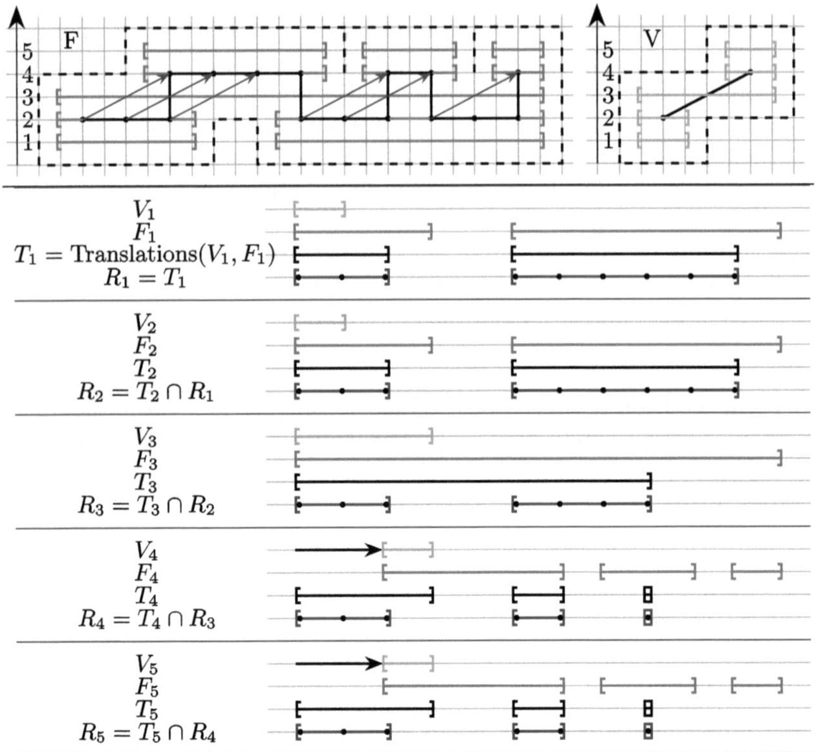

Fig. 4. Evolution of the visibility check algorithm for a (2, 1) vector. Green is the vector lattice map, black is the figure lattice map, red are the current intervals of positions where the visibility is still possible. The last red intervals are the visible positions. We travel the lattice maps from bottom-up and the found visibilities are drawn on the uppest figure. (Color figure online)

Algorithm 2 Given 2 lists of integer intervals K and L, returns $K \cap L$

---

function INTERSECTION $(\cap)$(K, L: Intervals)
    $R \leftarrow \emptyset$; $k \leftarrow 0$; $l \leftarrow 0$;
    while $k < K.nbIntervals$ & $l < L.nbIntervals$ **do**
        $[a, b] \leftarrow K[k]$; $[c, d] \leftarrow L[l]$
        $e \leftarrow \max(a, c)$; $f \leftarrow \min(b, d)$;
        if $e \leq f$ **then** $R.append([e, f])$;
        if $b \leq d$ **then** $k \leftarrow k + 1$;
        if $d \leq b$ **then** $l \leftarrow l + 1$;
    return R

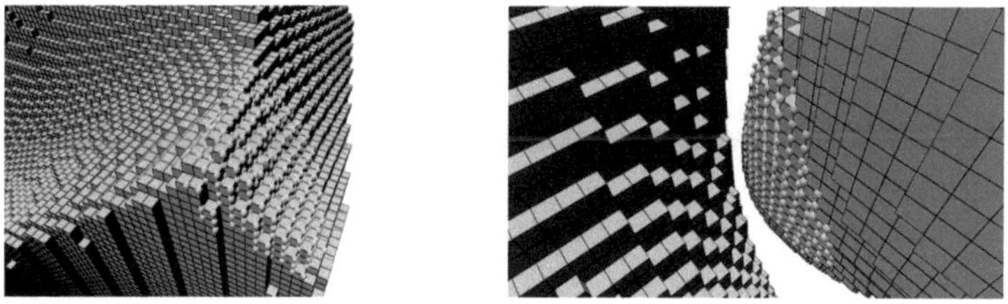

Fig. 5. Examples of visibilities from a source point: (left) the visibility is stopped at the sharp edge, (right) the visibility does not cross the gap between the two parts.

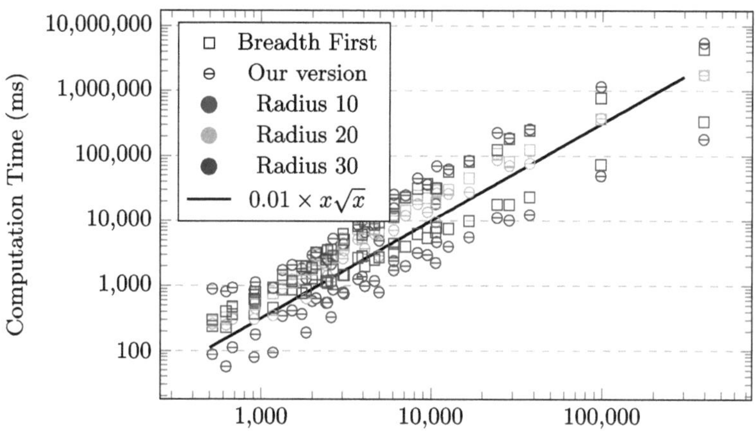

Fig. 6. Computation time of visibility as a function of the number of pointels. We compare the running time of the naive breadth first visibility algorithm against our algorithm using intervals. We input the same maximal radius to both versions (10, 20, 30) and test it on 6 different figures (goursat, torus, rcube, sphere9, leopold, "D20") with various gridsteps, (# pointels ranging from 520 to 390235).

We have plotted on Fig. 6 the running times to compute global visibility of both [13, Algorithm 3] and our algorithm. Timings are very similar. Our algorithm is faster for smaller maximal radii (up to 20), while the other is slightly faster for bigger radii. We have also measured the (discrete) average visibility distance along digitization of smooth shapes (Fig. 7). It follows a law proportionnal to $\sqrt{h}$. As shown in the next section, for real 3d images, a maximal radius below 10 is sufficient for normal and curvature estimation.

4 Saliency-Aware Normal and Curvature Estimation

Saliency-Aware Normal Estimator. We propose a new normal estimator on digital surfaces, which uses the visibility of the point of interest while taking into account a user-given scale σ. If σ is proportionnal to the average distance between visible points (so some $\Theta\left(\sqrt{h}\right)$), then this estimator is observed to be multigrid

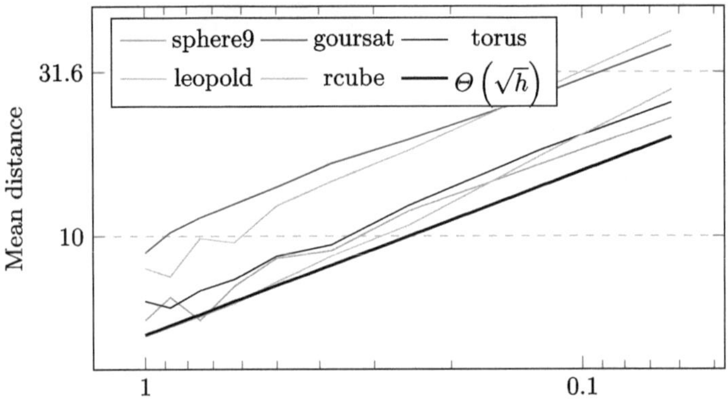

Fig. 7. Mean distance of visibility as a function of the gridstep h (see text).

convergent along digitization of smooth shapes. However, since the computation window is limited by the visibility, it better approximates the geometry near sharp or salient features.

More precisely, we choose a Gaussian function $w_\sigma(x) := e^{-\frac{x^2}{2\sigma^2}}$ as a weight kernel. Let V_p be the set of visible points from the point p. Let c_p be the weighted centroid of the visible points around p, i.e. $c_p := \frac{\sum_{q \in V_p} w_\sigma(\|q-p\|)q}{\sum_{q \in V_p} w_\sigma(\|q-p\|)}$. We form the weighted covariance matrix $\mathcal{V}_p$ of the points V_p as:

$$\mathcal{V}_p = \sum_{q \in V_p} w_\sigma(\|q-p\|)(q-c_p)(q-c_p)^T. \tag{1}$$

The *visibility normal* $\boldsymbol{n}(p)$ of point p at scale σ is defined as the first eigenvector of the covariance matrix $\mathcal{V}_p$ of the visible points, corresponding to its smallest eigenvalue. Its orientation is chosen to point in the same direction as the average of the trivial normals to the surfels touching the point p.

Experimental Validation. Figure 9 shows an example of our normals compared to normals computed with the integral invariant (II) estimator [10]. In order to compare quantitatively the different normal estimators, we compute the root-mean-square error (RMSE) and the maximum error $E_{\max}$ of the normals computed on a digital surface with respect to the normals computed on the continuous surface (Fig. 8). We use 4 different digital surfaces (ellipsoid, goursat, rcube and sphere9). We see that the RMSE and $E_{\max}$ errors of our estimator are convergent with respect to the grid resolution with a rate of convergence of $h^{\frac{2}{3}}$ for the RMSE and $h^{\frac{1}{2}}$ for $E_{\max}$.

Application to Curvature Estimation. In order to illustrate better that the visibility normals better take into account the sharp features of a digital surface while staying meaningfull in digitization of smooth parts, we compare the curvature estimates induced by different discrete normal estimators. We exploit

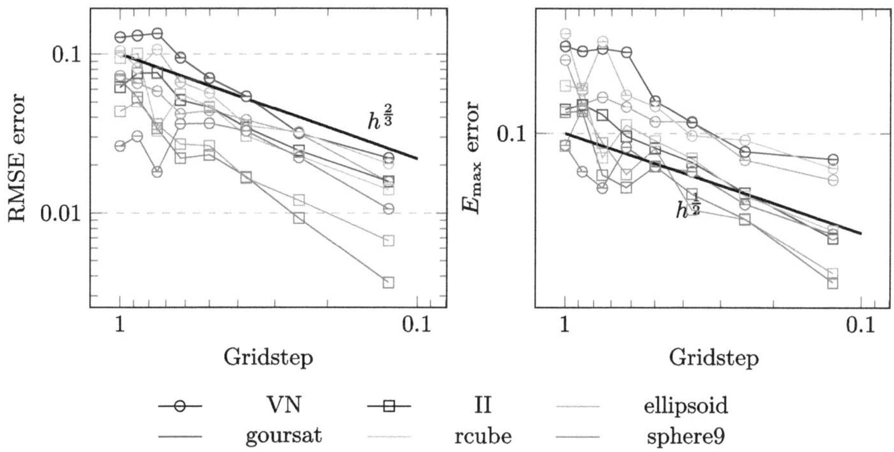

Fig. 8. Error RMSE and $E_{\max}$ as a function of grid resolution.

the *Corrected Normal Current (CNC) estimator*, whose theory is detailed in [15] and which can estimate all kinds of curvatures given positions and normals. It produces state-of-the-art curvature estimates on polygonal surfaces [16], point clouds [14], and even outperforms Integral Invariant curvature estimates [3] on digital surfaces [15]. Its implementation is also available in DGtal.

Figure 10 displays the results of CNC curvature estimations on the digitization of a piecewise smooth shape "Talking D20", taken from Thingi10K. We compare the differences obtained by just changing the discrete normal estimator (available in the DGtal library): (column II normals) Integral Invariant normals with $r = 6$, (column CTriv normals) convolved trivial normals with $r = 6$, (column Visibility normals) our proposed estimator (VN) with $\sigma = 4$ and maximal visibility distance 2σ. These parameters were chosen so that the respective computation windows of the different estimators are approximately the same.

The II estimator is good along digitization of smooth or flat parts of the dice, but presents curious artefacts near sharp features induced by the hollowed out numbers: this is due to the nature of II normal estimates, which computes a PCA of a ball centered on the point of interest and may include points on the other side of the saliencies.

The CTriv estimator only performs averaging of trivial normal directions along the surface. It behaves much better than II near features (although it tends to smooth curvatures) since it does not use information from across the gap. However, some oscillations of the curvatures are distinguishable especially along the flat parts, and some curvature estimates are erroneous on smoother parts (like the edge above 6 for κ_2 or several dice vertices).

The VN estimator takes the best of the two previous approaches. It remains precise and stable on smooth and flat regions, while perfectly delineating sharp features and holes, whatever the kind of estimated curvatures. Its drawback is the computation time, which is $52s$ for a maximal visibility distance 8, compared to

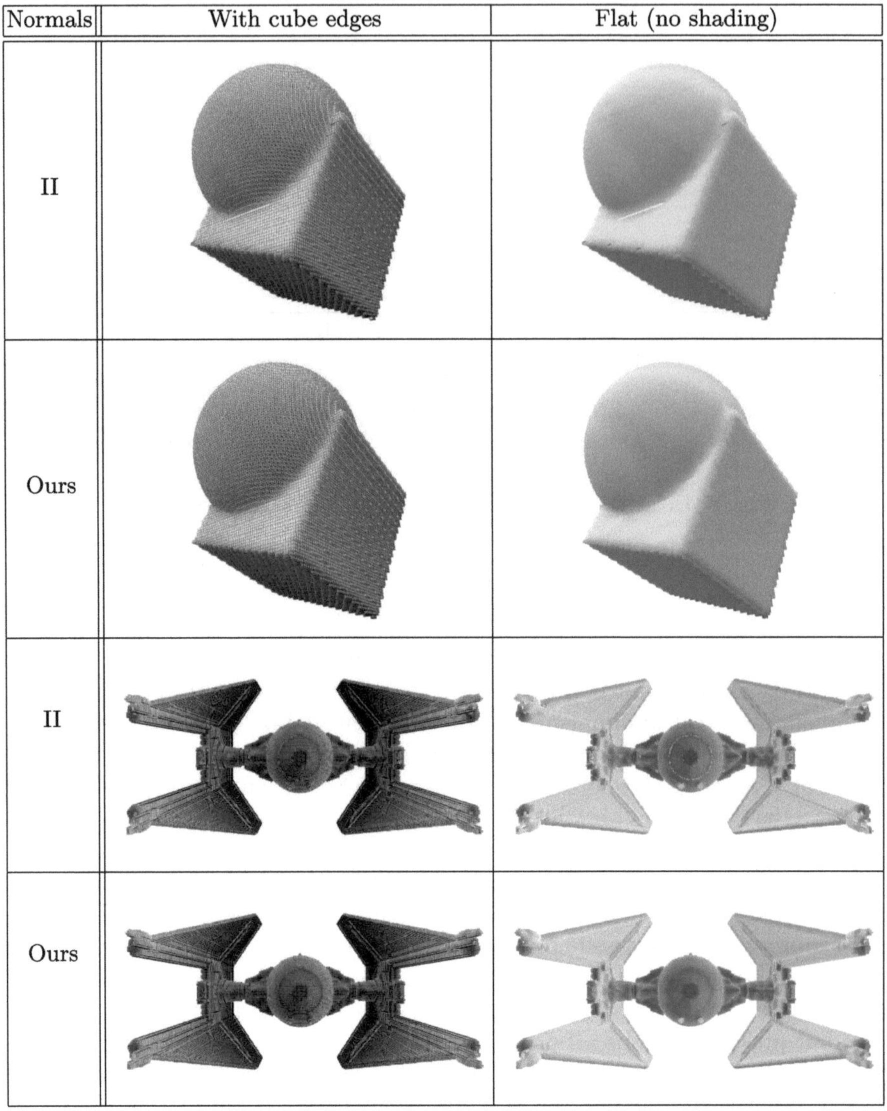

Fig. 9. Examples of normals computed on a cps, then on a tie as a color map. First set displays II normals, second set uses the visibility algorithm to compute them. The normals are colored according to their orientation as $0.5 \cdot (n_i + \mathbb{1})$. The right images are the smooth version of the left images. "Integral Invariant normals" are computed using a radius of $r = 4.5$, while our normal estimator are computed using a deviation of $\sigma = 4$, and so a radius of $r = 2\sigma = 8$.

2.3s for II and $0.1s$ for CTriv. Note that the computation time falls back to $33s$ for a maximal distance of 6 (and same $\sigma = 4$) while the result is visually indistinguishable.

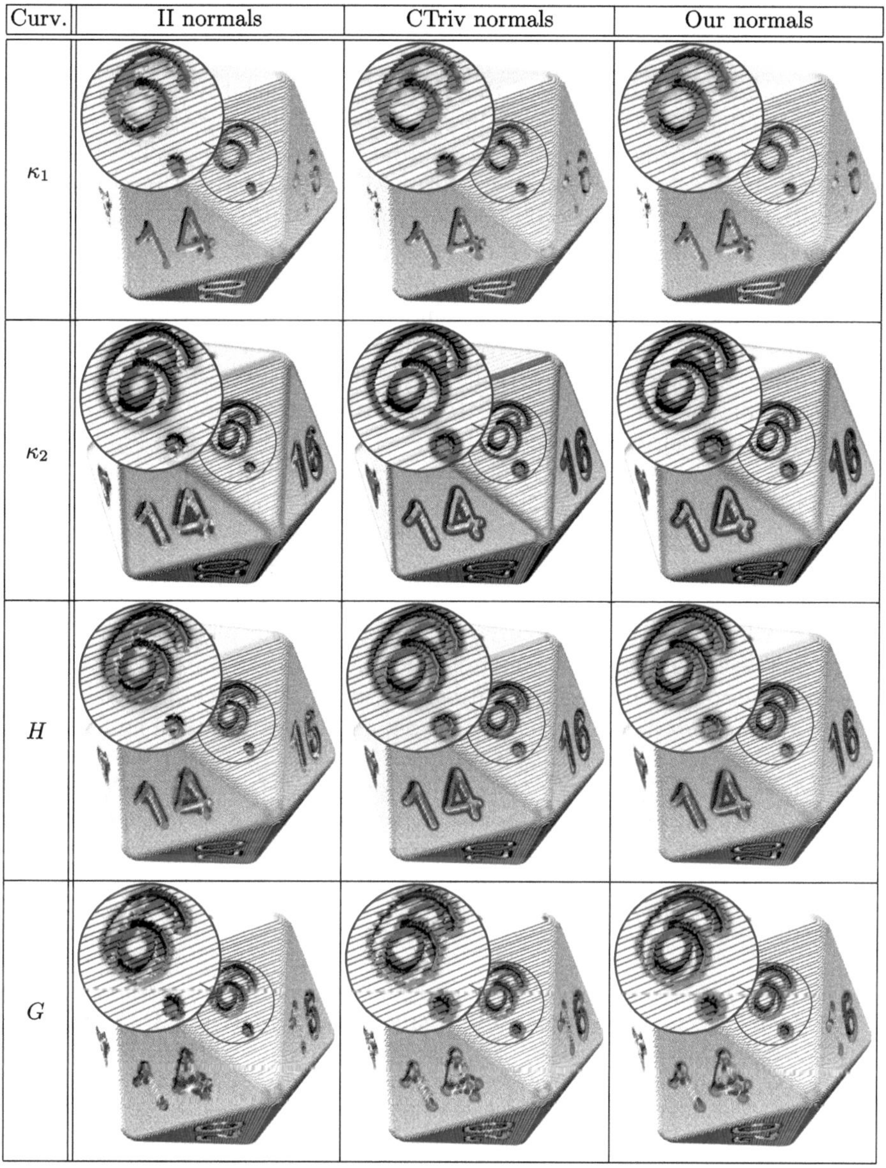

Fig. 10. Estimation of curvatures using Corrected Normal Current estimators [15] with a measure of radius 2 for the shape "Dice-20" of Thingi10K database at resolution 256^3. This estimator is parameterized by an input normal estimator: first column by "Integral Invariant normals" (radius $r = 6$), second column by "Convolved Trivial normals" (radius $k = 6$), third column by our presented normal estimator (deviation $\sigma = 4$, radius $r = 2\sigma = 8$). Per row are displayed the estimated curvatures in order: first and second principal curvatures κ_1 and κ_2, mean curvature H, Gaussian curvature G. The range of curvature between deep blue and deep red is $[-0.1, 0.1]$ with white color equal to 0. (Color figure online)

5 Conclusion

We have proposed a new algorithm that computes the visibility between all pairs of points on a digital surface. It uses the specificity of a structure to represent sets of lattice points or cells, a mapping from projected coordinates to list of integer intervals. Compared to a classical breadth-first approach to compute visibility, it provides the exact visibility result, even when the set of visible points from a source is disconnected. As shown by experiments the running times are comparable with the breadth-first algorithm, and even faster for small maximal visibility distance, which are typical when processing standard 3d images up to 512^3. We have used visibility to define a discrete tangent estimator that respects salient features while giving good approximations on digitization of smooth parts. The obtained normal field is more suitable to curvature estimation than the classical Integral Invariant normal estimator.

We have several ideas for possible future works. First, it is possible to speed up the identification of non visibility by a coarse-to-fine approach and a precomputation of a pyramid of covering cells. We would like to explore if we can build such a pyramid variant of our algorithm, considering the primitive vectors as a hierarchy. Second we would like to prove the muligrid convergence of our normal estimator, which is observable on experiments. Last, our algorithm is not limited to pairwise visibility, but is more an algorithm of exact pattern matching. We would like to explore new applications of pattern identification along digital surfaces (like corner detection, locally convex zones identification, etc.).

Acknowledgments. This work is partially supported by the French National Research Agency within the StableProxies project (ANR-22-CE46-0006).

References

1. Charrier, E., Lachaud, J.-O.: Maximal planes and multiscale tangential cover of 3D digital objects. In: Aggarwal, J.K., Barneva, R.P., Brimkov, V.E., Koroutchev, K.N., Korutcheva, E.R. (eds.) IWCIA 2011. LNCS, vol. 6636, pp. 132–143. Springer, Heidelberg (2011). https://doi.org/10.1007/978-3-642-21073-0_14
2. Chica, A.: Visibility-based feature extraction from discrete models. In: Proceedings of the 2008 ACM Symposium on Solid and Physical Modeling, pp. 347–352 (2008)
3. Coeurjolly, D., Lachaud, J.O., Levallois, J.: Multigrid convergent principal curvature estimators in digital geometry. Comput. Vis. Image Underst. **129**, 27–41 (2014)
4. Coeurjolly, D., Miguet, S., Tougne, L.: 2D and 3D visibility in discrete geometry: an application to discrete geodesic paths. Pattern Recogn. Lett. **25**(5), 561–570 (2004)
5. Cuel, L., Lachaud, J.-O., Thibert, B.: Voronoi-based geometry estimator for 3D digital surfaces. In: Barcucci, E., Frosini, A., Rinaldi, S. (eds.) DGCI 2014. LNCS, vol. 8668, pp. 134–149. Springer, Cham (2014). https://doi.org/10.1007/978-3-319-09955-2_12

6. Durand, F., Drettakis, G., Puech, C.: The 3D visibility complex. ACM Trans. Graph. (TOG) **21**(2), 176–206 (2002)
7. Feschet, F., Tougne, L.: Optimal time computation of the tangent of a discrete curve: application to the curvature. In: Bertrand, G., Couprie, M., Perroton, L. (eds.) DGCI 1999. LNCS, vol. 1568, pp. 31–40. Springer, Heidelberg (1999). https://doi.org/10.1007/3-540-49126-0_3
8. Fourey, S., Malgouyres, R.: Normals estimation for digital surfaces based on convolutions. Comput. Graph. **33**(1), 2–10 (2009)
9. Ghosh, S.K.: Visibility Algorithms in the Plane. Cambridge University Press (2007)
10. Lachaud, J.-O., Coeurjolly, D., Levallois, J.: Robust and convergent curvature and normal estimators with digital integral invariants. In: Najman, L., Romon, P. (eds.) Modern Approaches to Discrete Curvature. LNM, vol. 2184, pp. 293–348. Springer, Cham (2017). https://doi.org/10.1007/978-3-319-58002-9_9
11. Lachaud, J.O., Vialard, A., de Vieilleville, F.: Fast, accurate and convergent tangent estimation on digital contours. Image Vis. Comput. **25**(10), 1572–1587 (2007)
12. Lachaud, J.-O.: An alternative definition for digital convexity. In: Lindblad, J., Malmberg, F., Sladoje, N. (eds.) DGMM 2021. LNCS, vol. 12708, pp. 269–282. Springer, Cham (2021). https://doi.org/10.1007/978-3-030-76657-3_19
13. Lachaud, J.-O.: An alternative definition for digital convexity. J. Math. Imaging Vision 1–18 (2022). https://doi.org/10.1007/s10851-022-01076-0
14. Lachaud, J.O., Coeurjolly, D., Labart, C., and, P.R.: Lightweight curvature estimation on point clouds with randomized corrected curvature measures. Comput. Graph. Forum **42**(5), e14910 (2023)
15. Lachaud, J.O., Romon, P., Thibert, B.: Corrected curvature measures. Discret. Comput. Geom. **68**(2), 477–524 (2022). https://doi.org/10.1007/s00454-022-00399-4
16. Lachaud, J.O., Romon, P., Thibert, B., Coeurjolly, D.: Interpolated corrected curvature measures for polygonal surfaces. Comput. Graph. Forum **39**(5), 41–54 (2020). https://doi.org/10.1111/cgf.14067
17. Marêché, A., Debled-Rennesson, I., Feschet, F., Ngo, P.: A parameter-free normal estimator on digital surfaces. In: International Conference on Intelligent Systems and Pattern Recognition, pp. 130–142. Springer (2024)
18. Nguyen, T.P., Debled-Rennesson, I.: A discrete geometry approach for dominant point detection. Pattern Recogn. **44**(1), 32–44 (2011)
19. Nirenstein, S., Blake, E., Gain, J.: Exact from-region visibility culling. In: Proceedings of 13th Eurographics Workshop on Rendering (2002)
20. O Rourke, J.: Visibility. In: Handbook of Discrete and Computational Geometry, pp. 875–896. Chapman and Hall/CRC (2017)
21. Soille, P.: Generalized geodesy via geodesic time. Pattern Recogn. Lett. **15**(12), 1235–1240 (1994)
22. Soille, P., et al.: Morphological Image Analysis: Principles and Applications, vol. 2. Springer (1999)

Convex and Concave Decomposition of Digitized Shapes Using Plane Probing and Visibility

Jacques-Olivier Lachaud[1,2] and Tristan Roussillon[2(✉)]

[1] University Savoie Mont Blanc, CNRS, LAMA, 73000 Chambéry, France
jacques-olivier.lachaud@univ-smb.fr
[2] Claude Bernard Lyon 1 University, CNRS, INSA, LIRIS, 69000 Lyon, France
tristan.roussillon@liris.cnrs.fr

Abstract. In this paper, we consider the geometrical analysis of oriented digital surfaces, which form the boundary of connected voxel sets. We present a method for the detection and reconstruction of its convex and concave parts. The proposed method composes two tools: the first one is a probing approach for the extraction of locally extremal points, the second one, based on the notion of visibility, joins some pairs of points whenever the straight segment between them stays sufficiently close to the input surface without crossing it. Finally, outer and inner candidate segments are assembled separately into facets, thus building polyhedral approximations of the local convex and concave zones.

Keywords: surface approximation · digital geometry · digital convexity

1 Introduction

Numerous studies focus on capturing the geometry of 3D shapes based on sampled data, whether in the form of point clouds, polygonal meshes, or digital surfaces. In this paper, we consider digital surfaces, defined as boundaries of connected voxel sets. They may be seen as closed, oriented, quadrangular meshes with vertices of integer coordinates.

Given a digital surface, we present a method for the detection and reconstruction of its convex and concave parts. The geometry of the convex hull is exactly recovered when the input data is digitally convex. This is therefore a first step towards a polygonal reconstruction that respects the convex and concave parts, which would be useful for many tasks in geometry processing.

Several methods could be considered for our purposes. Global approaches, e.g., based on the α-shape [2,11] or on a decimation strategy [5], are not expected to better capture the local geometry of the surface than local approaches. In the field of digital geometry, numerous works focus on the recognition of planar parts (see, e.g., [1]). In particular, one may use a plane-probing algorithm [10] to extract some tangent triangles. However, even when the input data is

M. H. F. Wilkinson and J. Kosinka (Eds.): DGMM 2025, LNCS 16296, pp. 44–59, 2026.
https://doi.org/10.1007/978-3-032-09544-2_4

digitally convex, this approach is not able to recover all facets of the convex hull. Alternatively, we propose in this paper a probing algorithm to address the different problem of deciding whether a point is extremal or not.

In 2D, very good polygonal reconstructions of digital curves are obtained by local convex hull computations, done either explicitly [4,14,15] or implicitly [6,9,12]. We know only one attempt to extend this approach to 3D [13], but the proposed algorithm only works for a limited class of input. We go further in this direction, using the more recent notion of full convexity and tangency [7,8].

The outline is as follows. Section 2 defines the notion of convex and concave visibilities along a digital surface and establishes their links to convexity. Section 3 presents a probing approach to speed up the detection of convex and concave vertices. Section 4 explains the global reconstruction algorithm of convex and concave zones. Section 5 concludes the paper and offers some perspectives.

2 Convex and Concave Visibility

The input data is a set K of d-dimensional cells, i.e. *voxels*. Our objective is to identify the convex and concave zones of the digital boundary of K, which is an oriented digital surface.[1] We introduce the concept of *visibility* within exterior and interior parts of the digital surface, a kind of oriented alternative to the *tangency* to a digital set [8, Definition 4]. Remarkably, Theorem 2 shows that convex hull edges and facets of K are indeed visible when K is digitally convex.

Background Definitions. There is a natural grid cell decomposition of the Euclidean space $\mathbb{R}^d$ with open unit cells (segments, squares, cubes, etc.) such that their vertices are exactly the digital space $\mathbb{Z}^d$. We denote this cubical complex $\mathcal{C}^d$, while its k-dimensional cells are denoted by $\mathcal{C}^d_k$. For any subset Y of $\mathbb{R}^d$, the *star* of Y is $\mathrm{Star}\,(Y) := \{c \in \mathcal{C}^d, \bar{c} \cap Y \neq \emptyset\}$, where $\bar{c}$ is the topological closure of c. The *cover* of Y is $\mathrm{Cover}\,(Y) := \{c \in \mathcal{C}^d, c \cap Y \neq \emptyset\}$. Obviously the cover of a set is included in its star. The 0-dimensional cells of $\mathcal{C}^d$ are identified with $\mathbb{Z}^d$ and are called (lattice) points, or often *pointels* in the literature when thought of as vertices of higher dimensional cells. Given a subset Y of $\mathbb{R}^d$, its *convex hull* is denoted by $\mathrm{CvxH}\,(Y)$. The *set of extremal points* of a convex set C is denoted by $\mathrm{Extr}\,(C)$ (when C is a polytope, they are often called *vertices*). For a grid cell c, which is a convex set, we have $\mathrm{Extr}\,(c) = \bar{c} \cap \mathbb{Z}^d$. We recall one characterization of full convexity [7,8], and the related notion of tangency:

Definition 1. *A digital set* $X \subset \mathbb{Z}^d$ *is* digitally convex *iff* $\mathrm{CvxH}\,(X) \cap \mathbb{Z}^d = X$. *It is* fully convex *iff* $\mathrm{Star}\,(\mathrm{CvxH}\,(X)) = \mathrm{Star}\,(X)$. *A digital subset* $A \subset X \subset \mathbb{Z}^d$ *is* tangent *to* X *whenever* $\mathrm{Star}\,(\mathrm{CvxH}\,(A)) \subset \mathrm{Star}\,(X)$. *The elements of* A *are called* cotangent *in* X.

Otherwise said, if we consider two pointels $A = \{p, q\}$ of the digital surface, with X its set of pointels, they are cotangent in X whenever the straight line

[1] A more generic definition of oriented surface could be used [3], but the presentation is made easier if we consider the surface to be the boundary of a set of cells.

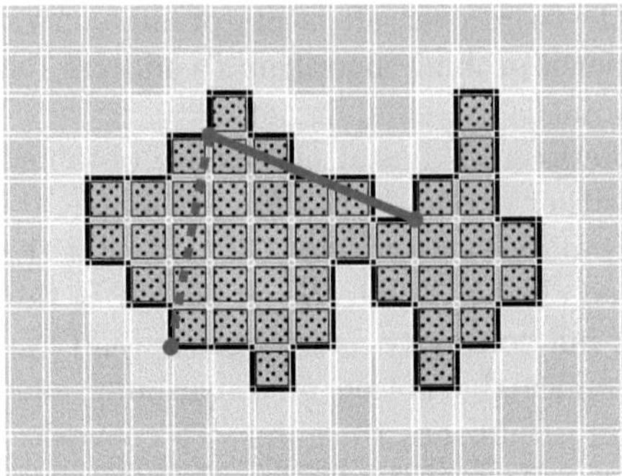

(a) Example of tangent edges in K (●—●) and non tangent edges in K (● -●)

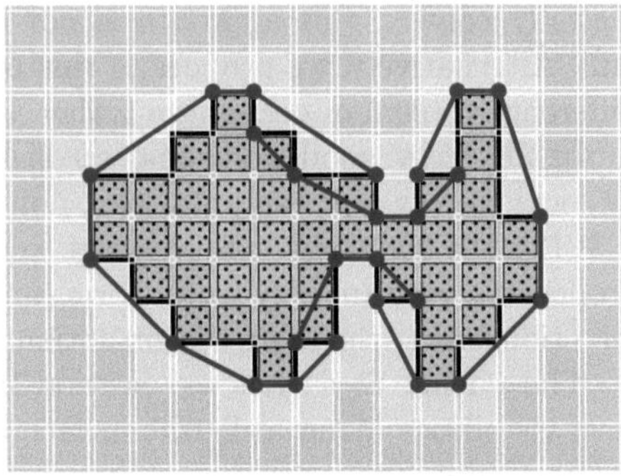

(b) Examples of convex K-visible edges (●—●) concave K-visible edges (●—●)

Fig. 1. Decomposition of the cellular grid $\mathcal{C}^d$ into disjoint set of cells and illustration of tangency and visibility. The input of set of voxels K are hatched as ⊠, and $\mathcal{C}^d = \text{Outer}(K) \cup \text{Out}(K) \cup \text{Bd}(K) \cup \text{In}(K) \cup \text{Inner}(K)$. The set X_K of 0-cells of $\text{Cl}(K)$ is decomposed into the inner pointels I_K (the 0-cells of $\text{Inner}(K)$) and the boundary pointels Z_K (the 0-cells ▪ of $\text{Bd}(K)$). (Color figure online)

segment $[pq]$ touches cells that are also touched by X. Tangency defines a kind of visibility, since p and q can "see" each other if and only if their line of sight stays close to the input digital surface. However this line of sight may cross the digital surface. Therefore we exploit the fact that we know where is the interior of the surface (the input set of interior voxels K), to keep this line of sights either exterior or interior. Convex vertices (resp. concave vertices) are visible to each other in the exterior (resp. interior) of the surface.

Let $K \subset \mathcal{C}^d_d$ be the input set of interior voxels to the surface. For any subset L of $\mathcal{C}^d$, its (combinatorial) *closure* $\text{Cl}(L)$ is $\{c \in \mathcal{C}^d, \text{Star}(c) \cap L \neq \emptyset\}$. For simplicity, the k-dimensional cells of a subset L of $\mathcal{C}^d$ are denoted by L_k. We denote by X_K the 0-cells of $\text{Cl}(K)$. The digital set X_K is subdivided into two disjoint subsets: the set of *inner pointels* I_K is defined as $\{x \in X_K, \text{Star}(x) \subset \text{Cl}(K)\}$ and the set of *boundary pointels* Z_K is $X_K \setminus I_K$. We subdivide the cubical complex $\mathcal{C}^d$ into several disjoint subsets (see Fig. 1).

$$\begin{aligned}
\text{Outer}(K) &:= \{c \in \mathcal{C}^d, \text{Extr}(c) \subset (\mathbb{Z}^d \setminus X_K)\},\\
\text{Out}(K) &:= \{c \in \mathcal{C}^d, \text{Extr}(c) \cap (\mathbb{Z}^d \setminus X_K) \neq \emptyset \text{ and } \text{Extr}(c) \cap (Z_K) \neq \emptyset\},\\
\text{Bd}(K) &:= \{c \in \mathcal{C}^d, \text{Extr}(c) \subset Z_K\},\\
\text{In}(K) &:= \{c \in \mathcal{C}^d, \text{Extr}(c) \cap (Z_K) \neq \emptyset \text{ and } \text{Extr}(c) \cap I_K \neq \emptyset\},\\
\text{Inner}(K) &:= \{c \in \mathcal{C}^d, \text{Extr}(c) \subset I_K\}.
\end{aligned}$$

The analogy of $\text{Bd}(K)$ with the topological boundary is clear, since it represents the cells whose star touches K and the complement of K.

Visibility and Local Convexity/Concavity. As said above, tangency is not restricted enough to capture local convex edges and facets of a digital set. We introduce convex/concave visibility for this more specific purpose (see also Fig. 1).

Definition 2. *The elements of a non-empty set of points* $A := \{p_1, \ldots, p_n\}$ *of* Z_K *are* convex (resp. concave) K-visible *if and only if* $\mathrm{Cover}(\mathrm{CvxH}(A)) \subset \mathrm{Out}(K) \cup \mathrm{Bd}(K)$ *(resp.* $\mathrm{Cover}(\mathrm{CvxH}(A)) \subset \mathrm{In}(K) \cup \mathrm{Bd}(K)$*).*

The convex hull of the points visible from a point x of Z_K tells if x is salient:

Definition 3. *Let* $x \in Z_K$. *The point* x *is a* convex (resp. concave) corner *iff exactly one cell of* K *(resp. exactly one cell of* $\mathcal{C}_d^d \setminus K$*) is incident to* x. *The* convex visibility cone $\mathcal{V}_K^{\mathrm{Out}}(x)$ *of* x *is the set of points* y *of* Z_K *such that* x, y *are convex* K*-visible. The* concave visibility cone $\mathcal{V}_K^{\mathrm{In}}(x)$ *of* x *is the set of points* y *of* Z_K *such that* x, y *are concave* K*-visible. The point* x *is* locally convex (resp. concave) on the boundary of K *iff* $x \in \mathrm{Extr}\left(\mathrm{CvxH}\left(\mathcal{V}_K^{\mathrm{Out}}(x)\right)\right)$ *and* x *is a convex corner, resp.* $x \in \mathrm{Extr}\left(\mathrm{CvxH}\left(\mathcal{V}_K^{\mathrm{In}}(x)\right)\right)$ *and* x *is a concave corner.*

Note that we require x to be a convex (resp. concave) corner in order to define it as a locally convex (resp. concave) corner. At first glance, this condition may seem unnecessary, since if it is not met, x cannot be locally convex or concave anyway. However, a concave corner may appear as a vertex of the convex hull of its convex K-visible points and vice versa. To eliminate this undesirable edge case, we only consider convex (resp. concave) K-visible points from a convex (resp. concave) corner.

From now on, we focus on convex definitions and properties, but analogous results hold by replacing convex with concave. We assemble the locally convex points into polyhedral objects as follows.

Definition 4. *Let* $A \subset Z_K$. *The set* $\mathrm{CvxH}(A)$ *is called a* locally convex face of K *whenever* $\mathrm{CvxH}(A)$ *is a face of* $\mathrm{CvxH}\left(\cup_{a \in A} \mathcal{V}_K^{\mathrm{Out}}(a)\right)$.

Lemma 2 (in Appendix A, page 13) shows that incident vertices, edges, etc., to a locally convex face are also locally convex. It is clear that the locally convex vertices and faces of K form a disjoint set with the locally concave vertices and faces of K, since their extrema must belong to either the convex corners or the concave corners. The following theorem justifies this definition of local convexity, since it coincides with the convex hull of Z_K when X_K is fully convex.

Theorem 1. *Let* $K \subset \mathcal{C}_d^d$ *and* X_K *be fully convex. The vertices and the faces of* $\mathrm{CvxH}(Z_K)$ *are locally convex vertices and locally convex faces of* K.

Proof. First it is clear that any vertex of $\mathrm{CvxH}(Z_K)$ must be a convex corner of K and is also convex K-visible (belonging to $\mathrm{Bd}(K)$). We use here Theorem 2 (Appendix A, page 14), which tells that any vertex and face of $\mathrm{CvxH}(Z_K)$ are convex K-visible. Now, let $A \subset Z_K$ be such that $F := \mathrm{CvxH}(A)$ is a face of $\mathrm{CvxH}(Z_K)$. So $F \subset \partial \mathrm{CvxH}(Z_K)$ and any $a \in A$ is a vertex of $\mathrm{CvxH}(Z_K)$. Clearly $\mathcal{V}_K^{\mathrm{Out}}(a) \subset Z_K$ so $\mathrm{CvxH}\left(\mathcal{V}_K^{\mathrm{Out}}(a)\right) \subset \mathrm{CvxH}(Z_K)$. If a is not locally convex, then it is a convex combination of points of $\mathcal{V}_K^{\mathrm{Out}}(a) \subset \mathrm{CvxH}\,(Z_K)$, and it cannot be a vertex of $\mathrm{CvxH}\,(Z_K)$, a contradiction. Thus a is locally convex.

The face $F = \mathrm{CvxH}\,(A)$ is a subset of each $\mathrm{CvxH}\,(\mathcal{V}_K^{\mathrm{Out}}(a))$ for $a \in A$ (since any subset of its vertices are convex visible according to Theorem 2). So $F \subset \cup_{a \in A} \mathrm{CvxH}\,(\mathcal{V}_K^{\mathrm{Out}}(a)) \subset \mathrm{CvxH}\,(\cup_{a \in A} \mathcal{V}_K^{\mathrm{Out}}(a))$.

If we assume that there is a $y \in F$ with $y \in \mathrm{int}(\mathrm{CvxH}\,(\cup_{a \in A} \mathcal{V}_K^{\mathrm{Out}}(a)))$, then it follows that $y \in \mathrm{int}(\mathrm{CvxH}\,(Z_K))$ since any point of $\mathcal{V}_K^{\mathrm{Out}}(a)$ belongs to Z_K. But this is a contradiction to F being a face of $\mathrm{CvxH}\,(Z_K)$. So either $y \in \partial\mathrm{CvxH}\,(\cup_{a \in A} \mathcal{V}_K^{\mathrm{Out}}(a))$ or y is outside this convex set. The second case is impossible since $F \subset \mathrm{CvxH}\,(\cup_{a \in A} \mathcal{V}_K^{\mathrm{Out}}(a))$ as shown above. So F lies in the boundary of $\mathrm{CvxH}\,(\cup_{a \in A} \mathcal{V}_K^{\mathrm{Out}}(a))$. A convex set lying on the boundary of a convex set Y such that each of its extremal points are extremal points of Y is by definition a face of Y. So F is a locally convex face of K. □

3 Extremal Point Detection

In this section, we consider a digital set $Z \subset \mathbb{Z}^3$. The dimension of the space is restricted to $d = 3$ to better align with the illustrations, but it can be extended to arbitrary d without difficulty. A point $\mathbf{z} \in Z$ is *extremal* if and only if $\mathbf{z} \in \mathrm{Extr}\,(\mathrm{CvxH}\,(Z))$. In addition, $\mathbf{z} \in Z$ is *locally extremal* if and only if it is extremal for the convex hull of some non-empty subset $Z' \subseteq Z$ containing $\mathbf{z}$. In theory, Z' could be equal to $\{\mathbf{z}\}$ or to Z. In the first case, the definition would be too local, as every point would be considered locally extremal. In the second case, it would not be local enough, as only extremal points would be considered locally extremal. The challenge is to find the right trade-off, which the approach described below manages remarkably well in practice.

We propose a probing algorithm that labels a point $\mathbf{z} \in Z$ as locally extremal or not extremal. Informally, the algorithm starts with an initial tetrahedron. The apex is placed at $\mathbf{z}$ and the opposite triangular face stands below it. The triangle is then iteratively updated with elementary vector operations using only queries "is $\mathbf{x}$ in Z?" (see Fig. 2). The algorithm stops if it detects a point alignment that makes $\mathbf{z}$ not extremal or if the normal vector to the triangle gives a direction for which $\mathbf{z}$ is extremal with respect to the set Z' of visited points, thus locally extremal. It is very similar to the H-algorithm [10] but solves a different problem.

In more details, by translation, we assume without loss of generality that the point to consider is the origin $\mathbf{0}$. The tetrahedron is then conveniently represented by a 3×3 matrix $\mathbf{M}$ with integer coefficients, where the three column vectors are denoted by $\mathbf{m}_1$, $\mathbf{m}_2$, $\mathbf{m}_3$. See Fig. 3(a) and Definition 5.

Definition 5 (Fig. 3(a)). *Let $Z \subset \mathbb{Z}^3$ be a digital set containing $\mathbf{0}$. A tetrahedron is a matrix $\mathbf{M} \in \mathbb{Z}^{3\times3}$, which is valid with respect to Z if*

$$-\mathbf{m}_1, -\mathbf{m}_2, -\mathbf{m}_3 \in Z, \quad \mathbf{m}_1, \mathbf{m}_2, \mathbf{m}_3 \notin Z, \quad \det(\mathbf{M}) = 1. \tag{1}$$

The update of the tetrahedron is conveniently represented by a 3×3 matrix with integer coefficients taken from a set of six matrices with determinant 1:

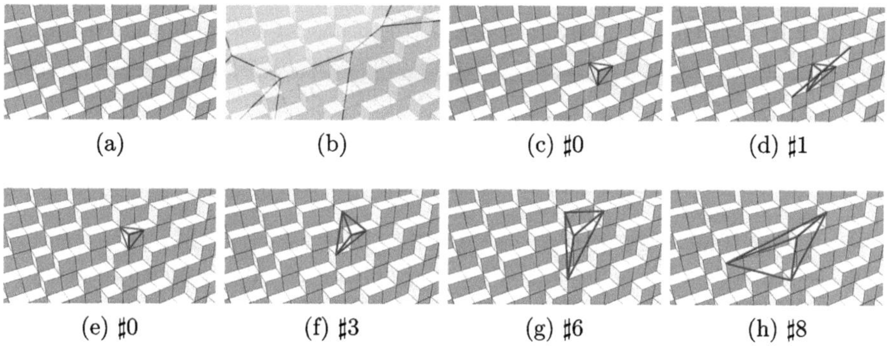

Fig. 2. Z is the set of pointels of a digital surface, denoted by Z_K in Sect. 2 and shown in gray colors in (a). CvxH(Z) is shown in (b). An initial tetrahedron is shown in (c). In this case, the algorithm finds a point alignment (in red) after one update and the starting point is thus labelled as not extremal. An execution from a vertex of CvxH(Z) is shown from (e) to (h). The base facet of the tetrahedron aligns with a supporting plane after eight iterations and the starting point is labelled as locally extremal. (Color figure online)

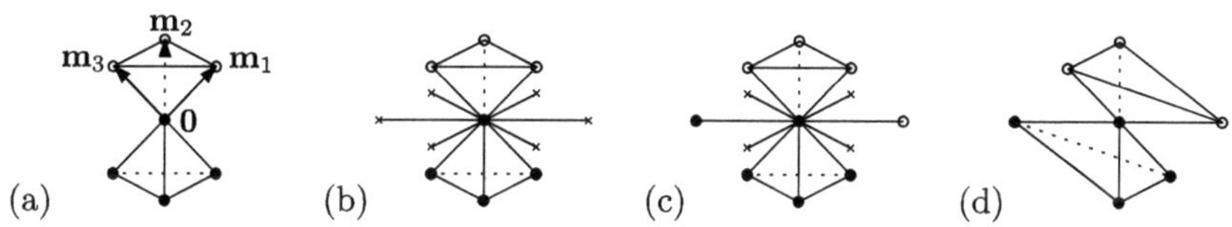

Fig. 3. Illustration of Definition 5 (a) and Definition 7 (b–d). Points in (resp. not in) Z are depicted with black (resp. white) disks. A valid tetrahedron is shown in (a). Among the six possible operations, depicted with crosses in (b), a valid one is shown in (c). Having applied this operation, the resulting tetrahedron is shown in (d). One can check that it is valid.

Definition 6. *Let Σ be the set of all permutations over $\{1, 2, 3\}$. For a permutation $\sigma \in \Sigma$, let $\boldsymbol{U}_\sigma$ be a 3×3 matrix having -1 at the intersection between the $\sigma(1)$-th column and the $\sigma(2)$-th row and the same entries as $\mathbf{I}$, the 3×3 identity matrix, elsewhere. Let $\mathbb{U}$ be the set $\{\boldsymbol{U}_\sigma \mid \sigma \in \Sigma\}$.*

Note that computing $\mathbf{M}\boldsymbol{U}_\sigma$ amounts to subtracting the $\sigma(2)$-th column of $\mathbf{M}$ to the $\sigma(1)$-th column. As shown in Fig. 3(b), the six possible results (determined by the six permutations $\sigma \in \Sigma$) form the vertices of an hexagon around the origin. Any such operation is possible if the resulting tetrahedron is valid.

Definition 7 (Fig. 3(c–d)). *Let $Z \subset \mathbb{Z}^3$ be a digital set containing* $\mathbf{0}$ *and let* $\mathbf{M} \in \mathbb{Z}^{3\times3}$ *be a valid tetrahedron. An operation is a function* $f : \mathbb{Z}^{3\times3} \mapsto \mathbb{Z}^{3\times3}$ *such that given* $\mathbf{M}' = f(\mathbf{M})$*, there exists a matrix* $\boldsymbol{U}_\sigma \in \mathbb{U}$ *that satisfies* $\mathbf{M}' = \mathbf{M}\boldsymbol{U}_\sigma$*. Moreover, f is valid if* $\mathbf{M}'$ *is valid, i.e., if*

$$-\mathbf{m}_{\sigma(1)} + \mathbf{m}_{\sigma(2)} \in Z, \quad \mathbf{m}_{\sigma(1)} - \mathbf{m}_{\sigma(2)} \notin Z. \tag{2}$$

The algorithms halts if there is no valid operation, but also whenever two points symmetric with respect to $\mathbf{0}$ are in Z. Otherwise it continues to probe:

Definition 8. *Let $Z \subset \mathbb{Z}^3$ be a digital set containing* $\mathbf{0}$ *and let* $\mathbf{M} \in \mathbb{Z}^{3\times3}$ *be a valid tetrahedron. The state* STATE($\mathbf{M}$) *of* $\mathbf{M}$ *is defined as:*

$$\begin{array}{lll} NO & \text{if } \exists\sigma \in \Sigma \text{ s.t. } -\mathbf{m}_{\sigma(1)} + \mathbf{m}_{\sigma(2)}, \mathbf{m}_{\sigma(1)} - \mathbf{m}_{\sigma(2)} \in Z, \\ YES & \text{if } \forall\sigma \in \Sigma,\ -\mathbf{m}_{\sigma(1)} + \mathbf{m}_{\sigma(2)}, \mathbf{m}_{\sigma(1)} - \mathbf{m}_{\sigma(2)} \notin Z, \\ PROBING & \text{otherwise} \end{array} \tag{3}$$

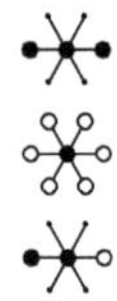

Finally, our algorithm IsExtremal reads as follows:
Input: a digital set Z containing $\mathbf{0}$, a valid tetrahedron $\mathbf{M}$.
while $STATE(\mathbf{M}) = PROBING$ **do**
  Find a valid operation f and apply it: $\mathbf{M} \leftarrow f(\mathbf{M})$
return STATE($\mathbf{M}$)

The initial tetrahedron is usually a (positively oriented) orthonormal basis. By rotations, we can assume it is $\mathbf{I}$ without loss of generality.

If the algorithm terminates, the output is either No or Yes. In the former case, $\mathbf{0}$ is not extremal because it lies in a segment joining two other points of Z. In the latter case, $\mathbf{0}$ is locally extremal and more precisely, extremal with respect to the set of visited points. We can show that the algorithm indeed terminates for any finite Z and that it returns Yes if $\mathbf{0}$ is a vertex of CvxH (Z) (see Theorem 3 and Theorem 4 in Appendix B, page 15).

Note that finding a valid operation is quite easy: we have to check if there exists $\sigma \in \Sigma$ such that $-\mathbf{m}_{\sigma(1)} + \mathbf{m}_{\sigma(2)} \in Z$ and $\mathbf{m}_{\sigma(1)} - \mathbf{m}_{\sigma(2)} \notin Z$. This can be done by at most six calls to a set-membership predicate. There are usually several valid operations. The above-mentioned termination results do not depend on a particular choice. In some cases, however, the way we choose the operations may change the final answer. In the experiments, we follow the approach previously used in the H-algorithm [10]: we use an in-sphere criterion and, if several points are in a cospherical position, we resort to a lexicographic order.

Algorithm 1: Given the interior voxels K, returns a surface meshing of the convex zones as a pair of vertices and faces joining them.

```
Function BuildLocallyConvexZones( In K: subset of C_d^d ) : Mesh;
begin
    E ← ∅ ;                                                        // (I)
    Z ← Cl(K) ∩ Z^d ;
    foreach convex corner z of K do
        M ← GetValidTetrahedron(z, Z) ;
        if IsExtremal(M, Z − z) then E ← E ∪ {z}
    V ← GetVisibilityPairs(K, E) ;                                 // (II)
    (C_v, C_e) ← GetLocallyConvexVerticesAndEdges(E, V) ;          // (III)
    C_f ← GetLocallyConvexFacets(C_v, C_e, V) ;                    // (IV)
    return (C_v, C_f)
```

Algorithm IsExtremal provides in practice a fast way to discard a large amount of non-extremal points, especially along digitizations of smooth surfaces. As an example, the ellipsoid $x^2 + 25y^2 + 25z^2 - 90 \leq 0$ is digitized at grid steps $0.5, 0.1, 0.05$ and 0.01 to obtain four sets of voxels. For each one, we have processed the set of boundary pointels Z after surface tracking. The results are reported in the table below, where n_c denotes the number of corners, i.e., points $\mathbf{z} \in Z$ such that $\forall k \in \{1,2,3\}, \mathbf{z} \pm \mathbf{e}_k \in Z, \mathbf{z} \pm \mathbf{e}_k \notin Z$, from where the algorithm is initialized, and n_e is the number of points detected as locally extremal. Note that $n_c \geq n_e$ by definition and $n_e \geq |\mathrm{CvxH}(Z)|$. Experimentally, n_e is equal or slightly greater than $|\mathrm{CvxH}(Z)|$, while it is far less than n_c.

grid step	$\|Z\|$	n_c	n_e	$\|\mathrm{CvxH}(Z)\|$
0.5	984	112	112	112
0.1	24.808	2.032	1.128	1.128
0.05	99.448	7.784	3.064	3.064
0.01	2.488.104	186.664	33.864	33.784

4 Reconstruction of Locally Convex and Concave Zones

General Presentation. Algorithm 1 provides the general algorithm for reconstructing the locally convex zones along the boundary of a set of interior voxels K. It is composed of 4 stages:

(I) Filtering the convex corners with the probing algorithm IsExtremal
(II) Computing the convex visibility between these points (Algorithm 2)
(III) Extracting the locally convex vertices and edges among them (Algorithm 3)
(IV) Extracting the locally convex facets between them (Algorithm 4)

Algorithm 2: Given the interior voxels K and extremal convex corners E, returns all the visible K-convex pairs.

```
Function GetVisibilityPairs( In K: voxels, In E: points ) : set of
(point,point);
begin
    R ← ∅;
    foreach (p, q) ∈ E × E do
        if p ≠ q and Cover (CvxH ({p, q})) ⊂ Bd (K) ∪ Out (K) then
            R ← R ∪ {(p, q)};
    return R
```

Algorithm 3: Given points E and their visibility V, filters them to return the locally convex vertices and edges.

```
Function GetLocallyConvexVerticesAndEdges( In E: points, In V: set
of (point,point) ) : (points, set of (point,point)) ;
Var M: map point to bool, N: map (point,point) to bool;
begin
    S ← ∅ ;                                  // set of candidate convex edges
    foreach p ∈ E do M[p] ← true foreach (p, q) ∈ V do N[(p, q)] ← false
    foreach p ∈ E do
        Q ← ConvexHull({p} ∪ {q ∈ E, (p, q) ∈ V});
        if Q is not full dimensional or p ∉ Extr (Q) then
            M[p] ← false;
            continue;
        foreach facet f = (p_0, ..., p_{k−1}) of Q do
            if ∃i ∈ {0, ..., k − 1}, p_i = p then
                S ← S ∪ {(p, p_{i−1})} ∪ {(p, p_{i+1})} ;   // indices modulo k
    foreach (p, q) ∈ V do
        if (p, q) ∉ S or (q, p) ∉ S then continue
        L ← {r ∈ E, (p, r) ∈ V and (p, r) ∈ S and (r, p) ∈ S};
        L ← L ∪ {r ∈ E, (q, r) ∈ V and (q, r) ∈ S and (r, q) ∈ S};
        Q ← ConvexHull(L);
        if (p, q) is an edge of Q then N[(p, q)] ← true
    C_v ← {p ∈ E, M[p] = true};
    C_e ← {(p, q) ∈ V, N[(p, q)] = true};
    return (C_v, C_e)
```

Computing the locally concave zones is a completely symmetric algorithm and is not detailed here (just starts from concave corners). Note that the true C++ implementation uses indices instead of sets to speed up computations.

Computational Considerations. Note that more than 95% of the computation time is related to stage (II), the computation of all the visible pairs. Its complex-

Algorithm 4: Given the locally convex vertices C_v and edges C_e, the visibility V, returns the locally convex facets joining them.

```
Function GetLocallyConvexFacets( In C_v: points, In C_e, V: set of
(point,point) ) : set of facets ;
begin
    F ← ∅;
    foreach p ∈ C_v do
        Q ← ConvexHull({p} ∪ {q ∈ C_v, (p, q) ∈ V});
        foreach facet f ∈ Q do
            if p ∈ Extr (f) and every edge of f is in C_e then
                // Some faces may be present several times
                f' ← LowestCircularPermutation(f);
                F ← F ∪ {f'} ;                    // (3, 1, 6) → (1, 6, 3)
    return F
```

ity grows quadratically with the number of candidate extremal points. This is why it is crucial to filter the convex corners as most as possible and this is exactly what does the probing algorithm IsExtremal. Furthermore, many of the visibility tests are done between candidate points that are very far away from each other and probably unrelated. Instead of using a heuristic (like assuming that points further away from a given distance are not visible), we use a coarse-to-fine pyramid approach to check covering and tangency, that can be proven to be exact. We do not have the place to develop it here, but we achieve between 10× to 200× speed-up with this technique.

Results and Timings. We have tested our method on a large amount of digital surfaces. Some results are shown in Fig. 4 and Table 1 (computations made on a Macbook Pro, apple M2 Max, 64 Go RAM). The convex and concave parts are accurately detected, and their polyhedral reconstruction closely matches the convex hull in the most salient regions as shown in Fig. 4(d) and (g). Remarkably, the columns n_e and n_v are identical, which means that the probing algorithm IsExtremal identifies all the locally convex and concave vertices. Thanks to algorithm IsExtremal and the above-mentioned additional optimization tricks, our method achieves low running times relative to the difficulty of the problem. Note that the concave parts are located inside the volume and are thus not visible on Fig. 4(c), (e) and (h). Many parts are of course neither convex nor concave and addressing them remains a key direction for future work.

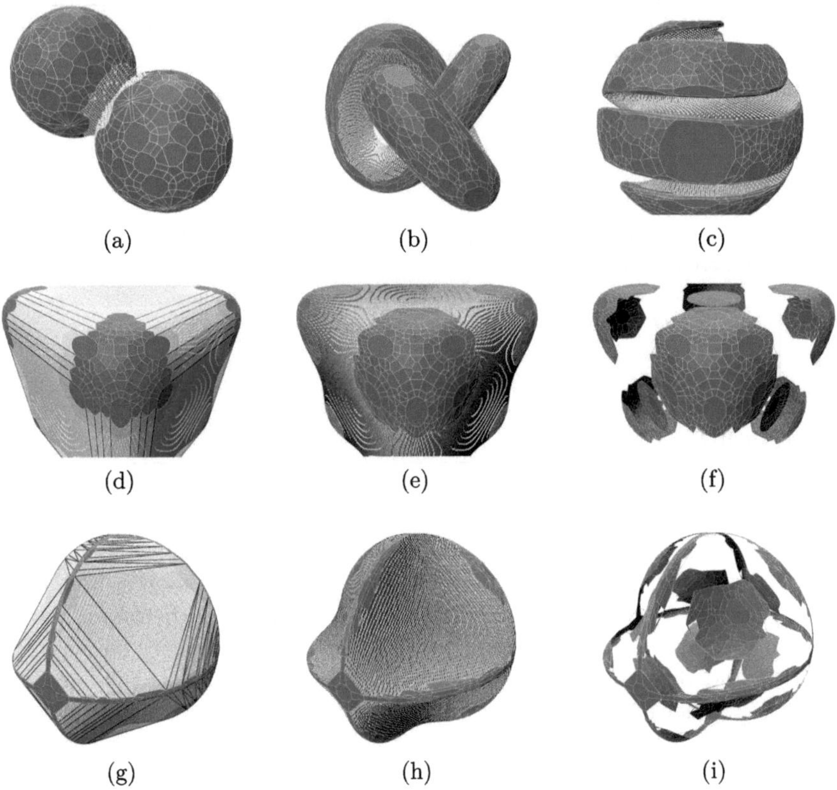

Fig. 4. Reconstructions of `twospheres`, `torusknot`, `sharpsphere`, `goursat` and `leopold` are respectively shown in (a), (b), (c), (d)–(f) and (g)–(i). The digital surface, the convex hull, the convex and concave parts are respectively colored in grey, yellow, blue and pink. Statistics and timings are collected in Table 1. (Color figure online)

Table 1. For each digital surface of Fig. 4, we provide the overall running time together with the number of input points ($|Z_k|$), corners (n_c), extremal points (n_e), locally convex/concave points and faces (n_v and n_f).

shape	$\lvert Z_k\rvert$	convex zones				concave zones				time (s)
		n_c	n_e	n_v	n_f	n_c	n_e	n_v	n_f	
`twospheres`	21.758	3.258	1.580	1.580	1.339	2.664	0	0	0	5
`torusknot`	96.622	15.196	2.930	2.930	2.765	13.347	5	5	0	31
`sharpsphere`	119.846	16.715	3.111	3.111	2.546	15.011	412	412	271	40
`goursat`	104.984	7.776	2.160	2.160	2.408	7.376	288	288	246	24
`leopold`	144.440	33.672	1.664	1.664	1.538	32.256	856	856	744	19

5 Conclusion

In this paper we have proposed an original definition of local convexity and concavity along a digital surface. The definition, based on the notion of visibility, is sound in the sense that it matches the convex hull when the digital surface is digitally (fully) convex. Computing local convexity/concavity requires the identification of extremal points. We have proposed a fast probing algorithm that quickly prune corners to retain extremal points. The computation of visibilities between extremal points is thus greatly reduced. Locally convex/concave edges and faces are then quickly determined with local convex hull computations. Experiments illustrate that the decomposition of the digital surface into convex and concave zones is exact and not approximate.

One of the limitations of the work is the computation of visibility between all pairs of extremal points. A pyramid approach is used to speed up these computations, but computing visibility remains the bottleneck of the method. We are currently investigating "compatibility criteria" between extremal points to prune further the number of tested pairs. Our final objective remains the full polyhedral decomposition of the digital surface into convex, concave, and saddle parts. A plane-probing variant could be able to detect potential saddle zones. It remains to understand how to combine this information to triangulate these zones in a deterministic way.

Acknowledgments. This work is partially supported by the French National Research Agency within the StableProxies project (ANR-22-CE46-0006).

A Properties Related to Visibility

We gather here some properties related to visibility and necessary to give some theoretical guarantees for local convexity.

We first state without proof a few elementary properties of the sets defined in Sect. 2.

Lemma 1. *For any* $K \subset \mathcal{C}^d_d$*, the set of cells* $\mathcal{C}^d$ *is the disjoint union of* $\mathrm{Outer}\,(K)$, $\mathrm{Out}\,(K)$, $\mathrm{Bd}\,(K)$, $\mathrm{In}\,(K)$, $\mathrm{Inner}\,(K)$. *Furthermore* $\mathrm{Outer}\,(K)$, $\mathrm{Bd}\,(K)$ *and* $\mathrm{Inner}\,(K)$ *are closed.* $\mathrm{Out}\,(K)$ *and* $\mathrm{In}\,(K)$ *are open.*

The next lemma tells that the definition of locally convex face is sound, i.e., all faces incident to a locally convex face are also locally convex.

Lemma 2. *For any* $\emptyset \neq A' \subset A \subset Z_K$*, if* $\mathrm{CvxH}\,(A)$ *is a locally convex face of* K*, then* $\mathrm{CvxH}\,(A')$ *is a locally convex face of* K *too.*

Proof. Assume $\mathrm{CvxH}\,(A)$ is a locally convex face of K. Let $C := \mathrm{CvxH}\left(\cup_{a \in A} \mathcal{V}^{\mathrm{Out}}_K(a)\right)$ and $C' := \mathrm{CvxH}\left(\cup_{a \in A'} \mathcal{V}^{\mathrm{Out}}_K(a)\right)$. We have $\mathrm{CvxH}\,(A') \subset C'$. If $\mathrm{CvxH}\,(A') \not\subset \partial C'$ then $\mathrm{CvxH}\,(A') \cap \mathrm{int}(C') \neq \emptyset$ (it must cross its interior). But obviously $C' \subset C$, which implies $\mathrm{int}(C') \subset \mathrm{int}(C)$. It follows that $\mathrm{CvxH}\,(A') \cap \mathrm{int}(C) \neq \emptyset$. This contradicts $\mathrm{CvxH}\,(A') \subset \mathrm{CvxH}\,(A) \subset \partial C$ ($\mathrm{CvxH}\,(A)$ is a face of C) since ∂C and $\mathrm{int}(C)$ are disjoint. □

We show below that the pointels I_K are indeed interior points as they belong to the interior of the convex hull of X_K.

Lemma 3. *Let $K \subset \mathcal{C}^d_d$ non-empty. Then X_K and Z_K have the same convex hull (i.e.* $\mathrm{CvxH}\,(Z_K) = \mathrm{CvxH}\,(X_K)$*). Furthermore* $\mathrm{CvxH}\,(Z_K)$ *is full dimensional, and I_K belongs to* $\mathrm{int}(\mathrm{CvxH}\,(Z_K))$*, which is not empty.*

Proof. Since the convex hull is an increasing operator and $Z_K \subset X_K$, we have $\mathrm{CvxH}\,(Z_K) \subset \mathrm{CvxH}\,(X_K)$. To show the reverse inclusion, X_K being finite, the extremal points V of $\mathrm{CvxH}\,(X_K)$ form a subset of X_K. Let $v \in V$. Let i be an arbitrary direction. If $v + \mathbf{e}_i \in X_K$ and $v - \mathbf{e}_i \in X_K$, it means that v is a linear combination of two points of X_K and cannot be an extremal point of $\mathrm{CvxH}\,(X_K)$. So either $v^+ := v + \mathbf{e}_i$ or $v^- := v - \mathbf{e}_i$ or both are not in X_K. Assume it is v^+ wlog. So the 1-cell c^+ of boundary $\{v, v^+\}$ must be in $\mathrm{Out}\,(K)$ and v must belong to Z_K. Therefore, $V \subset Z_K$ and we conclude that $\mathrm{CvxH}\,(X_K) = \mathrm{CvxH}\,(V) \subset \mathrm{CvxH}\,(Z_k)$.

Besides, if $x \in I_K$, it follows that all neighbors $N := (x \pm \mathbf{e}_i)_{i \in \{1,\ldots,d\}}$ are in X_K too. Since $\mathrm{CvxH}\,(N)$ is full dimensional, $x \in \mathrm{int}(\mathrm{CvxH}\,(N)) \subset \mathrm{int}(\mathrm{CvxH}\,(X_K)) = \mathrm{int}(\mathrm{CvxH}\,(Z_K))$. □

It is quite intuitive that the boundary of the convex hull of a digitally convex digital set should be close to the boundary points of this digital set. However this result if false if we consider classical digital convexity. For instance just pick two points separated by a primitive lattice vector of ∞-norm strictly greater than 1. The full convexity is thus necessary in the following property:

Theorem 2. *Let $K \subset \mathcal{C}^d_d$. If X_K is fully convex, then* $\mathrm{Cover}\,(\partial \mathrm{CvxH}\,(Z_K)) \subset \mathrm{Out}\,(K) \cup \mathrm{Bd}\,(K)$*. Hence any edge or facet of* $\mathrm{CvxH}\,(Z_K)$ *are convex K-visible.*

Proof. Let F be any facet of $\mathrm{CvxH}\,(X_K)$, which is also a facet of $\mathrm{CvxH}\,(Z_K)$ by Lemma 3. Hence $F \subset \partial \mathrm{CvxH}\,(Z_K)$. Let $c \in \mathrm{Cover}\,(F)$, so $c \cap F \neq \emptyset$, and let $y \in c \cap F$. Let $V := \mathrm{Extr}\,(c)$ be the vertices of c.

Since $y \in F$, we have $y \in \partial \mathrm{CvxH}\,(Z_K) \subset \mathrm{CvxH}\,(Z_K) = \mathrm{CvxH}\,(X_K)$. So $\mathrm{Star}\,(y) = \{c\} \subset \mathrm{Star}\,(\mathrm{CvxH}\,(X_K)) = \mathrm{Star}\,(X_K)$, the latter equality coming from full convexity of X_K. At least one of the extremal point of c must belong to X_K, and c cannot be in $\mathrm{Outer}\,(K)$.

We show now that no point of V can be in I_K. Assume there is some $x \in V$ with $x \in I_K$. Then $c \in \mathrm{Inner}\,(K)$ or $c \in \mathrm{In}\,(K)$. If $c \in \mathrm{Inner}\,(K)$, then $V \subset I_K$. By Lemma 3, we know that $I_K \subset \mathrm{int}(\mathrm{CvxH}\,(Z_K))$. Hence $\mathrm{CvxH}\,(V) \subset \mathrm{CvxH}\,(\mathrm{int}(\mathrm{CvxH}\,(Z_K))) = \mathrm{int}(\mathrm{CvxH}\,(Z_K))$. It follows that c lies entirely in the interior of $\mathrm{CvxH}\,(Z_K)$, which is a contradiction with $y \in \partial \mathrm{CvxH}\,(Z_K)$.

If $c \in \mathrm{In}\,(K)$, then $\mathrm{Star}\,(c) \subset \mathrm{In}\,(K)$ since $\mathrm{In}\,(K)$ is open (see Lemma 1). Now $\overline{\mathrm{In}\,(K)} \subset \mathrm{Cl}\,(K)$. It follows that:

$$
\begin{aligned}
& \mathrm{CvxH}\,(\mathrm{Star}\,(c)) \subset \mathrm{CvxH}\,(\mathrm{Cl}\,(K)) = \mathrm{CvxH}\,(Z_K) && (\mathrm{CvxH}\,(\cdot) \text{ is increasing}) \\
\Rightarrow\ & \mathrm{int}(\mathrm{CvxH}\,(\mathrm{Star}\,(c))) \subset \mathrm{int}(\mathrm{CvxH}\,(Z_K)) && (\text{Taking their interior}) \\
\Rightarrow\ & \mathrm{CvxH}\,(\mathrm{Star}\,(c)) \subset \mathrm{int}(\mathrm{CvxH}\,(Z_K)) && (\text{since } \mathrm{Star}\,(c) \text{ is open}) \\
\Rightarrow\ & c \subset \mathrm{int}(\mathrm{CvxH}\,(Z_K)) && (\text{since } c \in \mathrm{Star}\,(c))
\end{aligned}
$$

There is again a contradiction since $y \in c$ and $y \in \partial\text{CvxH}(Z_K)$.

We have just shown that $\text{Cover}(F) \subset (\mathcal{C}^d \setminus (\text{Outer}(K) \cup \text{In}(K) \cup \text{Inner}(K)))$, hence $\text{Cover}(F) \subset (\text{Out}(K) \cup \text{Bd}(K))$ by Lemma 1. Furthermore, $\partial\text{CvxH}(Z_K)$ is the union of finitely many convex facets (F_j). Thus $\text{Cover}(\partial\text{CvxH}(Z)) = \text{Cover}(\cup_j F_j) = \cup_j \text{Cover}(F_j) \subset \text{Out}(K) \cup \text{Bd}(K)$. □

It immediately implies the tangency of convex hull facets.

Corollary 1. *Let $K \subset \mathcal{C}^d_d$. If X_K is fully convex, then the vertices of any facet of $\text{CvxH}(Z_K)$ are cotangent in Z_K (i.e. $\text{Star}(\partial\text{CvxH}(Z_K)) \subset \text{Star}(Z_K)$).*

Proof. It suffices to notice than $\text{Star}(Z_K) = \text{Out}(K) \cup \text{Bd}(K) \cup \text{In}(K)$. □

B Properties Related to Probing

In this section, we focus on the termination of algorithm IsExtremal. We first show that it returns Yes if $\mathbf{0}$ is a vertex of $\text{CvxH}(Z)$ and then show that it terminates for any finite Z, whatever the relative position of $\mathbf{0}$.

Theorem 3. *Let $Z \subset \mathbb{Z}^3$ be a digital set containing $\mathbf{0}$ and let $\mathbf{I}$ be a valid initial tetrahedron. If $\mathbf{0}$ is a vertex of $\text{CvxH}(Z)$, Algorithm IsExtremal returns Yes after a number of iterations bounded from above by $2\sqrt{3}$ times the total area of the facets of $\text{CvxH}(Z)$ incident to $\mathbf{0}$.*

Proof. First, note that since $\mathbf{I}$ is assumed to be a valid tetrahedron, $-\mathbf{e}_1, -\mathbf{e}_2, -\mathbf{e}_3$ and $\mathbf{0}$ are all in Z by Definition 5, which means that $\text{CvxH}(Z)$ is full-dimensional.

Every facet $\mathcal{F}$ of $\text{CvxH}(Z)$ incident to $\mathbf{0}$ has an orthogonal vector $\mathbf{v}_\mathcal{F} \in \mathbb{Z}^3$ and an area $a_\mathcal{F}$ that satisfy $\|\mathbf{v}_\mathcal{F}\|_2 \leq 2a_\mathcal{F}$. Denoting by $\mathbf{v}$ (resp. a), the sum of $\mathbf{v}_\mathcal{F}$ (resp. $a_\mathcal{F}$) over all facets $\mathcal{F}$ incident to $\mathbf{0}$, we have again $\|\mathbf{v}\|_2 \leq 2a$. Therefore, $\|\mathbf{v}\|_1 \leq \sqrt{3}\|\mathbf{v}\|_2 \leq 2\sqrt{3}a$.

We denote by $\mathcal{H}(\mathbf{v})$ the half-space $\{\mathbf{x} \in \mathbb{R}^3 \mid \mathbf{x}\cdot\mathbf{v} < 0\}$. By construction, $\mathcal{H}(\mathbf{v})$ contains $\text{CvxH}(Z) \setminus \{\mathbf{0}\}$, therefore $Z \setminus \{\mathbf{0}\}$. Note that two nonzero points $\mathbf{x}, -\mathbf{x}$ cannot be both in $\mathcal{H}(\mathbf{v})$, therefore cannot be both in Z. Consequently, algorithm IsExtremal, if it terminates, cannot return No but only Yes in that case.

The algorithm will stop before crossing the boundary of $\mathcal{H}(\mathbf{v})$, while the maximum number of iterations to reach it can be determined through a decreasing sequence of integers (see also [10, Lemma 5, Lemma 6, Theorem 1]).

More formally, we denote by $\mathbf{M}$ the current tetrahedron and c_1, c_2, c_3 the entries of $\mathbf{v}^\top\mathbf{M}$. At initialization, $\mathbf{M} = \mathbf{I}$ and c_1, c_2, c_3 are just the coordinates of $\mathbf{v}$. Since $\mathbf{M}$ is a valid tetrahedron (1), we have for all $k \in \{1, 2, 3\}$, $-\mathbf{m}_k \in Z$, which implies $-\mathbf{m}_k \in \mathcal{H}(\mathbf{v})$. Therefore, $(-\mathbf{m}_k) \cdot \mathbf{v} < 0$ and, equivalently, $c_k > 0$.

If there is a valid operation f, we denote by $\mathbf{M}' := f(\mathbf{M})$ the new tetrahedron. There is a matrix $\boldsymbol{U}_\sigma$ such that $\mathbf{M}' = \mathbf{M}\boldsymbol{U}_\sigma$ and by permutations, we assume w.l.o.g. that σ is the identity. We thus have $\mathbf{v}^\top\mathbf{M}' = (c_1 - c_2, c_2, c_3)$. In addition, since f is valid (2), we have $-\mathbf{m}_1 + \mathbf{m}_2 \in Z$, which implies $-\mathbf{m}_1 + \mathbf{m}_2 \in \mathcal{H}(\mathbf{v})$. Therefore, $(-\mathbf{m}_1 + \mathbf{m}_2) \cdot \mathbf{v} < 0$ and, equivalently, $c_2 - c_1 > 0$. In other words, the entries c_1, c_2, c_3 are strictly positive integers and their sum is strictly decreasing from $\|\mathbf{v}\|_1$ at every iteration. As a consequence, the number of iterations is less than $\|\mathbf{v}\|_1$, which concludes the proof. □

Theorem 4. *Let $Z \subset \mathbb{Z}^3$ be a finite digital set containing $\mathbf{0}$ and let $\mathbf{M}$ be a valid initial tetrahedron. Algorithm ISEXTREMAL terminates after a finite number of iterations.*

Proof. Let $C_\mathbf{M}$ be the cone formed by all convex combinations of $-\mathbf{m}_1$, $-\mathbf{m}_2$, $-\mathbf{m}_3$, i.e., $\{-\sum_{k\in\{1,2,3\}} c_k\mathbf{m}_k \mid c_1, c_2, c_3 \in \mathbb{R}_{\geq 0}\}$. If there is a valid operation f, we denote by $\mathbf{M}' := f(\mathbf{M})$ the new tetrahedron. There is a matrix $\boldsymbol{U}_\sigma$ such that $\mathbf{M}' = \mathbf{M}\boldsymbol{U}_\sigma$ and by permutations, we assume w.l.o.g. that σ is the identity.

For all $\mathbf{x} \in C_\mathbf{M}$, we have by definition $\mathbf{x} = -c_1\mathbf{m}_1 - c_2\mathbf{m}_2 - c_3\mathbf{m}_3$ with nonnegative c_1, c_2, c_3. We can write also

$$\begin{aligned}\mathbf{x} = -c_1'\mathbf{m}_1' - c_2'\mathbf{m}_2' - c_3'\mathbf{m}_3' &= -c_1'(\mathbf{m}_1 - \mathbf{m}_2) - c_2'\mathbf{m}_2 - c_3'\mathbf{m}_3 \\ &= -c_1'\mathbf{m}_1 - (c_2' - c_1')\mathbf{m}_2 - c_3'\mathbf{m}_3,\end{aligned}$$

for any $c_1', c_2', c_3' \in \mathbb{R}$. The nonnegativity of c_1, c_2, c_3 implies $c_1', c_3' \geq 0$ and $c_2' \geq c_1' \geq 0$, i.e. the nonnegativity of c_1', c_2', c_3'. Thus, $\mathbf{x} \in C_{\mathbf{M}'}$.

As a result, $C_{\mathbf{M}'}$ contains $C_\mathbf{M}$. By induction, the new cone $C_{\mathbf{M}'}$ also contains all past cones and therefore, all vertices of past tetrahedra. Consequently, Algorithm IsExtremal selects at every iteration a point $z \in Z$ to be a vertex of the next tetrahedron and this point cannot be selected again afterwards. Since Z is finite, Algorithm IsExtremal terminates after a finite number of iterations. □

References

1. Brimkov, V., Coeurjolly, D., Klette, R.: Digital planarity - a review. Discret. Appl. Math. **155**(4), 468–495 (2007)
2. Edelsbrunner, H., Mücke, E.P.: Three-dimensional alpha shapes. ACM Trans. Graph. **13**(1), 43 72 (1994). https://doi.org/10.1145/174462.156635
3. Herman, G.T.: Oriented surfaces in digital spaces. CVGIP Graph. Models Image Process. **55**(5), 381–396 (1993)
4. Klette, G.: Recursive calculation of relative convex hulls. In: Debled-Rennesson, I., Domenjoud, E., Kerautret, B., Even, P. (eds.) DGCI 2011. LNCS, vol. 6607, pp. 260–271. Springer, Heidelberg (2011). https://doi.org/10.1007/978-3-642-19867-0_22
5. Kobbelt, L., Campagna, S., Seidel, H.P.: A general framework for mesh decimation. In: Proceedings of the Graphics Interface 1998 Conference, 18–20 June 1998, Vancouver, BC, Canada, pp. 43–50 (1998). http://graphicsinterface.org/wp-content/uploads/gi1998-6.pdf
6. Lachaud, J.O., Provençal, X.: Two linear-time algorithms for computing the minimum length polygon of a digital contour. Discret. Appl. Math. **159**(18), 2229–2250 (2011). https://doi.org/10.1016/j.dam.2011.08.002
7. Lachaud, J.-O.: An alternative definition for digital convexity. In: Lindblad, J., Malmberg, F., Sladoje, N. (eds.) DGMM 2021. LNCS, vol. 12708, pp. 269–282. Springer, Cham (2021). https://doi.org/10.1007/978-3-030-76657-3_19

8. Lachaud, J.O.: An alternative definition for digital convexity. J. Math. Imaging Vis. **64**(7), 718–735 (2022). https://doi.org/10.1007/s10851-022-01076-0
9. Lachaud, J.-O., Provençal, X.: Dynamic minimum length polygon. In: Aggarwal, J.K., Barneva, R.P., Brimkov, V.E., Koroutchev, K.N., Korutcheva, E.R. (eds.) IWCIA 2011. LNCS, vol. 6636, pp. 208–221. Springer, Heidelberg (2011). https://doi.org/10.1007/978-3-642-21073-0_20
10. Lachaud, J.O., Provençal, X., Roussillon, T.: Two plane-probing algorithms for the computation of the normal vector to a digital plane. J. Math. Imaging Vision **59**(1), 23–39 (2017). https://doi.org/10.1007/s10851-017-0704-x
11. Portaneri, C., Rouxel-Labbé, M., Hemmer, M., Cohen-Steiner, D., Alliez, P.: Alpha wrapping with an offset. ACM Trans. Graph. (TOG) **41**(4), 1–22 (2022)
12. Roussillon, T., Sivignon, I.: Faithful polygonal representation of the convex and concave parts of a digital curve. Pattern Recognit. **44**(10-11), 2693–2700 (2011). https://hal.archives-ouvertes.fr/hal-00643573
13. Schulz, H.: Polyhedral approximation and practical convex hull algorithm for certain classes of voxel sets. Discret. Appl. Math. **157**(16), 3485–3493 (2009). https://doi.org/10.1016/j.dam.2009.04.008
14. Sklansky, J., Chazin, R.L., Hansen, B.J.: Minimum-perimeter polygons of digitized silhouettes. IEEE Trans. Comput. **21**(3), 260–268 (1972)
15. Wiederhold, P.: Computing the minimal perimeter polygon for sets of rectangular tiles based on visibility cones. J. Math. Imaging Vision **66**(5), 873–903 (2024). https://doi.org/10.1007/s10851-024-01203-z

First Results on Locally Turn-Bounded Surfaces in 3D Euclidean Space

Lysandre Macke(✉), Étienne Baudrier, and Étienne Le Quentrec

University of Strasbourg, ICube, UMR 7357, Strasbourg, France
lmacke@unistra.fr

Abstract. Information loss induced by digitization is inevitable, and constitutes a classical problem in digital geometry. Indeed, preserving the geometric and topological features of an object is crucial for image processing tasks. To ensure this preservation when converting continuous objects to discrete representations, different classes of shapes have been introduced. However, the current state of the art still leaves room for the definition of non smooth classes that have digitization preservation properties. To that end, we focus on Locally Turn-Bounded (LTB) Curves, introduced by É. Le Quentrec et al. in 2019. While LTB curves has shown promising results in two dimensions, their extension to higher dimensions is not trivial. This paper introduces the Locally Turn-Bounded Surfaces, a generalization of LTB curves to the 3D Euclidean space. As first results, we not only show that our definition preserves basic properties from LTB curves to LTB surfaces, but also demonstrate that LTB surfaces impose conditions on topology, notably preventing the surface from containing small holes.

1 Introduction

Emerging alongside image computing, digital geometry deals with the study of discrete models. Unlike real-world continuous ones, these models can be adapted for computer processing and have a wide range of applications, including computer graphics, medical imaging, or computer vision. However, the digitization process comes at a cost. Indeed, when converting points from a continuous space into a finite number of pixels (or voxels), a lot of information is lost, and the geometric and topological properties of a shape may be altered. Being able to control this information loss from continuous to discrete is therefore a fundamental point.

Even though many digitization processes have been proposed to address this issue, the accuracy of a digitization is, eventually, always highly dependent on the regularity hypothesis of the original shape. To this end, several studies have been carried out to determine classes of shapes that would maintain a faithful digitization, by imposing conditions on the shape's border. For instance, one of the first and most widely used hypotheses for two dimensions, called *par(r)-regularity*, was introduced independently in 1982 by T. Pavlidis in [Pav82], and later by J.

M. H. F. Wilkinson and J. Kosinka (Eds.): DGMM 2025, LNCS 16296, pp. 60–73, 2026.
https://doi.org/10.1007/978-3-032-09544-2_5

Serra in [Ser82]. However, as its name implies, par(r)-regularity involves working with differentiable objects, thus discarding irregular shapes that are yet ubiquitous in real life (polygons, surfaces with edges and/or vertices, ...). In this setting, we aim to find a class of shapes that permits the geometric estimation of a shape from its discretization and the topological conservation between the two models, while allowing the inclusion of both regular and irregular shapes.

Some later hypotheses, such as *half-regularity* [Ste05], *r-stability* [MKS09], *quasi(r)-regularity* [NPKD19], *μ-reach* [CCL09] and *locally tun-boundedness* [LMBT20], took inspiration from the idea of controlling the border of the shape, and aim to generalize the notion of par(r)-regularity to also include shapes with spiky and angular corners such as polygons. Half regularity for example, allows this inclusion by preventing significant shrinkage only in the middle of the shape. However, even though this definition permits topological conservation and the inclusion of polygons, this class of shapes also includes shapes with large elongated spikes which can lead to a geometrical discrepancy between the continuous and discretized object. The class of r-stable shapes that generalizes the notion of half-regularity manages to have topological properties too, but still allows the presence of thin spikes which on one hand allow them to include a wide range of polygon but on the other hand cause them to fail to admit geometric estimation. Quasi-regular shapes also shows robustness towards topological preservation, allowing well-composedness for their discretization. However, quasi-regularity does not give control on the fine variations on the shape border, still making discrete geometric estimation complicated. The notion of μ-reach allows the construction of a large class of shapes with properties on their homotopy equivalence, but has not yet been studied in the cubic grid. Finally, previous studies focusing on *locally turn-bounded (LTB) curves* have demonstrated, among other properties, the connectivity preservation from continuous to discrete space, together with well-composedness for a Gauss digitization with a grid of step under a specific threshold. Concerning geometric estimation, results have also been presented regarding multigrid convergence for length estimation. However, while these results are well-established for planar curves, the notion of local turn-boundedness has not yet been studied in higher dimensions. In this paper, we investigate on the three-dimensional case, introducing *locally turn-bounded surfaces* as a generalization of LTB curves.

Following the general framework used for LTB curves (and par-regular shapes), the present paper aims to lay the ground for future discretization properties of LTB surfaces. We demonstrate, as a preliminary study, their first results in 3D Euclidean space: their local topological behavior and how their definition precludes topological artifacts that could be lost during discretization. Our main intention is to present propositions that will useful in future works to understand and control the behavior of LTB surfaces in the cubic grid, and to handle inconvenient configurations in it. The following section recalls preliminary concepts, along with basic notations and definitions related to the notion of turn. In Sect. 3, we define and present in a first time some examples of LTB surfaces, and then show initial properties on our 3D extension. Among them, we focus on results regarding local connectivity and topology, relying on the notion of distortion defined in [Gro07]. Finally, Sect. 4 depicts our last proposition, demonstrating that LTB surfaces cannot contain small holes.

Notations The Euclidean distance between two points a and b is denoted by $d(a, b)$. The length of a curve $\mathcal{C}$ is denoted by $\mathcal{L}(\mathcal{C})$ and the maximal length of the sides of a polygonal line or a polygon L by $\lambda(L)$.

2 Turn of a Curve

The definition of local turn-boundedness is based on the notion of *turn* introduced by Milnor [Mil50], which is recalled hereafter as useful definitions and properties. Let us first denote by $\angle(u, v)$ the *geometric angle* between two vectors u and v, defined as the absolute value $\angle(u, v) := |\phi|$ of the angle $\phi \in (-\pi, \pi]$ between u and v. This definition allows us to introduce the notion of turn:

Definition 1 (Turn of polygonal lines and polygons, [AR89, part 5.1.1]**).**

- *The turn $\kappa(L)$ of a polygonal line $L = [x_i]_{i=0}^{n-1}$ with $n \geq 3$ is defined as:*

$$\kappa(L) := \sum_{i=1}^{n-2} \angle(\overrightarrow{x_{i-1}x_i}, \overrightarrow{x_i x_{i+1}})$$

- *Respectively, the turn $\kappa(P)$ of a polygon $P = [x_i]_{i \in \mathbb{Z}/n\mathbb{Z}}$ is defined as (see Fig. 1):*

$$\kappa(P) := \sum_{i=1}^{n} \angle(\overrightarrow{x_{i-1}x_i}, \overrightarrow{x_i x_{i+1}})$$

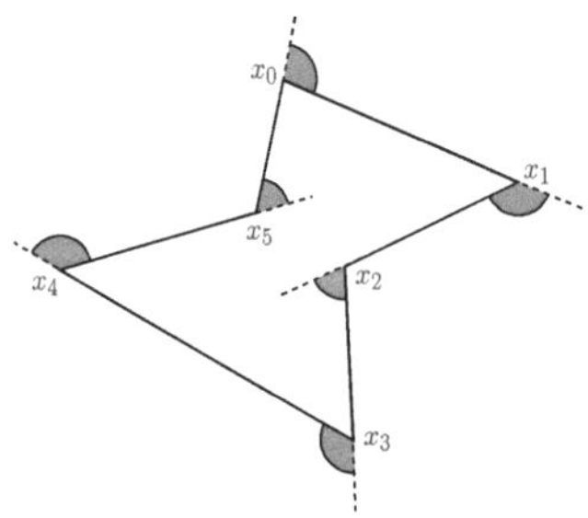

Fig. 1. The turn of the polygon is equal to the sum of the blue angles. (Color figure online)

Definition 2 (Turn of a curve, [Mil50, part 2]**).** *The turn $\kappa(\mathcal{C})$ of a curve $\mathcal{C}$ is the supremum of the turns of the polygonal lines inscribed in $\mathcal{C}$ (which are, any polygonal chain formed by points ordered along $\mathcal{C}$).*

Following this idea, another property that will be of great use for our final proposition is the convergence of the length of a sequence of polygonal lines, depicted in Property 3.

Property 3 ([AR89, Corollary 2.1.2])**.** *Let (X^n, d) be a n dimensional metric space. Let $\mathcal{C}$ be a curve in X^n, and let $(L_m)_{m\in\mathbb{Z}}$ be a sequence of polygonal lines inscribed in $\mathcal{C}$ such that $\lim_{m\to\infty} \lambda(L_m) = 0$. Then, we have $\lim_{m\to\infty} \mathcal{L}(L_m) = \mathcal{L}(\mathcal{C})$.*

We can notice that in their book [AR89], where most of our definitions come from, A. Alexandrov and Y. Reshetnyak mention the equivalence between the name "turn", and "*integral curvature*". In point of fact, the turn coincides, for the case of regular curves, with the integral of the curve's curvature with respect to an arc-length parameterization.

Property 4 ([AR89, part 5.3.1])**.** *Let $\mathcal{C}$ be a curve of the class C^2 and let $\gamma : [t_0, t_1] \to \mathcal{C}$ be a parameterization of $\mathcal{C}$. Then we have*

$$\kappa(\mathcal{C}) = \int_{t_0}^{t_1} |k(t)| \mathrm{d}t$$

where $k(t)$ is the curvature of $\mathcal{C}$ at point $\gamma(t)$.

3 Locally Turn-Bounded Surfaces

The notion of local turn-boundedness was first introduced in [LMBT20] for two dimensional spaces, as a local geometric feature based on the turn. A Jordan curve $\mathcal{C}$ is said to be *δ-locally-turn-bounded* if, for each pair of points a and b in $\mathcal{C}$ such that their Euclidean distance $d(a, b)$ is less than δ, the turn of one of the arcs of $\mathcal{C}$ delimited by a and b is less than or equal to $\frac{\pi}{2}$. More generally, (θ, δ)-locally-turn-bounded curves follow the same definition with a turn bounded by a given θ. For the sake of readability, we will shorten *locally turn-bounded* by *LTB* in the rest of the article.

By bounding locally the turn, LTB curves filter any curve containing parts that are too thin or that have a lot of local variations, while encompassing a large variety of shapes, including angular ones like rectangles and some other polygons.

Based on the work previously made on LTB curves, we propose an extension of local turn-boundedness to three dimensional spaces, introducing a new class of shapes called LTB surfaces.

Definition 5 (δ-LTB surface). *Let S be a topological surface possibly with boundary in $\mathbb{R}^3$. The surface S is said to be δ-locally turn-bounded (δ-LTB) if for each pair of points a and b in S such that $d(a, b) < \delta$, there exists at least one curve $\mathcal{C}_a^b$ lying on S with a turn that is less than or equal to $\frac{\pi}{2}$.*

Like for δ-LTB curves, it is possible to generalize δ-LTB surfaces to the class of (θ, δ)-LTB surfaces that have a turn bounded by $\theta \in [0, 2\pi]$ instead of $\frac{\pi}{2}$. However, even if (θ, δ)-LTB curves are more flexible and more inclusive for $\theta > \frac{\pi}{2}$, it has been shown in [LMBT20] that the discretization results obtained by setting a $\theta \leq \frac{\pi}{2}$ do not hold for a larger theta. For example, the Gauss digitization of LTB curves for a compatible grid step h is always a Jordan curve with a 4-connected interior only when $\theta \leq \frac{\pi}{2}$. Accordingly, we will essentially focus on δ-LTB surfaces rather than (θ, δ)-LTB ones.

With the Definition 5, it is apparent that verifying the local turn-boundedness of a surface S is not an easy operation, since contrary to planar curves, two points on S delimit not only two arcs on the shape, but an infinity of them. In consequence, we observe that two main challenges arise: the first being the choice of the two points a and b, and the second being the searching for a minimal path between them, with respect to the turn.

Below are presented a few elementary examples of δ-LTB surfaces.

Proposition 6. *A cube with edge size c is c-LTB.*

Proof: Let C be a cube with edges size c. Each pair of points a and b at distance strictly shorter than c is either of the same face of the cube, either on two adjacent faces. In the first case, it is possible to draw a straight line between a and b, with turn 0. In the second case, let a' be the orthogonal projection of a on the edge separating the two points. Then, we have aa' orthogonal to $a'b$, and the polygonal line $aa'b$ has a turn of exactly $\frac{\pi}{2}$. The cube of size c is thus c-LTB. An illustration of this proof is depicted in Fig. 2. □

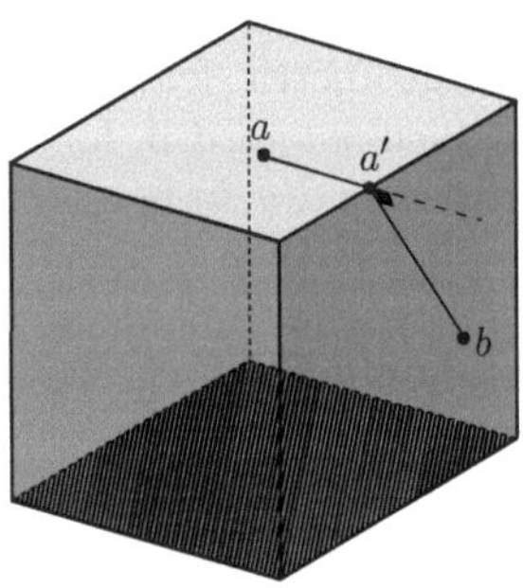

Fig. 2. The polygonal line $aa'b$ has a turn of $\frac{\pi}{2}$.

Proposition 7. *A sphere with radius r is $\sqrt{2}r$-LTB.*

Proof: Let S be a sphere with radius r. Let a and b be two points on S. Then, we know that there exists a great circle Γ (which is a cross section formed by the intersection of S and a plane passing through its center), and only one, passing through a and b. We consider $\mathcal{C}_a^b$ the shortest arc of Γ delimited by a and b, and denote θ its central angle. The notations are illustrated in Fig. 3. Since $\mathcal{C}_a^b$ is an arc of circle, we know that its curvature is equal to $\frac{1}{r}$ everywhere. Hence, by Property 4, the turn of $\mathcal{C}_a^b$ is

$$\kappa(\mathcal{C}_a^b) = \frac{1}{r}\,\mathcal{L}(\mathcal{C}_a^b) = \frac{1}{r}r\theta = \theta$$

Thus, we have $\kappa(\mathcal{C}_a^b) = \theta$. The sphere of radius r is thus $(\theta, 2r\sin(\frac{\theta}{2}))$-LTB, and in particular, if $\theta = \frac{\pi}{2}$, it is $\sqrt{2}r$-LTB.

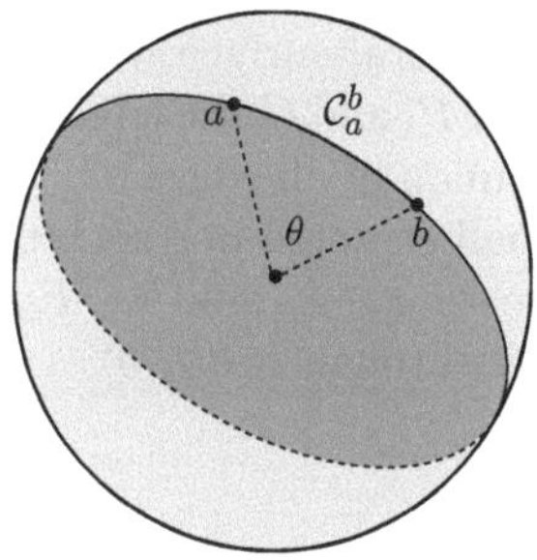

Fig. 3. The arc $\mathcal{C}_a^b$ is the shortest arc of Γ delimited by a and b.

Remark 8. *$\sqrt{2}r$ is the maximal value of δ for a sphere S of radius r to be δ-LTB.*

Proof: Set $a, b \in S$ such that $d(a,b) > \sqrt{2}r$. The length of the shortest arc S_a^b of S joining a and b is strictly greater than $\frac{\pi}{2}r$. Because the curvature of S is exactly $\frac{1}{r}$ everywhere, by Property 4 we have $\kappa(S_a^b) > \frac{\pi}{2}$.

In order to be able to further investigate with the notion of δ-LTB surfaces, we present from now on some primary propositions, often directly extended from the 2D case.

Proposition 9. *Let S be a δ-LTB surface. Let a and b be two points of S such that $d(a,b) < \delta$. The curve on S delimited by a and b and with minimal turn is included in the ball of diameter $[a,b]$.*

Proof: Since $d(a,b) < \delta$, by the definition of δ-LTB surfaces, the curve $\mathcal{C}_a^b$ of minimal turn delimited by a and b has a turn bounded by $\frac{\pi}{2}$. Let c be a point on $\mathcal{C}_a^b$. Then, by definition of the turn, we have $\kappa([a,c,b]) \leq \kappa(\mathcal{C}_a^b) \leq \frac{\pi}{2}$, thus $\widehat{acb} \geq \frac{\pi}{2}$.

Corollary 10. *A cube with edge size c is not δ-LTB if $\delta > c$.*

Proof. Let a and b the centers of two opposite faces of the cube. By 9, any arc of turn upper-bounded by $\frac{\pi}{2}$ is included in the ball of diameter $[a, b]$. The ball does not intersect the square delimiting the face containing a. The, there exist no arc of the cube whose turn is upper-bounded by $\frac{\pi}{2}$ connecting a and b. □

Thanks to the Proposition 9, we demonstrate through the following proposition that even though smoothness is not imposed (allowing the inclusion of polyhedrons), the locally turn-boundedness hypothesis controls not only the local curvature of certain curves but also the curvature of the whole surface.

Proposition 11. *Let P be a δ-LTB polyhedral surface with finite vertex set. Then, the dihedral angle between each pair of adjacent faces of P is obtuse.*

Proof: Let P be a δ-LTB polyhedral surface with finite vertex set. Let F_A and F_B denote two adjacent faces of P, and C the edge separating them. We set c the midpoint of C. Let E be the union of all edges of P, and set $r < d(c, E \setminus C)$. We denote by $\overline{B}(c, r)$ the closed ball centered in c such that $\overline{B}(c, r)$ cannot cross any other edge than C. Finally, we set $a \in F_A$ and $b \in F_B$ such that a and b are in the plane orthogonal to C passing through c, $d(a, b) < \delta$ and $d(c, a) = d(c, b) < \frac{r}{\sqrt{2}}$, as depicted in Fig. 4.

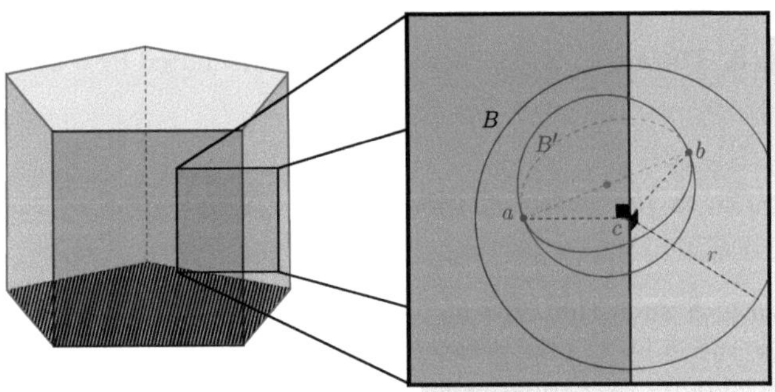

Fig. 4. A zoom on an edge of a pentagonal prism: the curve of minimal turn is included in the ball B'.

Then, we know that the ball B' of diameter $[a, b]$ is included in $\overline{B}(c, r)$ and does not intersect any other edge than C either. Moreover, the polygonal line acb is the curve joining a and b with a minimal turn, and corresponds exactly to the dihedral angle of F_A and F_B. Then, by definition of LTB surfaces, acb has a turn upper-bounded by $\frac{\pi}{2}$, and the angle between the two faces is obtuse. □

Proposition 12 hereafter lets us see that δ-LTB surfaces directly inherit the local connectivity property from their two-dimensional counterparts. This notably implies the lower bounding on the shape's overall thickness and more generally its stability through morphological dilation.

Proposition 12 (Local connectivity). *Let S be a δ-LTB surface. Let $p \in S$. The intersection of S with an open ball in $\mathbb{R}^3$ centered in p and of radius $\epsilon \leq \delta$ is path-connected.*

Proof: Let p be a point on S, and let $B(p, \epsilon)$ be the ball centered in p with radius $\epsilon \leq \delta$. Since S is δ-LTB, for all $q \in S \cap B(p, \epsilon)$, there exists an arc $\mathcal{C}_p^q \in S \cap B(p, \epsilon)$ with a turn less than or equal to $\frac{\pi}{2}$. By Proposition 9, the arc $\mathcal{C}_p^q$ is included in a ball of diameter $[p, q[$, itself included in $B(p, \epsilon)$. Thus, $S \cap B(p, \epsilon)$ is path connected. □

4 About Local Topology of LTB Surfaces

A major focus point of this study is made on local topology of LTB surfaces. Indeed, to favor topological preservation, we want to work with surfaces without topological artefacts. Moreover, three-dimensional surfaces are richer than planar curves when it comes to topological information, which in the other hand make their study much more complex. Accordingly, in order to be able to determine propositions on LTB surfaces, we make use in this section of concepts taken from Riemannian geometry and also algebraic topology detailed below.

Property 13 (Arc-length upper bound, [AR89, Theorem 5.8.1]**).** *Let $\mathcal{C}$ be an arc such that $\kappa(\mathcal{C}) < \pi$ and let d be the Euclidean distance between its two endpoints. Then, its length $\mathcal{L}(\mathcal{C})$ is upper bounded:*

$$\mathcal{L}(\mathcal{C}) \leq \frac{d}{\cos \frac{\kappa(\mathcal{C})}{2}}$$

Equality is reached if and only if $\mathcal{C}$ is a polygonal line formed by two segments of equal length.

Definition 14 (Distortion, [Gro07, section 1.14]**).** *Let X be a subset of a path metric space A (which is a set together with a notion of distance between its elements that is equal to the infimum of the lengths of the curves joining them). We denote d_X and d_A the respective induced metrics on X and A. The distortion of X, $\operatorname{distort}(X)$, denotes the dilation of the identity map on X with respect to the two induced metrics:*

$$\operatorname{distort}(X) = \sup_{x,y \in X} \frac{d_X(x, y)}{d_A(x, y)}$$

In our particular case, it is to note that the path metric space A of the Definition 14 corresponds to $\mathbb{R}^3$, and therefore that the induced metrics d_A corresponds to the Euclidean distance d. X is depicted as a curve lying on a surface S and is measured with the induced metric on S.

Remark 15 ([Gro07, Remark 1.13.b])**.** *Let (X, d) be a compact path metric space. Then, every free homotopy class (which is, any class of curves that are homotopic to each others) is represented by a length minimizing curve.*

In [Gro07, Sec. 1.14], M. Gromov proves that for any compact subset X of $\mathbb{R}^n$, if distort(X) is upper bounded by $\frac{\pi}{2}$, then X is simply connected. However, the original proof requires the surface to be differentiable, which may not be the case for δ-LTB surfaces. When we investigated how this could still apply in our case, experimenting with twisting Gromov's proof led us to Proposition 17. This proposition lets us observe that δ-LTB surfaces control the size of the holes that they may contain, forcing them to have a diameter at least equal to δ. The following lemma asserts an inequality on the distortion of a closed curve, which will be of use in Proposition 17.

Lemma 16. *Let c be a closed curve of image $\mathcal{C}$, and let l denote its length. By extending c to a periodic function on $\mathbb{R}$, set $x(s) := c(\frac{l}{2} + s) - c(s)$, $r(s) := d(c(\frac{l}{2} + s), c(s))$ and $u(s) := \frac{x(s)}{r(s)}$. Then,*

$$\mathcal{L}(u) \leq 4\,\mathrm{distort}(\mathcal{C})$$

For a better visualization, the notations of Lemma 16 are illustrated in Fig. 5.

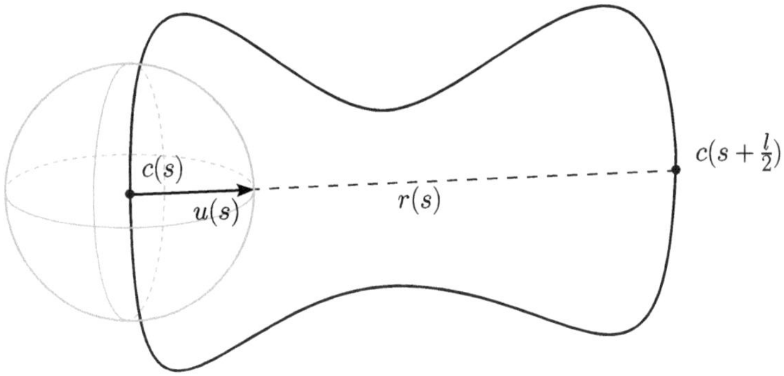

Fig. 5. $u(s)$ is the unit vector from $c(s)$ to $c(s+\frac{l}{2})$ lying within the unit sphere in $\mathbb{R}^n$.

Proof: Let $\epsilon > 0$. Since u is a unit vector along c, it traces out a certain curve on the unit sphere in $\mathbb{R}^n$. Let L_u be a polygonal line inscribed in this curve. Then by Property 3, we know that there exists a η_0 such that if $\lambda(L_u) < \eta_0$, then $\mathcal{L}(u) \leq \mathcal{L}(L_u) + \epsilon$. Let $k := \frac{l}{2\,\mathrm{distort}(\mathcal{C})}$. By continuity of u on the compact set $[0, l]$, u is uniformly continuous on $[0, l]$. Then there exists $\eta := \frac{l}{2N}$ with $N \in \mathbb{N}^*$ such that for any $s \in [0, l]$, $||u(s + \eta) - u(s)|| \leq \eta_0$ and $\eta \leq \frac{k}{2}\frac{(1+\epsilon)^2-1}{(1+\epsilon)^2}$. Hence,

$$\mathcal{L}(u) \leq \sum_{i=0}^{2N-1} ||u((i+1)\eta) - u(i\eta)|| + \epsilon \tag{1}$$

Let $\Delta f(s) := f(s + \eta) - f(s)$ for any f. As a first step, we have:

$$\begin{aligned}\langle \Delta u(s), u(s+\eta) + u(s)\rangle &= \|u(s+\eta)\|^2 - \|u(s)\|^2 \\ &= 0 \end{aligned} \tag{2}$$

Then $\langle \Delta u(s), u(s+\eta)\rangle = -\langle \Delta u(s), u(s)\rangle$, hence:

$$\begin{aligned}\|\Delta u(s)\|^2 &= \langle \Delta u(s), u(s+\eta)\rangle - \langle \Delta u(s), u(s)\rangle \\ &= -2\langle \Delta u(s), u(s)\rangle.\end{aligned}$$

Thus

$$\langle \Delta u(s), u(s)\rangle = -\frac{1}{2}\|\Delta u(s)\|^2. \tag{3}$$

Secondly, let $\rho(s) := \frac{\Delta r(s)}{r(s+\eta)}$. Then, we have:

$$\begin{aligned}\Delta u(s) &= \frac{x(s+\eta)}{r(s+\eta)} - \frac{x(s)}{r(s)} \\ &= \frac{x(s+\eta)r(s) - x(s)r(s+\eta)}{r(s+\eta)r(s)} \\ &= \frac{x(s+\eta)r(s) - x(s)r(s) + x(s)r(s) - x(s)r(s+\eta)}{r(s+\eta)r(s)} \\ &= \frac{\Delta x(s)}{r(s+\eta)} - \frac{x(s)}{r(s)}\frac{\Delta r(s)}{r(s+\eta)} \\ &= \frac{\Delta x(s)}{r(s+\eta)} - u(s)\rho(s)\end{aligned} \tag{4}$$

Thus

$$\begin{aligned}\|\Delta u(s)\|^2 &= \langle \Delta u(s), \Delta u(s)\rangle \\ &= \left\langle \frac{\Delta x(s)}{r(s+\eta)}, \Delta u(s)\right\rangle + \rho(s)\langle -u(s), \Delta u(s)\rangle && \text{by Eq. 4} \\ &= \left\langle \frac{\Delta x(s)}{r(s+\eta)}, \Delta u(s)\right\rangle + \frac{1}{2}\rho(s)\|\Delta u(s)\|^2 && \text{by Eq. 3} \\ &= \left\langle \frac{\Delta x(s)}{r(s+\eta)}, \frac{\Delta x(s)}{r(s+\eta)} - u(s)\rho(s)\right\rangle + \frac{1}{2}\rho(s)\|\Delta u(s)\|^2 && \text{by Eq. 4} \\ &= \left\|\frac{\Delta x(s)}{r(s+\eta)}\right\|^2 - \left\langle \frac{\Delta x(s)}{r(s+\eta)}, u(s)\rho(s)\right\rangle + \frac{1}{2}\rho(s)\|\Delta u(s)\|^2\end{aligned}$$

$$\|\Delta u(s)\|^2 = \left\|\frac{\Delta x(s)}{r(s+\eta)}\right\|^2 - \langle \Delta u(s) + u(s)\rho(s), u(s)\rho(s)\rangle + \frac{1}{2}\rho(s)\|\Delta u(s)\|^2$$

by Eq. 4

$$= \left\|\frac{\Delta x(s)}{r(s+\eta)}\right\|^2 - \rho(s)\langle \Delta u(s), u(s)\rangle - \rho(s)^2\|u(s)\|^2 + \frac{\rho(s)}{2}\|\Delta u(s)\|^2$$

$$= \left\|\frac{\Delta x(s)}{r(s+\eta)}\right\|^2 - \rho(s)^2 - \rho(s)\langle \Delta u(s), u(s)\rangle + \frac{\rho(s)}{2}\|\Delta u(s)\|^2$$

since u is a unit vector

$$= \left\|\frac{\Delta x(s)}{r(s+\eta)}\right\|^2 - \rho(s)^2 + \rho(s)\|\Delta u(s)\|^2$$

by Eq. 3

$$\leq \left\|\frac{\Delta x(s)}{r(s+\eta)}\right\|^2 + \rho(s)\|\Delta u(s)\|^2$$

By definition of $\mathrm{distort}(\mathcal{C})$ and $r(s)$, for any s we have $\frac{l/2}{r(s)} \leq \mathrm{distort}(\mathcal{C})$. Then, $k = \frac{l}{2\,\mathrm{distort}(\mathcal{C})} \leq r(s)$, hence $\frac{1}{r(s)} \leq \frac{1}{k}$ for any s. Then:

$$\rho(s) = \frac{\Delta r(s)}{r(s+\eta)} \leq \frac{\Delta r(s)}{k} \leq \frac{2\eta}{k}.$$

Hence:

$$\begin{aligned}
\|\Delta u(s)\|^2 &\leq \frac{1}{k^2}\|\Delta x(s)\|^2 + \frac{2\eta}{k}\|\Delta u(s)\|^2 \\
&\leq \frac{1}{k^2}\left\|c\left(\frac{l}{2}+s+\eta\right) - c(s+\eta) - c\left(\frac{l}{2}+s\right) + c(s)\right\|^2 + \frac{2\eta}{k}\|\Delta u(s)\|^2 \\
&\leq \frac{1}{k^2}\left\|\left(c\left(\frac{l}{2}+s+\eta\right) - c\left(\frac{l}{2}+s\right)\right) + (c(s) - c(s+\eta))\right\|^2 + \frac{2\eta}{k}\|\Delta u(s)\|^2 \\
&\leq \frac{1}{k^2}\left(\mathcal{L}\left(\mathcal{C}_{c(\frac{l}{2}+s+\eta)}^{c(\frac{l}{2}+s)}\right) + \mathcal{L}\left(\mathcal{C}_{c(s)}^{c(s+\eta)}\right)\right)^2 + \frac{2\eta}{k}\|\Delta u(s)\|^2 \\
&\leq \frac{1}{k^2}(\eta+\eta)^2 + \frac{2\eta}{k}\|\Delta u(s)\|^2 \qquad \text{since } c \text{ is parametrized by arc-length} \\
&\leq \frac{4\eta^2}{k^2} + \frac{2\eta}{k}\|\Delta u(s)\|^2
\end{aligned}$$

Since $\eta \leq \frac{k}{2}\frac{(1+\epsilon)^2-1}{(1+\epsilon)^2}$,

$$\begin{aligned}
\|\Delta u(s)\|^2 &\leq \frac{\left(\frac{2\eta}{k}\right)^2}{1-\frac{2\eta}{k}} \\
&\leq (1+\epsilon)^2\left(\frac{2\eta}{k}\right)^2
\end{aligned}$$

Then

$$\begin{aligned}\|\Delta u(s)\| &\leq (1+\epsilon)\frac{2\eta}{k} \\ &\leq (1+\epsilon)4\,\text{distort}(\mathcal{C})\frac{\eta}{l} && \text{by definition of } k\end{aligned}$$

Hence:

$$\begin{aligned}\mathcal{L}(u) &\leq \sum_{i=0}^{2N-1} ||\Delta u(i\eta)|| + \epsilon && \text{by Eq. 1} \\ &\leq 2N(1+\epsilon)4\,\text{distort}(\mathcal{C})\frac{\eta}{l} + \epsilon \\ &\leq (1+\epsilon)4\,\text{distort}(\mathcal{C}) + \epsilon && \text{by definition of } N \text{ and } \eta\end{aligned}$$

Finally, as ϵ tends to 0:

$$\mathcal{L}(u) \leq 4\,\text{distort}(\mathcal{C})$$

□

Proposition 17. *Let S be a compact δ-LTB surface possibly with boundary. Then, every non-trivial loop in S has diameter greater than or equal to δ.*

Proof: For the sake of contradiction, suppose that there exists a non-trivial loop c in S of minimal length l, and of image $\mathcal{C}$ such that $\text{diam}(\mathcal{C}) < \delta$. Because S is compact, the existence of such a curve is ensured by Remark 15.

We divide the proof in two main steps: first, the upper bounding of the distortion of $\mathcal{C}$. Then, by setting a unit vector u from one point of c to its antipode, we shall arrive to a contradiction regarding the length of the curve traced by u and the distortion hypothesis.

First let us prove that $\text{distort}(\mathcal{C}) \leq \sqrt{2}$. Let $a, b \in \mathcal{C}$. Then $d(a,b) < \delta$ and hence there exists an arc $S_a^b \subset S$ joining a and b and such that $\kappa(S_a^b) \leq \frac{\pi}{2}$. Hence by Property 13, $\mathcal{L}(S_a^b) \leq \sqrt{2}d(a,b)$. Moreover let $\mathcal{C}_a^b$ be the shortest arc of $\mathcal{C}$ between a and b.

- If $S_a^b \cup (\mathcal{C} \setminus \mathcal{C}_a^b)$ is a non-trivial loop, by minimality of $\mathcal{C}$, $\mathcal{L}(\mathcal{C}) \leq \mathcal{L}(S_a^b \cup (\mathcal{C} \setminus \mathcal{C}_a^b))$, then $\mathcal{L}(\mathcal{C}_a^b) \leq \mathcal{L}(S_a^b) \leq \sqrt{2}d(a,b)$ (see Fig. 6).
- Otherwise, $\mathcal{C}_a^b \cup S_a^b$ is a non-trivial loop. By minimality of $\mathcal{C}$, $\mathcal{L}(\mathcal{C}) \leq \mathcal{L}(\mathcal{C}_a^b \cup S_a^b)$. Then $\mathcal{L}(S_a^b) \geq \mathcal{L}(\mathcal{C} \setminus \mathcal{C}_a^b) \geq \mathcal{L}(\mathcal{C}_a^b)$, then $\mathcal{L}(\mathcal{C}_a^b) \leq \sqrt{2}d(a,b)$ (see Fig. 7).

Hence $\text{distort}(\mathcal{C}) \leq \sqrt{2}$.

Define the parametrized curve u as in Lemma 16,

$$u(s) := \frac{c(\frac{l}{2}+s) - c(s)}{d(c(\frac{l}{2}+s), c(s))}$$

By Lemma 16, $\mathcal{L}(u) \leq 4\,\text{distort}(\mathcal{C}) \leq 4\sqrt{2}$. Because we have $u(s+\frac{l}{2}) = -u(s)$, we have $\mathcal{L}(u) \geq 2\pi$. Thus, $2\pi \leq \mathcal{L}(u) \leq 4\sqrt{2} < 2\pi$, which is the desired contradiction. □

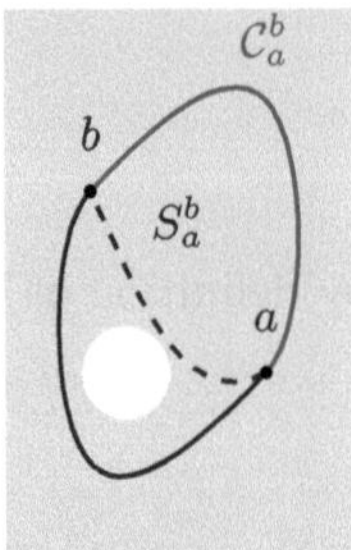

Fig. 6. $S_a^b \cup (\mathcal{C} \setminus \mathcal{C}_a^b)$ is a non-trivial loop

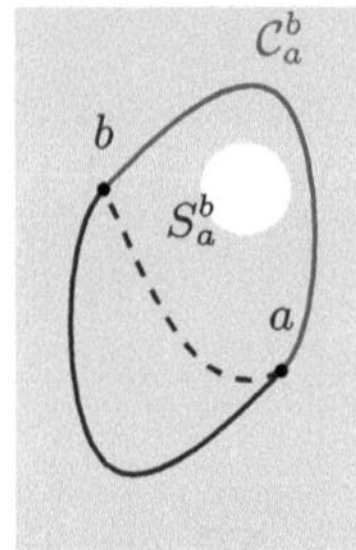

Fig. 7. $\mathcal{C}_a^b \cup S_a^b$ is a non-trivial loop

Figure 8 illustrates the last proposition with torus of different sizes: the torus containing two independent non-trivial loops running respectively around its central hole and around its tube, local turn-boundedness restricts it from shrinking or inflating as the process shortens either loop.

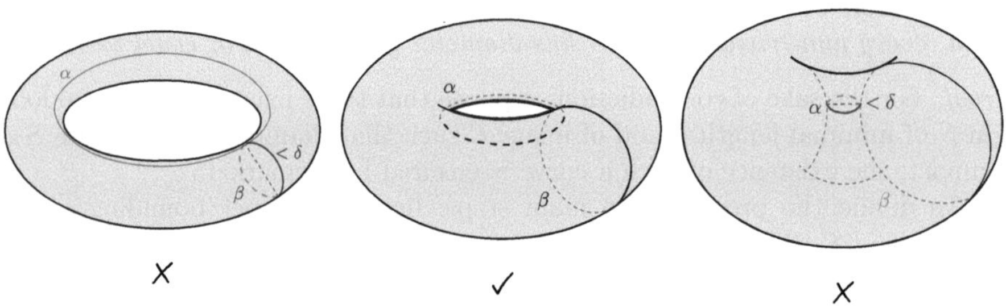

Fig. 8. The torus in the middle is δ-LTB and then has no non-trivial loop with diameter less than δ, contrary to the torus on the left and on the right.

5 Conclusion

In this article, we proposed an extension to the notion of local turn-boundedness to three dimensions through LTB surfaces. In particular, we presented some initial propositions on the topology of LTB surfaces, such as local connectivity, upper bounding of dihedral angles, and lower bounding of holes diameters. By demonstrating those fundamental properties, the results established in this study form a first proof of concept of our three dimensional extension, opening the way to further work regarding hypothesis on the continuous shape. Indeed, as topology in three dimensions is much more complex than in two dimensions, it is apparent that there is still work to do before moving on to discrete space. For example, simple connectivity of the intersections of a δ-LTB surface with an open ball of radius $\frac{\delta}{2}$ that has still not been proven yet, and will be a direct topic of interest of our further works. Moreover, another challenge arises in three dimension, namely the optimization (maximization) of the δ parameter. If this problem seems highly complex in the case of smooth surfaces, intuitions appears for regular ones, and while an algorithmic solution exists for 2D polygons, this problem merits no less attention. The result on the inclusion of the arc of minimal turn presented in this paper is, in particular, relevant to answer this problematic, and we hope that those first results will pave the way to an extension of the algorithm to polyhedrons.

References

[Mil50] Milnor, J.W.: On the total curvature of knots. Ann. Math. **52**(2), 248 (1950). ISSN: 0003486X

[Pav82] Pavlidis, T.: Algorithms for Graphics and Image Processing. Springer, Heidelberg (1982)

[Ser82] Serra, J.: Image analysis and mathematical morphology, vol. 1, 4th print. Academic Press, London (1982)

[AR89] Alexandrov, A.D., Reshetnyak, Y.G.: General Theory of Irregular Curves Ed. by M. Hazewinkel, vol. 29. Mathematics and Its Applications. Springer, Dordrecht (1989)

[Ste05] Stelldinger, P.: Digitization of non-regular shapes. In: Ronse, C., Najman, L., Decencière, E. (eds.) Mathematical Morphology: 40 Years On. Computational Imaging and Vision, vol. 30, pp. 269–278. Springer, Heidelberg (2005)

[Gro07] Gromov, M.: Metric structures for Riemannian and non-Riemannian spaces. Modern Birkhäuser classics. Birkhäuser, Boston (2007)

[CCL09] Chazal, F., Cohen-Steiner, D., Lieutier, A.: A sampling theory for compact sets in euclidean space. Discret. Comput. Geom. **41**(3), 461–479 (2009). https://doi.org/10.1007/s00454-009-9144-8

[MKS09] Meine, H., Köthe, U., Stelldinger, P.: A topological sampling theorem for Robust boundary reconstruction and image segmentation. Discret. Appl. Math. **157**(3), 524–541 (2009)

[NPKD19] Ngo, Phuc, et al.: Geometric preservation of 2D digital objects under rigid motions. J. Math. Imaging Vision **61**(2), 204–223 (2019)

[LMBT20] Quentrec, É.L., et al.: Local turn-boundedness: a curvature control for continuous curves with application to digitization. J. Math. Imaging Vision **62**(5), 673–692 (2020)

Computational Aspects of Discrete Structures and Tilings

How to Peel Fully Convex Digital Sets

Fabien Feschet[1] and Jacques-Olivier Lachaud[2](✉)

[1] Clermont Auvergne University, CNRS, ENSMSE, LIMOS, 63000 Clermont-Ferrand, France
fabien.feschet@u-auvergne.fr

[2] University Savoie Mont Blanc, CNRS, LAMA, 73000 Chambéry, France
jacques-olivier.lachaud@univ-smb.fr

Abstract. Full convexity is an interesting alternative to classical digital convexity since it guarantees connectedness and even simple connectedness in digital spaces $\mathbb{Z}^d$, for any dimension d. This paper aims at giving a better understanding of the monotonicity properties of fully convex digital sets, since earlier works showed that the question was difficult for thin fully convex sets. To decipher the hierarchy of fully convex sets ordered by inclusion, we study how we can peel a fully convex set progressively while keeping its full convexity. We provide a characterization of peelable points and fast algorithms to identify them. Furthermore, we show that fully convex set can be peeled one point at a time till reduced to the empty set, similarly to digitally convex sets in the classical sense. The peeling of a fully convex set can be seen as an analog to homotopic thinning processes, but with an additional geometric property.

Keywords: Digital convexity · Homotopic thinning · Digital geometry

1 Introduction

Convexity is a fundamental tool to analyse the geometry of shapes in Euclidean spaces, and it is the basis of many other fields like convex optimization. Our aim is to get a better understanding of the properties of full convexity. Full convexity is a recent alternative definition to digital convexity in the digital lattice $\mathbb{Z}^d$ [16,17], which has the considerable advantage of guaranteeing in arbitrary dimension the digital connectedness of its elements and even the simple connectedness of their geometric realization. This is in stark contrast with classical digital convexity and to most other adaptations of digital convexity (see [15] for a recent overview). Indeed, other geometric definitions of digital convexity add the digital connectedness as an external property [6,11–13,21], but this trick does not give any insight at what is a true digital analog to convexity and is not helpful in dimension greater or equal to three. Besides, axiomatic definitions of digital convexity [18,26] or more functional versions [14,19,22] lose the Euclidean geometric intuition of convexity and its relation to digital planarity.

Full convexity is already showing a strong potential in digital geometry analysis [7,8,16,17]: local characterizations of convex and concave parts, unambiguous

M. H. F. Wilkinson and J. Kosinka (Eds.): DGMM 2025, LNCS 16296, pp. 77–89, 2026.
https://doi.org/10.1007/978-3-032-09544-2_6

notion of visibility, exact geodesics, tangent planes, computation time comparable to classical convexity. Fully convex shapes also includes standard digital primitives like digital planes and balls. One can design a fully convex envelope operator idempotent on fully convex set, which gives a proper definition of polyhedral model. Full convexity also enjoys two new nice characterizations [9,10]. However these works have highlighted a difficulty that is proper to the fully convex envelope operator, and which is certainly related to the particular connectedness properties of fully convex sets: the envelope operator may not be increasing, and this may happen around "thin" digital sets.

In order to understand better this peculiar hierarchical structure of fully convex sets and regular digital sets in-between, we propose in this paper to study the *peeling* of a fully convex set. More precisely, we determine when the full convexity property is preserved after removing one point, or even removing several points in parallel. This can be seen as a generalization of the works of Tarsissi *et al.* [23–25], which addresses the case of convex polyominoes (hence 4-connected digital convex objects of $\mathbb{Z}^2$) and which focuses on the word combinatorics formulation of these objects [3]. There is also a strong relation with the fruitful works on simple points and homotopic thinning [2,4], which has led to many applications in 3D imaging [20], see also [1,5] for parallel homotopic thinning. In a sense, peeling is a more restrictive homotopic thinning, since it preserves homotopy and convexity, and that kind of property could be of interest in many skeletonization or deformation algorithms.

The paper is organized as follows. Section 2 presents the necessary background information, notations and results. Section 3 provides sufficient conditions for peelability, valid in a sequential or in a parallel process. Section 4 focuses on the characterization of peelability. It is shown to be locally decidable in $\mathbb{Z}^2$, but no more starting from $\mathbb{Z}^3$. We also show that any fully convex set is peelable one point at a time till reduced to the empty set, confirming the hierarchical (yet not arborescent) nature of the space of fully convex sets. Section 5 summarizes the results and outlines some perspectives to this work.

2 Basic Notions

We introduce here basic definitions and properties needed in the rest of the paper (they can be found in [8,10,17]). In the sequel, $\mathcal{C}^d$ is the cubical cell complex induced by $\mathbb{Z}^d$. Its 0-dimensional cells are identified to points of $\mathbb{Z}^d$. The set $\mathcal{C}^d$ is the set of open k-dimensional cells of $\mathcal{C}^d$.

The *(topological) boundary* ∂Y of a subset Y of $\mathbb{R}^d$ is the set of points in its closure but not in its interior. The star of a cell σ in $\mathcal{C}^{\mathrm{d}}$, denoted by $\mathrm{Star}\,(\sigma)$, is the set of cells of $\mathcal{C}^{\mathrm{d}}$ whose boundary contains σ, plus the cell σ itself. The closure $\mathrm{Cl}\,(\sigma)$ of σ contains σ and all the cells in its boundary. In this paper, the cell boundary operator, also denoted by ∂, maps a k-cell to all its proper faces, that is all its k'-cells, $0 \leqslant k' < k$, and not only its $(k-1)$-cells.

A subcomplex K of $\mathcal{C}^{\mathrm{d}}$ with $\mathrm{Star}\,(K) = K$ is *open*, while being *closed* when $\mathrm{Cl}\,(K) = K$. The *body* of a subcomplex K, i.e. the union of its cells in $\mathbb{R}^d$, is written $\|K\|$.

For any real subset Y of $\mathbb{R}^d$, we denote by $\bar{\mathcal{C}}^d_k[Y]$ the set of k-cells whose topological closure intersects Y, i.e. $\bar{\mathcal{C}}^d_k[Y] = \{c \in \mathcal{C}^d_k, \bar{c} \cap \ Y \neq \emptyset\}$, where $\bar{c} = \|\mathrm{Cl}\,(c)\,\|$ for any cell c. For any subset $Y \subset \mathbb{R}^d$, it is natural to define $\mathrm{Star}\,(Y) := \bar{\mathcal{C}}^d[Y] = \cup_{0 \leqslant k \leqslant d} \bar{\mathcal{C}}^d_k[Y]$. Last, the set $\mathrm{CvxH}\,(Y)$ is the *convex hull* of Y in $\mathbb{R}^d$.

Definition 1 (Full convexity). *A subset $X \subset \mathbb{Z}^d$ is* digitally k-convex *for $0 \leqslant k \leqslant d$ whenever*

$$\bar{\mathcal{C}}^d_k[X] = \bar{\mathcal{C}}^d_k[\mathrm{CvxH}\,(X)]. \tag{1}$$

Subset X is fully (digitally) convex *if it is digitally k-convex for all $k, 0 \leqslant k \leqslant d$.*

The digital 0-convexity is the classical digital convexity. The following characterization will be useful:

Lemma 1 [7, Lemma 4] *A digital set X is fully convex iff* $\mathrm{Star}\,(X) = \mathrm{Star}\,(\mathrm{CvxH}\,(X))$.

Finite convex sets Y in $\mathbb{R}^d$ are intended as $Y = \mathrm{CvxH}\,(\mathrm{Vtcs}\,(Y))$ where $\mathrm{Vtcs}\,(Y)$ is the set of extreme points of Y, also called *vertices* in this paper. By definition, $\mathrm{Vtcs}\,(Y) \subset Y$. Moreover, any one of the vertices of Y cannot be written as a convex linear combination of other points of Y. Since we are considering finite digital sets, the convex set $\mathrm{CvxH}\,(Z)$ is always a bounded polytope of $\mathbb{R}^d$ for all finite subsets Z of $\mathbb{Z}^d$. A vertex z of a convex set $\mathrm{CvxH}\,(Y)$ is such that z does not belong to any open segment $]u, w[$ with $u, w \in \mathrm{CvxH}\,(Y)$. We can specialize this property for digital sets. Let us denote by $z[\mathbf{e}]$ the line with direction $\mathbf{e}$ passing through z. It is the union z with two open rays $z^+[\mathbf{e}]$ and $z^-[\mathbf{e}]$ which are both semi-infinite in the direction of $+\mathbf{e}$ and resp. $-\mathbf{e}$. We then have the following property.

Lemma 2. *For $X \subset \mathbb{Z}^d$, $z \in \mathrm{Vtcs}\,(\mathrm{CvxH}\,(X))$ and $\varepsilon \in \{+, -\}$, $z^\varepsilon[\mathbf{e}] \cap X \neq \emptyset$ $\Rightarrow z^{-\varepsilon}[\mathbf{e}] \cap X = \emptyset$.*

Proof. Indeed, if both intersections are not empty then z is a convex combination of others points of X and cannot be a vertex of $\mathrm{CvxH}\,(X)$. □

Fully convex sets are stable with respect to orthogonal projections along axes in $\mathbb{R}^d$. Let $I_d = \{1, \ldots, d\}$ and for $j \in I_d$, let π_j denote the orthogonal projector associated to the j-th axis, which consists in omitting the j-th coordinates for all points of $\mathbb{R}^d$. The image of a point in $\mathbb{Z}^d$ by any axis projector is a point in $\mathbb{Z}^{d-1}$. Let us recall the definition of P-convexity [10] and its relation with full convexity.

Definition 2 (P-convexity). *Let $X \subset \mathbb{Z}^d$ be a digital set. The set X is P-*convex *if and only if X is digitally 0-convex (i.e.* $\mathrm{CvxH}\,(X) \cap \mathbb{Z}^d = X$*) and when $d > 1$, for any $j \in I_d$, $\pi_j(X)$ is P-convex in $\mathbb{Z}^{d-1}$.*

Theorem 1 ([10], theorem 5). *For arbitrary dimension $d \geqslant 1$, for any $X \subset \mathbb{Z}^d$, X is fully convex if and only if X is P-convex.*

The property below recalls that the vertices of an axis projected convex set are to be found among the projection of the vertices of the original convex set.

Proposition 1. *Let X be a 0-convex set. Then for each direction $i \in I_d$, $\pi_i(X)$ is 0-convex and $\forall z \in \mathrm{Vtcs}\,(\mathrm{CvxH}\,(\pi_i(X)))$, $\exists z' \in \mathrm{Vtcs}\,(\mathrm{CvxH}\,(X))$, $z = \pi_i(z')$.*

3 Sufficient Conditions for Sequential and Parallel Peelability

We present in this part two possible ways to peel fully convex sets until reaching a unique point. A first way considers the vertices of the convex hull of the digital set and their possible sequential and parallel elimination. A second way is to peel along the bounding box slice per slice.

The following lemma tells that we can peel a digitally convex set by its vertices. This does not however guarantee to maintain full convexity.

Lemma 3. *Let $X \subset \mathbb{Z}^d$ be a 0-convex set, i.e.* $\mathrm{CvxH}(X) \cap \mathbb{Z}^d = X$, *and* $V := \mathrm{Vtcs}(\mathrm{CvxH}(X))$. *Then for any subset of vertices $Z \subset V$, $X \setminus Z$ is 0-convex.*

Proof. Let $X' := X \setminus Z$. We have $\mathrm{CvxH}(X') \cap \mathbb{Z}^d \supset X'$ since $\mathrm{CvxH}(\cdot)$ is increasing and $X' \subset \mathbb{Z}^d$. Let us thus show that $\mathrm{CvxH}(X') \cap \mathbb{Z}^d \subset X'$. This is true when $X' = \emptyset$ so we can suppose that $X' \neq \emptyset$, which implies that $\mathrm{CvxH}(X') \cap \mathbb{Z}^d \neq \emptyset$.

We reason by contradiction: let suppose that $\exists x \in \mathrm{CvxH}(X') \cap \mathbb{Z}^d$, while $x \notin X'$. We have also $x \in \mathrm{CvxH}(X) \cap \mathbb{Z}^d$ (since $\mathrm{CvxH}(\cdot)$ is increasing and $X' \subset X$). The fact that X is 0-convex implies $x \in X$. It follows that $x \in X \setminus X' = Z \subset V$. Since $x \notin X'$ and since $\mathrm{Vtcs}(\mathrm{CvxH}(X')) \subset X'$, $x \notin \mathrm{Vtcs}(\mathrm{CvxH}(X'))$. So, $x \in \mathrm{CvxH}(X') \cap \mathbb{Z}^d$ implies that x is a convex combination of $\mathrm{Vtcs}(\mathrm{CvxH}(X'))$. But $X' \subset X$ implies that x is a convex combination of $\mathrm{Vtcs}(\mathrm{CvxH}(X))$, which is a contradiction with $x \in V = \mathrm{Vtcs}(\mathrm{CvxH}(X))$. □

It is necessary to peel convex hull vertices if we peel X one point at a time.

Lemma 4. *Let $X \subset \mathbb{Z}^d$ be a 0-convex set. Let $z \in X$. Then $X \setminus \{z\}$ is 0-convex if and only z is a vertex of* $\mathrm{CvxH}(X)$.

Proof. ($\Leftarrow$) by using Lemma 3.

($\Rightarrow$) Assume $X \setminus \{z\}$ is 0-convex, i.e. $\mathrm{CvxH}(X \setminus \{z\}) \cap \mathbb{Z}^d = X \setminus \{z\}$. Hence, $z \notin \mathrm{CvxH}(X \setminus \{z\})$ because $z \in \mathbb{Z}^d$. But since $\mathrm{CvxH}(X) = \mathrm{CvxH}(\mathrm{Vtcs}(X))$, $\mathrm{CvxH}(X \setminus \{z\}) \subsetneq \mathrm{CvxH}(X)$ implies that $\mathrm{Vtcs}(\mathrm{CvxH}(X \setminus \{z\})) \neq \mathrm{Vtcs}(\mathrm{CvxH}(X))$. So at least one point of $\mathrm{Vtcs}(\mathrm{CvxH}(X))$ has been removed. Only z was removed so z is a vertex of $\mathrm{CvxH}(X)$. □

We show below that a local test is sufficient to determine if removing a vertex of a fully convex set keeps the full convexity property.

Theorem 2. *Let X be a non empty finite fully convex set of $\mathbb{Z}^d$ and let $Z \subset \mathrm{Vtcs}(\mathrm{CvxH}(X))$ be a subset of its convex hull vertices. Assume that, for every direction $i \in I_d$, for every $z \in Z$, we have $\{z - \mathbf{e}_i, z + \mathbf{e}_i\} \cap (X \setminus Z) \neq \emptyset$. Then $X \setminus Z$ is fully convex.*

Proof. We prove that $X \setminus Z$ is P-convex. First of all, the result is trivial if Z is empty. Otherwise, the first requirement for P-convexity is that $X \setminus Z$ is 0-convex, which is a consequence of Lemma 3. This is enough to conclude if $d = 1$.

Otherwise, if $d \geqslant 2$, we have to prove that for every direction $i \in I_d$, $\pi_i(X \setminus Z)$ is P-convex. Let $X' := X \setminus Z$. Let $Y := \pi_i(Z)$. We consider the two subsets of X and X' that projects exactly onto Y, i.e. $C := \pi_i^{-1}(Y) \cap X$ and $C' := \pi_i^{-1}(Y) \cap X'$. It is clear that C is non empty since Z is included in C, and that $\pi_i(C) = Y$ ($Z \subset C$, $\pi_i(Z) \subset \pi_i(C) \subset Y$ and $\pi_i(Z) = Y$ by definition).

Looking at C', for any $y \in Y$, there is some $z \in Z$ with $z = \pi_i(z)$. By hypothesis $\{z - \mathbf{e}_i, z + \mathbf{e}_i\} \cap (X \setminus Z) \neq \emptyset$. Let $z' \in \{z - \mathbf{e}_i, z + \mathbf{e}_i\} \cap (X \setminus Z)$. Clearly $\pi_i(z') = \pi_i(z \pm \mathbf{e}_i) = y$. Furthermore $z' \in X \setminus Z = X'$ and projects onto Y. So $z' \in C'$. Since for any $y \in Y$, we have found some $z' \in C'$ that projects onto it, we have $\pi_i(C') \supset Y$. And obviously $\pi_i(C') = \pi_i(\pi_i^{-1}(Y) \cap X') \subset Y$. This proves $\pi_i(C') = Y$.

The sets $\pi_i(X \setminus C)$ and $\pi_i(C)$ are disjoint since C contains all the elements of X that projects onto Y. The same is true for $\pi_i(X' \setminus C')$ and $\pi_i(C')$. The equality $X \setminus C = X' \setminus C'$ holds (using $\pi_i(x) \notin Y = \pi_i(Z)$ implies $x \notin Z$) since:

$$x \in X' \setminus C' \Leftrightarrow x \in X, x \notin Z, \pi_i(x) \notin Y \Leftrightarrow x \in X, \pi_i(x) \notin Y \Leftrightarrow x \in X \setminus C.$$

$$\begin{aligned}
\text{Then}\quad \pi_i(X') &= \pi_i(X' \setminus C') \cup \pi_i(C') && \text{(disjoint union and } C' \subset X')\\
&= \pi_i(X' \setminus C') \cup \pi_i(C) && \text{(since } \pi_i(C') = \pi_i(C) = Y)\\
&= \pi_i(X \setminus C) \cup \pi_i(C) && (X \setminus C = X' \setminus C')\\
&= \pi_i(X). && \text{(since } C \subset X)
\end{aligned}$$

Since X was fully convex by hypothesis, it is also P-convex, so its projection $\pi_i(X)$ is P-convex. So $\pi_i(X')$ is P-convex and Theorem 1 concludes. □

The previous theorem induces a very simple algorithm to thin a fully convex set X sequentially up to a given constraint set Y: (i) extract vertices of the convex hull of X, (ii) check if one vertex that is not in Y satisfies the theorem hypothesis, (iii) remove it from X and loop to (i) till no such vertex can be found. However it requires to compute or update the convex hull of X at each step. The same theorem shows sufficient conditions under which we can peel X in parallel. The convex hull must still be updated but for generally less iterations.

It is also quite easy to remove points on the bounding box of a fully convex set X. We recall [17, Lemma 5]: if X is fully convex and $Y \subset \mathbb{R}^d$ is *stable*, then $X \cap Y$ is fully convex. Now [17, Lemma 6] states that any half-space of integer intercept and axis normal vector is stable, and any intersection of stable sets is itself stable. The result below is straightforward by choosing P as the half-space just touching one side of the bounding box.

Proposition 2. *Let $X \subset \mathbb{Z}^d$ fully convex, non-empty, finite. Let (l, u) be the lowest and highest points of its axis-aligned bounding box. If P is a $d-1$-hyperplane orthogonal to an axis, with l or u in P, then $X \setminus P$ is fully convex.*

We can thus remove any one of the side of X and yet keep its full convexity.

4 Peelability

Previous peelability results apply only in some circumstances and do not show that one can peel a set one point at a time. To provide a more precise answer on the deletion of points in fully convex sets while maintaining the full convexity, we therefore provide a definition of what peelable really means.

Definition 3. *For a fully convex set* X *and some* $z \in X$*, we say that* z *is* peelable in X *when* $z \in \mathrm{Vtcs}\,(\mathrm{CvxH}\,(X))$ *and, when* $d \geqslant 2$*, for every direction* $i \in I_d$*, either (i)* $z - \mathbf{e}_i \in X$ *or* $z + \mathbf{e}_i \in X$*, or (ii)* $\pi_i(z)$ *is peelable in* $\pi_i(X)$*.*

When $d = 1$, a point z is peelable as soon as it is a vertex of $\mathrm{CvxH}\,(X)$. Let us also remark that when using projections, a peelable point z must but projected onto a vertex of the convex hull of the projection such that the condition $z \in \mathrm{Vtcs}\,(\mathrm{CvxH}\,(X))$ must be checked again. We now explain why peelability is not a local notion. To do this, we first prove that, for $d \leqslant 2$, peelability of a vertex z is indeed local, and then provides a counter example in $d \geqslant 3$ of local decidability of peelability for z.

For some $z \in \mathbb{Z}^d$, we consider $\mathcal{N}_X(z) = \mathrm{Cl}\,(\mathrm{Star}\,(z)) \cap X$ the neighborhood of z in X. Let us show now that peelability is locally decidable in 2d (see Appendix A for the complete list of local configurations and a fast implementation).

Lemma 5. *Let* $X \subset \mathbb{Z}^2$*, finite and full convex. Then point* $z \in \mathrm{Vtcs}\,(\mathrm{CvxH}\,(X))$ *is peelable in* X *if and only if* z *is peelable in* $\mathcal{N}_X(z)$*.*

Proof. The cases depend on the number of intersections between X and rays $z[.]$ along directions $\mathbf{e}_1$ and $\mathbf{e}_2$ (rotations excluded), $N = \sharp\{X \cap (z[\mathbf{e}_1] \cup z[\mathbf{e}_2])\}$.

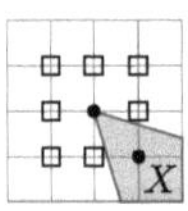

$N = 0$

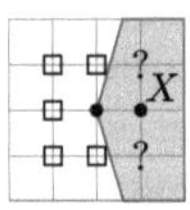

$N = 1$

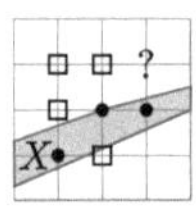

$N = 1$

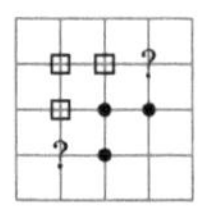

$N = 2$

(**$N = 0$**) z is peelable in X and in $\mathcal{N}_X(z)$ because z is projected at a vertex for any direction $\mathbf{e}_1$ and $\mathbf{e}_2$. (**$N = 1$ left**) z is peelable in X and in $\mathcal{N}_X(z)$ because z is only projected using $\mathbf{e}_2$ and it is a vertex both in projection for X and for $\mathcal{N}_X(z)$. (**$N = 1$ right**) z is not peelable along direction $\mathbf{e}_2$ because it is covered by an edge of $\mathrm{CvxH}\,(X)$, but it is also not peelable in $\mathcal{N}_X(z)$ along $\mathbf{e}_2$ because it is not a vertex in projection. Note that point $e = z - \mathbf{e}_1 - \mathbf{e}_2$ must belong to X as X is fully convex: otherwise $\mathrm{CvxH}\,(X)$ would intersect a 1d cell c not touching X so $c \notin \mathrm{Star}\,(X)$, but $c \in \mathrm{Star}\,(\mathrm{CvxH}\,(X))$ which is a contradiction to Lemma 1. (**$N = 2$**) Point z is peelable in both X and $\mathcal{N}_X(z)$ because it has a neighbor using both direction $\mathbf{e}_1$ and $\mathbf{e}_2$. □

Starting at $d = 3$, peelability cannot be locally decided as explicited in Fig. 1. It is quite easy to generalize this example so that it requires arbitrary large neighborhoods to decide the peelability.

We remark first that peelability imposes no specific ordering on projections.

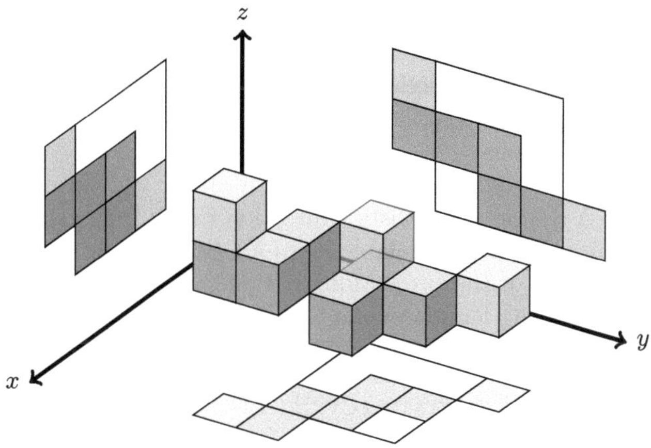

Fig. 1. Counter-example to local decidability of peeling in 3D. The fully convex set X is composed of the blue, green and red points (represented as cubes). The red point is the vertex p. As one can see, p is locally peelable along each projection (see Fig. 3 in appendix). However it is not a vertex of the convex hull of the projection of X along axis x into the plane (yz), since it lies in the middle of the edge formed by the projection of the two green points. So $X \setminus \{p\}$ is not P-convex, hence not full convex. (Color figure online)

Lemma 6. *For $J = (i_1, \cdots, i_p)$, let $\pi_J := \bigcirc_{k=1..p} \pi_{i_k}$ be the projection composing every axis projections in J, and let σ_J be a permutation of J. Then $\pi_{\sigma_J} = \pi_J$.*

Proof. It is sufficient to show it for two projections with directions $i \neq j$. Let us pick $z = (z_1, \cdots, z_d) \in \mathbb{Z}^d$. $\pi_k(z) = z_{\overline{k}}$ where $\overline{k}$ means omission of the coordinate of index k that is the set $I_d \setminus \{k\}$. So $(\pi_i \bigcirc \pi_j)(z) = z'_{\overline{i}}$ where $z' = z_{\overline{j}}$. Since $i \neq j$, $z'_{\overline{i}} = z_{\overline{i,j}}$. Thus, $(\pi_i \bigcirc \pi_j)(z) = z_{\overline{i,j}} = z_{\overline{j,i}} = (\pi_j \bigcirc \pi_i)(z)$. □

We start with an initial fully convex set X in $\mathbb{Z}^d$. A point $z \in \mathbb{Z}^d$ is peelable only if $z \in \mathrm{Vtcs}(\mathrm{CvxH}(X))$. If it is the case there exists a set of directions, called *easy directions*, defined by $\mathrm{Easy}_X[z] := \{i \in I_d : \exists \varepsilon_i \in \{+1, -1\}, z + \varepsilon_i \mathbf{e}_i \in X\}$. So to check peelability, projections will only be tested for *hard directions*, defined as $\mathrm{Hard}_X[z] := I_d \setminus \mathrm{Easy}_X[z]$. If $\mathrm{Hard}_X[z] = \emptyset$ then z is a peelable point and all such points are peelable.

Hard directions are stable in projection since emptyness of $z[\mathbf{e}_i]$ is stable under π_i and obviously emptyness in projection implies emptyness in π^{-1}. So the following property holds:

Lemma 7. *Suppose X is a non empty fully convex set in $\mathbb{Z}^d$, and let $z \in \mathrm{Vtcs}(\mathrm{CvxH}(X))$. Assume $i \neq j$ are two different directions of I_d with $\pi_i(z) \in \mathrm{Vtcs}(\mathrm{CvxH}(\pi_i(X)))$. We have $j \in \mathrm{Hard}_{\pi_i(X)}[\pi_i(z)] \Leftrightarrow j \in \mathrm{Hard}_X[z]$.*

Suppose that $\mathrm{Hard}_X[z] \neq \emptyset$. Let us remark that the set $\mathrm{Easy}_X[z]$ is preserved in the sense that for any projection $i \in \mathrm{Hard}_X[z]$ and any $\varepsilon_i \in \{+1, -1\}$, it holds that $z + \varepsilon_i \mathbf{e}_k \in X \Rightarrow \pi_i(z) + \varepsilon_i \pi_i(\mathbf{e}_k) \in \pi_i(X)$ for $k \in \mathrm{Easy}_X[z]$. In other words, when checking the peelability of z and taking projections in $\mathrm{Hard}_X[z]$, we preserve the fact that directions in $\mathrm{Easy}_X[z]$ do not have to be checked. Hence, we have to study what happened when peelability conditions are not verified for hard directions.

Point $z \in X$ is not peelable if and only if either $z \notin \mathrm{Vtcs}(\mathrm{CvxH}(X))$ or $z \in \mathrm{Vtcs}(\mathrm{CvxH}(X))$ and there exists a direction i such that: (i) $z - \mathbf{e}_i \notin X$ and $z + \mathbf{e}_i \notin X$ and (ii) $\pi_i(z)$ is not peelable in $\pi_i(X)$. So it is clear that the condition stopping the iteration of the projections is that $z \notin \mathrm{Vtcs}(\mathrm{CvxH}(X))$ in some projection. The following lemma explains this case geometrically.

Lemma 8. *Let X fully convex and let $z \in \mathrm{Vtcs}(\mathrm{CvxH}(X))$ and a direction $i \in \mathrm{Hard}_X[z]$. Then $\pi_i(z) \notin \mathrm{Vtcs}(\pi_i(\mathrm{CvxH}(X))) \Leftrightarrow \exists F^\varepsilon$ face of $\mathrm{CvxH}(X)$ such that $z^\varepsilon[\mathbf{e}_i] \cap \mathrm{relint}(F^\varepsilon) \neq \emptyset$ for some $\varepsilon \in \{+1, -1\}$.*

Proof. (Preamble) Since $z \in X$, $z + \mathbf{e}_i \notin X$ and $z - \mathbf{e}_i \notin X$, we have that $z + k\mathbf{e}_i \notin X$ for any $k \in \mathbb{Z} \setminus \{0\}$ since X is fully convex (0-convexity suffices).

($\Rightarrow$) If $\pi_i(z) \notin \mathrm{Vtcs}(\pi_i(\mathrm{CvxH}(X)))$ then there exists a segment $[u, v]$ in $\pi_i(\mathrm{CvxH}(X))$ such that $\pi_i(z) \in]u, v[$. Considering a segment of the form $]u', v'[$ with $u' \in \pi_i^{-1}(u)$, $v' \in \pi_i^{-1}(v)$ and $u', v' \in \mathrm{CvxH}(X)$, we note that $z[\mathbf{e}_i] \cap]u', v'[\neq \emptyset$ and that $\pi_i(z[\mathbf{e}_i] \cap]u', v'[) = \pi_i(z)$ for all possible u' and v'.

Suppose that $z[\mathbf{e}_i] \cap \mathrm{CvxH}(X) = \{z\}$. For any u' and v', there is a point of $]u', v'[$ in $z[\mathbf{e}_i]$. By hypothesis this point is z. Hence z is a convex combination of u' and v' which is a contradiction. So, $z[\mathbf{e}_i] \cap \mathrm{CvxH}(X)$ is a segment S containing z as a vertex. Consider the other vertex v of S and a facet F^ε containing it. Point v cannot be a vertex of F^ε since this would be a point in $\mathbb{Z}^d$ other than z and belonging to $z[\mathbf{e}_i]$ which is impossible by the preamble. Hence $v \in \mathrm{relint}(F^\varepsilon)$. We get that $z^\varepsilon[\mathbf{e}_i] \cap \mathrm{relint}(F^\varepsilon) \neq \emptyset$.

($\Leftarrow$) The point v in $z^\varepsilon[\mathbf{e}_i] \cap \mathrm{relint}(F^\varepsilon)$ is not in $\mathbb{Z}^d$ with $\varepsilon \in \{+1, -1\}$. So the projections of any vertex of the face F^ε are different from $\pi_i(z)$. Hence $\pi_i(z)$ belongs to the convex hull of the projections of $\mathrm{Vtcs}(\mathcal{F}^\varepsilon)$ because convex combinations are preserved by the linear operator π_i. This implies that $\pi_i(z) \notin \mathrm{Vtcs}(\pi_i(\mathrm{CvxH}(X)))$. □

Lemma 8 tells us that when the projection of a vertex z is covered by the projection of a face then z would not be peelable. This indicates that points on the boundary of X are the points which are the most difficult to cover using projections. We now give the main result for the usage of peelability which legitimates its definition.

Theorem 3. *Let X be a non empty finite fully convex set of $\mathbb{Z}^d$. Then, z is a peelable point of $X \Leftrightarrow X \setminus \{z\}$ is fully convex.*

Proof. ($\Rightarrow$) If z is a peelable point of X, then $z \in \mathrm{Vtcs}(\mathrm{CvxH}(X))$. We know that vertex removal preserves 0-convexity (Theorem 2) when condition (i) of peelability is verified for all directions. Else we can recursively use the peelability along projections, as in the definition of P-convexity. Noticing that 0-convexity is preserved by projections of full convex sets - while it is not true in general for arbitrary 0-convex sets - and using Theorem 1, it is clear that removing peelable points preserves full convexity.

($\Leftarrow$) Suppose that both X and $X \setminus \{z\}$ are full convex. We cannot have $\text{CvxH}(X) = \text{CvxH}(X \setminus \{z\})$ because it would imply that $\text{CvxH}(X) \cap \mathbb{Z}^d = \text{CvxH}(X \setminus \{z\}) \cap \mathbb{Z}^d$ which is impossible because z is in the former set but not in the later one. So since $\text{CvxH}(X) \neq \text{CvxH}(X \setminus \{z\})$ then $\exists z_0 \in \text{Vtcs}(\text{CvxH}(X))$ such that $z_0 \notin \text{Vtcs}(\text{CvxH}(X \setminus \{z\}))$. So in particular $z_0 \notin X \setminus \{z\}$. But $X \setminus (X \setminus \{z\}) = \{z\}$. Hence $z = z_0$ and $z \in \text{Vtcs}(\text{CvxH}(X))$.

If $\text{Easy}_X[z] = I_d$ then z is peelable. Otherwise $\text{Hard}_X[z]$ is not empty and let $i \in \text{Hard}_X[z]$. We consider $\mathcal{Z}_i = z[\mathbf{e}_i] \cap \text{CvxH}(X)$. Assume $\mathcal{Z}_i \neq \{z\}$. So we consider the farthest face F^ε such that $z[\mathbf{e}_i] \cap F^\varepsilon \neq \emptyset$ using direction $\varepsilon\mathbf{e}_i$ from z. By construction F^ε if also a face of $\text{CvxH}(X \setminus \{z\})$. But F^ε separates z from $z + \varepsilon e_i \notin X$. Consider now the 1d cell $c \in \mathcal{C}^{\mathrm{d}}$ whose realization is $\|c\| =]z; z + \varepsilon\mathbf{e}_i[$. It is clear that $\|c\| \cap F^\varepsilon \neq \emptyset$. Since F^ε is a face of $\text{CvxH}(X \setminus \{z\})$, we have $\|c\| \cap \text{CvxH}(X \setminus \{z\}) \neq \emptyset$. So since $X \setminus \{z\}$ is full convex, we must have that $\text{Vtcs}(\|c\|) \cap (X \setminus \{z\}) \neq \emptyset$. This is a contradiction since neither points are in $X \setminus \{z\}$.

So, for any $i \in \text{Hard}_X[z]$, we must have $\mathcal{Z}_i = z[\mathbf{e}_i] \cap \text{CvxH}(X) = \{z\}$. But using Lemma 8, this implies that $\pi_i(z) \in \text{Vtcs}(\pi_i(\text{CvxH}(X)))$ for direction i. Hence z is a peelable point. □

Thanks to Theorem 3, we know that peelable points are necessary and sufficient to peel fully convex sets while maintaining full convexity. However, we have to prove that such points exist for every non empty fully convex set. This is done by the following theorem (see Fig. 2 for an illustration).

Theorem 4. *Suppose X is a non empty fully convex set in $\mathbb{Z}^d$. Then, any facet of the bounding box of X contains a peelable point.*

Proof. We are going to prove the theorem by recursion on the dimension. To do this, we need to remark that for any direction $i \in I_d$ the bounding box facets for the projection $\pi_i(X)$ are precisely the projections of the facets of the bounding box of X. Thus, any facet f of the bounding box of $\pi_i(X)$ can be written as $\pi_i(F)$ where F is a facet of the bounding box of X. We can made this fact precise simply by looking at the facets of the bounding box of X whose normal vectors are the $\pm\mathbf{e}_j$ for $j \neq i$. Indeed, these are precisely the facets whose projections define the bounding box of $\pi_i(X)$.

Let us now consider the starting case: $d = 1$. As previously seen, a point z is peelable for $d = 1$ if and only if z is a vertex of X. So X has one peelable point if $X = \{z\}$ and two peelable points when X has two points or more. So the property is true for $d = 1$.

Let us suppose that for any dimension less than d, the property is true. We consider a direction $i \in I_d$. Let us denote by $X_i := \pi_i(X)$. Let us consider a facet f of the bounding box of X_i. As recalled previously, $f = \pi_i(F)$ where F is a facet of the bounding box of X. By hypothesis, the dimension of X_i is strictly lower than the dimension of X. Thus applying the hypothesis since $\pi_i(X)$ is fully

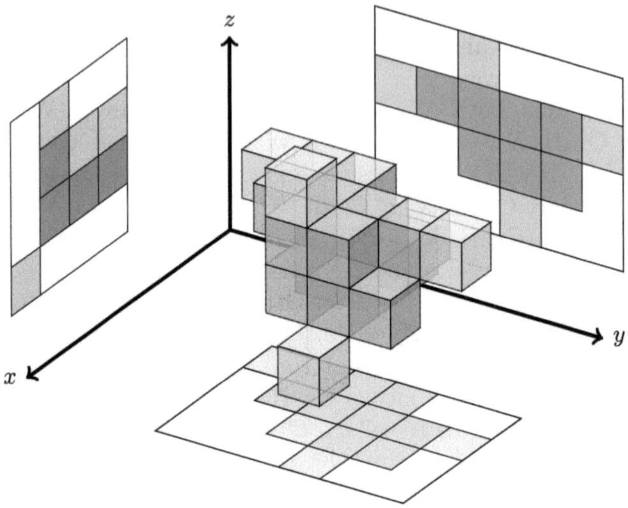

Fig. 2. Any facet of the bounding box of a fully convex set X contains a peelable point (displayed in green). (Color figure online)

convex, we get that in f, there exists a peelable point z_f. Using Lemma 1, we now that, z_f being a vertex of $\mathrm{CvxH}(X_i)$ in f, it is the projection of a vertex z_F of $\mathrm{CvxH}(X)$. By construction $z_F \in F$. But since we know that hard directions are stable by projection (Lemma 7), we know that all directions which are hard in projections behave well because z_f is peelable. The peelability of z_F is thus solely govern by the remaining direction i. There are two cases. If $i \in \mathrm{Easy}_X[z_F]$ then by definition z_F is peelable. If $i \in \mathrm{Hard}_X[z_F]$, then we note that $z_f \in \mathrm{Vtcs}(X_i)$ and thus direction i behave well in the sense that $\pi_i(z_F) = z_f$ is a vertex of X_i. Hence again z_F is peelable. Overall we obtain that the face F has a peelable point z_F. But, since the all the facets of the bounding box of X are constructible by the facets of the bounding box of X_i (by considering all the directions $i \in I_d$), we obtain that the property of true for dimension d. □

5 Conclusion and Perspectives

In this paper we have studied the inclusion hierarchy of full convex sets in $\mathbb{Z}^d$ by examining if and how one can peel a full convex set and keep this property throughout the process. Such sets must be peeled necessary at vertices. Using the equivalence of full convexity with P-convexity, we have exhibited easy sufficient conditions for parallel peeling. We have then characterized the peelability conditions through projections and proved that any full convex set is peelable at least one point at a time, which may be found lying on its bounding box.

Our next objective is to prove that a full convex set X is peelable under the constraint of leaving the full convex subset $Y \subset X$ unchanged. It could lead to a new method for computing the fully convex envelope of a digital set, by

Minkowski dilation, then peeling. The current envelope operator [7,8] is indeed an iterative finite growing process but with unclear bound: a peeling method would be much easier to bound. A related problem is defining a meaningful intersection for full convex sets, and peelability could help in this endeavour.

Acknowledgments. This work is partially supported by the French National Research Agency within the StableProxies project (ANR-22-CE46-0006). The authors also thank Gilles Bertrand for initiating discussions about thinning full convex sets.

A Peelable Configurations in 2D

```
// Check peelability of some point in a set X, given its neighborhood
// input:  cfg  the encoded neighborhood of the point (in {0,...,255})
// output: true iff the point is peelable wrt its neighborhood
// prerequisite: the point should be a vertex of CvxH(X)
bool isLocallyPeelable( unsigned int cfg )
{ // 45 configurations are peelable
  static const unsigned int p[ 8 ]
    = { 0x5151f1ff, 0x11110001, 0x00101000b, 0x01010003,
        0x000000ab, 0x00000000, 0x00001000b, 0x00010003 };
  return p[ cfg >> 5 ] & ( 1 << (cfg & 0x1f) );
}
```

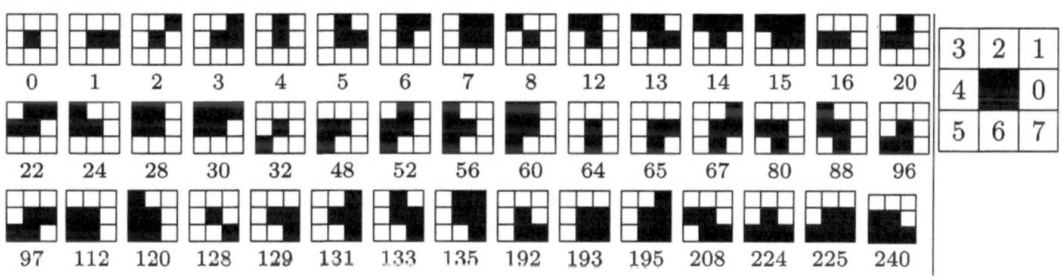

Fig. 3. The 45 configurations peelable in 2D. Each configuration is encoded as an 8-bit integer, each bit b being set to 1 whenever the b-th neighbor (as displayed on the right) is present in the digital set.

References

1. Bertrand, G., Couprie, M.: Powerful parallel and symmetric 3D thinning schemes based on critical kernels. J. Math. Imaging Vision **48**, 134–148 (2014)
2. Bertrand, G., Malandain, G.: A new characterization of three-dimensional simple points. Pattern Recogn. Lett. **15**(2), 169–175 (1994)
3. Brlek, S., Lachaud, J.O., Provençal, X., Reutenauer, C.: Lyndon+ christoffel= digitally convex. Pattern Recogn. **42**(10), 2239–2246 (2009)

4. Couprie, M., Bertrand, G.: New characterizations of simple points in 2D, 3D, and 4D discrete spaces. IEEE Trans. Pattern Anal. Mach. Intell. **31**(4), 637–648 (2008)
5. Couprie, M., Bertrand, G.: Asymmetric parallel 3D thinning scheme and algorithms based on isthmuses. Pattern Recogn. Lett. **76**, 22–31 (2016)
6. Eckhardt, U.: Digital lines and digital convexity. In: Bertrand, G., Imiya, A., Klette, R. (eds.) Digital and Image Geometry. LNCS, vol. 2243, pp. 209–228. Springer, Heidelberg (2001). https://doi.org/10.1007/3-540-45576-0_13
7. Feschet, F., Lachaud, J.O.: Full convexity for polyhedral models in digital spaces. In: Discrete Geometry and Mathematical Morphology, pp. 98–109. Springer, Cham (2022)
8. Feschet, F., Lachaud, J.O.: An envelope operator for full convexity to define polyhedral models in digital spaces. J. Math. Imaging Vis. **65**(5), 754–769 (2023). https://doi.org/10.1007/s10851-023-01155-w
9. Feschet, F., Lachaud, J.O.: New characterizations of full convexity. In: Brunetti, S., Frosini, A., Rinaldi, S. (eds.) Discrete Geometry and Mathematical Morphology, pp. 41–53. Springer, Cham (2024)
10. Feschet, F., Lachaud, J.O.: New properties for full convex sets and full convex hulls. J. Math. Imaging Vis. **67**(8) (2025). https://doi.org/10.1007/s10851-024-01225-7
11. Kim, C.E., Rosenfeld, A.: Convex digital solids. IEEE Trans. Pattern Anal. Machine Intel. **6**, 612–618 (1982)
12. Kim, C.E., Rosenfeld, A.: Digital straight lines and convexity of digital regions. IEEE Trans. Pattern Anal. Machine Intel. **2**, 149–153 (1982)
13. Kim, C.E., Sklansky, J.: Digital and cellular convexity. Pattern Recogn. **15**(5), 359–367 (1982). https://doi.org/10.1016/0031-3203(82)90038-3
14. Kiselman, C.O.: Convex functions on discrete sets. In: Combinatorial Image Analysis. 10th International Workshop, IWCIA. LNCS, vol. 3322, pp. 443–457. Springer (2004)
15. Kiselman, C.O.: Elements of digital geometry, mathematical morphology, and discrete optimization. World Sci. (2022). https://doi.org/10.1142/12584
16. Lachaud, J.-O.: An alternative definition for digital convexity. In: Lindblad, J., Malmberg, F., Sladoje, N. (eds.) DGMM 2021. LNCS, vol. 12708, pp. 269–282. Springer, Cham (2021). https://doi.org/10.1007/978-3-030-76657-3_19
17. Lachaud, J.-O.: An alternative definition for digital convexity. J. Math. Imaging Vision 1–18 (2022). https://doi.org/10.1007/s10851-022-01076-0
18. Llinares, J.V.: Abstract convexity, some relations and applications. Optimization **51**(6), 797–818 (2002)
19. Murota, K., Shioura, A.: Relationship of M/L-convex functions with discrete convex functions by Miller and Favati-Tardella. Discret. Appl. Math. **115**, 151–176 (2001)
20. Pudney, C.: Distance-ordered homotopic thinning: a skeletonization algorithm for 3D digital images. Comput. Vis. Image Underst. **72**(3), 404–413 (1998)
21. Ronse, C.: A bibliography on digital and computational convexity (1961–1988). IEEE Trans. Pattern Anal. Machine Intel. **11**(2), 181–190 (1989)
22. Roy, A.J., Stell, J.G.: Convexity in discrete space. In: Kuhn, W., Worboys, M.F., Timpf, S. (eds.) COSIT 2003. LNCS, vol. 2825, pp. 253–269. Springer, Heidelberg (2003). https://doi.org/10.1007/978-3-540-39923-0_17
23. Tarsissi, L., Coeurjolly, D., Kenmochi, Y., Romon, P.: Convexity preserving contraction of digital sets. In: Palaiahnakote, S., Sanniti di Baja, G., Wang, L., Yan, W.Q. (eds.) ACPR 2019. LNCS, vol. 12047, pp. 611–624. Springer, Cham (2020). https://doi.org/10.1007/978-3-030-41299-9_48

24. Tarsissi, L., Kenmochi, Y., Djerroumi, H., Coeurjolly, D., Romon, P., Borel, J.P.: Algorithms for pixelwise shape deformations preserving digital convexity. In: International Conference on Discrete Geometry and Mathematical Morphology, pp. 84–97. Springer (2022)
25. Tarsissi, L., Kenmochi, Y., Romon, P., Coeurjolly, D., Borel, J.P.: Convexity preserving deformations of digital sets: characterization of removable and insertable pixels. Discret. Appl. Math. **341**, 270–289 (2023)
26. Webster, J.: Cell complexes and digital convexity. In: Bertrand, G., Imiya, A., Klette, R. (eds.) Digital and Image Geometry. LNCS, vol. 2243, pp. 272–282. Springer, Heidelberg (2001). https://doi.org/10.1007/3-540-45576-0_16

Exact Polyominoes and Non-decomposability

Michela Ascolese(✉) and Andrea Frosini

Dipartimento di Matematica e Informatica "U. Dini",
Università degli Studi di Firenze, Firenze, Italy
michela.ascolese@unifi.it

Abstract. A configuration of the plane is a subset (both finite or infinite) A of points in $\mathbb{Z}^2$. Provided a finite window probe W, i.e., a finite set of positions of $\mathbb{Z}^2$, we shift W in all possible positions on $\mathbb{Z}^2$ and count the points of A that fit in its positions. In case all the values are equal, say h, then we say that A is h-homogeneous w.r.t. W. When W has a tile shape, i.e., a shape that can be used to tile the plane by translation, h-homogeneous configurations related to W can be decomposed into disjoint, 1-homogeneous ones, in some cases. Very few is known about the properties that a tile has to satisfy in order to allow such a decomposition. In this paper, we shed some light on the topic, by investigating the decomposability of a class of windows, single squares, i.e., those tiles that can be surrounded in one only possible way by four copies of themselves in order to produce a tiling of the plane, that constitute one of the simplest classes of tiles. We also introduce other classes of tiles that allow homogeneous, non-decomposable planar configurations. Finally, we define the notion of reducible configuration, that concerns the possibility of decomposing a configuration into smaller, non 1-homogeneous ones, moving toward the classification of tiles w.r.t. the decomposability property.

Keywords: Discrete Tomography · Discrete Geometry · Exact polyominoes · Tilings

1 Introduction

In Discrete Tomography we try to infer structural properties of a physical object, regarded as a discrete set of points A, from quantitative measures of its internal configuration, say projections, usually taken along discrete parallel lines. For an exhaustive introduction to the topic one can refer to [11,12]. Here we consider a different paradigm introduced by M. Nivat et al. in [9,13], and that proposes to inspect the set of points by moving on it a finite size window W and collecting the number of points that lie inside it. In case W contains the same number of points h for each of its possible placements on A, say the set is h-homogeneous, then A reveals some structural properties that include the presence of patterns

M. H. F. Wilkinson and J. Kosinka (Eds.): DGMM 2025, LNCS 16296, pp. 90–102, 2026.
https://doi.org/10.1007/978-3-032-09544-2_7

that can be detected and reconstructed [4,10], so intersecting the main topics of the wide area of Computerized and Discrete Tomography with the notion of planar tilings.

In [13] it is also observed that only exact polyominoes, those ones that tile the plane by translation, allow 1-homogeneous configurations. In case of rectangular windows, a decomposition theorem also holds, which allows to split each h-homogeneous configuration A into h disjoint sub-configurations that are 1-homogeneous. This result can be generalized including the class of Aztec diamonds and all those tiles that are balls in a generalized norm L^1 of $\mathbb{Z}^2$ [4]. On the other hand, planar configurations that are h-homogeneous w.r.t. a tile but cannot be decomposed into h 1-homogeneous ones have been discovered in [1].

Moving from this result, in this paper we consider the class of single square tiles, we recall some previous results on them and we provide an exhaustive computation of the smallest ones, up to perimeter 18, that are not decomposable w.r.t. some homogeneous configuration, together with the involved homogeneity values. Relying on that, we define the notion of irreducibility for a non-decomposable configuration, and we conjecture a connection between irreducible configurations and the area of the related single square tiles. Finally, we define some classes of single squares for which non-decomposable configurations always exist.

The paper is organized as follows: in Sect. 2, we introduce the basic notions of our research concerning tiles and planar configurations, in particular regarding their homogeneity and decomposability. In Sect. 3, we deepen the notion of non-decomposability and introduce the definition of irreducible configuration. We furthermore define some classes of single squares for which a non-decomposable, homogeneous configuration can be constructed, starting from the value of their areas and shape. Finally, we conclude introducing some open research lines in Sect. 4.

2 Notation and Preliminary Results

Let us consider *polyominoes*, i.e., finite, 4-connected discrete sets of points in $\mathbb{Z}^2$, usually represented, up to translation, as sets of cells on a squared surface, each cell representing a point. We further require that no holes are present.

Due to these conditions, we can characterize a polyomino by simply referring to its boundary: it can be coded, as a path, by means of a word on a four letters alphabet, according to the so called Freeman-chain coding [7,8]. More precisely, we consider as alphabet the set of symbols $\Sigma = \{1, 0, \overline{1}, \overline{0}\}$, whose letters correspond to the unit steps $\{\uparrow, \rightarrow, \downarrow, \leftarrow\}$, respectively, that we perform when traveling the boundary of a polyomino (see Example 1). Opposite directions are addressed as *complement* and pointed out with $\overline{x}$, where $x \in \Sigma$. It is clear that $\overline{\overline{1}} = 1$ and $\overline{\overline{0}} = 0$. The notation Σ^* indicates the set of all the words that are defined on the alphabet Σ, including the empty one ε.

The coding of a polyomino is ambiguous, in general, meaning that the same boundary can be coded using different words. Indeed, there is freedom of choice both for the starting point of the coding and for the traveling orientation, that can be clockwise or counterclockwise. In this work, we tackle such ambiguities by choosing the clockwise orientation and the bottom-left point of the boundary as the starting point of the coding.

Example 1. *According to our choices, the boundary word of the polyomino in Fig. 1 is*

$$P = 1\overline{0}10101010\overline{1}0\overline{11001010}.$$

The starting point of the coding is highlighted with a dot, as well as the chosen orientation (clockwise).

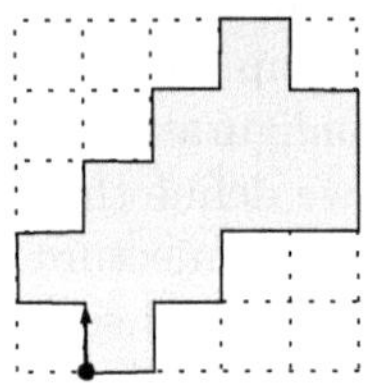

Fig. 1. A polyomino in $\mathbb{Z}^2$ represented as a set of cells on a squared surface. Its bottom-left corner is highlighted, as well as the clockwise orientation, that concur to the unambiguous coding of its boundary.

Provided a word $w = w_1 w_2 \ldots w_n \in \Sigma^*$, we consider the *hat* operator $\widehat{w} = \overline{w}_n \overline{w}_{n-1} \ldots \overline{w}_1$. The geometrical interpretation of $\widehat{w}$ is relevant: indeed, this word codes the same path coded by w, but traveled in the opposite direction.

2.1 Exact Polyominoes

A polyomino can also be used to tile the plane $\mathbb{Z}^2$ by translation, meaning that it is possible to join infinitely many copies of it to cover the whole discrete plane with no holes or intersections. In such a case, the polyomino is called *exact*, or *tile* (by the fact that it creates a tessellation of $\mathbb{Z}^2$). There are two ways to quickly check if a polyomino is exact:

Geometrically. It is sufficient to check if one copy of the polyomino can be *completely* surrounded with copies of itself. In case of positive answer, the surrounding can be iteratively repeated on its neighbors to create a tessellation of $\mathbb{Z}^2$.

Combinatorially. In 1991, Beauquier and Nivat characterized the boundary words of exact polyominoes by splitting them into four or six mutually related parts as specified in the following theorem.

Theorem 2 (Beauquier and Nivat [5]**).** *A polyomino P is exact if and only if there exist $A, B, C \in \Sigma^*$ such that*

$$P = ABC\widehat{A}\widehat{B}\widehat{C},$$

where at most one among A, B and C can be empty. Moreover, this factorization may not be unique.

By the authors' names, we address as *BN-factorization* each factorization of a boundary word as in Theorem 2. Furthermore, a geometrical interpretation is given: the BN-factors A and $\widehat{A}$ (B and $\widehat{B}$, C and $\widehat{C}$, respectively) actually identify the same path in $\mathbb{Z}^2$, thus defining the translations that have to be applied to an exact polyomino P to create a tessellation. Indeed, the part of the boundary coded by A will match in a tiling to the part of the boundary coded by $\widehat{A}$, in a neighbor copy of P. An example of such interpretation is shown in Fig. 2. In case one among the factors A, B or C is empty, we call the tile P a *pseudo-square*, and *pseudo-hexagon* otherwise. Since the same tile can have different BN-factorizations, it can be both a pseudo-square and a pseudo-hexagon at the same time. In [6], it has been proven that if two different BN-factorizations as a pseudo-square exist for the same tile, then it cannot be factorized as a pseudo-hexagon. Further studies have been carried on in [2,3]. Notice that different BN-factorizations produce different tilings, each one characterized by a different periodic behavior in the placement of its tiles.[1]

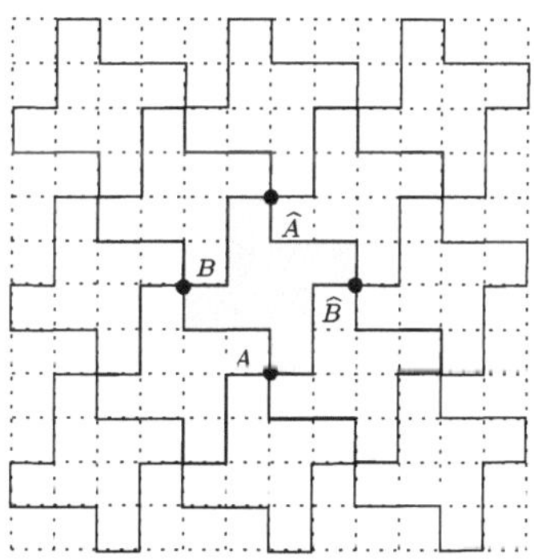

Fig. 2. A regular tiling induced by an exact polyomino. In this case, the tile is a single square. Its boundary word is (uniquely) factorized as $P = (1\overline{00}1)(0110)(\overline{1}00\overline{1})(\overline{0110})$.

[1] In this work, we only consider regular tilings, meaning that two directions of periodicity are always present in a tessellation. See [5] for more details concerning the case of semi-regular tilings.

2.2 Homogeneous Configurations and Decomposability

A *configuration of the plane* is a subset A of points in $\mathbb{Z}^2$. To represent it, we consider an infinite binary matrix, whose entries are equal to 1 if and only if the corresponding point of $\mathbb{Z}^2$ is in the set A. Actually, we will omit 0s to simplify the notation. Given a discrete direction v, i.e., a vector in $\mathbb{Z}^2 \setminus \{(0,0)\}$, the configuration A is *periodic with respect to* v if, for any point $a \in A$, also $a \pm v \in A$ holds, where $a \pm v$ stands for a translation of the point a along the direction v. Since we are in a bi-dimensional space, each configuration of the plane can have at most two directions of periodicity that are linearly independent.

We consider a polyomino W and use it to inspect a planar configuration, by moving it all over the discrete plane and revealing the points that fit inside, as a shifting window that reveals what hidden behind a wall. So, for each possible position of W in $\mathbb{Z}^2$, we count the number of points of A that lie inside the polyomino; these values range from 0 to ω, where ω indicates the area of W. In case the detected value is constant for each possible position of W in A, we say that A is *h-homogeneous with respect to* W, being h such a constant value.

Remark 3. *For a configuration A, the property of being homogeneous strictly depends on the choice of the polyomino W that is used to inspect it. Indeed, a configuration that is homogeneous w.r.t. a window W could be non-homogeneous w.r.t. a different one (see Fig. 3).*

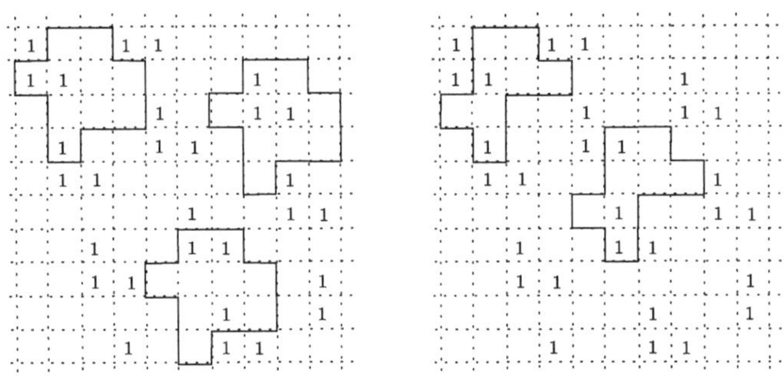

Fig. 3. A configuration A that is 3-homogeneous w.r.t. the polyomino on the left, but is not homogeneous w.r.t. the polyomino on the right.

An important result connecting homogeneous configurations with exact polyominoes is the following:

Theorem 4 (Nivat [13]). *Given a polyomino W, there exists a 1-homogeneous configuration w.r.t. W if and only if W is exact. Moreover, such a configuration is periodic w.r.t. the directions of periodicity of (the tiling induced by) W.*

Notice that the previous result provides a necessary and sufficient condition. On the other hand, h-homogeneous configurations (potentially) exist for any type of polyomino, not necessarily exact, on the condition that $h \geq 2$.

As a matter of fact, 1-homogeneous configurations are strictly connected to the periodic behavior that exact polyominoes induce in a tiling. Indeed, by exploiting this idea, M. Nivat proved the following result in 2002:

Theorem 5 (Nivat [13]). *Let us consider a configuration A of the plane that is h-homogeneous w.r.t. a rectangle W, with $2 \leq h \leq \omega$, with ω being the area of W. Then, there exist h disjoint configurations $A_1, \ldots, A_h$ such that A_i is 1-homogeneous w.r.t. W, for all $1 \leq i \leq h$, and $A = \bigcup_{i=1}^{h} A_i$.*

As a consequence, we introduce the notion of *decomposability.*

Definition 6. *Given an exact polyomino W of area ω and a homogeneity value $1 \leq h \leq \omega$, we say that W is h-*decomposable *if any configuration that is h-homogeneous w.r.t. W can be partitioned into h disjoint and 1-homogeneous ones. If a polyomino is h-decomposable for any possible homogeneity value, then it is addressed as* decomposable.

So, rectangles are decomposable (Theorem 5), as well as diamonds and all those polyominoes that are open sets in $\mathbb{Z}^2$ (see [4]). The decomposability property cannot be extended to the whole class of exact polyominoes, as shown in [1].

Starting from these results, the aim of this work is to deepen the notion of decomposability, thus characterizing new classes of tiles for which such a property holds or does not.

The following intuitive result allows to halve the homogeneity values we have to consider when studying planar configurations:

Theorem 7. *Let W be an exact polyomino of area ω that is h-decomposable for some $2 \leq h \leq \lfloor \frac{\omega}{2} \rfloor$. Then, W is $(\omega - h)$-decomposable.*

Proof. Let A be a $(\omega - h)$ homogeneous configuration w.r.t. W, and let us denote $\overline{A}$ the configuration obtained by exchanging 0s with 1s in A. The new configuration is clearly h-homogeneous w.r.t. W and so, by hypothesis, we can decompose it in h (periodic) 1-homogeneous configurations, $\overline{A}_1, \ldots, \overline{A}_h$.

Let us now consider $\mathcal{C}_1$ the set of all the possible ω configurations of the plane that are 1-homogeneous w.r.t. W, and let us denote $\{A_1, \ldots, A_{\omega-h}\} = \mathcal{C}_1 \setminus \{\overline{A}_1, \ldots, \overline{A}_h\}$. Since $A_i \cap A_j = \emptyset$ for all $1 \leq i \neq j \leq \omega - h$, the configuration $A = \bigcup_{i=1}^{\omega-h} A_i$ is such that A is $(\omega - h)$-homogeneous w.r.t. W. Moreover, the set $\{A_i\}_{1 \leq i \leq \omega-h}$ defines a decomposition of A in disjoint 1-homogeneous configurations, i.e., W is $(\omega - h)$-decomposable. □

As a consequence, from now on we will consider homogeneity values h ranging from 2 to $\lfloor \frac{\omega}{2} \rfloor$ only.

2.3 Single Squares and Their Properties

Theorem 4 suggests that periodicity properties become relevant when dealing with decomposability. So, in this work we restrict our investigation to a special class of tiles that allow one BN-factorization only, and, as a consequence, a unique tesselation of $\mathbb{Z}^2$ having one couple of directions of periodicity.

Definition 8. *A tile* P *is a* single square *when it admits a unique BN-factorization as* $P = AB\widehat{A}\widehat{B}$.

An example of single square is the tile depicted in Fig. 2. We underline that a single square induces a unique tiling of the plane, that is obtained by surrounding it with four copies of itself, and whose directions of periodicity can be easily obtained from the BN-factors A and B. Let us introduce the standard notation $|w|_x$ to denote the number of occurrences of the symbol $x \in \{0, 1, \overline{0}, \overline{1}\}$ in the word w.

Theorem 9 (A. and F. [1]**).** *Let us consider* P *a single square and* $P = AB\widehat{A}\widehat{B}$ *its (unique) BN-factorization. Each BN-factor defines a direction of periodicity in the induced regular tiling,*

$$v_1 = (|A|_0 - |A|_{\overline{0}}, |A|_1 - |A|_{\overline{1}}),$$
$$v_2 = (|B|_0 - |B|_{\overline{0}}, |B|_1 - |B|_{\overline{1}}).$$

As an example, the directions of periodicity of the single square in Fig. 2 are $v_1 = (-2, 2)$ and $v_2 = (2, 2)$.

We underline that, if we consider a cell of the polyomino and then move along one of the directions of periodicity in the tiling, we reach the cell in the same position of a neighbor tile. This trivial observation together with Theorem 4 is relevant in the study of decomposability.

Property 10. *Let us consider a single square* W *and a configuration of the plane* A *that is* 1*-homogeneous w.r.t.* W*. Then, in a tiling of the plane, the* 1*s of* A *that are present in the copies of* W *always appear in the same cell of the tile.*

The same property holds when considering a h-homogeneous configuration that is also decomposable, with $h \geq 2$. Indeed, the decomposability in 1-homogeneous configurations implies periodicity, meaning that, for the same tessellation, all the 1s appear in the same cells of W.

3 Non-decomposability

In Sect. 2.2 we observed that the property of being decomposable does not hold for all exact polyominoes, contrary to what was conjectured in [9]. The existence of non-decomposable tiles has been established in 2022, by exhaustively generating all the square tiles up to perimeter $n = 22$ and then all their 2-homogeneous configurations [1]. The possible decompositions of such configurations follow from the periodicity properties stated in Theorem 4, and a counterexample to the conjecture in [9] has been obtained. Still in [1], the authors

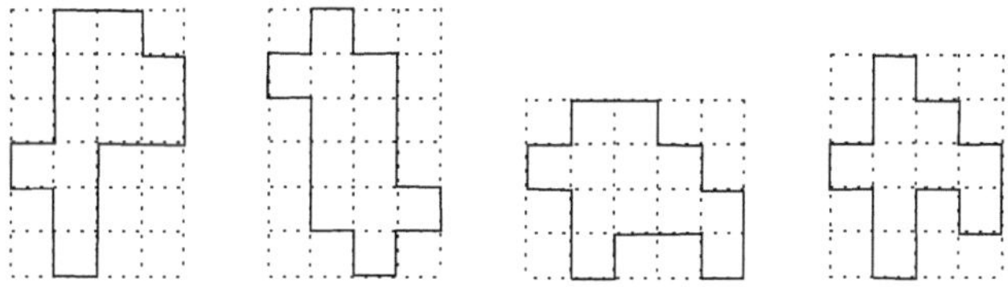

(a) (b) (c) (d)

Fig. 4. The single squares of perimeter $n = 20$ that allow non-decomposable 2-homogeneous configurations.

found four single squares, all of perimeter $n = 20$, that are not 2-decomposable (we report them in Fig. 4).

At the end of the paper, they also noticed that the single square (c) in Fig. 4 can be slightly modified to construct a new polyomino, that is a single square too, that is 2-decomposable but not decomposable when increasing the homogeneity value to $h = 3$. Starting from this observation, we want to deepen the study of non-decomposable tiles, with the aim of classifying them according to the homogeneity values for which the decomposability holds or does not.

3.1 Experimental Results and Some Non-decomposable Configurations

Following the strategy in [1], we exhaustively generate all the single squares up to perimeter $n = 18$ (we stop here due to the high computational cost) and, for each of them, we test the decomposability for any admissible homogeneity value. Table 1 shows the obtained results; we remark that tiles that are not 2-decomposable start appearing from perimeter $n = 20$ on.

Table 1. List of all the non-decomposable single squares, up to perimeter $n = 18$, together with the homogeneity values for which the decomposability does not hold.

ID	Boundary word	Perimeter	Area	Non-dec
1	1101010010110010	16	10	$h = 5$
2	$1\overline{0}1011001\overline{1}0\overline{100110}$	16	8	$h = 4$
3	$1\overline{00}10110\overline{1}00\overline{10110}$	16	8	$h = 4$
4	$1\overline{0}1011100\overline{1}0\overline{1001110}$	18	9	$h = 3$
5	$111\overline{00}1010\overline{1}00\overline{111010}$	18	10	$h = 5$
6	$11\overline{0}1\overline{0}1010\overline{1}0\overline{1}0\overline{11010}$	18	10	$h = 5$
7	$11\overline{00}11010\overline{11}00\overline{11010}$	18	10	$h = 5$
8	$11\overline{00}10110\overline{1}00\overline{110110}$	18	10	$h = 5$
9	$1\overline{00}101110\overline{1}00\overline{101110}$	18	10	$h = 5$
10	$11\overline{0}101110\overline{1}0\overline{1101110}$	18	9	$h = 3, 4$

A close inspection of Table 1 suggests that the non-decomposability seems to be related to the value of the area of the considered polyomino. Indeed, in most of the cases where the h-decomposability does *not* hold, the homogeneity value h is a divisor of the area of the tile. To go deep into this observation and to shed some light on the remaining cases, we introduce some definitions together with a result that actually characterizes those homogeneous configurations that cannot be decomposed.

Definition 11. *Let us consider a configuration A that is h-homogeneous w.r.t. a tile W. We call* pattern *the set of the positions of the elements of A that lie inside W when placing the window on the discrete plane.*

It is clear that there exist at most $\binom{\omega}{h}$ distinct patterns, with ω the area of W and h the homogeneity value of the configuration.

Then, Property 10 can be reformulated as follows:

Property 12. *If we fix a tiling of the plane induced by a single square W, a configuration A is 1-homogeneous w.r.t. W if and only if all the copies of W in the tiling show the same pattern.*

Property 12 leads to a necessary and sufficient condition for a homogeneous configuration to be decomposable, in case of single squares.

Theorem 13. *Let A be a h-homogeneous configuration w.r.t. a single square W of area ω, with $1 \leq h \leq \lfloor \frac{\omega}{2} \rfloor$. If we fix on $\mathbb{Z}^2$ a tiling induced by W, then, A can be decomposed into h disjoint 1-homogeneous configurations w.r.t. W if and only if all the copies of W in the tiling show the same pattern.*

Proof. We fix on $\mathbb{Z}^2$ a tiling induced by the exact polyomino W, and prove the statement.

Let us assume that A can be decomposed into h disjoint configurations $A_1, \ldots, A_h$, all 1-homogeneous w.r.t. W. By Theorem 4, each configuration A_i is periodic w.r.t. the directions of periodicity of W, and so the unique element that appears in each copy of the tile is always in the same position. By the disjointedness of the configurations, it follows that when considering the whole configuration A, the h elements in each tile always appear in the same positions, that is, all the tiles display the same pattern.

On the other hand, if each tile of the tiling displays the same pattern, the h elements that appear inside it follow the directions of periodicity of the tiling. As a consequence of Theorem 4, there are exactly h configurations all 1-homogeneous w.r.t. W, whose (disjoint) union gives all the elements in A, that so can be decomposed. □

Corollary 14. *Given W a single square of area ω and $2 \leq h \leq \lfloor \frac{\omega}{2} \rfloor$, W is not h-decomposable if and only if there exists a configuration A that is h-homogeneous w.r.t. W and such that any tiling of the plane induced by W contains at least two different patterns.*

As an example, in Fig. 5, on the left, we depict a 4-homogeneous configuration that cannot be decomposed in 1-homogeneous ones. Indeed, the reader can see that two different patterns appear in the same tiling, according to Corollary 14.

3.2 Irreducible Configurations

Let us focus on the non-decomposable, 4-homogeneous configuration in Fig. 5, on the left, w.r.t. the tile **10** in Table 1. We observe that the homogeneity value $h = 4$ does not divide the area of the tile, $\omega = 9$. However, such a configuration has a remarkable property: it can be obtained as the union of a 3-homogeneous, non-decomposable configuration, and a 1-homogeneous one (Fig. 5 on the right).

On the other hand, we can think of constructing bigger and non-decomposable configurations as the disjoint union of smaller ones, when intersections are not present. To formalize what we described above, we introduce the definition of *irreducible configuration.*

Definition 15. *Let us consider W an exact polyomino and A a h-homogeneous configuration w.r.t. W, for some value h. The configuration A is h-*irreducible *if it cannot be decomposed in two configurations A_1 and A_2 that preserve the homogeneity w.r.t. W. Otherwise, A is* reducible.

It is clear by definition that if a configuration A is h-decomposable, with $h \geq 2$, then it is reducible.

Property 16. *A 2-homogeneous configuration is irreducible if and only if it is not decomposable.*

Property 16 is not true for $h \geq 3$, in general. Indeed, a h-homogeneous configuration can be reducible even if it is not possible to decompose it in 1-homogeneous ones. Furthermore, the union of non-decomposable, homogeneous configurations can be decomposable.

Using the notion of irreducibility, we can state the following conjecture, that we experimentally tested for all single squares up to perimeter 18:

Conjecture 17. *Given a single square W of area ω, if there exists a h-irreducible configuration w.r.t. W, then h is a divisor of ω.*

3.3 Classes of Non-decomposable Tiles

Relying on Conjecture 17, we are now ready to define classes of non-decomposable tiles w.r.t. h-homogeneous configurations, where h is a divisor of their areas. We introduce the following notation: given W a single square of area ω, we define the vector $C = (c_1, \dots, c_p)$ such that c_i is the number of cells in the i-th column of the polyomino, with $1 \leq i \leq p$, counted from left to right, and the vector $R = (r_1, \dots, r_q)$ such that r_j is the number of cells in the j-th row of the polyomino, with $1 \leq j \leq q$, counted from the top to the bottom. We underline that p and q are the smallest indexes such that C and R have non-null entries.

As an example, the vectors C and R of the single square in Fig. 5 are $C = (1, 6, 2)$ and $R = (1, 2, 2, 2, 1, 1)$.

Starting from the vectors C and R we construct, if possible, irreducible configurations for the related single square. We describe the procedure basing on

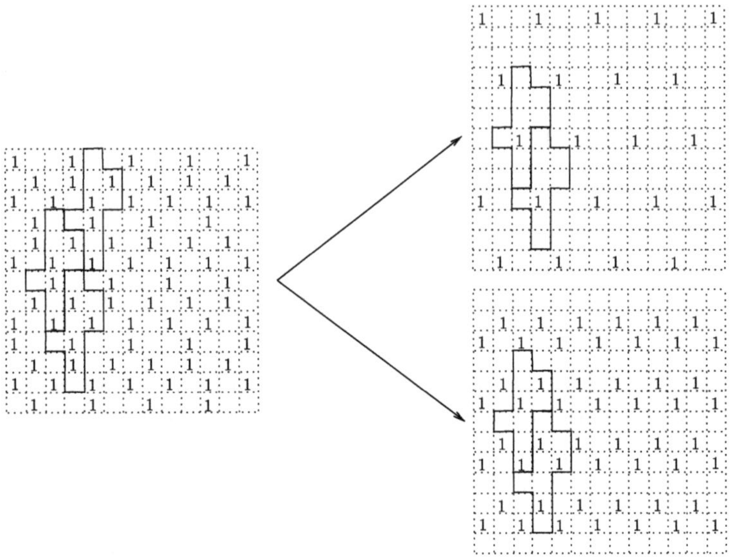

Fig. 5. A reducible 4-homogeneous configuration (on the left) and its decomposition in irreducible configurations (on the right).

the the values in R, being aware that the same method can rely on the vector C.

So, let W be a single square of area ω, with $R = (r_1, \ldots, r_q)$ the vector of its row cardinalities and $1 \leq h \leq \lfloor \frac{\omega}{2} \rfloor$ a divisor of ω. We proceed in partitioning, if possible, the values in R in $\lambda = \frac{\omega}{h}$ disjoint sets such that:

i) the sum of the values in each set is exactly h,
ii) the elements in each set can be expressed as

$$\{r_i, r_{i+\lambda}, r_{i+2\lambda}, \ldots, r_{i+k\lambda}\},$$

for $i = 1, \ldots, \lambda$, and $q \geq i + k\lambda$.

If such a decomposition exists, we consider the smallest configuration of the plane $A_{u,v}$, different from the empty one, that is periodic w.r.t. the directions $u = (1, 0)$ and $v = (0, \lambda)$. By construction, it is h-homogeneous w.r.t. W. Example 18, concerning the tile in Fig. 6, clarifies the construction.

Example 18. *The polyomino W in Fig. 6 has perimeter $n = 20$, area $\omega = 12$ and row cardinalities equal to $R = (1, 2, 3, 2, 2, 1, 1)$. Choosing $h = 4$, we can partition the values of R in three groups such that the sum of the elements in each group is four, and the distance between the rows in each group is constant and equal to $\lambda = 3$. We underline in the following vectors the elements belonging* *to each group, highlighted in Fig. 6 as well.*

$$R_1 = (\underline{1}, 2, 3, \underline{2}, 2, 1, \underline{1}),$$
$$R_2 = (1, \underline{2}, 3, 2, \underline{2}, 1, 1),$$
$$R_3 = (1, 2, \underline{3}, 2, 2, \underline{1}, 1).$$

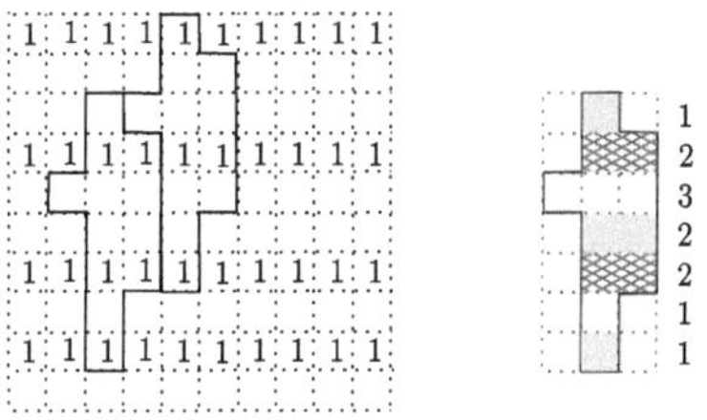

Fig. 6. A 4-homogeneous irreducible configuration that is constructed starting from the vector R of row cardinalities.

Now, we consider the directions $u = (1, 0)$ and $v = (0, 3)$, and the configuration of the plane $A_{u,v}$ that is periodic along them. It is clear from Fig. 6 that such a configuration is 4-homogeneous w.r.t. W. Moreover, it cannot be decomposed in 1-homogeneous disjoint configurations, since distinct patterns are present in the same tiling.

After the construction of $A_{u,v}$, two cases arise:

i) The directions $u = (1, 0)$ and $v = (0, \lambda)$ coincide with the directions of periodicity of W. In this case, $A_{u,v}$ can be decomposed in disjoint configurations that are 1-homogeneous w.r.t. W (Theorem 4), and the construction does not provide any hint concerning the h-decomposability of W.
ii) The directions $u = (1, 0)$ and $v = (0, \lambda)$ do not coincide with the directions of periodicity of W, as in case of Example 18, and so the polyomino is not h-decomposable. Moreover, in this case the configuration $A_{u,v}$ is also irreducible, by construction.

As already mentioned, a similar construction can be performed starting from the vector C that groups the column cardinalities.

4 Conclusion and Open Research Lines

In this paper, we deepen the notion of non-decomposability concerning homogeneous configurations and exact polyominoes, providing further counterexamples to a conjecture that was presented in [9] and then proved to be false in [1]. Moreover, we introduce the notion of irreducible configuration.

Focusing on the subclass of single square tiles, we describe a construction of possibly non-decomposable configurations when the area of the considered polyomino is not a prime number. The construction bases on the possibility of grouping its cells w.r.t. the rows or columns cardinalities. As a first step forward, we propose to inspect the possibility to generalize such a technique to any possible discrete direction, two for all, the diagonal ones.

As an immediate consequence, we conjectured (Conjecture 17) that the existence of irreducible configurations, and so the non-decomposability property, is strictly related to the values of the area of the involved tiles. In this direction, it could be interesting to explore the possibility that all non-decomposable single square tiles admit at least one irreducible configuration that can be obtained by grouping their cells in sets of equal cardinality, w.r.t. some discrete direction.

References

1. Ascolese, M., Frosini, A.: On the decomposability of homogeneous binary planar configurations with respect to a given exact polyomino. Lect. Notes Comput. Sci. **13493**, 139–152 (2022)
2. Ascolese, M., Frosini, A.: Proving a conjecture on prime double square tiles. Disc. Appl. Math. **350**, 71–83 (2024)
3. Ascolese, M., Frosini, A.: Setting the path to the combinatorial characterization of prime double square polyominoes. In: Proceedings of the 24th Italian Conference on Theoretical Computer Science, Palermo, Italy 13–15 September 2023 (2023). https://ceur-ws.org/Vol-3587/5918.pdf
4. Battaglino, D., Frosini, A., Rinaldi, S.: A decomposition theorem for homogeneous sets with respect to diamond probes. Comput. Vis. Image Underst. **17**, 319–325 (2013)
5. Beauquier, D., Nivat, M.: On translating one polyomino to tile the plane. Disc. Comput. Geom. **6**, 575–592 (1991)
6. Massé, A.B., Brlek, S., Labbé, S.: A parallelogram tile fills the plane by translation in at most two distinct ways. Disc. Appl. Math. **170**(7–8), 1011–1018 (2021)
7. Freeman, H.: Boundary encoding and processing. In: Lipkin, B., Rosenfeld, A. (eds.) Picture Processing and Psychopictorics, pp. 241-266. Academic Press, New York (1970)
8. Freeman, H.: On the encoding of arbitrary geometric configurations. IRE Trans. Electron. Comput. **10**, 260–268 (1961)
9. Frosini, A., Nivat, M.: On a tomographic equivalence between $(0, 1)$-matrices. Pure Math. Appl. **16**, 251–265 (2005)
10. Frosini, A., Nivat, M.: Binary matrices under the microscope: a tomographical problem. Theor. Comput. Sci. **370**, 201–217 (2007)
11. Herman, G.T., Kuba, A. (eds.): Discrete Tomography: Foundations Algorithms and Applications. Birkhäuser, Boston (1999)
12. Herman, G.T., Kuba, A. (eds.): Advances in Discrete Tomography and its Applications. Birkhäuser, Boston (2006)
13. Nivat, M.: Sous-ensembles homogènes de $\mathbb{Z}^2$ et pavages du plan (in French). C.R. Acad. Sci. Paris Ser. I **335**, 83–86 (2002)

Tilepaint and Aquarium Puzzles in Periodic Grids

Yan Gerard[1(✉)], Pascal Lafourcade[1], Lola-Baie Mallordy[1], and Léo Robert[2]

[1] University Clermont Auvergne, LIMOS, Clermont-Ferrand, France
{yan.gerard,pascal.lafourcade}@uca.fr
[2] University of Picardie Jules Verne, MIS, Amiens, France
leo.robert@u-picardie.fr

Abstract. We consider two classes of puzzles `Aquarium` and `Tilepaint` connected to the field of Discrete Tomography. The input mixes a tiling of a rectangular grid with projections prescribing the number of squared cells that have to be filled in each row and column. In `Tilepaint`, the tiles are either entirely filled or remain empty while in `Aquarium`, the tiles are filled like aquariums with an horizontal level of water. `Tilepaint` is known to be NP-complete with a proof showing that `Aquarium` is also NP-complete, even with tiles of size at most 3. In this paper, we investigate the complexity of the two puzzles in the special case of regular tilings with small tiles. We show that `Tilepaint` can be solved in polynomial time while `Aquarium` is NP-complete for regular triomino tilings and is polynomial time for some regular domino tilings. As `Aquarium` is NP-complete, we propose, in a second part, a physical zero-knowledge proof (ZKP) protocol for `Aquarium`, using decks of playing cards (using a total of $2mn + 2$ cards for an $m \times n$ grid).

1 Introduction

The aim of this work is to investigate two combinatorial problems at the intersection of Discrete Tomography and tilings. Discrete Tomography focuses on reconstructing a discrete set of points S based on its projections, which count the number of points in S along given straight lines [6,7]. Tilings, on the other hand, is the antique domain involving partitioning a space —here, a 2D grid— into smaller pieces called *tiles*, each with prescribed shapes.

Previous works have examined the relationship between these two fields by considering tilings of a rectangular grid using dominoes [10] or bars [3], where the goal was to compute a tiling that satisfies certain tomographic constraints. In contrast, our work investigates a different class of problems: given an existing tiling of a grid, the challenge is to select specific tiles or tile subsets that satisfy the prescribed tomographic constraints. As nonograms [13], these problems are variants of the foundational problem of Discrete Tomography i.e. the reconstruction of a binary matrix from its row and column sums or equivalently the reconstruction of a lattice set from its horizontal and vertical projections [5,12].

M. H. F. Wilkinson and J. Kosinka (Eds.): DGMM 2025, LNCS 16296, pp. 103–117, 2026.
https://doi.org/10.1007/978-3-032-09544-2_8

This variation originates from two recreational puzzles. The first, *Tilepaint*, is a puzzle published by Nikoli[1]. The second, *Aquarium*, was invented in 2004 by Naoki Inaba. For both problems, the input consists of a regular grid of $m \times n$ squares tiled by polyominoes called *tiles* and $m + n$ integers, one for each row and column of the grid.

The aim of *Tilepaint* is to select a set of tiles satisfying the tomographic constraint that the total number of selected squares in each row and column corresponds to the input projections (Fig. 1).

`Tilepaint`(A, H, V)

- `Input:` a tiling of the regular grid of $m \times n$ squares by a set A of polyominoes A_k called *tiles* and a pair of integer vectors $H = (h_i)_{1 \leq i \leq m} \in \mathbb{Z}^m$, $V = (v_j)_{1 \leq j \leq n} \in \mathbb{Z}^n$ called the horizontal and vertical *projections*.
- `Ouput:` A subset S of the tiles such that for all row and column indices i and j, the number of squares of the tiles of S in row i and column j are equal to h_i and v_j.

The problem *Aquarium* also involves filling certain squares of the tiles with the constraint of having the prescribed number of squares filled in each row and column. But unlike *Tilepaint* where either none or all the squares of each tile are filled, tiles are filled in *Aquarium* as aquariums. It means that for each tile, a water level is chosen below which all the squares of the tile are filled and above which no square of the tile is filled.

`Aquarium`(A, H, V)

- `Input:` a tiling of the regular grid of $m \times n$ squares by a set A of polyominoes A_k called *tiles* and a pair of integer vectors $H = (h_i)_{1 \leq i \leq m} \in \mathbb{Z}^m$, $V = (v_j)_{1 \leq j \leq n} \in \mathbb{Z}^n$ called the horizontal and vertical *projections*.
- `Ouput:` An horizontal water level l_k for each tile A_k so that for any indices i or j, the number of filled squares *i.e.*, under the water level of its tile is equal to h_i for the row i and to h_j for the column j.

The water level constraint makes the *Aquarium* problem asymmetric. Notice also that when tiles are of height 1, *Tilepaint* and *Aquarium* are equivalent.

In this paper, we consider the problem of reconstructing a solution of `Tilepaint` and `Aquarium` in the special case where the tiling is periodic and has small tiles of size 2 or 3 i.e. with dominoes or triominoes. The paper first presents the state of the art and the results. Then it provides the proofs with a polynomial-time algorithm for some domino tilings and some reductions of NP-complete problems for the triomino tilings. We end the paper with a zero-knowledge proof (ZKP) protocol for proving that a solution is known without leaking it.

[1] Nikoli is a well-known Japanese publisher of logic puzzles, famous for popularizing Sudoku.

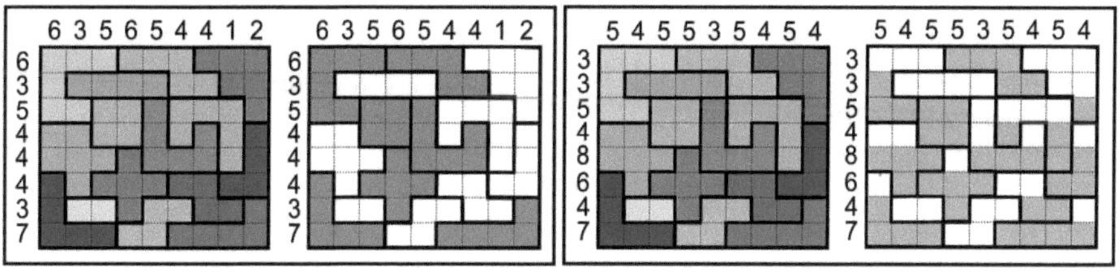

Fig. 1. Instances of Tilepaint(A, H, V) and Aquarium(A, H, V) and their solutions (left for *Tilepaint* and right for *Aquarium*).

2 State of the Art

The problem `Tilepaint`(A, H, V) has been proved NP-complete in 2022 [8]. The proof is a reduction of `3D-matching` and since it only uses tiles of height 1, it also proves that `Aquarium` is NP-complete. See Fig. 2 for the reduction of a small instance of `3D-matching` into a `Tilepaint` instance of height 1. However, this reduction slightly differs from the original in order to show the property that `Tilepaint` remains NP-complete with tiles of size at most 3.

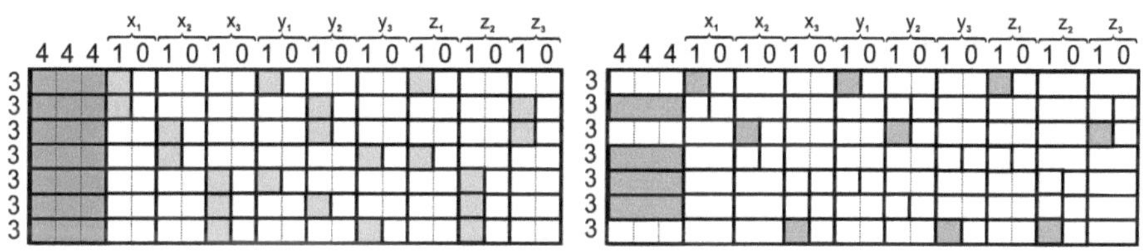

Fig. 2. Reduction of `3D-matching` to `Tilepaint` or `Aquarium`. Left, an instance of `Tilepaint`(A, H, V) encoding the instance of 3D-matching composed of the sets $p_1 = (x_1, y_1, z_1)$, $p_2 = (x_1, y_2, z_3)$, $p_3 = (x_2, y_2, z_3)$, $p_4 = (x_2, y_3, z_1)$, $p_5 = (x_3, y_1, z_2)$, $p_6 = (x_3, y_2, z_2)$, $p_7 = (x_3, y_3, z_2)$. Right, the solution of `Tilepaint` describes the solution $\{p_1, p_3, p_7\}$ of the 3d-matching instance.

The proof of NP-completeness of `Tilepaint` or `Aquarium` encodes the instance of 3d-matching in the tiling which is thus quite irregular, while the projections are fixed. By using a periodic tiling and only variable projections, we can expect different complexities. The first case to investigate is a periodic tiling by $m \times n$ tiles of unit size. In this case, solving `Aquarium` or `Tilepaint` means reconstructing a set of squares with prescribed projections. This problem is one of the milestones of Discrete Tomography. It has been independently proved to be sovable in polynomial time by D. Gale and R. Ryser in 1957 [5,12].

The Gale strategy is to reduce the problem to a problem of edge selection in a bipartite graph with unit capacities on the edges. Then the problem can be solved by the initial Gale algorithm or by more general graph algorithms such that Ford-Fulkerson or Hopcroft-Karp algorithms or with a more recent quasi-linear time flow algorithm [1]. It proves the next proposition:

Proposition 1. *Given horizontal, vertical projections and the tiling A of the region by $m \times n$ unit tiles,* `Aquarium` *and* `Tilepaint` *can be solved in time almost linear in the size of the grid ($O(mn)$).*

Let us now consider the extension of the problem to periodic domino and triomino tilings. Are `Tilepaint` and `Aquarium` still NP-complete in this restricted framework, or do there exist polynomial-time reconstruction algorithms?

3 Complexity Results

We consider four periodic domino tilings that we call the *stack*, the *jail*, the *wall* and the *traffic jam* (Fig. 3).

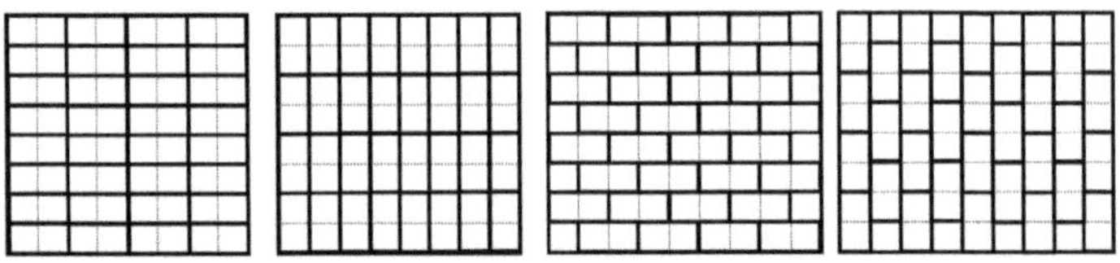

Fig. 3. Four domino tilings. From left to right, the *stack*, the *jail*, the *wall* and the *traffic jam.*

We also consider three periodic triomino tilings drawn in Fig. 4 and respectively called *E*, *Waves*, and *LL*.

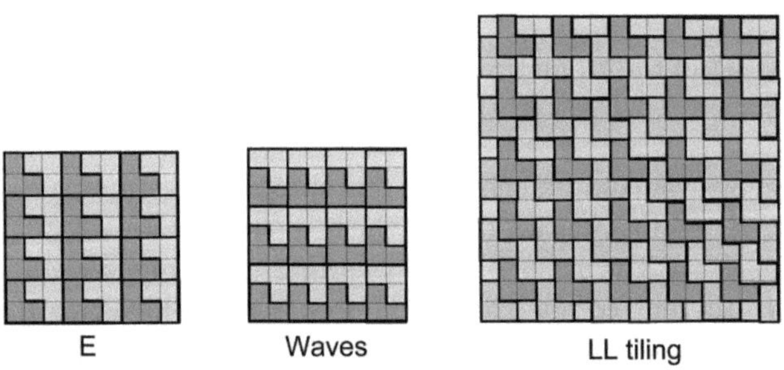

Fig. 4. Three Periodic tilings by triominoes. The *E*, *Waves*, and *LL* tilings.

We summarize our complexity results in Table 1, and express them in the next theorem.

Table 1. Complexities of the periodic instances of `Tilepaint` and `Aquarium`.

Tilings	Stack	Jail	Wall	Traffic	Waves	E	LL
`Tilepaint`	P	P	P	P	P	P	P
`Aquarium`	P	?	P	?	NP-complete	NP-complete	NP-complete

Theorem 1. *(i)* ***Tilepaint*** *can be solved in polynomial-time for the Stack, the Jail, the Wall, the Traffic Jam, the Waves, E and LL.*
(ii) ***Aquarium*** *can be solved in polynomial-time for the Stack and the Wall.*
(iii) ***Aquarium*** *is NP-complete for the Waves, the E and the LL triomino tilings.*

4 Periodic Domino Tilings

We now present the proofs of Theorem 1 for the four considered domino tilings. Our main contribution is to show that the wall instances can be solved in polynomial time.

The stack. It is the easiest case, since one vertical projection out of two is useless. Both instances of `Aquarium` and `Tilepaint` can be reduced to instances with unit tiles and thus can be solved in almost linear time.

The jail. Since `Tilepaint` is symmetrical (direction x and y are equivalent), `Tilepaint` makes no difference between the jail and the stack. Then `Tilepaint` is solvable in polynomial time for the jail. For `Aquarium` the jail and stack instances are different since the vertical tiles can be half-full. Surprisingly, `Aquarium` is challenging. We suspect that for the jail tiling, `Aquarium` is NP-complete but its complexity is an open question.

The wall. As the height of all the aquariums is 1, there is no difference between `Tilepaint` and `Aquarium`. We now show that the domino instances of `Aquarium` and `Tilepaint` with the wall domino tiling can be solved in polynomial time. We distinguish two cases depending on the parity of the number of columns.

(i) If the number of columns is odd, for instance, equal to $2k+1$, each row is made of k tiles of size 2 and a unit tile. It follows that the value of an horizontal projection is odd if and only if the unit tile of the corresponding row is filled. Then the parity of the values of the horizontal projections directly determine the unit tiles which have to be filled and the ones which have to remain empty. Now we remove the unit tiles from the picture and focus our attention on the dominoes. We call rank of a domino the index of the column of its left square so that the ranks of dominoes go from 1 to $2k$. We explain now how to determine the projections of the dominoes of each rank. First, we substract the number of unit squares in the first column from the initial projection v_0 of the first column. It provides the integer $v'_1 = v_1 - v_0$ counting the number of filled dominoes in the first rank. Then by subtracting v_1' to v_2, we obtain the integer $v_2' = v_2 - v_1'$ counting the number of dominoes of rank 2. By repeating this computation in the next ranks, we obtain the number of dominoes of each rank (see Fig. 5 for a detailed example of the algorithm). Knowing the number of dominoes of each rank, we decompose the problem into two independent sub-problems by considering the rows of even and odd indices. Each sub-problem is structured as the stack tiling and thus can be solved in polynomial time. In the case where the two stack instances, with the rows of even indices or the rows of odd indices are both feasible, we aggregate the two solutions to build the final solution. If one of the two instances with even or odd rows is not feasible, then the puzzle admits no solution.

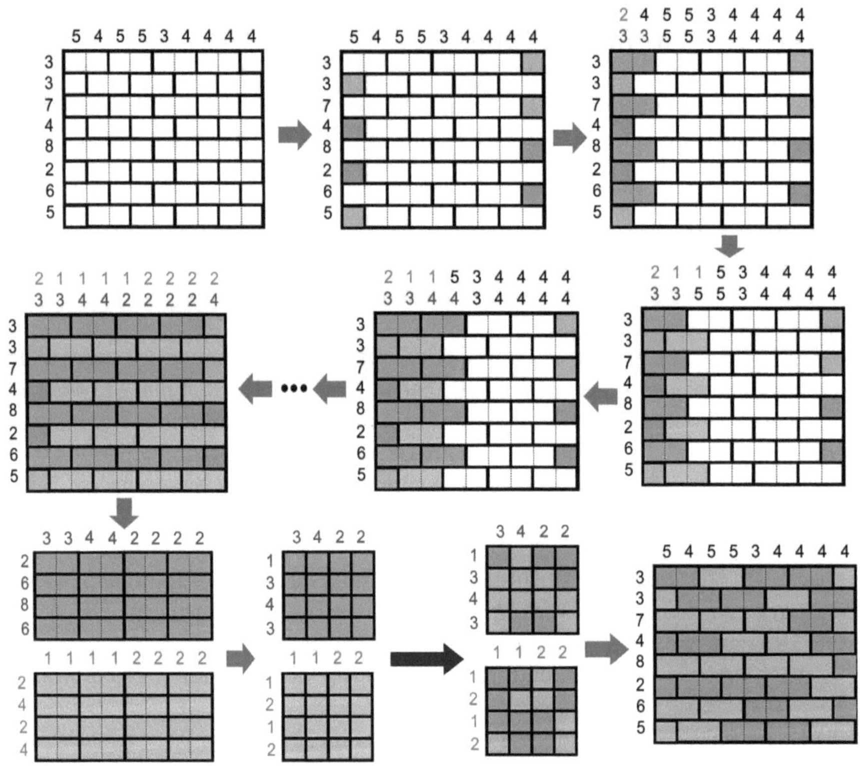

Fig. 5. Algorithm for solving the wall when the number of columns is odd.

(ii) If the number of columns $n = 2k$ is even, there are two unit tiles in some rows and zero unit tiles in others. The columns have indices going from 1 to $2k$. For a row with a pair of unit tiles and an odd value of projection, only one of the two unit tiles is filled. It only remains some uncertainty about the side, left or right, of the unit square which should be filled but it can be determined with two different ways of counting the filled dominoes. Let us denote v'_0 and v'_{2k} the a-priori unknown numbers of unit tiles that have to be filled respectively in the left-most and right-most columns. We also denote t the unknown number of dominoes which have to be filled. By computing the sum of the vertical projections with an odd index, we count the number of filled dominoes t plus the number of unit tiles in the left-most column. Then $\sum_{j=1}^{k} v_{2j-1} = t + v'_0$ while by symmetry, the sum of the vertical projection with an even index $\sum_{j=1}^{k} v_{2j} = t + v'_{2k}$ counts the number of filled dominoes plus the filled unit tiles of the right-most column. It follows that $\sum_{j=1}^{k} v_{2j} - \sum_{j=1}^{k} v_{2j-1} = v'_{2k} - v'_0$ (a). If we now consider a row with an even projection and two unit tiles, by parity, it might contain either two filled unit tiles or zero filled unit tiles. In both cases, the contribution of the row to the difference $v'_{2k} - v'_0$ is null. The only rows contributing to the difference $v'_{2k} - v'_0$ are the ones with an odd projection. They contribute either with a value $+1$ or with a value -1. The difference (a) tells us exactly how many unit tiles (from the rows with an odd projection) have to be filled on the left and how many tiles have to be filled on the right. Then the unit tiles from the rows with an odd projection can be addressed with some basic computations and fully removed from the picture by removing their squares and updating the projections of the remaining tiles. Now let us focus on the unit tiles in the rows having an even projection. We cannot fill one unit tile without filling the other one. Then these pairs of unit tiles are similar to dominoes. By making a bridge from the left column to the right column, the remaining combinatorial problem has a cylinder structure which does not seem to be of any help to solve the problem (Fig. 6).

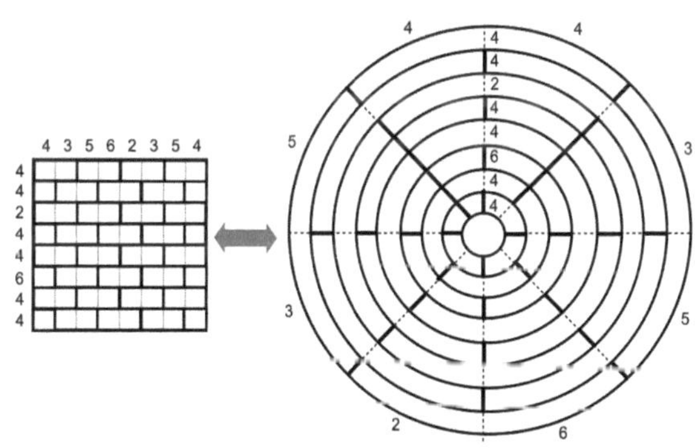

Fig. 6. Cylinder structure of the wall when the number of columns is even.

Our algorithm breaks the cylinder symmetry by assuming that the number of dominoes of a given rank, for instance of rank 1 is fixed (the rank of a domino is the index of the column of its left square). Then by difference with v_2, we obtain the number of dominoes of rank 2 and so on until the number of dominoes of rank $n-1$. As in the odd case (i), the knowledge of the number of dominoes of each rank allows us to decompose the problem in two sub-problems each one structured as a stack or a stack with some missing squares but it does stop the flow algorithms to solve it. The two instances with the odd or even rows can be solved independently in polynomial time. If one of them has no solution (as illustrated in the second line of Fig. 7), then the assumption on the number of filled dominoes of rank 1 is not valid. Then we test another value. If the two independent instances are feasible, the solutions are aggregated to provide a general solution. For each potential number of filled dominoes of rank 1, the reconstruction routine takes an almost linear time $O(mn)$. Since we have at most m numbers of filled dominoes of rank 1 to test, the total complexity of the reconstruction algorithm is $O(m^2n)$.

It follows that the reconstruction algorithm for the wall is $O(mn)$ if n is odd and $O(m^2n)$ if n is even.

The Traffic Jam. By symmetry of the wall, we know that `Tilepaint` is polynomial time solvable. For `Aquarium`, the fact that the tiles are vertical and can be half-filled makes the problem non trivial. If all the vertical projections with an even index are null, then the even columns are empty and the non empty columns are tiled as the jail. Then the jail can be reduced to the traffic jam, showing that the complexity for the traffic jam is at least the complexity for the jail.

5 Periodic Triomino Tilings

We now prove that `Tilepaint` is polynomial-time for the three considered triomino tilings while `Aquarium` becomes NP-complete.

Polynomial-Time Algorithms for Tilepaint. We start with the waves triomino tiling (by symmetry, it is equivalent to E, which is not true for Aquarium). We decompose the tiles in two classes denoted L if their shape looks like the letter L and $\bar{L}$ for their symmetric. Two horizontal projections out of three directly provide the number of filled L and $\bar{L}$ in each row. Then if they are consistent, we do not need to take into account the intermediary rows crossing L and $\bar{L}$ triominoes. Concerning the vertical projections, the computation is less trivial. The vertical projection v_1 is equal to $v_1 = 2|L_1| + 1|\bar{L}_1|$ where $|L_1|$ and $|\bar{L}_1|$ are respectively the numbers of filled L and $\bar{L}$ crossing the first column. With $v_2 = |L_1| + 2|\bar{L}_1|$, it leads to $|L_1| = \frac{2}{3}v_1 - \frac{1}{3}v_2$ and $|\bar{L}_1| = \frac{2}{3}v_2 - \frac{1}{3}v_1$. Equivalent expressions hold for the other columns. As we can compute the numbers of filled L and $\bar{L}$ in each row and column, we can reconstruct the sets of filled L and $\bar{L}$ independently as for unit tiles. It solves $\texttt{Tilepaint}(A, H, V)$ for the waves and E with time complexity $O(mn)$.

The strategy for LL is similar in that we decompose the tiling in three classes of tiles (all of them are L triominoes) as shown with the colors red, green and yellow in Fig. 4. Assume that the number of filled yellow tiles is known in the left-most column. By difference, we get the number of filled green tiles in this first column. Then the second vertical projection provides the number of filled

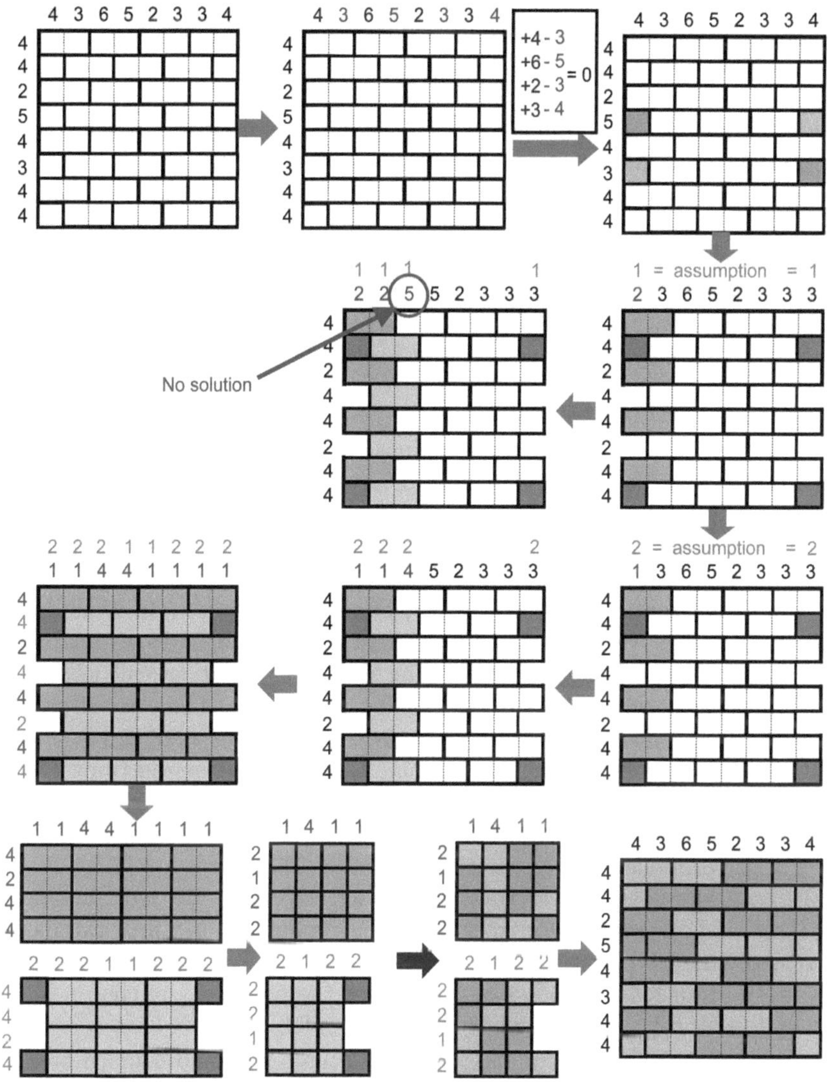

Fig. 7. Algorithm for solving the wall when the number of columns is even. Above, the difference between the vertical projections of the even and odd columns provides the difference between the filled unit cells in the left-most and rightmost columns by considering uniquely the rows of odd horizontal projections. It allows us to remove from the unit tiles of the rows with an odd projection (by updating the projections). Then the algorithm requires to fix the number of dominoes filled with a given rank *i.e.*, with a left square in a given column, here the left-most. It can either lead to a contradiction as in the second row of the figure or to a solution, through a decomposition into two sub-problems.

red tiles in the second column and so on. By this same reasoning, we obtain the number of filled tiles of each color in each column. It works in the same way for the rows: with an assumption on the number of filled green tiles in the lowest row, we obtain all the other projections for each set of colored tiles. Then the filled tiles of each color can be computed in almost linear time (Proposition 1) as performed for unit tilings. If there exists a solution for each color, then we obtain a solution. If a color is not feasible, then there is no global solution with the considered assumptions on the leftmost and lowest projections. In other words, by assuming the number of green filled tiles in the left-most column and lowest row, we can determine in polynomial time whether it provides solutions. As the number of values to test is at most m for the left-most column and n for the lowest row, it provides a polynomial time algorithm of time complexity $O(m^2n^2)$ for solving any instance in the LL tiling.

Aquarium is NP-Complete. We now prove that $\texttt{Aquarium}(A, H, V)$ is NP-complete with the regular triomino tilings. We reduce a problem of Discrete Tomography with three colors that we denote $\texttt{DT3}(H, H', V, V')$ [4]. The input of $\texttt{DT3}(H, H', V, V')$ consists of a rectangular region of $m \times n$ unit squares with two vertical projections $V \in \mathbb{Z}^n$ and $V' \in \mathbb{Z}^n$ and two horizontal projections $H \in \mathbb{Z}^m$ and $H' \in \mathbb{Z}^m$. The output consists of two disjoint subsets *Blue* and *Red* whose horizontal and vertical projections are respectively H and V for *Blue* and H' and V' for *Red*. The top rectangle of Fig. 8 shows an instance of $\texttt{DT3}(H, H', V, V')$ and its solution. The third color that we call *white* is the one of the squares which are neither in *Blue* nor in *Red*. The class of problems $\texttt{DT3}(H, H', V, V')$ has been proved to be NP-complete in 2012 in [4]. The two problems $\texttt{DT3}(H, H', V, V')$ and $\texttt{Aquarium}(A, H, V)$ on triomino tiling are connected by using the property that in $\texttt{Aquarium}$ a triomino can have three states: empty, half-filled, or full. The strategy of the reduction is to put these three states in correspondence with the three colors of $\texttt{DT3}(H, H', V, V')$.

We consider the wave triomino tiling. We encode an instance of $\texttt{DT3}(H, H', V, V')$ with m rows and n columns in an instance of $\texttt{Aquarium}(A, H^A, V^A)$ with $3m$ rows and $2n$ columns. With indices i going from 0 for the bottom rows to $m-1$ or $3m-1$ for the top rows, we choose $h^A_{3i} = 0$, $h^A_{3i+1} = h_i + h'_i$, $h^A_{3i+2} = 2h_i$, and for j going from 0 to $n-1$, we have $v^A_{2j} = v_j$ and $v^A_{2j+1} = 2v_j + v'_j$. The null rows require that all the L triominoes are empty. Then a solution of this aquarium instance has exactly $v^A_{2j} = v_j$ full tiles in the pair of columns of indices $2j$,$2j+1$. The difference $v^A_{2j+1} - 2v^A_{2j} = v'_j$ gives the number of half-filled tiles in the pair of columns $2j$,$2j+1$. For the rows, we have exactly $\frac{h_{3i+2}}{2} = h_i$ full tiles in the triplet of rows of indices $3i$, $3i+1$, $3i+2$. The number of half-full tiles is $h_{3i+1} - \frac{h_{3i+2}}{2} = h_i + h'_i - \frac{2h_i}{2} = h'_i$. Then if we consider the full tiles and the half-filled tiles, their number on each pair of columns and triplet of rows fits exactly with the number of expected blue and red squares of a solution of $\texttt{DT}(3, H, H', V, V')$. It shows that $\texttt{DT}(3, H, H', V, V')$ admits a solution if and only if $\texttt{Aquarium}(A, H^A, V^A)$ is feasible. The NP-hardness of $\texttt{DT}(3, H, H', V, V')$ implies the NP-hardness of $\texttt{Aquarium}(A, H^A, V^A)$.

By well choosing the rows and columns with null projections, a similar reduction can be replicated for the E and LL triomino tilings.

6 ZKP Protocol for Aquarium

Recall that a Zero-Knowledge Proof protocol describes the interactions between a prover Pv, who possesses a secret, and a verifier Vf, who wants to be convinced that Pv indeed has the solution without revealing the solution itself. In our context, we aim to construct a protocol that convinces Vf that Pv has the solution to a given `Aquarium` instance without leaking the solution.

Our protocol is split into two phases: (i) the construction of Pv's solution done by Vf where the level constraint is enforced; (ii) the verification of the tomographic constraints. In total, for a grid of size $n \times m$ and A its set of polyominos, our protocol needs $2mn + 2$ cards, and uses $\sum_{a \in A}(\text{height}(a) - 1)$ AND protocols, $\sum_{a \in A} \text{height}(a)$ copy protocols and $n + m$ shuffles. We formally describe the different phases of our protocol.

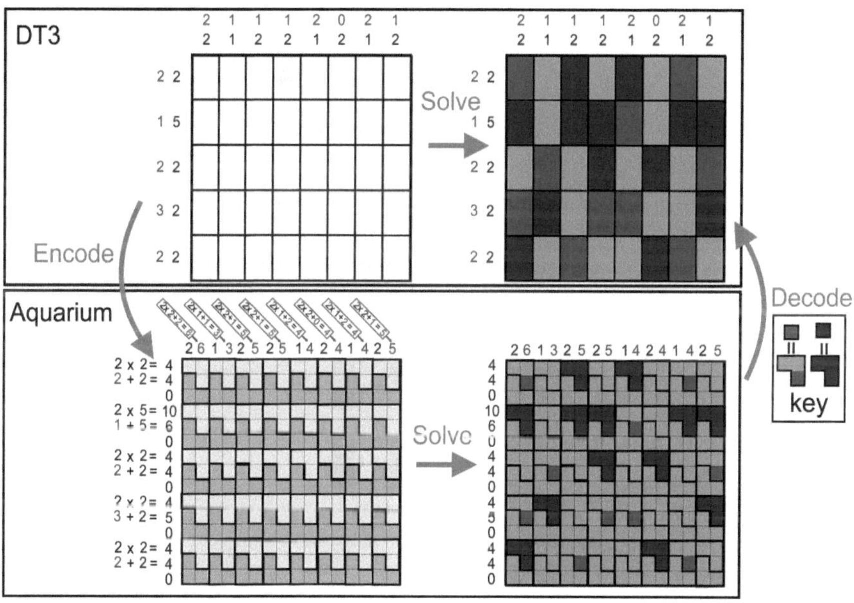

Fig. 8. An instance of $\texttt{DT}(3, H, H', V, V')$**, its encoding in** $\texttt{Aquarium}(A, H, V)$ **and the corresponding solutions.**

6.1 Setup Phase

Before getting involved in the protocol, Pv and Vf need to agree on public parameters. The grid of an $\texttt{Aquarium}(A, H, V)$ instance is composed of m rows and n

columns, with A, H and Vf being public values of the polyominoes (*e.g.*, the tiles), the numbers on the left and upper edges of the grid, respectively.

The grid has no commitment yet, the prover Pv will commit its solution in the *construction phase* (where Vf knows that all the tiles are correctly filled). The next step is to verify that the tomographic constraints are respected, if so, the protocol does not abort and Vf is convinced that Pv has the solution.

6.2 Construction Phase

The prover Pv wants to select some squares of the tiles. The selected squares are blackened while the remaining squares are left white. There are two steps, Pv indicates the level separating the black and white squares (without Vf knowing its height) then Vf will fill the region up to the given level.

Level Indicator

1. Pv and Vf consider the leftmost commitments of each row. Then, the prover Pv commits a black value on a given cell to represent the level; the other cells are committed to white. Note that if the region is empty (*i.e.*, no black cell), then Pv simply commits white value on each cell.
2. The verifier Vf will spread the level downward: 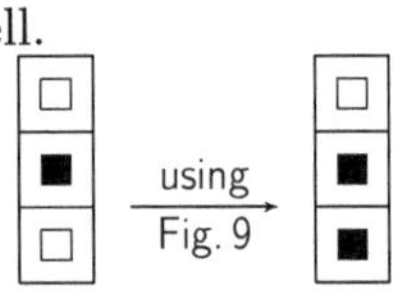

We use the technique of [11] to enforce that every cell under the level indicator will be coloured in black. Basically, for each adjacent commitments (from top to bot), Vf computes an AND operation with the Mizuki-Sone AND protocol [9] on the second commitment (and keeps the first commitment in its initial state). The general sub-routine is illustrated below in Fig. 9.

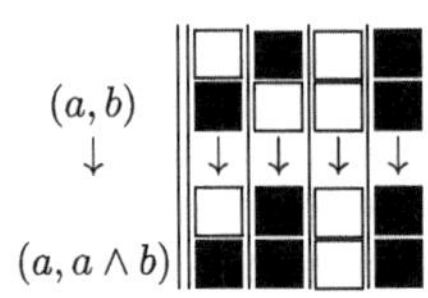

Fig. 9. Subroutine of [11] for modifying each consecutive cell where $a, b \in \{0, 1\}$ represents the colours of the leftmost and rightmost cells, respectively.

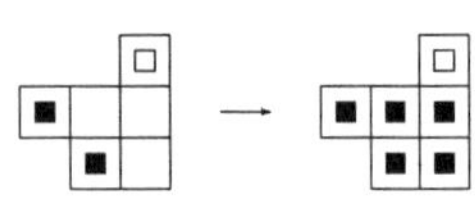

Fig. 10. Copies of right to leftmost commitment to fill each row.

Spreading the Level. The next step is to propagate the configuration of the leftmost cells to the rest of the tile (done via the Mizuki-Sone copy protocol [9]). Formally, for this step, For each row of length ℓ, apply the copy protocol of the leftmost commitment to output ℓ identical commitments to form the committed row as described in Fig. 10. The previous two steps are repeated for each tile.

6.3 Verification Phase

Verifying that Pv has indeed the solution is done by checking the numbers for each row are respected. This means that Vf needs to check that the numbers on each row/column is correct, to be convinced that Pv has committed a solution.

The verification of this rule is straightforward, each leftmost card of commitments will be used to verify the row constraints while the rightmost card will be used for column constraints. This approach has the advantage of using cards only once, so no input-preserving technique is needed (thus reducing the number of steps and shuffles). Formally, Pv and Vf apply the following steps:

1. For each row i, Vf picks the leftmost card of each commitment. Then Pv and Vf shuffle them. When Vf reveals the cards, if the number of $\boxed{\clubsuit}$ is equal to h_i (the number at the left of row i) then Vf continues, otherwise aborts.
2. For each column j, Vf picks the rightmost card of each commitment. Then Pv and Vf shuffle them. When Vf reveals the cards, if the number of $\boxed{\heartsuit}$ is equal to $m - v_j$ (v_j the number indicated at the top of column j) then Vf continues, otherwise aborts.

6.4 Security Proofs

Our protocol provides completeness, soundness and is zero-knowledge. Note that the sub-protocols used from the literature have been proven secure *i.e.*, they are correct, complete, sound and zero-knowledge.

Theorem 2 (Completeness). *If* Pv *knows the solution of an Aquarium instance, then* Pv *can convince* Vf.

Proof. Suppose that Pv has the solution, we want to show that the protocol will not abort, hence that Pv is able to prove the number rule. Let S be a solution, S is a valid grid respecting all the constraints.

During the construction phase, Pv indicates the water level of S for each room to Vf. As S is a valid grid, Pv can always choose the top left cell of the region (if the region is empty, then Pv simply puts white commitments) which indicates the level. Hence, when Vf spreads the level downward with the AND protocol and fills each row with the copy protocol, Vf ends up with a room coloured in exactly the same way as Pv from the correctness of the AND and copy protocols. As this routine is repeated for each room, Vf will colour each region following the "level" rule, so Pv can construct the solution S with this routine. Furthermore, since S is the solution, the "number" rule is enforced. Therefore, the protocol will not abort at the number rule verification.

Finally, the protocol will not abort so the completeness is verified.

Theorem 3 (Soundness). *If* Pv *does not provide a solution of the* $p \times q$ *Aquarium grid, then* Vf *always rejects.*

Proof. We prove the contrapositive of this statement. Suppose V accepts, meaning the verification passes.

When Vf checks the number of $\boxed{\clubsuit}$ for each row, this number corresponds to the number of black commitments in the row (a black commitment is represented as $\boxed{\clubsuit}\boxed{\heartsuit}$) since the leftmost cards are picked. For the columns, the complementary of each commitment is checked meaning that the number of $\boxed{\heartsuit}$ is considered. Since the number of $\boxed{\heartsuit}$ cards corresponds to the difference of the total number of rows in the grid and the column indicating number, the complement of each commitments is equal to the column indicating number. Since the rightmost cards are picked, this corresponds to the solution of the grid.

Notice that we omit the construction phase in this analysis. Indeed, since Vf constructs Pv's solution, then no invalid configuration can occur (*e.g.*, a filled black row with a white cell in it, or a white cell below a black cell in the same room). This claim is straightforward since the only input of the prover is during the level indicator phase. At this step, the prover chooses the values of the commitments (only one black value and the other white). Even if the prover puts multiple black values on the commitments, the sub-routine of [11] *smooths* the final values of the sequence. This comes from the fact that only the topmost black cell is taken into account in the final configuration. This last point allows us to let Pv making its own commitments and not using heavier sub-protocols, such as the chosen-pile protocol [2], that would result in worst complexity (as the number of shuffles and cards would increase).

Theorem 4 (Zero-knowledge). Vf *learns nothing about* Pv*'s solution of the given grid* G.

Proof. Suppose that Pv has a grid S, we want to show that Vf will not have learned anything about S at the end of the protocol. For this, we prove that the interaction between Pv and Vf can be simulated by a simulator Sim that does not know Pv's solution. It is sufficient to show that all distributions of face-up cards can be simulated by Sim.

The verification phase is the only step where cards are revealed. Since pile-scramble shuffle is used, on each row and each column, all outputs have the same probability to occur (given a specific initial number of $\boxed{\heartsuit}$ and $\boxed{\clubsuit}$ which corresponds to the row/column indicating number).

Thus, the verifier Vf learns nothing about the solution.

7 Conclusion

We focused our attention on instances of `Tilepaint` and `Aquarium` on regular tilings (see results in Table 1). As shown for domino and triomino tilings, the `Tilepaint` puzzles seem to be polynomial time solvable (for a fixed tiling) with a polynomial whose degree is linear in the size of the tiles. On the other hand, `Aquarium` remains NP-complete for periodic triomino tilings and the question of its complexity is open for some of the most simple domino tilings such as the jail and the traffic jam.

Finally, we proposed a ZKP protocol for `Aquarium` and proved its security. Our method could be extended to `Tilepaint`, since it is also NP-complete for irregular tiling and works in a similar way. An interesting future work would be to design ZKP protocols on other `Aquarium` tilings.

References

1. Chen, L., Kyng, R., Liu, Y.P., Peng, R., Gutenberg, M.P., Sachdeva, S.: Maximum flow and minimum-cost flow in almost-linear time. In: 2022 IEEE 63rd Annual Symposium on Foundations of Computer Science (FOCS), pp. 612–623 (2022). https://doi.org/10.1109/FOCS54457.2022.00064
2. Dumas, J.-G., Lafourcade, P., Miyahara, D., Mizuki, T., Sasaki, T., Sone, H.: Interactive physical zero-knowledge proof for norinori. In: Du, D.-Z., Duan, Z., Tian, C. (eds.) COCOON 2019. LNCS, vol. 11653, pp. 166–177. Springer, Cham (2019). https://doi.org/10.1007/978-3-030-26176-4_14
3. Dürr, C., Ch., E.G., Rapaport, I., Rémila, E.: Tiling with bars under tomographic constraints. CoRR (1999). https://arxiv.org/abs/cs/9903020
4. Dürr, C., Guiñez, F., Matamala, M.: Reconstructing 3-colored grids from horizontal and vertical projections is np-hard: a solution to the 2-atom problem in discrete tomography. SIAM J. Disc. Math. **26**(1), 330–352 (2012). https://doi.org/10.1137/100799733
5. Gale, D.: A theorem on flows in networks. Pac. J. Math. **7**, 1073–1082 (1957). https://doi.org/10.2140/pjm.1957.7.1073. https://msp.org/pjm/1957/7-2/pjm-v7-n2-p04-s.pdf
6. Gardner, R.J.: Geometric Tomography. Encyclopedia of Mathematics and its Applications. Cambridge University Press, Cambridge (1995)
7. Herman, G.T., Kuba, A.: Discrete Tomography - Foundations, Algorithms and Applications. Birkhauser, Basel (1999)
8. Iwamoto, C., Ide, T.: Five cells and tilepaint are np-complete. IEICE Trans. Inf. Syst. (2022). https://doi.org/10.1587/transinf.2021fcp0001
9. Mizuki, T., Sone, H.: Six-card secure AND and four-card secure XOR. In: Deng, X., Hopcroft, J.E., Xue, J. (eds.) FAW 2009. LNCS, vol. 5598, pp. 358–369. Springer, Heidelberg (2009). https://doi.org/10.1007/978-3-642-02270-8_36
10. Picouleau, C.: Reconstruction of domino tiling from its two orthogonal projections. Theor. Comput. Sci. **255**(1-2), 437–447 (2001). https://doi.org/10.1016/S0304-3975(99)00312-6
11. Robert, L., Miyahara, D., Lafourcade, P., Mizuki, T.: Physical ZKP protocols for nurimisaki and kurodoko. Theor. Comput. Sci. **972**, 114071 (2023). https://doi.org/10.1016/J.TCS.2023.114071
12. Ryser, H.J.: Combinatorial properties of matrices of zeros and ones. Canbadian J. Math. **9**, 371–377 (1957)
13. Ueda, N., Nagao, T.: Np-completeness results for nonogram via parsimonious reductions (1996). https://api.semanticscholar.org/CorpusID:3040166

Conversion of a Digital Object into a Finite Set of Balls: Complexity

Isabelle Sivignon(✉)

Univ. Grenoble Alpes, CNRS, Grenoble INP, GIPSA-Lab, 38000 Grenoble, France
isabelle.sivignon@gipsa-lab.grenoble-inp.fr

Abstract. This paper addresses the problem of converting a 2D digital object S (a finite set of points in $\mathbb{Z}^2$) into a finite set $\mathcal{B}$ of balls centered on $\mathbb{R}^2$, such that the balls of $\mathcal{B}$ cover all points of S and no point of $\mathbb{Z}^2 \backslash S$, and the cardinality of $\mathcal{B}$ is minimum. In a previous work, we showed that the problem was polynomial for the specific class of 2D hole-free digital objects. In this article, we show that the problem is NP-complete in the general case using a reduction from the 3-Planar Vertex Cover problem.

Keywords: digital object · set of balls · covering · complexity

1 Introduction

In this work, a 2D digital object S is a finite 4-connected subset of $\mathbb{Z}^2$. Given a finite set $\mathscr{B}$ of open balls, we note $\bigcup \mathscr{B} = \cup_{\mathscr{b} \in \mathscr{B}} \mathscr{b}$. We say that $\mathscr{B}$ covers S if $\bigcup \mathscr{B} \cap \mathbb{Z}^2 = S$. In particular, this means that $\bigcup \mathscr{B} \cap (\mathbb{Z}^2 \backslash S) = \emptyset$. We consider the following optimisation problem:

Problem 1. Given a 2D digital object S, compute a finite set $\mathscr{B}$ of balls that covers S and such that the cardinality of $\mathscr{B}$ is minimum.

A digital ball b is a subset of $\mathbb{Z}^2$ for which there exists a ball $\mathscr{b}$ such that $\mathscr{b} \cap \mathbb{Z}^2 = b$. Otherwise said, a digital ball is a subset of $\mathbb{Z}^2$ that can be covered using one ball, excluding any other point of $\mathbb{Z}^2$. In the rest of the article, calligraphic font $\mathscr{b}$ and $\mathscr{B}$ are used to denote open balls and sets of open balls, respectively, while mathematical font b and $\mathcal{B}$ are used to denote digital balls and sets of digital balls, respectively. A finite set $\mathcal{B}$ of digital balls covers a digital object S if $\bigcup \mathcal{B} = \cup_{b \in \mathcal{B}} b = S$. We say that $\mathcal{B}$ is a *digital ball covering* of S, its size is its cardinality $|\mathcal{B}|$ and it is minimum if $|\mathcal{B}|$ is minimum among all digital ball coverings of S. Note that the set of digital ball coverings of any digital object S is non-empty: the set of balls $\{b$ of center p and radius $\frac{1}{2}$, $p \in S\}$ is a digital ball covering of S. Then Problem 1 can be rephrased as Problem 2, and expressed as a decision problem in Problem 3:

Problem 2 (optimisation). Given a 2D digital object S, compute a minimum digital ball covering of S.

M. H. F. Wilkinson and J. Kosinka (Eds.): DGMM 2025, LNCS 16296, pp. 118–130, 2026.
https://doi.org/10.1007/978-3-032-09544-2_9

Problem 3 (decision). Given a 2D digital object S and an integer $k > 0$, does there exist a digital ball covering $\mathcal{B}$ of S with $|\mathcal{B}| \leqslant k$?

In addition to being a specific instance of the geometric set cover problem, this problem falls in the larger families of problems that consist in changing the computer representation of a shape. Many geometric models exists to represent either the surface or the volume of a solid shape, but the model that comes as output of the acquisition device is not always tailored for subsequent processing, whether it be creation, simulation, transmission for instance [14]. Therefore, the problem of converting one geometric model into another one with guarantees has been the subject of many studies, for a variety of pairs of geometric models. The problem we study in this work joins the family of problems where the output of the conversion algorithm is a finite set of balls. Many provably good algorithms have been proposed to solve specific problems in this family, when the input geometric model is a point cloud [1], a polygonal mesh [18,20], a digital shape [8,13] or even another finite set of balls [4,5,10,16]. These conversion algorithms are useful in a number of applications, as for instance in robotics [6,19], computer graphics [7], simulation of physical processes [13] or protein modelisation [5].

Problem 3 or close variants have been studied in previous works. On one side, it was shown in [8] that a variant of Problem 3, where the digital balls of $\mathcal{B}$ are restricted to the ones centered on $\mathbb{Z}^2$, is NP-complete. On the other side, a polynomial-time algorithm was proposed in [17] to solve Problem 2 when the digital object S has no hole (i.e. $\mathbb{Z}^2 \backslash S$ has exactly one 8-connected component). We continue to explore the frontier between polynomial and hard problems by proving the following theorem:

Theorem 1. *Problem 3 is NP-complete.*

2 Problem 3 is NP-Complete in Dimension 2

First, Problem 3 is in NP. Indeed, given a set $\mathcal{B}$ of digital balls and a digital object S, we can verify in polynomial time that $|\mathcal{B}| \leqslant k$ and that $\mathcal{B}$ covers S. This can be done by scanning all points p of each ball $b \in \mathcal{B}$ and either mark the point p as covered if $p \in S$, or end the verification algorithm with a negative answer if $p \notin S$. If all points of S are marked after scanning all balls, then the verification algorithm returns a positive answer. Scanning of the ball points is done in polynomial time in the size of the balls, and checking if a point is in S is done in polynomial time in the size of S.

The rest of the article is mainly dedicated to showing that Problem 3 is also NP-hard by exhibiting a reduction from (a variant of) the Vertex Cover problem. A vertex cover of a graph $G = (V, E)$ is a subset C of V such that for each edge $uv \in E$, either u or v (or both) belongs to C. The Vertex Cover problem is NP-complete even when restricted to 3-planar graphs (planar graphs where all vertices are of maximum degree 3) (see Lemma 1 in [9]). Without loss of generality, we can also assume that G is connected.

Problem 4 (3-Planar Vertex Cover). Given a connected 3-planar graph $G = (V, E)$ and an integer $k > 0$, does there exist a vertex cover C of G such that the cardinality of C is not greater than k ?

In the next sections, we show how to reduce Problem 4 to Problem 3: for any connected 3-planar graph G, we define a digital object $S(G)$ such that there exists a vertex cover C of G with $|C| \leqslant k$ if and only if there exists a digital ball covering $\mathcal{B}$ of $S(G)$ with $|\mathcal{B}| \leqslant k'$. For a 3-planar connected graph $G = (V, E)$ with $V = \{v_1, \dots v_n\}$ and $E = \{e_1, \dots e_m\}$, $S(G)$ will be the union of n vertex gadgets $\{S(v_i)\}_{i=1..n}$ and m edge gadgets $\{S(e_j)\}_{j=1..m}$: these gadgets are finite 4-connected subsets of $\mathbb{Z}^2$ which are precisely described in Sects. 2.2 to 2.4. We have $S(G) = \bigcup_{v \in V} S(v) \cup \bigcup_{e \in E} S(e)$.

The reduction is somehow an adaptation of the reduction used in [2], though new gadgets that are subsets of $\mathbb{Z}^2$ must be defined. In the following, the points of $\mathbb{Z}^2$ are called pixels. Given a pixel of coordinates $(x, y) \in \mathbb{Z}^2$, its *south neighbour* is the pixel of coordinates $(x, y-1)$ (similar definition for the other three cardinal directions). For each gadget, we have three kinds of pixels: gadget pixels (either *vertex pixels* or *edge pixels*), *out pixels* that must not belong to any gadget of $S(G)$, and *free pixels*. In the figures, gadget pixels are represented as black points (or circled black points), out pixels are depicted as white points, while free pixels are not marked.

2.1 Digital Ball or Not ?

The construction of vertex and edge gadgets is based on a family of small digital balls, and a family of small sets of pixels that are not digital balls. Figure 1 shows a set of 7 digital balls, the pixels of which are depicted in black: for each digital ball b, there exists a ball that covers exactly all the pixels of b (one is depicted in red).

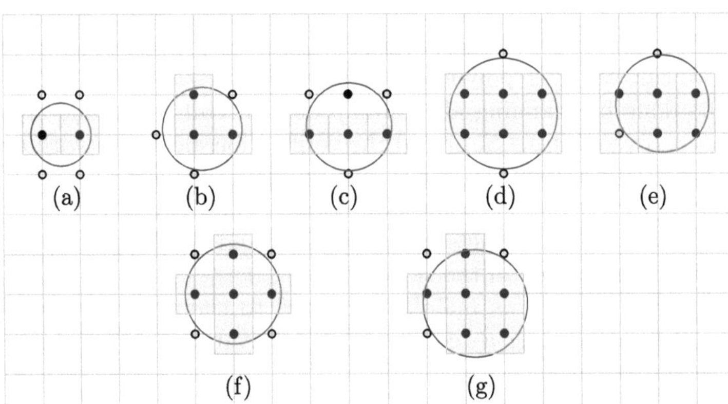

Fig. 1. Some useful digital balls. (Color figure online)

Lemma 1. *For each case depicted in Fig. 2, there is no ball that covers all black pixels while excluding all white pixels.*

Proof. For each case, the center of any ball that covers all black pixels and excludes all white pixels must belong to the intersection of the open halfplanes that are depicted in red, blue and green (for instance in (b)-top, the center must be closer to D than to A and C, and closer to B than to C). This intersection is empty in each case. ◻

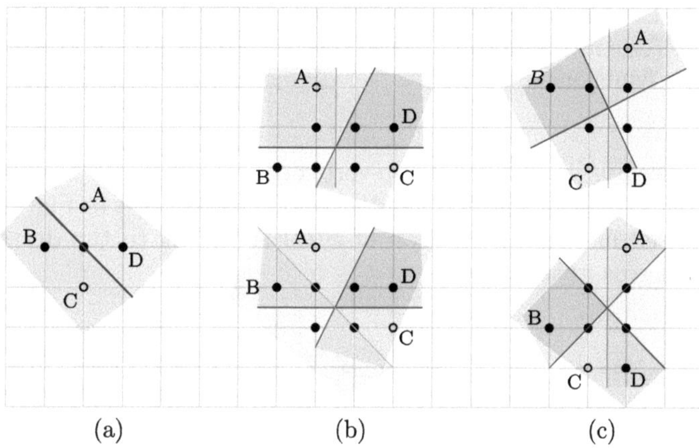

Fig. 2. Any ball that covers all black pixels also covers at least one white pixel: the subsets of black pixels are not digital balls.

2.2 Vertex Gadget

Since considering 3-planar graphs is enough, we define vertex gadgets for vertices of maximum degree 3.

For each vertex $u \in V$ of degree 3, the vertex gadget $S(u)$ (called vertex gadget of degree 3) is a translated and/or rotated and/or symmetric copy of the base vertex gadget $\dddot{V}$ defined as follows. $\dddot{V}$ is composed of 6 *regular pixels* that form a rectangle R of size 3×2, plus three *connecting pixels*, denoted by $\kappa_1, \kappa_2, \kappa_3$ and placed as depicted in Fig. 3(a). All pixels that are 4-connected to $R \cup \{\kappa_1, \kappa_2, \kappa_3\}$ are out pixels, except for the west neighbour of κ_1, the east neighbour of κ_2 and the south neighbour of κ_3. Finally, the lower left pixel of R is called *anchor pixel* and is denoted by α (see Fig. 3(a)). When needed, the connecting pixels of a vertex gadget $S(u)$, $u \in V$, will be denoted by $\kappa_i^u, i = 1 \ldots 3$, and the anchor pixel by α^u.

Gadgets for vertices of degree 2, denoted by $\ddot{V}$, and degree 1, denoted by $\dot{V}$, are simpler versions of the vertex gadget of degree 3: they are composed of the same set of regular pixels and have either two or one connecting pixels,

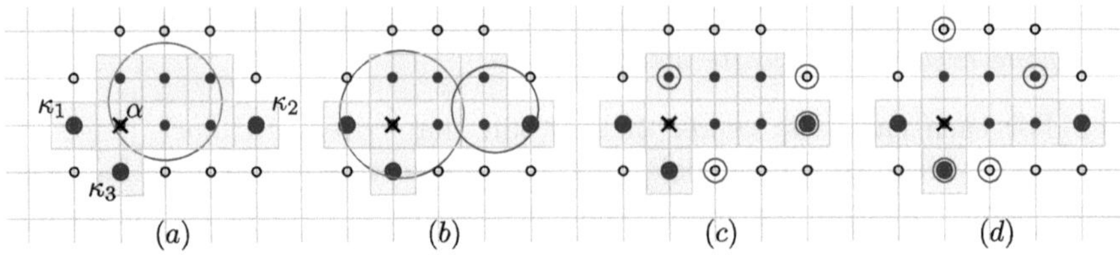

Fig. 3. The vertex gadget $\ddot{V}$ of degree 3: one ball is sufficient to cover all the regular pixels (a) while two balls are sufficient to cover all the regular pixels plus the three connecting pixels (b). Two balls are necessary to cover all regular pixels plus any connecting pixel: configurations that forbid a unique ball are highlighted in blue in (c) when pixel κ_2 (or equivalently κ_1) is also covered, and in (d) when pixel κ_3 is also covered. (Color figure online)

respectively (see Fig. 4). Note that two vertex gadgets of degree 2 are proposed: a straight vertex gadget in Fig. 4(a), and a bend vertex gadget in Fig. 4(b). Either one will be used in the construction of $S(G)$ according to the drawing of the edges adjacent to a given vertex.

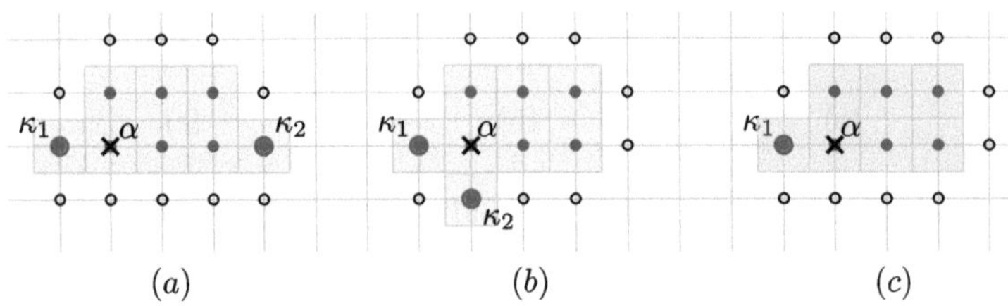

Fig. 4. (a) A straight vertex gadget of degree 2, (b) a bend vertex gadget of degree 2, and (c) a vertex gadget of degree 1.

The orientation $o(\kappa_i) \in \{\uparrow, \downarrow, \rightarrow, \leftarrow\}$ of a connecting pixel is the direction of the vector between κ_i and its only free 4-neighbour pixel. In $\ddot{V}$, we have $o(\kappa_1) = \leftarrow$, $o(\kappa_2) = \rightarrow$ and $o(\kappa_3) = \downarrow$. The orientation of a vertex gadget is the set of orientations of its connecting pixels. We have for instance $o(\ddot{V}) = \{\leftarrow, \rightarrow, \downarrow\}$.

Lemma 2. *For any vertex gadget $\tilde{V}$ of degree* 1 *to* 3*, we have the following properties:*

(i) the minimum size of a digital ball covering of the regular pixels of $\tilde{V}$ is one;
(ii) the minimum size of a digital ball covering of the regular pixels of $\tilde{V}$ plus one, two or three connecting pixels of $\tilde{V}$ is two.

Proof. Since a vertex gadget of degree 2 (resp. 1) is a vertex gadget of degree 3 minus one (resp. two) connecting pixel(s), it is enough to prove the property for a vertex gadget of degree 3. We use the digital balls described in Sect. 2.1

to exhibit digital ball coverings and thus prove the sufficient conditions. We refer to Fig. 3: one ball is enough to cover all regular pixels (a), and two balls are enough to cover all connecting pixels as well (b). The set of regular pixels being non empty, at least one digital ball is necessary to cover them. We use Lemma 1 to prove that two balls are necessary to cover all regular pixels plus any one of the connecting pixels (see Fig. 3(c)–(d)): in case (c) (resp. case (d)), the configuration of Fig. 2(b) (resp. (c)) prevents from using one ball only. ◻

2.3 Edge Gadget

We define two types of edge gadgets: straight edge gadgets, and one-bend edge gadgets. For an edge $e \in E$, $S(e)$ is a translated and/or symmetric and/or rotated copy of a base straight edge gadget $\bar{E}$ or a base one-bend edge gadget $\check{E}$.

An edge gadget $\tilde{E}$, whether it be $\bar{E}$ or $\check{E}$, connects two pixels $\gamma_1(x,y)$ and $\gamma_2(x',y')$, called *ghost pixels*, such that the length $l(\tilde{E})$ of the gadget, defined as $l(\tilde{E}) = d_1(\gamma_1,\gamma_2) = |x-x'|+|y-y'|$, verifies $l(\tilde{E}) \equiv_2 0$. We now refer to Fig. 5 to describe these gadgets. A straight edge gadget $\bar{E}$ (a) is a horizontal or vertical 4-connected path of an odd number of pixels. A one-bend edge gadget $\check{E}$ (b)-(c) is composed of one horizontal and one vertical 4-connected paths of pixels, one with an odd number of pixels, the other with an even number of pixels. These two paths are linked by a 3×2 rectangle R of 6 pixels, as depicted in Fig. 5.

For $\bar{E}$ and $\check{E}$, all pixels that are 4-connected to edge pixels are out pixels, except for the two ghost pixels (depicted in light gray in Fig. 5). Similarly to connecting pixels in vertex gadgets, the orientation $o(\gamma_i) \in \{\uparrow,\downarrow,\rightarrow,\leftarrow\}$ of a ghost pixel of an edge gadget is the direction of the vector between γ_i and its only 4-connected edge pixel in this edge gadget. The orientation of an edge gadget $\tilde{E}$ is $o(\tilde{E}) = \{o(\gamma_1), o(\gamma_2)\}$ where γ_1 and γ_2 are the ghost pixels of $\tilde{E}$.

Lemma 3. *The following properties hold for edge gadgets :*

(i) the minimum size of a digital ball covering of $\bar{E}$ plus at most one of its ghost pixels is equal to $L(\bar{E}) = \frac{l(\bar{E})}{2}$;
(ii) the minimum size of a digital ball covering of $\check{E}$ plus at most one of its ghost pixels is equal to $L(\check{E}) = \frac{l(\check{E})}{2} - 1$.

Proof. A straight edge gadget $\bar{E}$ of length $l(\bar{E}) \equiv_2 0$ is a path of $l(\bar{E})-1$ edge pixels. By Lemma 1 case (a), $\lceil \frac{l(\bar{E})-1}{2} \rceil = \frac{l(\bar{E})}{2}$ balls are necessary to cover all edge pixels. Figure 5 shows a digital ball covering of size $\frac{l(\bar{E})}{2}$, that also covers any of the two ghost pixels (up to symmetry).

A one-bend edge gadget of length $l(\check{E})$ is composed of an even path of length l, a 3×2 rectangle and an odd path of length $l' = l(\check{E}) - l - 5$. If one ball is used to cover all the rectangle, then by Lemma 1 case (b)-(c), no other pixel of the edge gadget can be covered by this ball: then the total number of balls needed to cover the gadget is equal to $1 + \frac{l}{2} + \lceil \frac{l'}{2} \rceil = \frac{l(\check{E})}{2} - 1$. Figure 5(b) illustrates this case, and moreover shows that there exists a digital ball covering of $\breve{E}$ that also covers the ghost pixel adjacent to the path of odd length.

Suppose now that k balls are used to cover the rectangle. Then, using case (a) of Lemma 1, these balls can at most also cover one pixel of each path. Then the total number of balls needed is equal to $k + \lceil \frac{l-1}{2} \rceil + \lceil \frac{l'-1}{2} \rceil = \frac{l(\breve{E})}{2} + k - 3$. Two balls are enough to cover the rectangle plus one pixel of each path as illustrated by Fig. 5(c), thus proving that $\frac{l(\breve{E})}{2} - 1$ balls are necessary and sufficient. Figure 5(c) illustrates this case and depicts a digital ball covering of $\breve{E}$ with $k = 2$, that also covers the ghost pixel adjacent to the path of even length. ◻

This lemma shows that for any given edge gadget $\tilde{E}$, two families of minimum digital ball coverings of size $L(\tilde{E})$ can be defined, according to the ghost pixel that is covered. These are called *minimum edge coverings*. Given a set of disjoint edge gadgets $F = \{\tilde{E}_j\}_{j \in 1...m}$, we set $L(F) = \sum_j L(\tilde{E}_j)$: a set of $L(F)$ balls is necessary and sufficient to cover the edge pixels of all $\tilde{E}_j \in F$ and exactly one ghost pixel of each $\tilde{E}_j \in F$.

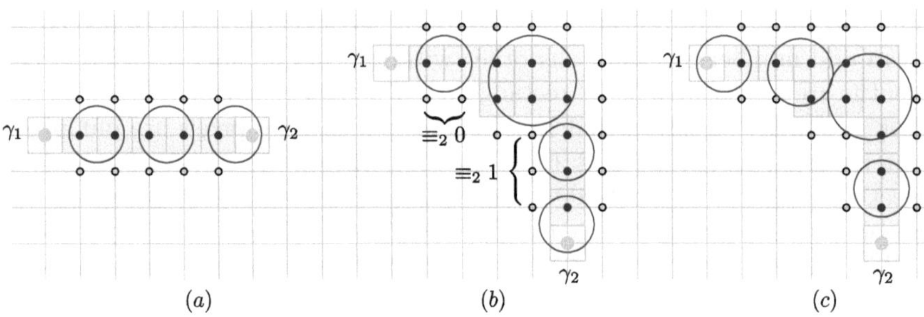

Fig. 5. (a) A straight edge gadget and (b)–(c) a one-bend edge gadget: in (b), one ball is used to cover the rectangle whereas in (c), two balls are used.

2.4 Building a Digital Object

In order to combine these gadgets to define $S(G)$ and have a reduction that is polynomial in space, we use the following result by [12].

Theorem 2 ([12]). *There is a linear time and space algorithm to draw any 3-planar graph with n vertices on an* $\lfloor \frac{n}{2} \rfloor \times \lfloor \frac{n}{2} \rfloor$ *grid with at most* $\lfloor \frac{n}{2} \rfloor + 1$ *bends, with the property that there is a spanning tree of* $n - 1$ *straight-line edges, while all non-tree edges have at most 1 bend (if* $n > 4$*).*

A drawing $D(G)$ of a graph G on a grid of size $s \times s$ associates coordinates $(x, y) \in \mathbb{Z}^2$ to each vertex $u \in V$ and a set of consecutive grid edges with at most one bend to each edge $e \in E$. From $D(G)$, we build $S(G)$ in a grid of size $8s \times 8s$.

Figure 6 gives an example of construction that uses all types of vertex and edge gadgets.

As said earler, for $u \in V$ of degree $d(u)$, $S(u)$ is a translated/rotated copy of a vertex gadget of degree $d(u)$. When $d(u) = 2$, $S(u)$ is a straight vertex gadget if the two edges adjacent to u in $D(G)$ are both horizontal or vertical, and a bend vertex gadget otherwise. Let us now precise how the gadgets are translated and/or rotated.

Given two adjacent vertices $u, v \in V$, we define the orientation $o(u,v) \in \{\uparrow, \downarrow, \leftarrow, \rightarrow\}$ in $D(G)$ as the direction of the vector between u and v in $D(G)$. The orientation $o(u)$ of u in $D(G)$ is the set $\{o(u,v), v \text{ neighbour of } u\}$. If u has coordinates (x, y) in $D(G)$, we set the coordinates of the anchor pixel α^u of $S(u)$ to $(8x, 8y)$. Once the anchor pixel of each vertex gadget is positionned, the vertex gadget $S(u)$ is rotated around α^u so that $o(S(u)) = o(u)$. If several transformations comply with this constraint, any one can be picked.

Notice that all anchor pixels have even-even coordinates (abscissa and ordinate are both even). By the definition of vertex gadget, this implies that all connecting pixels have either even-odd or odd-even coordinates. Therefore, given any two connecting pixels κ_k^u and $\kappa_{k'}^v$, $u, v \in V$, we have $d_1(\kappa_k^u, \kappa_{k'}^v) \equiv_2 0$. Using the fact that $D(G)$ has only straight-line edges and edges with at most one bend, edge gadgets defined in Sect. 2.3 can be used to complete the construction of $S(G)$ as follows. Let u and v be two adjacent vertices of G and $S(u)$ and $S(v)$ be the corresponding vertex gadgets placed in the grid as described before. Let κ (resp. κ') be the connecting pixel of $S(u)$ (resp. $S(v)$) such that $o(\kappa) = o(u,v)$ (resp. $o(\kappa') = o(v,u)$). If the edge (u,v) is a straight edge (resp. one-bend edge) in $D(G)$, we add to $S(G)$ a translated/rotated/symmetric copy of the straight edge gadget $\bar{E}$ (resp. one-bend edge gadget $\check{E}$). The length of the gadget is equal to $d_1(\kappa, \kappa')$, and it is rotated or symmetrized so that $o(\bar{E}) = \{o(\kappa), o(\kappa')\}$ (resp. $o(\check{E}) = \{o(\kappa), o(\kappa')\}$). Moreover, we set $\gamma_1 = \kappa$ and $\gamma_2 = \kappa'$ if $o(\kappa) = o(\gamma_1)$ and $o(\kappa') = o(\gamma_2)$ (the other way round otherwise).

Lemma 4. *We have the following properties:*

(i) any ball that covers vertex pixels from different vertex gadgets also covers out pixels;
(ii) any ball that covers any edge pixel and any regular pixel of any vertex gadget also covers out pixels;
(iii) any ball that covers edges pixels from different edge gadgets also covers out pixels.

Proof. Property (ii) is true by Lemma 1 case (a). The proof of property (i) follows from the size of the grid, the size of the vertex gadgets and the fact that vertex pixels are surrounded by out pixels, except where edge gagdets are connected. But these edge gadgets all contain at least one horizontal or vertical 4-connected path of length at least 3, and Lemma 1 case (a) enables to conclude. Property (iii) follows from the size of the grid, and the fact that edge gagdets are placed according to $D(G)$, in which edges do not cross. ◻

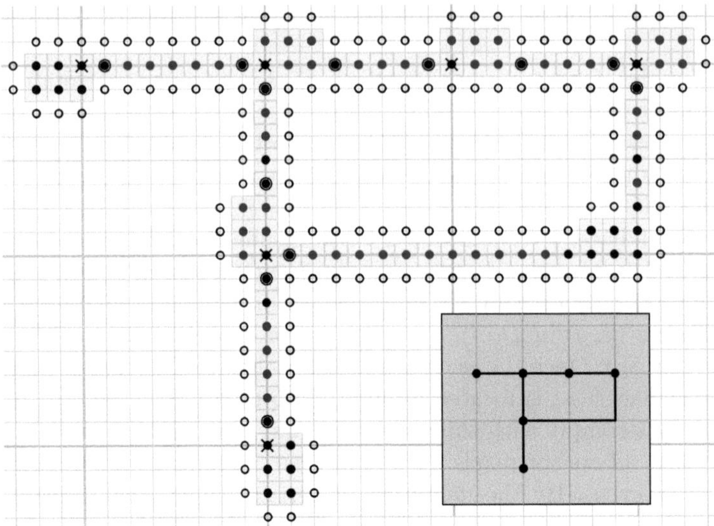

Fig. 6. An example of $D(G)$ and $S(G)$.

2.5 Proof of the Reduction

Now that the construction of $S(G)$ is set, let us first notice that, if G is a 3-planar connected graph, then $S(G)$ is a 2D (4-connected) digital object. The proof of the reduction is very similar to the proof in [2], and we follow the same reasoning. The two lemmas below complete the proof of Theorem 1.

Lemma 5. *Given $G = (V, E)$, if G has a vertex cover $C \subseteq V$, then $S(G)$ has a digital ball covering $\mathcal{B}$ with $|\mathcal{B}| = |C| + L + |V|$, where $L = L(\{S(e)|e \in E\})$.*

Proof. To build $\mathcal{B}$, we use Lemma 2 to select, for each vertex $u \in C$, two digital balls to cover all regular and connecting pixels of $S(u)$. This gives a set of $2|C|$ balls that covers all regular and connecting pixels of $S(u)$ for all $u \in C$. Moreover, since C is a vertex cover of G, these $2|C|$ balls also cover at least one ghost pixel of each edge. In order to cover all edge pixels, we use Lemma 3 and add in $\mathcal{B}$ a set of $L = L(\{S(e), e \in E\})$ digital balls by picking for each edge e, the minimum edge covering of $S(e)$ that covers the ghost pixel of $S(e)$ that is not already covered. This ensures that all connecting pixels of $S(G)$ are covered. Finally, we add $|V| - |C|$ digital balls to cover the regular pixels of all $S(v)$ with $v \in V \backslash C$. □

Lemma 6. *Given $G = (V, E)$, if $S(G)$ has a digital ball covering $\mathcal{B}$, then G has a vertex cover C with $|C| \leqslant |\mathcal{B}| - L - |V|$, where $L = L(\{S(e)|e \in E\})$.*

Proof. First, we build from $\mathcal{B}$ another ball covering $\mathcal{B}'$ of $S(G)$ such that $|\mathcal{B}'| \leqslant |\mathcal{B}|$. We initialize $\mathcal{B}'$ to $\mathcal{B}$ and discard from $\mathcal{B}'$ the balls that cover no pixel of $S(G)$. Then we change $\mathcal{B}'$ such that it contains a minimum edge covering of each edge $e \in E$. To do so, for $e \in E$, let $\mathcal{B}_e = \{b \in$

$\mathcal{B}' \mid b$ contains an edge pixel of $S(e)\}$. Note that, by Lemma 4(ii), balls of $\mathcal{B}_e$ do not contain regular pixels from vertex gadgets. We replace the set of balls $\mathcal{B}_e$ in $\mathcal{B}'$ as follows:

(i) if $\mathcal{B}_e$ covers exactly one ghost pixel γ of $S(e)$, then replace the set of balls $\mathcal{B}_e$ with the minimum edge covering of $S(e)$ that covers γ;
(ii) if $\mathcal{B}_e$ covers no ghost pixel of $S(e)$, then replace the set of balls $\mathcal{B}_e$ by any minimum edge covering of $S(e)$;
(iii) if $\mathcal{B}_e$ covers the two ghost pixels of $S(e)$, then $|\mathcal{B}_e| \geqslant L(S(e))+1$ by Lemma 3. Then we replace $\mathcal{B}_e$ by one minimum edge covering of $S(e)$ plus one ball b that covers only the uncovered ghost pixel (and no other pixel of $S(G)$).

Since by Lemma 4(iii), all $\mathcal{B}_e$ are disjoint and by Lemma 3 $|\mathcal{B}_e| \geqslant L(S(e))$, these substitutions do not increase the cardinality of $\mathcal{B}'$. Moreover, after these substitutions, $\mathcal{B}'$ contains two types of balls: (a) balls of minimum edge coverings of the edge gadgets, that all cover an edge pixel by definition of minimum edge covering ($\mathcal{B}'$ contains exactly L such balls) ; (b) balls that cover no edge pixel (some cover only a ghost pixel - see case (iii) above). So if we consider now the set of balls $\mathcal{C}$ of $\mathcal{B}'$ that cover no edge pixel, we have $|\mathcal{C}| = |\mathcal{B}'| - L$, and the property that the balls of $\mathcal{C}$ cover exactly one ghost pixel of each edge gadget $S(e)$ (the other ghost pixel is covered by a ball of the minimum edge covering of $S(e)$, that contains an edge pixel).

For a vertex $u \in V$, let $\mathcal{C}_u = \{b \in \mathcal{C} \mid b$ contains a regular pixel of $S(u)\}$. By Lemma 4(i), a ball cannot cover pixels (regular or connecting) of two different vertex gadgets, so that the sets $\mathcal{C}_u$ are disjoint, meaning that $\sum_{u\in V} |\mathcal{C}_u| = |\mathcal{C}|$. We replace each set $\mathcal{C}_u$ in $\mathcal{B}'$ by a smaller set $\mathcal{C}'_u$ as follows:

- if $\mathcal{C}_u$ covers no connecting pixel of $S(u)$, then by Lemma 2(i) $|\mathcal{C}| \geqslant 1$ and we define $\mathcal{C}'_u$ as a set containing a unique ball that covers all regulars pixels;
- if $\mathcal{C}_u$ covers at least one connecting pixel κ of $S(u)$, then by Lemma 2(ii) $|\mathcal{C}_u| \geqslant 2$ and we define $\mathcal{C}'_u$ as a set of two balls that cover all regular pixels and κ.

Again, these substitutions do not increase the cardinality of $\mathcal{B}'$. Let $C = \{u, |\mathcal{C}'_u| = 2)\}$. By definition, we have $\sum_{u\in V} |\mathcal{C}'_u| = |V| + |C|$. We claim that $|C| \leqslant |\mathcal{B}| - |V| - L$ and that C is a vertex cover of G.

First, we have $|V| + |C| = \sum_{u\in V} |\mathcal{C}'_u| \leqslant \sum_{u\in V} |\mathcal{C}_u| = |\mathcal{C}| = |\mathcal{B}'| - L$, thus $|C| \leqslant |\mathcal{B}'| - |V| - L \leqslant |\mathcal{B}| - |V| - L$. Second, suppose that C is not a vertex cover and let $e = (u, v)$ be an edge that is not covered by C. By definition of C, this means that $|\mathcal{C}_u| = |\mathcal{C}_v| = 1$, meaning that $\mathcal{C}_u$ and $\mathcal{C}_v$ cover no connecting pixel of $S(u)$ and $S(v)$ respectively. This implies that both ghost pixels of $S(e)$ are neither covered by $\mathcal{C}_u$ nor by $\mathcal{C}_v$. But ghost pixels of e can only be covered by $\mathcal{C}_u$, $\mathcal{C}_v$ or $\mathcal{B}_e$. Since $\mathcal{B}_e$ covers at most one ghost pixel by Lemma 3, we reach a contradiction with the fact that $\mathcal{B}$ is a digital ball covering of $S(G)$. □

3 Discussion

As a concluding section, we end with insights on two observations: the first one is about a naive upperbound for the optimisation problem (Problem 2), and the second is about the complexity of Problem 3 when S is a 6-connected 3D digital object and the balls of $\mathcal{B}$ are 3D digital balls.

For these two observations, we use a result from [17] that states that the boundary of the Voronoi diagram of $\mathbb{Z}^2 \backslash S$, denoted by $\partial \mathrm{Vor}(\mathbb{Z}^2 \backslash S)$, is a superset of the set of centers of the maximal digital balls of S. More precisely, let us denote by $\mathring{\mathrm{pixel}}(p)$ the open unit square centered on p for $p \in \mathbb{Z}^2$, and for $c \in \bigcup_{p \in S} \mathring{\mathrm{pixel}}(p)$, let $\mathscr{b}(c)$ denote the open ball centered on c and with radius equal to the distance between c and $\mathbb{Z}^2 \backslash S$. Then $b(c) = \mathscr{b}(c) \cap \mathbb{Z}^2$ is the maximal digital ball included in S centered on c and the set of balls $\mathcal{B} = \{b(c), c \in \mathrm{Vor}^{\sqcap}(S)\}$ is a digital ball covering of S.

A Naive Upper Bound. The first observation is that the simpler set $\mathcal{B}^v = \{b(c), c \text{ vertex of } \mathrm{Vor}^{\sqcap}(S)\}$ is also a digital ball covering of S. Indeed, let us first notice that each edge of $\mathrm{Vor}^{\sqcap}(S)$ defines a family of balls that go through a pair of points of $\mathbb{Z}^2 \backslash S$. Otherwise said, any edge $[c_1, c_2]$ can be described as a parabolic pencil of circles [3,15] defined by two points of $\mathbb{Z}^2 \backslash S$ and delimited by its two extremities $\mathscr{b}(c_1)$ and $\mathscr{b}(c_2)$. For a parabolic pencil $[c_1, c_2]$, we have $\bigcup_{c \in [c_1, c_2]} \mathscr{b}(c) = \mathscr{b}(c_1) \cup \mathscr{b}(c_2)$. This implies that $(\bigcup_{c \in [c_1, c_2]} \mathscr{b}(c)) \cap \mathbb{Z}^2 = (\mathscr{b}(c_1) \cup \mathscr{b}(c_2)) \cap \mathbb{Z}^2 = b(c_1) \cup b(c_2)$. Since $\mathrm{Vor}^{\sqcap}(S)$ contains the center of all maximal balls of S, for any $p \in S$, there exists $\mathscr{b}(c, r)$ with $c \in \mathrm{Vor}^{\sqcap}(S)$ such that $p \in \mathscr{b}$. Consider the edge $[c_1, c_2]$ that contains c: then $b(c_1) \cup b(c_2)$ contains p, showing that $\mathcal{B}^v$ is a digital ball covering of S. This upperbound can be further improved by computing a subset $\mathrm{Vor}^{\sqcap}_{\downarrow}(S)$ of $\mathrm{Vor}^{\sqcap}(S)$ such that $\mathcal{B}^v_{\downarrow} = \{b(c), c \text{ vertex of } \mathrm{Vor}^{\sqcap}_{\downarrow}(S)\}$ is still a digital ball covering of S. To compute it, we iteratively remove pending vertices of $\mathrm{Vor}^{\sqcap}(S)$ as follows: a pending vertex c of an edge $[c, c']$ is removed if $b(c) \subseteq b(c')$. Figure 7 gives an illustration where $\mathrm{Vor}^{\sqcap}_{\downarrow}(S)$ is depicted in green: in (a) the balls of $\mathcal{B}^v_{\downarrow}$ are drawn in red; in (b), the digital object S being hole-free, the set of balls in red is an optimal digital ball covering of S obtained using the algorithm from [17]. The upperbound does not make any assumption on the digital object S. However, as suggested by Fig. 7, it is not tight at all and further simplifications of $\mathrm{Vor}^{\sqcap}(S)$ could be considered in order to obtain a set of necessary edges.

Complexity in Dimension 3. Consider now the same problem as Problem 3 where S is a 3D 6-connected digital object. We give a sketch of a polynomial reduction that strongly - and unsurprisingly - suggests that the problem is still NP-hard. From any 2D digital object S, we build a 3D 6-connected digital object S' such that there exists a digital ball covering $\mathcal{B}$ of S with $|\mathcal{B}| \leqslant k$ if and only if there exists a digital ball covering $\mathcal{B}'$ of S' with $|\mathcal{B}'| \leqslant k'$. Given a maximal ball $\mathscr{b}(c)$ centered on $c = (x, y)$ vertex of $\mathrm{Vor}^{\sqcap}_{\downarrow}(S)$, and of radius r, let $\mathscr{b}^+(c)$ be the 3D ball of center $(x, y, 0)$ and radius r. Then let $b^+(c) = \mathscr{b}^+(c) \cap \mathbb{Z}^3$ and define $S' = \{b^+(c), \text{with } c \in \mathrm{Vor}^{\sqcap}_{\downarrow}(S)\}$. If there exists a digital ball covering $\mathcal{B}$ of S using k balls, then a digital ball covering of S' using k balls can be exhibited: for each ball b of $\mathcal{B}$ of center (x, y) and radius r, we add in $\mathcal{B}'$ the ball of center $(x, y, 0)$ and radius r. Conversely, suppose that there exists a digital ball covering $\mathcal{B}'$ of S' using k' balls, then a digital ball covering of S using $k \leqslant k'$ balls can be

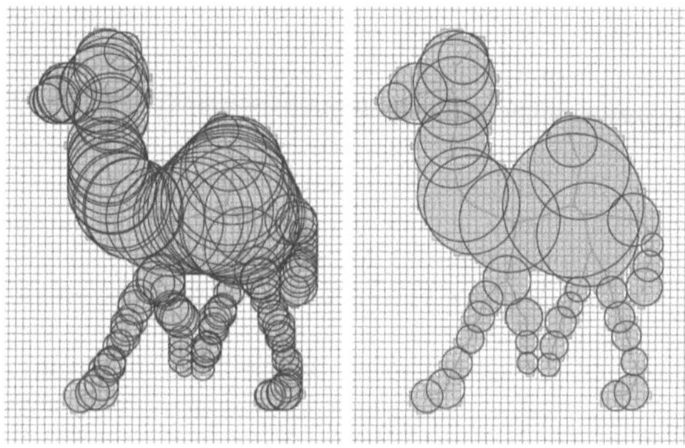

Fig. 7. The set of balls of $\mathcal{B}^{v}_{\downarrow}$ (a) contains 137 balls, while the optimal digital ball covering depicted in (b) has 40 balls.

found as follows: first, ignore the balls of $\mathcal{B}'$ that to not contain a pixel of the plane $z = 0$; second, for all the other balls $b' \in \mathcal{B}'$, consider the set of pixels of b' that lie in the plane $z = 0$ and add in $\mathcal{B}$ a maximal digital ball of S that contains this set of pixels. All in all, we have $|\mathcal{B}| \leqslant |\mathcal{B}'|$ which enables to conclude.

Several lines of research can however be explored. The first idea would be to follow the same line as in the 2D case to obtain an upperbound. The question is the following: is the 3D counterpart of $\mathrm{Vor}^{\sqcap}(S)$ still a superset of the centers of maximal balls ? If so, properties of parabolic pencils would still imply that the balls centered on its vertices define a digital ball covering. Another natural question is whether it is possible or not to simplify it using for instance contractions of edges, to obtain a tighter bound. A second idea is to reformulate the problem to use existing algorithms for geometric set cover problems, as for instance the problem of minimum stabbing of halfplanes by points [11].

References

1. Amenta, N., Choi, S., Kolluri, R.K.: The power crust. In: Proceedings of the Sixth ACM Symposium on Solid Modeling and Applications, pp. 249–266 (2001)
2. Attali, D., Nguyen, T.B., Sivignon, I.: (δ, ε)-ball approximation of a shape: definition and complexity. Discret. Comput. Geom. **61**, 595–625 (2019)
3. Boissonnat, J.D., Yvinec, M.: Géométrie algorithmique (1995)
4. Bradshaw, G., O'Sullivan, C.: Adaptive medial-axis approximation for sphere-tree construction. ACM Trans. Graph. **23**(1), 1–26 (2004)
5. Cazals, F., Dreyfus, T., Sachdeva, S., Shah, N.: Greedy geometric algorithms for collection of balls, with applications to geometric approximation and molecular coarse-graining. Comput. Graph. Forum **33**(6), 1–17 (2014)
6. Chiu, J.R., Sleiman, J.P., Mittal, M., Farshidian, F., Hutter, M.: A collision-free MPC for whole-body dynamic locomotion and manipulation. In: 2022 International Conference on Robotics and Automation (ICRA), pp. 4686–4693. IEEE (2022)
7. Choi, A.R., Sung, M.Y.: Performance improvement of haptic collision detection using subdivision surface and sphere clustering. PLoS ONE **12**(9), 1–17 (2017)

8. Coeurjolly, D., Hulin, J., Sivignon, I.: Finding a minimum medial axis of a discrete shape is NP-hard. Theoret. Comput. Sci. **406**(1), 72–79 (2008)
9. Garey, M.R., Johnson, D.S.: The rectilinear Steiner tree problem is NP-complete. SIAM J. Appl. Math. **32**(4), 826–834 (1977)
10. Giesen, J., Miklos, B., Pauly, M., Wormser, C.: The scale axis transform. In: Proceedings of the Twenty-Fifth Annual Symposium on Computational Geometry, pp. 106–115. Association for Computing Machinery (2009)
11. Har-Peled, S., Lee, M.: Weighted geometric set cover problems revisited. J. Comput. Geom. **3**(1), 65–85 (2012)
12. Kant, G.: Drawing planar graphs using the canonical ordering. Algorithmica **16**, 4–32 (1996)
13. Mede, T., Chambon, G., Hagenmuller, P., Nicot, F.: A medial axis based method for irregular grain shape representation in dem simulations. Granular Matter **20**(16) (2018)
14. O'Rourke, J., Badler, N.: Decomposition of three-dimensional objects into spheres. IEEE Trans. Pattern Anal. Mach. Intell. **3**, 295–305 (1979)
15. Schwerdtfeger, H.: Geometry of complex numbers: circle geometry, Moebius transformation, non-euclidean geometry. Courier Corporation (1979)
16. Shamir, A., Scharf, A., Cohen-Or, D.: Enhanced hierachical shape matching for shape transformation. Int. J. Shape Model. **09**(02), 203–222 (2003)
17. Sivignon, I.: Exact and optimal conversion of a hole-free 2D digital object into a union of balls in polynomial time. In: Discrete Geometry and Mathematical Morphology. LNCS, pp. 382–394. Springer (2022)
18. Stolpner, S., Kry, P., Siddiqi, K.: Medial spheres for shape approximation. IEEE Trans. Pattern Anal. Mach. Intell. **34**(6), 1234–1240 (2012)
19. Völz, A., Graichen, K.: Computation of collision distance and gradient using an automatic sphere approximation of the robot model with bounded error. In: ISR 2018; 50th International Symposium on Robotics, pp. 1–8 (2018)
20. Wang, R., et al.: Variational sphere set approximation for solid objects. Vis. Comput. **22**, 612–621 (2006)

Discrete and Combinatorial Topology

The Mid-sphere Cousin of the Medial Axis Transform

Herbert Edelsbrunner[1], Elizabeth Stephenson[1,2](✉), and Martin Hafskjold Thoresen[3]

[1] ISTA, 3400 Klosterneuburg, Austria
edels@ist.ac.at, elizasteprene@gmail.com
[2] Orteliu, Oslo, Norway
[3] Independent Researcher, Drammen, Norway
m@mht.wtf

Abstract. The *medial axis* of a smoothly embedded surface in $\mathbb{R}^3$ consists of all points for which the Euclidean distance function on the surface has at least two global minima. We generalize this notion to the *mid-sphere axis*, which consists of all points for which the Euclidean distance function has two interchanging saddles that swap their partners in the pairing by persistent homology. It offers a discrete-algebraic multi-scale approach to computing ridge-like structures on the surface. As a proof of concept, an algorithm that computes stair-case approximations of the mid-sphere axis is provided.

Keywords: Medial axes · Morse functions · persistent homology · vineyards · computation

1 Introduction

In the late 1960s, Harry Blum [4] revolutionized the automated classification of shapes that typically arise in biology and related fields by suggesting the medial axis as a skeleton fit to describe amorphous blobs. The concept was introduced in the smooth setting by Federer [11, Definition 4.1 on page 432] almost a decade earlier, but his contribution was largely overlooked. Since then, the medial axis has been used as a tool in shape classification, image analysis, animation, computer graphics, and other fields that benefit from the feature-preserving simplification of unwieldy objects; see e.g. [3,15,17].

Given a closed surface, $M \subseteq \mathbb{R}^3$, the *medial axis* is the set of centers of spheres that touch M in at least two points and do otherwise not intersect M. Slightly more elaborately, the *medial axis transform* is the same set of points paired with the radii at which the spheres touch M. Requiring instead that M lies inside the spheres, we get an apparently less useful complementary concept, which for the purposes of this paper we refer to as the *circum-sphere axis transform* of M. We observe that there is an appealing third option: the *mid-sphere axis transform*, which consists of centers and radii of spheres that touch M in at least two points,

M. H. F. Wilkinson and J. Kosinka (Eds.): DGMM 2025, LNCS 16296, pp. 133–147, 2026.
https://doi.org/10.1007/978-3-032-09544-2_10

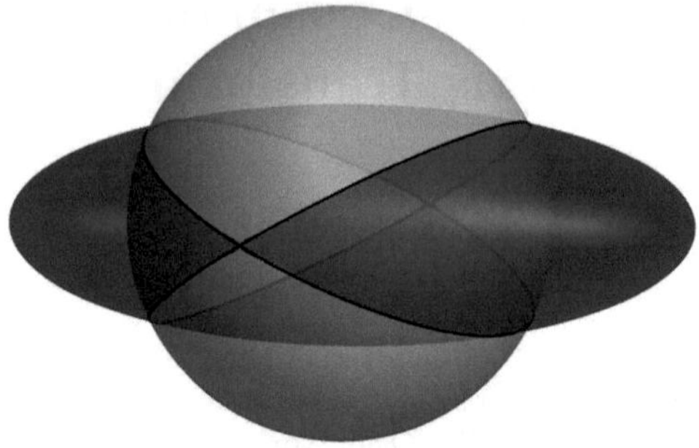

Fig. 1. A mid-sphere touching an ellipsoid in two antipodal points. The vertical plane that passes through these two points intersects the ellipsoid in a curve that lies inside the sphere, and the horizontal plane that passes through the same two points intersects the sphere in a great-circle filled by a disk that both lie inside the ellipsoid.

and for any two there is a closed curve on M inside the sphere, and a closed curve on the sphere together with a filling surface both inside M; see Fig. 1 for an illustration.[1]

All three axes are subsets of the *symmetry set* of M defined as the set of centers of spheres that touch M in at least two points. Compared to this "hopelessly complicated object" [5], the three axes are sparse skeleta that specialize in bringing out particular types of features. For example, the mid-sphere axis is sensitive to ridge-like features, in which a point $x \in M$ belongs to a curve called a *ridge* if it locally minimizes or maximizes the curvature along a principle curvature direction; see e.g. Chap. 11 on "Ridges and Ribs" in [14]. This definition is in terms of derivatives and thus hyper-sensitive to local noise. We propose that complementing it with more global topological requirements, we get more reliably computable access to ridge-like structures.

Outline. Section 2 formally defines the mid-sphere axis transform. Section 3 reviews the Morse theory of the squared distance function from a point. Section 4 introduces the notion of Faustian interchanges, which characterize the mid-sphere axis transform. Section 5 explains the pairing mechanism of persistent homology and its connection to Faustian interchanges. Section 6 exploits the vineyard algorithm to compute an approximation of the mid-sphere axis transform. Section 7 concludes the paper.

[1] We borrow the terminology from 3-dimensional geometry, where the *mid-sphere* of a convex polytope is a sphere that touches every one of its edges. The sphere thus intersects every facet in an inscribed circle. By the Circle Packing Theorem of Koebe [12], Andreev [1], and Thurston [16, Chapter 13.6], every 2-connected planar graph has a geometric realization as the edge skeleton of a convex polytope that has a mid-sphere. Curiously, this is not true if we require an in-sphere or a circum-sphere.

2 The Mid-sphere Axis Transform

As illustrated in Fig. 2, a point may belong to the mid-sphere axis for more than one radius. In other words, the projection of the mid-sphere axis transform in $\mathbb{R}^4$ to the mid-sphere axis in $\mathbb{R}^3$ is not necessarily injective, which is in contrast to the medial and the circum-sphere axes.

Let $M \subseteq \mathbb{R}^3$ be a smoothly embedded connected closed surface, and write $S = S(x,r)$ for the sphere with center $x \in \mathbb{R}^3$ and radius $r > 0$. The set $\mathbb{R}^3 \setminus M$ consists of two open components, a bounded one called the *inside* of M, and an unbounded one called the *outside* of M. Similarly, we distinguish between the *inside* of S, which is an open ball, and the *outside* of S, which is the complement of a closed ball. The sphere *touches* the surface in a point $b \in M \cap S$ if the tangent planes of M and S at b agree, and it touches M *generically* if $1/r$ is different from the two principal curvatures of M at b.

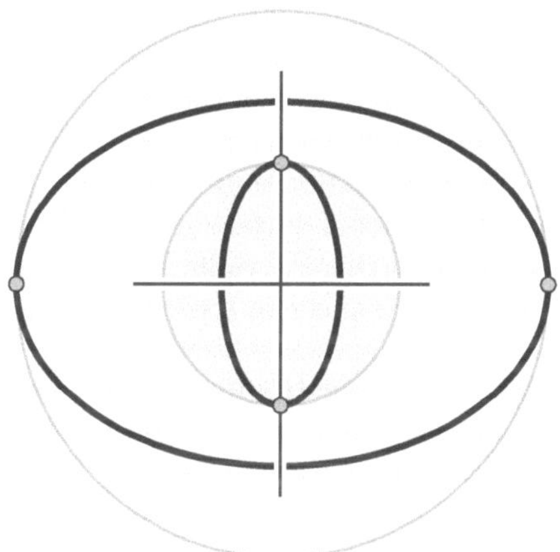

Fig. 2. Two locally cylindrical pieces of a *blue* surface with elliptic cross-sections as shown. The smaller cylinder runs inside the wider cylinder, and both share an axis normal to the drawing plane. The two *yellow* spheres touch the two cylinders in two points each, and because these spheres are concentric but have different radii, their common center belongs to two different *red* sheets of the mid-sphere axis, which intersect in a line that runs along the common central axis of the two cylinders. (Color figure online)

Definition 1. *Let $M \subseteq \mathbb{R}^3$ be a smoothly embedded connected closed surface. A point-radius pair $(x,r) \in \mathbb{R}^3 \times \mathbb{R}$ belongs to the* mid-sphere axis transform *of M if*

(i) $S = S(x,r)$ touches M generically in at least two points, $b \neq c$;
(ii) there is a smooth closed curve, $\mu \subseteq M$, that passes through b and c and lies otherwise inside S;
(iii) there is a smooth closed curve, $\sigma \subseteq S$, that passes through b and c, and a filling surface, τ, with boundary σ, that both lie otherwise inside M (Fig. 3).

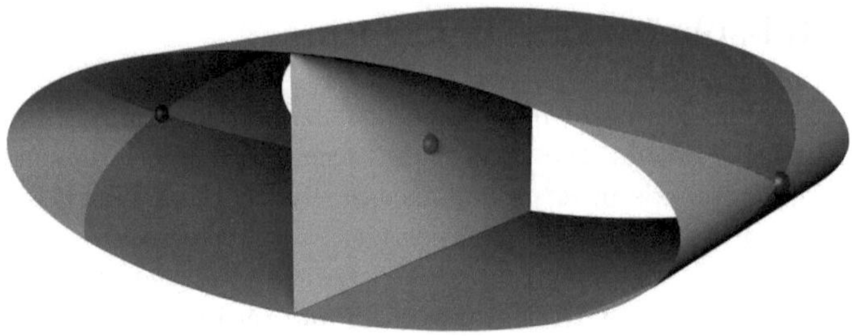

Fig. 3. Another example of $S \cap M$, in which M is a flattened cylinder and S has its center marked in red on the yellow mid-sphere axis. The points b, c at which the sphere touches the surface are seen in dark blue. (Color figure online)

See Fig. 1 for an intuitive example of the closed curves, μ and σ and the filling surface, τ. In this example, τ is a disk, but more generally it may be an orientable 2-manifold with boundary σ. Its purpose is to separate the two open curves obtained by removing points b and c from μ. It is not necessarily simply-connected and thus may have handles, like the torus. If M is homeomorphic to a sphere, then Condition (iii) can be weakened because the existence of τ follows from the existence of σ. We write $\mathcal{M}id(M)$ for the mid-sphere axis transform of M. Generically, it is a 2-dimensional surface in $\mathbb{R}^4$. The *mid-sphere axis* is the projection to $\mathbb{R}^3$ defined by forgetting the radii. This projection is not necessarily an embedding, so the mid-sphere axis can have self-intersections.

Remark. The topological requirements on the point-radius pairs of the mid-sphere axis transform can be strengthened by requiring that b and c be antipodal on S, or that σ be a circle on S and τ the disk whose boundary is this circle. We could also require that μ be part of the intersection of M with a plane that passes through b and c, and even that this plane be orthogonal to the disk τ. We will not pursue any of these modifications to the definition in this paper, primarily because the geometric conditions are more difficult to recognize algorithmically.

3 Morse Theory of Squared Distance Function

The interaction of a surface, M, with spheres centered at $x \in \mathbb{R}^3$ is conveniently described using the Euclidean distance of the points on M from x. This is a smooth function on M, unless $x \in M$, so it is preferable to use the square of this distance, which is always smooth. We thus introduce $f_x \colon M \to \mathbb{R}$ defined by $f_x(y) = \|y - x\|^2$. Writing $\nabla f_x(y)$ and $H_x(y)$ for the gradient and the Hessian of f_x at $y \in M$, a point $y \in M$ is a *critical point* of f_x if $\nabla f_x(y) = 0$, and it is *non-degenerate* if $H_x(y)$ is invertible. The values of critical points are called *critical values* of f_x, and all other values are *non-critical values*. Non-degenerate critical points are necessarily isolated, so if all critical points are non-degenerate, then there are only finitely many of them. Indeed, f_x is generically a *Morse function* [13], which is slightly stronger and requires that f_x satisfies two conditions:

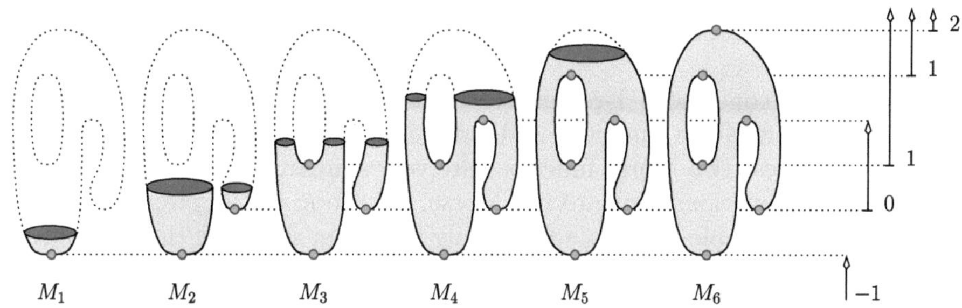

Fig. 4. The six non-empty sublevel sets at non-critical values interleaved between the critical values of the height function on the torus-with-a-nose. The lowest and highest of the three saddles both give birth, while the middle saddle gives death. On the *right*, we see the barcode that shows for which value a gap, loop, or closed surface exists in the sublevel set (see Sect. 5 for definitions).

- all critical points are non-degenerate;
- the values of the critical points are distinct.

Assuming $f_x \colon M \to \mathbb{R}$ is Morse, there are only three types of critical points: *minima*, *saddles*, *maxima*, which can be distinguished by the number of negative eigenvalues of the Hessian, which is 0, 1, 2, in this sequence. This number is traditionally referred to as the *index* of the critical point; see Fig. 4 where we see 2 minima, 3 saddles, and 1 maximum of the displayed height function.

Besides the three *local* types, we also distinguish between two *global* types of critical points. For this, we write $M_r = f_x^{-1}[0, r^2]$ for the *sublevel set* of f_x at r^2, which contains all points whose Euclidean distance from x is at most r. A fundamental result in Morse theory [13] is that M_r and M_s have the same homotopy type if there is no critical values in the open interval (r, s). Furthermore, whenever (r, s) contains a single critical value, then M_r and M_s have different homotopy types. To quantify this difference, we use the ranks of the p-th reduced homology groups, which we denote $\beta_p(M_r)$. Whenever we pass the value of a minimum, β_0 increases by 1, unless this is the first minimum, in which case β_{-1} decreases from 1 to 0. Whenever we pass the value of a saddle, either two components merge, in which case β_0 decreases by 1, or a component gains a loop, in which case β_1 increases by 1. Finally, whenever we pass the value of a maximum, β_1 decreases by 1, unless this is the last maximum, in which case β_2 increases from 0 to 1. This exhausts all cases and we note that each non-degenerate critical point changes only one rank by 1. We say the critical point gives *birth* if the rank increases by 1, and it gives *death* if the rank decreases by 1.

Both the local and the global types of the critical points of the squared distance function will be important to determine whether a point, x, with associated radius, r belongs to the mid-sphere axis transform.

4 Faustian Interchanges

Consider now a continuous 1-parameter family (a *curve*) of smooth functions that starts with a Morse function on M and ends with another Morse function on M. Unless these two Morse functions are very similar, the family necessarily passes through functions that are not Morse. By the *essential property* of Cerf theory [6], it is possible to choose the family such that it passes through only a finite number of functions that are not Morse, and each such function has only one violation of the two conditions that characterize a smooth function as Morse:

I. two critical points that share the same value;
II. one degenerate critical point with simple local neighborhood.

A violation II is the transient configuration that appears when two critical points collide and in the process annihilate each other. We refer to it as a *cancellation* of the two critical points. Its inverse is an *anti-cancellation*, which creates two critical points where there were none. We are however more interested in a violation I, which is the transient configuration that appears when two critical points swap the order of their values. We refer to it as an *interchange*.

An interchange of two minima does not affect their global type, unless they are the first two minima (the ones with smallest value). Then the first minimum becomes second and changes from giving death to giving birth, and the other way around for the second minimum. This happens where the curve of squared distance functions crosses the medial axis, because the point at the crossing has two closest points on M or, equivalently, the squared distance function from the crossing has two minima with smallest value.

Definition 2. *A* Faustian interchange *is an interchange of two critical points during which they swap their global types, one from giving death to giving birth and other from giving birth to giving death.*

Critical points that swap their global types when they interchange necessarily have the same index; see Sect. 5. As described above, the Faustian interchanges of two minima correspond to points of the medial axis, and by a similar argument, the Faustian interchanges of two maxima correspond to points of the circumsphere axis. Most important for this paper are the Faustian interchanges of two saddles, which correspond to points of the mid-sphere axis.

Theorem 1. *A point $x \in \mathbb{R}^3$ with radius $r > 0$ belongs to the mid-sphere axis transform of F iff f_x has a Faustian interchange of two saddles at r^2.*

Proof. " $\Longrightarrow$ ": Assuming $(x, r) \in \mathcal{M}id(M)$, let $b, c \in M$ be the points at which $S = S(x, r)$ touches M, and let μ, σ be the two closed curves and τ the surface that fills σ whose existence is guaranteed by Conditions (ii) and (iii) of Definition 1. Let X be the connected component of M_r that contains b and c. Since μ exists, $X \setminus \{b\}$ is connected, and so is $X \setminus \{c\}$. It is also clear that b and c cannot be minima or maxima, so they are saddles. If $X \setminus \{b, c\}$ is connected, then there is a curve that starts at one branch of $\mu \setminus \{b, c\}$ and ends at the other whose points lie on or inside S. But this curve contradicts the existence of the curve σ and the surface τ that fills σ. Hence, if we add b to $X \setminus \{b, c\}$ before c, this first gives death to a component and then birth to a loop, and if we add c to $X \setminus \{b, c\}$ before b, we observe the same pattern. Hence, f_x has a Faustian interchange of b and c at r^2.

"$\Longleftarrow$": Let b and c be the two saddles of the Faustian interchange of f_x at r^2. Since b and c swap their global types, they must belong to the same component of the corresponding sublevel set, $X \subseteq M_r$, this component contains a loop that passes through b and c, and $X \setminus \{b, c\}$ is not connected. This loop is the closed curve μ required by Condition (ii), and that $X \setminus \{b, c\}$ is not connected implies the existence of the curve $\sigma \subseteq S$ and the surface τ that fills σ inside M, as required by Condition (iii).

5 Pairing of Critical Points

The idea of persistent homology is that the critical points of a function can be paired to delimit the ranges in which the corresponding features appear in the sublevel sets of the function; see e.g. [10]. This additional structure is useful in two ways. First, it provides a means to assess how robust a Faustian interchange is, and we will introduce several ways to quantify this notion. Second, it leads to an algorithm to construct the mid-sphere axis transform.

To explain the pairing, we recall that each critical point gives either birth or death, namely to a homology class of the sublevel set at the corresponding critical value. Since we use reduced homology, the empty sublevel set has a (-1)-dimensional class, which dies when the first minimum is encountered. Later minima give birth to 0-dimensional classes, which die at the hand of the saddles that merge them with other 0-dimensional classes born before them. This is the *elder rule*, which prescribes that the older class survives by absorbing the younger class. Alternatively, a saddle may give birth to a 1-dimensional class, which later dies at the hand of a maximum. Indeed, every maximum gives death to such a class, except the last maximum, which gives birth to a 2-dimensional class, namely that of the entire surface. Besides the last maximum, there are two saddles per genus of the closed surface that remain unpaired in this process, and they represent the homology of M.

To visualize this structure, we list the critical points from left to right in the order of their values and mark a pair (a, b) by the interval that begins at a and ends at b, writing $a = \mathrm{bth}(b)$ and $b = \mathrm{dth}(a)$. The index of b is necessarily one higher than that of a, so there are only three kinds of (finite) intervals for a function on a 2-dimensional surface: those with endpoints of index $-1, 0$, of index $0, 1$, and of index $1, 2$. Letting b, c be two saddles of a Faustian interchange, we therefore have an interval from $a = \mathrm{bth}(b)$ to $b = \mathrm{dth}(a)$ and another interval from $c = \mathrm{bth}(d)$ to $d = \mathrm{dth}(c)$, as illustrated in Fig. 5. Since b and c swap their global types, they also swap their partners in the pairing; that is: $a = \mathrm{bth}(c), c = \mathrm{dth}(a)$ and $b = \mathrm{bth}(d), d = \mathrm{dth}(b)$ after the interchange. Note however that interchanging two saddles does not necessarily make a Faustian

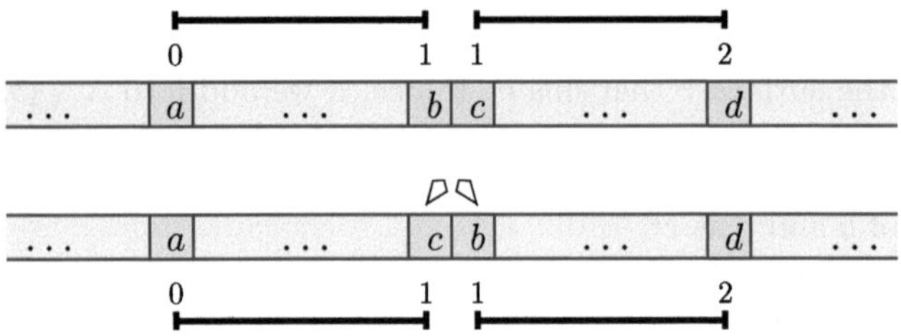

Fig. 5. The ordered sequences of critical points before and after a Faustian interchange of two saddles. Note that the two saddles swap their partners under the pairing.

interchange. Assuming a Faustian interchange as illustrated in Fig. 5, we consider the two intervals and call

$$\mathrm{Pre}_x(b,c) = f_x(b) - f_x(a) = r - \|x-a\|, \tag{1}$$
$$\mathrm{Post}_x(b,c) = f_x(d) - f_x(c) = \|x-d\| - r \tag{2}$$

the *pre-meditation* and *post-meditation* of the Faustian interchange, in which $r = \|x-b\| = \|x-c\|$, a is the first point of the second component of $X \setminus \{b,c\}$ encountered by the growing sphere centered at x, and d is the last point of the first component of $M \setminus \mu$ encountered by that sphere. We will return to these notions when we discuss the pruning on the mid-sphere axis in the next section.

6 Computation and Examples

This section uses the characterization of the mid-sphere axis in terms of Faustian interchanges (Theorem 1) to describe an algorithm that constructs a staircase approximation of this axis. To simplify the discussion, we ignore the radii, which may be computed without much extra effort. The algorithm is discrete and extends to the construction of the medial and the circum-sphere axes.

The Algorithm. We assume the input surface is given as a triangulation; that is: a 2-dimensional simplicial complex whose vertices are locations in $\mathbb{R}^3$ such that each edge belongs to exactly two triangles and each vertex belongs to an alternating cyclic sequence of edges and triangles. Given a point $x \in \mathbb{R}^3$, we assign $f_x(u) = \|u-x\|^2$ to each vertex, u, and the maximum value of the two or three incident vertices to each edge and triangle. Sorting the vertices, edges, and triangles by value and breaking ties by dimension, we get the filter for which we compute the pairing as defined by persistent homology; see e.g. [10, Chapter VII]. In this setting, the vertices, edges, and triangles play the roles of the minima, saddles, and maxima of the squared distance, respectively, and the transposition of two edges—one paired with a vertex and the other with a triangle—is a Faustian interchange if they replace each other in their respective pairs and thus alter their global types.

However, only a measure-zero set of points x in $\mathbb{R}^3$ induce squared distance functions that catch a Faustian interchange in the act. We therefore use the vineyard algorithm introduced in [9] to probe space with short segments of filters to find points of the mid-sphere axis. Given the endpoints $x', x'' \in \mathbb{R}^3$ of such a segment, this algorithm maintains the filter and the pairing of simplices while continuously moving x from x' to x''. The elementary operations in this sweep are transpositions of contiguous simplices in the filter, which are scheduled according to the linear interpolation between their initial and final values. This amounts to running bubble sort on the filter at x' using the filter at x'' as the target ordering. Whenever a transposition swaps two edges, we check whether one gives birth, the other gives death, and the swap changes their global types. If yes, then we have found a point whose squared distance function witnesses a Faustian interchange.

We use segments connecting neighboring points of the integer lattice, $\mathbb{Z}^3 \subseteq \mathbb{R}^3$. For example, we run the vineyard algorithm for the segment from $x' = (i, j, k)$ to $x'' = (i+1, j, k)$, and add the square $(i + \frac{1}{2}) \times [j \pm \frac{1}{2}] \times [k \pm \frac{1}{2}]$ to the staircase approximation of the mid-sphere axis iff there is at least one Faustian interchange along this segment. We begin the hunt for Faustian interchanges at an arbitrary point of $\mathbb{Z}^3$ and reduce the boundary matrix of the corresponding filter from scratch. Thereafter, we run the vineyard algorithm along the segments to the six neighboring integer points. When we arrive at the end of one of these segments, we repeat while reusing the final reduced matrix instead of computing it from scratch and making sure that no segment is swept twice. Sample results of this algorithm are shown in Fig. 6.

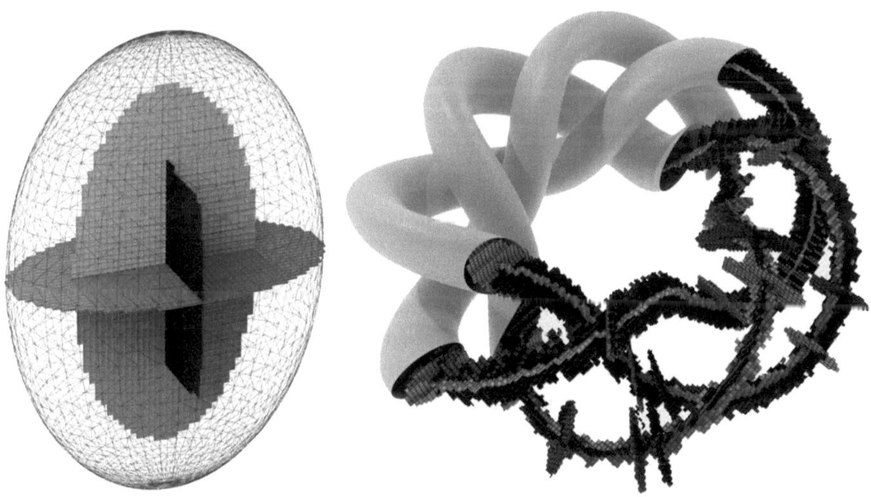

Fig. 6. The *pink* medial, *black* mid-sphere, and *blue* circum-sphere axes of an ellipsoid on the *left* and an elliptically thickened $(3, 7)$-torus knot on the *right*. To improve the visualization, we restrict ourselves to points inside the surfaces, which explains why the mid- and circum-sphere axes do not extend outside the two surfaces. (Color figure online)

Two or more segments can be swept in parallel, so we get some mileage out of partitioning the segments connecting neighboring integer points into a few sets, which are then processed in parallel. However, no matter how we schedule the segments, the number of swaps remains large. To see this note that two vertices swap along a segment iff it intersects the perpendicular bisector of the pair. Assuming we explore all segments connecting neighboring integer points in $[0, g]^3$, the number of such segments is $O(g^2)$. Letting n be the number of vertices in the triangulation of the surface, this amounts to a total of $O(g^2n^2)$ swaps. However, no matter how we schedule the segments, the number of interchanges is large. Focusing on the vertices, and letting n be their number and $[0, g]^3$ the domain, we can naïvely bound the number of interchanges by $O(g^3n^2)$, where every pair of vertices is interchanged along every segment. To improve this bound, we count the number of segments along which a fixed pair is interchanged. This happens when their positions trade place in the ordering, so any segment that crosses the perpendicular bisector of the pair contributes to exactly one such interchange. There are $O(g^2)$ such segments, so summing over all pairs of vertices gives at most $O(g^2n^2)$ interchanges.

The Matrices. The efficient implementation of the algorithm is challenging, in particular the computation of the reduced boundary matrices. Assuming familiarity with the vineyard algorithm described in [9], we sketch some of the more important design decisions that help in this respect. Following [9], we use a sparse matrix representation tailored to efficiently support the maintenance operations in the vineyard algorithm. For convenience, we use $\mathbb{Z}/2\mathbb{Z}$ coefficients so homology can be computed using boolean matrices. To reduce the boundary matrix, the algorithm swaps rows and columns, and it adds columns to each other. We thus store the matrix as a sequence of sparse columns, and we maintain permutations π_R and π_C for the rows and columns, respectively. With these permutations, it is easy to swap two rows or two columns, but they add an indirection when we access entries: the entry in row i and column j is physically stored in row $\pi_R(i)$ and column $\pi_C(j)$.

Each column is an ordered list of non-zero entries. Adding two columns means computing the `xor` function, and since the columns are sorted, this can be done by walking along the two lists in parallel and ignoring rows that appear in both. To decide which columns to add, the algorithm needs access to the maximum row index of the non-zero entries in a column. Because of the permutation π_R, this row is not necessarily the last in the list. Profiling an earlier version of our software revealed that finding this row index in a linear scan was the main bottleneck, so we decided to explicitly record this row index for each column and to update the record whenever it changes. We mention that the vineyard algorithm also adds rows to each other in a second matrix, which we support by storing the transpose of that matrix in the aforementioned format.

In spite of using sparse versions, the amount of memory needed to remember reduced matrices is significant, so we dispose of them as soon as we can while still avoiding the reduction of the boundary matrix from scratch. This is where breadth-first search in scheduling the segments shines, as its front of unfinished integer points is much smaller than for example for depth-first search.

Pruning Strategies. Similar to the medial axis, the mid-sphere axis is sensitive to small fluctuations in the data, and since our algorithm is approximate, this tends to give rise to artifacts at small scales. We therefore measure scale and prune artifacts if their scale is found to be below a fixed threshold. We distinguish between two strategies: the first uses a notion of *distance* between the two interchanging simplices, and the second assesses their *persistence*.

In *Euclidean pruning*, we check the Euclidean distance between the two interchanging simplices. For edges and triangles, we use the Euclidean distance between their midpoints and barycenters, respectively. See Fig. 7 for an illustration of the effect for different thresholds. A likely improvement would be using the geodesic distance along the surface, but we get satisfactory results with the Euclidean distance, which we use for all three types of axes.

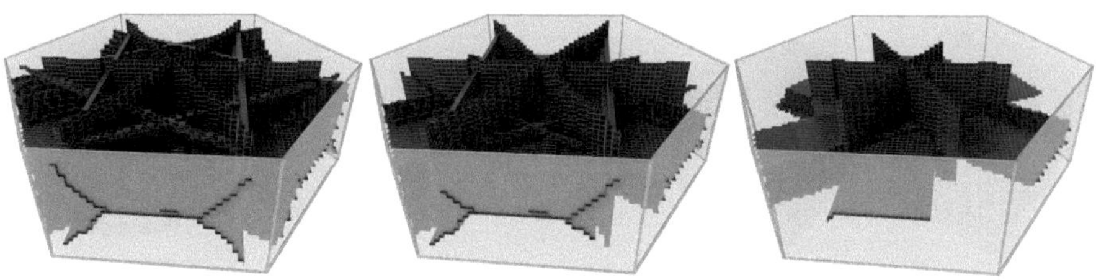

Fig. 7. A hexagonal solid with its mid-sphere axis progressively modified by Euclidean pruning. The top and bottom faces of the solid are removed for improved visibility.

Alternatively, we may measure the distance combinatorially, by counting the simplices needed to connect the interchanging simplices. We use only the simplest incarnation of this idea. In *face pruning*, we make sure that the interchanging simplices do not have a common face, and we apply this constraint to the construction of the mid-sphere and the circum-sphere axes. Likewise, in *coface pruning*, we make sure that they do not have a common coface, and we apply this to the construction of the medial and the mid-sphere axes.

In *persistence pruning*, we consider the persistence of the two interchanging simplices; that is: their pre- and post-meditations, as defined in Sect. 5. We note that they have different geometric meaning: for the mid-sphere axis, the pre-meditation measures the two components inside the mid-sphere before they merge due to adding the first of the two edges, while the post-meditation measures the loop formed by adding both edges simultaneously. In our current implementation, we require that both meditations exceed a fixed threshold. We find persistence pruning indispensable for computing the mid-sphere axis, as it detects many spurious loops that form the boundaries of triangles and are filled right after being formed. More generally, we use it for all three types of axes (Fig. 8).

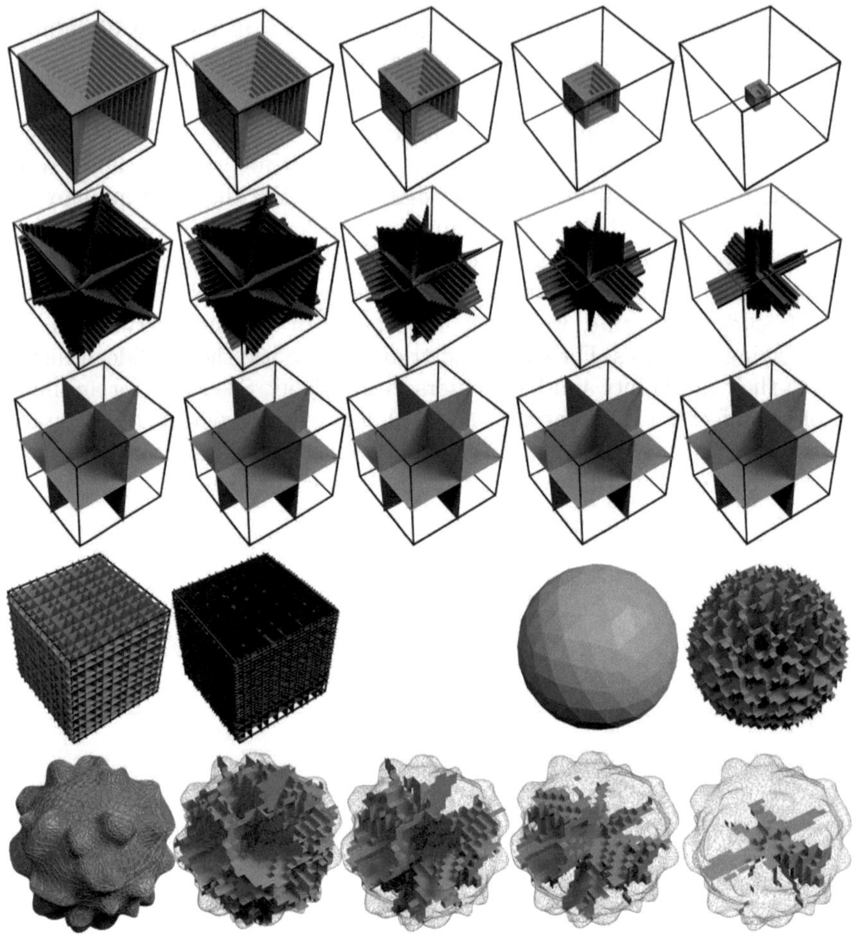

Fig. 8. Here we demonstrate some effects of pruning on the medial (pink), midsphere (black), and circumsphere (blue) axes. The top three rows demonstrate the effect of Euclidean pruning on the axes resulting from a cube. We also see that pruning is sometimes not needed for the circumsphere axis. In the fourth row, we demonstrate the necessity of the other types of pruning. Without coface pruning, the medial axis of the cube blows up, as does the midsphere axis without persistence pruning. To show the circumsphere axis behaving similarly, we needed a different input than the cube, and so used the sphere (in yellow). Without face pruning, it too becomes unwieldy. In the bottom row, we demonstrate an input (the purple virus-like shape) for which Euclidean pruning has an impact on the circumsphere axis. (Color figure online)

Code Availability. We have published the code used to compute these examples on GitHub[2]. It can be compiled and run locally, or sampled via a live demo we have running via GitHub Pages[3].

7 Discussion

The main contribution of this paper is the introduction of the mid-sphere axis transform of a smoothly embedded connected closed surface in $\mathbb{R}^3$, which is sensitive to ridge-like structures on the surface. The definition is in terms of local and global topological aspects of the family of squared distance functions on the surface from a point. We also introduce the notion of a Faustian interchange, which offers a unified framework in which the medial and mid-sphere axis transforms are particular cases. This framework extends beyond three dimensions and leads to d axis transforms for any smoothly embedded hypersurface in $\mathbb{R}^d$.

The presented algorithm serves as a proof of concept, and improved algorithms are desirable. One promising idea is the extension of the Voronoi heuristic to the mid-sphere axis. Note that the Voronoi tessellation is already used to compute the standard medial axis [2], and, by analogy, the furthest-point Voronoi tessellation yields the circum-sphere axis. To extend this heuristic to the mid-sphere axis, we will have to collect and combine substructures from various higher order Voronoi tessellations of the same set of points.

Another interesting direction is the analysis of effective pruning strategies that lower the sensitivity to noise in the data by defining mid-sphere axes on different scale levels. For example, can the λ-medial axis introduced by Chazal and Lieutier [7] be extended to the mid-sphere axis?

Appendix A Pseudocode

Here we present pseudocode of our algorithm. This is written per homology dimension and per component of the grid. In our implementation, we run different grid components in parallel.

[2] https://github.com/medial-ax/medial-ax.

[3] https://medial-ax.github.io/medial-ax/.

Algorithm 1. Computing the mid-sphere axis

```
function INITIALIZE(v)
    π_v ← ORDERING(v)    ▷ Order simplices by distance function from grid point v
    D_v ← boundary matrix ordered by π_v
    R_v, U_v ← REDUCE(D_v)    ▷ Following [9] we have D_v = R_v U_v
function PROCESSSEGMENT(u, v)
    F_{u,v} ← {}
    D_v ← D_u, R_v ← R_u, U_v ← U_u    ▷ Copy matrices. These are about to be updated
    π_u ← ORDERING(u), π_v ← ORDERING(v)    ▷ Simplex ordering wrt. u and v
    I ← BUBBLEINTERCHANGES(π_u, π_v)    ▷ Interchanges required to go from π_u to π_v
    for (i, j) in I do    ▷ Handle interchange of simplices σ_i and σ_j
        VINEYARDS(i, j, D_v, R_v, U_v)    ▷ Adjust matrices for v following [9]
        if was Faustian then    ▷ See Definition 2
            F_{u,v} ← F_{u,v} ∪ {(i, j)}    ▷ Register (i, j) as a Faustian interchange
function MAIN(v_0)    ▷ We start reducing from any vertex v_0
    INITIALIZE(v_0)
    for (u, w) ∈ BFS(v_0) do    ▷ Breadth-first traversal
        PROCESSSEGMENT(u, w)
        if {u, v} is visited for the last time then    ▷ Eager cleanup to save memory
            Delete D_w, R_w, U_w where w = u or w = v (or both)
    for (u, v) with entries in F_{u,v} do
        output PRUNE(F_{u,v})
```

References

1. Andreev, E.M.: Convex polyhedra in Lobačevskiĭ spaces. Mat. Sb., New Series**81**, 445–478 (1970)
2. Attali, D., Boissonnat, J-D., Edelsbrunner, H.:Stability and computation of medial axes—a state-of-the-art report. In: Moeller, T., Hamann, B., Russell, R. (eds.) Mathematical Foundations of Scientific Visualization, Computer Graphics, and Massive Data Exploration, pp. 109–125. Springer, Berlin (2009)
3. Bittar, E., Tsingos, N., Gascuel, M.: Automatic reconstruction of unstructured 3D data: combining a medial axis and implicit surfaces. Comput. Graph. Forum **14**, 457–468 (1995)
4. Blum, H.: A transformation for extracting new descriptors of shape. In: Dunn, W.W. (ed.) Proceedings of the Symposium on Models for the Perception of Speech and Visual Form, pp. 362–389. MIT Press, Cambridge (1967)
5. Bruce, J.W., Giblin, P.J., Gibson, C.G.: Symmetry sets. Proc. Royal Soc. Edinburgh Sect. A: Math.**101**, 163–186 (1985)
6. Cerf, J.: Sur les difféomorphismes de la sphère de dimension trois ($\Gamma_4 = 0$). In: Lecture Notes in Mathematics, vol. 53. Springer-Verlag, Berlin (1968)
7. Chazal, F., Lieutier, A.: The λ-medial axis. Graph. Models **67**, 304–331 (2005)
8. Cohen-Steiner, D., Edelsbrunner, H., Harer, J.: Stability of persistence diagrams. Disc. Comput. Geom. **37**, 103–120 (2007)
9. Cohen-Steiner, D., Edelsbrunner, H., Morozov, D.: Vines and vineyards by updating persistence in linear time. In: Proceedings of 22nd Annual Symposium Computer Geomentry, pp. 119–126 (2006)
10. Edelsbrunner, H., Harer, J.L.: Computational topology: an introduction. Am. Math. Soc., Providence (2010)

11. Federer, H.: Curvature measures. Trans. Amer. Math. Soc. **93**, 418–491 (1959)
12. Koebe, P.: Kontaktprobleme der konformen Abbildung. Ber. Sächs. Akad. Wiss. Leipzig, Math.-Phys. Kl.**88**, 141–164 (1936)
13. Milnor, J.: Morse Theory. Princeton Univ. Press, Princeton (1963)
14. Porteous, I.R.: Geometric Differentiation: For the Intelligence of Curves and Surfaces, 2nd edn. Cambridge Univ. Press, Cambridge (2001)
15. Tam, R. Heidrich, W.: Shape simplification based on the medial axis transform. In: Proceedings of 14th IEEE Visualization, pp. 63–71 (2003)
16. Thurston, W.: The Geometry and Topology of Three-manifolds. Electronic version, MSRI, Berkely, California (2002)
17. Yang, B., Yao, J.: DMAT: deformable medial axis transform for animated mesh approximation. Comput. Graph. Forum **37**, 301–311 (2018)

Subfield-Based 3D Curve-Thinning on (18, 12) Pictures of the FCC Grid

Kálmán Palágyi(✉)

Department of Image Processing and Computer Graphics, University of Szeged, Szeged, Hungary
palagyi@inf.u-szeged.hu

Abstract. Thinning is an iterative object reduction to extract skeleton-like shape features from binary objects. Parallel 3D curve-thinning algorithms are composed of parallel reductions to produce centerlines. In subfield-based parallel thinning the digital space is partitioned into $s \geq 2$ subfields which are alternatively activated, and only certain points in the active subfield can be deleted. This paper presents the first two subfield-based parallel 3D curve-thinning algorithms acting on the unconventional face-centered cubic (FCC) grid. The proposed algorithms use partitionings of the the FCC grid into four and eight subfields.

Keywords: Discrete shape representation · Unusual grids · Discrete topology · Centerlines · Curve-thinning · Face-centered cubic (FCC) grid · Topology preservation

1 Introduction

Parallel reductions transform *binary pictures* (containing black and white points) by changing only certain black points to white simultaneously (which is called *deletion*) [6].

Skeletonization plays a key role in a broad range of problems in image processing and computer vision [19]. *Thinning* is a frequently applied skeletonization method: certain border points are deleted in iteration steps, and the entire process is repeated until stability is reached. *Parallel curve-thinning algorithms* are composed of parallel reductions to produce *centerlines* (i.e., stick-like 1D representations) of objects in a *topology-preserving* way [11,12].

3D digital images are usually sampled on the regular cubic grid that is dual to $\mathbb{Z}^3$. In this paper, our attention is focussed on an unconventional 3D grid, the *face-centered cubic (FCC) grid* [12]. The point $(x, y, z) \in \mathbb{Z}^3$ is an element of the FCC grid if

$$(x + y + z) \; mod2 = 0\,.$$

Herman pointed out some disadvantages of the cubic grid [7], Gau and Kong reported three advantageous properties of the FCC grid over the cubic grid [5], and Edelsbrunner et al. showed that the FCC grid provides the densest sphere packing [3]. Due to its advantageous geometrical and topological properties, the importance of the FCC grid is on the rise [1,2,4,9,13,15,18,20].

M. H. F. Wilkinson and J. Kosinka (Eds.): DGMM 2025, LNCS 16296, pp. 148–162, 2026.
https://doi.org/10.1007/978-3-032-09544-2_11

This also justifies the development of skeletonization methods for this unconventional grid.

This paper presents the first two subfield-based 3D parallel curve-thinning algorithms on the FCC grid. They use partitionings of the the FCC grid into four and eight subfields. Both algorithms preserves the topology, and they are capable of producing one point thin centerlines.

2 Basic Notions and Results

Here, we recall the fundamental concepts of digital topology that apply specifically to the FCC grid [5,11,12].

Let us denote $\mathbb{F} \subset \mathbb{Z}^3$ the set of points in the FCC grid. A (18, 12) *picture* on $\mathbb{F}$ is a quadruple $(\mathbb{F}, 18, 12, B)$, where $B \subseteq \mathbb{F}$ denotes the set of *black points*, each point in $\mathbb{F} \setminus B$ is a *white point*, 18-*adjacency* is assigned to B, and 12-*adjacency* belongs to $\mathbb{F} \setminus B$, see Fig. 1.

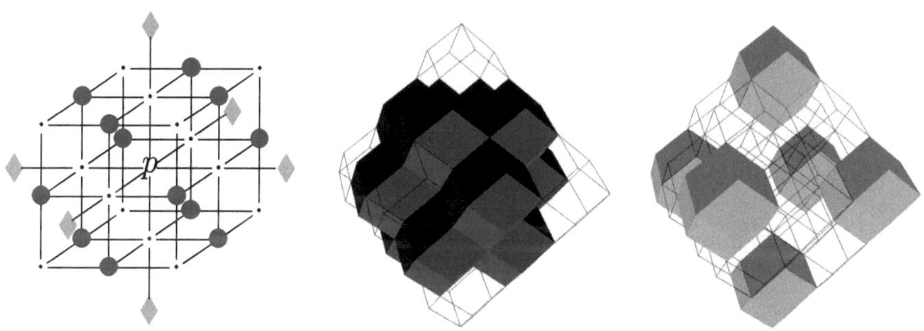

Fig. 1. The twelve points marked'●' are 12-adjacent to the central point p, and form the set $N_{12}(p)$; $N_{18}(p)$ contains $N_{12}(p)$ and the six further points marked'◆' that are 18-adjacent to p (left). (Note that points marked '·' are in $\mathbb{Z}^3 \setminus \mathbb{F}$.) The voxel representations of $N_{12}(p)$ (middle) and $N_{18}(p) \setminus N_{12}(p)$ (right) in which each voxel is a rhombic dodecahedron (i.e., a convex polyhedron with 12 congruent rhombic faces).

Since the studied adjacency relations are symmetric, their reflexive-transitive closure induces two equivalence relations on B and $\mathbb{F} \setminus B$, and the generated equivalence classes are called *black* 18-*components* (or *objects*) and *white* 12-*components*, respectively. It is assumed that B contains finitely many points, therefore there is a unique infinite white component which is said to be the *background*, and a finite white component is called a *cavity*.

A black point $p \in B$ is a *border point* if at least one point being 12-adjacent to p is white, see Fig. 2*a*-*e*; it is an *interior point* if it is not a border point (i.e., all the twelve points being 12-adjacent to p are black), see Fig. 2*f*; it is said to be *isolated* if all the eighteen points being adjacent to p are white, see Fig. 2*c*; and it is an *endpoint* if p is 18-adjacent to exactly one black point, see Fig. 2*a*.

Topology preservation is a major concern in thinning algorithms. It is crucial to guarantee that all parallel reductions of the thinning process are topology-preserving for all possible pictures. A parallel reduction acting on 2D pictures is *topology-preserving* if it never splits an object into two or more, never deletes completely any object, never merges a cavity with the background or another cavity, and never creates a cavity [11]. Since 3D objects may contain *holes* or *tunnels* (like solid tori or American donuts), it is generally agreed that a topology-preserving reduction in 3D neither creates nor eliminates a hole, and it does not merge any holes with each other [11].

Conditions for topology preservation were obtained with the help of the concept a *simple point* [11,12]. A single black point is simple if its deletion preserves the topology. The simpleness of a point p in $(18, 12)$ pictures depends only on $N_{18}(p)$ as it is stated by the following theorem:

Theorem 1. [5] *A black point $p \in B$ in picture $(\mathbb{F}, 18, 12, B)$ is simple if and only if both of the following conditions hold:*

1. *The set $N_{18}(p) \cap B$ is not empty and forms exactly one 18-component.*
2. *The set $N_{12}(p) \setminus B$ is not empty and forms exactly one 12-component.*

Based on Theorem 1, it is obvious that only non-isolated border points may be simple. Figure 2 also gives examples of simple and non-simple points.

Parallel reductions can delete a set of black point at the same time and not just a single point. Therefore, we need to consider what is meant by topology preservation when a number of black points are deleted simultaneously. Now we will make use the following result:

Theorem 2. [9] *A parallel reduction $\mathcal{R}$ preserves the topology for an arbitrary picture $(\mathbb{F}, 18, 12, B)$ if all of the following conditions hold:*

1. *Only simple points for B are deleted by $\mathcal{R}$.*
2. *If two 12-adjacent points p and q are deleted by $\mathcal{R}$, p is simple for $B \setminus \{q\}$. (Note that in this case q is simple for $B \setminus \{p\}$.)*
3. *If three mutually 12-adjacent points p, q, and r are deleted by $\mathcal{R}$,*
   p is simple for $B \setminus \{q, r\}$, or
   q is simple for $B \setminus \{p, r\}$, or
   r is simple for $B \setminus \{p, q\}$.
4. *$\mathcal{R}$ does not delete completely any object of B composed of four mutually 12-adjacent points, see Fig. 3a-b.*
5. *$\mathcal{R}$ does not delete completely any object composed of mutually 18-adjacent points in which the points are not mutually 12-adjacent (i.e., in such an object, there are two points that are not 12-adjacent), see Fig. 3c.*

Note that all the conditions of Theorem 2 can be checked locally. It is important to point out that Theorem 2 provides a method of verifying that a parallel reduction preserves the topology, rather than a methodology, for constructing topology-preserving reductions.

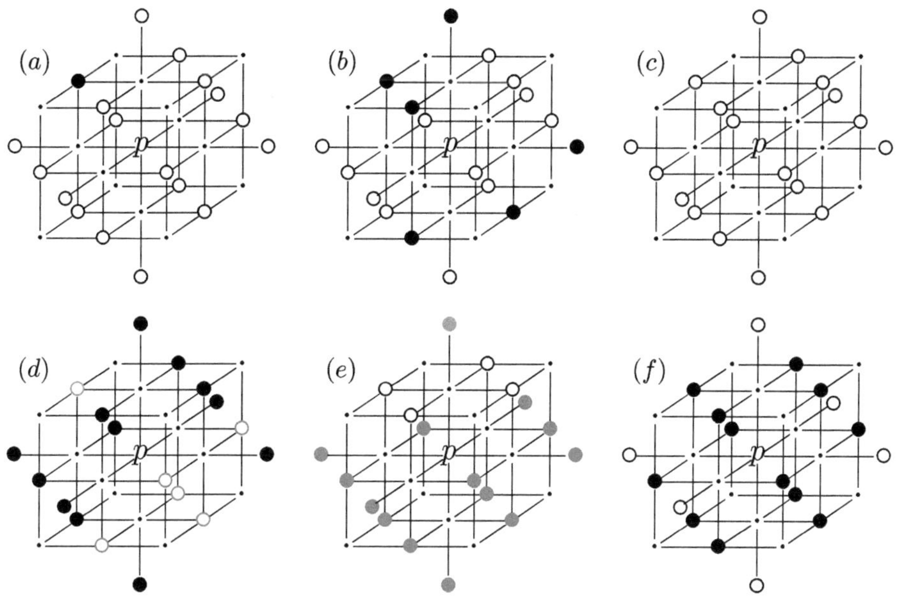

Fig. 2. Examples of border points (a)-(e), an interior point (f), an isolated point (c), simple points (a)-(b), and non-simple points (c)-(f) in $(18, 12)$ pictures, where the central point p is black. Points marked'●' and '○', respectively, refer to black and white points, and points denoted by '·' are not in $\mathbb{F}$. By Theorem 1, a black point p is simple only for the first two configurations (a) and (b), since both conditions of the theorem are satisfied; isolated point p is not simple in (c), since condition 1 is violated (i.e., there is no black 18-component in $N_{18}(p)$); p is not simple in (d), since condition 2 is not met (i.e., there are two white 12-components in $N_{12}(p)$, as depicted in different colors); p is not simple in (e), since condition 1 is violated (i.e., there are two black 18-components in $N_{18}(p)$, as depicted in different colors); interior point p is not simple in (f), since condition 2 is not met (i.e., there is no white 12-component in $N_{12}(p)$.

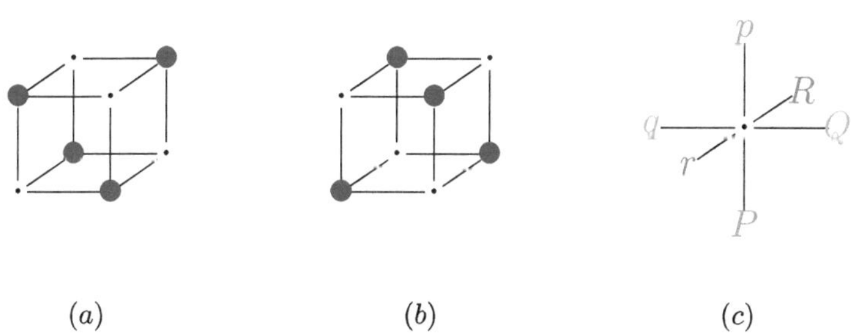

Fig. 3. The two possible sets of four mutually 12-adjacent points marked'●' (a)-(b), see condition 4 of Theorem 2. The set of six points$\mathcal{X} = \{p, P, q, Q, r, R\}$ is the maximal set of mutually 18-adjacent points (c). If there are two points that are not 12-adjacent in a set $\mathcal{Y} \subseteq \mathcal{X}$, then$\{p, P\} \subseteq \mathcal{Y}$, or$\{q, Q\} \subseteq \mathcal{Y}$, or$\{r, R\} \subseteq \mathcal{Y}$. One can check that condition 5 of Theorem 2 applies to 37 possible such sets. In all three figures, the points marked '·' are not elements of the FCC grid.

3 The Proposed Algorithms

In this section, we present two subfield-based 3D parallel curve-thinning algorithms acting on (18, 12) pictures of the $\mathbb{F}$ grid. These algorithms, respectively, use four and eight subfields. The subfields in the proposed partitionings are denoted by $SF_s(i)$ ($s \in \{4, 8\}$; $i \in \{0, \ldots, s-1\}$), see Fig. 4.

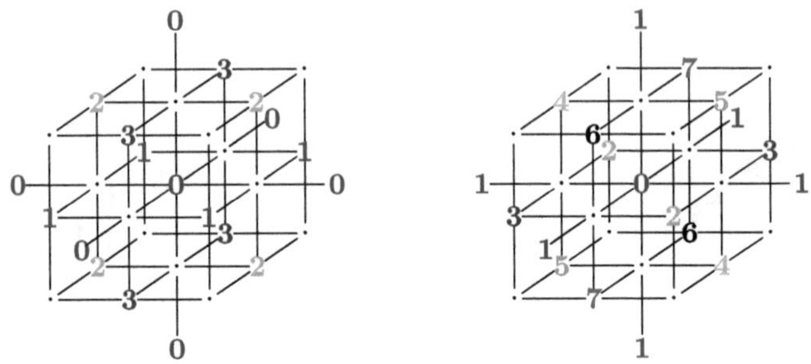

Fig. 4. Partitionings of $\mathbb{F}$ into four (left) and eight (right) subfields. All points marked 'i' belong to the i-th subfield $SF_s(i)$ ($s \in \{4, 8\}$; $i \in \{0, \ldots, s-1\}$) and points marked '·' are in $\mathbb{Z}^3 \setminus \mathbb{F}$.

It is easy to see the following relationship between the two partitionings:

$$SF_4(i) = SF_8(2i) \cup SF_8(2i+1) \quad (i \in \{0, 1, 2, 3\}).$$

For reasons of scope, only the partitioning of $\mathbb{F}$ into four subfields is given formally:

$$\begin{aligned} SF_4(0) &= \{\ (x, y, z) \mid x, y \text{ and } z \text{ are all even}\ \} \\ SF_4(1) &= \{\ (x, y, z) \mid x \text{ and } y \text{ are both odd and } z \text{ is even}\ \} \\ SF_4(2) &= \{\ (x, y, z) \mid y \text{ and } z \text{ are both odd and } x \text{ is even}\ \} \\ SF_4(3) &= \{\ (x, y, z) \mid x \text{ and } z \text{ are both odd and } y \text{ is even}\ \} \end{aligned}$$

Note that both the conventional 3D cubic and the body-centered cubic (BCC) grids can be partitioned into two, four and eight subfields [8,10,17]. Unfortunately, we could not manage to propose a reasonable partitioning of $\mathbb{F}$ into two subfields.

Algorithm 1 presents our subfield-based 3D parallel curve-thinning algorithms $\mathbf{SF}_4(m, n)$ and $\mathbf{SF}_8(m, n)$ ($m, n \in \{1, 2, \ldots\}$). Note that both parameters m and n are positive integers, and they are to control the *endpoint re-checking phase* with m *endpoint-oriented* parallel reductions of the iteration steps (see lines 15-17 in Algorithm 1).

Algorithm 1 uses two concepts to be defined:

Definition 1. *Let* $(\mathbb{F}, 18, 12, Y)$ *be the current picture and* $Z \subseteq Y$ *be the set of black (border) points in the active subfield. Then a point* $p \in Z$ *is said to be* subfield-deletable for (Y, Z) *if*

Algorithm 1: algorithm $\mathbf{SF}_s(m,n)$ $(s \in \{4,8\};\ m,n \in \{1,2,\ldots\})$

1 **Input**: picture $(\mathbb{F},18,12,X)$
2 **Output**: picture $(\mathbb{F},18,12,Y)$
3 $Y \leftarrow X$
4 **repeat**
5 $\quad D \leftarrow \emptyset$
6 $\quad$ `// iteration-level partitioning of border points`
7 $\quad$ **for** $i = 0$ **to** $s-1$ **do**
8 $\quad\quad B(i) \leftarrow \{\ p \mid p$ is a border point in $Y \cap SF_s(i)\ \}$
9 $\quad$ `//` s `subfield-oriented parallel reductions`
10 $\quad$ **for** $i = 0$ **to** $s-1$ **do**
11 $\quad\quad D(i) \leftarrow \{\ p \mid p$ is *subfield-deletable* for $(Y, B(i))\ \}$
12 $\quad\quad Y \leftarrow Y \setminus D(i)$
13 $\quad\quad D \leftarrow D \cup D(i)$
14 $\quad$ `//` m `endpoint-oriented parallel reductions`
15 $\quad$ **for** $j = 0$ **to** $m-1$ **do**
16 $\quad\quad E \leftarrow \{\ p \mid p$ is *end-deletable*(n) for $Y\}$
17 $\quad\quad Y \leftarrow Y \setminus E$
18 **until** $D = \emptyset$;

- *point p is simple for Y and*
- *p is not an endpoint in the current picture.*

Definition 2. *A black point p is* end-deletable(n) $(n \in \{1,2,\ldots\})$ *if*

- *p is an endpoint in the current picture* $(\mathbb{F},18,12,Y)$,
- *the only (black) point in $N_{18}(p) \cap Y$ is not an endpoint, and*
- *at least n point/points in $N_{18}(p)$ was/were deleted by the subfield-oriented reductions of the current iteration.*

In Algorithm 1, line 3 initializes the set of black points in the output picture (i.e., elements of the produced centerline). The kernel of the **repeat** cycle (see lines 5-17) corresponds to one iteration step that comprises $s+m$ $(s \in \{4,8\};\ m \in \{1,2,\ldots\})$ parallel reductions in which the first s reductions are *subfield-oriented*, and the remaining m ones are *endpoint-oriented*.

In the i-th $(i \in \{0,\ldots,s-1\})$ subfield-oriented reduction, subfield $SF_s(i)$ is activated and all black points in the current picture $(\mathbb{F},18,12,Y)$ that are *subfield-deletable* for $(Y, B(i))$ are deleted simultaneously (see lines 11 and 12).

A well-known property of subfield-based thinning algorithms is that they tend to dig 'potholes' that cause unwanted side branches to grow on the centerlines. That is why Ma et al. introduced the concept of a *twig voxel* in their 2-subfield 3D parallel curve-thinning algorithm acting on the conventional cubic grid [14]. Unfortunately, the detection of deletable twig voxels is a rather time-consuming process, and their use is not promising for more than two subfields. In order to reduce the count of parasitic side branches in the produced centerlines, we used the following two tricks:

- The subfield-oriented reductions in an iteration step can only delete points that were already border points at the beginning of the current iteration step (see lines 8 and 11).
- The s subfield-oriented reductions (see lines 10-13) are followed by an *endpoint re-checking* phase that contains m endpoint-oriented reductions (see lines 15-17) that are capable of deleting *end-deletable*(n) points (see line 16 and Definition 2).

The proposed algorithms terminate if no points are deleted in the current iteration (see line 18). Since all considered input pictures contain finitely many black points, algorithms $\mathbf{SF}_4(m,n)$ and $\mathbf{SF}_8(m,n)$ always terminate. It can be readily seen that interior points at the beginning of an iteration step cannot be transformed into endpoints by the s subfield-oriented reductions. In addition, if $D = \emptyset$, then $E = \emptyset$ (see line 16). That is why we could write $D = \emptyset$ in line 18.

Note that the endpoint re-checking with m endpoint-oriented parallel reductions can truncate at most m points from a one point thin line segment that was created in the current iteration step. Since $N_{18}(p)$ may contain at most 17 black points for a border point p, there is no *end-deletable*(n) points in a picture if $n \geq 17$. Thus the reasonable parameter setting is $n \in \{1, 2, \ldots, 16\}$. In Fig. 5, the effect of iteration-level endpoint re-checking is illustrated for the 4-subfield algorithm $\mathbf{SF}_4(m,n)$.

4 Implementation Details and Results

The author proposed a general and computationally efficient implementation scheme for arbitrary parallel thinning algorithms [16]. It utilizes the following properties of topology-preserving thinning:

- Only border points in the current picture are examined in each thinning phase (i.e., we do not have to evaluate the deletion rules for interior points). We can use a linked list *border_list* for storing the border points in the current picture, thus the repeated scans/traverses of the entire array (that stores the picture) are avoided.
- A second list called *deletable_list* stores the 'deletable' points of the current parallel reduction. In order to ensure that both the input and the output pictures of a thinning phase can be stored in a single array, the evaluation and the deletion phases are separated: first all 'deletable' points are added to the *deletable_list*, then they are deleted and *border_list* is updated accordingly. All interior points that are adjacent (in our case: 12-adjacent) to 'deletable' point become border points. These brand new border points of the resulted picture are added to the *border_list*.
- Only some simple points in the current picture may be deleted (since all existing sufficient conditions for topology-preserving reductions require that).

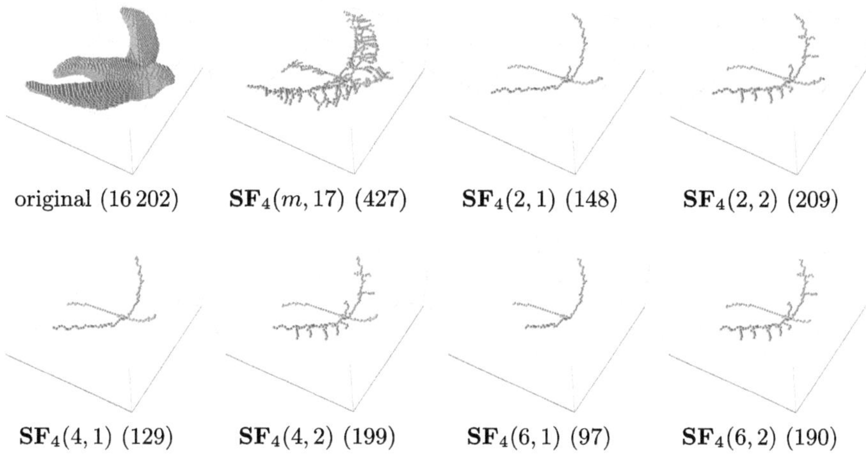

Fig. 5. Centerlines produced by algorithm $\mathbf{SF}_4(m,n)$ with variable parameter settings for a $104\times36\times94$ image of a bird. Numbers in parenthesis are the counts of points in the centerlines. No endpoint re-checking is performed in case $\mathbf{SF}_4(m,17)$, since if $n=17$ is chosen, we cannot delete an endpoint (for any $m\in\{1,2,\ldots\}$). We can state that by increasing the values of m, the centerlines will contain fewer and fewer unwanted side branches, while by increasing the value of n, the number of branches increases. The choice of $(m,n)=(4,1)$ seems reasonable, while $(m,n)=(6,1)$ 'overprunes' the centerline.

Since simpleness of a point p is a local property (it depends on $N_{18}(p)$ for $(18,12)$ pictures of the FCC grid), the corresponding Boolean function can be evaluated for all possible (in our case: 2^{18}) configurations, and the results can be stored in a (pre-calculated) look-up-table. This trick involves a remarkable speed up.

The proposed curve-thinning algorithms $\mathbf{SF}_4(m,n)$ and $\mathbf{SF}_8(m,n)$ were tested for various shapes. For reasons of scope, we could select just eight illustrative examples shown in Figs. 6-13. Based on the lessons learned from Fig. 5, the choice of parameters for all test images is: $(m,n)=(4,1)$. The triplet (o,i,t) gives the count of object/black points in the centerline, the required numbers of iteration steps, and the computation time (in msec) on a usual PC.

We can state that the produced centerlines are well-positioned, they contain all the essential branches and are free from unwanted side branches. It can be concluded that the 4-subfield and the 8-subfield algorithms resulted 'similar' centerlines (with comparable running time).

5 Topology Preservation and Maximality

Topology preservation is a key issue in skeletonization approaches, including thinning. Thus, there is a fairly general agreement that a thinning algorithm

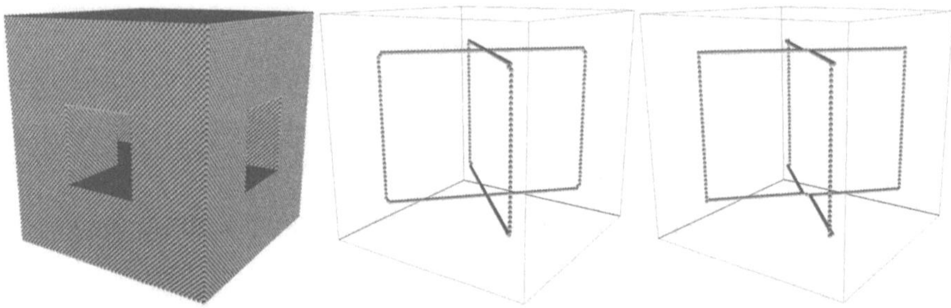

original (395 306, N/A, N/A) $\mathbf{SF}_4(4,1)$ (394, 33, 88.11) $\mathbf{SF}_8(4,1)$ (390, 33, 99.17)

Fig. 6. A 99 × 99 × 99 image and its centerlines produced by the proposed algorithms.

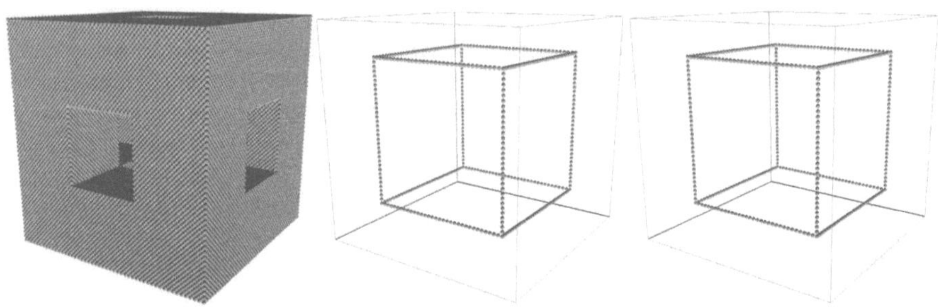

original (359 368, N/A, N/A) $\mathbf{SF}_4(4,1)$ (392, 17, 74.85) $\mathbf{SF}_8(4,1)$ (392, 17, 85.74)

Fig. 7. A 99 × 99 × 99 image and its centerlines produced by the proposed algorithms.

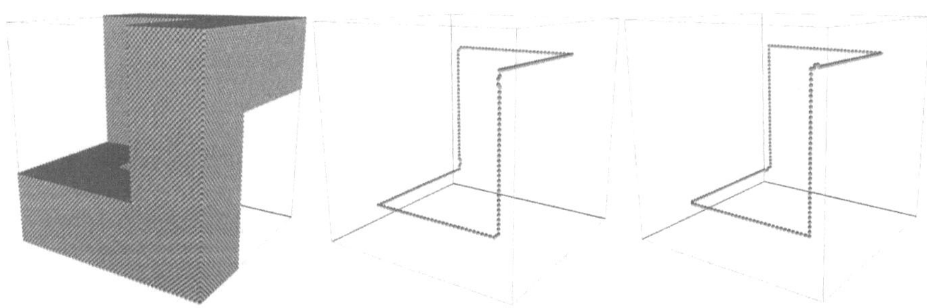

original (216 622, N/A, N/A) $\mathbf{SF}_4(4,1)$ (200, 17, 45.43) $\mathbf{SF}_8(4,1)$ (200, 17, 52.01)

Fig. 8. A 99 × 99 × 99 image and its centerlines produced by the proposed algorithms.

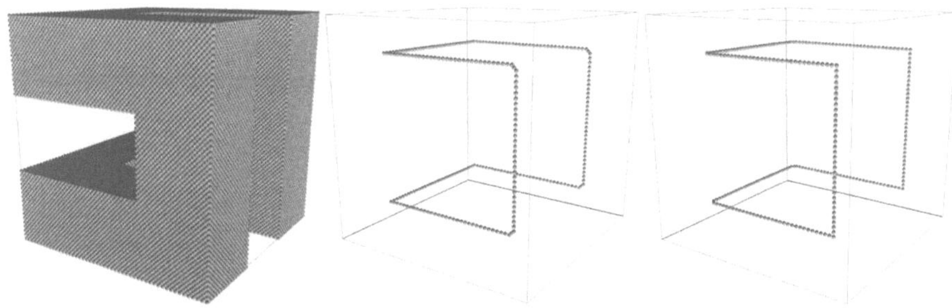

original (287 496, N/A, N/A) $\mathbf{SF}_4(4,1)$ (264, 17, 56.71) $\mathbf{SF}_8(4,1)$ (264, 17, 65.82)

Fig. 9. A $99 \times 99 \times 99$ image and its centerlines produced by the proposed algorithms.

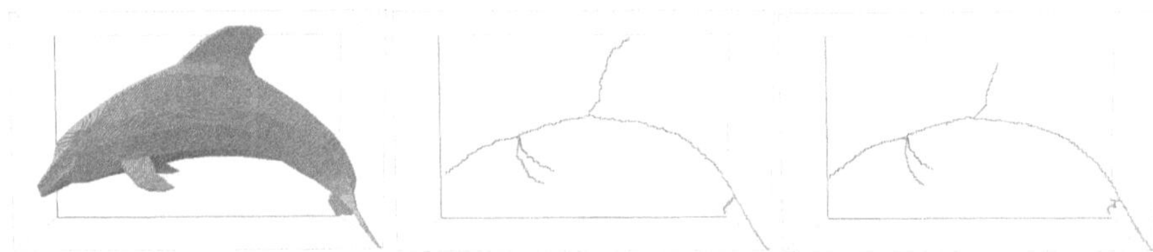

original (601 457, N/A, N/A) $\mathbf{SF}_4(4,1)$ (550, 32, 214.81) $\mathbf{SF}_8(4,1)$ (533, 32, 229.06)

Fig. 10. A $348 \times 130 \times 215$ image and its centerlines produced by the proposed algorithms.

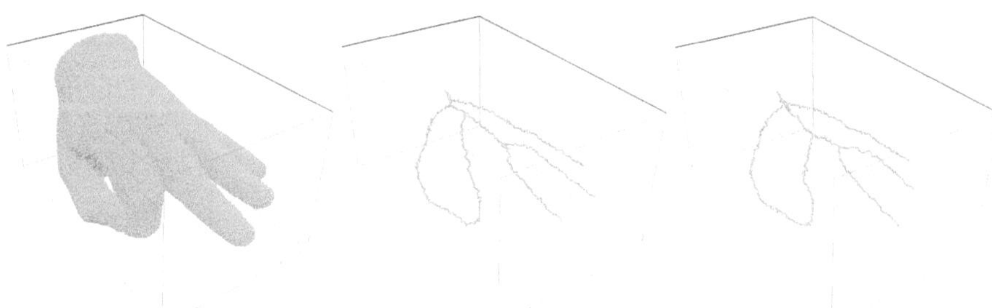

original (227 988, N/A, N/A) $\mathbf{SF}_4(4,1)$ (410, 20, 89.74) $\mathbf{SF}_8(4,1)$ (376, 26, 92.88)

Fig. 11. A $191 \times 96 \times 114$ image and its centerlines produced by the proposed algorithms.

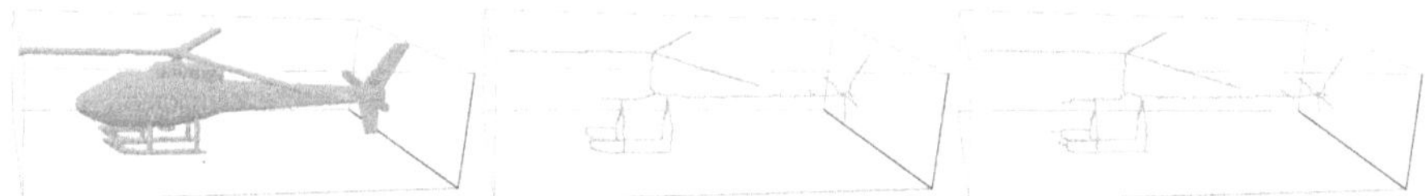

original (136 948, N/A, N/A) $\mathbf{SF}_4(4,1)$ (768, 18, 61.22) $\mathbf{SF}_8(4,1)$ (799, 23, 69.08)

Fig. 12. A $103 \times 381 \times 255$ image and its centerlines produced by the proposed algorithms.

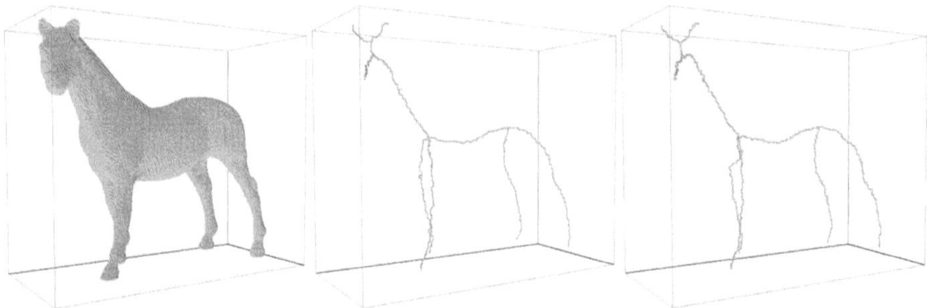

original (494 282, N/A, N/A) $\mathbf{SF}_4(4,1)$ (680, 29, 199.94) $\mathbf{SF}_8(4,1)$ (686, 29, 211.12)

Fig. 13. A $285 \times 238 \times 131$ image and its centerlines produced by the proposed algorithms.

is absolutely useless if it cannot preserve the topology for all possible pictures. Now we will show that the proposed algorithms meet this requirement.

We present first some straightforward properties concerning simple points (see Theorem 1) and the proposed partitionings (see Fig. 4):

Proposition 1. *Endpoints are simple points.*

Proposition 2. *Both points of a two-element object are (simple) endpoints.*

Proposition 3. *In both considered partitionings, two, three, and four mutually 12-adjacent points are in different subfields.*

Proposition 4. *Let us assume that an object $\mathcal{O}$ is composed of mutually 18-adjacent points in which the points are not mutually 12-adjacent. In the case of $s = 4$, if all the elements of $\mathcal{O}$ belong to the same subfield, then $\mathcal{O} = \{p, q\}$ (and $q \in N_{18}(p) \setminus N_{12}(p)$).*

Proposition 5. *In the case of $s = 8$, each element of an object composed of mutually 18-adjacent points is in a different subfield. (Note that such an object may contain at most six elements, see Fig. 3c.)*

We are now ready to state our key theorem:

Theorem 3. *Both algorithms $\mathbf{SF}_4(m,n)$ and $\mathbf{SF}_8(m,n)$ are topology-preserving.*

Proof. The following two cases need to be considered:

- The subfield-oriented parallel reduction assigned to the active subfield $SF_s(i)$ ($s \in \{4,8\}$; $i \in \{0,\ldots,s-1\}$) (see lines 11-13 in Algorithm 1) is topology-preserving.
- Endpoint-oriented parallel reductions (see lines 16 and 17 in Algorithm 1) are also topology-preserving.

We need to verify whether all the conditions of Theorem 2 are satisfied for both of the above-mentioned parallel reductions. Let us first examine the subfield-oriented case (i.e., the simultaneous deletion of *subfield-deletable* points, see Definition 1):

- Since only simple points are deleted, condition 1 of Theorem 2 holds.
- Conditions 2, 3, and 4 of Theorem 2 are absolutely irrelevant by Proposition 3. (Since two, three, and four mutually 12-adjacent black points cannot all be in the same subfield, they cannot be deleted at a time.)
- Let us now investigate the two partitionings separately:
  - If $s = 4$, by Proposition 4, the only object to be considered contains two 18-adjacent points. By Proposition 2, both points in that object are endpoints. Since endpoints are not *subfield-deletable*, condition 5 of Theorem 2 is also satisfied.
  - In the $s = 8$ case, by Proposition 5, condition 5 of Theorem 2 is irrelevant (i.e., there is no object in a subfield that are composed of mutually 18-adjacent points in which the points are not mutually 12-adjacent).

Next we examine the endpoint-oriented parallel reductions (i.e., the simultaneous deletion of *end-deletable*(n) points, see Definition 2):

- By Proposition 1, endpoints are simple points. Thus condition 1 of Theorem 2 is satisfied.
- Let us assume that p is an *end-deletable*(n) point, and let q be the only black point that is 12-adjacent to p. Then the following two cases are to be investigated:
  - If q is not an endpoint, it is not *end-deletable*(n). Thus condition 2 of Theorem 2 holds.
  - If q is an endpoint, p is not an *end-deletable*(n) point, and we arrive at a contradiction.
- It is obvious that three or four endpoints cannot be mutually 12-adjacent. Thus conditions 3 and 4 of Theorem 2 are irrelevant.
- Let us assume that an object $\mathcal{O}$ is composed of mutually 18-adjacent points in which the points are not mutually 12-adjacent. It is easy to see that if all points of $\mathcal{O}$ are endpoints, then $\mathcal{O} = \{p, q\}$ (i.e., $\mathcal{O}$ is composed of two endpoints). Since both points p and q are not *end-deletable*(n), condition 5 of Theorem 2 is also satisfied.

Note that the topology preservation of the endpoint-oriented reductions for the endpoints does not depend on the parameter 'n'.

Since each thinning phase (i.e., parallel reduction) is topology-preserving, both algorithms preserve the topology. □

Lastly, we investigate an important property of the proposed algorithms: A (2D or 3D) curve-thinning algorithm is said to be *maximal* if the centerlines it produces contain only non-simple points and (simple) endpoints. (If a curve-thinning algorithm is not maximal, it is unable to produce one point thin centerlines for all possible objects.)

An advantageous property of maximal curve-thinning algorithms is that the resulted centerlines can only contain three types of points: endpoints (which have only one black 'neighbor'), line-points (that are 'adjacent' to exactly two black points), and branch-points (that have more than two black 'neighbors') that form junctions (bifurcations, trifurcations, etc.). Hence, a complex descriptor named *skeletal graph* can be derived from such centerlines: the set of vertices of a skeletal graph is formed by the endpoints and the branch-points, and there is an edge between two vertices if they are connected via 'adjacent' line-points.

Theorem 4. *Both algorithms* $\mathbf{SF}_4(m,n)$ *and* $\mathbf{SF}_8(m,n)$ *are maximal.*

Proof. Both algorithms terminate if no points are deleted in a subfield-oriented phase of an iteration (see lines 10-13 in Algorithm 1). Then the subsequent endpoint-oriented phase of this (last) iteration (see lines 15-17 in Algorithm 1) cannot delete any more points.

Since a subfield-oriented parallel reduction deletes all simple points that are not endpoints in the active subfield (see line 11 in Algorithm 1 and Definition 1), the produced centerlines cannot contain simple points other than endpoints. Thus both proposed algorithms are maximal. □

6 Conclusions

In this work, the first two 3D parallel subfield-based curve-thinning algorithms on (18, 12) pictures of the unconventional FCC grid are reported. The proposed algorithms use four and eight subfields with iteration-level border detection and a novel endpoint re-checking. It is proved that both algorithms preserve the topology and produce one point thin centerlines for all possible objects.

Acknowledgements. This research was supported by project no. TKP2021-NVA-09 provided by the Ministry of Culture and Innovation of Hungary from the National Research, Development and Innovation Fund.

References

1. Biswas, R., Largeteau-Skapin, G., Zrour, R., Andres, E.: Rhombic dodecahedron grid – coordinate system and 3D digital object definitions. In: Proc. 21st IAPR Int. Conf. on Discrete Geometry for Computer Imagery, pp. 27–37 (2019). https://doi.org/10.1007/978-3-030-14085-4_3
2. Čomić, L., Magillo, P.: On hamiltonian cycles in the FCC grid. Comput. Graphics **89**, 88–93 (2020). https://doi.org/10.1016/j.cag.2020.05.015
3. Edelsbrunner, H., Iglesias-Ham, M., Kurlin, V.: Relaxed disk packing. In: Proc. 27th Canadian Conf. on Computational Geometry, pp. 128–135 (2015). https://doi.org/10.48550/arXiv.1505.03402

4. Gastineau, N., Togni, O.: Coloring of the dth power of the face-centered cubic grid. Discussiones Mathematicae - Graph Theory **41**, 1001–1020 (2021). https://doi.org/10.7151/dmgt.2257
5. Gau, C.J., Kong, T.Y.: Minimal nonsimple sets of voxels in binary images on a face-centered cubic grid. Int. J. Patt. Recogn. Artif. Intell. **13**, 485–502 (1999). https://doi.org/10.1142/S021800149900029X
6. Hall, R.W.: Parallel connectivity-preserving thinning algorithms. In: Kong, T.Y., Rosenfeld, A. (eds.) topological algorithms for digital image processing, pp. 145–179. Elsevier Science, Amsterdam (1996). https://doi.org/10.1016/S0923-0459(96)80014-0
7. Herman, G.T.: Geometry of digital spaces. Birkhäuser, Boston (1998). https://doi.org/10.1007/978-1-4612-4136-2
8. Karai, G., Kardos, P., Palágyi, K.: Subfield-based parallel kernel-thinning algorithms on the BCC grid. In: Proc. 11th Int. Conf. Pattern Recognition Applications and Methods, pp. 288–295 (2022). https://doi.org/10.5220/0010831000003122
9. Karai, G., Kardos, P., Palágyi, K.: Sufficient conditions for topology-preserving parallel reductions on the face-centered cubic grid. J. Math. Imaging Vision **66**, 271–292 (2024). https://doi.org/10.1007/s10851-024-01177-y
10. Kardos, P.: Topology preservation on the BCC grid. J. Comb. Optim., 1–15 (2021). https://doi.org/10.1007/s10878-021-00828-9
11. Kong, T.Y.: On topology preservation in 2-d and 3-d thinning. Int. J. Patt. Recogn. Artif. Intell. **9**, 813–844 (1995). https://doi.org/10.1142/S0218001495000341
12. Kong, T.Y., Rosenfeld, A.: Digital topology: introduction and survey. Comput. Vision Graph. Image Process. **48**, 357–393 (1989). https://doi.org/10.1016/0734-189X(89)90147-3
13. Koshti, G., Biswas, R., Largeteau-Skapin, G., Zrour, R., Andres, E., Bhowmick, P.: Sphere construction on the FCC grid interpreted as layered hexagonal grids in 3D. In: Proc. 19th Int. Workshop on Combinatorial Image Analysis, pp. 82–96 (2018). https://doi.org/10.1007/978-3-030-05288-1_7
14. Ma, C.M., Wan, S.Y., Chang, H.K.: Extracting medial curves on 3D images. Pattern Recogn. Lett. **23**, 895–904 (2002). https://doi.org/10.1016/S0167-8655(01)00162-3
15. Magillo, P.: Non-square grids: a new trend in imaging and modeling? Comput. Sci. Rev. **56**, 100695 (2025). https://doi.org/10.1016/j.cosrev.2024.100695
16. Palágyi, K.: A 3D fully parallel surface-thinning algorithm. Theoret. Comput. Sci. **406**, 119–135 (2008). https://doi.org/10.1016/j.tcs.2008.06.041
17. Palágyi, K., Németh, G., Kardos, P.: Topology preserving parallel 3D thinning algorithms. In: Brimkov, V.E., Barneva, R.P. (eds.) Digital Geometry Algorithms, pp. 165–188. Springer (2012). https://doi.org/10.1007/978-94-007-4174-4_6
18. Rácz, G.F., Csébfalvi, B.: Cosine-weighted B-spline interpolation on the face-centered cubic lattice. Comput. Graph. Forum **37**, 503–511 (2018). https://doi.org/10.1111/cgf.13437

19. Saha, P.K., Borgefors, G., Sanniti di Baja, G.: Skeletonization: theory, methods and applications. Academic Press, London (2017). https://doi.org/10.1016/B978-0-08-101291-8.00017-1
20. Strand, R., Stelldinger, P.: Topology preserving marching cubes-like algorithms on the face-centered cubic grid. In: 14th Int. Conf. on Image Analysis & Processing, pp. 781–786 (2007). https://doi.org/10.1109/ICIAP.2007.4362871

Characterization of the Computed Homology and Cohomology Bases

Yann-Situ Gazull(✉), Aldo Gonzalez-Lorenzo, and Alexandra Bac

Aix Marseille Univ, CNRS, LIS, Marseille, France
{yann-situ.gazull,aldo.gonzalez-lorenzo,alexandra.bac}@univ-amu.fr

Abstract. Computing homology and cohomology is at the heart of many recent works and a key issue for topological data analysis. Among homological objects, homology generators are useful to locate or understand holes (especially for geometric objects). The present paper provides a characterization of the class of homology bases that are computed by standard algorithmic methods. The proof of this characterization relies on the Homological Discrete Vector Field, a combinatorial structure for computing homology, which encompasses several standard methods (persistent homology, tri-partitions, Smith Normal Form, discrete Morse theory). These results refine the combinatorial homology theory and provide novel ideas to gain more control over the computation of homology generators.

Keywords: Computational Homology · Combinatorial Topology · Homological Discrete Vector Field · Persistent Homology

1 Introduction

After having been a branch of mathematics from early 20th century, homology and cohomology have become a key issue in computer science since the 80's to study holes of data independently of geometry. Later in the 90's, persistent homology even became a central tool for topological data analysis. Indeed, it formalizes "hole changes" along a time-varying sequence of objects and computes a notion of "hole importance" or "duration", stable against noise.

(Co)homology computation can be carried out in several ways. Algebraic computation, as a pseudo-diagonalization (Smith Normal Form) is widely used, for instance, for persistent homology. On the other hand, effective homology provides a "categorical" approach (see [17]), while Discrete Morse Theory proposes a combinatorial approach (see [7]). In 2017, Aldo-Gonzalez Lorenzo et al. [12] introduced Homological Discrete Vector Field (which we refer to as *HDVF* in the present paper), which provides a framework in between combinatorial approaches (such as discrete Morse theory) and algebraic approaches (such as Smith normal form), encompassing these state-of-art approaches but distinct from them.

It has long been known that not all homology basis can be computed by standard algorithmic approaches (Smith Normal Form, persistent homology,

M. H. F. Wilkinson and J. Kosinka (Eds.): DGMM 2025, LNCS 16296, pp. 163–177, 2026.
https://doi.org/10.1007/978-3-032-09544-2_12

tri-partitions, Discrete Morse Theory). For instance, 1-dimensional computed generators are elementary cycles. In the present paper, we introduce a precise characterization of homology bases computed by HDVFs, which we call *explicit bases*. We then show (or recall) the equivalence between specific HDVFs and other algorithmic methods and thus deduce that all of them actually compute such explicit bases.

The paper is organised as follows: Sect. 2 introduces homology and HDVFs. Section 3 contains our main contribution; we define and discuss the notion of explicit homology bases, and prove that being explicit is equivalent to being induced by a perfect HDVF. In Sect. 4, we explain how this characterization applies to homology bases computed by other standard methods. In particular, we prove that tri-partitions (defined in [5]) can be seen as layers of perfect HDVFs (and vice versa) and show that a homology basis is explicit if and only if it is induced by a tri-partition. We also explain how this characterization applies to persistent homology bases.

Technical and lengthy proofs are not provided in this paper, but can be found in a companion technical report [8].

2 Preliminaries

2.1 Homology, Complexes and Notations

The present work considers homology over a field $\mathcal{F}$. In the rest of the paper, we consider that $\mathcal{F} = \mathbb{Z}/2\mathbb{Z}$ but results hold for any field. In this section, we quickly introduce the main concepts of homology theory. For more information on homology theory, see [13].

A cellular complex $\mathscr{C}$ is a topological structure that consists of cells of different dimensions ($\mathscr{C}_q$ the set of cells of dimension q) attached together by "gluing" maps (for a more precise definition, see [13]). Depending on the type of cells, cellular complexes are categorized into a variety of combinatorial structures: simplicial complexes and delta-complexes (triangles, tetrahedra, etc.), cubical complexes (squares, cubes, etc.), or general cellular complexes. In this article, we mainly use cubical complexes, but the results apply to any cellular complex.

The definition of homology for these structures is based on the algebraic version of the concept of boundary. The boundary morphism ∂_q must associate to a q-cell a linear combination of $(q-1)$-cells corresponding to its geometric faces, while respecting the constraint that $\partial_q \circ \partial_{q+1} = 0$. When such a definition is possible, we obtain a chain complex, an abstract algebraic structure on which we can compute homology:

Definition 1 (chain complex [13]). *A* chain complex (K, ∂) *is a sequence of* $\mathcal{F}$*-vector fields* K_q *with linear maps* $\partial_q : K_q \to K_{q-1}$ *satisfying* $\partial_q \circ \partial_{q+1} = 0$*:*

$$K_n \xrightarrow{\partial_n} \dots \xrightarrow{\partial_{q+1}} K_q \xrightarrow{\partial_q} K_{q-1} \xrightarrow{\partial_{q-1}} \dots \xrightarrow{\partial_1} K_0 \xrightarrow{\partial_0} 0$$

K_q is the space of q-chains *and ∂_q is the* q-boundary map. *When (K, ∂) is derived from a cellular complex $\mathscr{C}$, the q-chains are defined as the linear combinations of q-cells: $K_q = \operatorname{Span} \mathscr{C}_q$. The boundary map ∂ is defined as the algebraization of the geometric boundary (see [13]).*

The notion of HDVF is deeply related to the relation between cells and chains. That is why this paper only considers chain complexes that are derived from cellular complexes, which we simply refer to as *complexes.*

Given a complex (K, ∂), we denote $K := \bigoplus_{q=0}^{n} K_q$ the whole chain space and $\partial := \bigoplus_{q=0}^{n} \partial_q$ the boundary operator. We also denote $\overline{K_q} := \mathscr{C}_q$ the set of q-cells and $\overline{K} := \bigcup_{q=0}^{n} \overline{K_q}$ the set of every cell in the associated cellular complex. We respectively call *q-cycles* and *q-boundaries* the chains in $\ker \partial_q$ and in $\operatorname{im} \partial_{q+1}$.

Definition 2 (homology group [13]**).** *Given a complex (K, ∂), its* q-homology group *is defined as the quotient space of q-cycles over q-boundaries:*

$$H_q(K) = \ker \partial_q / \operatorname{im} \partial_{q+1}$$

As $\mathcal{F}$ is a field, there exists an integer β_q (called the q-th Betti number) such that $H_q(K)$ is isomorphic to $\mathcal{F}^{\beta_q}$. Intuitively, β_q counts the number of holes of dimension q in the associated cellular complex.

We denote $[\cdot]_q^K : \ker \partial \to H_q(K)$ the map that associates a q-cycle to its corresponding q-homology class: $[x]_q^K = [y]_q^K \iff x - y \in \operatorname{im} \partial_{q+1}$.

Definition 3 (homology bases and generators). *A family $(\mathtt{g}_i)_{i \leq \beta}$ of q-cycles is a* q-homology basis *of (K, ∂) if and only if $\left([\mathtt{g}_i]_q^K\right)_{i \leq \beta}$ is a basis of $H_q(K)$. The cycles $\mathtt{g}_i$ are called* q-homology generators.

Notations and Conventions: given two complexes (K, ∂) and (K', ∂'), we will write $(K, \partial) \subseteq (K', \partial')$ if (K, ∂) is a sub-complex of (K', ∂') i.e. if $K \subseteq K'$ and ∂' agrees with ∂ on K. For simplicity, we will often identify ∂ and ∂'. If A is a set of cells, we will write $\mathtt{i}_A$ the inclusion map from the subspace $\operatorname{Span} A$ into the full chain space K, $\mathtt{j}_A$ the corresponding projection map from K to $\operatorname{Span} A$ and id_A the identity map on $\operatorname{Span} A$. Given two sets of cells A and B, we also denote $\partial_{AB} := \mathtt{j}_B \circ \partial \circ \mathtt{i}_A$ the boundary induced by $\operatorname{Span} A$ restricted to $\operatorname{Span} B$.

2.2 Reductions and HDVFs

Effective homology theory was introduced in the 90's in [17]. It defines the notion of *reduction* between two chain complexes (usually from a large to a small complex), which ensures that their homology groups are isomorphic. Reductions play an important role in understanding the homology of complexes.

Definition 4 (reduction [17]**).** *A* reduction *from a complex (K, ∂) to a complex (K', d) is a triplet of maps (f, g, h) with $f_q : K_q \to K'_q$, $g_q : K'_q \to K_q$ and $h_q : K_q \to K_{q+1}$ satisfying the following properties:*

1. $f\partial = df$ 2. $\partial g = gd$ 3. $fh, hg, hh = 0$ 4. $fg = id_{K'}$ 5. $gf = id_K - \partial h - h\partial$

$$\begin{array}{ccccccc}
\cdots & \underset{h_q}{\overset{\partial_{q+1}}{\rightleftarrows}} & K_q & \underset{h_{q-1}}{\overset{\partial_q}{\rightleftarrows}} & K_{q-1} & \underset{h_{q-2}}{\overset{\partial_{q-1}}{\rightleftarrows}} & \cdots \\
 & & g_q \big\uparrow\big\downarrow f_q & & g_{q-1} \big\uparrow\big\downarrow f_{q-1} & & \\
\cdots & \xrightarrow{d_{q+1}} & K'_q & \xrightarrow{d_q} & K'_{q-1} & \xrightarrow{d_{q-1}} & \cdots
\end{array}$$

Then we have $\forall q,\ H_q(K) \simeq H_q(K')$.

Definition 5 (perfect reduction [17]**).** *A reduction from* (K, ∂) *to* (K', d) *is perfect iff* $d = 0$*. In this case, the q-homology group of both complexes is isomorphic to* K'_q*. In addition,* $(g_q(\gamma))_{\gamma \in \overline{K'_q}}$ *is a q-homology basis.*

Homological discrete vector fields emerged as a generalization of Discrete Morse Theory [7]. Discrete Morse theory uses acyclic graphs (called *discrete gradient vector fields*) defined on the cells of the complex and provides bounds on Betti numbers. However, it was shown that it fails to compute homology, for instance, for well-known contractible but not collapsible complexes such as the Bing's house. HDVFs were first developed to bypass this limitation.

Definition 6 (homological discrete vector field [10] **(HDVF)).** *Given a complex* (K, ∂)*, a* homological discrete vector field *is a triplet* $X = (P, S, C)$ *with* P*,* S*, and* C *forming a partition of* $\overline{K}$ *such that* ∂_{SP} *is invertible. The cells of* P*,* S *and* C *are respectively called* primary, secondary *and* critical *cells.*
It is proved in [10] that the invertibility of ∂_{SP} *implies that the following* (f, g, h) *is a well defined reduction from* (K, ∂) *to* $(\mathrm{Span}\, C, d)$*:*

$$g := \begin{array}{c} C \\ \begin{pmatrix} 0 \\ G \\ id \end{pmatrix} \end{array}\!\!\begin{array}{l} \\ P \\ S \\ C \end{array} \qquad f := \begin{array}{c} P\ S\ C \\ (F\ 0\ id) \end{array}\, C \qquad h := \begin{array}{c} P\ S\ C \\ \begin{pmatrix} 0 & 0 & 0 \\ H & 0 & 0 \\ 0 & 0 & 0 \end{pmatrix} \end{array}\!\!\begin{array}{l} \\ P \\ S \\ C \end{array} \qquad d := \begin{array}{c} C \\ (D) \end{array}\, C$$

with $H = (\partial_{SP})^{-1}$*,* $F = -\partial_{SC} \cdot H$*,* $G = -H \cdot \partial_{CP}$ *and* $D = \partial_{CC} - \partial_{SC} \cdot H \cdot \partial_{CP}$*. Given an HDVF this reduction is called the associated reduction, an example is detailed in [8].*

In figures, primary, secondary and critical cells are respectively colourized in green, red and blue (see Figs. 1 and 3 for some examples).

Let us point out that HDVFs are singular objects, in between algebra and combinatorics, interleaving the notions of cells and chains. Indeed, HDVFs are not mere reductions as they rely on a cell partition and not on a space decomposition; but they are not purely combinatorial objects either, as HDVF validity condition is algebraic (matrix invertibility).

Definition 7 (perfect HDVF [10]**).** *A HDVF is perfect if its induced reduction is perfect, i.e. if the reduced boundary* $d = 0$*. In this case,* $\mathrm{Span}\, C_q$ *is isomorphic to* $H_q(K)$ *and* g *maps every critical cell to its associated homology generator:*

$(g_q(\gamma))_{\gamma \in C_q}$ is the q-homology basis associated to (P, S, C). Similarly, f^* (the transpose of f) maps every critical cell to its associated cohomology generator: $(f_q^*(\gamma))_{\gamma \in C_q}$ is the q-cohomology basis associated to (P, S, C). See [8] for more details on duality and cohomology, and see Fig. 2 for an illustration.

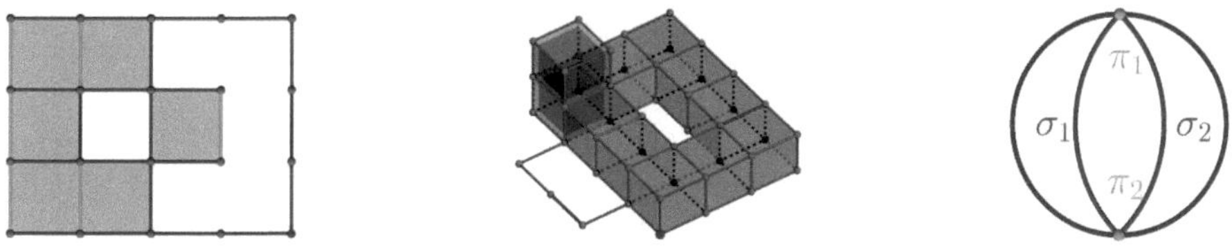

Fig. 1. Examples of three complexes with partitions of the cells between primary (in green), secondary (in red) and critical (in blue) cells. The first two partitions are HDVFs whereas the last one is not a HDVF: ∂_{SP}, the boundary of the secondary cells restricted to the primary cells, is not invertible because $\partial_{SP}(\sigma_1 + \sigma_2) = 0$. (Color figure online)

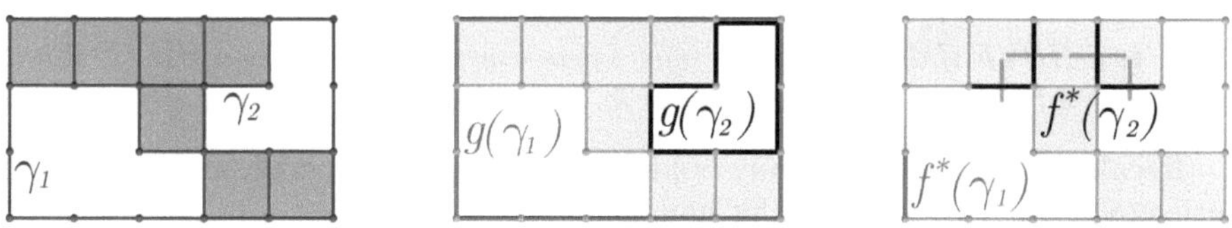

Fig. 2. A perfect HDVF (P, S, C) with $C_1 = \{\gamma_1, \gamma_2\}$ (left), its associated 1-homology basis $(g(\gamma_1), g(\gamma_2))$ (middle) and 1-cohomology basis $(f^*(\gamma_1), f^*(\gamma_2))$ (right).

Inspired by [5], we define two maps which we call the *canonical cycle* and *co-cycle* maps. Given a perfect HDVF (P, S, C) with reduction (f, g, h), they are respectively defined as follows:

$$z_q = id - h_{q-1}\partial_q \qquad z^q = id - h_q^* \partial_{q+1}^*$$

In particular, if $\gamma \in C_q$, the maps z_q and z^q coincide with the homology and cohomology generators: $z_q(\gamma) = g_q(\gamma)$ and $z^q(\gamma) = f_q^*(\gamma)$. This is a consequence of Definition 4(5.) combined with the action of f, g and h on critical cells (see Definition 6). Now let us state our first contribution; the following lemma will play a central role in what follows:

Lemma 1. *In a perfect HDVF (P, S, C):*

$$\forall x \in K_q \qquad (x + \operatorname{Span} S) \cap \ker \partial = \{z_q(x)\} \qquad (x + \operatorname{Span} P) \cap \ker \partial^* = \{z^q(x)\}$$

In particular, if $c \in \operatorname{Span} C_q$ we have $(c + \operatorname{Span} S) \cap \ker \partial = \{g_q(c)\}$.
In addition, we get $z_q \circ \mathtt{i}_{S_q} = 0$ and $z^q \circ \mathtt{i}_{P_q} = 0$.

Proof. The proof can be found in the technical report [8]. □

A core result of [10] is the following: in a non perfect HDVF, it is always possible to pair some critical cells (i.e. transform two critical cells to one primary and one secondary cell) in order to obtain a perfect HDVF. In particular, this proves the existence of a perfect HDVF for every cellular complex. This process is called *the completion* of a HDVF:

Proposition 1 (completion of a HDVF [10]**).** *Given two complexes* $(K, \partial) \subseteq (K', \partial')$ *and a HDVF* $X = (P, S, C)$ *for* (K, ∂)*, there exists a perfect HDVF* $X' = (P', S', C')$ *for* (K', ∂') *such that* $P \subseteq P'$ *and* $S \subseteq S'$*.*

Three operations were defined in [10] to locally modify a HDVF (and hence homology an cohomology generators) by swapping the status of two set of cells. We present two of them, that are used in our proofs:

Definition 8 (HDVF operations [10]**).** *Given a homological discrete vector field* $X = (P, S, C)$ *with induced reduction* (f, g, h) *and* $\Pi \subseteq P$*,* $\Sigma \subseteq S$ *and* $\Gamma \subseteq C$*. The* W *operation swaps secondary cells with critical cells whereas the* M *operation swaps primary cells with critical cells:*

$$\mathtt{W}_{\Sigma}^{\Gamma} X = (P,\ (S \cup \Gamma) \backslash \Sigma,\ (C \cup \Sigma) \backslash \Gamma) \qquad \mathtt{M}_{\Pi}^{\Gamma} X = ((P \cup \Gamma) \backslash \Pi,\ S,\ (C \cup \Pi) \backslash \Gamma)$$

$\mathtt{W}_{\Sigma}^{\Gamma} X$ *is a HDVF if and only if* $\mathtt{j}_{\Sigma} \circ g \circ \mathtt{i}_{\Gamma}$ *is invertible.* $\mathtt{M}_{\Pi}^{\Gamma} X$ *is a HDVF if and only if* $\mathtt{j}_{\Gamma} \circ f \circ \mathtt{i}_{\Pi}$ *is invertible*[1]*. See Fig. 3 for illustrations.*

Recall that these results only hold for homology over a field, and some limitations appear when we consider homology over a ring.

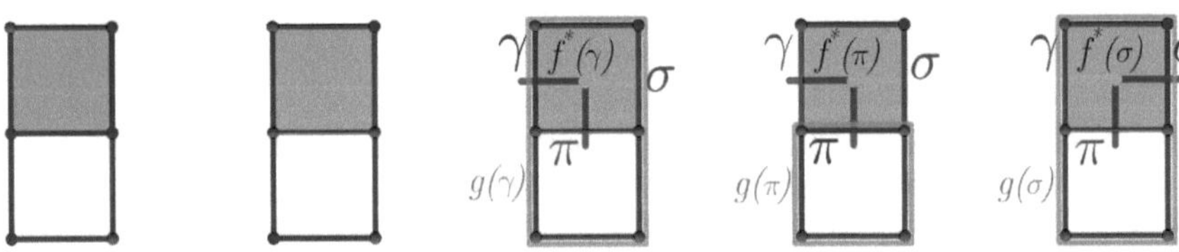

Fig. 3. Five HDVFs on the same complex: $X_1, \ldots, X_5$. X_1 and X_2 are not perfect as the number of critical q-cells is larger than the corresponding q-th Betti number. X_2 is a completion of X_1 (three pairs of secondary-primary cells were added) and X_3 is a completion of X_2. X_3, X_4 and X_5 are perfect. Using Definition 8 we can write $X_4 = \mathtt{M}_{\{\pi\}}^{\{\gamma\}} X_3$ and $X_5 = \mathtt{W}_{\{\sigma\}}^{\{\gamma\}} X_3$. Note that W and M respectively preserves the homology generators (in orange) and cohomology generators (in purple) of the cells involved in the operation.

3 HDVF and Homology Bases

Given a complex, a number of homology bases can be obtained using a perfect HDVF, while others cannot (see Fig. 4 for examples). This section is our main

[1] By a cardinality argument on critical cells, valid W and M operations transform a perfect HDVF into another perfect HDVF.

contribution. We first define and characterize bases induced by HDVFs (we call them "explicit" bases). In Theorem 1 we show that the homology basis induced by a HDVF is explicit. Then, in Theorem 2, we derive the construction of a HDVF from an explicit homology basis. As a result, we prove that a homology basis is explicit if and only if it is induced by a HDVF.

Actually, this result goes beyond HDVFs and we show in Sect. 4 that it extends to other standard approaches (such as persistent homology, Smith Normal Form or tri-partitions).

3.1 Explicit Homology Bases

In this section we introduce *explicit homology bases* and their properties. In what follows, (K, ∂) is a chain complex derived from a cellular complex $\mathscr{C}$. The following results tightly intricate combinatorics (cells) and algebra (chains), which is consistent with the nature of HDVFs. To avoid confusion between chains and cells, we use lowercase Greek letters for cells and lowercase Latin letters for chains. Cells are also considered as chains, therefore we identify a cell τ with the chain consisting of one τ. Given a chain $x \in K$ we denote $\bar{x}$ the set of cells that appear in it: $\bar{x} := \left\{\tau \in \overline{K} \;/\; \langle \tau, x\rangle \neq 0\right\}$ (with $\langle \cdot,\cdot\rangle$ the standard inner product on K). As a consequence, if A is a set of cells we have $a \in \operatorname{Span} A \iff \bar{a} \subseteq A$.

Proposition 2 gives various equivalent definitions of explicit bases and Proposition 3 shows that the generators of an explicit basis are elementary. First, let us introduce the notion of sub-complex induced by a set of cells:

Definition 9 (induced complex). *Given a set of cells A in a complex (K, ∂), we denote $(K(A), \partial)$ the minimum sub-chain complex containing A. Precisely, $K(A)_q$ is generated by the faces of dimension q of cells in A.*

Given a q-homology basis $(\mathbf{g}_i)_{i\leq\beta}$ for a complex (K, ∂) and $J \subseteq \{1, \dots, \beta\}$ a set of indices, we denote $K^J := K\left(\bigcup_{i\in J} \overline{\mathbf{g}_i}\right)$ the space of chains generated by the cells in the generators[2] indexed by J. We denote ι_q^J the homology map induced by the inclusion $(K^J, \partial) \subset (K, \partial)$. With this notation, we also denote $(K^\beta, \partial) := (K^{\{1,\dots,\beta\}}, \partial)$ the sub-complex induced by the whole basis (see Fig. 4) and ι_q^β its associated homology map.

Definition 10 (Explicit homology bases). *A q-homology basis $(\mathbf{g}_i)_{i\leq\beta}$ for a complex (K, ∂) is said to be* explicit *if it satisfies the following property:*

$$\forall k,\ \overline{\mathbf{g}_k}\setminus\bigcup_{i\neq k}\overline{\mathbf{g}_i} \neq \emptyset \qquad \textit{and} \qquad \mathit{dim}\left(H_q\left(K^\beta\right)\right) = \beta$$

Intuitively, a homology basis is explicit if and only if no generator is included in the others and the sub-complex induced by the generators have no more than β holes of dimension q (see Fig. 4 for examples of non explicit homology bases).

---

[2] Note that K_q^J is the span of a combinatorial object $(\cup_{i\in J}\overline{\mathbf{g}_i})$; it is not necessarily equal to Span $((\mathbf{g}_i)_{i\in J})$ (the "natural" algebraic object defined from generators).

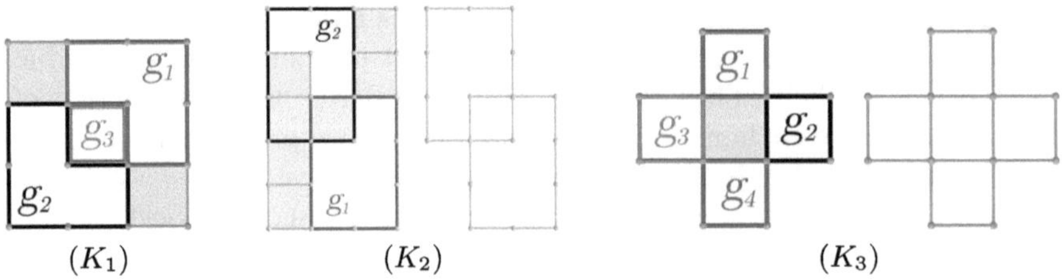

Fig. 4. Examples of 1-homology bases which are not explicit (right images: subcomplexes induced by the basis). By Theorem 1 they cannot be obtained from a HDVF. In (K_1), $\overline{g_3}\backslash(\overline{g_1} \cup \overline{g_2}) = \emptyset$ so the basis is not explicit. In (K_2), the complex (K^β, ∂) induced by the basis has three 1-holes instead of two: $3 = dim\left(H_1\left(K^\beta\right)\right) \neq \beta = 2$. Similarly, in (K_3), the complex induced by the basis has five 1-holes instead of four.

Proposition 2 (Explicit bases equivalent definitions). *Given a q-homology basis* $(g_i)_{i \leq \beta}$ *for a complex* (K, ∂)*, the three following properties are equivalent:*

1. $\forall k,\ \overline{g_k}\backslash\bigcup_{i\neq k}\overline{g_i} \neq \emptyset$ *and* $\forall J,\ \iota_q^J : H_q\left(K^J\right) \to H_q(K)$ *injective of rank* $|J|$
2. $\forall J \subseteq \{1, \ldots, \beta\},\quad \ker\partial \cap K_q^J \subseteq \mathrm{Span}\left((g_i)_{i\in J}\right)$
3. $\forall k,\ \overline{g_k}\backslash\bigcup_{i\neq k}\overline{g_i} \neq \emptyset$ *and* $dim\left(H_q\left(K^\beta\right)\right) = \beta$

Proof. The proof is lengthy and can be found in the technical report [8]. □

Interestingly, all these definitions combine combinatorics (cells) and algebra (chains), but cannot be fully captured by either. Even 2. of Proposition 2 entails an intersection with K_q^J, which definition relies on cells. This dual nature of bases computed by standard approaches is an important result of the present article. Moreover, we have the following property:

Proposition 3. *The generators of an explicit homology basis are* elementary. *A cycle is said to be* elementary *if it is not composed of smaller (non-zero) cycles.*

Proof. Let $(g_i)_{i\leq\beta}$ be an explicit q-homology basis of (K, ∂) and a, b be cycles such that $\bar{a} \sqcup \bar{b} = \overline{g_k}$ (and so $a + b = g_k$). Denote $J := \{k\}$. We get that a and b are in $\ker\partial \cap K_q^J$ so by 2(2.) they are in $\mathrm{Span}\left(g_k\right) = \{0, g_k\}$. Hence, either a or b is 0, otherwise we would have $g_k = g_k + g_k = 0$. □

3.2 Characterization of HDVF Homology Bases

We now prove that the homology bases obtained from HDVFs are explicit.

Theorem 1. *Let* $X = (P, S, C)$ *be a perfect HDVF for* (K, ∂) *with reduction* (f, g, h)*. Then, for all* q*, its associated* q*-homology basis* $\left(g_q(\gamma)\right)_{\gamma\in C_q}$ *is explicit.*

Proof. This proof uses the characterization 2. of Proposition 2 and can be found in the technical report [8]. □

As stated by the following theorem, the converse also holds:

Theorem 2. *Given* $(\mathrm{g}_i)_{i\leq\beta_q}$ *an explicit q-homology basis for* (K,∂), *there exists a perfect HDVF* $X = (P,S,C)$ *for* (K,∂) *with reduction* (f,g,h) *such that* $\left(g_q(\gamma)\right)_{\gamma\in C_q} = (\mathrm{g}_i)_{i\leq\beta_q}$.

We prove Theorem 2 by building a perfect HDVF from an explicit basis. To do so, we need the following lemmas about existence and completion of HDVFs. Firstly, given $(K,\partial)\subseteq(K',\partial)$, it is possible to complete a perfect HDVF for (K,∂) to a perfect HDVF for (K',∂) while preserving some critical cells (specifically the ones associated with holes of (K,∂) that were not filled in (K',∂)):

Lemma 2 (critical preservation during completion). *Let* $(K,\partial)\subseteq(K',\partial)$ *and let* $X=(P,S,C)$ *be a perfect HDVF for* (K,∂) *with reduction* (f,g,h).

Suppose that there is $C^*\subseteq C_q$ *such that* $([g_q(\gamma)]^{K'})_{\gamma\in C^*}$ *is linearly independent in* $H_q(K')$. *Then there exists* $X'=(P',S',C')$ *a perfect HDVF for* (K',∂) *such that* $P\subseteq P'$, $S\subseteq S'$ *and* $C^*\subseteq C'$.

Proof. The proof is involved and can be found in the technical report [8]. □

Secondly, the critical cells preserved in a completion process keep their associated homology generators:

Lemma 3 (generator preservation during completion). *Suppose* $(K,\partial)\subseteq(K',\partial)$ *and let* $X=(P,S,C)$ *and* $X'=(P',S',C')$ *be perfect HDVFs for* (K,∂) *and* (K',∂) *such that and* $S\subseteq S'$. *Then* $\forall x\in K_q$, $z_q(x)=z'_q(x)$.
In particular, $\forall\gamma\in C\cap C'$, $g(\gamma)=g'(\gamma)$.

Proof. Let $x\in K\subseteq K'$. X and X' are perfect so Lemma 1 gives:

$$z_q(x)\in\ker\partial\cap(x+\operatorname{Span}S)\subseteq\ker\partial\cap(x+\operatorname{Span}S')=\{z'_q(x)\}$$ □

Finally, given an explicit homology basis, it is possible to build a perfect HDVF X^β for the sub-complex $\left(K^\beta,\partial\right)$ induced by the basis, such that X^β is still associated to the same basis:

Lemma 4 (HDVF of the basis complex). *Let* $(\mathrm{g}_i)_{i\leq\beta}$ *be an explicit q-homology basis for* (K,∂). *Then there exists* $X^\beta=(P^\beta,S^\beta,C^\beta)$ *a perfect HDVF for* $\left(K^\beta,\partial\right)$ *with reduction* $(f^\beta,g^\beta,h^\beta)$ *such that* $\left(g^\beta_q(\gamma)\right)_{\gamma\in C^\beta_q}=(\mathrm{g}_i)_{i\leq\beta}$.

Proof. The construction of this HDVF is done by induction on $k\in\{0,\dots\beta\}$ and can be found in the technical report [8]. □

The full proof of Theorem 2 is technical and can be found the technical report [8]. However, once previous lemmas are known, it becomes intuitive and Fig. 5 illustrates it.

As a result, Theorems 1 and 2 imply that a homology basis is explicit if and only if it is induced by a perfect HDVF. Note that the proof of Theorem 2 provides an algorithm to build a perfect HDVF given an explicit homology basis.

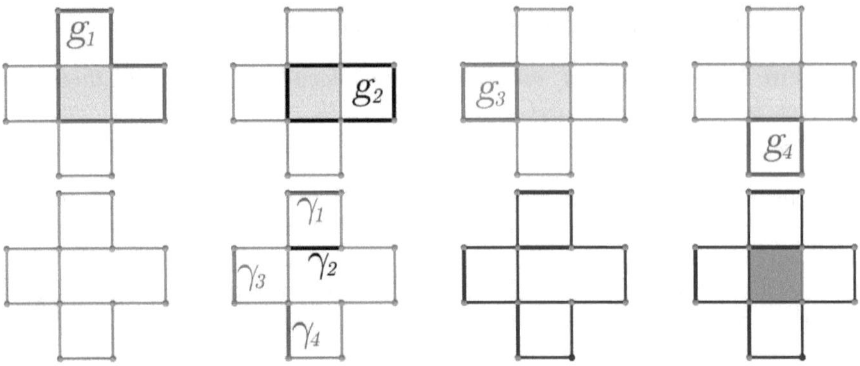

(a) (b) (c) (d)

Fig. 5. Illustration of the constructions involved in the proof of Theorem 2. The first row represents four generators forming an explicit 1-homology basis of a complex (K, ∂). The second row shows steps of the proof. It starts (a) with $(K(\bigcup \overline{\mathbf{g}_i}), \partial)$ the complex induced by this basis (which has four 1-holes as the basis is explicit). (b) represents a choice of cells γ_k that are respectively in $\overline{\mathbf{g}_k} \setminus \bigcup_{i \neq k} \overline{\mathbf{g}_i}$ (which is non-empty as the basis is explicit). (c) is the perfect HDVF X^β obtained with Lemma 4. By construction, we get $g^\beta(\gamma_i) = \mathbf{g}_i$. Finally, (d) is X the perfect HDVF for (K, ∂) obtained by completing X^β using Lemma 2. The 1-homology basis induced by X is $(\mathbf{g}_1, \mathbf{g}_2, \mathbf{g}_3, \mathbf{g}_4)$ by construction using Lemma 3.

4 Characterization of Homology Bases Obtained by Other Methods

In this section we discuss how the characterization of Sect. 3 transfers to bases computed by other methods.

First, it is important to distinguish computational homology approaches (such as HDVFs, but also Smith Normal Form, persistent homology and tri-partitions) and heuristics for computing small homology generators (the question of computing minimal ones is NP-difficult, see for instance [1]). Methods related to this latter question usually combine former approaches together with solvers and heuristics to minimize generators, hence homology bases may not be explicit (see for instance Fig. 4, K_3). Now, the remainder of this section examines the links between HDVFs and other approaches and deduces properties on the bases they compute.

Discrete Morse Theory. It has been proven in [12] that HDVFs generalize Discrete Morse Theory, in particular, a *discrete gradient vector field* is a HDVF. Hence, Theorem 1 applies, and Discrete Morse Theory computes only explicit bases.

Smith Normal Form (SNF). Homology is commonly computed by reducing the boundary matrix of a complex to the Smith-Normal-Form (SNF). It has been shown in [12] that a HDVF can emulate this SNF computation. Precisely, pairing secondary and primary cells is equivalent to performing a boundary matrix reduction step. Some works, such as [15], deduce homology bases from the SNF of the boundary matrix (or a modified version of it). Therefore, Theorem 1 applies, and such methods (when homology is computed over a field) actually compute explicit bases.

4.1 Characterization of Homology Bases Computed by Tri-Partitions

In a recent paper [5], the authors defined the notion of tri-partition. A tri-partition is a combinatorial object that provides homological information using the notion of tree and cotree:

Definition 11 (trees and cotrees). *A* q-tree *is a set of* q*-cells* A *that does not generate non-zero cycles, i.e.* $\operatorname{Span} A \cap \ker \partial = \{0\}$*. Similarly, a* q-cotree *is a set of* q*-cells* A *that does not generate non-zero cocycles, i.e.* $\operatorname{Span} A \cap \ker \partial^* = \{0\}$*. A* q*-tree* A *(resp.* q*-cotree) is maximal if there is no other* q*-tree (resp.* q*-cotree)* A' *such that* $A \subset A'$*.*

Definition 12 (tri-partition [5]). *Given a complex* (K, ∂)*, a* tri-partition *of dimension* q *is a triplet* (A^q, A_q, E_q) *forming a partition of the* q*-cells of* K *such that* A^q *is a maximal* q*-cotree,* A_q *is a maximal* q*-tree and* $|E_q| = \beta_q$*.*

It has been proved that every complex admits tri-partitions for all dimensions (Theorem 3.1 in [5]). Tri-partitions also provide homology bases.

Definition 13 (canonical cycle [5]). *Let* (A^q, A_q, E_q) *be a tri-partition for a complex* (K, ∂)*. If* $\tau \in A^q \sqcup E_q$*, there is a unique* q*-cycle* $z_q(\tau)$ *in* $\tau + \operatorname{Span} A_q$*. By duality, if* $\tau \in A_q \sqcup E_q$*, there is a unique* q*-cocycle* $z^q(\tau)$ *in* $\tau + \operatorname{Span} A^q$*.*

It was proved that $(z_q(\epsilon))_{\epsilon \in E_q}$ is a q-homology basis and $(z^q(\epsilon))_{\epsilon \in E_q}$ is a q-cohomology basis (see Theorem 4.3 in [5]). Proposition 4 states that tri-partitions are closely related to perfect HDVFs. Precisely, they can be viewed as "dimensional layers" of perfect HDVF. Conversely, a perfect HDVF can be viewed as a "stack" of tri-partitions.

Proposition 4. *If* (P, S, C) *is a perfect HDVF for* K*, then* (P_q, S_q, C_q) *is a* q*-tri-partition for every dimension* q*. Conversely, if for every* q*,* (A^q, A_q, E_q) *is a* q*-tri-partition of* K*, then* $(\bigcup A^q, \bigcup A_q, \bigcup E_q)$ *is a perfect HDVF for* K*.*

Proof. The proof can be found in the technical report [8]. □

This proposition underlines the fact that perfect HDVFs and tri-partitions represent the same object, but the former uses a "vertical" point of view whereas the latter uses a "horizontal" point of view. As a consequence, Definition 13 coincides with the canonical (co)cycle maps defined for HDVF. This is a consequence of Definition 4 and Lemma 1. Finally, the characterization from Sect. 3 and the link between tri-partitions and HDVF implies the following result:

Theorem 3. *A* q*-homology basis is explicit iff it is induced by a* q*-tri-partition.*

4.2 Characterization of Persistent Homology Bases

Persistent homology studies changes in the homology of a filtration. Formally, a filtration is a growing sequence of complexes $K^0 \subseteq K^1 \subseteq \cdots \subseteq K^N$. The inclusions between complexes induces maps between the homology groups associated with the complexes. For each dimension q we have:

$$H_q(K^0) \xrightarrow{\iota_q^0} H_q(K^1) \xrightarrow{\iota_q^1} \cdots \xrightarrow{\iota_q^{N-1}} H_q(K^N)$$

Given $i < j$, the image of $\iota_q^i \circ \cdots \circ \iota_q^{j-1}$ represents the q-homology classes of K^i that remain in the filtration until K^j. Intuitively, it represents the q-holes of K^i that were not filled between K^i and K^j. It is then possible to define a *birth* and a *death* index for each homology class that appears in the filtration. Thus, every homology class in the filtration is associated with a *lifetime* interval $[i, j]$ (j eventually equals to ∞ if the homology class never dies). These information can be summed up in a *persistence diagram*, which can be defined as a multiset over $\mathbb{R}^2$. See [4,20] for a detailed explanation of persistent homology.

It has been shown in [12] that persistent homology can be computed by tweaking the algorithm that computes a perfect HDVF. Given a filtration $\emptyset = K^0 \subseteq \cdots \subseteq K^N = K$, the method starts with an empty HDVF and completes it into a perfect HDVF $X^i = (P^i, S^i, C^i)$ for (K^i, ∂) at the i-th iteration. Precisely, when a cell τ_i is added to the filtration at step i, it is paired (as a secondary-primary pair) with the youngest possible cell τ_j already present. If no such τ_j exists, τ_i is left as a critical cell. This algorithm actually reproduces how positive-negative pairs are chosen in [3]. Its complexity is the same as for computing a perfect HDVF – at most cubic in the number of cells.

This method not only computes persistence intervals, but also HDVFs and reductions at each step of the filtration. In particular, it computes a q-homology basis for each (K^i, ∂), which is called the *i-th persistent q-homology basis* and whose elements are called *persistent q-homology generators.* Formally, if (f^i, g^i, h^i) is the reduction associated to X^i, the i-th persistent q-homology basis is given by $\left(g_q^i(\gamma)\right)_{\gamma \in C_q^i}$. Note that it is a *persistent basis* in the sense of [2,14,19].

A direct consequence of Lemma 3 is that persistent generators are preserved during the lifetime of the homology class they represent. Equivalently, for all $\gamma \in C_q^i$, we have $z_q(\gamma) = g_q^i(\gamma)$, where z_q is the canonical cycle map of the final HDVF X^N. Note that in [9], the authors prove similar results and define a similar "tweaking" of the algorithm computing *AT-models* (i.e. perfect reductions) to emulate persistent homology.

As a result, every point (i, j) in the persistence diagram can be associated to a unique cell γ (which was critical between step i and j) and to a unique persistent homology generator $z_q(\gamma)$. Theorem 1 implies that the persistent homology bases of a filtration are explicit. An illustration of persistent homology bases computed by HDVF is given in Fig. 6. Note that other works such as [16,18] deduce persistent homology bases from boundary matrix reductions. Again, as boundary matrix reductions can be reproduced by a HDVF completion, the persistent homology bases computed by these methods are explicit.

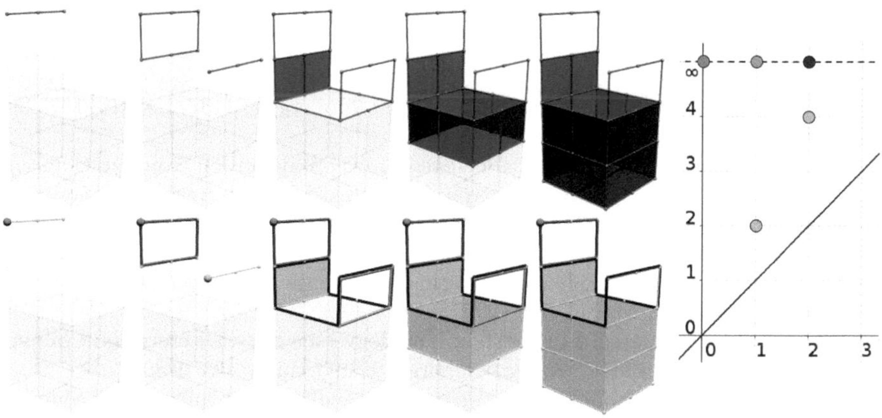

Fig. 6. Example of persistent homology computed by HDVF. First row: the filtration $K_0 \subseteq \cdots \subseteq K_4$ and its HDVFs $X^0, \ldots, X^4$. Second row: the ith-persistent homology bases for $i = 0, \ldots, 4$ (dimensions 0 and 1 are represented). On the right: the persistence diagram of the filtration, each point (i, j) corresponds to a homology class whose lifetime is $[i, j]$, its color matches the color of the corresponding persistent generator.

5 Conclusion and Future Works

This paper introduced a characterization of the homology bases computed by standard computational homology methods. We first proved that a class of homology bases called *explicit* is equivalent to being induced by a HDVF. We then showed how this result applies to other methods, in particular to tri-partitions (which we proved are closely related to perfect HDVFs), Discrete Morse Theory and boundary matrix reduction methods (and thus standard persistent homology algorithms). Note that a similar result (yet less intuitive) holds for cohomology bases (see [8] for more details).

This work suggests that the HDVF framework provides a way to study computational homology methods. In particular, these results might give some insights on the isomorphisms computed in [11] and on the hole closing and opening operations defined in [6]. In addition, they also raise computational homology questions. For instance, can the set of explicit bases over a complex give information about the topology? Can we use HDVF operations to adapt a perfect HDVF to a given explicit basis? Are minimal explicit bases easier to compute than standard minimal homology bases?

Another open area is the study of HDVFs when homology is computed over a ring. In this case, the existence of a perfect reduction is not guaranteed due to the presence of torsion. However, it might be possible to define a reduction that is somehow "optimal" and investigate the HDVFs related to it.

Disclosure of Interests. On behalf of all authors, the corresponding author declares that there is no conflict of interest.

References

1. Dey, T., Hou, T., Mandal, S.: Computing minimal persistent cycles: polynomial and hard cases, pp. 2587–2606 (2020). https://doi.org/10.1137/1.9781611975994.158
2. Dey, T.K., Hou, T., Mandal, S.: Persistent 1-cycles: definition, computation, and its application. In: Marfil, R., Calderón, M., Díaz del Río, F., Real, P., Bandera, A. (eds.) CTIC 2019. LNCS, vol. 11382, pp. 123–136. Springer, Cham (2019). https://doi.org/10.1007/978-3-030-10828-1_10
3. Edelsbrunner, L.: Zomorodian: topological persistence and simplification. Discrete Comput. Geom. **28**(4), 511–533 (2002)
4. Edelsbrunner, H., Harer, J.: Persistent homology—a survey. Discrete and Comput. Geom. - DCG **453** (2008). https://doi.org/10.1090/conm/453/08802
5. Edelsbrunner, H., Ölsböck, K.: Tri-partitions and bases of an ordered complex. Discrete Comput. Geom. **64**, 759–775 (2020). https://api.semanticscholar.org/CorpusID:210173079
6. Edelsbrunner, H., Ölsböck, K.: Holes and dependences in an ordered complex. Comput. Aided Geom. Design **73**, 1–15 (2019)
7. Forman, R.: Morse theory for cell complexes. Adv. Math. **134**(1), 90–145 (1998). https://doi.org/10.1006/aima.1997.1650
8. Gazull, Y.S., Gonzalez-Lorenzo, A., Bac, A.: Characterization of the computed homology and cohomology bases – technical report (2025). https://arxiv.org/abs/2509.09350
9. Gonzalez-Diaz, R., Ion, A., Jimenez, M.J., Poyatos, R.: Incremental-decremental algorithm for computing AT-models and persistent homology. In: Real, P., Diaz-Pernil, D., Molina-Abril, H., Berciano, A., Kropatsch, W. (eds.) CAIP 2011. LNCS, vol. 6854, pp. 286–293. Springer, Heidelberg (2011). https://doi.org/10.1007/978-3-642-23672-3_35
10. Gonzalez-Lorenzo, A.: Computational homology applied to discrete objects. Theses, Aix-Marseille Universite; Universidad de Sevilla (2016). https://hal.archives-ouvertes.fr/tel-01477399
11. Gonzalez-Lorenzo, A., Bac, A., Gazull, Y.S.: A constructive approach of Alexander duality. J. Appl. Comput. Topol. **9**(2) (2025). https://doi.org/10.1007/s41468-024-00198-1
12. Gonzalez-Lorenzo, A., Bac, A., Mari, J.L., Real, P.: Allowing cycles in discrete Morse theory. Topol. Appl. **228**, 1–35 (2017). https://doi.org/10.1016/j.topol.2017.05.008
13. Hatcher, A., Press, C.U., of Mathematics, C.U.D.: Algebraic Topology. Algebraic Topology, Cambridge University Press (2002). https://books.google.fr/books?id=BjKs86kosqgC
14. Obayashi, I.: Volume-optimal cycle: tightest representative cycle of a generator in persistent homology. SIAM J. Appl. Algebra Geom. **2** (2017). https://doi.org/10.1137/17M1159439
15. Peltier, S., Alayrangues, S., Fuchs, L., Lachaud, J.O.: Computation of homology groups and generators. Comput. Graph. **30**(1), 62–69 (2006)
16. Pham, T., Wagner, H.: Computing representatives of persistent homology generators with a double twist (2024). https://arxiv.org/abs/2403.04100
17. Rubio, J., Sergeraert, F.: 1. Homologie effective. Cours de l'institut Fourier **20**(17), 15–38 (1986). http://eudml.org/doc/272555

18. Čufar, M., Virk, v.: Fast computation of persistent homology representatives with involuted persistent homology. Found. Data Sci. **5**(4), 466–479 (2023https://doi.org/10.3934/fods.2023006
19. Wu, P., et al.: Optimal topological cycles and their application in cardiac trabeculae restoration. In: Niethamme, M., et al. (eds.) IPMI 2017. LNCS, vol. 10265, pp. 80–92. Springer, Cham (2017). https://doi.org/10.1007/978-3-319-59050-9_7
20. Zomorodian, A., Carlsson, G.: Computing persistent homology. Discrete Comput. Geom. **33**, 249–274 (2005). https://doi.org/10.1007/s00454-004-1146-y

Computing Gradient Vector Fields with Morse Sequences

Gilles Bertrand[1] and Laurent Najman[1,2](✉)

[1] Univ Gustave Eiffel, CNRS, LIGM, 77454 Marne-la-Vallée, France
{gilles.bertrand,laurent.najman}@esiee.fr
[2] Department of Mathematics, Khalifa University, Abu Dhabi, UAE

Abstract. We rely on the framework of Morse sequences to enable the direct computation of gradient vector fields on simplicial complexes. A Morse sequence is a filtration from a subcomplex L to a complex K via elementary expansions and fillings, naturally encoding critical and regular simplexes. Maximal increasing and minimal decreasing schemes allow constructing these sequences, and are linked to algorithms like Random Discrete Morse and Coreduction. Extending the approach to cosimplicial complexes ($S = K \setminus L$) allows for efficient computation using reductions, perforations, coreductions, and coperforations. We further generalize to F-sequences, which are Morse sequences weighted by an arbitrary stack function F, and provide algorithms to compute maximal and minimal sequences. A particular case is when the stack function is given through a vertex map, common in topological data analysis. For injective maps, the complex decomposes into lower stars, recovering established methods and enabling parallel computation; for non-injective maps, our approach applies directly without requiring perturbations. Thus, the paper adopts Morse sequences as a framework that simplifies and connects some important existing propagation-based methods, while also introducing new schemes that extend their scope and practical applicability.

Keywords: Discrete Morse theory · Expansions and collapses · Fillings and perforations · Simplicial complex

1 Introduction

A fundamental concept in discrete Morse theory [10] is the one of discrete gradient vector field. We rely on the framework of Morse sequences [4] to compute gradient vector fields on simplicial complexes. A Morse sequence is defined as a sequence of simplicial complexes transitioning from a subcomplex L to a complex K through elementary operations: expansions (adding free pairs) and fillings (adding critical simplexes). These sequences naturally yield a gradient vector field, with regular pairs being free and critical simplexes marking topological features.

The aim of this paper is not primarily to introduce algorithms that surpass existing ones in efficiency or in the number of critical simplexes. Rather, our

M. H. F. Wilkinson and J. Kosinka (Eds.): DGMM 2025, LNCS 16296, pp. 178–191, 2026.
https://doi.org/10.1007/978-3-032-09544-2_13

contribution is to adopt the framework of Morse sequences as a pedagogical and coherent way to describe several of the most important propagation-based methods. This perspective provides a simple language in which these techniques can be reformulated and their relationships made clearer. While we do not attempt an exhaustive survey, our presentation helps to organize part of the literature in a structured manner. At the same time, working consistently within this framework also leads to new algorithmic ideas and extensions, which we develop in the sequel of this paper.

In Sect. 2, we describe two construction schemes: a maximal increasing scheme (building from L to K by prioritizing expansions) and a minimal decreasing scheme (reducing from K to L by prioritizing collapses), linking these to existing algorithms like Random Discrete Morse [1] and Coreduction [17]. These approaches align with propagation-based methods that aim to minimize the number of critical simplexes, a key objective in many computational topology applications. In Sect. 3, we extend this framework to cosimplicial complexes (sets $S = K \setminus L$) and apply operations such as reductions, perforations, coreductions, and coperforations to efficiently compute Morse sequences.

In Sect. 4, we further generalize Morse sequences to F-sequences, weighted by a stack F (a map on simplexes), ensuring topological consistency across filtration levels. In Sect. 5, we provide some generic schemes to compute maximal and minimal F-sequences, and we give in Sect. 6 an algorithm for computing a maximal F-sequence. (See Appendix A for an algorithm that computes a minimal F-sequence).

In Sect. 7, we apply our framework to vertex maps, common in topological data analysis. For injective vertex maps, the complex is partitioned into lower stars, enabling parallel computation. In this case, we show that our approach enable us to retrieve established methods [11,18]. Thus, our algorithms offer flexibility for real-world data, by handling non-injective maps directly, without the need of a total order or of a perturbation.

In the conclusion (Sect. 8), we highlight that a Morse complex can be constructed while computing a Morse sequence, and we propose some future research directions.

2 Morse Sequences

Let K be a finite family composed of non-empty finite sets, called *simplexes*. The family K is a *(simplicial) complex* if $\sigma \in K$ whenever $\sigma \neq \emptyset$ and $\sigma \subseteq \tau$ for some $\tau \in K$. An element of a simplicial complex K is *a face of* K. A *facet of* K is a face of K that is maximal for inclusion. The *dimension* of $\sigma \in K$, written $dim(\sigma)$, is the number of its elements minus one. If $\sigma \in K$, we write:

- $\partial(\sigma, K) = \{\nu \in K \mid \nu \subseteq \sigma \text{ and } dim(\nu) = dim(\sigma) - 1\}$, and
- $\delta(\sigma, K) = \{\nu \in K \mid \sigma \subseteq \nu \text{ and } dim(\nu) = dim(\sigma) + 1\}$,

which are, respectively, the *boundary* and the *coboundary of* σ *in* K.
A *subcomplex of* K is a set $L \subseteq K$ which is a simplicial complex.

We recall the definitions of simplicial collapses and simplicial expansions [20]. Let K be a simplicial complex and let $\sigma, \tau \in K$. The couple (σ, τ) is a *free pair for* K, if τ is the only face of K that contains σ. If (σ, τ) is a free pair for K, then the simplicial complex $L = K \setminus \{\sigma, \tau\}$ is *an elementary collapse of* K, and K is *an elementary expansion of* L. We say that K *collapses onto* L, or that L *expands onto* K, if there exists a sequence $\langle K = K_0, \ldots, K_k = L \rangle$, such that K_i is an elementary collapse of K_{i-1}, $i \in [1, k]$.

We also recall the definitions of perforations and fillings [20].
Let K, L be simplicial complexes. If $\nu \in K$ is a facet of K and if $L = K \setminus \{\nu\}$, we say that L is *an elementary perforation of* K, and that K is *an elementary filling of* L.

We now introduce the notion of a "Morse sequence" by simply considering expansions and fillings of a simplicial complex [4].

Definition 1. *Let* $L \subseteq K$ *be two simplicial complexes. A* Morse sequence (from L to K) *is a sequence* $\overrightarrow{W} = \langle L = K_0, \ldots, K_k = K \rangle$ *of simplicial complexes such that, for each* $i \in [1, k]$, K_i *is either an elementary expansion or an elementary filling of* K_{i-1}. *If* $L = \emptyset$, *we say that* $\overrightarrow{W}$ *is a* Morse sequence on K.

Thus, any Morse sequence $\overrightarrow{W}$ on K is a *filtration on* K, that is a sequence of nested complexes $\langle \emptyset = K_0, ..., K_k = K \rangle$ such that, for each $i \in [0, k-1]$, we have $K_i \subseteq K_{i+1}$; see [8,19].

Let $\overrightarrow{W} = \langle L = K_0, \ldots, K_k = K \rangle$ be a Morse sequence. We write $\diamond\overrightarrow{W}$ for the sequence $\diamond\overrightarrow{W} = \langle \kappa_1, \ldots, \kappa_k \rangle$ such that, for each $i \in [1, k]$:

- If K_i is an elementary filling of K_{i-1}, then κ_i is the simplex such that $K_i = K_{i-1} \cup \{\kappa_i\}$; we say that the face κ_i is *critical for* $\overrightarrow{W}$.
- If K_i is an elementary expansion of K_{i-1}, then κ_i is the pair (σ, τ) such that $K_i = K_{i-1} \cup \{\sigma, \tau\}$; we say that κ_i, σ, τ, are *regular for* $\overrightarrow{W}$.

We say that $\diamond\overrightarrow{W}$ is a *simplex-wise (Morse) sequence (from L to K)*. Note that $\diamond\overrightarrow{W}$ is a sequence of faces and pairs.

Observe that, if $\overrightarrow{W} = \langle \emptyset = K_0, ..., K_k = K \rangle$ is a Morse sequence on K, with $k \geq 1$, then K_1 is necessarily a filling of $\emptyset$. That is, K_1 is made of a single face (a vertex) that is critical for $\overrightarrow{W}$.

Definition 2. *Let* $\overrightarrow{W}$ *be a Morse sequence. The* gradient vector field of $\overrightarrow{W}$ *is the set composed of all regular pairs for* $\overrightarrow{W}$. *We say that two Morse sequences* $\overrightarrow{V}$ *and* $\overrightarrow{W}$ *from* L *to* K *are* equivalent *if they have the same gradient vector field.*

Building a gradient vector field from a complex is a fundamental issue in discrete Morse theory. It is worth mentioning that using Morse sequences for computing gradient vector fields entails no loss of generality, see [4].

The two following schemes are two basic ways to build a Morse sequence $\overrightarrow{W}$ from L to K.

1. *The increasing scheme.* We build $\overrightarrow{W}$ from the left to the right. Starting from L, we obtain K by iterative expansions and fillings. We say that this scheme is *maximal* if we make a filling only if no expansion can be made.
2. *The decreasing scheme.* We build $\overrightarrow{W}$ from the right to the left. Starting from K, we obtain L by iterative collapses and perforations. We say that this scheme is *minimal* if we make a perforation only if no collapse can be made.

Clearly, any Morse sequence may be obtained by each of these two schemes whenever the condition of maximality or minimality is not imposed.

The purpose of maximal and minimal schemes is to try to minimize the number of critical simplexes. This problem is, in general, NP-hard [14]. Therefore, these methods do not, in general, give optimal results. Note that there may exist some differences between the maximal increasing scheme and the minimal decreasing one. See [11] and [3] for examples which illustrate this difference.

In the next section we will see that:

- There is a link between the minimal decreasing scheme and the scheme of the algorithm *Random Discrete Morse*, proposed by Benedetti and Lutz in [1]. See also Sect. 2.3 and Algorithm 1 in [19].
- There is a link between the maximal increasing scheme and the scheme of the algorithm *Coreduction* proposed by Mrozek and Batko in [17]. See also [11] and Algorithm 3.6 in [13].

Note that a distinctive feature of our approach is the use of these schemes to compute Morse sequences. A Morse sequence on a complex K not only defines a gradient vector field on K but also imposes a specific structure on it.

3 Cosimplicial Complexes

The computation of a Morse sequence from L to K, where L and K are simplicial complexes, can be carried out by inductively performing elementary operations restricted to the set $S = K \setminus L$. This set S is not a simplicial complex, but it possesses the following specific structure, that we refer to as a *cosimplicial complex*; see also [5] for more details.

Definition 3. *Let S be a finite set of simplexes. The set S is a* cosimplicial complex *if, for any $\sigma, \tau \in S$, we have $\nu \in S$ whenever $\sigma \subseteq \nu \subseteq \tau$.*

Note that each simplicial complex is also a cosimplicial complex.

Let S be a finite set of simplexes. We write $\overline{S}$ for the set of simplexes such that $\sigma \in \overline{S}$ if and only if there exists $\tau \in S$ with $\sigma \subseteq \tau$. Thus, we have $S \subseteq \overline{S}$. We observe that $\overline{S}$ is a simplicial complex, this complex is the smallest simplicial complex that contains S. It follows that S is a simplicial complex if and only if $\overline{S} = S$. We write $\underline{S} = \overline{S} \setminus S$. The following result is straightforward:

Proposition 1. *Let S be a finite set of simplexes. The set S is a cosimplicial complex if and only if $\underline{S}$ is a simplicial complex.*

If S is a cosimplicial complex, then the sets $L = \underline{S}$ and $K = \overline{S}$ are two simplicial complexes such that $L \subseteq K$ and $S = K \setminus L$.

In fact, a set S is a cosimplicial complex if and only if there exist two simplicial complexes L and K such that $S = K \setminus L$. This formulation corresponds to the definition of an *open simplicial complex*, see [15].

The following definition is a simple extension of the notion of a Morse sequence to cosimplicial complexes. Proposition 2 provides another way to describe a Morse sequence from L to K.

Definition 4. *Let S be a cosimplicial complex. We say that a sequence $\overrightarrow{W}$ is a* Morse sequence on *S if $\overrightarrow{W}$ is a Morse sequence from $\underline{S}$ to $\overline{S}$.*

Proposition 2. *Let (L, K) be a pair of simplicial complexes such that $L \subseteq K$, and let $S = K \setminus L$. A sequence $\overrightarrow{W}$ is a Morse sequence from L to K if and only if $\overrightarrow{W}$ is a Morse sequence on S.*

Let S be a cosimplicial complex. For each $\nu \in S$, we write:

$$\partial(\nu, S) = \partial(\nu, \overline{S}) \cap S \text{ and } \delta(\nu, S) = \delta(\nu, \overline{S}).$$

We have $\partial(\nu, S) \subseteq S$ by construction, we also note that $\delta(\nu, S) \subseteq S$.

We now introduce four operations which operate only on the set S.
Let S be a cosimplicial complex, and let $\sigma, \tau, \nu \in S$, with $\sigma \subseteq \tau$. We say that:

- the complex $S \setminus \{\sigma, \tau\}$ is a *reduction of S* if $\delta(\sigma, S) = \{\tau\}$,
- the complex $S \setminus \{\nu\}$ is a *perforation of S* if $\delta(\nu, S) = \emptyset$,
- the complex $S \setminus \{\sigma, \tau\}$ is a *coreduction of S* if $\partial(\tau, S) = \{\sigma\}$,
- the complex $S \setminus \{\nu\}$ is a *coperforation of S* if $\partial(\nu, S) = \emptyset$.

Reductions and perforations are used in the algorithm Random Discrete Morse [1]; in this algorithm S is a simplicial complex. Coreductions and coperforations have been introduced in [17] with the algorithm Coreduction; in this algorithm S is a complex which is more general than a cosimplicial complex. See also [11] and [13] for other algorithms based on coreductions. The link between these four operations and operations on the simplicial complexes $\underline{S}$ and $\overline{S}$ is the following.

Proposition 3. *Let S be a cosimplicial complex, and let $\sigma, \tau, \nu \in S$, with $\sigma \subseteq \tau$.*

- *$\overline{S} \setminus \{\sigma, \tau\}$ is a collapse of $\overline{S}$ iff $S \setminus \{\sigma, \tau\}$ is a reduction of S.*
- *$\overline{S} \setminus \{\nu\}$ is a perforation of $\overline{S}$ iff $S \setminus \{\nu\}$ is a perforation of S.*
- *$\underline{S} \cup \{\sigma, \tau\}$ is an expansion of $\underline{S}$ iff $S \setminus \{\sigma, \tau\}$ is a coreduction of S.*
- *$\underline{S} \cup \{\nu\}$ is a filling of $\underline{S}$ iff $S \setminus \{\nu\}$ is a coperforation of S.*

Thus, by Propositions 2 and 3:

- A Morse sequence $\overrightarrow{W}$ from L to K can be built with the decreasing scheme by iterative reductions and perforations on the set $S = K \setminus L$.
- A Morse sequence $\overrightarrow{W}$ from L to K can be built with the increasing scheme by iterative coreductions and coperforations on the set $S = K \setminus L$.

4 *F*-Sequences

In this section, we introduce Morse sequences weighted by a map. We first give some basic definitions relative to these maps, see also [5].

Let F be a map from a cosimplicial complex S to $\mathbb{Z}$. We say that F is a *stack on* S if we have $F(\sigma) \leq F(\tau)$ whenever $\sigma, \tau \in S$ and $\sigma \subseteq \tau$.

Let F be a map from a cosimplicial complex S to $\mathbb{Z}$. For any $\lambda \in \mathbb{Z}$, we write:

$$F_\lambda = \{\nu \in S \mid F(\nu) \leq \lambda\} \text{ and } F[\lambda] = \{\nu \in S \mid F(\nu) = \lambda\}.$$

The sets F_λ and $F[\lambda]$ are, respectively, *the cut and the section of* F *at level* λ.

Note that if K is a simplicial complex, then the indexed family $(F_\lambda)_{\lambda \in \mathbb{Z}}$ is a filtration on K.

The two following properties are straightforward.

Proposition 4. *Let S be a cosimplicial complex and F be a map from S to $\mathbb{Z}$. If F is a stack on S, then any cut and any section of F is a cosimplicial complex.*

Proposition 5. *Let K be a simplicial complex and F be a map from K to $\mathbb{Z}$. The map F is a stack on K if and only if any cut of F is a simplicial complex.*

Now, we extend the notion of a Morse sequence to an arbitrary stack F on a simplicial complex K. In such a sequence, each expansion is an *F-expansion*, that is, an expansion that preserves the topology of all cuts of F. See also [6], which establishes a connection between F-expansions and watersheds.

In the sequel of this paper, L and K will denote simplicial complexes.

Definition 5. *Let F be a stack on K and let L be a subcomplex of K. Let $\kappa = (\sigma, \tau)$ be a free pair for L. We say that κ is* a free pair for F *if $F(\sigma) = F(\tau)$. If κ is a free pair for F, we say that $L' = L \setminus \{\sigma, \tau\}$ is an* (elementary) F-collapse *of L and L is an* (elementary) F-expansion of L'. *We write $F(\kappa) = F(\sigma) = F(\tau)$.*

Let F be a stack on K and let L be a subcomplex of K. Let (σ, τ) be a free pair for L which is also a free pair for F. Then, it can be easily checked that:

- For each $\lambda < F(\sigma)$, we have $\sigma \notin L \cap F_\lambda$ and $\tau \notin L \cap F_\lambda$,
- For each $\lambda \geq F(\sigma)$, the pair (σ, τ) is a free pair for $L \cap F_\lambda$.

Definition 6. *Let $\overrightarrow{W}$ be a Morse sequence from L to K and let F be a stack on K. We say that $\overrightarrow{W}$ is a* Morse sequence on F *or an* F-sequence *(from L to K), if we have $F(\sigma) = F(\tau)$ whenever (σ, τ) is a regular pair for $\overrightarrow{W}$. We also say that an F-sequence from $\emptyset$ to K is an* F-sequence on K.

Let $\overrightarrow{W} = \langle L = K_0, ..., K_k = K \rangle$ be an F-sequence. Then $\overrightarrow{W}$ induces a "double filtration": the indexed family $(K_i)_{i \in [0,k]}$ is a filtration where each K_i induces the filtration $(K_i \cap F_\lambda)_{\lambda \in \mathbb{Z}}$. Also, $\overrightarrow{W}$ induces the sequence $\langle F_0, ..., F_k \rangle$ where each F_i is the stack on K_i which is the restriction of F to K_i.

Scheme 1: computing a sequence $\overrightarrow{W}(L, K, F)$ that is maximal for F.

$I := L$; $X := \emptyset$; $\overrightarrow{W} := \langle L \rangle$;
while $I \neq K$ **do**
  if *there exists* C, *with* $I \subseteq C \subseteq K$, *such that* C *is an* F*-expansion of* I **then** $X := C$;
  if $X = \emptyset$ **then** compute a complex C, with $I \subseteq C \subseteq K$, such that C is an elementary filling of I, and set $X := C$;
  $\overrightarrow{W} := \overrightarrow{W} \cdot X$; $I := X$; $X := \emptyset$;

Scheme 2: computing a sequence $\overrightarrow{W}(L, K, F)$ that is minimal for F.

$I := K$; $X := \emptyset$; $\overrightarrow{W} := \langle K \rangle$;
while $I \neq L$ **do**
  if *there exists* C, *with* $L \subseteq C \subseteq I$, *such that* C *is an* F*-collapse of* I **then** $X := C$;
  if $X = \emptyset$ **then** compute a complex C, with $L \subseteq C \subseteq I$, such that C is an elementary perforation of I, and set $X := C$;
  $\overrightarrow{W} := X \cdot \overrightarrow{W}$; $I := X$; $X := \emptyset$;

5 Maximal and Minimal *F*-Sequences

We extend to F-sequences the maximal and minimal schemes presented in Sect. 2. These schemes may be formalized with the following definition.

Let F be a stack on K and let $\overrightarrow{W} = \langle L = K_0, ..., K_k = K \rangle$ be an F-sequence.

For any $i \in [0, k]$, we say that K_i is *maximal* (resp. *minimal) (for F)* if no F-expansion (resp. F-collapse) of K_i is a subset of K (resp includes L).

We say that $\overrightarrow{W}$ or $\diamond\overrightarrow{W}$ is *maximal (for F)* if, for any $i \in [1, k]$, the complex X_{i-1} is maximal for F whenever X_i is critical for $\overrightarrow{W}$.

We say that $\overrightarrow{W}$ or $\diamond\overrightarrow{W}$ is *minimal (for F)* if, for any $i \in [0, k-1]$, the complex X_{i+1} is minimal for F whenever X_i is critical for $\overrightarrow{W}$.

Maximal F-sequences fit with the strategy underlying the algorithm *Morse-Reduce* [16], which operates on filtrations of Lefschetz complexes; here, we work with stacks rather than filtrations, and we focus specifically on simplicial complexes. In addition, minimal F-sequences can be seen as extending the construction used in the algorithm *Random Discrete Morse* [1] to the setting of stacks.

Some pseudocode for computing a maximal (*resp.* minimal) F-sequence is provided in Scheme 1 (*resp.* Scheme 2).

Figure 1b illustrates an example of a maximal F-sequence on K, where K is a triangulation of a square, and F is the stack depicted Fig. 1a. Such a sequence begins from either one of the two possible points, both assigned a weight of 0, as shown in Fig. 1a. Using Scheme 1 with $L = \emptyset$, and processing simplexes according to their weights, we initiate the sequence with a critical 0-simplex, denoted **a**.

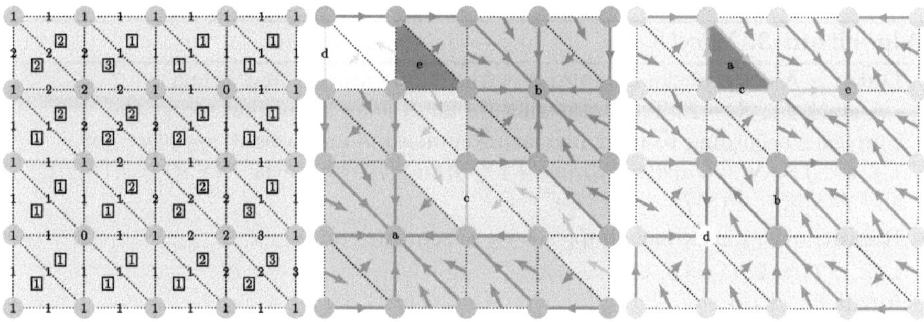

(a) A simplicial stack F (b) A maximal F-sequence (c) A minimal F-sequence

Fig. 1. (a) A simplicial stack F on a triangulation K of a square. (b) A maximal F-sequence on K. (c) A minimal F-sequence on K.

We then perform all feasible expansions with 1-simplices, followed by all possible 2-simplices, resulting in the red region depicted in Fig. 1b. Next, we introduce a second critical point, **b**, and repeat the expansion process, yielding the blue region. Subsequently, a critical 1-simplex (an edge), labeled **c**, is added, and its expansion produces the green region. A further critical 1-simplex, labeled **d**, is incorporated, leading to the yellow region after expansion. Finally, the sequence is completed by adding a critical 2-simplex, a triangle labeled **e**.

Figure 1c illustrates an example of a minimal F-sequence on the same stack as in Fig. 1a. Scheme 2 begins by performing all possible collapses, shown in gray. Next, it introduces the first critical 2-simplex—the triangle labeled **a**. This is followed by the introduction of the first critical 1-simplex—the edge labeled **b** (highlighted in blue). After performing all possible collapses from this step, we obtain the region composed of blue edges. The scheme then introduces a second critical 1-simplex—the edge labeled **c** (in green)—leading, after further collapses, to the region of green edges. Finally, the scheme terminates with two critical 0-simplices: the points labeled **e** and **d**.

6 An Algorithm for Computing Maximal F-Sequences

We now give a description of an algorithm for computing a maximal (simplex-wise) F-sequence from L to K. The input of Algorithm 3 is the set $S = K \setminus L$ and the map F, which is restricted to S. The output $\mathbf{Max}(S, F)$ is a *maximal F-sequence on* S, that is, a maximal F-sequence from $\underline{S}$ to $\overline{S}$. By Proposition 2, this gives the desired sequence from L to K. By Proposition 3, $\mathbf{Max}(S, F)$ can be computed using operations limited to the set S. Consequently, the sets $\underline{S}$ to $\overline{S}$ are not needed for this computation.

The following facts lead directly to the soundness of the algorithm; note that the two sets $T \cup \underline{S}$ and $S \setminus T$ are not explicitly considered in the algorithm.

1. At the beginning of the algorithm we have $T \cup \underline{S} = \underline{S}$, and at the end we have $T \cup \underline{S} = \overline{S}$.

Algorithm 3: Max(S, F)

Data: - A cosimplicial complex S with its operators ∂ and δ; and
- A stack $F : S \to \mathbb{Z}$. The datastructure for S is an array that stores the simplexes according to their increasing dimension and weight: we have, for $1 \leq i < j \leq N = Card(S)$: $dim(S[i]) \leq dim(S[j])$ whenever $F(S[i]) = F(S[j])$; and $F[S[i]] < F(S[j])$ otherwise.

Result: $\diamond\overrightarrow{W}$, a maximal simplex-wise F-sequence on S.

1 $i := 1$; $T := \emptyset$; $U := \emptyset$; $\diamond\overrightarrow{W} := \langle\,\rangle$;
2 **forall** $\tau \in S$ **do**
3     $\rho(\tau) = Card(\partial(\tau, S))$;
4   **if** $\rho(\tau) = 1$ **then** $U := U \cup \{\tau\}$;
5 **while** $i \leq N$ **do**
6   **while** $U \neq \emptyset$ **do**
7   Extract $\tau \in U$;
8   **if** $\rho(\tau) = 1$ **then**
9   Find out the simplex $\sigma \in \partial(\tau, S)$ such that $\sigma \notin T$;
10   **if** $F(\tau) = F(\sigma)$ **then**
11                 $\diamond\overrightarrow{W} := \diamond\overrightarrow{W} \cdot (\sigma, \tau)$; $T := T \cup \{\sigma, \tau\}$;
12   **forall** $\mu \in \delta(\sigma, S) \cup \delta(\tau, S)$ **do**
13                     $\rho(\mu) := \rho(\mu) - 1$;
14   **if** $\rho(\mu) = 1$ **then** $U := U \cup \{\mu\}$;
15   **while** $S[i] \in T$ *and* $i \leq N$ **do** $i := i + 1$;
16   **if** $i \leq N$ **then**
17         $\sigma := S[i]$; $T := T \cup \{\sigma\}$; $\diamond\overrightarrow{W} := \diamond\overrightarrow{W} \cdot \sigma$;
18   **forall** $\tau \in \delta(\sigma, S)$ **do**
19             $\rho(\tau) := \rho(\tau) - 1$;
20   **if** $\rho(\tau) = 1$ **then** $U := U \cup \{\tau\}$;
21 **return** $\diamond\overrightarrow{W}$

2. At each step of the algorithm, the set $T \cup \underline{S}$ is a simplicial complex and the set $S \setminus T$ is a cosimplicial complex.
3. If $\nu \in S \setminus T$, we have $\rho(\nu) = Card(\partial(\nu, S) \cap (S \setminus T))$.
4. At line 11, the complex $T \cup \underline{S} \cup \{\sigma, \tau\}$ is an elementary F-expansion of $T \cup \underline{S}$.
5. At line 17, the complex $T \cup \underline{S} \cup \{\sigma\}$ is an elementary filling of $T \cup \underline{S}$.

Observe that, when $S = K$, that is, when $L = \emptyset$, $\underline{S} = \emptyset$, and $\overline{S} = K$, Algorithm 3 provides a maximal F-sequence on the simplicial complex K. This case corresponds to the one given Fig. 1b.

Proposition 6. *If S is a cosimplicial complex, then* **Max**(S, F) *is a maximal simplex-wise F-sequence on S.*

The complexity of the algorithm depends on the data structure used to access the simplicial complex (see [11] for a description of several such data structures).

As long as we can compute $\partial(.,.)$ and $\delta(.,.)$ in $\mathcal{O}(d)$ time, where d is the dimension of the complex (*e.g.*, for example with a graph, or with cubical complexes and the use of a mask to check the neighborhood of a simplex), the complexity of $\mathbf{Max}(K, F)$ is $\mathcal{O}(dn)$, where n the number of its simplexes.

In a dual way, we can derive an algorithm for a Morse sequence that is minimal for F, see Algorithm 5 in Appendix A.

Another important case is when a simplicial complex is decomposed into a disjoint union of cosimplicial complexes. In certain cases, Algorithm 3 allows processing each cosimplicial complex independently, possibly in parallel. This fact is used in the next section.

7 Maximal Sequences and Vertex Maps

In topological data analysis, functions often act on the vertex set of a complex, especially for data from digital images or point clouds [8]. We now derive an instance of our algorithm for this class of functions.

Let K be a simplicial complex. A *vertex map on* K is a map $f : V(K) \to \mathbb{Z}$, where $V(K)$ is *the vertex set of* K, that is, the set composed of all simplexes $\sigma \in K$ such that $dim(\sigma) = 0$.

Let f be a vertex map on K. The map f may be extended to all simplexes of K as follows. For any $\tau \in K$, we write:

$$F(\tau) = \max\{f(\sigma) \mid \sigma \in V(K) \text{ and } \sigma \subseteq \tau\}.$$

Clearly, the map F is a stack on K. We say that F is the *stack induced by* f.

We say that a vertex map f on K is a ϑ*-map* if f maps distinct elements to distinct elements, that is, if f is injective. If f is a ϑ-map, we say that the stack induced by f is a ϑ*-stack*. Also, if $\sigma_1, ..., \sigma_N$ is a total ordering of the vertices of K, we say that this ordering is the *ordering induced by* f if we have $f(\sigma_i) < f(\sigma_j)$ whenever $i < j$.

Let F be a ϑ-stack on K. If $\sigma \in V(K)$, the *lower star of* σ is the set:

$$\hat{\delta}(\sigma) = \{\tau \in K \mid \sigma \subseteq \tau \text{ and } F(\tau) = F(\sigma)\}.$$

Since a ϑ-map is injective, we immediately derive the following.

Proposition 7. *Let F be a ϑ-stack on K. For each $\sigma \in V(K)$, we have:*

$$\hat{\delta}(\sigma) = \{\tau \in K \mid F(\tau) = F(\sigma)\}.$$

Thus, each lower star is a section of F, and each non-empty section of F is a lower star. By Proposition 4, it means that each lower star is a cosimplicial complex. Also, it means that the set of all lower stars constitutes a partition of the complex K.

An F-sequence can thus be obtained by treating each lower star independently. Since F is constant on a lower star, we introduce the following.

Algorithm 4: Max$(V, \hat{\delta})$

Data: A vertex set V of a simplicial complex K with its lower star operator $\hat{\delta}$. An array $V[i]$, $i \in [1, N]$, stores the vertices according to a total order.

Result: $\diamond\overrightarrow{W}$, a Morse sequence on K which is maximal for each lower star.

for $i \in [1, N]$ **do in parallel**
  $S := \hat{\delta}(V[i])$; $\diamond\overrightarrow{W_i}$:=**Max**(S)
$\diamond\overrightarrow{W} := \langle\,\rangle$;
for $i \in [1, N]$ **do** $\diamond\overrightarrow{W} := (\diamond\overrightarrow{W}) \cdot (\diamond\overrightarrow{W_i})$;
return $\diamond\overrightarrow{W}$

If S is cosimplicial complex, we write $\mathbb{1}_S$ for the function $S \to \mathbb{Z}$ such that, for each $x \in S$, we have $\mathbb{1}_S(x) = 1$. This $\mathbb{1}_S$ is a constant function, which is trivially a stack on S. If $\overrightarrow{W}$ is a Morse sequence on a simplicial complex K, we say that $\overrightarrow{W}$ or $\diamond\overrightarrow{W}$ is *maximal* if $\overrightarrow{W}$ is maximal for the stack $\mathbb{1}_K$.

Algorithm 4 allows to compute an F-sequence, where F is a stack on K induced by a ϑ-map f. The algorithm uses a function **Max**(S) which is a particular case of the function **Max**(S, F) introduced in the previous section with Algorithm 3. More precisely, we have **Max**(S) = **Max**$(S, \mathbb{1}_S)$.

The output of the algorithm is written **Max**$(V, \hat{\delta})$. The set V is the vertex set $V(K)$ of the simplicial complex K. This set is given by an array where the vertices are ordered by the total order induced by f. Thus the map f and the stack F are not explicitly used. The lower star $\hat{\delta}(\sigma)$ of each vertex σ is encoded in order to fulfill the conditions of Algorithm 3.

By construction, we have the following:

Proposition 8. *The result* **Max**$(V, \hat{\delta})$ *of Algorithm 4 is a simplex-wise Morse sequence on* K *that is maximal for each lower star* $\hat{\delta}(\sigma)$*, with* $\sigma \in V = V(K)$.

In [18], V. Robins *et al.* proposed an algorithm for computing a gradient vector field of a cubical complex, see algorithm ProcessLowerStars of this paper. As mentioned in [9], several aspects "candidate the algorithm to be one of the best, topologically correct algorithms for computing a Forman gradient". It is worthwhile to observe that the core of this algorithm and Algorithm 4 are the same. See Figs. 1 and 2 in [18] which can be used for illustrating Algorithm 4, if we replace simplicial complexes by cubical ones.

In [11], U. Fugacci *et al.* also proposed an algorithm for computing a Forman gradient. The input of the algorithm is a totally ordered set of the vertices of a simplicial complex which agrees with the input of Algorithm 4. Here again, it may be seen that the core of this algorithm and Algorithm 4 are the same. Observe that a total ordering $\sigma_1, ..., \sigma_N$ of the vertices of K induces the ϑ-map f such that $f(\sigma_i) = i$, which itself induces the ϑ-stack F as defined above.

A vertex map f given by real data almost never fulfills the condition of a ϑ-map where all values are distinct. In [18], the map f is perturbed with a tie-breaking scheme to ensure unique values. An alternative approach to this issue consists simply in computing the Morse sequence $\mathbf{Max}(K, F)$ with Algorithm 3, where F is the vertex stack induced by f.

8 Conclusion

In this paper, we have introduced Morse sequences as a simple and pedagogical framework within discrete Morse theory, supporting the construction of gradient vector fields on simplicial and cosimplicial complexes. Defined through elementary expansions and fillings, Morse sequences provide a coherent way to encode the topological structure of a space. Building on this, our maximal and minimal schemes, together with algorithms such as $\mathbf{Max}(K, F)$ and $\mathbf{Min}(K, F)$, extend naturally to F-sequences and vertex maps, demonstrating their usefulness in applications including digital image analysis and point cloud processing. The alignment with existing methods shows that this perspective is compatible with well-established approaches, while parallel processing of cosimplicial complexes enhances computational efficiency. Beyond these specific contributions, the broader value of our work lies in showing how Morse sequences can serve as a clear and consistent language for presenting several important propagation-based methods, highlighting their connections, and motivating new algorithmic directions. In this sense, Morse sequences provide both a convenient tool for understanding the literature and a promising basis for further developments.

A distinctive strength of Morse sequences lies in their direct connection to *Morse references* [4,7]. A Morse reference Υ is an application that maps each simplex to a set of critical simplices of the same dimension via gradient path parity: a critical simplex ν belongs to $\Upsilon(\sigma)$ if and only if the number of gradient paths from σ to $\pm\nu$ is odd. Hence, the Morse reference straightforwardly relates to the *Morse complex* (also known as the *critical complex.*) The Morse reference Υ of a Morse sequence $\diamond\overrightarrow{W} = \langle \kappa_1, \ldots, \kappa_k \rangle$ can be computed by traversing the sequence $\diamond\overrightarrow{W}$ from left to right. Thus, it is possible, based on an algorithm such as $\mathbf{Max}(K, F)$, to compute both the gradient vector field and the Morse complex in a single pass. Since the Morse complex is itself a central structure in computational topology, it can then be directly leveraged for efficient computation of Betti numbers and persistent homology [11,18], which are key tools in topological data analysis and related applications.

An interesting direction for future research is the systematic use of critical kernels [2] in the context of parallel algorithms. Critical kernels provide a symmetric and homotopy-preserving way of collapsing complexes, and can therefore serve as natural building blocks for parallel reductions. Integrating them into the Morse sequence framework may yield new n-dimensional strategies distributing computations across independent substructures, while maintaining global topological consistency (see [12] for a review of existing strategies). This perspective opens promising avenues for scaling up to very large complexes, such as those arising in 3D imaging or high-dimensional data analysis.

Python code implementing the algorithms of this paper for illustration purpose is available at https://github.com/lnajman/MorseSequences. All these algorithms can straightforwardly be extended to cubical complexes of arbitrary dimension.

A An Algorithm for Minimal Morse Sequences

In this appendix, we give an algorithm for computing a Morse sequence from L to K that is minimal for F. The input of Algorithm 5 is the set $S = K \setminus L$ and the map F, which is restricted to S. The output $\mathbf{Min}(S, F)$ is a *minimal F-sequence on S*, that is, a minimal F-sequence from $\underline{S}$ to $\overline{S}$.

The same notations as for Algorithm 3 are used. We derive in the same manner the soundness of the algorithm. Here again, the sets $\underline{S}$ to $\overline{S}$ are not needed for computing $\mathbf{Min}(S, F)$.

Algorithm 5: Min(S, F)

Data: - A cosimplicial complex S with its operators ∂ and δ; and
- A stack $F : S \to \mathbb{Z}$. The datastructure for S is an array that stores the simplexes according to their decreasing dimension and weight: we have, for $1 \leq i < j \leq N = Card(S)$: $dim(S[i]) \geq dim(S[j])$ whenever $F(S[i]) = F(S[j])$; and $F[S[i]] > F(S[j])$ otherwise.

Result: $\diamond\overrightarrow{W}$, a minimal simplex-wise F-sequence on S.

$i := 1$; $T := \emptyset$; $U := \emptyset$; $\diamond\overrightarrow{W} := \langle\,\rangle$;
forall $\sigma \in S$ **do**
    $\rho(\sigma) := \mathrm{Card}(\delta(\sigma, S))$;
    if $\rho(\sigma) = 1$ **then** $U := U \cup \{\sigma\}$;
while $i \leq N$ **do**
    while $U \neq \emptyset$ **do**
        Extract $\sigma \in U$;
        if $\rho(\sigma) = 1$ **then**
            Find out the simplex τ such that $\tau \in \delta(\sigma, S)$ and $\tau \notin T$;
            if $F(\tau) = F(\sigma)$ **then**
                $\diamond\overrightarrow{W} := (\sigma, \tau) \cdot (\diamond\overrightarrow{W})$; $T := T \cup \{\sigma, \tau\}$;
                forall $\mu \in \partial(\sigma, S) \cup \partial(\tau, S)$ **do**
                    $\rho(\mu) := \rho(\mu) - 1$;
                    if $\rho(\mu) = 1$ **then** $U := U \cup \{\mu\}$;
    while $S[i] \in S$ *and* $i \leq N$ **do** $i := i + 1$;
    if $i \leq N$ **then**
        $\tau := S[i]$; $T := T \cup \{\tau\}$; $\diamond\overrightarrow{W} := \tau \cdot (\diamond\overrightarrow{W})$;
        forall $\sigma \in \partial(\tau)$ **do**
            $\rho(\sigma) := \rho(\sigma) - 1$;
            if $\rho(\sigma) = 1$ **then** $U := U \cup \{\sigma\}$;
return $\diamond\overrightarrow{W}$;

Proposition 9. *If S is a cosimplicial complex, then* $\mathbf{Min}(S, F)$ *is a minimal simplex-wise F-sequence on S.*

References

1. Benedetti, B., Lutz, F.H.: Random discrete Morse theory and a new library of triangulations. Exp. Math. **23**(1), 66–94 (2014)
2. Bertrand, G.: On critical kernels. Comptes rendus de l'Acad. Sci. Sér. I Math. **1**(345), 363–367 (2007)
3. Bertrand, G.: Morse sequences. In: Brunetti, S., Frosini, A., Rinaldi, S. (eds.) DGMM 2024. LNCS, vol. 14605, pp. 377–389. Springer, Cham (2024). https://doi.org/10.1007/978-3-031-57793-2_29
4. Bertrand, G.: Morse sequences: a simple approach to discrete Morse theory. J. Math. Imaging Vision **67**(16), 1–22 (2025)
5. Bertrand, G.: Morse sequences on stacks and flooding sequences. In: International Workshop on Combinatorial Image Analysis. LNCS, Springer (2025). https://arxiv.org/abs/2509.01384
6. Bertrand, G., Boutry, N., Najman, L.: Discrete Morse functions and watersheds. J. Math. Imaging Vision **65**(5), 787–801 (2023)
7. Bertrand, G., Najman, L.: Morse frames. In: Brunetti, S., Frosini, A., Rinaldi, S. (eds.) DGMM 2024. LNCS, vol. 14605, pp. 364–376. Springer, Cham (2024). https://doi.org/10.1007/978-3-031-57793-2_28
8. Edelsbrunner, H., Morozov, D.: Persistent homology: theory and practice. In: European Congress of Mathematics, pp. 31–50. European Mathematical Society (2013)
9. Floriani, L.D., Fugacci, U., Iuricich, F., Magillo, P.: Morse complexes for shape segmentation and homological analysis: discrete models and algorithms. Comput. Graph. Forum **34**(2), 761–785 (2015)
10. Forman, R.: Morse theory for cell complexes. Adv. Math. **134**, 90–145 (1998)
11. Fugacci, U., Iuricich, F., De Floriani, L.: Computing discrete Morse complexes from simplicial complexes. Graph. Models **103**, 101023 (2019)
12. Guillou, P., Vidal, J., Tierny, J.: Discrete Morse sandwich: fast computation of persistence diagrams for scalar data-an algorithm and a benchmark. IEEE Trans. Visual Comput. Graphics **30**(4), 1897–1915 (2023)
13. Harker, S., Mischaikow, K., Mrozek, M., Nanda, V.: Discrete Morse theoretic algorithms for computing homology of complexes and maps. Found. Comput. Math. **14**, 151–184 (2014)
14. Joswig, M., Pfetsch, M.E.: Computing optimal Morse matchings. SIAM J. Discrete Math. **20**, 11–25 (2006)
15. Knudson, K.P., Scoville, N.A.: Discrete Morse theory for open complexes (2024). https://arxiv.org/abs/2402.12116. Working paper or preprint
16. Mischaikow, K., Nanda, V.: Morse theory for filtrations and efficient computation of persistent homology. Discrete Comput. Geom. **50**(2), 330–353 (2013)
17. Mrozek, M., Batko, B.: Coreduction homology algorithm. Discrete Comput. Geom. **41**(1), 96–118 (2009)
18. Robins, V., Wood, P.J., Sheppard, A.P.: Theory and algorithms for constructing discrete Morse complexes from grayscale digital images. IEEE Trans. Pattern Anal. Mach. Intell. **33**(8), 1646–1658 (2011)
19. Scoville, N.A.: Discrete Morse Theory, vol. 90. American Mathematical Society (2019)
20. Whitehead, J.H.C.: Simplicial spaces, nuclei and m-groups. Proc. Lond. Math. Soc. **2**(1), 243–327 (1939)

Hierarchical and Graph-Based Models, Analysis and Segmentation

Designing Connected Operators Using the Topological Tree of Shapes

Julien Mendes Forte[1(✉)], Nicolas Passat[2(✉)], and Yukiko Kenmochi[1(✉)]

[1] Normandie Université, UNICAEN, ENSICAEN, CNRS, GREYC, 14050 Caen, France
{julien.mendes-forte,yukiko.kenmochi}@unicaen.fr
[2] Université de Reims Champagne-Ardenne, CRESTIC, Reims, France
nicolas.passat@univ-reims.fr

Abstract. We recently introduced the topological tree of shapes as a new structure in the family of morphological trees. The topological tree of shapes is both a topological invariant of grey-level images and a hierarchical model of these images. Building upon these two properties, we explain how the topological tree of shapes can be used to relevantly design connected operators. In this theoretical article, we show that such operators present algebraic properties related to openings/closings, while also guaranteeing the preservation of the topological structure of the processed images. Software implementation available at https://github.com/jmendesf/TopologicalTreeOfShapes.

Keywords: Topological tree of shapes · Connected operators · Opening/closing · Topological invariance · Grey-level images

1 Introduction

Mathematical morphology provides a rich family of graph-based structures [17], and in particular trees, which can be used to model, analyze and/or process images. These trees are essentially defined from sequences of partitions and the way they can be ordered by the refinement relation. In the case where these partitions are total, the induced trees belong to the family of binary partition trees [28] and watershed trees [18] and they are mostly used for segmentation purpose. In the case where these partitions are partial, the induced trees belong to the family of the component trees [29] and trees of shapes [6] and they are mostly dedicated to model and/or process grey-level images.

In this article, we focus on this second family. In a recent study [21], we formalized the links that exist between the component trees (min-tree and max-tree) and the tree of shapes. To this end, we introduced the new notion of a complete tree of shapes, a tree composed by the nodes of the min- and max-trees and structured with respect to the two order relations inherited from the component tree and the adjacency tree [26], respectively. By clustering the nodes of the complete tree of shapes with respect to well-chosen equivalence relations, we can retrieve the tree of shapes, but also a new tree

This work was supported by *Région Normandie* (thesis grant RIN) and *Agence Nationale de la Recherche* (Grants ANR-22-CE45-0034 and ANR-23-CE45-0015).

M. H. F. Wilkinson and J. Kosinka (Eds.): DGMM 2025, LNCS 16296, pp. 195–209, 2026
https://doi.org/10.1007/978-3-032-09544-2_14

called the topological tree of shapes [20]. Similarly to the tree of shapes, the topological tree of shapes is an image model, i.e. it provides a way to hierarchically represent a grey-level image in a lossless way. It is also a topological invariant of this image, that characterizes its structure with respect to connectedness and strong deletability [24].

In previous works, we investigated how to build the topological tree of shapes of an image [14,21]. In this article, we now investigate how the topological tree of shapes can be involved in the design of connected operators [30,31]. In particular, we show that such connected operators exhibit various algebraic properties of openings and closings [25] while remaining topologically invariant, in the sense where the topological trees of shapes of the initial and final images are isomorphic.

This theoretical article is organized as follows. In Sect. 2, we recall related works on morphological trees. In Sect. 3, we provide definitions related to the topological tree of shapes. In Sect. 4, we show how an image can be reconstructed from its topological tree of shapes. In Sect. 5, we explain how to build connected operators from the topological tree of shapes while preserving topological and algebraic properties. In Sect. 6, we illustrate the behaviour of such connected operators. In Sect. 7, we provide concluding remarks. The proofs of the properties and propositions are provided in the research report [15].

2 Related Works

The partial partition trees can be used to model grey-level images. These trees are subdivided into two families, on the one hand the component tree [29] and its variants [12,23], on the other hand the tree of shapes [6] and its variants [3,32]. Beyond the scope of mathematical morphology, these trees are also considered under different names. In particular, in computational geometry, the max-tree is referred to as join tree, the min-tree as split-tree and the tree of shapes as contour tree [4]. The component tree family, which builds upon the (region-based) inclusion relation of the connected components of the threshold sets of the image, only requires that a grey-level image be defined in a space supporting connectedness. This weak hypothesis allows to use component trees not only on grey-level images, but more generally on valued graphs. The tree of shapes family, which builds upon the (contour-based) nested relation of external borders of the threshold sets of the image, requires to fulfill stronger topological hypotheses [6, Chapter 4]. In practice, such hypotheses are satisfied on standard models of digital topology [11,27] and efforts were also geared towards making the tree of shapes self-dual despite the usual open/closed duality of topological spaces [8].

Trees of the component tree and tree of shapes families are image models. This means that the images represented by these trees can be reconstructed in a lossless fashion from them. Such trees are then efficient structures for processing images. In particular, they allow to design connected operators [30,31] that do not create artificial boundaries in the resulting images. Such kinds of operators were proposed by following two main paradigms: optimal cut computation [9] (mostly for segmentation purpose) and node selection guided by attributes [2,10] (mostly for filtering purpose). In particular, these connected operators generally act as image simplification / quantification processes [5,16].

Partial partition trees as defined in the framework of mathematical morphology also provide a relevant way of describing the topological structure of grey-level images. On the one hand, the tree of shapes is the grey-level generalization of the adjacency tree [26], a popular topological invariant that describes the nesting of the foreground and background connected components of binary images. On the other hand, the component trees allow to retrieve the persistence diagram [7] of a 2-dimensional grey-level image, while more generally, persistent homology and morphological dynamics are deeply related in any dimension [1]. These considerations have notably motivated the involvement of component trees in deep learning procedures. In [22], a loss function is based on component trees and proved to be differentiable and then usable in neural network training [19]. In [33], self-supervised learning strategies are based on component trees so that neural networks learn the tree structure in pretext tasks.

In [21], we proposed a theoretical framework that describes the continuum between component trees and trees of shapes. In this context, we defined the unifying notion of a complete tree of shapes. The complete tree of shapes contains exactly the nodes of both the min- and the max-trees. By partitioning these nodes based on adequate equivalence classes, we can retrieve other trees. For instance, if the equivalence is related to the equality of the external borders of the connected components, then the complete tree of shapes can be turned into the tree of shapes by a decreasing homeomorphism. Alternatively, if the equivalence is related to the notion of strong deletability [24] between the connected components, the complete tree of shapes can be turned into the topological tree of shapes [20] by a decreasing homeomorphism. The tree of shapes and the topological tree of shapes are then strongly related, yet distinct trees. Similarly to the tree of shapes, the topological tree of shapes is an image model and then allows to design connected operators [5,16]. It is also a rich topological invariant. Based on these two virtues, our purpose is to study the connected operators that can be designed from the topological tree of shapes, and the ability of these operators to satisfy algebraic properties of morphological operators while preserving topological properties of images.

3 Topological Tree of Shapes

In this section, we provide the minimal set of definitions and notations required to make this article self-contained. More details are available in [21].

Let $\mathbb{U}$ be a set endowed with a topological structure that satisfies the hypotheses of [6, Chapter 4]. For the sake of readability, we will consider here that $\mathbb{U} = \mathbb{Z}^2$ endowed with the digital topology framework [27], and that the only authorized images are well-composed [13]. When $\mathbb{U} = \mathbb{Z}^2$, the elements of $\mathbb{U}$ are generally called pixels.

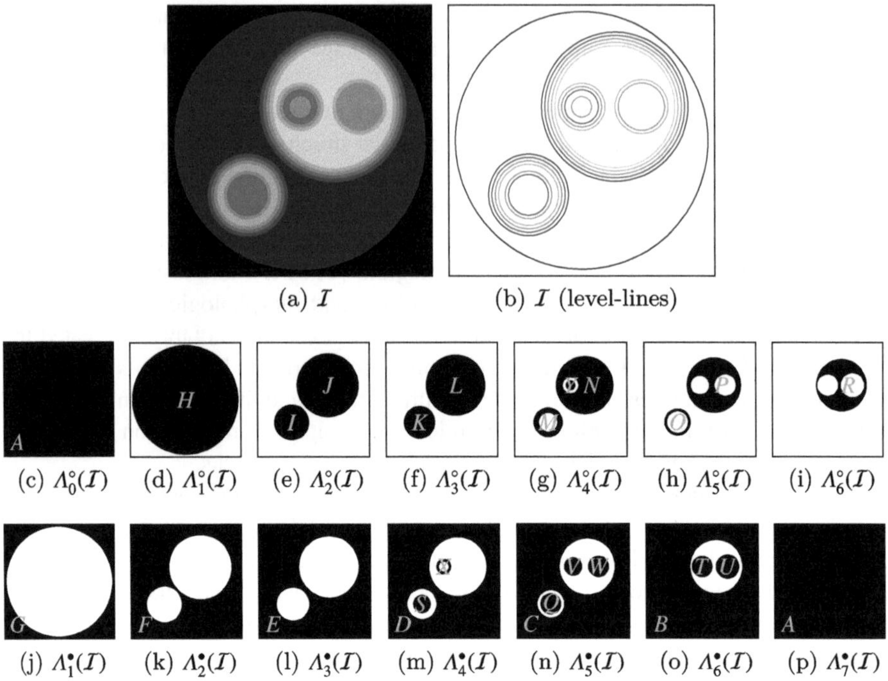

Fig. 1. (a) An image $\mathcal{I} : \mathbb{U} \to \mathbb{V}$ with $\mathbb{V} = [\![0, 7]\!]$ (from 0 in black to 7 in white). (b) The image $\mathcal{I}$ viewed by its level lines. (c–i) Upper threshold sets $\Lambda_v^\circ(\mathcal{I})$ (in black) of $\mathcal{I}$, from $v = 0$ (c) to 6 (i). (j–p) Lower threshold sets $\Lambda_v^\bullet(\mathcal{I})$ (in black) of $\mathcal{I}$, from $v = 1$ (j) to 7 (p).

Let $\mathbb{V}$ be a set endowed with a total order relation. Without loss of generality, we consider that $\mathbb{V} = [\![\bot, \top]\!] \subset \mathbb{Z}$ endowed with the canonical order $\leqslant$, with $\bot = 0$. In practice, $\mathbb{V}$ can be seen as a finite set of grey-level values.

Let $\mathcal{I} : \mathbb{U} \to \mathbb{V}$ (Fig. 1(a)). We assume that $\mathcal{I}^{-1}(\{\top\}) = \emptyset$. We also assume that $\mathcal{I}^{-1}(\mathbb{V} \setminus \{\bot\})$ is finite. In other words, no pixels in the image are assigned the maximal value $\top$ (which is useful when considering the complement of the image) and there is a finite number of pixels of non-minimal value (these pixels are lying in the finite window of the image).

Let $v \in \mathbb{V}$. The upper- and lower-threshold sets of $\mathcal{I}$ at value v (Fig. 1(c–p)) are the subsets of $\mathbb{U}$ defined as

$$\begin{aligned} \Lambda_v^\circ(\mathcal{I}) &= \{\mathbf{x} \in \mathbb{U} \mid v \leqslant \mathcal{I}(\mathbf{x})\} \\ \Lambda_v^\bullet(\mathcal{I}) &= \{\mathbf{x} \in \mathbb{U} \mid v > \mathcal{I}(\mathbf{x})\} \end{aligned} \tag{1}$$

Let $X \subseteq \mathbb{U}$. We denote by $\Pi[X] \subseteq 2^{\mathbb{U}}$ the set of the connected components of X. For all $v \in \mathbb{V}$, we define the following sets

$$\begin{aligned} \Theta_v^\circ &= \Pi[\Lambda_v^\circ(\mathcal{I})] \\ \Theta_v^\bullet &= \Pi[\Lambda_v^\bullet(\mathcal{I})] \\ \Theta_v &= \Theta_v^\circ \cup \Theta_v^\bullet \end{aligned} \quad \text{and} \quad \begin{aligned} \Theta^\circ &= \textstyle\bigcup_{v \in \mathbb{V}} \Theta_v^\circ \\ \Theta^\bullet &= \textstyle\bigcup_{v \in \mathbb{V}} \Theta_v^\bullet \\ \Theta &= \Theta^\circ \cup \Theta^\bullet = \textstyle\bigcup_{v \in \mathbb{V}} \Theta_v \end{aligned} \tag{2}$$

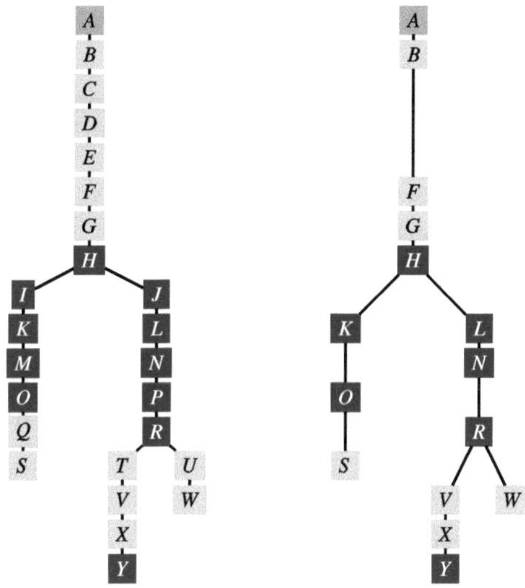

(a) $\mathfrak{T}_\Theta = (\Theta, \lhd)$ (b) $\mathfrak{T}_\Xi = (\Xi, \blacktriangleleft)$

Fig. 2. (a) The complete tree of shapes and (b) the topological tree of shapes of the image $\mathcal{I}$ of Fig. 1. The nodes in blue (resp. yellow) correspond to connected components of upper- (resp. lower) threshold sets. The node A (in green) is both a connected component of upper- (resp. lower) threshold sets. In the topological tree of shapes (b), each node is an equivalence class represented here (arbitrarily) by its minimum node. (Color figure online)

Let $X \subset \mathbb{U}$ be a (connected) set. We denote by $X^\tau \supseteq X$ the set obtained by saturating (i.e. closing the holes of) X.

Let $\sqsubseteq_\Theta$ be the partial order relation defined on Θ by

$$X \sqsubseteq_\Theta Y \iff (X \subseteq Y) \vee (X^\tau \subseteq Y^\tau) \tag{3}$$

We denote by $\lhd$ the reflexive-transitive reduction of $\sqsubseteq_\Theta$ on Θ.

Definition 1 (Complete tree of shapes [21]**).** *The complete tree of shapes of $\mathcal{I}$ is the tree $\mathfrak{T}_\Theta = (\Theta, \lhd)$. See Fig. 2(a).*

Let $D \subset \Lambda \subseteq \mathbb{U}$. Let $\iota : \Pi[\Lambda \setminus D] \to \Pi[\Lambda]$ and $\bar{\iota} : \Pi[\overline{\Lambda}] \to \Pi[\overline{\Lambda \setminus D}]$ be the two functions defined by $X \subseteq \iota(X)$ and $Y \subseteq \bar{\iota}(Y)$. In other words, $\iota(X)$ is the connected component of Λ that includes the connected component X of $\Lambda \setminus D$, while $\bar{\iota}(Y)$ is the connected component of the complement of $\Lambda \setminus D$ that includes the connected component Y of the complement of Λ. We say that D is a strongly deletable set (of Λ) if ι and $\bar{\iota}$ are bijective [24], meaning that the set D is topologically neutral, in the sense where it neither splits/removes nor joins/creates connected components in Λ or its complement. Let $X, Y \in \Theta$ be such that $X \lhd Y$. If X is unique for this property and $Y \setminus X$ is a strongly deletable set of Y, then we write $Y \searrow X$. We denote by $\sim$ the equivalence relation on Θ derived from $\searrow$. We denote by $\Xi = \Theta/{\sim}$ the set of all the equivalence classes of Θ with respect to $\sim$, i.e. the quotient set of Θ induced by $\sim$. Let $\sqsubseteq_\Xi$ be the order relation on Ξ

defined by

$$J \sqsubseteq_\Xi K \Longleftrightarrow \bigwedge^{\sqsubseteq_\Theta} J \sqsubseteq_\Theta \bigwedge^{\sqsubseteq_\Theta} K \tag{4}$$

We denote by ◂ the reflexive-transitive reduction of $\sqsubseteq_\Xi$ on Ξ.

Definition 2 (Topological tree of shapes [20]**).** *The topological tree of shapes of I is the tree $\mathfrak{T}_\Xi = (\Xi, ◂)$. See Fig. 2(b).*

Proposition 3 ([21])**.** *The equivalence $\sim$ induces a decreasing homeomorphism from the complete tree of shapes to the topological tree of shapes.*

This derives from the fact that a given node in the topological tree of shapes is a collection of node(s) from the complete tree of shapes along a branch (with no bifurcation), with respect to strong deletability.

4 Image Reconstruction from the Topological Tree of Shapes

The usual partial partition trees (component trees, trees of shapes) are image models. In particular, we can reconstruct their associated image from them in a lossless way. This can also be the case for the topological tree of shapes if we rely on the underlying complete tree of shapes.

4.1 The Case of the Complete Tree of Shapes

In order to reconstruct an image from a partial partition tree, it is mandatory to store two kinds of information at each node X: (1) its proper part $\rho(X) \subseteq X$ that corresponds to the part of X that spatially contributes to the definition of the image, and (2) its valuation $\eta(X) \in \mathbb{V}$ that spectrally contributes to the definition of the image. In the case of the complete tree of shapes, these information are defined as follows.

Definition 4 (Proper part of a node in the complete tree of shapes). *Let $X \in \Theta$. The proper part of X in $\mathfrak{T}_\Theta = (\Theta, \lhd)$ is defined by*

$$\rho(X) = X \setminus \bigcup_{Y \lhd X} Y^\tau \tag{5}$$

Less formally, the proper part of X is obtained by removing from X its saturated children nodes. We have $\bigcup_{X \in \Theta} \rho(X) = \mathbb{U}$ and $\rho(X) \cap \rho(Y) = \emptyset$ for any distinct $X, Y \in \Theta$. However, $\rho(\Theta)$ may not be a partition of $\mathbb{U}$ since we may have $\rho(X) = \emptyset$ for some $X \in \Theta$ [21]. We set $\Theta_\rho = \{X \in \Theta \mid \rho(X) \neq \emptyset\}$. Then, $\rho(\Theta_\rho)$ is a partition of $\mathbb{U}$.

Definition 5 (Valuation of a node in the complete tree of shapes). *Let $X \in \Theta$. The valuation $\eta(X)$ of X in $\mathfrak{T}_\Theta = (\Theta, \lhd)$ is defined by*

$$\eta(X) = \begin{cases} \bigvee^{\leqslant} \mathbb{I}(X) & \text{if } X \in \Theta^\circ \\ \bigwedge^{\leqslant} \mathbb{I}(X) - 1 & \text{if } X \in \Theta^\bullet \end{cases} \tag{6}$$

where $\mathbb{I}(X) = \{v \in \mathbb{V} \mid X \in \Theta_v\}$.

The non-symmetric formulation of Eq. (6) induced by the "−1" term derives from the non-symmetric formulation of the threshold sets in Eq. (1) induced by the $\leqslant$ and $>$ relations. Note that $\eta(X)$ is also commonly referred to as the grey-level value of the node X.

Following the same reconstruction paradigm as proposed in [10] for the component tree, it is possible to reconstruct the image $\mathcal{I}$ from its complete tree of shapes $\mathfrak{T}_\Theta$ as follows, for any $\mathbf{x} \in \mathbb{U}$

$$\mathcal{I}(\mathbf{x}) = \bigvee_{X \in \Theta}^{\leqslant} \eta(X) \cdot \mathbf{1}_{\rho(X)}(\mathbf{x}) = \bigvee_{X \in \Theta_\rho}^{\leqslant} \eta(X) \cdot \mathbf{1}_{\rho(X)}(\mathbf{x}) \tag{7}$$

where $\mathbf{1}_A : \mathbb{U} \to \{0, 1\}$ is the indicatrix function of the set $A \subseteq \mathbb{U}$.

4.2 The Case of the Topological Tree of Shapes

A node $K \in \Xi$ of the topological tree of shapes $\mathfrak{T}_\Xi = (\Xi, \blacktriangleleft)$ is an equivalence class $K \subseteq \Theta$ of nodes of the complete tree of shapes $\mathfrak{T}_\Theta = (\Theta, \lhd)$. The proper part of a node $K \in \Xi$ directly derives from the proper parts of the nodes $X \in K$ that compose it.

Definition 6 (Proper part of a node in the topological tree of shapes). *Let $K \in \Xi$. The proper part of K in $\mathfrak{T}_\Xi = (\Xi, \blacktriangleleft)$ is defined by*

$$\rho(K) = \bigcup_{X \in K} \rho(X) = \bigcup_{X \in K \cap \Theta_\rho} \rho(X) \tag{8}$$

We have $\bigcup_{K \in \Xi} \rho(K) = \mathbb{U}$ and $\rho(J) \cap \rho(K) = \emptyset$ for any distinct $J, K \in \Xi$. However, $\rho(\Xi)$ may not be a partition of $\mathbb{U}$ since we may have $\rho(K) = \emptyset$ for some $K \in \Xi$. We set $\Xi_\rho = \{K \in \Xi \mid \rho(K) \neq \emptyset\}$. Then, $\rho(\Xi_\rho)$ is a partition of $\mathbb{U}$. For any $K \in \Xi$, we set $K_\rho = K \cap \Theta_\rho$.

By contrast with a node $X \in \Theta_\rho$ of a complete tree of shapes, that contributes to the image by exactly one value $\eta(X)$, a node $K \in \Xi_\rho$ of a topological tree of shapes may contribute to the image by many values, i.e. one value $\eta(X)$ per node $X \in K_\rho$. It follows that the valuation of a node in the topological tree of shapes is defined as a function instead of a value.

Definition 7 (Valuation of a node in the topological tree of shapes). *Let $K \in \Xi$. The valuation $\eta(K)$, also denoted by η_K, of K in $\mathfrak{T}_\Xi = (\Xi, \blacktriangleleft)$ is defined by the function*

$$\left| \begin{array}{rcl} \eta_K : K & \longrightarrow & \mathbb{V} \\ X & \longmapsto & \eta(X) \end{array} \right. \tag{9}$$

where $\eta(X)$ is the valuation of X in the complete tree of shapes (Eq. (6)).

It is then possible to reconstruct the image $\mathcal{I}$ from its topological tree of shapes $\mathfrak{T}_\Xi$ as follows, for any $\mathbf{x} \in \mathbb{U}$

$$\mathcal{I}(\mathbf{x}) = \bigvee_{K\in\Xi}^{\leqslant} \bigvee_{X\in K}^{\leqslant} \eta_K(X) \cdot \mathbf{1}_{\rho(X)}(\mathbf{x}) = \bigvee_{K\in\Xi_\rho}^{\leqslant} \bigvee_{X\in K_\rho}^{\leqslant} \eta_K(X) \cdot \mathbf{1}_{\rho(X)}(\mathbf{x}) \tag{10}$$

5 Building Connected Operators from the Topological Tree of Shapes

5.1 Standard Connected Operators

From Eq. (10), we can reconstruct an image $\mathcal{I}$ from its topological tree of shapes $\mathfrak{T}_\Xi$. It is also possible to design morphological operators Γ_υ by substituting a valuation functional υ to η in Eq. (10). (Note that Γ_η is then the identity operator.)

Let us consider such a functional υ. For each $K \in \Xi$, we have $\upsilon(K) = \upsilon_K : K \to \mathbb{V}$. Let $\mathcal{F} : \mathbb{U} \to \mathbb{V}$ be the image obtained as $\mathcal{F} = \Gamma_\upsilon(\mathcal{I})$. The morphological operator Γ_υ is a connected operator.

Property 8 (Connectedness). *Let* $\mathbf{x}, \mathbf{y} \in \mathbb{U}$. *We have*

$$((\mathbf{x} \frown \mathbf{y}) \wedge (\mathcal{I}(\mathbf{x}) = \mathcal{I}(\mathbf{y}))) \Longrightarrow (\mathcal{F}(\mathbf{x}) = \mathcal{F}(\mathbf{y})) \tag{11}$$

with $\frown$ *the adjacency relation on* $\mathbb{U}$.

In other words, no new contours are created by applying Γ_υ. However, there is no guarantee that Γ_υ preserve the spectral integrity of the processed image.

First, we can consider value integrity, to ensure that no new value will be locally created by application of Γ_υ. Let us consider the following property

($\mathscr{P}_1$) $\forall K \in \Xi, \upsilon_K(K_\rho) \subseteq \eta_K(K_\rho)$

We have the following result.

Property 9 (Local value preservation). *If* ($\mathscr{P}_1$) *holds, then we have*

$$\forall \mathbf{x} \in \rho(K), \exists \mathbf{y} \in \rho(K), \mathcal{F}(\mathbf{x}) = \mathcal{I}(\mathbf{y}) \tag{12}$$

Second, we can consider differential integrity, to ensure that the sign of the gradients will not be altered by application of Γ_υ. Let $K \in \Xi \setminus \{\mathbb{U}\}$. There exists $J \in \Xi$ such that $K \blacktriangleleft J$. We set $\widetilde{K} = K \cup \bigwedge^{\sqsubseteq_\Theta} J$. We observe that K and $\widetilde{K}$ are totally ordered by $\sqsubseteq_\Theta$. We also have $K \subseteq \Theta^\circ$ or $K \subseteq \Theta^\bullet$. We set $\eta_{\widetilde{K}} : \widetilde{K} \to \mathbb{V}$ as $\eta_{\widetilde{K}}(\mathbf{x}) = \eta_K(\mathbf{x})$ if $\mathbf{x} \in K$ and $\eta_{\widetilde{K}}(\bigwedge^{\sqsubseteq_\Theta} J) = \eta_J(\bigwedge^{\sqsubseteq_\Theta} J)$. If $K \subseteq \Theta^\circ$ (resp. $K \subseteq \Theta^\bullet$) then η_K induces an isomorphism between $(K, \sqsubseteq_\Theta)$ and $(\eta_K(K), \geqslant)$ (resp. $(\eta_K(K), \leqslant)$). It follows that $\eta_{\widetilde{K}}$ also induces an isomorphism between $(\widetilde{K}, \sqsubseteq_\Theta)$ and $(\eta_{\widetilde{K}}(\widetilde{K}), \geqslant)$ (resp. $(\eta_{\widetilde{K}}(\widetilde{K}), \leqslant)$), i.e.

$$\forall X, Y \in \widetilde{K}, (X \sqsubseteq_\Theta Y) \Longleftrightarrow \left(\eta_{\widetilde{K}}(X) \geqslant \eta_{\widetilde{K}}(Y)\right) \quad (\text{resp. } \left(\eta_{\widetilde{K}}(X) \leqslant \eta_{\widetilde{K}}(Y)\right)) \tag{13}$$

Less formally, within a node of the topological tree of shapes, the structural ordering of nodes (via $\sqsubseteq_\Theta$) is given by the grey-level ordering of their valuation: progressing along the inclusion chain corresponds to a monotonic evolution in values. Let us consider the following property

$(\mathscr{P}_2)$ $\forall K \in \Xi \setminus \{\mathbb{U}\}, \forall X, Y \in \widetilde{K}, \left(\eta_{\widetilde{K}}(X) \leqslant \eta_{\widetilde{K}}(Y)\right) \Longrightarrow \left(\upsilon_{\widetilde{K}}(X) \leqslant \upsilon_{\widetilde{K}}(Y)\right)$

where $\upsilon_{\widetilde{K}}$ is defined from υ_K analogously to $\eta_{\widetilde{K}}$ and η_K. We then have the following result.

Property 10 (Gradient sign preservation). *If $(\mathscr{P}_2)$ holds, then we have*

$$\forall \mathbf{x}, \mathbf{y} \in \mathbb{U}, ((\mathbf{x} \frown \mathbf{y}) \wedge (\mathcal{I}(\mathbf{x}) \leqslant \mathcal{I}(\mathbf{y}))) \Longrightarrow (\mathcal{F}(\mathbf{x}) \leqslant \mathcal{F}(\mathbf{y})) \tag{14}$$

5.2 Topology-Preserving Connected Operators

In the previous section, we proposed sufficient conditions for guaranteeing basic properties (local value and gradient sign preservation) of connected operators built upon the topological tree of shapes. We now go one step further by proposing sufficient conditions for guaranteeing topological properties of connected operators built upon the topological tree of shapes.

We consider the following property, that guarantees that the minimal elements of the nodes $K \in \Xi$ preserve their valuation:

$(\mathscr{P}_3)$ $\forall K \in \Xi, \upsilon_K(\bigwedge^{\sqsubseteq_\theta} K) = \eta_K(\bigwedge^{\sqsubseteq_\theta} K)$

We then have the following result.

Proposition 11 (Topological tree of shapes preservation). *Let $\mathcal{I}$ be an image and $\mathfrak{T}_\mathcal{I} = (\Xi_\mathcal{I}, \blacktriangleleft_\mathcal{I})$ its topological tree of shapes. Let υ be a valuation functional. Let $\mathcal{F}$ be the image obtained from $\mathcal{I}$ by application of the operator induced by υ, and $\mathfrak{T}_\mathcal{F} = (\Xi_\mathcal{F}, \blacktriangleleft_\mathcal{F})$ its topological tree of shapes. If $(\mathscr{P}_2)$ and $(\mathscr{P}_3)$ hold, then there exists a bijection $f : K \in \Xi_\mathcal{I} \mapsto \widehat{K} \in \Xi_\mathcal{F}$ that induces an isomorphism between $\mathfrak{T}_\mathcal{I}$ and $\mathfrak{T}_\mathcal{F}$, i.e.*

$$\forall J, K \in \Xi_\mathcal{I}, (K \blacktriangleleft_\mathcal{I} J) \Longleftrightarrow \left(\widehat{K} \blacktriangleleft_\mathcal{F} \widehat{J}\right) \tag{15}$$

5.3 Towards Openings and Closings

We finally investigate a family of connected operators based on the topological tree of shapes which present algebraic properties related to openings and closings.

Let υ be a valuation functional that satisfies $(\mathscr{P}_3)$. Let $K \in \Xi$. We have to define the values $\upsilon_K(X)$ for any $X \in K \setminus \{\bigwedge^{\sqsubseteq_\theta} K\}$. We propose, in particular, four ways to design functionals υ for any $X \in K \setminus \{\bigwedge^{\sqsubseteq_\theta} K\}$

$$\upsilon_K^{\wedge\wedge}(X) = \eta_K\left(\bigwedge^{\sqsubseteq_\theta} K\right) \tag{16}$$

$$\upsilon_K^{\wedge\vee}(X) = \begin{cases} \eta_K\left(\bigwedge^{\sqsubseteq_\theta} K\right) & \text{if } K \subseteq \Theta^\circ \\ \eta_J\left(\bigwedge^{\sqsubseteq_\theta} J\right) & \text{if } K \subseteq \Theta^\bullet \end{cases} \tag{17}$$

$$\upsilon_K^{\vee\wedge}(X) = \begin{cases} \eta_J\left(\bigwedge^{\sqsubseteq_\theta} J\right) & \text{if } K \subseteq \Theta^\circ \\ \eta_K\left(\bigwedge^{\sqsubseteq_\theta} K\right) & \text{if } K \subseteq \Theta^\bullet \end{cases} \tag{18}$$

$$\upsilon_K^{\vee\vee}(X) = \eta_J\left(\bigwedge^{\sqsubseteq_\theta} J\right) \tag{19}$$

with $K \blacktriangleleft_{\mathcal{I}} J$. It is plain that for these four functionals, $(\mathscr{P}_2)$ holds. It follows that they all preserve gradient sign (Proposition 10). Since $(\mathscr{P}_3)$ also holds by definition, it follows that they all preserve the topological tree of shapes structure (Proposition 11).

In addition, they also present the following algebraic properties. Let Γ_υ be the operator induced by the functional υ.

Property 12 (Idempotence). *For any functional υ defined in Eqs. (16–19), we have*

$$\Gamma_\upsilon \circ \Gamma_\upsilon = \Gamma_\upsilon \tag{20}$$

Property 13 ((Anti-)extensivity). *Let $\mathbf{x}^\circ \in \bigcup_{X \in \Theta^\circ} \rho(X)$ and $\mathbf{x}^\bullet \in \bigcup_{X \in \Theta^\bullet} \rho(X)$. For any functional υ defined in Eqs. (16–19), we have*

$$\Gamma_{\upsilon^{\wedge\star}}(\mathcal{I})(\mathbf{x}^\circ) \geqslant \mathcal{I}(\mathbf{x}^\circ) \tag{21}$$

$$\Gamma_{\upsilon^{\vee\star}}(\mathcal{I})(\mathbf{x}^\circ) \leqslant \mathcal{I}(\mathbf{x}^\circ) \tag{22}$$

$$\Gamma_{\upsilon^{\star\vee}}(\mathcal{I})(\mathbf{x}^\bullet) \geqslant \mathcal{I}(\mathbf{x}^\bullet) \tag{23}$$

$$\Gamma_{\upsilon^{\star\wedge}}(\mathcal{I})(\mathbf{x}^\bullet) \leqslant \mathcal{I}(\mathbf{x}^\bullet) \tag{24}$$

where $\star$ can be $\wedge$ or $\vee$. It follows in particular that $\Gamma_{\upsilon^{\wedge\vee}}$ is extensive while $\Gamma_{\upsilon^{\vee\wedge}}$ is anti-extensive.

Remark 14 (Non-increasingness). *In general, for any functional υ defined in Eqs. (16–19), Γ_υ is not increasing. Indeed, for two images $\mathcal{I}_1$ and $\mathcal{I}_2$ we may have*

$$(\forall \mathbf{x} \in \mathbb{U}, \mathcal{I}_1(\mathbf{x}) \leqslant \mathcal{I}_2(\mathbf{x})) \wedge (\exists \mathbf{y} \in \mathbb{U}, \Gamma_\upsilon(\mathcal{I}_1)(\mathbf{y}) > \Gamma_\upsilon(\mathcal{I}_2)(\mathbf{y})) \tag{25}$$

This default of increasingness derives from the fact that the nodes of the topological tree of shapes gather nodes of the complete tree of shapes based on properties which are not correlated with the grey-level values. For comparable reasons, this is also the case of other connected operators based on morphological trees, see e.g. [34].

An opening (resp a closing) is an idempotent, anti-extensive (resp. extensive) and increasing operator. The operator $\Gamma_{\upsilon^{\wedge\vee}}$ (Eq. (17)) is idempotent and extensive. The operator $\Gamma_{\upsilon^{\vee\wedge}}$ (Eq. (18)) is idempotent and anti-extensive. Except the increasingness property, these two operators present the algebraic properties required by a closing and an opening, respectively. In addition, they present the advantage to preserve the topology of the processed image, which is generally not the case of usual opening and closing operators. Indeed the topological tree of shapes is not altered by these operators.

The operators $\Gamma_{\upsilon^{\wedge\wedge}}$ (Eq. (16)) and $\Gamma_{\upsilon^{\vee\vee}}$ (Eq. (19)) are neither extensive nor anti-extensive. However, they present extensive/anti-extensive properties when considered on the "ascending" or "descending" parts of the image.

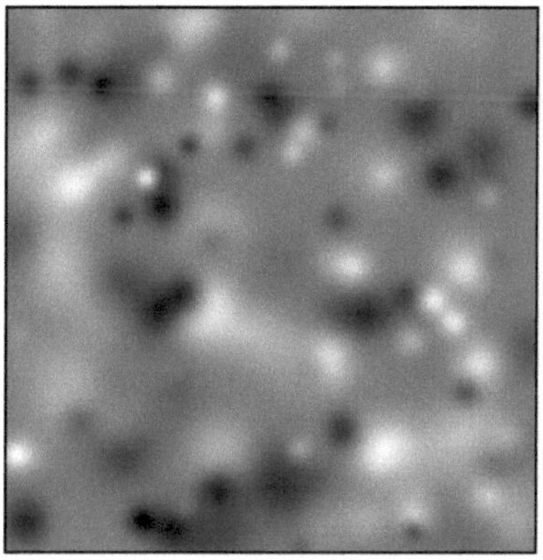

(a) $\mathcal{I}$

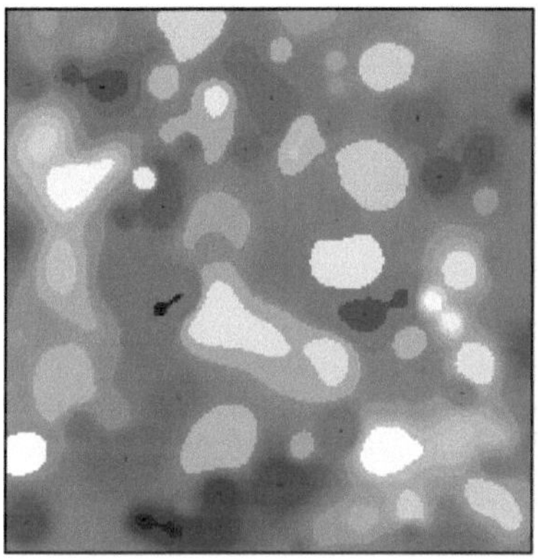

(b) $\Gamma_{\upsilon^{\wedge\vee}}(\mathcal{I})$ ("quasi-closing")

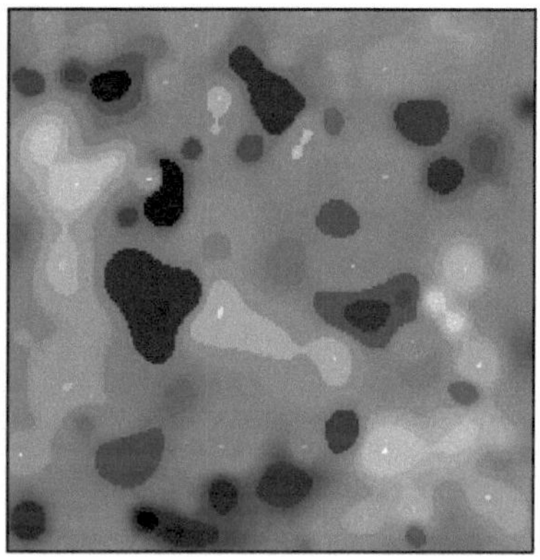

(c) $\Gamma_{\upsilon^{\vee\wedge}}(\mathcal{I})$ ("quasi-opening")

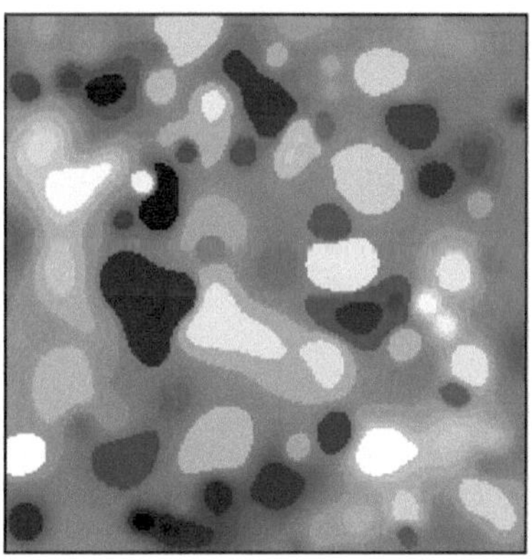

(d) $\Gamma_{\upsilon^{\wedge\wedge}}(\mathcal{I})$

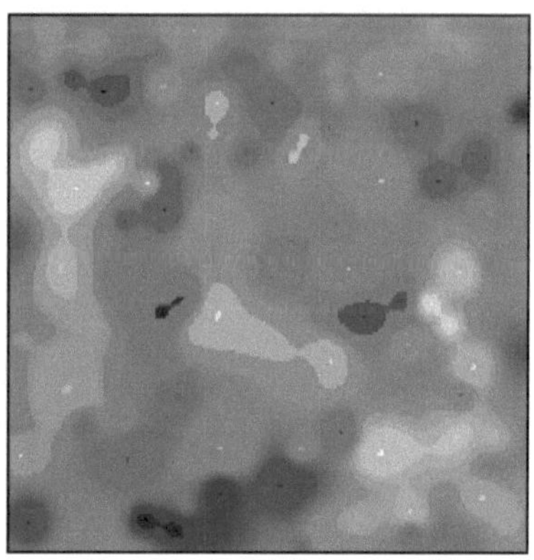

(e) $\Gamma_{\upsilon^{\vee\vee}}(\mathcal{I})$

Fig. 3. (a) Initial image $\mathcal{I}$ of a synthetic Gaussian landscape. (b–e) Images obtained from $\mathcal{I}$ by applying the operators $\Gamma_{\upsilon^{\star\star}}$ built from valuation functionals $\upsilon^{\star\star}$ of Eqs. (16–19).

(a) I

(b) $\Gamma_{\upsilon^{\wedge\vee}}(I)$ (“quasi-closing”)

(c) $\Gamma_{\upsilon^{\vee\wedge}}(I)$ (“quasi-opening”)

(d) $\Gamma_{\upsilon^{\wedge\wedge}}(I)$

(e) $\Gamma_{\upsilon^{\vee\vee}}(I)$

Fig. 4. (a) Initial image I (Landsat-7 image courtesy of the U.S. Geological Survey). (b–e) Images obtained from I by applying the operators $\Gamma_{\upsilon^{\star\star}}$ built from valuation functionals $\upsilon^{\star\star}$ of Eqs. (16–19).

6 Examples

In Figs. 3–4, we illustrate the behaviour of the four operators $\Gamma_{\upsilon^{\star\star}}$ defined in Eqs. (16–19). The operator $\Gamma_{\upsilon^{\wedge\vee}}$ is idempotent and extensive. It then acts as a "quasi-closing" (Figs. 3–4(b)). The operator $\Gamma_{\upsilon^{\vee\wedge}}$ is idempotent and antiextensive. It then acts as a "quasi-opening" (Figs. 3–4(c)). The operator $\Gamma_{\upsilon^{\wedge\wedge}}$ is idempotent, extensive in the ascending parts and anti-extensive in the descending parts of the image. It then increases the surface of the "peaks" and "craters" of the image (Figs. 3–4(d)). The operator $\Gamma_{\upsilon^{\vee\vee}}$ is idempotent, anti-extensive in the ascending parts and extensive in the descending parts of the image. It then decreases the surface of the "peaks" and "craters", generating an image as flat as possible (Figs. 3–4(e)). Beyond their respective behaviours, these operator have the ability to preserve the topological tree of shapes between the initial and the final image. Codes available at https://github.com/jmendesf/TopologicalTreeOfShapes.

7 Conclusion

In this article, we started to investigate how connected operators could be built from the topological tree of shapes. We observed that it is possible to ensure relevant basic properties (no creation of new values, gradient sign preservation), but also to guarantee the preservation the topological tree of shapes between the initial and the final image. This last property is important, as it opens the way to the efficient design of topology-aware morphological operators. We also observed that in a context of topology preservation, it was possible to build connected operators satisfying some algebraic properties (idempotence, (anti-)extensivity) required by the class of openings and closings. At this stage, the designed operators do not satisfy the increasingness property, as a side effect of gathering the nodes into equivalence classes induced by the chosen topolocal invariant. Another side effect of considering/preserving equivalence classes of nodes is that for real (e.g. natural) images, the noise and high level of detail of the image generally leads to a complex topological structure, i.e. small equivalence classes. Under these hypotheses, the effects of a connected operator that guarantees the preservation of the topological tree of shapes may be limited (see e.g. Fig. 4 vs. Fig. 3), and even null in the worst cases (for singleton equivalence classes). This problem sheds light on the necessity to discriminate topological information from topological noise in the images. In our next works, we will investigate how these various issues may be tackled. Since the proposed topological tree of shapes relies on the notion of strong deletability, we may also investigate the effect of considering alternative topological invariance paradigms (e.g. homology, homotopy type).

References

1. Boutry, N., Najman, L., Géraud, T.: Some equivalence relation between persistent homology and morphological dynamics. J. Math. Imaging Vision **64**, 807–824 (2022)
2. Breen, E.J., Jones, R.: Attribute openings, thinnings, and granulometries. Comput. Vis. Image Underst. **64**, 377–389 (1996)

3. Carlinet, E., Géraud, T.: MToS: a tree of shapes for multivariate images. IEEE Trans. Image Process. **24**, 5330–5342 (2015)
4. Carr, H., Snoeyink, J., Axen, U.: Computing contour trees in all dimensions. Comput. Geom. **24**, 75–94 (2003)
5. Caselles, V., Monasse, P.: Grain filters. J. Math. Imaging Vision **17**, 249–270 (2002)
6. Caselles, V., Monasse, P.: Geometric Description of Images as Topographic Maps. Lecture Notes in Mathematics. vol 1984, Springer, Cham (2010)
7. Edelsbrunner, H., Harrer, J.: Persistent homology - a survey. Contemp. Math. **453**, 257–282 (2008)
8. Géraud, T., Carlinet, E., Crozet, S.: Self-duality and digital topology: links between the morphological tree of shapes and well-composed gray-level images. In: Mathematical Morphology and Its Applications to Signal and Image Processing. ISMM 2015. LNCS, vol 9082. Springer., pp. 573–584 (2015)
9. Guigues, L., Cocquerez, J.P., Le Men, H.: Scale-sets image analysis. Int. J. Comput. Vision **68**, 289–317 (2006)
10. Jones, R.: Connected filtering and segmentation using component trees. Comput. Vis. Image Underst. **75**, 215–228 (1999)
11. Kovalevsky, V.A.: Finite topology as applied to image analysis. Comput. Vision Graph. Image Process. **46**, 141–161 (1989)
12. Kurtz, C., Naegel, B., Passat, N.: Connected filtering based on multivalued component-trees. IEEE Trans. Image Process. **23**, 5152–5164 (2014)
13. Latecki, L.J., Eckhardt, U., Rosenfeld, A.: Well-composed sets. Comput. Vis. Image Underst. **61**, 70–83 (1995)
14. Mendes Forte, J., Passat, N., Kenmochi, Y.: Building the topological tree of shapes from the tree of shapes. In: Discrete Geometry and Mathematical Morphology. DGMM 2024. LNCS, vol 14605. Springer., pp. 271–285 (2024)
15. Mendes Forte, J., Passat, N., Kenmochi, Y.: Connected filters from the topological tree of shapes. Technical report (2025). https://hal.science/hal-04994612
16. Mendes Forte, J., Passat, N., Shimizu, A., Kenmochi, Y.: Consistent connected operators based on trees of shapes. SIAM J. Imaging Sci. In press
17. Najman, L., Cousty, J.: A graph-based mathematical morphology reader. Pattern Recogn. Lett. **47**, 3–17 (2014)
18. Najman, L., Schmitt, M.: Geodesic saliency of watershed contours and hierarchical segmentation. IEEE Trans. Pattern Anal. Mach. Intell. **18**, 1163–1173 (1996)
19. Nguyen, A.Q., Cousty, J., Kenmochi, Y., Higashiyama, S., Kawabe, J., Shimizu, A.: Multifactor component tree loss function: a topology-preserving method for skeleton segmentation from bone scintigrams. In: Topology- and Graph-Informed Imaging Informatics. TGI3 2024. LNCS, vol 15239. Springer., pp. 66–75 (2024)
20. Passat, N., Kenmochi, Y.: A topological tree of shapes. In: DGMM, pp. 221–235 (2022)
21. Passat, N., Mendes Forte, J., Kenmochi, Y.: Morphological hierarchies: a unifying framework with new trees. J. Math. Imaging Vision **65**, 718–753 (2023)
22. Perret, B., Cousty, J.: Component tree loss function: definition and optimization. In: Discrete Geometry and Mathematical Morphology. DGMM 2022. LNCS, vol 13493. Springer., pp. 248–260 (2022)
23. Perret, B., Lefèvre, S., Collet, C., Slezak, É.: Hyperconnections and hierarchical representations for grayscale and multiband image processing. IEEE Trans. Image Process. **21**, 14–27 (2012)
24. Ronse, C.: A topological characterization of thinning. Theoret. Comput. Sci. **43**, 31–41 (1986)
25. Ronse, C., Heijmans, H.J.A.M.: The algebraic basis of mathematical morphology: II. Openings and closings. GIP Image Underst. **54**, 74–97 (1991)

26. Rosenfeld, A.: Adjacency in digital pictures. Inf. Control **26**, 24–33 (1974)
27. Rosenfeld, A.: Digital topology. Am. Math. Mon. **86**, 621–630 (1979)
28. Salembier, P., Garrido, L.: Binary partition tree as an efficient representation for image processing, segmentation, and information retrieval. IEEE Trans. Image Process. **9**, 561–576 (2000)
29. Salembier, P., Oliveras, A., Garrido, L.: Anti-extensive connected operators for image and sequence processing. IEEE Trans. Image Process. **7**, 555–570 (1998)
30. Salembier, P., Serra, J.: Flat zones filtering, connected operators, and filters by reconstruction. IEEE Trans. Image Process. **4**, 1153–1160 (1995)
31. Salembier, P., Wilkinson, M.H.F.: Connected operators. IEEE Signal Process. Mag. **26**, 136–157 (2009)
32. Song, Y., Zhang, A.: Monotonic tree. In: Discrete Geometry for Computer Imagery. DGCI 2002. LNCS, vol 2301. Springer., pp. 114–123 (2002)
33. Tang, Q., Du, B., Xu, Y.: Self-supervised learning based on max-tree representation for medical image segmentation. In: International Joint Conference on Neural Networks. IJCNN 2022., pp. 1–6 (2022)
34. Urbach, E.R., Roerdink, J.B.T.M., Wilkinson, M.H.: Connected shape-size pattern spectra for rotation and scale-invariant classification of gray-scale images. IEEE Trans. Pattern Anal. Mach. Intell. **29**, 272–285 (2007)

Efficient Connected Alternating Sequential Filters Based on Component Trees

Wonder A. L. Alves[1], Nicolas Passat[2], Dennis J. Silva[2,3,4(✉)], Alexandre Morimitsu[4], and Ronaldo F. Hashimoto[4]

[1] Informatics and Knowledge Management Graduate Program, Universidade Nove de Julho, São Paulo, Brazil

[2] Université de Reims Champagne Ardenne, CRESTIC, Reims, France
dennis-jose.da-silva@univ-reims.fr

[3] Université de Reims Champagne Ardenne, MEDyC, UMR CNRS 7369, Reims, France

[4] Department of Computer Science, Institute of Mathematics and Statistics, Universidade de São Paulo, São Paulo, Brazil

Abstract. Alternating Sequential Filters (ASFs) rely on the iterated application of openings and closings of increasing strength. Since connected openings and closings can be designed from the component tree, it is possible to define Connected ASFs (CASFs) by alternatively pruning the branches of the dual max- and min-trees of an image. The main drawback of this approach is that pruning one of the trees modifies not only the image but also the dual tree. Recomputing a component tree at each iteration of the process is computationally expensive. In this article, we show how a component tree can be efficiently updated (avoiding reconstruction) when its dual tree is pruned. We build upon this algorithmic scheme to propose a computationally efficient CASF algorithm.

Keywords: Alternating Sequential Filters · Connected operators · Component Tree · Algorithmics

1 Introduction

The component tree was popularized in the field of mathematical morphology at the end of the 90's [15]. This morphological tree is especially well fitted for modeling grey-level images. By encoding the inclusion relation on the connected components over the successive threshold sets of a grey-level image, it allows to represent in a lossless way the information carried by the image in a mixed spatial/spectral space. The strength of the component tree derives, on the one hand, from its low computational cost. This assertion holds both for its construction [1,3,6,8] and for its processing, in particular based on attribute filtering [2,9,19]. On the other hand, by pruning a max-tree (resp. min-tree), one can easily compute openings (resp. closings) [12] in the framework of connected operators [14].

Alternating Sequential Filters (ASFs), are a class of morphological operators that apply successive, alternated operations of openings and closings with increasing strength [10,11,16,17]. These openings/closings can be designed from the classical

M. H. F. Wilkinson and J. Kosinka (Eds.): DGMM 2025, LNCS 16296, pp. 210–223, 2026.
https://doi.org/10.1007/978-3-032-09544-2_15

paradigm of structuring element. However, any valid opening/closing can be involved, and in particular those designed from component trees. Building ASFs from component trees leads in particular to Connected ASFs (CASFs), which gather the behaviours of ASFs and connected operators. Building a CASF from component trees requires to process, at each iteration, either the max- or the min-tree of the current image and to prune that tree. This pruning modifies the image but also its dual tree, involved at the next iteration. Recomputing one of the trees at each step is indeed tractable, but this leads to low computational performance, as highlighted in [5].

In this article, we address the computational bottleneck of CASFs built from component trees. To reach that goal, we propose an algorithmic scheme that allows to update a component tree when its dual tree is pruned, thus avoiding its whole reconstruction. We then embed this scheme in the design of CASFs and we experimentally compare their computational cost to CASFs that reconstruct ex nihilo one tree at each iteration.

The remainder of this article is organized as follows. Section 2 recalls the basics of component trees. Section 3 presents the theoretical aspects of the update of the component tree, the induced component tree updating algorithms, and the CASF approach based on these updating algorithms. Section 4 presents experimental results. Section 5 concludes the article.

2 Basics on the Component Tree

A grey-level image is defined as a mapping $f : \mathcal{D} \to \mathbb{K}$. The domain $\mathcal{D}$ and the co-domain $\mathbb{K}$ are defined as finite sets. The elements of $\mathcal{D}$ are called points. The elements of $\mathbb{K}$ are called values. Without loss of generality, we set $\mathbb{K} = [\![0, K]\!] \subset \mathbb{Z}$, namely the interval of $K + 1$ integers from 0 to K and we assume that $f(p) \neq K$ for any $p \in \mathcal{D}$. The definition of a component tree only requires that the domain $\mathcal{D}$ be endowed with a non-directed graph structure. However, since we will simultaneously work with the max- and min-trees, we consider here stronger hypotheses. In particular, we assume that $\mathcal{D}$ is endowed with a topological structure. For the sake of readability, in this article, we consider $\mathcal{D}$ as a connected subset of $\mathbb{Z}^2$ endowed with the digital topology [13]. In particular, the points of $\mathbb{Z}^2$ are linked by the dual 4- and 8-adjacencies, denoted by $\frown_4$ and $\frown_8$, respectively. If α denotes an adjacency, we denote its dual as $\overline{\alpha}$. These α-adjacencies induce the notions of α-connectedness. (Note that for the sake of concision, the illustrative figures will be proposed for $\mathcal{D} \subset \mathbb{Z}$ instead of $\mathbb{Z}^2$).

A tree is an acyclic, connected graph $\mathcal{T} = (V(\mathcal{T}), E(\mathcal{T}))$. The elements of $V(\mathcal{T})$ are called nodes. The (directed) elements of $E(\mathcal{T})$ are called edges. A tree $\mathcal{T}$ supports two basic operations on any node $\tau \in V(\mathcal{T})$:

$$\texttt{parent}(\mathcal{T}, \tau) = \tau_P \iff (\tau, \tau_P) \in E(\mathcal{T}) \tag{1}$$

$$\texttt{children}(\mathcal{T}, \tau) = \{\tau_C \in V(\mathcal{T}) \mid (\tau_C, \tau) \in E(\mathcal{T})\} \tag{2}$$

From these two operations, we define the root and the leaves of the tree $\mathcal{T}$ as:

$$\texttt{root}(\mathcal{T}) = \tau_R \iff \forall \tau \in V(\mathcal{T}), \tau \neq \texttt{parent}(\mathcal{T}, \tau_R) \tag{3}$$

$$\texttt{leaves}(\mathcal{T}) = \{\tau_L \in V(\mathcal{T}) \mid \texttt{children}(\mathcal{T}, \tau_L) = \emptyset\} \tag{4}$$

The component tree is a tree obtained from the connected components (CCs) of the threshold sets of f and the inclusion relations between these CCs. Let $k \in \mathbb{K}$. We define the upper and lower threshold sets at value k as:

$$\mathcal{X}_k(f) = \{p \in \mathcal{D} \mid f(p) \geqslant k\} \tag{5}$$

$$\mathcal{X}^k(f) = \{p \in \mathcal{D} \mid f(p) < k\} \tag{6}$$

The image f can be reconstructed from either the family of lower or upper level sets, i.e. $\forall p \in \mathcal{D}$ we have:

$$f(p) = \sup\{k \mid p \in \mathcal{X}_k(f)\} = \inf\{k - 1 \mid p \in \mathcal{X}^k(f)\} \tag{7}$$

Each nonempty threshold set of f is partitioned into CCs. By gathering all these CCs, we define:

$$\mathcal{U}(f) = \{\tau \in CC(\mathcal{X}_k(f), \alpha) \mid k \in \mathbb{K}\} \tag{8}$$

$$\mathcal{L}(f) = \{\tau \in CC(\mathcal{X}^k(f), \overline{\alpha}) \mid k \in \mathbb{K}\} \tag{9}$$

where $CC(\mathcal{X}, \alpha)$ denotes the set of CCs of $\mathcal{X}$ induced with respect to the chosen adjacency α.

The CCs of upper and lower level sets and the inclusion relation $\subseteq$ on sets induce two dual trees [9, 15] called component trees. Formally a component tree of an image f is defined as a tree $\mathcal{T}_f = (V(\mathcal{T}_f), E(\mathcal{T}_f))$ such that:

$$V(\mathcal{T}_f) = \mathcal{U}(f) \text{ or } \mathcal{L}(f) \tag{10}$$

$$E(\mathcal{T}_f) = \{(\tau_C, \tau_P) \in V(\mathcal{T}_f) \times V(\mathcal{T}_f) \mid \tau_C \subset \tau_P \ \wedge \nexists \tau \in V(\mathcal{T}_f), \tau_C \subset \tau \subset \tau_P\} \tag{11}$$

A component tree $\mathcal{T}_f$ of the image f can be completely represented by a compact, non-redundant data structure [4]. In particular, we denote by $\texttt{flat}(\mathcal{T}_f, p)$ the smallest CC of $V(\mathcal{T}_f)$ containing p. We also say that $p \in \mathcal{D}$ is a compact node point (CNP) of $\texttt{flat}(\mathcal{T}_f, p)$. For any node $\tau \in V(\mathcal{T}_f)$, we set:

$$\texttt{compact}(\mathcal{T}_f, \tau) = \{p \in \mathcal{D} \mid \tau = \texttt{flat}(\mathcal{T}_f, p)\} \tag{12}$$

the set composed of the CNPs of τ. Note that, from Eq. (12), for any node $\tau \in V(\mathcal{T}_f)$,

$$p \in \texttt{compact}(\mathcal{T}_f, \tau) \iff \tau = \texttt{flat}(\mathcal{T}_f, p). \tag{13}$$

The compact representation of a component tree $\mathcal{T}_f = (V(\mathcal{T}_f), E(\mathcal{T}_f))$ is denoted by $\widehat{\mathcal{T}_f} = (\widehat{V}(\mathcal{T}_f), \widehat{E}(\mathcal{T}_f))$ with:

$$\widehat{V}(\mathcal{T}_f) = \{\texttt{compact}(\mathcal{T}_f, \tau) \mid \tau \in V(\mathcal{T}_f)\} \tag{14}$$

$$\widehat{E}(\mathcal{T}_f) = \{(\texttt{compact}(\mathcal{T}_f, \tau_C), \texttt{compact}(\mathcal{T}_f, \tau_P)) \mid (\tau_C, \tau_P) \in E(\mathcal{T}_f)\} \tag{15}$$

Since $\texttt{compact}(\mathcal{T}_f, \tau) \neq \emptyset$ for any $\tau \in V(\mathcal{T}_f)$, the sets $\widehat{V}(\mathcal{T}_f)$ and $V(\mathcal{T}_f)$ are in bijection. This bijection induces a trivial isomorphism between $\mathcal{T}_f = (V(\mathcal{T}_f), E(\mathcal{T}_f))$ and $\widehat{\mathcal{T}_f} = (\widehat{V}(\mathcal{T}_f), \widehat{E}(\mathcal{T}_f))$.

In practice, our algorithms will act on compact component trees $\widehat{\mathcal{T}_f} = (\widehat{V}(\mathcal{T}_f), \widehat{E}(\mathcal{T}_f))$. These compact representations of component trees for upper and lower level sets are called max-tree and min-tree [15] and denoted by $\widehat{\mathcal{T}}_f^{\max}$ and $\widehat{\mathcal{T}}_f^{\min}$, respectively. For the sake of simplicity, we will also refer to the max-tree (resp. min-tree) as the component tree of upper (resp. lower) level sets.

Based on Eq. (7), we consider the mapping $\mathtt{level}(\mathcal{T}_f, \cdot) : V(\mathcal{T}_f) \to \mathbb{K}$ from CCs of a component tree $\mathcal{T}_f$ to the grey-level values. This mapping is defined so that, for any node $\tau \in V(\mathcal{T}_f)$, we have $\mathtt{level}(\mathcal{T}_f, \tau) = f(p)$ with $p \in \mathtt{compact}(\mathcal{T}_f, \tau)$. From Eq. (13), we have $\tau = \mathtt{flat}(\mathcal{T}_f, p)$. Based on this mapping, the image f can be reconstructed from its tree $\mathcal{T}_f$ by setting, for all $p \in \mathcal{D}$:

$$f(p) = \mathtt{level}(\mathcal{T}_f, \mathtt{flat}(\mathcal{T}_f, p)) \tag{16}$$

3 Method

In this section, we describe our approach to efficiently building CASFs using component trees. To this end, we first build the max- and min-trees of the image. Then, we alternatively remove a leaf from one of them and update the other, repeating this process until no further leaf needs to be removed. We first present theoretical results in Subsect. 3.1. Then we describe the induced algorithms in Subsect. 3.2.

3.1 Theory

We first investigate how the max-tree evolves when removing a leaf node from the min-tree, as illustrated in Fig. 1, which serves as a running example throughout this theoretical section. For interactive exploration and a clearer illustration of the method, we provide a dedicated notebook[1].

For a given grey-level image $f : \mathcal{D} \to \mathbb{K}$, we assume that its max-tree $\mathcal{T}_f^{\max}$ is built for a given α-adjacency ($\alpha = 4$ or 8), while its min-tree $\mathcal{T}_f^{\min}$ is built for the dual $\overline{\alpha}$-adjacency. We set:

- $L \in V(\mathcal{T}_f^{\min})$, a leaf node in the min-tree $\mathcal{T}_f^{\min}$;
- $\mathcal{T}_g^{\min}$, the min-tree obtained after pruning the leaf node L;
- $g : \mathcal{D} \to \mathbb{K}$, the image reconstructed from the pruned min tree $\mathcal{T}_g^{\min}$;
- $\mathcal{T}_g^{\max}$ the max-tree of the image g.

The problem that we consider is formulated as follows.

Problem 1. *How to modify $\mathcal{T}_f^{\max}$ to obtain $\mathcal{T}_g^{\max}$?*

By definition, the leaf node L is $\overline{\alpha}$-connected. If $\overline{\alpha} = 4$, then L is 4-connected and then also 8-connected. It follows that L is included in a unique node in both max-trees $\mathcal{T}_f^{\max}$ and $\mathcal{T}_g^{\max}$.

If $\overline{\alpha} = 8$, then L may be composed of many 4-connected components. If so, let us choose L_1, L_2 two distinct 4-connected components of L which are 8-connected. Then,

[1] Code available at https://github.com/wonderalexandre/MorphoTreeAdjust.

there exists $p_1 \in L_1$ and $p_2 \in L_2$ such that $p_1 \frown_8 p_2$. Let q be such that $p_1 \frown_4 q \frown_4 p_2$. We have $q \notin L$, otherwise L_1, L_2 would be 4-connected. By definition, we have $q \notin L$, and, since L is a leaf node of the min-tree, we have $f(q) > f(p_1) = f(p_2)$. It follows that p_1, p_2, q belong to a same node in the max-trees $\mathcal{T}_f^{\max}$ and $\mathcal{T}_g^{\max}$. By induction, it comes that L is included in a unique node in both max-trees $\mathcal{T}_f^{\max}$ and $\mathcal{T}_g^{\max}$. This justifies the following result.

Proposition 1. *All the points in the node L are compact node points of the two nodes* $\texttt{flat}(\mathcal{T}_f^{\max}, p)$ *and* $\texttt{flat}(\mathcal{T}_g^{\max}, p)$ *with* $p \in L$.

Let τ_L denote the node $\texttt{flat}(\mathcal{T}_f^{\max}, p)$. By removing the leaf node L from the min-tree $\mathcal{T}_f^{\min}$, the points $p \in L$ are transferred from the node L to the node $\texttt{parent}(\mathcal{T}_f^{\min}, L)$. Doing so, they are assigned the new value $g(p) = \texttt{level}(\mathcal{T}_f^{\min}, \texttt{parent}(\mathcal{T}_f^{\min}, L)) > \texttt{level}(\mathcal{T}_f^{\min}, L) = f(p)$. Let $a = f(p)$ and $b = g(p)$.

Node Updating

We first investigate the differences between the nodes of the max-trees $\mathcal{T}_f^{\max}$ and $\mathcal{T}_g^{\max}$.

Let τ be a node of $\mathcal{T}_f^{\max}$ such that (i) $\texttt{level}(\mathcal{T}_f^{\max}, \tau) \notin [\![a, b]\!]$ or (ii) τ is not α-connected to L. If (i) holds, then the CC associated to τ is not affected by modifying the value of L from a to b, since the threshold sets of f and g are the same outside the interval $[\![a, b]\!]$ (see nodes τ_R and τ_I in Fig. 1(c)). If (ii) holds, then the CC associated to τ is not affected by modifying the value of L, since L is not part of this CC (see nodes τ_{III} and τ_V in Fig. 1(c)). This justifies the following result.

Proposition 2. *Let* $\tau \in V(\mathcal{T}_f^{\max})$. *If* $\texttt{level}(\mathcal{T}_f^{\max}, \tau) \notin [\![a, b]\!]$ *or* τ *is not* α*-connected to L, then we have* $\tau \in V(\mathcal{T}_g^{\max})$, *with* $\texttt{compact}(\mathcal{T}_g^{\max}, \tau) = \texttt{compact}(\mathcal{T}_f^{\max}, \tau)$ *and* $\texttt{level}(\mathcal{T}_g^{\max}, \tau) = \texttt{level}(\mathcal{T}_f^{\max}, \tau)$.

Let us now focus on the other nodes of $\mathcal{T}_f^{\max}$. We denote by $\mathcal{F}$ the set of all the nodes τ of $\mathcal{T}_f^{\max}$ such that $\texttt{level}(\mathcal{T}_f^{\max}, \tau) \in [\![a, b]\!]$ and τ is α-connected to L. The set $\mathcal{F}$ is not empty, since $\texttt{level}(\mathcal{T}_f^{\max}, \tau_L) = a$ and $L \subseteq \tau_L$. The set $\mathcal{F}$ can be subdivided into subsets $F_\lambda = \{\tau \in \mathcal{F} \mid \texttt{level}(\mathcal{T}_f^{\max}, \tau) = \lambda\}$ for $\lambda \in [\![a, b]\!]$ (see sets F_2, F_3 and F_5 in Fig. 1(c)). We have $\bigcup_{\lambda \in [\![a,b]\!]} F_\lambda = \mathcal{F}$ and, for any distinct $k, \ell \in [\![a, b]\!]$, we have $F_k \cap F_\ell = \emptyset$. However, $\{F_\lambda\}_{\lambda \in [\![a,b]\!]}$ is not necessarily a partition of $\mathcal{F}$, since we may have $F_\lambda = \emptyset$ for certain $\lambda \in [\![a + 1, b]\!]$ (see F_4 in Fig. 1(c)).

We also consider the additional set $F_{>b}$ composed of nodes τ of $\mathcal{T}_f^{\max}$ such that $\texttt{level}(\mathcal{T}_f^{\max}, \tau) > b$, τ is α-connected to L and $\texttt{parent}(\mathcal{T}_f^{\max}, \tau) \in \mathcal{F} \setminus F_b$ (see node τ_I in Fig. 1(c)).

We consider the following three (mutually exclusive) cases.

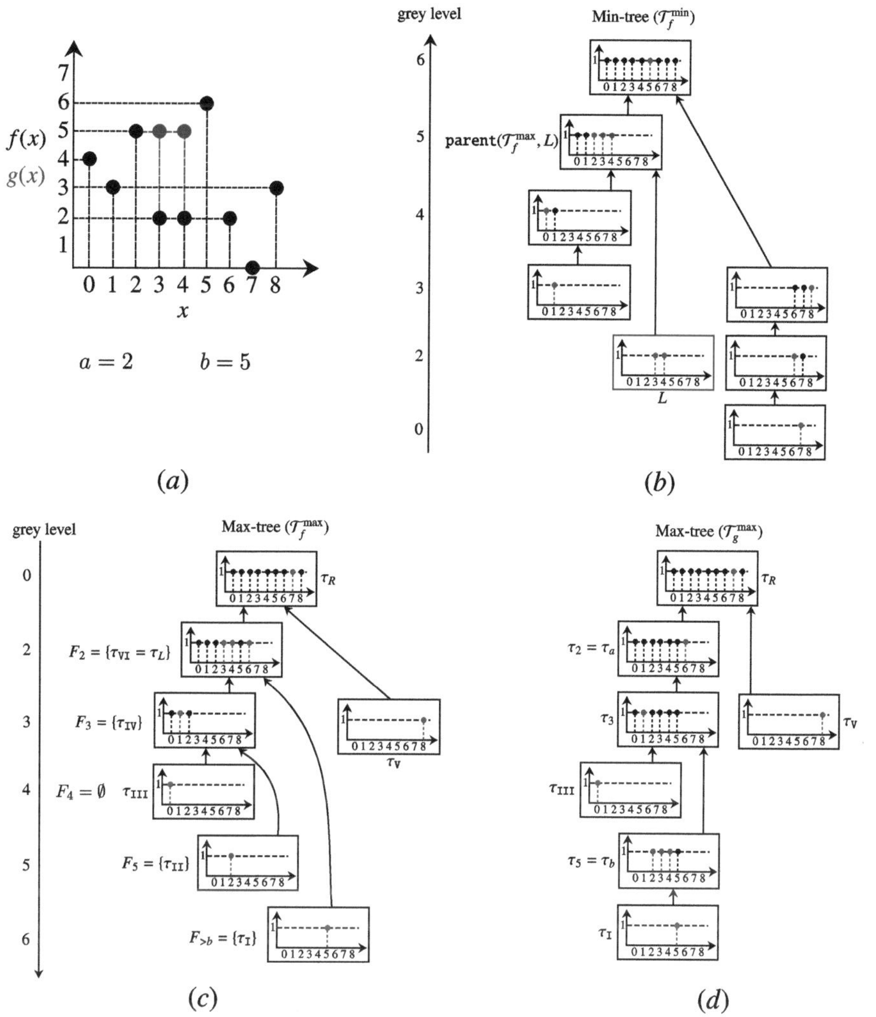

Fig. 1. Illustration of a 1D image transformation and its associated trees, with a grey-level bar indicating the level of each node. The CNPs are shown in red, while the non-CNPs are shown in black. (a) The original image f and the resulting image g, obtained by reconstructing the min-tree of f after pruning node L (highlighted with a blue box in (b)). Pixels affected by the pruning of the min-tree are shown in blue. (b) The min-tree $\mathcal{T}_f^{\min}$ of the image f, with the leaf node L highlighted by a blue box. (c) The max-tree $\mathcal{T}_f^{\max}$ of the image f, highlighting the node τ_L and the sets F_2, F_3, F_5, and $F_{>b}$, which are affected by the pruning of L in the min-tree. (d) The max-tree $\mathcal{T}_g^{\max}$ of the reconstructed image g, highlighting the new nodes τ_2, τ_3, and τ_5, which are defined from the sets F_2, F_3, and F_5. (Color figure online)

– $\lambda = b$. Each $\tau \in F_b \cup F_{>b}$ is an α-connected component of $\mathcal{X}_b(f)$. Since the value of L is modified so that $g(p) = b$ for any $p \in L$, we have the following:

$$\tau_b = L \cup \bigcup F_b \cup \bigcup F_{>b} \tag{17}$$

is an α-connected component of $\mathcal{X}_b(g)$, and thus it is a node of the max-tree $\mathcal{T}_g^{\max}$. We have $\texttt{compact}(\mathcal{T}_g^{\max}, \tau_b) = L \cup \bigcup_{\tau \in F_b} \texttt{compact}(\mathcal{T}_f^{\max}, \tau)$ and $\texttt{level}(\mathcal{T}_g^{\max}, \tau_b) = b$ (see τ_5 in Fig. 1(d)).

- $\lambda \in [\![a+1, b-1]\!]$ with $F_\lambda \neq \emptyset$. Each $\tau \in F_\lambda$ is an α-connected component of $\mathcal{X}_\lambda(f)$. Let $\sigma = \min\{\widetilde{\lambda} \in [\![\lambda+1, b]\!] \mid F_{\widetilde{\lambda}} \neq \emptyset\}$. By induction, it follows that

$$\tau_\lambda = \tau_\sigma \cup \bigcup F_\lambda \tag{18}$$

is an α-connected component of $\mathcal{X}_\lambda(g)$, and thus it is a node of the max-tree $\mathcal{T}_g^{\max}$. We have $\texttt{compact}(\mathcal{T}_g^{\max}, \tau_\lambda) = \bigcup_{\tau \in F_\lambda} \texttt{compact}(\mathcal{T}_f^{\max}, \tau)$ and $\texttt{level}(\mathcal{T}_g^{\max}, \tau_\lambda) = \lambda$ (see τ_3 in Fig. 1(d)).

- $\lambda = a$. We have $F_a = \{\tau_L\}$. By definition,

$$\tau_a = \tau_L \tag{19}$$

is an α-connected component of $\mathcal{X}_a(g)$, and thus it is a node of the max-tree $\mathcal{T}_g^{\max}$. Two (mutually exclusive) subcases can occur.
  - $\texttt{compact}(\mathcal{T}_f^{\max}, \tau_L) \supset L$. In that case, τ_a is not an α-connected component of $\mathcal{X}_{a+1}(g)$, and we then have $\texttt{compact}(\mathcal{T}_g^{\max}, \tau_a) = \texttt{compact}(\mathcal{T}_f^{\max}, \tau_L) \setminus L$ and $\texttt{level}(\mathcal{T}_g^{\max}, \tau_a) = a$ (see τ_2 in Fig. 1(d)).
  - $\texttt{compact}(\mathcal{T}_f^{\max}, \tau_L) = L$. In that case, τ_a is an α-connected component of $\mathcal{X}_\lambda(g)$ for any $a \leqslant \lambda \leqslant \lambda^+$ with $\lambda^+ = \min\{\widetilde{\lambda} \in [\![a+1, b]\!] \mid F_{\widetilde{\lambda}} \neq \emptyset\}$, and we then have $\tau_a = \tau_{\lambda^+} = \tau_L$, $\texttt{compact}(\mathcal{T}_g^{\max}, \tau_a) = \texttt{compact}(\mathcal{T}_g^{\max}, \tau_{\lambda^+})$, and $\texttt{level}(\mathcal{T}_g^{\max}, \tau_a) = \texttt{level}(\mathcal{T}_g^{\max}, \tau_{\lambda^+})$.

Let $\widetilde{\mathcal{F}} = \{\tau_\lambda \mid \lambda \in [\![a, b]\!] \wedge F_\lambda \neq \emptyset\}$ (see τ_5, τ_3, and τ_2 in Fig. 1(d)). The previous discussion justifies the following result.

Proposition 3. *The set $\mathcal{F}$ in the max-tree $\mathcal{T}_f^{\max}$ is replaced with the set $\widetilde{\mathcal{F}}$ in the max-tree $\mathcal{T}_g^{\max}$.*

From Propositions 2–3, we can determine the set of nodes of the new max-tree $\mathcal{T}_g^{\max}$ from the former max-tree $\mathcal{T}_f^{\max}$.

Edge updating

We now deal with the definition of the $\texttt{parent}$ relation in $\mathcal{T}_g^{\max}$ from the $\texttt{parent}$ relation in $\mathcal{T}_f^{\max}$.

Let τ_C be a node of $\mathcal{T}_g^{\max}$ such that $\tau_C \neq \texttt{root}(\mathcal{T}_g^{\max})$. Hence, we have $\tau_C \neq \texttt{root}(\mathcal{T}_f^{\max})$. The following two cases (mutually exclusive) can occur.

- $\tau_C \notin \widetilde{\mathcal{F}}$. This implies $\tau_C \in V(\mathcal{T}_f^{\max})$. Let $\tau_P = \texttt{parent}(\mathcal{T}_f^{\max}, \tau_C)$. Two subcases (mutually exclusive) can occur.
  - $\tau_P \notin \mathcal{F}$. It follows from Proposition 2 that we have $\texttt{parent}(\mathcal{T}_g^{\max}, \tau_C) = \tau_P$ (see node τ_V in Fig. 1(d)).
  - $\tau_P \in \mathcal{F}$. Two subcases (mutually exclusive) can occur.
    - $\tau_C \in F_{>b}$. From Proposition 3 and the related discussion, we have $\tau_C \subset \tau_b$. It follows that $\texttt{parent}(\mathcal{T}_g^{\max}, \tau_C) = \tau_b$ (see node τ_I in Fig. 1(c–d)).

- * $\tau_C \notin F_{>b}$. We have $\tau_C \subset \tau_{\texttt{level}(\mathcal{T}_f^{\max},\tau_P)}$. Let us suppose that there exists $\tau \in V(\mathcal{T}_g^{\max})$ such that $\tau_C \subset \tau \subset \tau_P$. We cannot have $\tau \in V(\mathcal{T}_f^{\max})$, otherwise $\tau_P \neq \texttt{parent}(\mathcal{T}_f^{\max}, \tau_C)$. Then we have $\tau = \tau_\lambda \in \widetilde{\mathcal{F}}$ with $\texttt{level}(\mathcal{T}_g^{\max}, \tau_P) < \lambda < \texttt{level}(\mathcal{T}_g^{\max}, \tau_C)$, which implies that there exists $\tau' \in V(\mathcal{T}_f^{\max})$ such that $\texttt{level}(\mathcal{T}_g^{\max}, \tau') = \lambda$ and then $\tau_C \subset \tau' \subset \tau_P$, in contradiction with $\tau_P = \texttt{parent}(\mathcal{T}_f^{\max}, \tau_C)$. It follows that we have $\texttt{parent}(\mathcal{T}_g^{\max}, \tau_C) = \tau_{\texttt{level}(\mathcal{T}_f^{\max},\tau_P)}$ (see node τ_{III} in Fig. 1(c–d)).

- $\tau_C \in \widetilde{\mathcal{F}}$. Two subcases (mutually exclusive) can occur.
  - $\tau_C \neq \tau_a$. We have $\texttt{parent}(\mathcal{T}_g^{\max}, \tau_C) = \tau_{\lambda^-}$ with $\lambda^- = \max\{\widetilde{\lambda} \in [\![a, \lambda - 1]\!] \mid F_{\widetilde{\lambda}} \neq \emptyset\}$ (see nodes τ_3 and τ_5 in Fig. 1(d)).
  - $\tau_C = \tau_a$. This implies $\tau_C \in V(\mathcal{T}_f^{\max})$. With the same notations as above, we have $\texttt{parent}(\mathcal{T}_g^{\max}, \tau_C) = \tau_P$ (see node τ_2 in Fig. 1(d)).

The previous discussion justifies the following result.

Proposition 4. *Let* $\tau_C \in V(\mathcal{T}_g^{\max})$ *such that* $\tau_C \neq \texttt{root}(\mathcal{T}_g^{\max})$. *The relation* parent *is unchanged from* $\mathcal{T}_f^{\max}$ *to* $\mathcal{T}_g^{\max}$, *except in the following three cases:*

$$(\tau_C \in F_{>b}) \wedge (\tau_P \in \mathcal{F}) \Longrightarrow \texttt{parent}(\mathcal{T}_g^{\max}, \tau_C) = \tau_b \tag{20}$$

$$(\tau_C \notin F_{>b} \cup \widetilde{\mathcal{F}}) \wedge (\tau_P \in \mathcal{F}) \Longrightarrow \texttt{parent}(\mathcal{T}_g^{\max}, \tau_C) = \tau_{\texttt{level}(\mathcal{T}_f^{\max},\tau_P)} \tag{21}$$

$$(\tau_C \in \widetilde{\mathcal{F}} \setminus \{\tau_a\}) \Longrightarrow \texttt{parent}(\mathcal{T}_g^{\max}, \tau_C) = \tau_{\lambda^-} \tag{22}$$

3.2 Algorithms

We decompose the proposed method in two parts: first, the updating of the trees after a leaf removal operation; second the computation of the CASF.

Tree Updating

We propose an algorithmic scheme to tackle Problem 1. For the sake of readability, this scheme is split into three algorithms, which are direct transcriptions of the results established in Sect. 3.1:

- the determination of the nodes to update (Algorithm 1);
- the update of these nodes (Algorithm 2);
- the induced update the edges (Algorithm 3).

In the algorithms, we adopt the following notation: "$a := b$" means that b is assigned to a; "$a \leftarrow b$" means that b is added to a; "$a \rightarrow b$" means that b is removed from a.

The tree updating procedure is designed for the computation of CASF, that relies on the removal of regions based on size criterion. It is therefore relevant to compute and maintain an up-to-date area attribute for each node in the two component trees. The initial computation incurs no additional cost when building the component trees. The update from $\mathcal{T}_f^{\max}$ to $\mathcal{T}_g^{\max}$ only affects the new nodes τ_λ in $\mathcal{T}_g^{\max}$, which can be handled in a bottom-up fashion, in linear time.

Algorithm 1: Building $\mathcal{F}$ and $F_{>b}$

Input: $f : \mathcal{D} \to \mathbb{K}$
Input: $\mathcal{T}_f^{\max} = (V(\mathcal{T}_f^{\max}), E(\mathcal{T}_f^{\max}))$ denoted by $\mathcal{T}_f$ (considered wrt α-adjacency)
Input: $\mathcal{T}_f^{\min} = (V(\mathcal{T}_f^{\min}), E(\mathcal{T}_f^{\min}))$ and $L \in \mathtt{leaves}(\mathcal{T}_f^{\min})$
Notation: $a := \mathtt{level}(\mathcal{T}_f^{\min}, L)$ and $b := \mathtt{level}(\mathcal{T}_f^{\min}, \mathtt{parent}(\mathcal{T}_f^{\min}, L))$
Output: $\mathcal{F} = \{F_\lambda \mid a \leqslant \lambda \leqslant b\}$ and $F_{>b}$

foreach $\lambda \in [\![a, b]\!]$ **do** $F_\lambda := \emptyset$
$F_{>b} := \emptyset$
$Q := \{q \in \mathcal{D} \setminus L \mid \exists p \in L, q \frown_\alpha p\}$
foreach $q \in Q$ **do**
  if $a \leqslant f(q) \leqslant b$ **then** $F_{f(q)} \leftarrow \mathtt{flat}(\mathcal{T}_f, q)$
  if $b < f(q)$ **then** $F_{>b} \leftarrow \mathtt{flat}(\mathcal{T}_f, q)$
while $b \leqslant \max\{\mathtt{level}(\mathcal{T}_f, \mathtt{parent}(\mathcal{T}_f, \tau)) \mid \tau \in F_{>b}\}$ **do**
  $\tau := \arg_{\tau \in F_{>b}} \max\{\mathtt{level}(\mathcal{T}_f, \mathtt{parent}(\mathcal{T}_f, \tau))\}$
  $F_{>b} \to \tau$
  if $b < \mathtt{level}(\mathcal{T}_f, \mathtt{parent}(\mathcal{T}_f, \tau))$ **then** $F_{>b} \leftarrow \mathtt{parent}(\mathcal{T}_f, \tau)$
  if $b = \mathtt{level}(\mathcal{T}_f, \mathtt{parent}(\mathcal{T}_f, \tau))$ **then** $F_b \leftarrow \mathtt{parent}(\mathcal{T}_f, \tau)$
foreach $\tau \in F_{>b}$ **do**
  $F_{\mathtt{level}(\mathtt{parent}(\mathcal{T}_f, \tau))} \leftarrow \mathtt{parent}(\mathcal{T}_f, \tau)$
for λ **from** b **to** $a + 1$ **do**
  foreach $\tau \in F_\lambda$ **do**
    $F_{\mathtt{level}(\mathcal{T}_f, \mathtt{parent}(\mathcal{T}_f, \tau))} \leftarrow \mathtt{parent}(\mathcal{T}_f, \tau)$

Algorithm 2: Building $V(\mathcal{T}_g^{\max})$

Input: $V(\mathcal{T}_f^{\max})$ with $\mathcal{T}_f^{\max}$ denoted by $\mathcal{T}_f$
Input: $\mathcal{F} = \{F_\lambda \mid a \leqslant \lambda \leqslant b\}$
Output: $V(\mathcal{T}_g^{\max})$ with $\mathcal{T}_g^{\max}$ denoted by $\mathcal{T}_g$

$V(\mathcal{T}_g) := V(\mathcal{T}_f) \setminus \bigcup_{\lambda=a}^{b} F_\lambda$
$V(\mathcal{T}_g) \leftarrow \tau_b$
$\mathtt{level}(\mathcal{T}_g, \tau_b) := b$
$\mathtt{compact}(\mathcal{T}_g, \tau_b) := L$
foreach $\tau \in F_b$ **do**
  $\mathtt{compact}(\mathcal{T}_g, \tau_b) := \mathtt{compact}(\mathcal{T}_g, \tau_b) \cup \mathtt{compact}(\mathcal{T}_f, \tau)$
foreach $\lambda \in [\![a + 1, b - 1]\!]$ **such that** $F_\lambda \neq \emptyset$ **do**
  $V(\mathcal{T}_g) \leftarrow \tau_\lambda$
  $\mathtt{level}(\mathcal{T}_g, \tau_\lambda) := \lambda$
  $\mathtt{compact}(\mathcal{T}_g, \tau_\lambda) := \emptyset$
  foreach $\tau \in F_\lambda$ **do**
    $\mathtt{compact}(\mathcal{T}_g, \tau_\lambda) := \mathtt{compact}(\mathcal{T}_g, \tau_\lambda) \cup \mathtt{compact}(\mathcal{T}_f, \tau)$
$F_a \to \tau_L$
if $\mathtt{compact}(\mathcal{T}_f, \tau_L) \supset L$ **then**
  $V(\mathcal{T}_g) \leftarrow \tau_a$
  $\mathtt{level}(\mathcal{T}_g, \tau_a) := a$
  $\mathtt{compact}(\mathcal{T}_g, \tau_a) := \mathtt{compact}(\mathcal{T}_f, \tau_L) \setminus L$

Algorithm 3: Building $E(\mathcal{T}_g^{\max})$

Input: $\mathcal{T}_f^{\max} = (V(\mathcal{T}_f^{\max}), E(\mathcal{T}_f^{\max}))$ denoted by $\mathcal{T}_f$ with L and τ_L known
Input: $\mathcal{F} = \{F_\lambda \mid a \leqslant \lambda \leqslant b\}$ and $F_{>b}$
Output: $E(\mathcal{T}_g^{\max})$ with $\mathcal{T}_g^{\max}$ denoted by $\mathcal{T}_g$

$E(\mathcal{T}_g) := E(\mathcal{T}_f)$
for λ **from** b **to** $a+1$ **such that** $F_\lambda \neq \emptyset$ **do**
  $\lambda^- := \max\{\tilde{\lambda} \in [\![a, \lambda - 1]\!] \mid F_{\tilde{\lambda}} \neq \emptyset\}$
  if $\lambda^- = a$ **and** $\texttt{compact}(\mathcal{T}_f, \tau_L) = L$ **then**
    $\texttt{parent}(\mathcal{T}_g, \tau_\lambda) := \texttt{parent}(\mathcal{T}_f, \tau_L)$
  else
    $\texttt{parent}(\mathcal{T}_g, \tau_\lambda) := \tau_{\lambda^-}$
foreach $\lambda \in [\![a, b]\!]$ **do**
  foreach $\tau_P \in F_\lambda$ **do**
    foreach $\tau_C \in \texttt{children}(\mathcal{T}_f, \tau_P)$ **do**
      if $\tau_C \notin \bigcup_{\lambda=a}^{b} F_\lambda$ **then**
        if $\tau_C \in F_{>b}$ **then**
          $\texttt{parent}(\mathcal{T}_g, \tau_C) := \tau_b$
        else
          $\texttt{parent}(\mathcal{T}_g, \tau_C) := \tau_{\texttt{level}(\mathcal{T}_f^{\max}, \tau_P)}$

Algorithm 4: Building CASF by updating the component trees

Input: $f : \mathcal{D} \to \mathbb{K}$ (initial image)
Parameter: κ (node removal criterion)
Output: $g : \mathcal{D} \to \mathbb{K}$ (filtered image)

build $\mathcal{T}_f^{\max}$ from f, denoted by $\mathcal{T}^{\max}$
build $\mathcal{T}_f^{\min}$ from f, denoted by $\mathcal{T}^{\min}$
initiate κ
while κ **is valid do**
  while $\exists L \in \texttt{leaves}(\mathcal{T}^{\max})$ **such that** $\kappa(L) = True$ **do**
    update $\mathcal{T}^{\max}$ by removing L
    update $\mathcal{T}^{\min}$ from Algs. 1–3
  while $\exists L \in \texttt{leaves}(\mathcal{T}^{\min})$ **such that** $\kappa(L) = True$ **do**
    update $\mathcal{T}^{\min}$ by removing L
    update $\mathcal{T}^{\max}$ from Algs. 1–3
  update κ
reconstruct g from $\mathcal{T}^{\max}$ or $\mathcal{T}^{\min}$

Efficient Connected Alternating Sequential Filters

We are now ready to use the component tree updating strategy established above in order to build efficient CASFs. Our proposed approach is formalized in Algorithm 4.

As in the naive approach, this algorithm considers a given state of the node removal criterion κ at each iteration of the while loop (Line 4), and the two component trees are pruned sequentially (Lines 5–7 and 8–10). However, in contrast to a naive approach that would (1) prune all necessary leaves from the max-tree (resp. min-tree), (2) reconstruct the filtered image, and (3) build the corresponding min-tree (resp. max-tree), here Algorithm 4 iteratively removes each leaf from the max-tree (resp. min-tree) (Line 6/9) and updates the min-tree (resp. max-tree) accordingly (Line 7/10) without the need to reconstruct the image or fully rebuild a tree.

4 Results

To assess the performance of our method, we implemented two versions of CASFs: (1) the naive version and (2) our proposed approach described in Algorithm 4.

We conducted run-time experiments on the Occluded RoadText Challenge[2] which contains 1 019 driving-scene images (817 in the test set and 202 in the validation set) originally sized $1\,980 \times 1\,080$ pixels. These images, featuring natural elements, urban infrastructure, text, and crowded scenes, were rescaled to 576×324 pixels for processing efficiency. For each image, we ran the CASFs with increasing values of area for openings/closings, from 10 to 500 pixels, in steps of 10 (i.e. $10 \times i$ pixels at iteration i of the process). We then computed the average computation time of the both methods over these images.

The time costs for each iteration are given in Fig. 2(a). We can observe that the costs associated with the naive method remains mostly constant across successive iterations. This is because they are primarily driven by the construction of the component trees and the reconstruction of the image, both of which mainly depend on the image size. In contrast, our method incurs a higher cost in the initial iterations but quickly

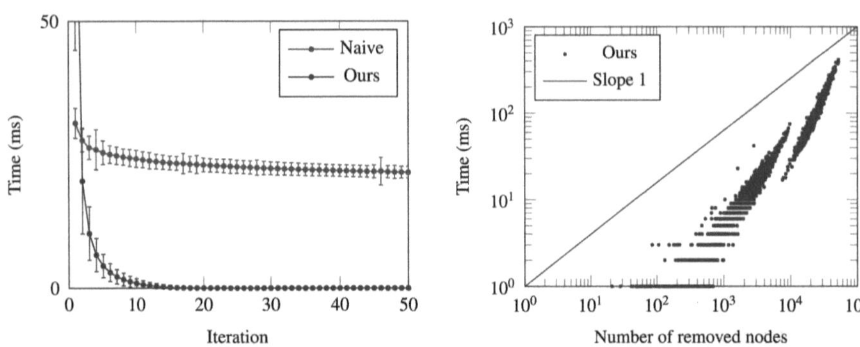

(a) Time cost per iteration (b) Time cost per number of removed nodes

Fig. 2. (a) Time cost of each iteration of the two CASF methods (naive, ours). (b) Correlation (log–log scale) between the time cost of each iteration (y-axis) and the number of removed nodes during the iteration (x-axis) for our method.

[2] https://rrc.cvc.uab.es/?ch=29.

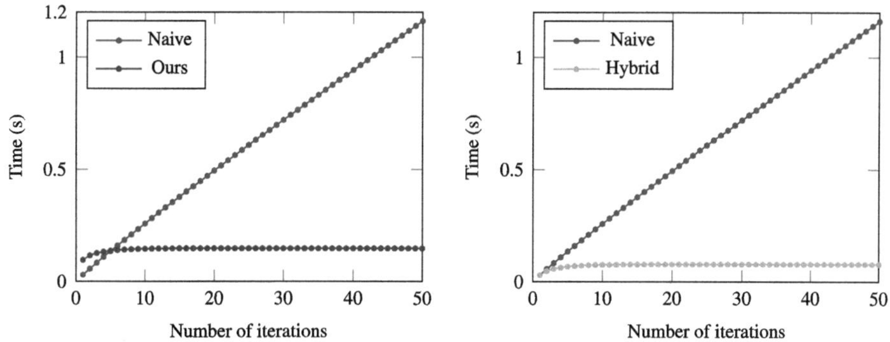

(a) Overall time cost: naive vs. ours (b) Overall time cost: naive vs. hybrid

Fig. 3. Time cost of CASFs. Naive approach (a–b, in red). Our approach (a, in blue). Hybrid approach obtained by considering, at each iteration, the most efficient approach (b, in green). (Color figure online)

decreases in subsequent ones, ultimately becoming significantly lower than that of the naive approach. This behavior is explained by the (sub)linear dependence of the computational cost per iteration on the number of removed nodes, as experimentally shown in Fig. 2(b). The reduction in cost observed during the iterations in Fig. 2(a) derives from the fact that the number of leaf nodes to be removed progressively decreases as the area increases.

The overall time costs of the two methods, obtained by summing the costs across all iterations, are shown in Fig. 3. The cost of the naive approach (red curve) increases linearly with the number of iterations, since each iteration has a similar time cost (see Fig. 2(a)). The cost of our method (blue curve) is initially higher than the naive one, due to the greater cost of the first few iterations. However, it then increases quasi-linearly, with a lower slope than the naive approach, and then becomes lower than the naive approach (here, around the 6th iteration). Our standalone approach thus becomes more efficient than the naive method when the number of iterations is sufficiently high. In addition, our approach can also be combined with the naive approach to form a hybrid strategy: the naive approach is used for the initial steps, and our approach is applied for the remaining steps. The cost of this hybrid method (Fig. 3(b), green curve) is significantly lower than that of the naive approach (Fig. 3(b), red curve).

In addition to the iteration-wise analysis, we evaluated the overall performance improvement by comparing the relative computational gains. Our approach achieved a speedup of approximately 5.55× compared to the naive method, while the hybrid method provided an even greater speedup of 8.61×. Moreover, the hybrid strategy was 1.55× faster than our standalone approach. These improvements correspond to relative reductions in computational cost of 81.97% and 88.38% for our method and the hybrid method, respectively, compared to the naive baseline. Furthermore, the hybrid method achieved a 41.5% reduction compared to our standalone method. Altogether, these findings support the hypothesis that updating the component trees (rather than reconstructing them at each iteration) not only improves CASF efficiency, but also demonstrates the benefit of integrating both strategies in a hybrid approach for optimal performance.

5 Conclusion

In this article, we investigated how to optimize the computational efficiency of Connected ASFs based on component tree openings and closings. To this end, we proposed an algorithmic approach for updating a max- (resp. min-) tree when a leaf node is removed from its dual min- (resp. max-) tree. This process avoids the complete reconstruction of the component tree at each iteration of a CASF. We observed that this strategy has a linear time cost with respect to the number of removed nodes. This implies that in the initial iterations, when many nodes need to be removed, our method is less efficient than fully reconstructing the trees. However, it quickly becomes more computationally efficient than full reconstruction. This observation suggests that hybrid CASF algorithms could be beneficial, initially relying on full tree reconstruction and transitioning to tree updating in later iterations.

The perspective for future work on this topic are manifold. We plan to study how to update a component tree when an entire distal subtree is removed from its dual tree. This could significantly reduce the computational cost of the early iterations in the CASF sequence. We also intend to conduct a complexity analysis of the proposed method to identify a cut-off point where updates become more efficient than full reconstructions. More generally, we aim to explore how to update a max- (resp. min-) tree when its dual min- (resp. max-) tree is modified by increasing (resp. decreasing) a subset of a flat zone. This may pave the way for a broader class of efficient image processing techniques based on component trees. Finally, we will investigate the efficient integration of flat and non-flat attribute updates during incremental tree updates [7, 18], enabling CASFs based on a wider range of attributes, such as perimeter, bounding box, height, among others.

Acknowledgements. This work was funded by the *Agence Nationale de la Recherche* (Grants ANR-22-CE45-0034 and ANR-23-CE45-0015), *CNPq – Conselho Nacional de Desenvolvimento Científico e Tecnológico* (Grants 141422/2018-1, 428720/2018-7 and 407242/2021-0) and *FAPESP – Fundação de Amparo a Pesquisa do Estado de São Paulo* (Grants 2015/22308 – 2, 2018/15652 – 7, and 2023/07656 – 0).

References

1. Blin, N., Carlinet, E., Lemaitre, F., Lacassagne, L., Géraud, T.: Max-tree computation on GPUs. IEEE Trans. Parallel Distrib. Syst. **33**, 3520–3531 (2022)
2. Breen, E.J., Jones, R.: Attribute openings, thinnings, and granulometries. Comput. Vis. Image Underst. **64**, 377–389 (1996)
3. Carlinet, E., Géraud, T.: A comparative review of component tree computation algorithms. IEEE Trans. Image Process. **23**, 3885–3895 (2014)
4. Caselles, V., Monasse, P.: Geometric Description of Images as Topographic Maps. Springer, Heidelberg (2010)

5. Dalla Mura, M., Benediktsson, J.A., Bruzzone, L.: Alternating sequential filters with morphological attribute operators for the analysis of remote sensing images. In: ISPRS Processing, pp. 43–50 (2010)
6. Gazagnes, S., Wilkinson, M.H.F.: Distributed connected component filtering and analysis in 2D and 3D tera-scale data sets. IEEE Trans. Image Process. **30**, 3664–3675 (2021)
7. Gobber, C.F., Hashimoto, R.F., Alves, W.A.: An efficient algorithm to update non-flat and incremental attributes in morphological trees. Pattern Recogn. Lett. **163**, 47–54 (2022)
8. Götz, M., Cavallaro, G., Géraud, T., Book, M., Riedel, M.: Parallel computation of component trees on distributed memory machines. IEEE Trans. Parallel Distrib. Syst. **29**, 2582–2598 (2018)
9. Jones, R.: Connected filtering and segmentation using component trees. Comput. Vis. Image Underst. **75**, 215–228 (1999)
10. Morales, A.W., Acharya, R., Ko, S.: Morphological pyramids with alternating sequential filters. IEEE Trans. Image Process. **5**, 965–977 (1995)
11. Pei, S.C., Lai, C.L., Shih, F.Y.: An efficient class of alternating sequential filters in morphology. CVGIP: Graph. Models Image Process. **59**, 109–116 (1997)
12. Ronse, C., Heijmans, H.J.A.M.: The algebraic basis of mathematical morphology: II. Openings and closings. CVGIP Image Underst. **54**, 74–97 (1991)
13. Rosenfeld, A.: Digital topology. Am. Math. Mon. **86**, 621–630 (1979)
14. Salembier, P., Wilkinson, M.: Connected operators. IEEE Signal Process. Mag. **26**, 136–157 (2009)
15. Salembier, P., Oliveras, A., Garrido, L.: Anti-extensive connected operators for image and sequence processing. IEEE Trans. Image Process. **7**, 555–570 (1998)
16. Serra, J. (ed.): Image Analysis and Mathematical Morphology. Theoretical Advances, vol. II. Academic Press, London (1988)
17. Sternberg, S.R.: Grayscale morphology. Comput. Vision Graph. Image Process. **35**, 333–355 (1986)
18. Wilkinson, M.H.F., Gao, H., Hesselink, W.H., Jonker, J.E., Meijster, A.: Concurrent computation of attribute filters on shared memory parallel machines. IEEE Trans. Pattern Anal. Mach. Intell. **30**, 1800–1813 (2008)
19. Xu, Y., Géraud, T., Najman, L.: Connected filtering on tree-based shape-spaces. IEEE Trans. Pattern Anal. Mach. Intell. **38**, 1126–1140 (2016)

3 Fundamental Structures

3.1 Spectrograms

We adopt a continuous framework for defining spectrograms, following [8], in order to maintain mathematical clarity. In practice, however, the resulting time-frequency representations are sampled on a uniform grid, with a fixed time step and frequency resolution.

Let $x \in \mathcal{L}_\infty(\mathbb{R};\mathbb{C})$ denote the input signal and $g \in \mathcal{L}_1(\mathbb{R};\mathbb{C})$ a window function. The Short-Time Fourier Transform (STFT) of x with respect to g is defined as:

$$\begin{aligned} V_g[x] &: \mathbb{R} \times \mathbb{R} \to \mathbb{C} \\ (t, f) &\mapsto \int_{\mathbb{R}} x(\tau)\, \overline{g(\tau - t)}\, e^{-2\pi i \tau f}\, d\tau \end{aligned} . \tag{1}$$

The STFT is a time-frequency transform whose output is a complex-valued function with domain $\mathbb{R} \times \mathbb{R}$. In analytical applications, the phase is discarded, and only the magnitude is considered. The corresponding *spectrogram* is the squared modulus of the STFT:

$$S_g[x](t, f) = |V_g[x](t, f)|^2 . \tag{2}$$

For interpretability, the spectrogram is often expressed in decibels (dB):

$$S_g^{\mathrm{dB}}[x](t, f) = 10 \log_{10}\left(|S_g[x](t, f)|\right) . \tag{3}$$

Since the STFT can be written as $V_g[x](t, f) = e^{-2\pi i t f}(x * M_f g^*)(t)$ (where $M_f g(t) = e^{2\pi i t f} g(t)$ and $g^*(t) = \overline{g(-t)}$), we have the norm inequality

$$||V_g[x](t, f)||_\infty = ||x * M_f g^*||_\infty \leq ||x||_\infty ||M_f g^*||_1 = ||x||_\infty ||g||_1 . \tag{4}$$

In real world applications, our input signals x will be bounded by 1 (or can be rescaled to be so), so the normalization $||g||_1 = 1$ ensures that the spectrogram will also be bounded by 1, that is

$$0 \leq S_g[x] \leq 1 \qquad \text{and} \qquad -\infty \leq S_g^{\mathrm{dB}}[x] \leq 0 . \tag{5}$$

In order to obtain a discretized grayscale image from a spectrogram, we quantize time and frequency with time resolution $\Delta t = 0.01s$ and frequency resolution $\Delta f = 10\,\mathrm{Hz}$. Bounds for the grayscale range are necessary, so we define a lower bound of $-120\,\mathrm{dB}$ and the upper bound of 0 dB is given by the Eq. (5). Next, we quantize the intensity values using 1 dB intervals, obtaining a range $Q = \{-120, -119, \ldots, 0\}$. The desired grayscale image is then

$$(\mathbf{S}_{kn})_{n=0,\ldots,N}^{k=0,\ldots,K} \in Q^{K \times N} , \tag{6}$$

with $f_k = k \cdot \Delta f$ and $t_n = n \cdot \Delta t$.

Here, $k = 0, \ldots, K-1$ indexes frequency bins and $n = 0, \ldots, N-1$ indexes time frames; K and N denote the total numbers of frequency and time samples, respectively.

This representation enables us to apply image processing techniques, and in particular mathematical morphological ones. The geometry of a spectrogram is constrained by time-frequency properties and implies that the resulting image only contains specific shapes: horizontal structures correspond to sinusoids, vertical ones to transients, and noisy components appear as irregular patterns where valleys and hills alternate. This is illustrated in Fig. 6a.

Since the image can be interpreted as a topographic map where the signals appear as mountains, it makes sense to apply morphology techniques based on level sets and hierarchical structures. We opted for the max-tree representation and we aim to apply successive filters in a pipeline to separate the mountains from the irregular landscape.

3.2 Max-Tree

Component trees are data structures that organize the objects in an image and represent their inclusion relationships. They are useful for image segmentation and analysis, as they allow a hierarchical representation of objects. These trees have received particular attention regarding their robustness to noise. For example, in [11], the authors studied the robustness of attribute profiles computed from these trees for the classification of hyperspectral image pixels. In [3], the stability of the tree of shapes is investigated for region of interest detection, which improves by 20 to 30% when using this structure. This robustness has even been studied more formally in [4], and some even use these trees to estimate the noise level in an image [7].

The max-tree, introduced by [17], is based on the notion of upper level sets of an image. Let u be a grayscale image; the upper level sets of u form the family $\mathcal{U} = [u \geq \lambda]_{\lambda \in \mathbb{R}}$, where $[u \geq \lambda]$ denotes the set of pixels with an intensity greater than or equal to λ. The max-tree is simply the set of connected components of the elements of $\mathcal{U}$. Since any two components are either nested or disjoint, this set of components forms a tree structure, as illustrated in Fig. 1.

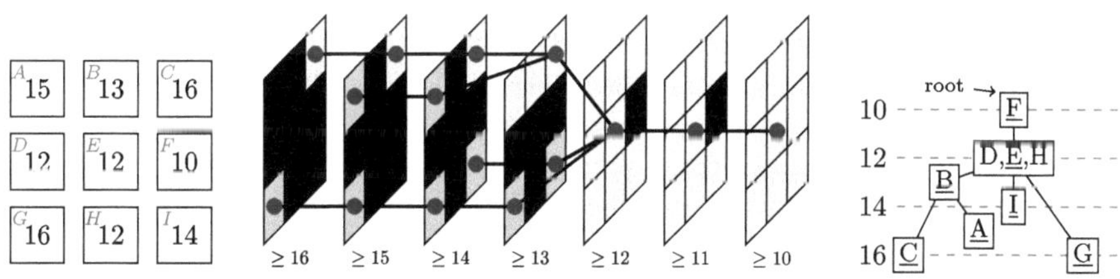

(a) Original image (b) Inclusion of upper-level-set decomposition (c) Max-tree

Fig. 1. Illustration of max-tree construction: (a) original image, (b) inclusion of upper-level-set decomposition, and (c) corresponding max-tree. Each node in the tree represents a connected component of the level-sets.

The max-tree applied on a spectrogram can contain different kinds of components: some components only contain noise, others only contain signal, and some (usually closer to the tree root) contain both noise and signal. Some examples of these components are illustrated in Fig. 2.

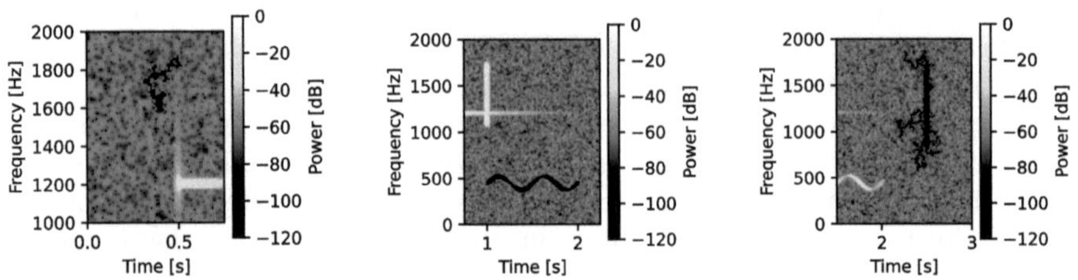

(a) Component containing only noise

(b) Component containing only signal

(c) Mixed noise and signal component

Fig. 2. Examples of max-tree components: (a) noise-only, (b) signal-only, and (c) mixed noise and signal. The component of interest is shown in black.

Our goal now is to select the components that only contain signal and filter out the ones that contain noise. Once this is achieved, we can obtain a mask by taking the union of the selected components. This will allow us to produce a new spectrogram by masking the original STFT and applying the inverse STFT [1].

4 Processing Pipeline

We propose a three-stage pipeline to denoise an audio signal by applying morphological filters to the max-tree of its spectrogram:

1. **Pre-processing:** We compute the STFT $V_g[x]$ with resolutions Δt, Δf, take its squared modulus, convert to dB, and quantize into Q levels to obtain the input spectrogram image $\mathbf{S}_{kn}$.
2. **Morphological filtering:** We build the max-tree $\mathcal{T}$ of the quantized spectrogram $\mathbf{S}_{kn}$ and apply, in any order, the following filters:
   - *Area filter*: remove components C_i with area $A_i > A_{\max}$,
   - *Height filter*: remove components C_i with height $H_i < H_{\min}$,
   - *Shape filter*: remove components C_i with average 8-neighbor count per pixel $R_i > R_{\max}$.

   Then, we form the binary mask M as the union of all remaining components.
3. **Reconstruction:** We mask the STFT coefficients, i.e. compute $V_g[x] \odot M$ and then perform the inverse STFT to obtain the denoised audio.

We apply *area* first to quickly remove near-root components and shrink the tree, then *height* to discard low-contrast structures, and finally *shape* on the few survivors. This minimizes cost while preserving recall.

This pipeline is illustrated in Fig. 3. Now, let us discuss in detail each step and expose the motivations and effects they have.

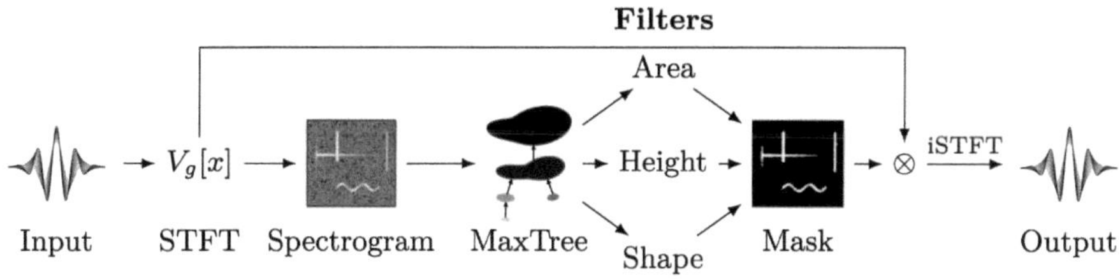

Fig. 3. Spectrogram processing pipeline.

4.1 Max-Tree Construction

We compute the max-tree $\mathcal{T}$ of the image $\mathbf{S} \in Q^{K \times N}$, seen as a function from the discrete time-frequency domain $\Omega = \{0, \ldots, K-1\} \times \{0, \ldots, N-1\}$ to the set of quantized values Q. The max-tree is defined as a tree where the nodes (indexed by I) are the connected components $C_i \subseteq \Omega$ and the edges are given by $C_i \to C_j \Leftrightarrow C_i \subset C_j$ and $\{i' \in I : C_i \subset C_{i'} \subset C_j\} = \emptyset$. We can then encode the max-tree as the triplet

$$\mathcal{T} = (\mathcal{C}, p, \Lambda) \tag{7}$$

where $\mathcal{C} := \{C_i\}_{i \in I}$ is the set of components, $p : I \to I \cup \{\emptyset\}$ is the function that associates to each component index the index of its parent (and $\emptyset$ for the root), and $\Lambda := \{\lambda_i\}_{i \in I} \subseteq Q$ is the set of values.

4.2 Filtering by Area

First, we apply a filter by area as a heuristic to remove components that are too large to only contain signal. The area of a component C_i is defined as $A_i = |C_i|$, i.e. the number of elements of the component.

This filter has additional benefits: it eliminates edge cases and reduces computation time for the following filters. We choose a value of $A_{\max} = 75\%$ of the image size[1]. Our experiments show this value produces a very conservative filter, ensuring that the probability of losing signal is negligible.

4.3 Filtering by Height

This is arguably the most critical filter. When applying the max-tree to noised signals, the noise is split into several components. Several of these display line-like morphological attributes locally (horizontal, vertical, or curved), although they do not correspond to signal. This phenomenon happens because their values are only slightly higher than those of the adjacent components.

Although these components do not contain signal and do not appear clearly as mountains, they are hard to distinguish from the rest of the noise: they are surrounded by a shallow ditch and thus remain disconnected from neighboring structures.

[1] Here image size means the total number of pixels $K \times N$.

To eliminate them we introduce a new attribute, the *height*. For a component C_i, its height H_i is the maximum elevation of any child component relative to C_i. Formally,

$$H_i = \max_{j \in I,\, C_j \subseteq C_i} \{\lambda_j - \lambda_i\} \geq 0\,. \tag{8}$$

Figure 4 illustrates this attribute.

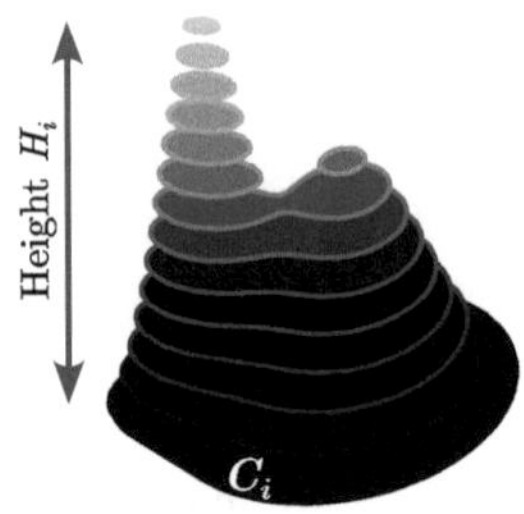

Fig. 4. Height attribute H_i of the component C_i.

In a sense, the height encodes the local signal-to-noise ratio. Thus, we discard all components whose height falls below a threshold $H_{\min}$. In our experiments, we set $H_{\min} = 6\,\text{dB}$ which is a value that preserves most signal components. This parameter should be tuned to the specific application.

4.4 Filtering by Shape

The last stage decides whether a surviving component has an acceptable *shape*. After the height filter, most components containing noise are removed. A few persist though, typically containing a mixture of signal with noise (see Fig. 5).

We perform the following computations to remove the noisy components based on their shape:

Thinning: Each candidate component is thinned with the parallel algorithm of Guo and Hall [9], which iteratively erodes contour pixels while preserving eight-connectedness and every 2×2 block [12]. Let T denote the binary image of the resulting skeleton.

Neighbor Count: For every pixel $p \in T$ we count its eight neighbors inside T:

$$N(p) \;=\; \sum_{q \in \mathcal{N}_8(p)} T(q) \quad \in \{0, \ldots, 8\}. \tag{9}$$

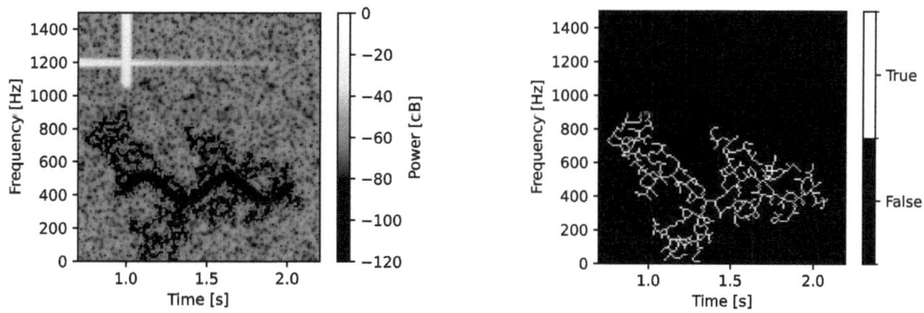

(a) Component to thin

(b) Thinning of the component

Fig. 5. Thinning of a component that contains signal and noise.

In practice, this is a convolution of T with a 3×3 kernel whose entries are all equal to 1 except for the central one which is equal to 0, followed by a point-wise multiplication with T. Pixels at line ends have $N(p) = 1$, interior line pixels give $N(p) = 2$, and bifurcations yield $N(p) \geq 3$.

Neighbor Ratio: To characterize the global shape, we average the neighbor count *only over pixels with* $N(p) \geq 2$, avoiding the bias introduced by the endpoints of a line:

$$R_i = \frac{\sum_{p \in C_i^2} N(p)}{|C_i^2|} > 2 \quad \text{with} \quad C_i^2 = \{p \in C_i : N(p) \geq 2\}. \tag{10}$$

For a perfectly thinned line, we have $R_i = 2$. Components whose skeleton is not a single line give a larger value (for Fig. 5, $R_i \approx 2.52$).

We filter by keeping the components whose neighbor ratio is below a certain threshold $R_{\max}$. In our experiments, we choose a value of $R_{\max} = 2.1$ that allows for a small tolerance to the components that are not exactly lines.

Note that, for mixed components, this criterion may remove some signal or retain residual noise; we regard this as a tunable trade-off set by $R_{\max}$.

4.5 Mask Construction

With the component-wise filters in place, we can now build the binary mask. A component is accepted only if it passes *all* the tests. Formally,

$$\mathcal{F}_i = (A_i \leq A_{\max}) \wedge (H_i \geq H_{\min}) \wedge (R_i \leq R_{\max}) \in \{\texttt{true}, \texttt{false}\}. \tag{11}$$

Let $I^\star = \{i \in I : \mathcal{F}_i = \texttt{true}\}$ be the indices of the surviving components. The final mask is the union,

$$M = \bigcup_{i \in I^\star} C_i. \tag{12}$$

5 Experimental Results

We evaluate the proposed method on both synthetic and real audio signals. The synthetic data are built from a mixture of sinusoids, transients, and additive white noise, which is the signal already used for illustration in previous sections. For real data, we select several isolated musical-instrument notes and contaminate them with artificial noise.

It is difficult to use a quantitative measure for evaluating the quality of the denoised signal. In order to evaluate it visually, we show the original spectrograms and the mask we retrieve. Auditory perception was also used as a qualitative method in the experimental setup to assess the quality of the denoising, but it is not included in this work due to the limitations of the written medium.

All experiments use a Gaussian analysis window $g(t) = \exp\left(-\frac{1}{2}\frac{t^2}{\sigma^2}\right)$, $t \in [-T/2, T/2]$ with duration T and relative width $\sigma = \sqrt{\frac{\Delta t}{2\pi \Delta f}} \approx 0.0126\,\text{s}$; this value produces an isometric time–frequency plane [8], which in particular means that the horizontal and vertical lines of same power have the same width. The window is normalized so that the upper bound of Eq. (5) holds, i.e. divided by $\|g\|_1$. Table 1 lists all analysis and filter parameters.

Table 1. Parameters used in the pipeline.

Parameter	Symbol	Value	Unit
Sampling rate	f_s	20 000	Hz
Time resolution	Δt	0.01	s
Frequency resolution	Δf	10	Hz
Window duration	T	0.1	s
Gaussian width	σ	0.0126	s

(a) Spectrogram parameters.

Filter	Symbol	Value	Unit
Height	H_{min}	6	dB
Neighbor ratio	R_{max}	2.1	n.u.
Area	A_{max}	75	%

(b) Filter parameters.

5.1 Synthetic Example

Figure 6 shows a synthetic signal composed of an AM/FM sinusoid, a decaying sinusoid, and two transients (one intersecting the decaying sinusoid) mixed with white noise at SNR = 3 dB. The recovered mask (Fig. 6b) closely matches the ground-truth signal components.

5.2 Violin with Vibrato

To assess robustness on non-stationary harmonics, we process a violin note with natural vibrato, down-sampled to 20 kHz and mixed with pink noise at -6 dB

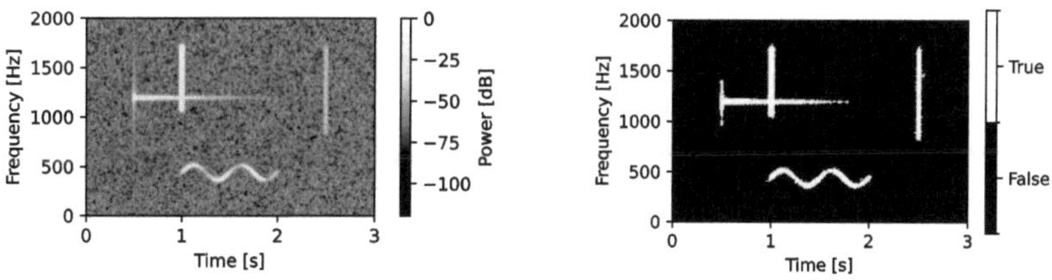

(a) Input spectrogram

(b) Resulting mask

Fig. 6. Spectrogram computed from a synthetic audio signal and the resulting binary mask M.

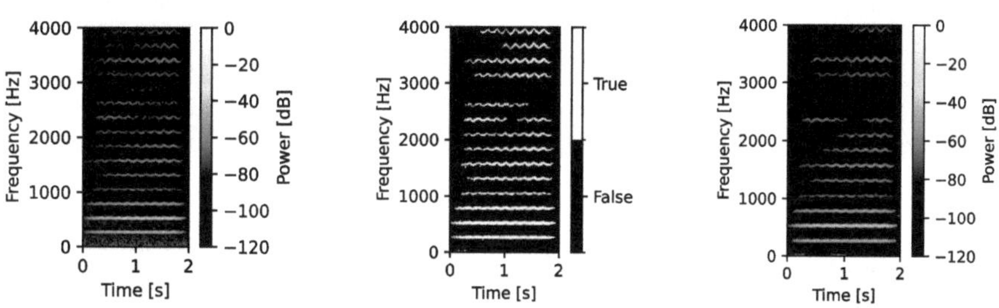

(a) Spectrogram of a violin signal with vibrato

(b) Mask obtained after filtering

(c) Result after applying the denoising pipeline

Fig. 7. Example of a violin signal with vibrato with added pink noise. (Color figure online)

SNR.[2] Fig. 7 confirms that the pipeline preserves the undulating harmonics while largely removing the broadband noise.

5.3 Parameter-Sensitivity Analysis

Finally, we investigate the effect of the two main thresholds, height H_{min} and neighbor ratio R_{max}. Figure 8 sweeps each parameter independently while holding the other constant, demonstrating that the method remains effective over a broad range and can be fine-tuned for specific applications.

[2] Pink noise exhibits a $1/f$ power spectrum.

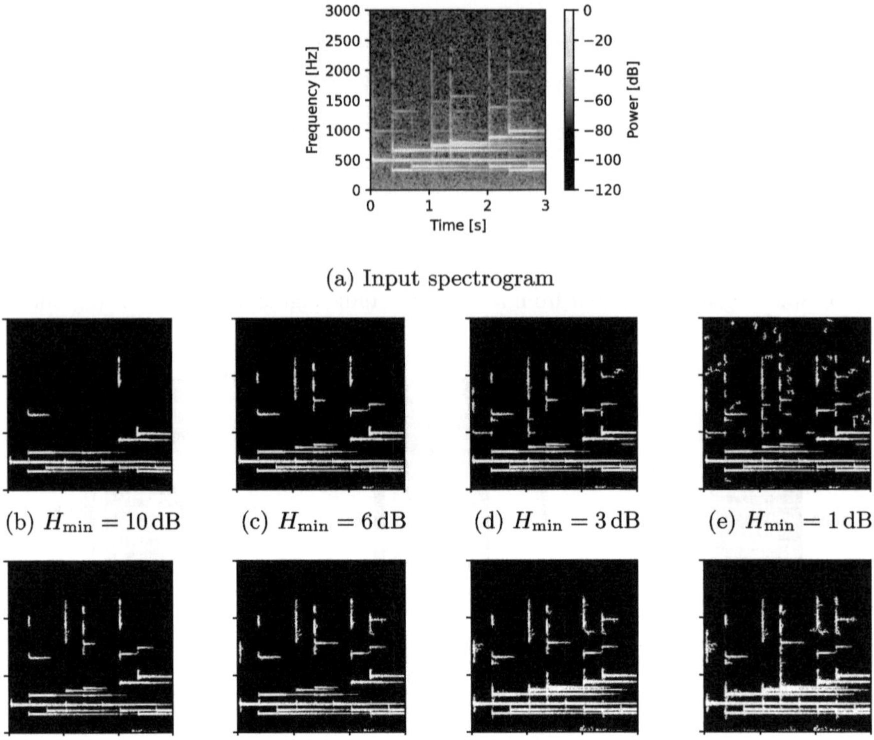

(a) Input spectrogram

(b) $H_{\min} = 10\,\text{dB}$ (c) $H_{\min} = 6\,\text{dB}$ (d) $H_{\min} = 3\,\text{dB}$ (e) $H_{\min} = 1\,\text{dB}$

(f) $R_{\max} = 2.01$ (g) $R_{\max} = 2.2$ (h) $R_{\max} = 2.3$ (i) $R_{\max} = 2.4$

Fig. 8. Influence of the two main threshold parameters on the binary mask: increasing $H_{\min}$ removes low-contrast components; increasing $R_{\max}$ tolerates more bifurcated skeletons, and thus more complex shapes.
Top row: input spectrogram.
Middle row: height threshold $H_{\min}$ is swept from 10 to 1 dB while the neighbor-ratio threshold is fixed at $R_{\max} = 2.1$.
Bottom row: $R_{\max}$ is swept from 2.01 to 2.4 with $H_{\min} = 6\,\text{dB}$ kept constant.

6 Conclusions and Future Work

We have presented a morphological spectrogram–denoising pipeline that treats the magnitude spectrogram as a grayscale image and exploits the max-tree representation to isolate signal components. By combining simple, interpretable attributes (area, height, and a shape-based neighbor ratio) our method removes broadband noise while preserving the harmonic and transient structures corresponding to signal. Our approach relies solely on morphological operations, so it is lightweight, does not require training data, and runs in a few seconds on a standard laptop CPU for spectrograms of typical size ($\approx 400 \times 300$).

Our technique is explainable and interpretable, as it is based on the geometry of spectrograms and the properties of the max-tree.

One of the main limitations of our approach is that it relies on hand-tuned thresholds that were calibrated on a limited set of signals; their transferability to very different recording conditions (*e.g.* low-bit-rate speech, ultrasound) remains to be established. Moreover, our evaluation is mostly qualitative; the absence of perceptual metrics (*e.g.* PESQ, STOI) and large-scale statistical tests prevents a rigorous comparison with recent deep-learning baselines.

The future work will focus on benchmarking the method against classical and modern denoising techniques: it remains to be determined the soundness of our approach in comparison to the state of the art. One of the main goals will be to determine the robustness of the method to different types of noise and to the choice of parameters.

Beyond denoising, we believe that the proposed morphological framework can serve as a generic *pre-processing front-end* for a variety of downstream tasks—including onset detection, harmonic/percussive source separation, and acoustic-event classification. We therefore plan to release an open-source implementation together with a small benchmark corpus and reference masks, so as to foster reproducible research and encourage the integration of hierarchical morphology with data-driven (e.g. neural) approaches in the broader audio-signal-processing community.

References

1. Allen, J.: Short term spectral analysis, synthesis, and modification by discrete Fourier transform. IEEE Trans. Acoust. Speech Signal Process. **25**(3), 235–238 (1977)
2. Boll, S.: Suppression of acoustic noise in speech using spectral subtraction. IEEE Trans. Acoust. Speech Signal Process. **27**(2), 113–120 (1979)
3. Bosilj, P., Kijak, E., Lefèvre, S.: Beyond MSER: maximally stable regions using tree of shapes. In: British Machine Vision Conference (2015)
4. Boutry, N., Tochon, G.: Stability of the tree of shapes to additive noise. In: Lindblad, J., Malmberg, F., Sladoje, N. (eds.) DGMM 2021. LNCS, vol. 12708, pp. 365–377. Springer, Cham (2021). https://doi.org/10.1007/978-3-030-76657-3_26
5. Cadore, J., Gallardo-Antolín, A., Peláez-Moreno, C.: Morphological processing of spectrograms for speech enhancement. In: Travieso-González, C.M., Alonso-Hernández, J.B. (eds.) NOLISP 2011. LNCS (LNAI), vol. 7015, pp. 224–231. Springer, Heidelberg (2011). https://doi.org/10.1007/978-3-642-25020-0_29
6. Cadore, J., Valverde-Albacete, F.J., Gallardo-Antolín, A., Peláez-Moreno, C.: Auditory-inspired morphological processing of speech spectrograms: applications in automatic speech recognition and speech enhancement. Cogn. Comput. **5**(4), 426–441 (2013)
7. Esteban, B., Tochon, G., Géraud, T.: Estimating the noise level function with the tree of shapes and non-parametric statistics. In: Vento, M., Percannella, G. (eds.) CAIP 2019. LNCS, vol. 11679, pp. 377–388. Springer, Cham (2019). https://doi.org/10.1007/978-3-030-29891-3_33
8. Gröchenig, K.: Foundations of Time-Frequency Analysis. Birkhäuser, Boston (2001)
9. Guo, Z., Hall, R.W.: Parallel thinning with two-subiteration algorithms. Commun. ACM **32**(3), 359–373 (1989)

10. Hussein, W.: Spectrogram enhancement by edge detection approach applied to bioacoustics calls classification. Signal Image Process.: Int. J. **3**(2), 1–20 (2012)
11. Koç, S.G., Aptoula, E., Bosilj, P., Damodaran, B.B., Dalla Mura, M., Lefevre, S.: A comparative noise robustness study of tree representations for attribute profile construction. In: 2017 25th Signal Processing and Communications Applications Conference (SIU), pp. 1–4. IEEE (2017)
12. Lam, L., Lee, S.W., Suen, C.: Thinning methodologies-a comprehensive survey. IEEE Trans. Pattern Anal. Mach. Intell. **14**(9), 869–885 (1992)
13. Lin, Z., Goubran, R.: Musical noise reduction in speech using two-dimensional spectrogram enhancement. In: Proceedings of the 2nd IEEE Internatioal Workshop on Haptic, Audio and Visual Environments and Their Applications, HAVE 2003, pp. 61–64 (2003)
14. de Oliveira, A.G., et al.: Bird acoustic activity detection based on morphological filtering of the spectrogram. Appl. Acoust. **98**, 34–42 (2015)
15. Romero-García, G., Bloch, I., Agón, C.: Mathematical Morphology Applied to Feature Extraction in Music Spectrograms. In: Brunetti, S., Frosini, A., Rinaldi, S. (eds.) DGMM 2024. LNCS, vol. 14605, pp. 431–442. Springer, Cham (2024). https://doi.org/10.1007/978-3-031-57793-2_33
16. Saldanha, J.C., R, S.O.: Reduction of noise for speech signal enhancement using Spectral Subtraction method. In: 2016 International Conference on Information Science (ICIS), pp. 44–47 (2016)
17. Salembier, P., Oliveras, A., Garrido, L.: Antiextensive connected operators for image and sequence processing. IEEE Trans. Image Process. **7**(4), 555–570 (1998)
18. Sharan, R.V., Moir, T.J.: Noise robust audio surveillance using reduced spectrogram image feature and one-against-all SVM. Neurocomputing **158**, 90–99 (2015)
19. Uchiyama, R., Tanabe, N.: Real-time noise suppression using harmonic/percussive separation with morphological operations for hammering test. In: 2023 Asia Pacific Signal and Information Processing Association Annual Summit and Conference (APSIPA ASC), pp. 1100–1106 (2023)

On the Constrained Maximization of Influence Over a Directed Tree Structure

Niccolò Di Marco[1(✉)], Andrea Frosini[2], and Shinichi Nakano[3]

[1] University of Tuscia, Viterbo, Italy
niccolo.dimarco@unitus.it
[2] University of Florence, Florence, Italy
[3] Gunma University, Maebashi City, Japan

Abstract. Maximizing influence in networks is a widely studied problem with applications across various domains. In this paper, we approach this problem by considering influence maximization in directed tree structures. Given a tree T, each node is associated with a binary label, obtaining 1−nodes and 0−nodes. The influence of 1−nodes over T is then computed as the total number of 0−nodes in their neighbourhood. Given an integer k, our objective is to identify a subset of k 1−nodes that maximizes the influence over T. After analysing heuristic approaches and their limitations, we present a dynamic programming algorithm for solving this problem exactly when the maximum degree and k are fixed. Our findings provide new insights into influence maximization in directed tree structures and establish a foundation for further exploration of constrained optimization problems.

1 Introduction

Maximizing influence in network structures is a fundamental problem in various domains, including graph theory [1,2], epidemiology [3] and information dissemination [4–7].

Generally speaking, the problem asks to find a set of nodes (also called *seed nodes*) that maximize the expected influence over the tree, where influence is usually computed as the number of infected nodes reached by the seed ones.

Notably, this problem has been extensively studied, with prior works proving its NP-hardness in the general setting and establishing that the best approximation algorithm is the greedy one [8,9].

In this work, we focus on a related problem: influence maximization over directed tree structures, where nodes are labeled with a binary state, and influence is measured in terms of the number of affected neighbors by one class in another. Given a tree structure and a constraint on the number of influential nodes, our goal is to determine the labeling that maximizes the overall influence.

Notably, previous studies [4] have explored related optimization problems in tree-based structures, finding a linear-time algorithm to solve the uncon-

M. H. F. Wilkinson and J. Kosinka (Eds.): DGMM 2025, LNCS 16296, pp. 237–248, 2026.
https://doi.org/10.1007/978-3-032-09544-2_17

strained version of this problem, i.e., where any number of influential nodes can be selected, through dynamic programming techniques.

However, the constrained variant, in which exactly k influential nodes must be chosen, presents additional computational challenges and remains an open problem.

In this work, given a directed tree T, we formally define and approach the *Constrained Influence Maximization Problem* $CIMP(T,k)$. We propose several heuristics and recursive methods for its solution, analyzing their theoretical properties and practical performance.

The work is organized as follows: first, we provide a family of greedy heuristics, proving its inefficacy to find the optimal solution in the general case. Then, we provide a dynamic programming approach to find the optimal labeling in $O(|V|(k+d)^{\min\{k,d\}})$ time, being d the maximum degree of the nodes of the tree. Through theoretical analysis and empirical validation, we provide insights into the efficiency and effectiveness of our proposed approaches.

2 Definitions

We consider a directed tree $T=(V,E)$, where V is the set of nodes and E is the set of directed edges. In general, each tree will be rooted in a single root r and $|T|=|V|$. For simplicity, we denote the out-neighbourhood of a node v (without v itself) as $N(v)$ and its out-degree as $d(v)$. We suppose each node v is associated with a binary label l_v. To keep it general, we say that v is a 1−node if $l_v=1$ and a 0−node otherwise. Therefore, the labeling determines a partition of the nodes $V=V_1^l\cup V_0^l$, where $V_i^l=\{v\in V\mid l_v=i\}$ is the subset of i−nodes. Finally, we refer to the number k_l of 1−nodes in the labeling l as its cardinality, i.e., $k_l=|V_1^l|$ (if the related label is obvious, we will omit it).

Here, we are interested in defining a measure of influence of 1−nodes over 0−nodes. To do that, we can specialize the out-degree of a node using $d_0^l(v)=|\{w\in N(v)\mid l_w=0\}|$, i.e. the number of 0−nodes in the out-neighbourhood of v. Please note that, when the context is clear, we will omit the l apex. Then, the influence of a configuration (T,l) is computed as

$$I(T,l)=\sum_{v\in V_1^l} d_0^l(v) \tag{1}$$

In this work, we study the following problem:

Constrained Influence Maximization Problem ($CIMP(T,k)$): Provided a directed tree T and an integer $0\le k\le |T|$, find a labelling $\bar{l}$ such that $I(T,\bar{l})=\max_l I(T,l)$, with $k=|V_1^l|$.

Example 1. *Figure 1 shows two labelings on the same tree: the left one has six (red-colored) 1−nodes. In that case, $I(T,l)=3$. On the right, three 1-nodes provide $I(T,l)=5$.*

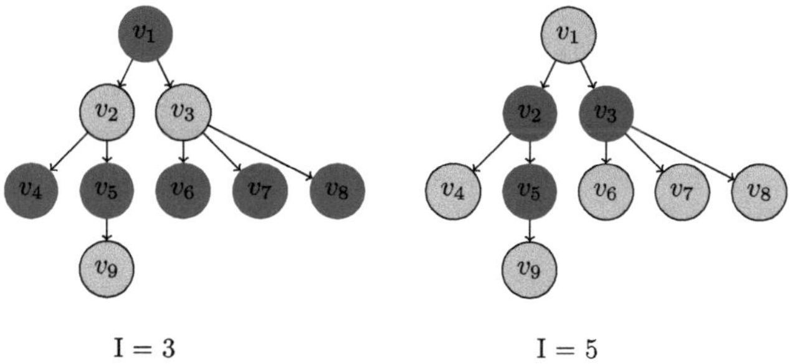

Fig. 1. Two examples of trees and their labellings. From here on, the red nodes represent 1−nodes. In more details, on the left v_1, v_4, v_5, v_6, v_7 and v_8 are 1−nodes. On the right, v_2, v_3 and v_5 are 1−nodes. (Color figure online)

Remark 1. *For a given tree T there may exist different labellings, possibly with a different number of 1-nodes, providing the same (maximal) influence. For example, consider the tree depicted in Fig. 1. The labelling $l = \{v_2, v_3, v_5\}$ and $l' = \{v_2, v_3\}$ provide the same influence $I(T, l) = I(T, l') = 5$.*

Remark 2. *We point out that, when $k \in \{0, 1, |V|\}$; $CIMP(T, k)$ admits only trivial solutions: in the first case no 1-nodes are present in T, so the influence is 0, in the second case the only 1-node of T is chosen among those having maximum degree, while, in the latter, all the nodes are 1-nodes and $I(T, l) = 0$. Therefore, the problem acquires relevance when $1 < k < |T|$.*

3 An Heuristic Approach to $CIMP(T, k)$

To start our study of $CIMP(T, k)$, we define the so-called *switch* operator. Let l be a labelling of the nodes of a given directed tree T and let v and w be two of its nodes. The operator $switch(v, w)$ creates a new label $\bar{l}$ such that $\bar{l}_v = l_w$, $\bar{l}_w = l_v$, and $\bar{l}_u = l_u$ for each $u \in V$, $u \neq v, w$.

The switch operator modifies a label l by exchanging the values of two nodes v and w.

To understand its functionality, consider two nodes $v \in V_1^l$ and $w \in V \setminus V_1^l$. We formally define the concept of *increment* $\Delta(v, w)$ as:

$$\Delta(v, w) := d_0^{\bar{l}}(w) - d_0^l(v) - p_{\bar{l}}(w) + p_l(v), \tag{2}$$

where $\bar{l}$ is the label resulting from $switch(v, w)$ and $p_l(x)$ is a binary variable equal to one if and only if the parent of x is a 1−node, according to label l. Equation (2) defines the increment obtained in the influence after one application of *switch*.

The motivation is easily obtained: $d_0^{\bar{l}}(w)$ counts the number of 0−nodes in the out-neighbour of w that has to be added to the total influence of $\bar{l}$, while $d_0^l(v)$ are the out-neighbour nodes of v that have to be subtracted when it changes into a 0-node. Moreover, we must add 1 to the influence if the parent of v is a 1-node, i.e., if $p_l(v) = 1$. In fact, when v is changed to a 0-node, its parent gains a new 0-node in its neighborhood. Similarly, we subtract 1 from the influence if the parent of w is a 1-node, i.e., if $p_{\bar{l}}(w) = 1$.

Therefore, if $\Delta(v, w) > 0$, then $switch(v, w)$ assures that $I_{\bar{l}} > I_l$. Figure 2 shows an example of an application of *switch* that leads to an increment in the influence.

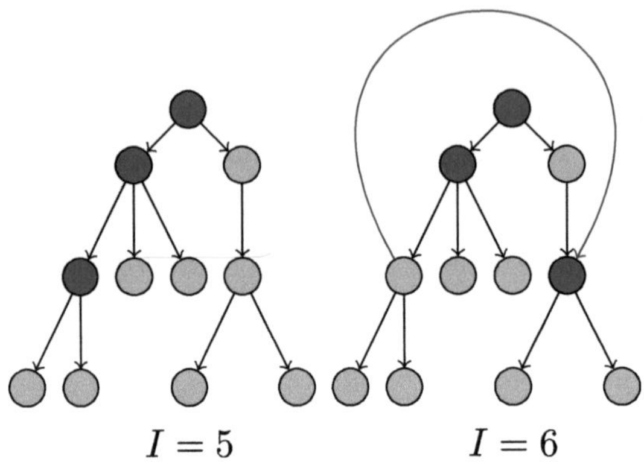

Fig. 2. The operator switch exchanges the labels of two nodes producing an increment in the influence of the new label.

3.1 First Heuristic: *TreeLabelling*

We propose the heuristic *TreeLabeling* (Algorithm 1) that relies on *TrySwitch* (Algorithm 2) to obtain an approximation of an optimal labeling l having cardinality k.

Algorithm 1 *TreeLabeling*

Require: $T = (V, E)$, k;
Ensure: a labeling l of T having cardinality k;
  for $i = 1 : k$ **do** ▷ Place the first k nodes
    $W = \{w \in V \setminus V_1^l \ : \ d_0(w) - p_l(w) = max_{u \in V}\,(d_0(u) - p_l(u))\}$;
    Select randomly a node $v \in W$;
    $l_v = 1$;
  end for
  Return $l = TrySwitch(T, l)$;

Algorithm 2 *TrySwitch*

Require: $T = (V, E)$ and one of its labels l;
Ensure: a new label $\bar{l}$;
  $check = TRUE$
  while $check$ **do** ▷ Try to apply *switch*
    $check = FALSE$;
    $S = \emptyset$;
    for (u, v) such that $u \in V_1^l$ and $v \in V \setminus V_1^l$ **do**
      if $\Delta(u, v) > 0$ **then**
        $check = TRUE$;
        $S = S \cup \{(u, v)\}$;
      end if
      choose randomly $(u, v) \in S$ such that $\Delta(u, v) \geq \Delta(u', v')$, for each $(u', v') \in S$;
      $l = switch(u, v)$;
    end for
  end while
  Return l;

Indeed, the heuristic performs a greedy approach to the problem: Algorithm 1 first labels with 1 the k nodes with maximal influence, then, it applies Algorithm 2 to the obtained label to increase the influence through a sequence of *switch* applications until no further increment occur. Obviously, in case of pairs of nodes with equal maximal increment, the algorithm randomly selects one of them to be switched. Therefore, it is possible that two calls on the same tree could return different labels.

Unfortunately, this simple heuristic does not provide an optimal label, as witnessed by the following example.

Example 2. *Consider the directed tree depicted in Fig. 3, and set* $k = 3$. *The label in (a) shows the (only) optimal solution while the label in (b) shows the output of the heuristic TreeLabeling.*

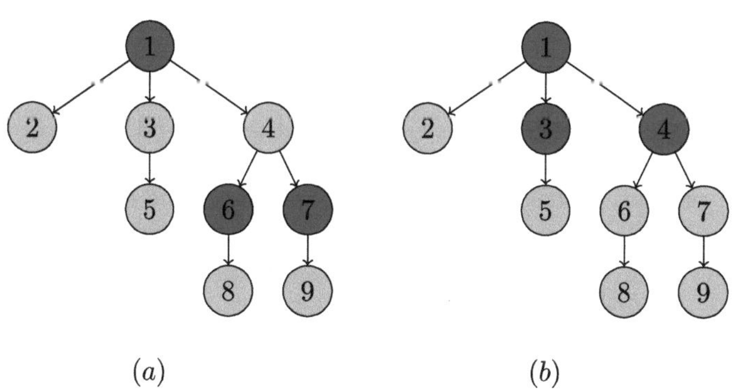

Fig. 3. (*a*) optimal labeling. (*b*) labeling obtained using *TreeLabeling*.

In particular, the algorithm follows these steps:

1. $l_v = 0$ *for each* $v \in V$ $(I = 0)$;
2. $l_1 = 1$ $(I = 3)$;
3. $l_4 = 1$ $(I = 4)$;
4. $l_3 = 1$ $(I = 4)$.
5. *Since applying switch does not increase the influence, the algorithm stops.*

Configuration (a) *obtains an influence equal to* 5*, while* (b) *gets only* 4*, providing a counterexample to the heuristic optimality.*

3.2 A Slightly Extended Version of *TreeLabeling*

To extend the class of the instances solved by (Algorithm 1), we propose a second heuristic *ExtTreeLabeling* (Algorithm 3), that acts after an initial call of *TreeLabeling* whose output is a labeling l of cardinality k and such that $\Delta(u, v) \leq 0$, for all $(v, u) \in V_1^l \times (V \setminus V_1^l)$.

Algorithm 3 *ExtTreeLabeling*

```
Require: T = (V, E), k;
Ensure: a labeling l of T of cardinality k;
  l = TreeLabeling(T, k);
  for (u, v) such that u ∈ V_1^l and v ∈ V \ V_1^l and Δ(v, u) = 0 do
     l_1 = switch(u, v);
     l_2 = TrySwitch(T, l_1);
     if l_1 ≠ l_2 then;
        l = l_2;                          ▷ the influence has been increased
     end if
  end for
  Return l;
```

For each couple (v, u) of nodes of T such that $\Delta(v, u) = 0$, the operator $switch(v, u)$ produces an equivalent (from the influence point of view) label, indicated as l_1 in Algorithm 3. If the call $TrySwitch(T, l_1)$ returns a different labeling l_2 with increased influence, i.e., $l_1 \neq l_2$, then l is updated to l_1, otherwise, l is preserved and another couple of tree nodes is considered until no further couples are present. Note that Algorithm 3 produces the optimal label in the situation depicted in Fig. 3, allowing us to move from the labeling (b) to the labeling (a).

Although Algorithm 3 produces the optimal label in a wider class of instances than Algorithm 1, there still exists some cases where this strategy fails, as witnessed in the next example.

Example 3. *Consider the directed graph* T *depicted in Fig. 4. We consider* $ExtTreeLabeling(T, 5)$.

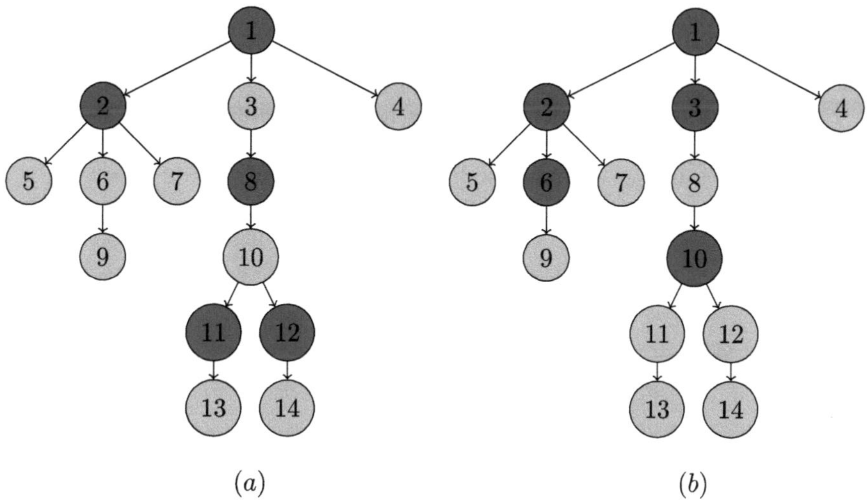

Fig. 4. A directed tree with two different labelings: in (a) the (only) optimal one ($I = 8$) obtained by *ExtTreeLabeling*$(T, 5)$, while in (b) the labelling obtained by *TreeLabeling*$(T, 5)$ ($I = 7$).

This call first initializes $l_v = 0$ *for each* $v \in V$ *(*$I = 0$*). Then it sets* $l_1 = 1$ *(*$I = 3$*),* $l_2 = 1$ *(*$I = 5$*), and* $l_{10} = 1$ *(*$I = 7$*). Since the addition of further 1-nodes does not increase the influence, the algorithm randomly chooses two nodes to reach a labeling* l *of cardinality* 5*, say* $l_6 = 1$ *and* $l_3 = 1$*. Notice that any other applications of switch on two nodes of* T *does not increase the influence (see Fig. 4, (b)).*

On the other hand, two successive applications of switch lead to the desired improvement: the first one is performed by $l_1 = switch(3, 11)$ *and the last one, i.e.,* $switch(6, 12)$*, is performed by the call* $l_2 = TrySwitch(T, l_1)$*. The optimal solution with* $I = 8$ *is finally reached (see Fig. 4, (a)).*

3.3 A Counterexample for the General Heuristic

The two previous heuristics can be generalized (with a relevant increase in complexity) to the case in which k applications of the *switch* operator are performed before the final call of Algorithm 2. In the next property, we show a general tree structure, depending on k, in which $k - 2$ applications of *switch* are needed to obtain the optimum influence.

Proposition 1. *For each* $k \geq 1$*, there exists a directed tree* T *such that* $k - 2$ *applications of the switch operator with* 0 *increment are required to move from a greedy labeling of dimension* k *(provided by Algorithm 1) to the optimal labeling of the same dimension.*

Proof. For a given k, let us consider the directed tree depicted in Fig. 5 having $|T| = 2(k-1) + 2(k-2) + 2 = 4k-4$. Algorithm 1 outputs the left labeling of dimension k whose nonoptimal influence is $I = 2k-3$.

The only way to further increase I is to apply $k-2$ times the operator *switch* and reach the labeling on the right, whose influence is still $I = 2k-3$. After that, *TrySwitch* applies a further *switch*, indicated with the red arrow in Fig. 5, on the right, and it increases the influence by one unit, leading to the optimal one $I = 2k-2$.

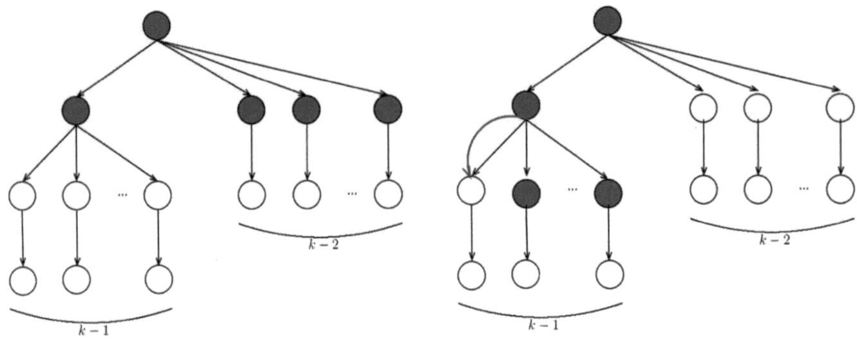

Fig. 5. A greedy labeling of a directed tree that requires $k-2$ applications of *switch* before finding a way to obtain the optimal label.

4 Dynamic Programming Solution

In this section, we show that $CIMP(T,k)$ can be solved in polynomial time with respect to the number of nodes of the tree through a dynamic programming approach, when the maximum degree of the nodes and k are constants. Let T be a directed tree and v one of its nodes. We indicate as T_v the (directed) subtree of T whose root is v, and with $CIMP_0(T_v,k)$ [resp. $CIMP_1(T_v,k)$] a solution l of $CIMP(T_v,k)$ with the constraint that $l_v = 0$ [resp. $l_v = 1$].

We associate to the node v two $(k+1)$-length vectors, $S_{0,v}$ and $S_{1,v}$ whose positions vary from 0 to k and whose i-th entries $S_{0,v}(i)$ and $S_{1,v}(i)$ are defined as $I(T_v,l)$, with $l = CIMP_0(T_v,i)$ and $l = CIMP_1(T_v,i)$, respectively. We assume the values $S_{0,v}(i)$ with $i > |V(T_v)|$ are set to 0, and the same holds for the values $S_{1,v}(i)$. So, if v is a leaf, it turns out to be $S_{0,v} = S_{1,v} = \overline{\mathbf{0}}$, the zero constant vector.

The vector $S_{0,v}$ stores the maximum influence values obtained in the subtree T_v by the labels whose cardinalities range from 0 to k, when the node v is a 0-node, while the vector $S_{1,v}$ stores the same values when the node v is a 1-node.

By definition, if v is the root of T, then $\max\{S_{0,v}(i), S_{1,v}(i)\}$ stores the cardinality of one of the labeling of $CIMP(T, i)$, and its k-th entry provides the required solution to $CIMP(T, k)$.

Now, we provide the recurrence to compute the two vectors $S_{0,v}$ and $S_{1,v}$, for each non-leaf node v in T: to lighten the notation, we denote $d = d(v)$ and let $v_1, \ldots, v_d$ be the children of v

$$\begin{cases} S_{0,v}(i) = \max_{i_1+i_2+\cdots+i_t=i} \{\max\{S_{0,v_1}(i_1), S_{1,v_1}(i_1)\} + \cdots + \max\{S_{0,v_d}(i_d), S_{1,v_d}(i_d)\}\} \\ S_{1,v}(i) = \left(\max_{i_1+i_2+\cdots+i_t=i-1} \{\max\{S_{0,v_1}(i_1), S_{1,v_1}(i_1)\} + \cdots + \max\{S_{0,v_d}(i_t), V_{1,v_d}(i_d)\}\} \right) + d_0(v) \end{cases} \tag{3}$$

where $d_0(v)$ is computed using the number of $S_{0,v_i}, 1 \leq i \leq d$ vectors involved in the sum.

In words, the maximum is taken over all indices satisfying $i_1 + \ldots + i_d = i$ [resp. $i_1 + \ldots + i_d = i - 1$] because the solution must contain exactly i [resp. $i - 1$] 1−nodes. The solution for T_v is constructed from the solutions obtained for each subtree rooted in each of the children of v, considering the optimal case, i.e. if the root must be included or not in each case.

Example 4. *Let us consider the directed tree in Fig. 6. Setting $k = 3$, we list the vectors S_0 and S_1 for the ten nodes. First of all, note that:*

1. $S_{m,v} = (0, 0, 0, 0)$ *for* $m = \{0, 1\}$,$v = \{5, 6, 7, 8, 9, 10\}$;
2. $S_{0,3} = S_{0,4} = (0, 0, 0, 0)$;
3. $S_{1,3} = (0, 3, 2, 1)$;
4. $S_{1,4} = (0, 2, 1, 0)$.

Concerning node 2, $S_{0,2}(1) = 3$ is obtained by choosing, as an example, the sequence $S_{1,3}(1) = 3, S_{0,4}(0) = 0, S_{0,5}(0) = 0$ related to the vectors of its children. Similarly, we get $S_{0,2}(2) = S_{0,2}(3) = 5$. Similarly, we obtain the vector $S_{1,2} = (0, 3, 5, 6)$.

To understand its computation, let's consider, as an example, $S_{1,2}(3) = 6$. That value is obtained by summing the sequence of values $S_{1,3}(1) = 3$, $S_{1,4}(1) = 2$, $S_{1,5}(0) = 0$ and $d_0(1) = 1$. Finally, we reach node 1 that has vectors $S_{0,1} = (0, 3, 5, 6)$ and $S_{1,1} = (0, 1, 4, 6)$, so providing the two solutions in Fig. 6 with maximal influence six.

The following theorem provides the correctness of the recurrence:

Lemma 1. *For each node v of a directed tree T, the values $S_{0,v}(i)$ and $S_{1,v}(i)$, store the influences of the labels that are solutions of $CIMP_0(T_v, i)$ and $CIMP_1(T_v, i)$, respectively.*

Proof. Let us proceed by induction on the height of the tree T_v.

Base: if v is a leaf, then the maximal influence of any of the two possible labels $l_v = 1$ or $l_v = 0$ is 0, and both the vectors $V_{0,n}$ and $V_{1,n}$ are the 0-constant vector;

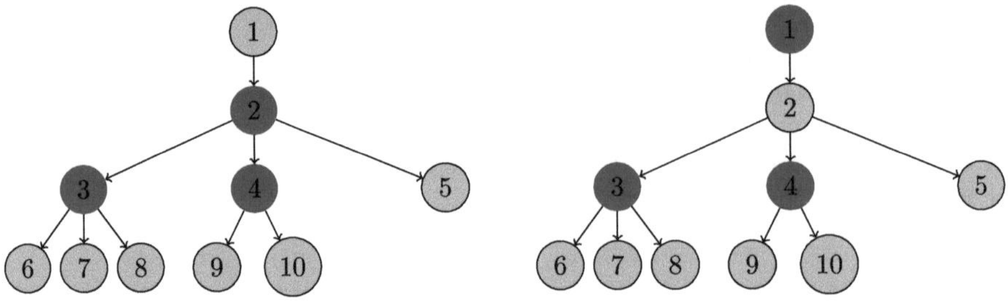

Fig. 6. The two solutions of $CIMP(T, 3)$. The labels of the nodes can be obtained from the computation of maximum values of the vectors $V_{0,1}$ and $V_{1,1}$, being node 1 the root of the tree.

Inductive step: let $v_1, \ldots v_d$ be the children of v. By inductive hypothesis, the vectors S_{0,v_j} and S_{1,v_j} store the influences of the solutions of $CIMP_0(T_{v_j}, i)$ and $CIMP_1(T_{v_j}, i)$, with $0 \leq i \leq k$ and $1 \leq j \leq d$, respectively. Let us focus on the vector $S_{0,v}$ whose i-th entry is computed using the maximal (according to the influence value) sequence of indexes $i_1 + i_2 + \cdots + i_d = i$. Assume by contradiction that

$$I(T_v, l) > S_{0,v}(i), \tag{4}$$

with $l = CIMP_0(T_v, i)$. In that case, the label l produces $l^{(1)}, \ldots, l^{(d)}$ labels of each subtree $T_{v_1}, \ldots, T_{v_d}$ such that $|l^{(1)}| + \cdots + |l^{(d)}| = i$. So, if (4) holds, it means that one of the labels V_{0,v_j} and V_{1,v_j} were not optimal, contradicting their definitions. A similar reasoning holds for $S_{1,v}$, reaching the thesis. □

Theorem 1. Given a directed tree $T = (V, E)$ and an integer $0 \leq k \leq |T|$, the problem $CIMP(T, k)$ can be solved in $O(|V|(k + \bar{d})^{\min\{k,\bar{d}\}})$ time, with $\bar{d} = \max_{v \in V} d(v)$.

Proof. Lemma 1 assures that the value $\max\{S_{0,v}(k), S_{1,v}(k)\}$ is the influence of one of the labeling l that is a solution of $CIMP(T, k)$, when n is the root of T. Furthermore, the sequences of indices used to compute such a value recursively provide l.

The complexity of the computation of l follows from the definition of the vectors $S_{0,v}$ and $S_{1,v}$. In particular, the heaviest computational step concerns the search of the indexes' sequences $i_1, \ldots, i_d$ that provide the maximum influences for each internal node v of T, and for each cardinality of the labels (recall that $d = d(v)$). The computations of the entries $S_{0,v}(i)$ and $S_{1,v}(i)$ require to scan all the $\binom{i+d-1}{i}$ indexes' sequences of the vectors S_{0,v_i} and S_{1,v_i}, with v_i ranging among the children of v, to find those indexes summing to i and producing the maximum influence. To hit the aim, the final complexity reaches $O(|V|(k + \bar{d})^{\min\{k,\bar{d}\}})$. □

The following example shows how the vectors S_0 and S_1 lead to the solutions of $CIMP(T, k)$ related to the tree T in Fig. 5 and provided as a counterexample to the greedy heuristic with a bounded number of switch operations to the problem.

Example 5. *Let us consider the problem $CIMP(T, 5)$, where T is the directed tree in Fig. 5. Let us number the nodes of T starting from node 1, the root, and proceeding to its children from left to right, using the labels $2, \ldots, 5$. We compute the vectors $S_{0,2} = (0, 1, 2, 3, 4, 3)$, $S_{1,2} = (0, 4, 4, 4, 4, 4)$, $S_{0,i} = \overline{\mathbf{0}}$ and $S_{1,i} = (0, 1, 0, 0, 0, 0)$, with $i \in \{3, 4, 5\}$. The computation of the vectors related to node 1 rely on their values obtaining $S_{0,1} = (0, 4, 5, 6, 7, 7)$, and $S_{1,1} = (0, 4, 7, 7, 7, 8)$. The maximal value $S_{1,1}(5) = 8$ is the (unique) solution of $CIMP(T, 5)$ and it is obtained by the sequence $S_{0,2}(4) = 4$, $S_{0,i}(0) = 0$, with $i \in \{3, 4, 5\}$ and $d_0(v) = 4$.*

5 Conclusions

In this work, we introduced and analyzed the Constrained Influence Maximization Problem ($CIMP$) on directed tree structures, extending the classical formulation of the influence maximization framework.

After presenting possible heuristics, we introduced a dynamic programming algorithm that computes the exact solution to CIMP in polynomial time, provided that k and the maximum degree of the tree are fixed. Our approach systematically builds optimal labelings for subtrees, storing influence values under the constraint of fixed cardinality and root label.

However, there is still room for improvement of the proposed algorithms. In fact, our approach led to a combinatorial explosion in k and d. Therefore, future works will focus on different strategies to reduce its complexity and, if possible, to extend these techniques to broader classes of graphs, incorporate probabilistic influence models, or consider additional constraints and real-world applications.

References

1. Mouravski, A.: Influence maximization on families of graphs. Rochester Institute of Technology, 2011
2. Gautam, R.K., Kare, A.S., Durga Bhavani, S.: Heuristics for influence maximization with tiered influence and activation thresholds. In: Verma, A., Verma, P., Pattanaik, K.K., Dhurandher, S.K., Woungang, I. (eds.) Advanced Network Technologies and Intelligent Computing. ANTIC 2023. CCIS, vol. 2092, pp. 134–148. Springer, Cham (2024). https://doi.org/10.1007/978-3-031-64070-4_8
3. Kitsak, M., et al.: Identification of influential spreaders in complex networks. Nat. Phys. **6**(11), 888–893 (2010)
4. Di Marco, N., Brunetti, S., Cinelli, M., Quattrociocchi, W.: Post-hoc evaluation of nodes influence in information cascades: the case of coordinated accounts. ACM Trans. Web (2024). Just Accepted
5. Shao, C., Ciampaglia, G.L., Varol, O., Yang, K.C., Flammini, A., Menczer, F.: The spread of low-credibility content by social bots. Nat. Commun. **9**(1), 1–9 (2018)
6. Bakshy, E., Rosenn, I., Marlow, C., Adamic, L.: The role of social networks in information diffusion. In: Proceedings of the 21st International Conference on World Wide Web, pp. 519–528, 2012

7. Agha Mohammad Ali Kermani, M., Ghesmati, R., Pishvaee, M.S.: A robust optimization model for influence maximization in social networks with heterogeneous nodes. Comput. Soc. Netw. **8**(1), 17 (2021)
8. Kempe, D., Kleinberg, J., Tardos, É.: Maximizing the spread of influence through a social network. In: Proceedings of the ninth ACM SIGKDD International Conference on Knowledge Discovery and Data Mining, KDD03. ACM, August 2003
9. Mossel, E., Roch, S.: On the submodularity of influence in social networks. In: Proceedings of the Thirty-Ninth Annual ACM Symposium on Theory of Computing, STOC07, pp. 128–134. ACM, June 2007

Discrete Spiral and Max-Link, Complementary Tools for Vision

Walter G. Kropatsch(✉)

Institute of Visual Computing and Human-Centered Technology, Virtual and Augmented Reality, E193-03, Technische Universität Wien, Vienna 1040, Austria
krw@prip.tuwien.ac.at

Abstract. Human vision is dynamic: body, head, and eye movements change the view of the actual environment. Regions and features close to the boundary are likely to disappear while the probability in the center is high that they re-appear after the movements. We therefore exploit a re-arrangement of the pixels that orders the pixels such that the center of the image is on one end of the sequence and the boundary pixels are on the other end. A discrete spiral that connects a corner of the image with the center and contains all the pixels satisfies these constraints. In addition it is a strict total order. This order enables a max-link strategy generating efficiently a spanning forest of the image. It allows the efficient contraction for the levels of an irregular pyramid. The max-link strategy on the spiral pushes the surviving vertices geometrically towards the center of the image. We explore the properties of the proposed combination and their integration into the contrast pyramid.

Keywords: Spiral total order · Max-Link strategy · Irregular Image Pyramids

1 Introduction

A spiral is a frequent concept appearing also in nature: spiral galaxies, young spiral ferns, or simply hose carts[1] The Stochastic pyramid of Peter Meer [13] was the first image pyramid where the levels of the pyramid were irregular graphs instead of arrays. The selection of survivors was done by choosing the local maxima of random numbers associated with every pixel. In this way the distance between surviving vertices in the next higher pyramid level was at least two. Then all the non-surviving vertices chose one survivor in their neighborhood. If there were gaps the process was repeated for the undecided pixels until every non-survivor had a surviving neighbor. Since the random numbers are not necessarily all different, decisions at such (rare cases) enhance the influence of noise and led to different pyramidal structures although the inputs were not substantially different.

[1] i.e., https://en.wikipedia.org/wiki/NGC_1300, https://www.shutterstock.com/search/ferns-spiral.

M. H. F. Wilkinson and J. Kosinka (Eds.): DGMM 2025, LNCS 16296, pp. 249–260, 2026.
https://doi.org/10.1007/978-3-032-09544-2_18

Jolion and Montanvert extended Meer's concept in the adaptive pyramid [8] by replacing the random selection by a data-dependent adaptation of the graphs at higher pyramid levels. Also the adaptive pyramid suffers from the fact that some of the decisions are very weak and hence also influenced by noise.

In the current paper we make only a small modification to the above concepts: We replace the random numbers by a strict total order of ranks that are derived from a spiral covering the image. Experience with the total order of a column major show that pyramids with identical input always create the same structure but representative pixels of homogeneous regions tend to move towards a corner of the image [1,3]. The spiral connects the center of the image with one of the corners by a Hamiltonian path. This allows to push vertices that represent relevant regions of the image towards the center. The center of a bounded image is more likely to re-appear in a sub-sequent view than a corner even by a slight shift of successive image captures. For the selection of survivors we use a max-link strategy that is somehow related to the principle of Peter Meer but it generates a spanning forest that can be used to construct the irregular pyramid. It also allows to adapt the structure to the image's salient regions.

We first recall the basic principles of the irregular graph pyramid (Sect. 2). Then the strict total order is introduced together with the max-link strategy for efficiently deriving a spanning forest (Sect. 3). Section 4 introduces the discrete spirals, how they can be computed from a template and how to combine them efficiently with the max-link strategy. The parallel computational complexity is addressed in Sect. 5. Section 6 proposes how to combine data adaptivity with the spiral total order in areas of low saliency. Section 7 reports some preliminary results with the contrast pyramid that is the first pyramid that is able to propagate high spatial frequency to the higher pyramid levels and enables top-down reconstructions which are hard to distinguish from the original images although only a few percent of the original colors are used. The Conclusion summarizes the paper and enumerates a few future research issues.

2 Recall on the Irregular Graph Pyramid

The irregular graph pyramid [10] consists of several levels of plane graphs $G_i(V_i, E_i)$, $i = 0, 1, \ldots t$ where the base level G_0 is the 4-neighborhood graph of an image. The pixel values $g(v), v \in V_0$, are the attributes of the vertices V_0. Two operations generate the higher pyramid levels: **edge contractions** and **edge removals** [4]. The operation of contracting an edge deletes the edge and one of its end points (the other "**survives**") while the removal of an edge deletes the edge and merges its adjacent (dual) faces. Contracting an edge has the important property of preserving the connectivity of the graph while removing an edge merges the two adjacent faces. A set of edges that are contracted with the same surviving vertex are called **contraction kernels (CK)**. The result of contracting graph G by a contraction kernel CK is denoted by G/CK.

The concept of **equivalent contraction kernels (ECK)** [9] allows to combine multiple contraction kernels resulting in the same graph

$$G_n = (((G_0/CK_0)/CK_1 \dots /CK_t) = G_0/ECK \tag{1}$$

as after several successive contractions. The **receptive field** $RF(v_t) \subset V_0$ of a higher level vertex $v_t \in V_t, t > 0$, can be derived directly from the base by the ECK(v_t) that is a forest spanning the receptive fields of all top vertices $v_t \in V_t$. A tree is an acyclic connected graph with one less edges $T \subset E$ than vertices: $|T| = |V| - 1$. The receptive fields of a graph $G_t = (V_t, E_t)$ at a (top) level $t > 0$ partition the base vertices,

$$V_0 = \bigcup_{v_t \in V_t} RF(v_t), \tag{2}$$

the receptive fields are the connected components of the spanning forest of the base graph G_0 corresponding to the ECK.

3 Total Order and Max-Link

The ECKs of a high level of the pyramid form spanning forests of the receptive fields of the roots [9]. In regular pyramids [5] the receptive fields are determined by the position of the surviving pixel, the reduction factor (in ML[2] called stride) and the shape of the reduction window. In irregular pyramids they are the result of the contraction process with selected contraction kernels. This selection can be related to properties of the data and, hence, the contraction adapts to the data like in the contrast pyramid [10]. But this may, in certain cases, lead to ambiguous selections that may also be influenced by noise. A simple example are large homogeneous regions where the surviving representative vertex should be in the middle of the region. To avoid that these representative vertices are located along the boundary of the region or of the image, and disappear by small shifts of the camera, we introduce a strict total order as additional attribute of the vertices.

Definition 1 (Strict Total Vertex Order (TO)). *Given a plane graph $G(V, E)$ we define the* **rank** *of a vertex as a function $\rho : V \mapsto \mathbb{N}$ with binary relations $\rho(v)$ satisfying $\forall v \neq w \in V : \rho(v) \neq \rho(w)$.*

The (strict) total order of the vertices determines unique contraction kernels and, in addition, enable a large number of independent contraction kernels that can be contracted simultaneously with low parallel complexity.

Definition 2 (Independence). *Two contraction kernels CK_a and CK_b are* **independent** *if the intersection of their receptive fields $RF(CK_a) \cap RF(CK_b) = \emptyset$ is empty.*

[2] machine learning.

3.1 Deriving Spanning Trees and Spanning Forests

$ECK \subset E_0$ are spanning forests of the base graph [9].

Definition 3. *Let $G(V, E)$ denote the plane graph and $\rho(v) : V \mapsto \mathbb{N}$ be the* **rank** *of a vertex in the TO. The local neighborhood $\Gamma(v) \subset V$ of a vertex $v \in V$ is defined by all vertices that are related by an edge in E:*

$\Gamma(v) = \{w \in V | (v, w) \in E$ or $(w, v) \in E\}^3$. With **max-link(TO)**, *every vertex $w \in V \setminus R_{max}$ chooses the vertex $v \in \Gamma(w) \subset V$ that has the highest rank in the TO. The link $(w, v) \in T_\oplus$ is then an edge of the ECK:*

$$T_\oplus = \{(w, v) \in E | w \in V \setminus R_{max}, v = \arg \max_{u \in \Gamma(w)} \rho(u)\}. \quad (3)$$

The roots $R_{max} \subset V$ of each tree of the spanning forest $T_\oplus \subset E$ are the local maxima of the TO (see Fig. 4a).

Proposition 1. *The edges $T_\oplus \subset E$ computed by (3) can be computed in parallel by all vertices of the graph. The parallel complexity depends only on the degree of the vertices and NOT on the number of vertices.*

4 Spirals

In polar coordinates (r, θ) an Archimedean spiral (Fig. 1(a)) can be described by the equation $r = b \cdot \theta$ with real number b. The constant parameter b controls the distance between the loops. The continuous spiral can be sub-sampled (Fig. 1(b)) with constant arc length between successive sampling points. After sub-sampling by a factor of 8 the first sampling point after the center covers nearly the complete first loop of the spiral. This creates complicated configurations around the center in all further sub-samplings.

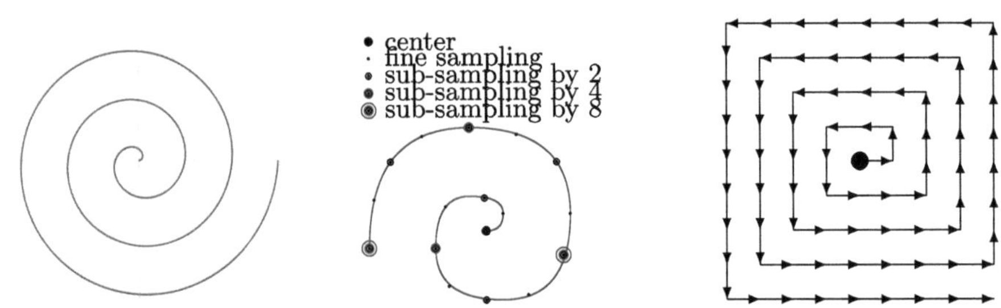

(a) Archimedean spiral (b) samplings (c) discrete spiral

Fig. 1. Continuous to discrete spirals.

[3] In images 4-neighborhood is $\Gamma_4(x, y) = \{(p, q)| \; |x - p| + |y - q| \leq 1\}$ and 8-neighborhood $\Gamma_8(x, y) = \{(p, q)| \max(|x - p|, |y - q|) \leq 1\}$.

A discrete version is shown in Fig. 1(c). Choosing $b = 1$ the discrete spiral is image-filling i.e. covering all the pixels of the square array of the image. Pixels along the Hamiltonian[4] path of the spiral receive an additional attribute: This **rank** ρ measures the number of discrete steps of the path from the starting pixel.

Spirals for different images and even shapes can be cut out from a pre-computed large template. Figure 2(a) shows the onion-like relations between spirals of different sizes. Larger versions can be created without re-computing the central part and smaller versions simply remove the respective differences.

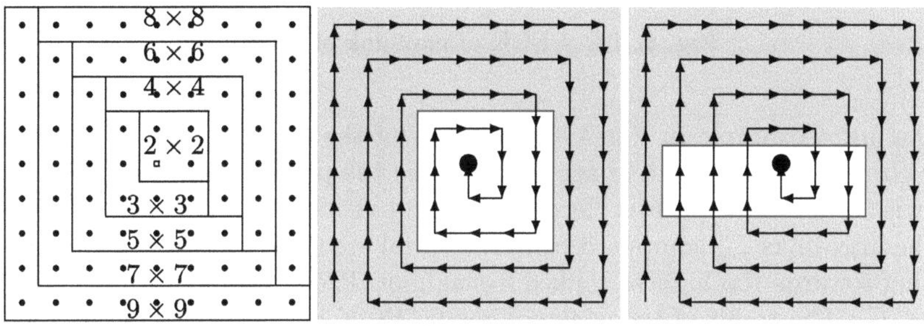

(a) $2 \times 2 \ldots 9 \times 9$ layers (b) $4 \times 4 \subset 9 \times 9$ (c) $2 \times 6 \subset 9 \times 9$

Fig. 2. Related smaller and larger sub-windows

Smaller sub-regions can be cut out from a larger template: sub-windows as centered squares in Fig. 2(a,b). The ranks of the smaller window correspond to the ranks of the template. Centered rectangular sub-windows can also be cut out from the template (Fig. 2(c)). In these cases the ranks are no more contiguous and contain gaps. However they still form a strict total order.

Virtual links (Fig. 3) in the same orientation as the spirals of the template connect the created gaps of the 2×6 subwindow. The selected ranks in Fig. 3 are cut out from Fig. 7(b). Non-rectangular shapes can also be cut-off from the template but they may create further local extrema. Consequently the resulting ranks may form only a spanning forest. How to resolve such situations is shown in Sect. 6.

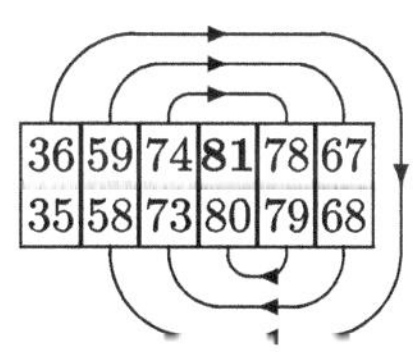

Fig. 3. Virtual links.

4.1 Max-Link of the Spiral Total Order

We consider now the use of the max-link strategy as defined in Definition 3 for generating a spanning forest of the image from the spiral ranks with the purpose of using the spanning forest as ECK for constructing the irregular pyramid efficiently.

[4] A Hamiltonian path is a path visiting every vertex of the graph exactly once.

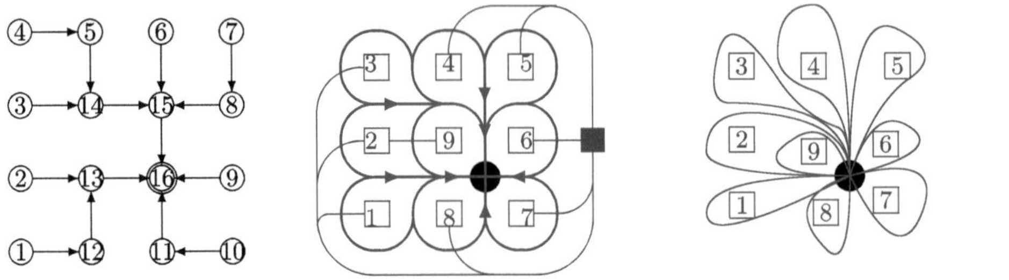

(a) Spiral ECK= $T_\oplus$ (b) G/ECK (red), $\overline{G/ECK}$ (green) (c) structure of self-loops

Fig. 4. 4×4 ECK of max-link of spiral.

The 4×4 example in Fig. 4 shows ranks of the spiral that start in the left lower corner ($\rho = 1$) of the image, and follows the pixels along a spiral curve in clockwise order. This path is Hamiltonian and visits all vertices once.

The max-links $T_\oplus$ computed from the spiral are shown in Fig. 4(a) by arrows pointing towards the highest ranked 4-neighbor. It constructs the pyramid with ECK= $T_\oplus$. The result of the contractions G/ECK is the graph $(\{\bullet\}, E_t)$ with a single vertex (Fig. 4(b)) and rank $\rho(\bullet) = 16$ in the base. The nine surviving edges form nine self-loops E_t directly attached to the root $\bullet$. Only the self-loop around $\boxed{2}$ includes self-loop $\boxed{9}$ in Fig. 4(c). This ECK spans the complete image if the spiral ranks are used for max-link and the branches of the spanning tree connect the root with 4-connected paths of length not longer than $\sqrt{|V|} = n = 4$ that are well suited for constructing the pyramid efficiently.

4.2 Properties of the Spanning Forests $T_\oplus$

Figure 5 shows a 4-connected 9×9 neighborhood graph $G(V, E)$, with $|V| = n^2 = 81, |E| = 2n(n-1) = 144$, a 9×9 discrete spiral $G_@ = (V, E_@), E_@ \subset E$ (in green). The 80 edges of the spiral cover all vertices of the image and form a Hamiltonian 4-path that connects a corner of the image with its center. The 80 max-links $G_\oplus = (V, T_\oplus)$ are shown in red. The spiral and the max-links together contain all edges: $E_@ \cup T_\oplus = E$. Both the spiral and the max-link $|E_@| = |T_\oplus| = n^2 - 1$ are trees spanning V. There are only $|E_@ \cap T_\oplus| = 2(n^2-1) - 2n^2 + 2n = 2(n-1) = 16$ common edges (red line inside green rectangle).

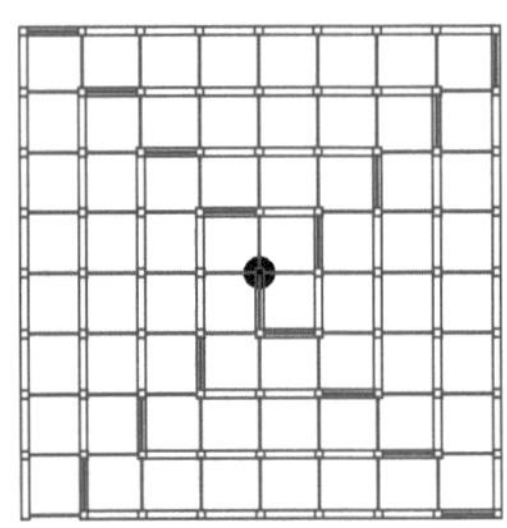

Fig. 5. Spiral and Max-links

The edges common to $E_@$ and $T_\oplus$ appear exclusively along the diagonal staircases and are independent.

Proposition 2. *Let ρ denote the strict total order of the vertices of graph $G(V, E)$ with $|V| = n^2$, $E_@ \subset E$ denotes the edges of the spiral and $T_\oplus \subset E$ the max-links of the spiral ranks.*

1. *In general max-link (3) produces a* **spanning forest** $SF \subset E$ *of the graph* G *where each connected component contains one local maximum of the TO. It is the root of the tree spanning the connected component.*
2. *If we require the TO to form a slope*[5] $T_\oplus$ *remains connected and forms a* **spanning tree** *of the graph* G *with* $|T_\oplus| = n^2 - 1$ *links.*
3. **Independence of edges:** *All 4-neighbors* $(v, w) \in \Gamma_4 \subset E$ *of a spiral TO connect a vertex with even* $\rho(v)$ *and odd* $\rho(w)$*. Then all edges of the subset* $SF_0 = \{(v, w) \in SF | \rho(v) \text{ is even}, \rho(w) \text{ is odd}\}$ *are pairwise independent except at branching points of* SF*.*
4. *All boundary vertices are connected to one of the four diagonal stair cases of* $T_\oplus$ *by a sequence of horizontal or of vertical edges.*
5. *orthogonality:* $\forall e_@ = (x, y) \in E_@ \exists e_\oplus = (x, z) \in T_\oplus : e_@ \perp e_\oplus$.

Proof. (Proposition 1) Every vertex $v \in V$ creates only one link to its highest neighbor. Since there are no identical ranks in the TO no cycle can be created. Any local extremum cannot link to a neighbor, following the created link backwards determines the receptive field of the extremum that is a connected component.

(Proposition 2) A slope cannot contain more than one maximum and one minimum [7,10]. Consequently a TO that is a slope has no other extrema than the global extrema and (3) produces a spanning tree.

(Proposition 3) Since the boundary of an $m \times n$ image has always an even length $2(m + n - 2)$ the last vertex of any loop has the opposite parity to the start vertex of the loop. This property propagates along the spiral (Fig. 4(a) and Fig. 7(b)). After contracting all vertices with even rank (except the root) only the edges with odd ranks $\rho_0 = \rho$ survive. Compacting $\rho_{i+1} = (1 + \rho_i)/2$ in $G/SF_{i+1}, i = 1, 2, \ldots$ creates again even ranks that can be contracted into their odd max-link neighbors iteratively until one root is left. The compaction of ranks reduces the number of ranks by a factor of two at each iteration and, hence, it has logarithmic parallel complexity.

(Proposition 4) Starting the spiral from a corner and incrementing the ranks along the spirals makes all ranks of the respective inner loops higher than the adjacent outer loop. Consequently max-link at vertices of the outer loop create 4-connected links to the inner loop. The only exceptions occur at the corners and when the spiral proceeds to the next inner loop.

(Proposition 5) This property is related to Property 4: links from the outer loop to the next inner loop are clearly orthogonal to the orientation of the spiral edge. Since the spiral edges at the corners form a right angle the max-link at the corners and the end-of-spiral-loop edges, the claim is satisfied. □

5 Parallel Computational Complexity

Independence of edges is a pre-condition for contracting a set of edges in parallel (Definition 2). If both end points of an edge $(x, y) \in T_\oplus$ have degrees < 3 then

[5] In a slope region, every pair of vertices is connected by a monotonic path [7].

maxima of T' and the surviving edges re-connect the different components of $T_A \subset E_A, T_B \subset E_B, T_C \subset E_C$. But the connections may not be acyclic and with multiple edges. In such case we re-apply max-link on the contracted graphs $(A/T_A, E_A), (B/T_B, E_B), (C/T_C, E_C)$. Since the contracted graphs contain only the local maxima but still form a TO, the result is a spanning forest with less local maxima. The contraction of the sub-trees and re-linking with max-tree can be repeated until reaching a spanning tree for all regions[6]. Our example shows the result in Fig. 8(a) where the new connections are marked by double edges. Because the original regions are connected and contraction preserves the connectivity, the process converges to the resulting spanning trees which can be down-projected to the ECK of the original graph using the unique ranks (Fig. 8(b)). The result of contracting $G/(T_A \cup T_B \cup T_C)$ is shown in Fig. 8(c). Redundant edges have been removed, the self-loop around region C is preserved. All three top vertices are at the closest position within their regions A, B, C.

General Algorithm for constructing the combined ECK
Input (Fig. 7(a,b): a) region partition $G = \bigcup_{\lambda \in \Lambda} R_\lambda$ with distinct labels $\lambda \in \Lambda$ for every vertex; and b) ranks ρ of the spiral TO.

1. For all $\lambda \in \Lambda : T_\oplus(\lambda) =$ max-link of all connected regions R_λ.
2. While links of any label have more than 1 component do:
   - Re-run and re-compute max-links on R_λ / T_λ inside each region.
3. Expand the top spanning trees to the base following the ranks: ECK.
4. Contract G/ECK with the resulting ECK.
5. Simplify the final graph by removing redundant parallel edges and redundant self-loops.

7 Max-Link with Spiral TO for the Contrast Pyramid

The basic strategy (see [10] for more details) is to contract edges with increasing contrast up to a certain level that is determined by the contrast histogram [11].

In real images there are many edges with low contrast (Fig. 9(b)). Hence we use the TO of vertices to deterministically select the edges in connected regions with same contrast by the max-link of the spiral. To process several contractions in parallel one further criterium is to group edges that are independent of each other. We observed in Proposition 2 that the spiral TO provides independent subsets of edges by the parity of the ranks of the edge's end points.

Figure 9 shows the example Fish of the Berkeley database [12] with its contrast histogram Fig. 9(b). Contrasts higher than 4 are shown in red and identify the contrast of the top level edges connecting vertices with 6% of the original colors (more examples and algorithmic details can be found in [6]). The reconstruction propagates the top colors down to the base image.

[6] The number of repetitions is the logarithm of the longest branch in the reduced graph.

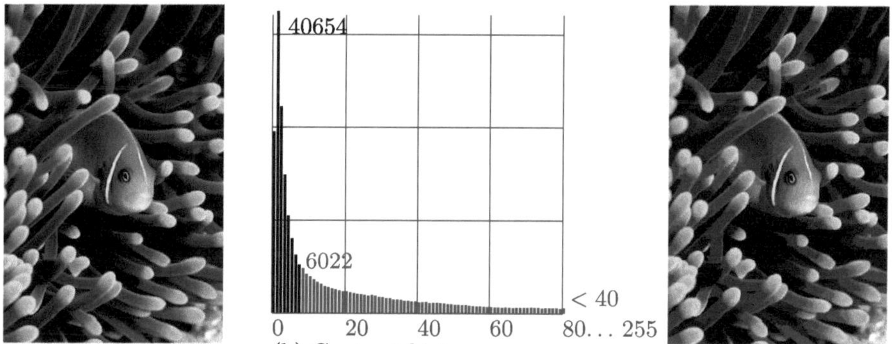

(a) original Fish (b) Contrast histogram (c) reconstruction

Fig. 9. Original image, contrast histogram, and reconstruction with 6% of colors.

Table 1 lists the highest contrasts c_t satisfying $\sum_{c=0}^{c_t} h_0(c) < |V_0|$ for some images from the Berkeley data base [12]. All images have the same size $|V_0| = 154401$, $|V_t|$ is the number of vertices at the top and # crit is the number of critical points (local max, local min, saddle) in the base. Therefore the mean error of the reconstructed pixels has a brightness difference smaller than 8 and 12 for images Fish and Pheasant, respectively, but is still below 30 for the other images. Empirical tests have shown (see section Gallery in [6]) that this is below the noticeable deviation for human observers, that can hardly distinguish the reconstruction from the original!

Table 1. Maximally contractible edges

Berkeley#	Picture	c_t	$\|V_t\|$	# crit
210088	Fish	8	9264	18019
43074	Pheasant	12	4632	19788
160068	Cat	15	9264	20541
156065	Coral	16	12352	45915
295087	arc	16	12352	
55073	stone statue	21	15440	
41069	squirrel	25	9264	

8 Conclusion and Outlook

This paper proposes complementary tools for building the contrast pyramids: the spiral total order together with the max-link strategy determine the ECK for maximally contracting a given image before actually constructing the pyramid. The shown properties of the combination of the discrete spiral together with the max-link strategy not only push the surviving vertices as close as possible towards the center of the spiral, they have the potential to allow a fast parallel implementation. Since the content is actual research we expect further experimental results soon.

There is a wide range of potential applications beyond the efficient derivation of the region adjacency graph in [2]. For the computation of the max-links also the 8-neighborhood can be used. The inside of a bounding box of an image object can be linearized by a spiral and contrast-contracted as a signature of the object. The total order for voxel images in 3D could enable efficient processing of volumetric data like CT or MRI in medical imaging.

Acknowledgments. Great thanks go to Noriane Petit and Victor Laurin who produced the experimental results with the contrast histogram during their internships in Wien, 2024–2025.

Disclosure of Interests. The author has no competing interest.

References

1. Banaeyan, M., Kropatsch, W.G.: Fast labeled spanning tree in binary irregular graph pyramids. J. Eng. Res. Sci. **1**(10), 69–78 (2022)
2. Banaeyan, M., Kropatsch, W.G., Hladuvka, J.: Redundant 1-cells in Multi-labeled 2-Gmap Irregular Pyramids. In: Bürstmayr, H., Gronauer, A., Holzinger, A., Roth, P.M., Stampfer, K. (eds.) Proc. OAGM Workshop, Digitalization for Smart Farming and Forestry, pp. 5–22. Technische Universität Graz (September 2022)
3. Banaeyan, M., Kropatsch, W.G.: Adapting a total vertex order to the geometry of a connected component. Pattern Recogn. Lett. **190**, 8–14 (2025)
4. Brun, L., Kropatsch, W.G.: Dual contraction of combinatorial maps. In: Kropatsch, W.G., Jolion, J.M. (eds.) 2nd IAPR-TC-15 Workshop on Graph-based Representation, pp. 145–154. OCG-Schriftenreihe, Österreichische Computer Gesellschaft (1999). band 126
5. Burt, P.J.: A pyramid framework for real-time computer vision. In: Davis, L.S. (ed.) Foundations of Image Understanding, chap. 12, pp. 349–380. Kluwer (2001)
6. Cerman, M.: Structurally Correct Image Segmentation using Local Binary Patterns and the Combinatorial Pyramid. Technical report PRIP-TR-133, PRIP, TU Wien (2015)
7. Gonzalez-Diaz, R., Batavia, D., Casablanca, R.M., Kropatsch, W.G.: Characterizing slope regions. J. Comb. Optim. 1–20 (2021)
8. Jolion, J.M., Montanvert, A.: The adaptive pyramid, a framework for 2D image analysis. Comput. Vis. Graph. Image Process. Image Underst. **55**(3), 339–348 (1992)
9. Kropatsch, W.G.: Equivalent contraction kernels to build dual irregular pyramids. In: Solina, F., Kropatsch, W.G., Klette, R., Bajcsy, R. (eds) Advances in Computer Vision. Advances in Computing Science, pp. 99–107. Springer, Vienna (1997). https://doi.org/10.1007/978-3-7091-6867-7_11
10. Kropatsch, W.G., Banaeyan, M., Gonzalez-Diaz, R.: Controlling topology preserving graph pyramids. In: A.El-Yacoubi, M., Vincent, N., Kurtz, C. (eds.) Emerging Topics in Pattern Recognition and Artificial Intelligence, Language Processing, Pattern Recognition, and Intelligent Systems, vol. 9, chap. 2, pp. 27–76. World Scientific (2024)
11. Kropatsch, W.G.: Lifting some secrets about contrast pyramids. In: Brun, L. (ed.) Graph-Based Representations in Pattern Recognition. submitted. (2025)
12. Martin, D.R., Fowlkes, C., Tal, D., Malik, J.: A database of human segmented natural images and its application to evaluating segmentation algorithms and measuring ecological statistics. In: 8th International Conference on Computer Vision, ICCV 2001, vol. 2, pp. 416–423 (2001)
13. Meer, P.: Stochastic image pyramids. Comput. Vis. Graph. Image Process. **45**(3), 269–294 (1989)

Numerical Implementation of Variational Morphological Operators on Graphs

Miguel Amorim, Jean-Christophe Pesquet, Tristan Portugues, Antonio Silveti-Falls, and Hugues Talbot(✉)

CentraleSupélec, Inria, Université Paris-Saclay, Paris, France
{miguel.amorim,jean.pesquet,tristan.portugues,antonio.falls, hgues.talbot}@centralesupelec.fr

Abstract. In this work, we study several implementations of the variational formulation of classical morphological operators such as erosion, dilation, opening, and closing, as well as arbitrary rank filters using convex optimization over graphs. Since the structuring element can naturally vary along nodes in the graph, this formulation allows for an effective implementation of spatially-variant mathematical morphology. It is particularly efficient on massively parallel architectures. We focus on solving the associated optimization problems using various algorithms from convex analysis. The problem being non-smooth, we consider algorithms based on the proximal operator, such as ADMM and PPXA+. We tested these approaches on both images and graph data. Our results show that proximal algorithms can handle the problems efficiently. We believe that this variational framework is a promising direction to combine mathematical morphology with modern computational tools, and is a direct avenue for solving inverse problems with mathematical morphology operators in the formulation.

1 Introduction

Mathematical morphology (MM) is a fundamental framework in image and signal processing, providing nonlinear operators such as erosion, dilation, opening and closing [7,12]. A recent variational framework has reformulated these operators, as well as arbitrary rank filters using convex optimization principles, making them well defined on arbitrary graphs [8]. This formulation opens the door to new computational strategies, parameter tuning, and potential integration with machine learning architectures, e.g. to solve inverse problems in a different way than with classical linear operators. In this work, we investigate the practical resolution of this variational formulation using methods suited to the structure of the problem.

In Sect. 2, we propose zeroth-order methods for solving the variational erosion/dilation problem. Section 3 addresses the more challenging task of implementing variational opening and closing. Two proximal splitting approaches are shown to yield effective solutions. Finally, conclusions and future works are presented in Sect. 4.

M. H. F. Wilkinson and J. Kosinka (Eds.): DGMM 2025, LNCS 16296, pp. 261–273, 2026.
https://doi.org/10.1007/978-3-032-09544-2_19

Notation and Graph Framework. Let $G = (V, E)$ be a directed, weighted, reflexive graph with $|V| = n$. Each node $v_i \in V$ indexed by $i \in \{1, \dots, n\}$, is associated with a scalar value $y_i \in \mathbb{R}$. The neighborhood of node i is $N_i = \{j \mid (i,j) \in E\}$ and its reciprocal neighborhood is $\check{N}_i = \{j \mid (j,i) \in E\}$.

2 Variational Formulation of Erosion and Dilation

We consider the variational formulation of the morphological erosion and dilation operators, as introduced in [8], which we will now describe. Given an input graph with node values $y \in \mathbb{R}^n$, the variational erosion/dilation are defined by the following convex optimization problem

$$\min_{x \in \mathbb{R}^n} \sum_{i=1}^{n} \sum_{j \in U_i} \varphi_i \left(\sigma_{\Omega_i} (x_i - y_j - w_{ij}) \right) \tag{1}$$

where U_i, a nonempty subset of V, is the set of neighbors of node i and $(w_{ij})_{j \in U_i} \in \mathbb{R}^{|U_i|}$ is a vector of shift parameters. The support function σ_{Ω_i} is defined for an interval $\Omega_i = [\alpha_i^-, \alpha_i^+]$, $\alpha_i^- < 0 < \alpha_i^+$, as

$$(\forall x \in \mathbb{R}) \quad \sigma_{\Omega_i}(x) = \begin{cases} \alpha_i^- x & \text{if } x < 0 \\ 0 & \text{if } x = 0 \\ \alpha_i^+ x & \text{if } x > 0. \end{cases}$$

We define on $\mathbb{R}_+$ the convex function $\varphi_i \colon x \mapsto x^p$ for $p \in [1, 2]$. The parameters of these two functions σ_{Ω_i} and φ_i can depend on the node i and therefore be different at each node of the graph. This property of the variational formulation allows for adaptive morphology operations. For $p = 1$ and $p = 2$, closed-form solutions can be derived from [8] for the formulation of erosion and dilation.

The choice of neighborhood U_i for each node i is totally flexible. The solution of the optimization problem remains well defined for arbitrary neighborhoods, although varying choices of neighborhoods can lead to significantly different filtering behaviors, as it is well-known in MM. Despite these differences, the computational process for solving the variational problem remains unchanged, making the approach both general and adaptable. Like the parameters of the support and convex cost function, the neighborhoods can differ from node to node, allowing the implementation of structuring functions.

2.1 Closed-Form Results from the Variational Formulation

First results found in [8], give closed-form solutions to the specific cases of the variational problem. The morphological erosion and dilation can be recovered by setting α_i^- / α_i^+ take infinity values, while a rank filter and a weighted mean are found for $p = 1$ and $p = 2$. For other values of $p \in]1, 2[$, we have an interpolation between the non-differentiable solution of $p = 1$ and the continuous and differentiable solution of $p = 2$.

Let $U_i = N_i$ be the neighborhood of node i and $(z_{ij})_{1 \le j \le |U_i|}$ the coefficients $(y_j + w_{ij})_{j \in U_i}$ indexed in increasing order.

For $\alpha_i^+ = +\infty$, we have

$$x_i = \min_{j \in U_i} y_j + w_{ij} \qquad \text{(Erosion)}.$$

For $\alpha_i^- = -\infty$, we have

$$x_i = \max_{j \in U_i} y_j + w_{ij} \qquad \text{(Dilation)}.$$

For $p_i = 1$, let $j_i = \left\lfloor \frac{|U_i|}{1-\alpha_i^+/\alpha_i^-} \right\rfloor + 1$, the optimal x_i is given by

$$\hat{x}_i = z_{ij_i}$$

if $j_i - 1 \notin \mathbb{N}$, otherwise $\hat{x}_i$ can be chosen arbitrarily in $[z_{ij_{i-1}}, z_{ij_i}]$. This corresponds to a weighted rank filter (as defined in [8, Prop. 3]).

For $p_i = 2$, the optimal x_i is given by

$$x_i = \frac{(\alpha_i^+)^2 \sum\limits_{j=1}^{j_i-1} z_j + (\alpha_i^-)^2 \sum\limits_{j=j_i}^{|U_i|} z_j}{(j_i - 1)(\alpha_i^+)^2 + (|U_i| - j_i + 1)(\alpha_i^-)^2}$$

where j_i is the rank index at which the minimizer lies (as defined in [8, Prop. 2]).

2.2 General Case for $p > 1$: Root-Finding Approach

For most values $p \in]1, 2[$, except for a few special cases, the variational formulation of erosion and dilation does not admit a closed-form solution. However, due to the separable structure of the objective, the problem reduces to a set of independent scalar convex minimization problems, one for each pixel or graph node $i \in \{1, \ldots, n\}$.

For each pixel value x_i, we must solve the following scalar convex minimization problem

$$\min_{x_i \in \mathbb{R}} F_i(x) = \min_{x_i \in \mathbb{R}} \sum_{j \in U_i} \varphi_i\left(\sigma_{\Omega_i}(x_i - y_j - w_{ij})\right) \tag{2}$$

For finite values of α_i^- and α_i^+, the derivative of the objective function F_i at $x \in \mathbb{R}$ can be computed directly by

$$F_i'(x) = \sum_{j \in U_i} \begin{cases} p\alpha_i^-(\alpha_i^-(x - y_j - w_{ij}))^{p-1} & \text{if } x < y_j + w_{ij}, \\ 0 & \text{if } x = y_j + w_{ij}, \\ p(\alpha_i^+)^p(x - y_j - w_{ij})^{p-1} & \text{if } x > y_j + w_{ij}. \end{cases} \tag{3}$$

This derivative F_i' is continuous and strictly increasing, F_i is strictly convex. Hence the minimizer $x_i^\star$ is unique and must satisfy

$$F_i'(x_i^\star) = 0.$$

We solve this root-finding problem directly for each node i. Due to nonsmoothness and lack of global Lipchitz continuity, classical first or second-order solvers are not efficient root finders. Instead, zeroth-order methods can be applied to find the unique zero of F'. Due to the convexity and strict monotonicity of the derivative, our root-finding approach is guaranteed to solve the variational erosion and dilation problem for all $p \in [1, 2]$. Since the problem is separable and each output node can be solved in parallel, the proposed solution is efficient and numerically stable, and well-suited for implementation on large graphs or image grids.

We tested different zeroth-order methods to solve the problem such as the Bisection, Golden search, Ternary search and Interpolate-Truncate-Project methods [2]. While all these methods share the same worst-case complexity, the Golden search and bisection methods show better results over larger images (see Fig. 1).

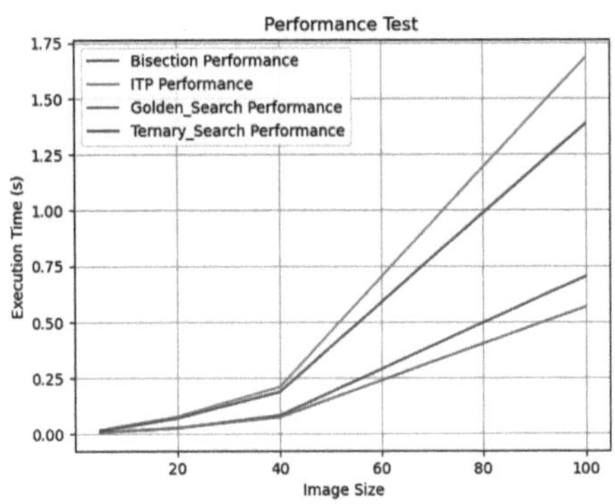

Fig. 1. Performance against image size for different zeroth-order methods.

The results applied to pictures and 3-D meshes show a clear interpolation between the two traditional morphological operators (Erosion and Dilation) when $-\alpha_+/\alpha_-$ goes from zero to large values (see Figs. 2 and 3). When using different neighborhood structures (also known as structuring elements), the results show a similar interpolation behavior while producing images with properties that follow the neighborhood shape. We show the resulting images using a diagonal cross neighborhood instead of a square of radius 1 (see Fig. 4). One can see the effect of the neighborhood shape by recognizing the diagonal cross or square over the solutions.

3 Variational Formulation of Opening and Closing

In this section, we consider the numerical resolution of the variational formulation of the morphological opening and closing operators proposed in [8]:

$$\min_{(x,t)\in(\mathbb{R}^n)^2} \sum_{i=1}^{n} \left(\sum_{j\in N_i} \varphi_i\left(\sigma_{\Omega_i}(t_i - y_j - w_{ij})\right) + \sum_{j\in \check{N}_i} \psi_i\left(\sigma_{\Lambda_i}(x_i - t_j - \nu_{ij})\right) \right). \tag{4}$$

This formulation is inspired by classical operators opening and closing, which are compositions of the classical erosion and dilation. A naive solution to problem

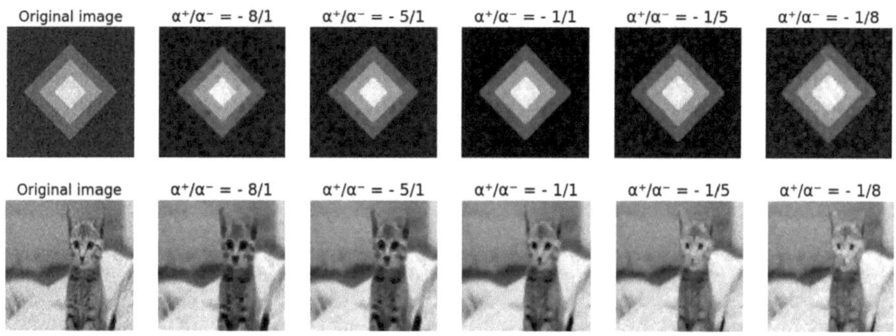

Fig. 2. Results of variational morphology for different ratios of α^+/α^- and $p = 1.2$.

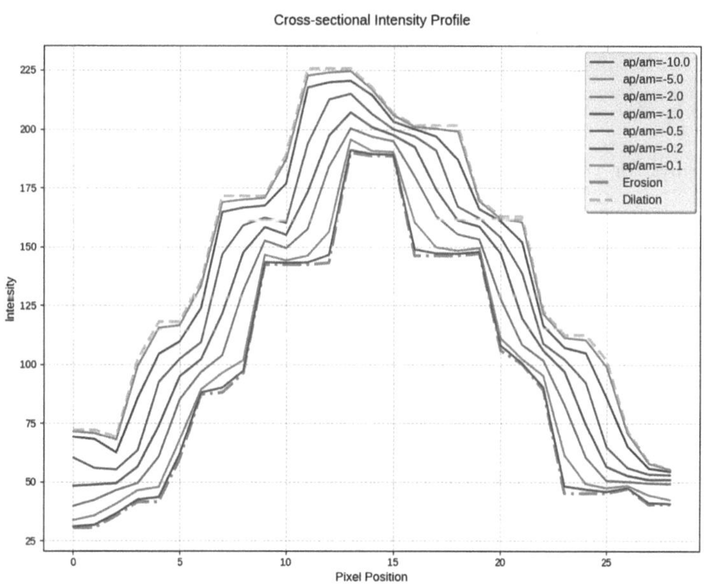

Fig. 3. Slice of the first result images for different ratios α^+/α^-.

Fig. 4. Results for variational erosion and dilation for 3D graph meshes (Top left: original image, Top right: median filter ($-\alpha^+/\alpha^- = 1$), Bottom left: variational erosion, Bottom right: variational dilation.

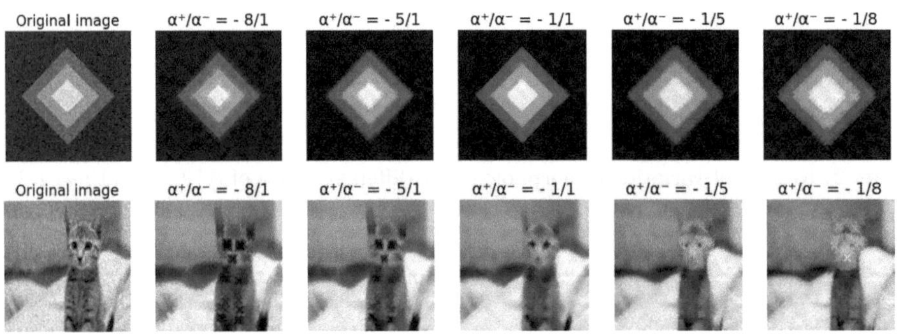

Fig. 5. Results for variational erosion and dilation with a diagonal cross structuring element.

(4) was first used in [8] composing the result of variational erosion and dilation to find an approximation of variational opening and closing. This solution first solves for t, then for x ignoring the coupling of the variables. While this approach gives a result close to the optimal solution and is theoretically founded under some asymptotic conditions [8], it cannot be considered a proper variational opening/closing. We will refer to this naive approach to compare the solutions computed in this section (see Figs. 5 and 6).

The operators are defined through a more complex convex problem involving two variables per node. Specifically, we solve the following optimization problem for variables $x = (x_i)_{i=1}^n$ and $t = (t_i)_{i=1}^n$. We remark that this formulation's objective function, while convex, couples the variables x and t, in contrast to (1). We can rewrite our problem in the form:

$$\min_{(x,t)\in(\mathbb{R}^n)^2} \bar{F}(t) + \bar{G}(x,t) \tag{5}$$

with

$$\bar{F}(t) = \sum_{i=1}^{n} \hat{F}_i(t_i)$$
$$(\forall i \in \{1,\ldots,n\})\ \hat{F}_i(t_i) = \sum_{j\in N_i} \underbrace{\varphi_i(\sigma_{\Omega_i}(t_i - y_j - w_{ij}))}_{F_{ij}(t_i)}$$

and

$$\bar{G}(x,t) = \sum_{i=1}^{n} \sum_{j\in \check{N}_i} \psi_i(\sigma_{\Lambda_i}(x_i - t_j - \nu_{ij})).$$

While the variational formulation of erosion/dilation (1) leads to an optimality condition which is completely separable, the structure of the objective function in (5) is not as simple. Nonetheless, each function F and G is a sum that is amenable to proximal splitting methods [9]. Indeed, due to the nature of our choice for Φ_i and σ_{Ω_i}, we can compute the proximal operator associated to the sum over each neighborhood or reciprocal neighborhood. This allows us to consider proximal splitting algorithms [4,5] such as Alternating Direction Method of Multipliers (ADMM) [11] and Parallel Proximal Point Algorithm (PPXA+) [10] to solve (4). The theory of monotone operators which encompasses optimisation with proximal operators can be found in [1]. A comprehensive repository of proximal operators can be found in [3].

Since each $\hat{F}_i$ is a function of only t_i, the proximal operator associated to $\bar{F}$ can be computed by computing the proximal operator associated to each $\hat{F}_i$ node by node,

$$(\forall i \in \{1,\ldots,n\}) \quad [\text{prox}_{\lambda\bar{F}}(t)]_i = \text{prox}_{\lambda\hat{F}_i}(t_i).$$

The proximal operator associated to each $\hat{F}_i$ can itself be computed by solving a scalar optimization problem, which amounts to solving the following inclusion problem

$$\begin{aligned}\text{prox}_{\lambda\hat{F}_i}(t_i) = z_i &\iff 0 \in \lambda\partial\hat{F}_i(z_i) + z_i - t_i = 0\\ &\iff 0 \in \lambda\left(\sum_{j\in\check{N}_i} \partial F_{ij}(z_i)\right) + z_i - t_i = 0.\end{aligned} \tag{6}$$

For $\bar{G}$, such a splitting is no longer possible because of the coupling between the variables t and x. In light of this, we can define, for all $i \in \{1,\ldots,n\}$ and $j \in \check{N}_i$, $u_{ij} = x_i - t_j$ and

$$G_{ij}(u_{ij}) = \psi_i(\sigma_{\Lambda_i}(u_{ij} - \nu_{ij})).$$

Since each G_{ij} only depends on u_{ij}, we can rewrite $\bar{G}$ in the following, separable way,

$$\bar{G}(x,t) = G(u) = \sum_{i=1}^{n} \sum_{j\in \check{N}_i} G_{ij}(u_{ij}).$$

From this, we can deduce the following expression of the proximal operator associated to G given componentwise, for all $i \in \{1,\dots,n\}$ and $j \in \check{N}_i$, by

$$[\mathrm{prox}_{\lambda G}(u)]_{ij} = \mathrm{prox}_{\lambda G_{ij}}(u_{ij}).$$

As is the case for $\mathrm{prox}_{\lambda \hat{F}_i}$, each proximal operator $\mathrm{prox}_{\lambda G_{ij}}(u)$ can be computed by solving the following inclusion problem,

$$\mathrm{prox}_{\lambda G_{ij}}(u_{ij}) = z_{ij} \iff 0 \in \lambda \partial G_{ij}(u_{ij}) + u_{ij} - z_{ij}, \tag{7}$$

which we will describe in more detail later in the paper.

As we have remarked earlier, these different splitting strategies are necessary because of the coupling induced by the function $\bar{G}$. We can rewrite (4) with respect to the lifted variables $(X,U) \in (\mathbb{R}^n)^2 \times \mathbb{R}^{n\times m}$,

$$\min_{X\in(\mathbb{R}^n)^2} F(X) + G(\underbrace{LX}_{U}), \tag{8}$$

where $X := (x,t) \in (\mathbb{R}^n)^2$, $F(X) = \bar{F}(t)$, $U = (u_{ij})_{i,j} \in \mathbb{R}^{n\times m}$, $L : (\mathbb{R}^n)^2 \to \mathbb{R}^{n\times m}$ is defined by the differences $x_i - t_j$ for every $i \in \{1,\dots,n\}$ and $j \in \check{N}_i$, and $m = \max_{i\in\{1,\dots,n\}} |\check{N}_i|$. The proximal splitting algorithms [4] that we employ will make use of the adjoint operator associated to the linear operator L. This adjoint operator $L^* : \mathbb{R}^{n\times m} \to (\mathbb{R}^n)^2$ maps a vector $u = (u_{ij})_{i,j} \in \mathbb{R}^{n\times m}$ to a pair $(x^\star, t^\star) \in (\mathbb{R}^n)^2$ as

$$(\forall i \in \{1,\dots,n\}) \quad \begin{cases} x_i^\star = \sum_{j\in\check{N}_i} u_{ij} \\ t_j^\star = -\sum_{i|j\in\check{N}_i} u_{ij}. \end{cases} \tag{9}$$

Once the problem is rewritten in this way, applying proximal splitting methods like ADMM or PPXA+ becomes straightforward as long as we can compute the linear operators L and L^*, and the proximal operators $\mathrm{prox}_{\lambda G_{ij}}$ and $\mathrm{prox}_{\lambda \hat{F}_i}$, which we will describe in the following subsection.

3.1 Explicit Form of the Proximal Operator Computation

We have seen that using the proximal splitting methods requires two different splitting strategies on $\bar{F}$ and $\bar{G}$. At the end, the computation of the proximal operators $\mathrm{prox}_{\lambda G_{ij}}(u_{ij})$ and $\mathrm{prox}_{\lambda \hat{F}_i}(v_i)$ reduces to solving two independent root-finding equations, in the case when $p > 1$, $\psi_i = \varphi_i$, and $\Lambda_i = \Omega_i$.

Using (7), we have

$$x = \mathrm{prox}_{\lambda G_{ij}}(u_{ij}) \iff 0 = \frac{1}{\lambda}(x - u_{ij}) + \begin{cases} p_i\alpha_i^-(\alpha_i^-(x-\nu_{ij}))^{p-1} & \text{if } x < \nu_{ij} \\ p_i(\alpha_i^+)^p(x-\nu_{ij})^{p-1} & \text{if } x > \nu_{ij} \\ 0 & \text{if } x = w_{ij}. \end{cases}$$

Using (6), we have

$$x = \mathrm{prox}_{\lambda \hat{F}_i}(t_i) \iff 0 = \frac{1}{\lambda}(x - t_i) + \sum_{j \in \check{N}_i} \begin{cases} p_i \alpha_i^-(\alpha_i^-(x - w_{ij}))^{p-1} & \text{if } x < w_{ij} \\ p_i(\alpha_i^+)^p (x - w_{ij})^{p-1} & \text{if } x > w_{ij} \\ 0 & \text{if } x = w_{ij}. \end{cases}$$

Since G_{ij} and $\hat{F}_i$ are differentiable and convex for $p > 1$, these two problems admit a unique minimizer. They are similar in form to the subproblems solved for the variational erosion/dilation in (1). These root-finding problems will be handled via a zeroth-order method, exactly as described in the computation of the variational formulation of erosion and dilation, since the functions encountered are of the same form. The proximal operators being computed term-by-term for the functions $\bar{F}$ and G, we can now make use of proximal splitting algorithms to solve the variational opening/closing problem (4).

3.2 ADMM Algorithm

To solve the variational opening/closing problem (8) using ADMM, we introduce the auxiliary variable $U \in \mathbb{R}^{n \times m}$ and reformulate the problem as

$$\min_{\substack{X \in (\mathbb{R}^n)^2 \\ U \in \mathbb{R}^{n \times m}}} F(X) + G(U) \quad \text{subject to} \quad LX - U = 0. \tag{10}$$

This is now a linearly constrained convex optimization problem to which we apply the linearized-ADMM algorithm as detailed in [11]; for simplicity we will refer to the linearized-ADMM as ADMM for the rest of the paper. We present the algorithm in its entirety below.

Algorithm 1 ADMM for Opening/Closing

Initialize: $X_0 = (x_0, t_0) \in (\mathbb{R}^n)^2,\ U_0 \in \mathbb{R}^{n \times m},\ z_0 \in \mathbb{R}^{n \times m}$
for $k - 0, 1, \dots$ **do**
  $X_{k+1} = \mathrm{prox}_{\mu F}\left(X_k - \frac{\mu}{\lambda} L^*(z_k + LX_k - U_k)\right)$
  $U_{k+1} = \mathrm{prox}_{\lambda G}(LX_{k+1} + z_k)$
  $z_{k+1} = z_k + LX_{k+1} - U_{k+1}.$
end for

Given step size parameters $\mu > 0$ and $\lambda > 0$ and a dual variable $z_k \in \mathbb{R}^{n \times m}$, the algorithm updates the variables X and U by applying, at each iteration, the proximal operator associated to either F or G, respectively. This algorithm is guaranteed to converge under standard assumptions; more specifically, we can ensure convergence under appropriate step sizes $\mu > 0$ and $\lambda > 0$ since the constraint $LX - U = 0$ is linear and nonempty and the functions F and G are proper, lower semicontinuous, and convex.

The choice of step sizes λ and μ influences the convergence speed. One choice for which convergence is guaranteed is given by

$$0 < \mu < \frac{\lambda}{\|L\|^2}$$

where $\|L\|$ is the operator norm of L. In our case, L is a sparse difference operator coupling the x_i and t_j variables. While computing the exact norm is costly, an upper bound can be estimated from the maximum number of neighbors for a node.

ADMM offers a flexible approach to solving the variational formulation of opening/closing. Its convergence is stable under our hypotheses, and its iterations are simple to implement due to the proximal calculus and separable structure of the functions involved. When the graph domain is fixed, such as in the case of images of a given size, the linear operator L and its adjoint L^* depend only on the chosen neighborhood (or structuring element), so that these operators (and potentially their factorizations) can be precomputed and stored to facilitate their use in ADMM.

Algorithm 2 PPXA+ (as in [10, Prop. 5.2])

Initialize: $x_0 \in (\mathbb{R}^n)^2$, $z_{0,1} \in (\mathbb{R}^n)^2$, $z_{0,2} \in \mathbb{R}^{n\times m}$
Set:

$$x_0 \in \underset{y\in(\mathbb{R}^n)^2}{\operatorname{argmin}} \sum_{i=1}^{2} \omega_i \|L_i y - z_{0,i}\|^2$$

for $n = 0, 1, 2, \ldots$ **do**
  $p_{n,1} = \operatorname{prox}_{F/\omega_1}(z_{n,1})$
  $p_{n,2} = \operatorname{prox}_{G/\omega_2}(z_{n,2})$
  Compute:

$$c_n \in \underset{y\in(\mathbb{R}^n)^2}{\operatorname{argmin}} \sum_{i=1}^{2} \omega_i \|L_i y - p_{n,i}\|^2$$

  for $i = 1, 2$ **do**
    $z_{n+1,i} = z_{n,i} + \lambda_n(L_i(2c_n - x_n) - p_{n,i})$
  end for
  $x_{n+1} = x_n + \lambda_n(c_n - x_n)$
end for

3.3 PPXA+ Algorithm

Another proximal splitting algorithm that is amenable to our problem is the so-called PPXA+ algorithm. This algorithm, introduced in [10], is designed to solve convex optimization problems of the form

$$\min_{X\in(\mathbb{R}^n)^2} F(L_1 X) + G(L_2 X) \tag{11}$$

where F, G are proper lower-semicontinuous convex functions, and L_1, L_2 are bounded linear operators. The method computes proximal operators of each term in parallel and aggregates their results via an update step. To apply PPXA+ to (4), we must first rewrite it by using the lifted variable $X := (x, t) \in (\mathbb{R}^n)^2$ and introducing the bounded linear operators $L_1 := \mathrm{Id}$ and $L_2 := L$. Then, the variational formulation for opening and closing in (8) can be rewritten as (11).

Given a sequence of relaxation parameters $(\lambda_n)_{n\in\mathbb{N}}$, with $\lambda_n \in (0, 2)$, satisfying $\lambda_{n+1} \leq \lambda_n$ and strictly positive weights ω_1, ω_2, with $\omega_1 + \omega_2 = 1$, the algorithm computes the prox for F and G separately at each iteration and then performs a sort of averaging step by solving a least-squares problem.

One of the main differences between ADMM and PPXA+ iterations is the computation of c_n that must be performed in PPXA+. This step requires solving a least squares problem of the form

$$c_n = \underset{z\in(\mathbb{R}^n)^2}{\operatorname{argmin}} \ \omega_1\|z - p_{n,1}\|^2 + \omega_2\|Lz - p_{n,2}\|^2.$$

This is a strictly convex quadratic problem, so its solution satisfies the normal equations

$$(\omega_1 \,\mathrm{Id} + \omega_2 L^* L)c_n = \omega_1 p_{n,1} + \omega_2 L^* p_{n,2}.$$

We solve this system using a linear solver (e.g., conjugate gradient) or, in smaller-scale cases, a direct method using sparse linear algebra. The operator $\omega_1 \,\mathrm{Id} + \omega_2 L^* L$ is symmetric positive definite by construction, ensuring that the system is well conditioned and uniquely solvable. The complexity of solving this linear system can worsen the performance of PPXA+ relatively to ADMM. However, we have found that this difference is not substantial for large images since the time required by both algorithms is dominated by the cost of the proximal operators involved.

In PPXA+, all proximal subproblems are decoupled and computed independently, making the algorithm naturally parallelizable between pixels or graph elements. Furthermore, PPXA+ offers a way to parallelize both proximal computations at the same time and can potentially be faster than ADMM.

Convergence for PPXA+ is guaranteed under the considered assumptions. Since L is sparse and L^* has a known structure (see (9)), the matrix vector products involved in the solver are efficient. Moreover, for fixed graph/neighborhood structure, $L^* L$ can be precomputed (and possible factorized) or reused across iterations.

Now, we show a comparison in performance for both algorithms and with the naive solution. We can see that both algorithms converge to an objective value that is lower than, but still close to, the naive solution. However the processed images are visually similar for both solutions of (4), although different from the naive image (see Figs. 5 and 6). This shows that even though the naive method can come really quickly to an almost optimal objective value, it produces images that are distinct from the optimal solution. We observe particularly that the naive method images are less smooth and noisier than the ones generated by ADMM and PPXA+ algorithms.

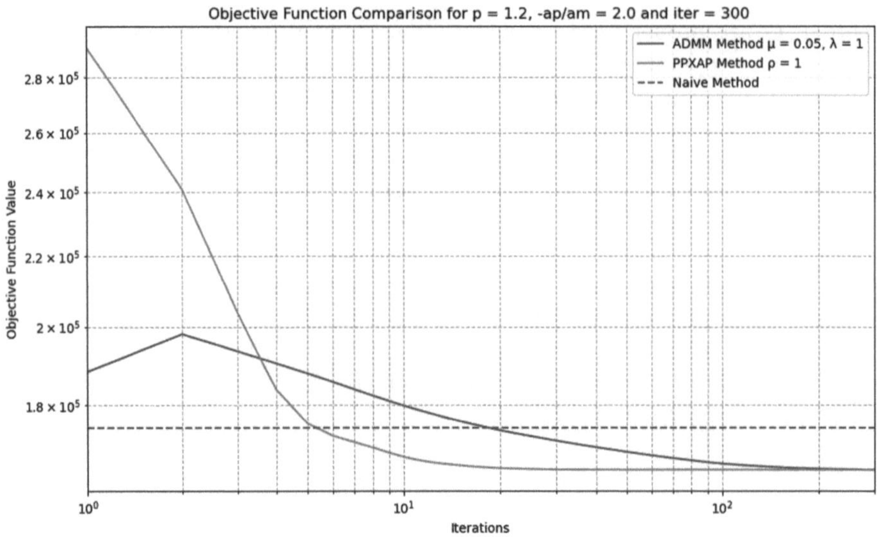

Fig. 6. Objective functions in the resolution of a variational closing processing for ADMM, PPXA+, and naive algorithms.

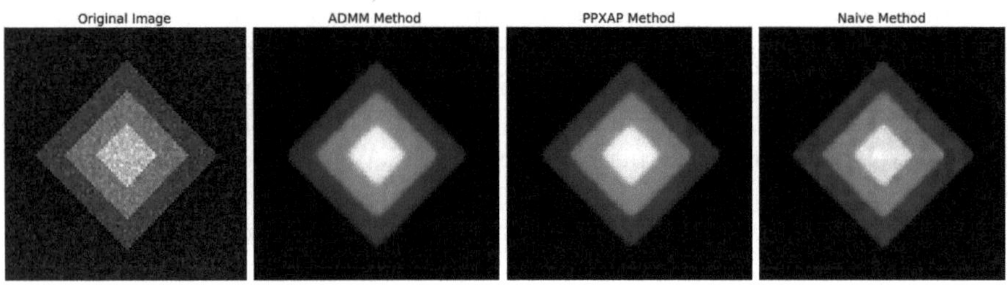

Fig. 7. Results of variational opening for ADMM, PPXA+, and naive algorithms.

4 Conclusion

We have detailed the implementation and numerical resolution of a variational framework for morphological operators on graphs. For the variational erosion and dilation, we use the separability of the variational formulation to decompose its solution into the solutions of scalar convex problems, solved efficiently using zeroth-order root-finding methods. For the variational opening and closing, we show that proximal operator splitting techniques are an effective approach, specifically demonstrating the PPXA+ and ADMM algorithms, as they are adapted to the structure of the problem. The methods presented here provide flexible and efficient tools for using variational morphological operators on graphs. They can be easily implemented as layers in deep neural networks, and can thus open up possibilities for further research, such as integrating adaptive or learned parameters. Another interesting perspective is the possibility that this approach be used as part of more sophisticated processing pipeline, where parameters can be adjusted or learned depending on the context. In future work we will also use these implementations of morphological operators for solving inverse problems using MM operators via variational methods [6].

References

1. Bauschke, H.H., Combettes, P.L.: Convex Analysis and Monotone Operator Theory in Hilbert Spaces,, 2 edn. Springer, Cham (2017). https://doi.org/10.1007/978-3-319-48311-5
2. Burden, R.L., Faires, J.D.: Numerical Analysis. The Prindle, Weber and Schmidt Series in Mathematics, fourth edn. PWS-Kent Publishing Company, Boston (1989)
3. Chierchia, G., Chouzenoux, E., Combettes, P.L., Pesquet, J.C.: The proximity operator repository (2025). http://proximity-operator.net/download/guide.pdf
4. Combettes, P.L., Pesquet, J.C.: Proximal splitting methods in signal processing. Fixed-point algorithms for inverse problems in science and engineering, pp. 185–212 (2011)
5. Combettes, P.L., Pesquet, J.C.: Fixed point strategies in data science. IEEE Trans. Signal Process. 3878–3905 (2021)
6. Gharbi, M., Villa, S., Chouzenoux, E., Pesquet, J.C., Duval, L.: Unrolled deep networks for sparse signal restoration in analytical chemistry. In: 2024 IEEE 34th International Workshop on Machine Learning for Signal Processing (MLSP), pp. 1–6. IEEE (2024)
7. Najman, L., Talbot, H. (eds.): Mathematical Morphology: From Theory to Applications. ISTE-Wiley, London, UK (September 2010). ISBN 978-1848212152
8. Najman, L., Pesquet, J.-C., Talbot, H.: When convex analysis meets mathematical morphology on graphs. In: Benediktsson, J.A., Chanussot, J., Najman, L., Talbot, H. (eds.) ISMM 2015. LNCS, vol. 9082, pp. 473–484. Springer, Cham (2015). https://doi.org/10.1007/978-3-319-18720-4_40
9. Parikh, N., Boyd, S.: Proximal algorithms. Found. Trends Optim. **1**(3), 127–239 (2014)
10. Pesquet, J.C., Pustelnik, N.: A parallel inertial proximal optimization method. Pac. J. Optim. **7**(2), 273–305 (2011)
11. Ryu, E.K., Yin, W.: Large-Scale Convex Optimization: Algorithms and Analyses via Monotone Operators. Cambridge University Press, Cambridge (2022)
12. Serra, J.: Image Analysis and Mathematical Morphology. Academic Press, Inc., Cambridge (1983)

Algorithms

Shape Filtering and Max-Tree Attribute Computation on a GPU

Paul D. Teeninga(✉) and Michael H. F. Wilkinson

Bernoulli Institute, University of Groningen, Groningen, The Netherlands
{p.d.teeninga,m.h.f.wilkinson}@rug.nl

Abstract. Max-trees are hierarchical structures for anti-extensive attribute filtering. A node attribute is a function of the connected component of a threshold set represented by the node. Filtering is carried out by removing nodes based on their attributes. In this paper we present a new algorithm for parallel attribute computation and filtering using the subtractive rule and image reconstruction on a GPU. Results show up to 30.4× speed-up compared to sequential algorithms, and increased numerical stability by several orders of magnitude.

Keywords: Mathematical morphology · Connected filters · Parallel algorithms · GPU processing

1 Introduction

In mathematical morphology, *connected filters* work by merging *flat zones*, which in a grayscale image are connected regions with equal value of maximal extent [14]. A *max-tree* is a hierarchical representation of (connected) components in an image for all possible thresholds, where the root node represents the set of all pixels [13]. Max-trees are particularly effective in implementing attribute filters [2]. A *max-tree attribute*, for some node, is a function of the pixels in the associated component. An example is the *area*, which is the number of pixels in a component. Connected filtering with a max-tree can be divided into four steps: max-tree construction, attribute computation, node filtering, and image reconstruction. One application is to detect and segment objects in an image, where node attributes are used for object detection, and the last two steps are used for segmentation [3,15].

In many cases, attribute computation is done concurrently with tree construction, but this is not always desirable, e.g., in cases where the application might require switching one attribute function with another without requiring complete rebuilding of the max-tree [20]. Separate, sequential computation of attributes on pre-existing trees is simple, but parallel computation is harder in such cases, especially in high bit depths.

Therefore, development of max-tree algorithms has been an area of active research [4,10], and in particular, parallel methods [1,9,21]. A first parallel CPU

M. H. F. Wilkinson and J. Kosinka (Eds.): DGMM 2025, LNCS 16296, pp. 277–290, 2026.
https://doi.org/10.1007/978-3-032-09544-2_20

algorithms for max-tree construction, for any bit depth, was given in [9], although the filtering solution was not applicable to the so-called subtractive rule, used in shape filtering [18]. In [17] a new algorithm without these restrictions is presented. Parallel CPU algorithms for attribute computation, node filtering, and image reconstruction are given in [16], based on algorithms in [8]. GPU algorithms for more general problems are given in [5], although they seem to require a non-trivial pre-processing step that creates an ordered node array by position in an Euler Tour.

In this paper, we implement attribute computation, node filtering, and image reconstruction on the GPU. The key motivation for this is that Max-trees are used on extreme-dynamic-range (XDR) images in, for example, astronomy [15]. While low-bit-depth algorithms on the GPU [1] and efficient parallel CPU algorithms [16,17] exist, efficient, subtractive max-tree filtering of XDR images on the GPU is an open problem, as is parallel attribute computation. These problems are not straightforward to parallelize, and are important since attribute computation in particular can take up a significant part of the compute time [13,20].

The max-tree is assumed to be constructed from an image with unique values, such that the weight function is injective. If an image does not have unique values, the combination of value and position (as tuple) can be used as unique value, since positions are unique. Two classes of max-tree attributes are defined, based on definitions in [7]: rootfix and leaffix attributes. Rootfix attributes can implement image reconstruction with filter rules. Leaffix attributes implement typical max-tree attributes, such as those based on moments [13,20], that reduce pixel values in a component to a single value. The rootfix and leaffix GPU implementations are compared to CPU implementations, including numerical stability.

The rest of the paper is organized as follows. Section 2 contains mathematical preliminaries, algorithms in Sect. 3, GPU implementations in Sect. 4, results in Sect. 5, discussion in Sect. 6, and finally conclusion in Sect. 7.

2 Preliminaries

An image is a node-weighted graph (V, E, w), where V is the finite set of image elements, E is the set of symmetric neighbor relations, and $w : V \to W$ is the weight function. We assume that W has a total order, neighbor count is $\mathcal{O}(D)$, where D is the number of dimensions, and w is injective. If w is not injective, an injective function w' can be constructed, where $w'(x) = (w(x), x)$, and the order is lexical.

A *connected component*, or simply *component*, of I at some threshold $t \in W$ is a maximal subset of $V_t = \{x \in V \mid w(x) \geq t\}$, where each distinct pair of elements is connected via a path in $E \cap V_t \times V_t$.

The *parent component* of a component $C \neq \emptyset$ is the first component, when decreasing the threshold, that contains more elements than C. For practical reasons, the parent of component $C = V \neq \emptyset$ is defined as C itself. For $C \neq \emptyset$, let $\text{root}(C) = \text{argmin}\{w(x) \mid x \in C\}$, and let component$(x)$, for $x \in V$, be the component at threshold $w(x)$ containing x. Function component(x) is the inverse of root(C), since w is injective. Let parent(x) be the root of the parent component of component(x).

Definition 1 (Max-tree). *Given image $I = (V, E, w)$, where w is injective,*

$$\text{maxtree}(I) = (V, \{(x, \text{parent}(x)) \mid x \in V \wedge x \neq \text{parent}(x)\}).$$

Proposition 1 (Bounded in-degree). *Given image (V, E, w) with at most k neighbors per node, any node is the max-tree parent of at most k nodes.*

Proof. An *ancestor* of node x is any other reachable node in the max-tree from x. An *origin* of a directed edge is the first element. Suppose that $x \in V$ has more than k incoming edges in the max-tree. For some image component at threshold $w(x)$, at least two different origins of edges to x in the max-tree, which have values greater than x, are connected to x via the same neighbor with weight greater than x. Let c be the minimum value of the two origins. The origins are in the same image component at threshold c. One of the origins is an ancestor of the other origin in the max-tree, and x cannot be the parent of both origins. Therefore, x has at most k incoming edges.

A max-tree (Definition 1) is a directed tree with edges towards the image root. The in-degree of every node is bounded by k (Proposition 1). A definition where w can map to the same value for multiple nodes is given in [21]. In this case, the in-degree is not necessarily bounded. When constructing injective weight functions $w'(x)$ that maintain the order between different weights, the set of possible max-trees for $w'(x)$ is a subset of possible max-trees for original weight function $w(x)$.

The utility of max-trees is to efficiently compute certain statistics of components, also referred to as *attributes*. We define two classes of attributes that can be efficiently computed, based on definitions in [7]. The leaffix definition here is slightly different, because the order of branches in a max-tree can be arbitrary.

Definition 2 (Rootfix attribute). *Let $z : V \rightarrow Z$, and let $\oplus : Z \times Z \rightarrow Z$ be a binary associative function. For any node $x \in V$, the result of $z(\text{root}(V)) \oplus \ldots \oplus z(\text{parent}(x)) \oplus z(x)$ is a rootfix attribute of x.*

Rootfix attributes can be used to implement filter rules. For example, to implement a *direct filter* on $f : V \rightarrow Z$, with criterion $m : V \rightarrow \{0, 1\}$, function z can be

$$z(x) = \begin{cases} f(x), & \text{if } x = \text{root}(V) \vee m(x) = 1 \\ \text{infimum}(Z), & \text{otherwise} \end{cases} \tag{1}$$

and $\oplus = \max$.

Definition 3 (Leaffix attribute). *Let $z : V \rightarrow Z$ be a function, let $\oplus : Z \times Z \rightarrow Z$ be a binary associative and commutative function, and let S be the set of max-tree descendants of $x \in V$. The result of $\bigoplus(\{x\} \cup S)$, where $\{x\} \cup S$ is reduced by $\oplus$ to a single element, is a leaffix attribute of x.*

Attributes that are typically computed during max-tree construction are leaf-fix attributes. For example, the *area* of node x is the number of elements in component(x). The area can be computed by letting $z(x) = 1$ and $\oplus = +$. To implement an *area filter* [19] with direct filter rule (Eq. 1), to remove components with a small area, $f(x) = w(x)$, and the criterion function $m(x)$ is equal to $(\text{area}(x) \geq \lambda)$, where λ is the minimum area.

3 Attribute Computation

The computation of max-tree attributes is vectorized by creating a list of edge sets, where the expected length of the list is $\mathcal{O}(\log n)$, and the expected total number of edges is $\mathcal{O}(n)$. We refer to combining partially computed attributes of nodes with some binary operator $\oplus$ as merging nodes, and the result is stored for the merge target. Attributes are computed by iterating over the list of edge sets in order, and merging node pairs in parallel for each edge set. In each iteration, if a node is merged to, this occurs at most once, and that node is not merged to another node. Fundamental parallel algorithms to contract a tree are introduced in [8].

Algorithm 1: Rootfix Merge List

input : max-tree (V, E) of an image
output : list of edge sets

$E' \leftarrow \{(r, x) \mid (x, r) \in E\}$ // *Order each pair by distance to image root*
$(\texttt{merge\_first}, \texttt{merge\_last}) \leftarrow$ (empty list, empty list)
while $\#E' > 0$ **do**
  let $hash(x)$ be a random 1-bit universal hash // *Example: Eq. 4*
  $\texttt{contracting} \leftarrow \{(r, x) \in E' \mid hash(x) = 1 \wedge hash(r) = 0\}$
  $\texttt{first} \leftarrow \{(x, y) \in E' \mid \exists r : (r, x) \in \texttt{contracting}\}$
  $\texttt{merge\_first} \leftarrow \texttt{merge\_first} + [\texttt{first}]$ // *List concatenation*
  $\texttt{merge\_last} \leftarrow [\texttt{contracting}] + \texttt{merge\_last}$
  $\texttt{update} \leftarrow \{(r, y) \mid (r, x) \in \texttt{contracting} \wedge (x, y) \in E'\}$
  $E' \leftarrow ((E' - \texttt{contracting}) - \texttt{first}) \cup \texttt{update}$
end
return $\texttt{merge\_first} + \texttt{merge\_last}$

Algorithm 1 constructs edge sets to compute rootfix attributes. Edges from a max-tree are iteratively contracted. In each edge, we refer to the first element, closer to the root, as the parent node. In each iteration, nodes are marked with 0 or 1 randomly with a universal hash function. Edges contract when the parent node is marked with 0, and the child node is marked with 1. For any removed node x with parent r, edge (r, x) is added to an edge set in `merge_last`. List `merge_last` is used to update the attribute of x with the final attribute of r in the last stage. For nodes y with removed parent x, edge (x, y) is added to an

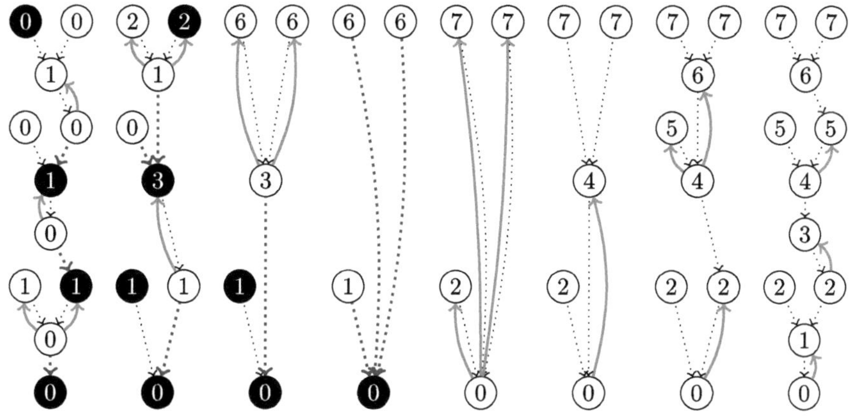

Fig. 1. Constructing (green) edges to compute rootfix attributes. Example attribute: root distance shown as node labels, where $z(x) = 0$, and $\oplus$ is the sum plus 1. For the first (green) edge sets E_0 to E_3, nodes are (pseudo)randomly marked with 0 (= black) or 1 (= white), and white-to-black edges are contracted, shown in red. If a node has an edge to a deleted node, then the attribute is updated. For the last (green) edge sets E_4 to E_7, contractions are reversed, and attributes of previously deleted nodes are updated to the final values. Edge sets are iterated over in order, starting from E_0. Attributes are updated at most once per node in an iteration, allowing parallelism. (Color figure online)

edge set in `merge_first`. List `merge_first` is used to update the attribute of y with the partial attribute of x in the first stage.

Given edge sets $E_0, E_1, \ldots, E_{l-1}$ from Algorithm 1, function $z : V \to Z$, and binary associative operator $\oplus : Z \times Z \to Z$, let $\text{rootfix}_0(x) = z(x)$, and let

$$\text{rootfix}_k(x) = \begin{cases} \text{rootfix}_{k-1}(r) \oplus \text{rootfix}_{k-1}(x), & \text{if } \exists (r, x) \in E_{k-1} \\ \text{rootfix}_{k-1}(x). & \text{otherwise} \end{cases} \tag{2}$$

The rootfix attribute of node x is $\text{rootfix}_l(x)$. Nodes closer to the root are always on the left side of $\oplus$. Figure 1 illustrates rootfix attribute computation.

Algorithm 2 creates edge sets to compute leaffix attributes. Similar to rootfix attributes, edges are iteratively contracted. A node is mergeable if it is not the image root, and it is either a leaf or it has one branch and the hash is 1. Edges are contractable if the left node is not mergeable and the right node is mergeable. Since the left node can be the merge target in multiple edges, at most one contractable edge (x, y) is selected arbitrarily per node x, per iteration. Edge (x, y) is added to an edge set in `merge_first`, and is used to update the attribute of x with the partial attribute of y. Any removed non-leaf node x has exactly one child node y, and (x, y) is added to an edge set in `merge_last`. List `merge_last` updates the attribute of x with the final attribute of y.

Algorithm 2: Leaffix Merge List

input : max-tree (V, E) of an image
output : list of edge sets

$E' \leftarrow \{(r,x) \mid (x,r) \in E\}$ // *Order each pair by distance to image root*
$\texttt{bc}(r) \leftarrow \#\{(r,x) \mid (r,x) \in E'\}$ // *Branch count*
$(\texttt{merge\_first}, \texttt{merge\_last}) \leftarrow$ (empty list, empty list)
while $\#E' > 0$ **do**
  let $hash(x)$ be a random 1-bit universal hash // *Example: Eq. 4*
  let $\texttt{mergeable}(x) = (\texttt{bc}(x) = 0 \vee (\texttt{bc}(x) = 1 \wedge hash(x) = 1))$
  $\texttt{contractable} \leftarrow \{(r,x) \in E' \mid \texttt{mergeable}(x) \wedge \neg\texttt{mergeable}(r)\}$
  // *Function* I *is the inverse of an arbitrary enumeration of* E'
  $\texttt{first} \leftarrow \{(r,x) \in \texttt{contractable} \mid \forall y : (r,y) \in \texttt{contractable} \rightarrow \texttt{I}(r,x) \geq \texttt{I}(r,y)\}$
  $\texttt{merge\_first} \leftarrow \texttt{merge\_first} + [\texttt{first}]$ // *List concatenation*
  $\texttt{last} \leftarrow \{(x,y) \in E' \mid \exists r : (r,x) \in \texttt{first}\}$
  $\texttt{merge\_last} \leftarrow [\texttt{last}] + \texttt{merge\_last}$
  // *Remove edges to merged leaves, update edges to other merged nodes*
  $\texttt{update} \leftarrow \{(r,y) \mid (r,x) \in \texttt{first} \wedge (x,y) \in E'\}$
  $E' \leftarrow (E' - \texttt{first}) \cup \texttt{update}$
end
return $\texttt{merge\_first} + \texttt{merge\_last}$

Given $[E_0, E_1, \ldots, E_{l-1}]$ as output from Algorithm 2, function $z : V \rightarrow Z$, and binary associative and commutative operator $\oplus : Z \rightarrow Z$, let $\text{leaffix}_0(x) = z(x)$, and let

$$\text{leaffix}_k(x) = \begin{cases} \text{leaffix}_{k-1}(x) \oplus \text{leaffix}_{k-1}(y), & \text{if } \exists (x,y) \in E_{k-1} \\ \text{leaffix}_{k-1}(x). & \text{otherwise} \end{cases} \tag{3}$$

The leaffix attribute of node x is $\text{leaffix}_l(x)$. There are some differences compared to Eq. 2: the merge target is the first element of an edge, and the $\oplus$ operator is commutative. Figure 2 illustrates leaffix attribute computation.

The parallelization strategy is to use parallel implementations of array operations. Code is executed by a single processor outside of array operations. The time complexities of both rootfix and leaffix algorithms are $\mathcal{O}(n/p + \log(n)\log(p))$ with high probability [16], where n is the pixel count and p is the processor count.

4 GPU Implementation

Rootfix and leaffix algorithms are implemented in (CUDA) C++, available at https://github.com/pauldavt/max-tree-attributes-gpu. The input is a max-tree (V, E) with enumerated nodes. Edge sets are stored as arrays and contain pairs of node numbers. An option is to process max-trees in batches, as a forest of max-trees, and nodes in the forest are enumerated instead. A potential benefit is a reduced fraction of time spent on host function calls for GPU operations.

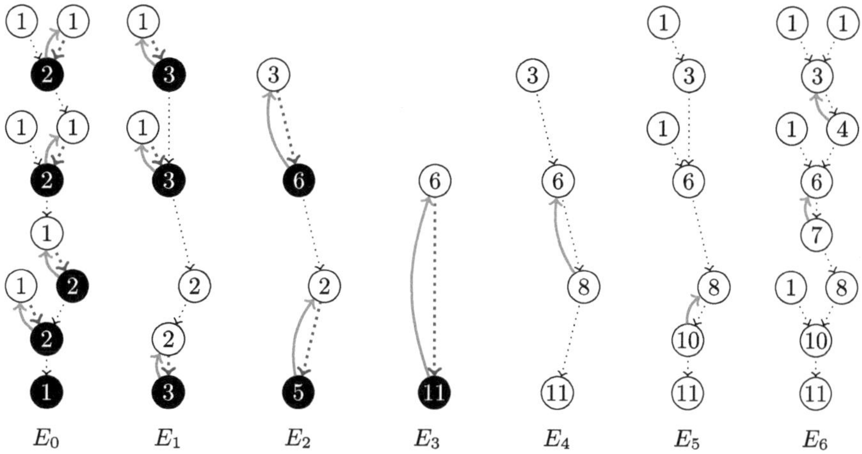

Fig. 2. Constructing (green) edges to compute leaffix attributes. Example attribute: area shown as node labels, where $z(x) = 1$, and $\oplus$ is the sum. For the first (green) edge sets E_0 to E_3, nodes with exactly in-degree 1 are (pseudo)randomly marked with 0 (= black) or 1 (= white). Nodes with in-degree 0 and in-degree > 1 are marked with 1 and 0 respectively. For each black node with at least one incoming edge from a white node, exactly one such edge is contracted, shown in red. For the last (green) edge sets E_4 to E_6, contractions are reversed, and attributes of previously deleted nodes are updated to the final values. Edge sets are iterated over in order, starting from E_0. Attributes are updated at most once per node in an iteration, allowing parallelism. (Color figure online)

The 1-bit hash function (from [22]) for node numbers stored as 32-bit unsigned integers is

$$hash(x) = \left([a(x+1) + b] \bmod 2^{32}\right) \gg 31, \tag{4}$$

where a is a pseudo-random odd 32-bit unsigned integer, b is a pseudo-random 31-bit unsigned integer, and $\gg$ is bitwise right shift. The values of a and b are generated at the beginning of each loop iteration in the host code.

An array operation applied to every element in an array is implemented as CUDA kernel. The CUB framework [11] is used for *select*, to select array items using a boolean function. Another useful operation in the framework is the *partition* operation, to partition an array into up to 3 subarrays using boolean functions. Data races in merge operations are avoided because the algorithms ensure that any node in an iteration is a merge target at most once, and written values are not read in the same array operation by a different thread.

If merge lists are not necessary, then the first merge steps can be integrated without the `merge_first` list, to improve throughput.

4.1 Rootfix Attributes

The implementation of rootfix attributes is described. Given a max-tree as input, active nodes, excluding roots, are stored in array `active`, and current parent nodes are stored in `roots`, such that for any active node x, edge $(\texttt{roots}[x], x) \in E'$. Array `aux` is used as auxilary array for active nodes, and boolean array `merging` indicates if a node is merging. The loop contains the following steps:

1. Partition `active` into non-merging and merging nodes x, with output `aux`. Non-merging nodes are the active nodes in the next iteration. Contracting edges $(\texttt{roots}[x], x) \in E'$ are copied, and a reference to the created array is added to the front of `merge_last`.
2. Select non-merging nodes y with merging parent x. Edges $(x, y) \in E'$ are copied, and a reference to the new array is added to the back of `merge_first`.

In step 1, boolean array `merging` is updated. In step 2, `roots` is updated for nodes with merging parent. After the while loop, the edge lists can be combined into a single list. When computing rootfix attributes (Eq. 2), a single attribute array is enough, since for each edge array, every node is a merge target at most once.

4.2 Leaffix Attributes

The implementation of leaffix attributes is described. Given a max-tree as input, edges are stored as $(\text{parent}(x), x)$, and radix sorted by parent, such that all edges with the same parent are next to each other. This simplifies determining branch counts and selecting which edges to contract. Because the CUB framework has an efficient radix sort implementation, this choice seems justified, although theoretically using a sorting step is not optimal.

Recall that the number of incoming edges for each node is bounded by the number of neighbors k (Proposition 1). The initial branch count is determined in $\mathcal{O}(k)$ time per node and stored in a byte array. This array is necessary to determine if a child node is mergeable. The loop has the following steps, where each step contains at most one iteration over current edges:

1. Determine contracting edges and distinguish between merging leaves and merging nodes in linked list segments. This information is stored per edge in a byte array. If a node x with exactly one branch is mergeable, store the endpoint in $\texttt{child}[x]$.
2. Select contracting edges with output `first`. Add `first` reference to the back of `merge_first`. Mark merging nodes in linked list segments as removed using the branch count array. Update contracting edges (r, x) in E', where x is not a leaf, to $(r, \texttt{child}[x])$, to maintain a sorted array.
3. Let `remove` be the edges that merge leaves or contain root self-loops. Let `to_update` be the edges where the first node merged. Let `retain` be $(E' - \texttt{remove}) - \texttt{to\_update}$. Next, partition E' into `retain` (next iteration edges), `to_update`, and `remove` (to be discarded). Add `to_update` reference to the front of `merge_last`. Decrease branch count for merge targets of leaves.

In step 3, self-loops are removed to support processing a forest of max-trees. Alternatively, these self-loops can be removed before the loop. Similar to rootfix attributes, the edge lists can be combined into a single list, and a single attribute array is enough.

```
def merge_fn(attributes, idx_targets, idx_sources):
  attributes[idx_targets] += attributes[idx_sources]

def leaffix(edge_array_list, merge_fn, attributes):
  # attributes have been initialized with function z.
  for edges in edge_array_list:
    idx_targets = edges[:, 0] # edges[:, 1] for rootfix attributes
    idx_sources = edges[:, 1] # edges[:, 0] -
    merge_fn(attributes, idx_targets, idx_sources)
```

Listing 1: Area attribute example in Python

4.3 User-Defined Attributes in Python

To make the algorithms accessible in Python, user-defined attributes within the family of rootfix and leaffix attributes can be efficiently computed with array operations that have a GPU backend. More specifically, the edge array lists are provided by the GPU implementation, arrays are stored in GPU memory, and array operations such as element-wise addition are implemented by some framework to run on the GPU. Listing 1 shows an example of a possible implementation of the area attribute. After initialization of `attributes` with user-defined z, function `leaffix` iterates over the list of edge arrays and calls user-defined `merge_fn`. While the leaffix merge function needs to be commutative, the GPU implementation ensures that leaffix targets are closer to the root. This can simplify the implementation of some leaffix merge functions.

4.4 Numerical Stability

Because attributes can contain floating-point values, numerical stability is important to consider. Robust formulas to compute statistical moments with a reduction operator are given in [12]. As an example, to compute the first and second moments with a leaffix attribute, for some function f, the attribute vector is a tuple containing respectively the area, first, and second moment. Function $z(x) = (1, f(x), 0)$ in Definition 3. Let $v_0 = (a_0, \mu_0, M_{2,0})$ and $v_1 = (a_1, \mu_1, M_{2,1})$ be tuples in Z. Then,

$$\begin{aligned} a &= a_0 + a_1, \\ \delta &= \mu_1 - \mu_0, \\ \mu &= \mu_0 + a_1\delta/a, \\ M_2 &= M_{2,0} + M_{2,1} + a_0 a_1 \delta^2/a, \text{ and merge function} \\ \oplus(v_0, v_1) &= (a, \mu, M_2). \end{aligned}$$

The subtractive filter rule, and possibly other filter rules, may introduce numerical errors depending on the range of values. In the commonly used IEEE 754 standard, filter operations on floating-point values are exact when differences can be exactly represented, for example when the set of 32-bit floating-point values contains the representable values in $[1, 2)$, or the set is $\{0, 1, \ldots, 2^{24} - 1\}$.

5 Results

In the following, throughput is shown versus image size and bit depth, for simple rootfix and leaffix attributes, on a desktop computer. With max-trees and image values in GPU memory, a host timer is started after device synchronization, and the timer is ended after attribute computation and device synchronization. GPU results are compared to CPU results. Additionally, numerical errors are shown for the first statistical moments. The sequential CPU rootfix algorithm uses a union-find method [10], and the sequential CPU leaffix algorithm iterates over nodes in decreasing order by distance to the root. The parallel CPU algorithms are similar to the GPU algorithms. The desktop computer contains an AMD 7800X3D processor and a NVIDIA RTX 4070 graphics card.

Throughput Versus Image Size: Throughputs for simple rootfix and leaffix attributes are shown in Fig. 3. Results are similar for leaffix and rootfix attributes. Batched GPU processing has the best performance for every image size, with throughputs up to 1 gigapixel/s. The batch size is the maximum image size in pixels divided by the current image size. When processing single images, the GPU implementation is fastest for megapixel images or larger. The median GPU speed-up is up to 6.8$\times$ compared to the parallel CPU implementations, and up to 30.4$\times$ compared to sequential CPU implementations. With merge lists, throughput decreases by 10% and 14% for respectively the rootfix and leaffix attributes (not shown).

Throughput Versus Image bit Depth: We test how different bit depths impact the throughput for simple leaffix and rootfix attributes. The injective weight function in this situation is $w(x) = (f(x), x)$, where $f(x)$ is a pseudo-random value, and the weight order is lexical. Memory accesses are potentially more efficient for max-trees of images with low bit depth, because parent nodes are more likely to be the left pixel in the image. Figure 4 shows that throughput indeed increases at lower bit depths, by 59% for the rootfix attribute, and 51% for the leaffix attribute. The parallel CPU implementations have similar changes in throughput. The maximum GPU speed-up compared to the parallel CPU implementation is 8.5$\times$, for the rootfix attribute and 19-bit values. The minimum GPU speed-up is 6$\times$, for the leaffix attribute and 32-bit values.

Numerical Stability: The mean image value and variance at the max-tree root is computed for floating-point values in the range $[-1, 1]$. Figure 5 shows the median absolute error. The GPU implementation has better numerical stability compared to the sequential CPU implementation. It is not obvious to us why this is the case. The parallel CPU implementation has similar results (not shown) compared to the GPU implementation.

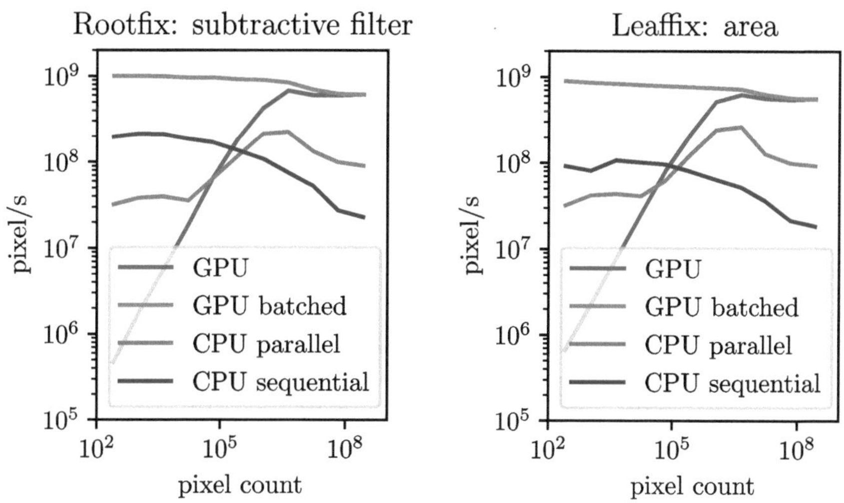

Fig. 3. Median throughput for a rootfix attribute (left) and leaffix attribute (right) versus image size. Square images with 32-bit pseudo-random values.

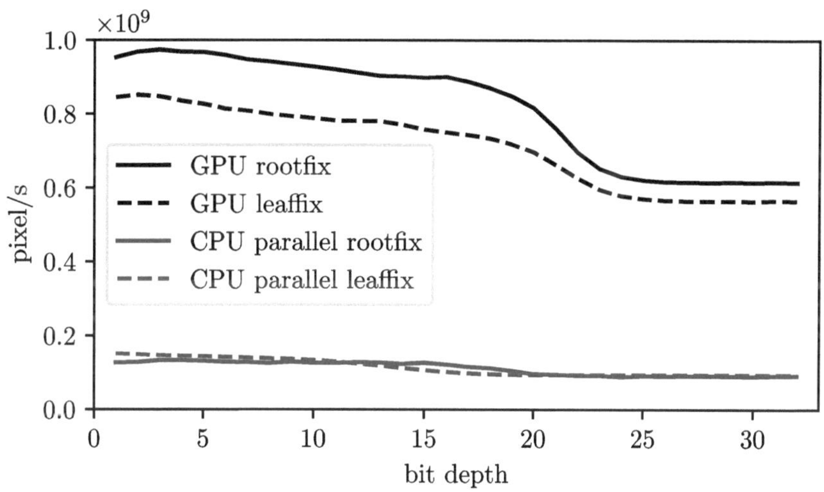

Fig. 4. Median throughput versus bit depth. Square images with 256×2^{20} pixels and pseudo-random values. The rootfix attribute is a subtractive filter, and the leaffix attribute is the area.

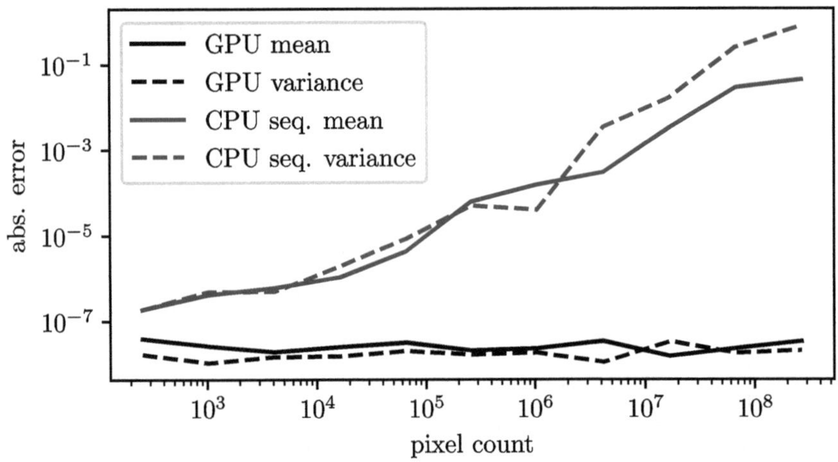

Fig. 5. Median absolute error for the mean value and variance at the max-tree root versus image size. Square images with pseudo-random 32-bit floating-point values in the range $[-1, 1)$.

6 Discussion and Future Work

While the GPU has $4888/8 = 611$ times more processing units compared to the CPU, a possible explanation for the seemingly modest speed-up is that the algorithm is memory bound, and a desktop GPU typically has an order of magnitude greater memory bandwidth.

An efficient GPU algorithm for max-tree construction of images with high bit depth does not seem to be available yet. An efficient GPU algorithm for low bit depth images is given in [1], although the max-tree in this case is canonical, and the in-degree of nodes is not bounded by the neighbor count. Algorithms in [5] could be used in this case.

A more thorough analysis of numerical stability in max-tree attributes is necessary, to know for which values the algorithms are numerically unstable, and understand why.

7 Conclusion

To compute max-tree attributes, filter nodes, and reconstruct images as part of a connected filter pipeline, we have implemented rootfix and leaffix attributes on the GPU. Results show up to 8.5× and 30.4× more throughput for the GPU algorithms compared to respectively parallel and sequential CPU algorithms. The absolute numerical error in the mean value and variance at the max-tree root remains low for the GPU implementation of the leaffix attribute, when the image size increases, while in the sequential CPU implementation the error increases significantly by several orders of magnitude. In future work, we aim to extend this to other tree representations, such as trees-of-shapes [6].

Acknowledgment. This work was in part funded by the Dutch Ministry of Economic Affairs, the "Samenwerkingsverband Noord-Nederland (SNN)" and the European Union through a subsidy from the Just Transition Fund for the DigiAgro3 Project, project number JTFN-00052.

References

1. Blin, N., Carlinet, E., Lemaitre, F., Lacassagne, L., Géraud, T.: Max-tree computation on GPUs. IEEE Trans. Parallel Distrib. Syst. **33**(12), 3520–3531 (2022). https://doi.org/10.1109/TPDS.2022.3158488
2. Breen, E.J., Jones, R.: Attribute openings, thinnings and granulometries. Comput. Vis. Image Underst. **64**(3), 377–389 (1996). https://doi.org/10.1006/cviu.1996.0066
3. Caldairou, B., Naegel, B., Passat, N.: Segmentation of complex images based on component-trees: methodological tools. In: Wilkinson, M.H.F., Roerdink, J.B.T.M. (eds.) ISMM 2009. LNCS, vol. 5720, pp. 171–180. Springer, Heidelberg (2009). https://doi.org/10.1007/978-3-642-03613-2_16
4. Carlinet, E., Géraud, T.: A comparative review of component tree computation algorithms. IEEE Trans. Image Process. **23**(9), 3885–3895 (2014). https://doi.org/10.1109/TIP.2014.2336551
5. Defour, D., Marin, M.: Regularity versus load-balancing on GPU for treefix computations. Procedia Comput. Sci. **18**, 309–318 (2013). https://doi.org/10.1016/j.procs.2013.05.194
6. Géraud, T., Carlinet, E., Crozet, S., Najman, L.: A quasi-linear algorithm to compute the tree of shapes of nD images. In: Hendriks, C.L.L., Borgefors, G., Strand, R. (eds.) ISMM 2013. LNCS, vol. 7883, pp. 98–110. Springer, Heidelberg (2013). https://doi.org/10.1007/978-3-642-38294-9_9
7. Leiserson, C.E., Maggs, B.M.: Communication-efficient parallel algorithms for distributed random-access machines. Algorithmica **3**(1), 53–77 (1988). https://doi.org/10.1007/BF01762110
8. Miller, G.L., Reif, J.H.: Parallel tree contraction and its application. In: FOCS, vol. 26, pp. 478–489 (1985). https://doi.org/10.1109/SFCS.1985.43
9. Moschini, U., Meijster, A., Wilkinson, M.H.F.: A hybrid shared-memory parallel max-tree algorithm for extreme dynamic-range images. IEEE Trans. Pattern Anal. Mach. Intell. **40**(3), 513–526 (2017). https://doi.org/10.1109/TPAMI.2017.2680765
10. Najman, L., Couprie, M.: Building the component tree in quasi-linear time. IEEE Trans. Image Process. **15**, 3531–3539 (2006). https://doi.org/10.1109/TIP.2006.877518
11. NVIDIA Corporation: CUB framework. https://docs.nvidia.com/cuda/cub/index.html. Accessed 20 Mar 2025
12. Pebay, P.P.: Formulas for robust, one-pass parallel computation of covariances and arbitrary-order statistical moments. Technical report, Sandia National Laboratories (SNL), Albuquerque, NM, and Livermore, CA, USA (2008)
13. Salembier, P., Oliveras, A., Garrido, L.: Antiextensive connected operators for image and sequence processing. IEEE Trans. Image Process. **7**(4), 555–570 (1998). https://doi.org/10.1109/83.663500

14. Salembier, P., Serra, J.: Flat zones filtering, connected operators, and filters by reconstruction. IEEE Trans. Image Process. **4**(8), 1153–1160 (1995). https://doi.org/10.1109/83.403422
15. Teeninga, P., Moschini, U., Trager, S.C., Wilkinson, M.H.F.: Statistical attribute filtering to detect faint extended astronomical sources. Math. Morphol. Theory Appl. **1**(1), 100–115 (2016). https://doi.org/10.1515/mathm-2016-0006
16. Teeninga, P.: Parallel max-tree attributes and shape filtering in polylogarithmic time. University of Groningen, Technical report (2024)
17. Teeninga, P.: Parallel Max-Tree Construction in Polylogarithmic Expected Time. Master's thesis, University of Groningen (2024)
18. Urbach, E.R., Roerdink, J.B.T.M., Wilkinson, M.H.F.: Connected shape-size pattern spectra for rotation and scale-invariant classification of gray-scale images. IEEE Trans. Pattern Anal. Mach. Intell. **29**, 272–285 (2007). https://doi.org/10.1109/TPAMI.2007.28
19. Vincent, L.: Morphological area openings and closings for grey-scale images. In: O, Y.L., Toet, A., Foster, D., Heijmans, H.J.A.M., Meer, P. (eds.) Shape in Picture: Mathematical Description of Shape in Grey-level Images, pp. 197–208. NATO (1993)
20. Westenberg, M.A., Roerdink, J.B.T.M., Wilkinson, M.H.F.: Volumetric attribute filtering and interactive visualization using the max-tree representation. IEEE Trans. Image Process. **16**, 2943–2952 (2007). https://doi.org/10.1109/TIP.2007.909317
21. Wilkinson, M.H.F., Gao, H., Hesselink, W.H., Jonker, J.E., Meijster, A.: Concurrent computation of attribute filters on shared memory parallel machines. IEEE Trans. Pattern Anal. Mach. Intell. **30**(10), 1800–1813 (2008). https://doi.org/10.1109/TPAMI.2007.70836
22. Wölfel, P.: Über die Komplexität der Multiplikation in eingeschränkten Branchingprogrammmodellen. Ph.D. thesis, Technical University of Dortmund, Germany (2003)

On the Properties of a Class of Greedy Algorithms for Multicriteria Optimization

Filip Malmberg[1(✉)], Zoe Dumoulin[1], and María Andreína Francisco Rodríguez[2]

[1] Centre for Image Analysis, Department of Information Technology, Uppsala University, Uppsala, Sweden
{filip.malmberg,zoe.dumoulin}@it.uu.se

[2] Division of Computing Science, Department of Information Technology, Uppsala University, Uppsala, Sweden
maria.andreina.francisco@it.uu.se

Abstract. We study a certain class of greedy algorithms for combinatorial optimization, and present a generalized version of this algorithmic pattern encompassing several previously published methods. We analyze the properties of the solutions produced by such algorithms, and provide proofs of their optimality through the concept of *lexicographic max-ordering*. By presenting a unified formulation of this class of greedy algorithms, we hope to facilitate the development of new such algorithms, and to provide a deeper understanding on the properties of existing algorithms.

Keywords: Combinatorial optimization · Multicriteria optimization · Pareto optimality · Lexicographic max-ordering

1 Introduction

In the field of image processing and computer vision, several greedy algorithms have been proposed for solving combinatorial optimization problems on graphs. Examples include the work of Cousty et al. [4,5], Wolf et al. [14,15], and Malmberg et al. [10–13]. We observe that these algorithms all follow very similar algorithmic patterns, yet their properties and optimality have generally been studied for each method in isolation.

Here, we formulate a generalized version of this algorithmic pattern that encompasses the methods cited above, and study the properties of the solutions computed with this approach. To characterize the optimality of such solutions, we study them through the lens of *lexicographic max-ordering* (Lex-MO) multicriteria optimization. The notion of Lex-MO optimality for multicriteria optimization problems was introduced by Ehrgott [6]. Our key insight is that under certain conditions, the problem of finding a Lex-MO optimal solution to a multi-objective optimization problem can be transformed into a sequence of constraint

M. H. F. Wilkinson and J. Kosinka (Eds.): DGMM 2025, LNCS 16296, pp. 291–302, 2026.
https://doi.org/10.1007/978-3-032-09544-2_21

satisfaction problems. If we can solve the associated constraint satisfaction problems, we can find a Lex-MO optimal solution. Even when the conditions for the solution to be exactly Lex-MO optimal are not met, we show that this algorithmic approach will still lead to solutions that satisfy two very desirable properties of Lex-MO optimal solutions, namely *max-ordering* optimality and *Pareto* optimality.

We do not propose any novel algorithms for solving specific classes of optimization problems – instead the general method presented here shows how previously published algorithms share similar characteristics, and provide insights into the properties of the solutions they produce. By identifying a common algorithmic pattern for combinatorial Lex-MO optimization, we hope to facilitate the development of new such algorithms, and shed light on how and why existing algorithms work.

2 Multicriteria Optimization

A multicriteria optimization problem can be defined as consisting of a set of solutions $\mathcal{X}$ together with a set of n criteria $\{f_1, f_2, \ldots, f_n\}$ [8]. Each criterion $f_i : \mathcal{X} \to \mathbb{R}, i \in \{1, \ldots, n\}$ maps every candidate solution $x \in \mathcal{X}$ to a real number. Here, we are only considering *combinatorial* optimization problems, in which the set of solutions $\mathcal{X}$ is finite.

Without loss of generality, we assume here that we seek solutions where the values attained by the criteria are as small as possible, and we will thus use the terms *optimal* and *minimal* interchangeably throughout the paper. We are thus interested in finding a solution $x \in \mathcal{X}$ that in some sense "minimizes" the criteria $\{f_1, f_2, \ldots, f_n\}$ among all possible solutions in $\mathcal{X}$. A central question is how to formally define what a "minimal" solution is in the presence of multiple, possibly conflicting, criteria? In the following subsections, we review three different optimality concepts for multicriteria optimization: *Pareto* optimality, *max-ordering* optimality, and *lexicographic max-ordering* optimality. We also discuss the relations between these different concepts.

2.1 Pareto Optimality

A fundamental notion when it comes to defining optimality for multicriteria optimization is *Pareto* optimality, defined as follows.

Definition 1. *A solution $x \in \mathcal{X}$ is* Pareto optimal *if there is no solution $x' \in \mathcal{X}$ such that*

1. *$f_i(x') \leq f_i(x)$ for all $i = 1, \ldots n$ and,*
2. *$f_i(x') < f_i(x)$ for at least one $i \in \{1, \ldots, n\}$.*

It can be argued that a solution that is not Pareto optimal should not be considered an optimal solution, since it is then possible to find another solution that improves one or more criteria without making any other criterion worse. On the other hand, the set of Pareto optimal solutions to a given multicriteria problem is often large, so a finer characterization of the solution set is often desirable.

2.2 Scalarization and Max-Ordering Optimality

Another common approach to defining optimality for multicriteria optimization is *scalarization*, which means that we define some function F that aggregates the values of $f_1, f_2, \ldots, f_n$ to a single scalar value. An optimal solution x is then defined as any solution that satisfies

$$F(x) = \min_{x' \in \mathcal{X}} F(x') \; . \tag{1}$$

In this way, the multicriteria optimization problem is transformed into a single criterion problem, for which the notion of a minimal solution is well defined.

Here, we will specifically consider the function

$$F_\infty(x) := \max_i f_i(x) \; . \tag{2}$$

Following the terminology of Ehrgott [8], we refer to a solution that minimizes F_∞ as a *max-ordering* solution. A max-ordering solution is not necessarily Pareto-optimal [8].

If we assume for a moment that all criteria f_i are non-negative, we can observe that the function F_∞ is the limit of a more general function

$$F_p(x) = \left(\sum_{i=1}^{n} f_i(x)^p \right)^{1/p} \; , \tag{3}$$

if we let $p \in \mathbb{R}$ tend to infinity. For non-negative criteria, F_p can be recognized as the *p-norm* of the vector containing all criteria.

2.3 Lexicographic Max-Ordering Optimality

Another alternative concept for optimality of multicriteria problems, proposed by Ehrgott [6], is *lexicographic max-ordering* (Lex-MO) optimality.[1]

For any element $v \in \mathbb{R}^Q$ we define $\Theta(v) = (\Theta_1(v), \ldots, \Theta_Q(v))$ to be the vector in $\mathbb{R}^Q$ containing the components of v in non-increasing order, so that $\Theta_1(v) \geq \ldots \geq \Theta_Q(v)$. We observe that even if the element v contains multiple components with the same value, the vector $\Theta(v)$ is uniquely defined. Let v and v' be elements in $\mathbb{R}^Q$. We say that *v precedes v' lexicographically*, and denote this $v \prec v'$, provided that there exists a j such that $\Theta_j(v) \neq \Theta_j(v')$, and for the smallest such j we have $\Theta_j(v) < \Theta_j(v')$. Also, we write $v \preceq v'$ provided either $v \prec v'$ or $\Theta_j(v) = \Theta_j(v')$ for all $j \in \{1, \ldots, Q\}$.

For multicriteria optimization, the values of the criteria $f_1, \ldots, f_n$ for a given solution can be viewed as a vector in $\mathbb{R}^n$. For a solution $x \in \mathcal{X}$, we denote this vector by $f_x = (f_1(x), \ldots, f_n(x))$.

Definition 2. *A solution x is said to be a Lex-MO solution if $f_x \preceq f_{x'}$ for all $x' \in \mathcal{X}$.*

---

[1] The same concept was also later independently proposed by Levi and Zorin [9], who used the term *strict minimizers*.

As shown by Ehrgott [7], Lex-MO optimal solutions have the following favorable properties:

- They are Pareto optimal, i.e., there is no other solution that improves one criterion without simultaneously worsening another one.
- They are max-ordering solutions, i,e., they minimize F_∞.

Figure 1 illustrates the relationship between Pareto optimality, max-ordering optimality, and Lex-MO optimality.

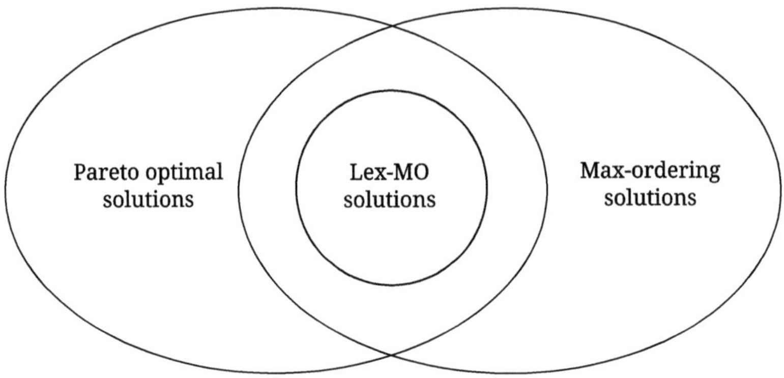

Fig. 1. The relation between Pareto optimality, Max-ordering optimality, and Lex-MO optimality.

Similarly to max-ordering optimality, Lex-MO optimality can also be motivated in terms of p-norm optimization. Consider the optimal solutions to the scalarization function F_p, as p increases. Levi and Zorin [9] observe that for non-negative criteria these solutions will converge to a Lex-MO optimal solution as p approaches infinity. In fact, for a given multicriteria optimization problem, there exists a finite $\hat{p} \in \mathbb{R}$ such that any solution optimizing F_p is also a Lex-MO optimal solution for $p \geq \hat{p}$. Wolf et al. [14] refer to such a value $\hat{p}$ as a *dominant power*, and Malmberg and Ciesielski [10] provide an example of how such a dominant power can be determined for a specific optimization problem.

2.4 A Brief Discussion About Scalarization in Relation to Multi and Single Criterion Optimization

As outlined in Sect. 2.2, scalarization provides a bridge between multicriteria and single criterion optimization. The scalarization function aggregates all criteria f_i into a single scalar value. Levi and Zorin [9] observe that in the context of computational geometry, many commonly occurring single criterion optimization problems have objective functions that can be viewed as consisting of

- A set of "local" error measures or cost function.
- A "global" error measure that aggregates the local costs into a single scalar value.

The same observation can be made for many optimization problems commonly occurring in image processing and computer vision.

Alternatively, we could also view this through the lens of multicriteria optimization, and interpret the local error measures as set of criteria, and the global error measure as a scalarization function. Consider for example the *minimal st-cut* problem, that is an important primitive for solving optimization problems in image processing and computer vision [2,3]. Given an edge weighted graph, where two special vertices are designated as s and t, the minimal st-cut problem consists of finding a partition of the graph vertices such that the weight sum of the edges spanning the boundary of the partition – the cut – is minimal.

If we instead seek to minimize the maximum edge weight along the cut, rather than the sum of the edge weights, then the resulting optimization problem can be solved via *watershed cuts* [1,4].

While these two optimization problems are typically presented as single criterion problems, the multicriteria optimization view highlights the similarity between them. Both problems have the same set of criteria (the individual edge weights) but use different scalarization functions (sum versus maximum). Additionally, the multicriteria optimization view allows for the use of *non-scalarizing* optimality concepts, such as the Lex-MO optimality considered here.

3 Preliminaries: Partitions and Equivalence Relations

We now introduce some concepts that will be required for the exposition of our proposed method.

Let $\mathcal{S}$ be a set. A *partition* of $\mathcal{S}$ is a set of mutually disjoint subsets of $\mathcal{S}$, whose union is $\mathcal{S}$. We refer to an element of a partition as a *region* of that partition. Let $\mathcal{P}$ and $\mathcal{Q}$ be partitions of $\mathcal{S}$. We say that $\mathcal{Q}$ is a *refinement* of $\mathcal{P}$ if every region of $\mathcal{Q}$ is included in a region of $\mathcal{P}$. Note that by this definition any partition is a refinement of itself.

Let $\sim$ be a binary relation on the set $\mathcal{S}$ (a subset of $\mathcal{S} \times \mathcal{S}$). We recall that $\sim$ is an *equivalence relation* if and only if the following properties hold for all $a, b, c \in \mathcal{S}$:

1. $a \sim a$ (*reflexivity*).
2. $a \sim b$ if and only if $b \sim a$ (*symmetry*).
3. If $a \sim b$ and $b \sim c$ then $a \sim c$ (*transitivity*).

The *equivalence class* of an element $a \in \mathcal{S}$ under the relation $\sim$ is defined as $\{b \in S | b \sim a\}$. The equivalence classes of a set under an equivalence relation are disjoint, and their union is the set itself. Thus, every equivalence relation on a set provides a partition of that set into its equivalence classes.

In the context of multicriteria optimization, we observe that for each individual criterion f_i, the relation "equal with respect to f_i" is an equivalence relation on the set of solutions $\mathcal{X}$. We denote this relation by $\underset{f_i}{=}$.

4 Proposed Method

Consider a combinatorial optimization problem with a set of solutions $\mathcal{X}$ and a set of n criteria $\{f_1, f_2, \ldots, f_n\}$. We are interested in finding Lex-MO optimal solutions to this multicriteria optimization problem. We assume that $|\mathcal{X}|$ is large, so that finding optimal solutions by exhaustive search across the set of solutions is not feasible. On the other hand, as observed by Zorin and Levi, it is often the case that the criteria are "local", in the sense that they only consider some particular aspect of each solution in $\mathcal{X}$. Here, we take this to mean that the set of distinct values attained by each criterion across $\mathcal{X}$ is "very small" compared to $|\mathcal{X}|$. For this scenario, we propose a general method for Lex-MO optimization, based on iteration over the equivalence classes of all $\underset{f_i}{=}$.

For every criterion f_i, let P_i be a partition of $\mathcal{X}$, such that for any two solutions x and x' it holds that if x and x' belong to the same subset of P_i then $f_i(x) = f_i(x')$. In other words, P_i must be a refinement of the partition given by the equivalence classes of $\underset{f_i}{=}$. The number of subsets of P_i is denoted $|P_i|$. Generally, we will assume that each $|P_i|$ is "very small" compared to $|\mathcal{X}|$.

By the requirement that each partition P_i must be a refinement of the partition induced by the equivalence relation $\underset{f_i}{=}$, each region A of every P_i can be associated with a real number, namely $f_i(x), x \in A$. Each region A can also be associated with a constraint $x \notin A$. We denote by $\mathcal{C}$ the set of constraints associated with all regions of all P_i for $i \in \{1, \ldots, m\}$. By extension, each constraint $x \notin A$ in $\mathcal{C}$ can also be associated with the real number associated with A. We define $\Phi : \mathcal{C} \to \mathbb{R}$ to be a mapping from the set of constraints $\mathcal{C}$ to the associated real values. In other words, a region A of any of the partitions P_i is associated with the value $f_i(x), x \in A$. Let's denote this value f_A. The constraint $c = x \notin A$ is then part of the set of constraints $\mathcal{C}$, and we have $\Phi(c) = f_A$.

Our proposed algorithmic procedure for solving Lex-MO optimization problems is outlined in Algorithm 1.

Algorithm 1 Proposed algorithmic pattern

```
Let Γ be an empty set of constraints.
for each constraint c ∈ C, in decreasing order by their value Φ(c) do
    if there exists a solution x ∈ X satisfying the constraints Γ ∪ {c} then
        Let Γ ← Γ ∪ {c}
    end if
end for
Return a solution that satisfies all constraints in Γ
```

Throughout the algorithm, we maintain a set Γ of *active* constraints. Initially, this set is empty. The algorithm iterates over all the constraints in $\mathcal{C}$ in order of decreasing value. At each iteration we check if the considered constraint can be added to Γ without making the set of solutions satisfying Γ empty. Answering this question amounts to solving a constraint satisfaction problem, and for the algorithm to be computationally feasible each such constraint satisfaction problem must be possible to solve in reasonable time. This, however, will depend on the specific optimization problem and solution set. At the termination of the algorithm we return a solution satisfying all constraints in Γ. Here, we assume that the constraint satisfaction problem is such that in addition to determining whether a solution exists, we can also find a concrete satisfying solution.

4.1 Optimality of the Solutions Produced by the Proposed Algorithmic Pattern

We will now study the properties of the solutions produced by the proposed algorithmic pattern. We present three theorems, that provide successively stronger optimality guarantees on the solutions produced by the algorithm, under different assumptions on the criteria f_i and partitions P_i.

We start by showing that under the assumptions already stated in the formulation of the algorithmic pattern, it indeed produces a feasible solution that is also a max-ordering solution.

Theorem 1. *If each P_i is a refinement of the partition induced by $\underset{f_i}{=}$, then the proposed algorithm returns a feasible, max-ordering solution.*

Proof. Since the set Γ is initially empty, and constraints are only added to Γ if that does not result in the set of feasible solutions becoming empty, there is, at the completion of the algorithm, at least one solution that satisfies all constraints in Γ and can be returned by the algorithm.

To see that the returned feasible solution is a max-ordering solution, we perform a proof by contradiction. Assume that the algorithm returns a solution $x \in \mathcal{X}$, and that there exists a solution $x' \in \mathcal{X}$ such that $F_\infty(x) > F_\infty(x')$. Consider the set of all constraints in $\mathcal{C}$ whose value is greater than $F_\infty(x')$. By the existence of x', the union of all such constraints must admit at least one feasible solution, namely x'. During the algorithm, all such constraints will be considered before all other constraints, and since their union does not render the set of feasible solutions empty, all such constraints will be added to Γ during the algorithm. Thus, x will satisfy all such constraints and the cost $F_\infty(x)$ of the solution returned by the algorithm can therefore be at most $F_\infty(x')$, contradicting our assumption. □

Next, we show that under slightly more strict conditions on the partitions P_i, the solutions produced by the algorithm will also be Pareto optimal.

Theorem 2. *If each P_i is equal to the partition induced by $\underset{f_i}{=}$, then the solution returned by the proposed algorithm is Pareto optimal.*

Proof. Let x be the solution returned by the algorithm. Assume, by way of contradiction, that there exists another solution x' such that for all criteria f_i

we have $f_i(x') \le f_i(x)$, and that there is at least one criterion f_k for which this inequality is strict. Let A be the region of P_k containing x, and let A' be the region of P_k containing x'. Since $f_k(x') < f_k(x)$, the constraint $x \notin A$ was evaluated before the constraint $x' \notin A'$ during the execution of the algorithm. Let Γ' be the state of the set Γ just prior to considering the constraint $x \notin A$. Since the algorithm produced a solution in which $x \in A$, it must hold that the set of solutions satisfying $\Gamma' \cup \{x \notin A\}$ is empty. This, in turn, means that x' must violate at least one constraint in Γ'. Let B be the region and f_j be the criterion associated with such a constraint. Since this constraint is in Γ', it follows that for any $x'' \notin B$ we have $f_j(x'') \ge f_j(x)$. Since each P_i is equal to the partition induced by $\underset{f_i}{=}$, this inequality must be strict. Since $x' \notin B$, it follows that $f_j(x') > f_j(x)$, contradicting our assumption that x is not Pareto optimal. □

We argue that satisfying the condition stated in Theorem 2 is not a problem in practice. If our partitionings are such that there are two constraints for the same criterion that have the same cost, we can merge them into a single constraint without violating the condition in Theorem 1. By repeating this merging procedure, the set of constraints $\mathcal{C}$ will eventually be equal to the constraint set of the partitionings induced by $\underset{f_i}{=}$ for $i \in \{1, \ldots, n\}$.

Finally, we will show that under certain conditions, the solutions produced by the algorithm are Lex-MO solutions. This holds if each of the constraints are associated with distinct values, i.e., no two constraints are associated with the same value. More formally, we will show that the algorithm produces Lex-MO optimal solutions if the mapping Φ between the constraints in $\mathcal{C}$ and their associated values is injective. The proof of this statement relies on the following Lemma.

Lemma 1. *Let $x \in \mathcal{X}$ be a solution and let $c \in C$ be a constraint. If the mapping Φ between the constraints in C and their associated values is injective, then x satisfies c if and only if $f_i(x) \neq \Phi(c)$ for all $i \in \{1, \ldots n\}$.*

Proof. First, let A and f_k be the region and criterion associated with the constraint c. If x does not satisfy c, then $x \in A$ and so $f_k(x) = \Phi(c)$. Thus

$$f_i \neq \Phi(c), \forall i \in \{1, \ldots, n\} \Rightarrow x \text{ satisfies } c\,. \tag{4}$$

Under the assumption that the mapping Φ is injective, there exists a mapping $\Psi : \mathbb{R} \to \mathcal{C}$ that is a *left inverse* of Φ, i.e., $\Psi(\Phi(c)) = c$ for every $c \in \mathcal{C}$. If $f_i(x) = \Phi(c)$ for some $i \in \{1, \ldots, n\}$, it must therefore hold that x does not satisfy the constraint $\Psi(f_i(x)) = \Psi(\Phi(c)) = c$. Thus

$$x \text{ satisfies } c \Rightarrow f_i \neq \Phi(c), \forall i \in \{1, \ldots, n\}\,. \tag{5}$$

□

Theorem 3. *If the mapping Φ between the constraints in $\mathcal{C}$ and their associated values is injective, the proposed algorithm returns a Lex-MO optimal solution.*

Proof. Assume that Φ is injective, and let Ψ be a left inverse of Φ. Let $x \in \mathcal{X}$ be the solution returned by the proposed algorithm, and let $x' \in \mathcal{X}$ be a Lex-MO optimal solution. We will now show by induction that $\Theta(f_x) = \Theta(f_{x'})$, and thereby show that x is also Lex-MO optimal.

Let k be an integer such that $1 \leq k < n$. Assume (†) that for all i such that $1 \leq i \leq k$ it holds that $\Theta_i(f_x) = \Theta_i(f_{x'})$. We will now show that this also implies that $\Theta_{k+1}(f_x) = \Theta_{k+1}(f_{x'})$.

Let $\mathcal{C}'$ be the set of all constraints in $\mathcal{C}$ with a value greater than $\Theta_{k+1}(f_x)$ and let $m = |\mathcal{C}'|$. We observe that $\mathcal{C}'$ is precisely the set of constraints in $\mathcal{C}$ that have been considered by the algorithm after m iterations. Let c be the constraint $\Psi(\Theta_{k+1}(f_x))$, i.e., c is the constraint considered by the algorithm at the $m+1$: th iteration. Let $\Gamma' \subseteq \mathcal{C}'$ be the state of the set Γ after m iterations. By definition, Γ' is precisely the subset of $\mathcal{C}'$ that is satisfied by x. Now, consider the subset of $\mathcal{C}'$ that is satisfied by x'. By Lemma 1, this is the set of all constraints in $\mathcal{C}'$ whose value is not attained by any $\Theta_i(f_{x'})$ for $i \in \{1, \ldots, k\}$. By the assumption (†), the subset of $\mathcal{C}'$ that is satisfied by x' is equal to Γ'.

Since we have $\Theta_{k+1}(f_x) = \Phi(c)$ for the solution x produced by the algorithm, it must hold that the set of solutions in $\mathcal{X}$ satisfying the constraints $\Gamma' \cup \{c\}$ is empty. Since x' is Lex-MO optimal, we must have $\Theta_{k+1}(f_x) \geq \Theta_{k+1}(f_{x'})$. Since Φ is injective, neither $\Theta(f_x)$ nor $\Theta(f_{x'})$ can have repeated entries. Thus $\Theta_i(f_{x'}) \neq \Phi(c)$ for all $i \in \{1, \ldots, k\}$. If $\Theta_{k+1}(f_x) > \Theta_{k+1}(f_{x'})$, then it also holds that $f_i(x') \neq \Phi(c)$ for all $i \in \{1, \ldots, n\}$, since by definition $\Theta_{k+1}(f_{x'}) > \Theta_j(f_{x'})$ for all $j \in \{k+1, \ldots, n\}$. Thus, if $\Theta_{k+1}(f_x) > \Theta_{k+1}(f_{x'})$ then $f_i(x') \neq \Phi(c)$ for all $i \in \{1, \ldots, n\}$. By Lemma 1, x' would then satisfy c and since we have already concluded that x' satisfies all constraints in Γ' it follows that x' would then satisfy all constraints in $\Gamma' \cup \{c\}$. This contradicts our conclusion that the set of solutions satisfying $\Gamma' \cup \{c\}$ is empty, and so it cannot hold that $\Theta_{k+1}(f_x) > \Theta_{k+1}(f_{x'})$.

In consequence, we have that the assumption (†) implies $\Theta_{k+1}(f_x) = \Theta_{k+1}(f_{x'})$.

By Theorem 1, we know that the assumption (†) holds for $k = 1$. By induction, it therefore also holds for all k. □

Let us now examine under what circumstances the conditions for Theorem 3 are fulfilled. We observe that Φ can only be injective if the conditions for Theorem 2 are met, i.e., if each partition P_i is equal to the unique partition induced by $\underset{f_i}{=}$. Otherwise, at least one partition has a pair of regions with the same value, and so Φ cannot be injective. If we thus fix each partition P_i to the partition induced by $\underset{f_i}{=}$, the question of whether Φ is injective or not will depend only on the criteria $f_1, \ldots, f_n$.

For each criterion f_i we consider the set of values that this criterion attains across the set of solutions $\mathcal{X}$, i.e., the image $f_i^{\rightarrow}(\mathcal{X})$ of $\mathcal{X}$ under f_i. If the sets $f_1^{\rightarrow}(\mathcal{X}), \ldots, f_n^{\rightarrow}(\mathcal{X})$ are pairwise disjoint, we say that the criteria have *unique values.* We observe that with the partitions fixed as stated above, Φ will be injective if and only if the criteria have unique values.

4.2 Computational Optimizations

While our proposed algorithmic pattern is abstract, and not a concrete algorithm for solving a specific optimization problem, we will still discuss some variations on the basic algorithm that may lead to improved computational performance in concrete algorithms.

First, we observe that for each criterion f_i, the constraints that correspond to regions of P_i that attain the lowest value for f_i can be omitted from the set of constraints $\mathcal{C}$ without affecting the outcome. To see this, we first observe that if there are multiple constraints corresponding to f_i that attain this lowest value, then they can be merged into a single region/constraint without violating the conditions of Theorems 1-3 as outlined in Sect. 4.1. Thus, we need only consider the case where there is a single such constraint. Next, we observe that by the time that this constraint is considered during the execution of the algorithm, all other constraints for that criterion have already been considered. There are then two possible cases to consider:

1. All of those constraints have been added to Γ, in which case the added constraints must imply that the solution is within the region associated with the lowest-value constraint.
2. Exactly one of the higher-value constraints has not been added to Γ. In this case, the solution must be in the region corresponding to the constraint that was not added.

In both these cases, the region that the solution belongs to is already determined by the higher-value constraints already added to Γ during previous iterations, and so we may omit the lowest-value constraint from the computations without affecting the result. If the number of regions of each P_k is very small, avoiding sorting the elements with minimal value may be substantially more computationally efficient.

The second optimization is based on the observation that if a constraint is not added to Γ during the iteration at which it is considered, then the solution produced by the algorithm must lie within the region corresponding to that constraint. Therefore, any remaining lower-value constraints corresponding to the same criterion are guaranteed to be satisfied by the solution, and so they can be added to Γ without evaluating the **if**-statement. For problems where solving the associated constraint satisfaction problem is computationally expensive, this may lead to a substantial reduction in computation time.

Table 1. Optimality properties of the solutions produced by the proposed algorithmic pattern, under different assumptions on P_i and f_i.

Assumption on P_i and f_i	Optimality of solutions		
	Max-ordering optimality	Pareto optimality	Lex-MO optimality
Each P_i is a refinement of the partition induced by $\underset{f_i}{=}$. (Theorem 1)	✓		
Each P_i is equal to the partition induced by $\underset{f_i}{=}$. (Theorem 2)	✓	✓	
The mapping Φ between the constraints in $\mathcal{C}$ and their associated values is injective. (Theorem 3)	✓	✓	✓

5 Conclusions

We have presented a generalized version of a class of greedy algorithms for combinatorial multicriteria optimization. In Sect. 4.1 we provide three theorems that give successively stronger guarantees on the optimality of the solutions produced by our proposed algorithmic pattern, under different assumptions on the underlying optimization problem. These results are summarized in Table 1.

Our general conclusion is that concrete implementations of the proposed algorithmic pattern can be designed so that they are guaranteed to return solutions that are max-ordering and Pareto-optimal, i.e. they satisfy two of the main desirable properties of Lex-MO optimal solutions. If the criteria of the optimization problem we are trying to solve have unique values, in the sense defined in Sect. 4.1, the solutions are also guaranteed to be Lex-MO optimal.

By identifying a common algorithmic pattern for combinatorial Lex-MO optimization, we hope to facilitate the development of new such algorithms. With the results presented here, the optimality properties of new such algorithms do not need to be derived independently for each algorithm. Instead, it is sufficient to show that a new algorithm is an instance of the general algorithmic pattern described here.

As hinted in the introduction, we postulate that several previously known algorithms for max-norm optimization can indeed be seen as instances of the general algorithmic approach presented here. This includes watershed cuts[2] [4], the *mutex watershed* algorithm by Wolf et al. [14] and the binary labeling algorithm by Malmberg and Ciesielski [10]. A detailed motivation of this claim, however, is beyond the scope of this conference paper and we plan to expand on these ideas in future publications.

[2] When implemented using Kruskal's algorithm, as described by Cousty et al. [4].

Acknowledgements. The authors wish to thank Ingela Nyström and Gunilla Borgefors for valuable feedback on draft versions of this manuscript. This work was supported by the Swedish Science Council, grant number 2023-03943.

References

1. Allène, C., Audibert, J.Y., Couprie, M., Keriven, R.: Some links between extremum spanning forests, watersheds and min-cuts. Image Vis. Comput. **28**(10), 1460–1471 (2010)
2. Boykov, Y., Funka-Lea, G.: Graph cuts and efficient nd image segmentation. Int. J. Comput. Vis. **70**(2), 109–131 (2006)
3. Boykov, Y., Veksler, O., Zabih, R.: Fast approximate energy minimization via graph cuts. IEEE Trans. Pattern Anal. Mach. Intell. **23**(11), 1222–1239 (2001)
4. Cousty, J., Bertrand, G., Najman, L., Couprie, M.: Watershed cuts: Minimum spanning forests and the drop of water principle. IEEE Trans. Pattern Anal. Mach. Intell. **31**(8), 1362–1374 (2009)
5. Cousty, J., Bertrand, G., Najman, L., Couprie, M.: Watershed cuts: thinnings, shortest path forests, and topological watersheds. IEEE Trans. Pattern Anal. Mach. Intell. **32**(5), 925–939 (2009)
6. Ehrgott, M.: Lexicographic max-ordering-a solution concept for multicriteria combinatorial optimization (1995)
7. Ehrgott, M.: A characterization of lexicographic max-ordering solutions (1999). http://nbn-resolving.de/urn:nbn:de:hbz:386-kluedo-4531
8. Ehrgott, M.: Multicriteria Optimization, vol. 491. Springer, Berlin, Heidelberg (2005). https://doi.org/10.1007/3-540-27659-9
9. Levi, Z., Zorin, D.: Strict minimizers for geometric optimization. ACM Trans. Graph. (TOG) **33**(6), 185 (2014)
10. Malmberg, F., Ciesielski, K.C.: Two polynomial time graph labeling algorithms optimizing max-norm-based objective functions. J. Math. Imaging Vis. **62**(5), 737–750 (2020)
11. Malmberg, F., Ciesielski, K.C., Strand, R.: Optimization of max-norm objective functions in image processing and computer vision. In: Couprie, M., Cousty, J., Kenmochi, Y., Mustafa, N. (eds.) Discrete Geometry for Computer Imagery. DGCI 2019. LNCS, vol. 11414, p. 206. Springer, Cham (2019). https://doi.org/10.1007/978-3-030-14085-4_17
12. Malmberg, F., Falcão, A.X.: A practical algorithm for max-norm optimal binary labeling of graphs. In: Vento, M., Foggia, P., Conte, D., Carletti, V. (eds.) Graph-Based Representations in Pattern Recognition. GbRPR 2023. LNCS, vol. 14121, pp. 35–45. Springer, Cham (2023). https://doi.org/10.1007/978-3-031-42795-4_4
13. Malmberg, F., Strand, R., Nyström, I.: Generalized hard constraints for graph segmentation. In: Heyden, A., Kahl, F. (eds.) Image Analysis. SCIA 2011. LNCS, vol. 6688, pp. 36–47. Springer, Berlin, Heidelberg (2011). https://doi.org/10.1007/978-3-642-21227-7_4
14. Wolf, S., et al.: The mutex watershed and its objective: efficient, parameter-free graph partitioning. IEEE Trans. Pattern Anal. Mach. Intell. **43**(10), 3724–3738 (2020)
15. Wolf, S., et al.: The mutex watershed: efficient, parameter-free image partitioning. In: Ferrari, V., Hebert, M., Sminchisescu, C., Weiss, Y. (eds.) Computer Vision – ECCV 2018. ECCV 2018. LNCS, vol. 11208, pp. 546–562. Springer, Cham (2018). https://doi.org/10.1007/978-3-030-01225-0_34

The Tree of Shapes as a Distance Transform: Building the ToS on High-Dynamic Range Images

Edwin Carlinet(✉) and Baptiste Esteban

EPITA Research Laboratory, 14-16 rue Voltaire, 94270 Le Kremlin-Bicêtre, France
edwin1.carlinet@epita.fr

Abstract. The tree of shapes (ToS) is a self-dual and contrast invariant hierarchical representation of images, making it highly suitable for various image processing tasks such as filtering, segmentation, and object detection. The existing linear algorithm for computing the ToS is effective for low-quantized images, but exhibits increased complexity for high-quantized images due to the reliance on data structures like red-black trees, which have logarithmic complexity for insertion and deletion operations. This paper introduces a novel algorithm based on the fast marching method to compute a distance map from which the ToS is extracted. This method takes advantage of the distribution of the gradient values of the image pixels to make use of an efficient priority queue structure that lowers the computational complexity of the ToS on high-dynamic range images.

Keywords: Tree of Shapes · Mathematical Morphology · Distance transforms

1 Introduction

Hierarchical representations of image [3] are extensively employed in image processing. These representations enable the description of objects of interest at multiple levels of detail, based on the inclusion relationships of image regions. Among them, the tree of shapes (ToS) encodes the inclusion relationships of an image's connected components and, consequently, its level lines. It is utilized in various tasks, such as image simplification [22], visual saliency [14], feature extraction [2], and image classification [7]. Figure 1 demonstrates a simple ToS-based application, where a grain filter is used as a pre-filtering step to accelerate text detection. Components are selected according to their size, resulting in a residual image that contains text characters.

The Fast Level Line Transform (FLLT) [12] introduced an algorithm to construct the Tree of Shapes (ToS) by merging the min-tree and max-tree [17]. Subsequently, the Fast Level Set Transform (FLST) [6] employed a region-growing approach to flood an image from its extrema, thereby organizing its level sets.

M. H. F. Wilkinson and J. Kosinka (Eds.): DGMM 2025, LNCS 16296, pp. 303–314, 2026.
https://doi.org/10.1007/978-3-032-09544-2_22

In [19], the author presented a top-down method that initiates from the image border and follows the level lines. Géraud et al. [10] first sorted the level lines of an image based on their inclusion levels and then applied a Union-Find procedure to derive the ToS. Similarly, [4] and [8] adapted the algorithm from [10] by initially constructing an order map and then extracting the ToS from this map using a max-tree, resulting in the first linear and parallel ToS algorithms.

Fig. 1. Illustration of a grain filters with the ToS. Left: original image. Middle: image simplified with a grain filter. Right: residual image after filtering.

The algorithm presented in [4] achieves linear time complexity under specific conditions. The quantization of image pixel values must be sufficiently low to enable the use of an efficient data structure, such as a hierarchical queue with constant insertion and deletion complexities, during the propagation phase. Additionally, the maximum number of connected components must not exceed a certain threshold to allow the application of a linear max-tree algorithm for computing the ToS from the order map [5]. Thus, in this paper, we introduce an algorithm for computing the ToS based on a distance transform. Specifically, rather than determining an ordering relationship, our approach computes a distance map that measures the distance between the level lines of an image and its border. Additionally, we leverage the distance transform to employ an efficient heap structure, enhancing the overall computational efficiency, particularly when processing high-dynamic range images.

This paper is structured as follows: Sect. 2 provides a review of the definition and construction algorithm of the Tree of Shapes (ToS). Section 3 details the computation of the order map as a distance map. Section 4 explains the management of the priority queue when dealing with XDR images. Section 5 compares the performance of computing the ToS using the order map versus the distance map. Finally, Sect. 6 presents the conclusions and the perspectives.

2 Sequential Tree of Shapes Computation

2.1 Definitions

Let $f : \Omega \rightarrow \mathcal{V}$ be an image defined on a domain $\Omega \subset \mathbb{Z}^n$ and whose value set $\mathcal{V}$ is endowed with an order relationship $<$. Let $[f \geq \lambda] = \{p \in \Omega \mid f(p) \geq \lambda\}$ be the upper thresholds set and $[f < \lambda] = \{p \in \Omega \mid f(p) < \lambda\}$ be the lower thresholds set, with $\lambda \in \mathcal{V}$. The set of upper connected components is denoted by $\mathfrak{C}^{\geq} = \cup_{\lambda \in \mathcal{V}}\{X \mid X \in \mathcal{CC}([f \geq \lambda])\}$ and the set of lower connected components by $\mathfrak{C}^{<} = \cup_{\lambda \in \mathcal{V}}\{X \mid X \in \mathcal{CC}([f < \lambda])\}$, with the operator $\mathcal{CC}(X)$ returning the set of connected components in a set X. The max-tree $\mathcal{T}_{\geq}$ (resp. the min-tree $\mathcal{T}_{<}$) encodes the inclusion relationship of the components in $\mathfrak{C}^{\geq}$ (resp. $\mathfrak{C}^{<}$).

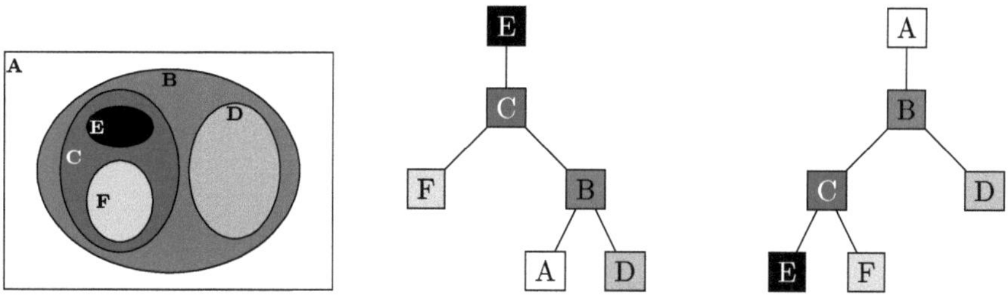

(a) An image f (b) Max-tree $\mathcal{T}_{\geq}(f)$ (c) ToS $\mathcal{T}_{\circ}(f)$

Fig. 2. An image and its max-tree and tree of shapes.

A shape $\mathcal{C}$ is a connected component of f with its holes filled and belongs to the set $\mathfrak{C}^{\circ} = \{\text{Sat}(X) \mid X \in \mathfrak{C}^{<}\} \cup \{\text{Sat}(X) \mid X \in \mathfrak{C}^{\geq}\}$ with $\text{Sat}(X)$ the hole-filling (saturation) operator. The tree of shapes $\mathcal{T}_{\circ}$ encodes the inclusion relationship of the shapes of $\mathfrak{C}^{\circ}$, and thus the level lines of f. Figure 2 illustrates a max-tree $\mathcal{T}_{\geq}$ and a tree of shapes $\mathcal{T}_{\circ}$ computed on an image.

2.2 Construction Algorithm

This section describes the different steps to compute the tree of shapes based on the algorithms presented in [10] and [4]. To ensure the root of the tree encloses the whole image, the image is padded with a border whose value is usually set to the median value of the image bordering pixels so that the self-duality of the resulting tree is maintained.

Immersion. The *immersion* step involves embedding the image into a set-valued map on a Khalimsky grid. This map models the level lines passing between the image pixels by introducing intermediate pixels (0-faces and 1-faces). As illustrated in Fig. 3, original pixels are depicted as blue squares, 1-faces as rectangles, and 0-faces as circles. The 1-faces and 0-faces use intervals to represent all possible level lines passing between adjacent original pixels. For example, in the case of the saddle point $\begin{smallmatrix}3&1\\0&3\end{smallmatrix}$, level lines from 0 to 3 may pass through the center (depending on the interpolation), while only level lines from 1 to 3 can pass between the two top pixels. This interval-valued map forms the basis of the ToS algorithm, enabling the *physical* tracing of level lines within the image.

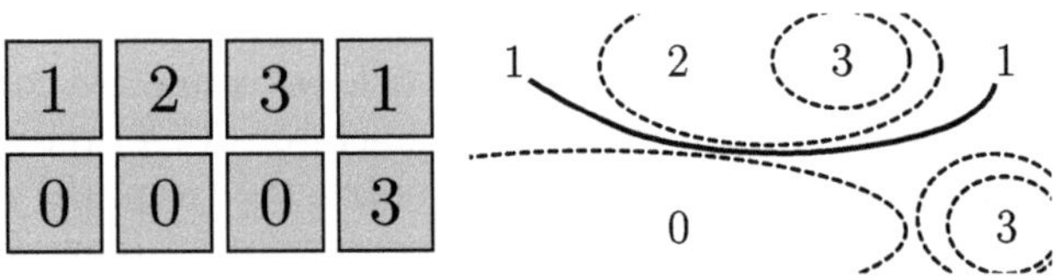

(a) Example image and its level lines

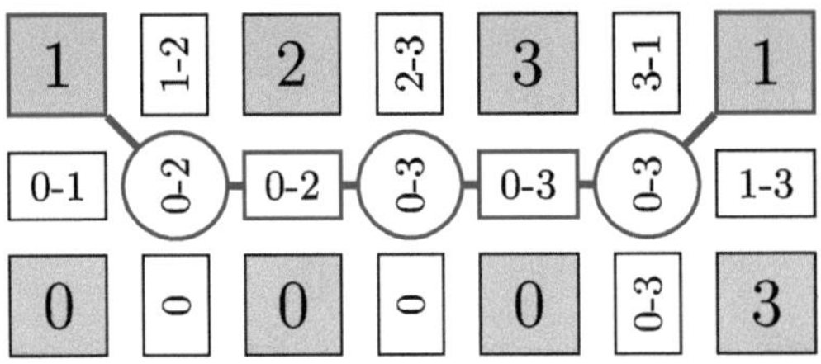

(b) Interval-valued map after immersion

Fig. 3. Illustration of the immersion and level set distances. The top-left and top-right "one" pixels are not connected in (a) but they belong to the same level set. In the interval-valued map (b), the level line "One" can be materialized and connects "physically" the two pixels "One": the distance between these two pixels is thus zero.

Propagation. The propagation depicted in Fig. 5 aims at sorting the pixels of the image. It processes the previously computed set-valued map F, starting from a root point p_∞ with an associated value v_∞. In [10], this step outputs a sorted array of pixels S while in [4], it outputs a mapping ORD that associates for each pixel p the wavefront step number (somewhat related to the depth of the pixel in the final ToS). The two algorithms are very similar and differ in the way the "sorting" is stored. In these algorithms, λ denotes the current wavefront level, and d is the wavefront counter. Each time the wavefront level is updated, d is incremented, and the current wavefront level is stored in the order map ORD. Both algorithms rely on a priority queue Q whose efficiency is crucial for the overall performance. The queue is implemented as a hierarchical queue, which is a set of queues $Q_0, Q_1, \ldots, Q_n$ where n is the number of levels in the image.

- PUSH(Q, v, p) appends a point p to the queue at level v
- POP(Q, v) retrieves the point p at the nearest level λ of v. It involves traversing the priority queue until a non-empty queue is found. The point p is then removed from the queue, and the level λ is updated to the index of the queue.

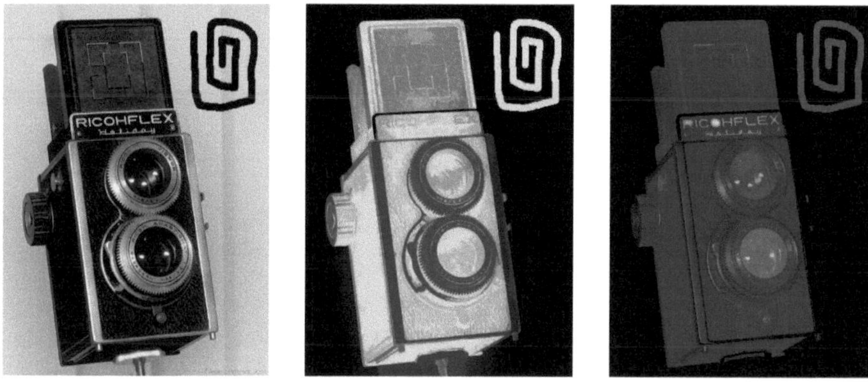

(a) (b) (c)

Fig. 4. The order map (b) computed on the camera image using Fig. 5 and the distance transform image (c) using alg. 3. The order map values range from 0 to 5700, while the distance transform values range from 0 to 600.

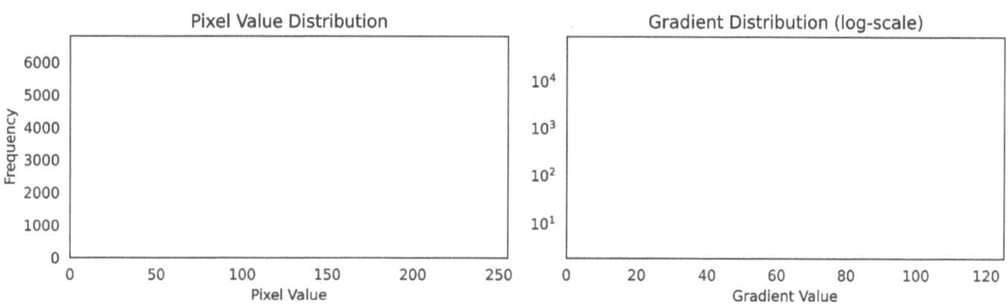

Fig. 5. Propagation algorithm from [10] (left) and order map algorithm from [4] (right). [10] outputs a pair $(\mathcal{F}, S)$ suitable for the Union-Find procedure while [4] outputs the order map ORD suitable for any max-tree algorithm.

The resulting order map ORD, computed during the propagation on the *camera* image, is illustrated in Fig. 4b, with values ranging from 0 to 5700.

Union-find Tree or Max-Tree Computation. In [10], the sorted array S is used to build the ToS by immersion using the Union-Find algorithm [1]. In [4], a max-tree is constructed from the order map using either a linear or quasi-linear algorithm, depending on the maximum value of the order map. After constructing the tree, post-processing steps are typically performed, such as "un-immersion" to return to the original domain and associating the nodes with their original grayscale values.

To adapt this process to H/XDR images, the challenging part is to adapt the propagation algorithm. Indeed, efficient max-tree algorithms have been proposed for High Dynamic Range images [13,21], but the propagation step consists of continuously traversing a hierarchical queue back and forth to order the pixels. One could obviously tackle this issue by using an ordered set instead of a hierar-

Algorithm 3: Our Distance Transform Algorithm

```
Function dt(F, p∞, v∞):
    Initialize Q
    D(p) ← +∞ for all p
    𝓕(p∞) ← v∞
    Push(Q, p∞, 0)
    while Q not empty :
        (p, d) ← Pop(Q)
        foreach q ∈ 𝒩(p) such that D(q) = +∞ do
            (m, M) ← F(q)
            𝓕(q) ← clamp 𝓕(p) in [m, M]
            d_pq ← |𝓕(p) − 𝓕(q)|
            Push(Q, q, d_pq)
            D(q) ← d + d_pq
    return F, D
```

chical queue, but this would not be efficient (all operations would be $O(\log n)$). In the next section, we propose a workaround to this issue by proposing an alternative to the order map for which efficient data structures exist and can handle high-dynamic range priorities.

3 Ordering with a Distance Transform

The order map computed during the propagation step of the ToS is not unique. Any image levelling (that preserves the ordering relationship between any pair of neighboring pixels) can be used as an order map. Here we propose to use the distance transform from the border on an interval-valued image. Instead of counting the number of level-lines (as done in Fig. 5), we instead sum the differences between the level-lines on the path. This computes the gray-weighted distance transform [11] on the interval-valued image F. More formally, let $\Pi(x)$ be the set of paths from the border to the pixel x in the image, the distance of x is defined as:

$$D(x) = \min_{\mathcal{F}, \pi \in \Pi(x)} \sum_{i=1}^{n-1} |\mathcal{F}(\pi_i) - \mathcal{F}(\pi_{i+1})|$$

where π_i is the i-th pixel of the path π and $\mathcal{F}$ is a projection of the interval-valued image F (a projection $\mathcal{F}$ of the interval-valued image F being any scalar image for which $\mathcal{F}(p) \in F(p)$). Note that in this distance, marching on the pixels of the same level set does not contribute to the distance as illustrated in Fig. 3 where the distance between the two pixels "One" is zero.

This distance could be implemented using an iterative algorithm [20] or the Fast Marching Method [18]. The first is more suited for parallelization but requires many passes over the images to converge.

The Fast Marching methods were initially designed to solve the Eikonal equations that model a wavefront propagation in a medium. In the context of image processing, the medium is a landscape where the gray levels are the elevations, and we aim at computing the distance from a set of source pixels (the image border) to all other pixels. The Fast Marching Method is essentially a Dijkstra-like algorithm for continuous domains. It works on a grid and maintains a priority queue of front points sorted by their current best-known distance. Algorithm 3 shows our adaptation of the algorithm to compute the distance. It outputs the distance map D and the level line map $\mathcal{F}$. $D(p)$ holds the distance to the border, and $\mathcal{F}(p)$ holds the value chosen in the interval $F(p)$ to reach this distance. It begins by initializing D where each point is assigned a tentative distance value: zero for source points (p_∞ or the image border) and infinity for all others. Source points are then added to the priority queue Q. The algorithm then iteratively pops the point with the smallest distance from the queue and enqueues the neighbors that are still *Far* ($D(q) = +\infty$).

There is nothing fancy in this algorithm, it is a classical distance transform very similar to Fig. 5. The only twist is that we do not push absolute distances into the heap, but distances relative to the current wavefront. Those distances are bounded as the difference between two level lines is bounded to 255 on 8-bit grayscale images. A priority queue can thus be implemented with a circular array of queues to store the pixels as in [9] and enables O(1)-amortized time queue operations. Q stores an active state d that is the current wavefront distance.

- PUSH(Q, p, δ) inserts p into the queue $(d+\delta) \bmod 256$, where δ is the relative distance to the current wavefront.
- POP$(Q) \rightarrow (p, d)$ iterates (circularly) over the queues until it finds a non-empty queue. The internal wavefront distance d is updated accordingly. It then pops the first pixel p from the active queue.

PUSH and POP operations are shown in alg. 4. The distance map computed with this algorithm is illustrated in Fig. 4 with values ranging from 0 to 600. One may wonder why it is preferable to use the distance transform instead of the order map. The answer is simple: *almost none* for low-dynamic range images (except for the fact the distance transform image has generally a lower dynamic and max-tree algorithms run faster on LDR images). However, the distance transform is suitable for high-dynamic range as we will see in the next section.

4 The Distance Transform on XDR Images

As seen in the previous section, there is a *minor* difference between the propagation algorithms (that compute the order map) and the distance transform algorithm: the first one uses a priority queue to enqueue the levels while the second enqueues the distances - actually, the differences between the distances. This *minor* difference is actually a major one when it comes to HDR images. Indeed, while grayscale levels could be uniformly distributed, the distribution

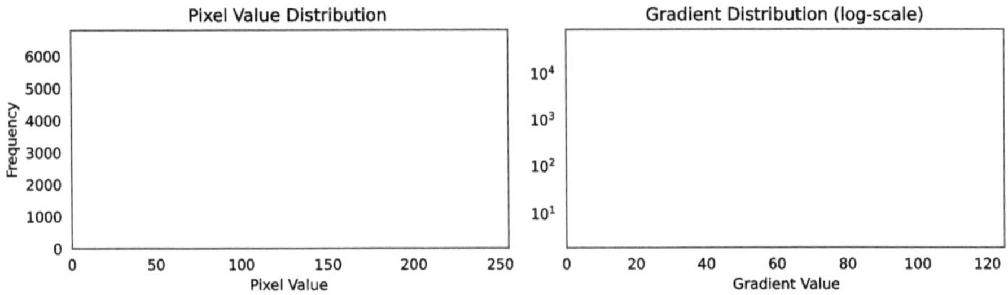

Fig. 6. Distribution of the grayscale values and distribution of gradients of the camera image. The histogram is not uniform and has a long tail: most edges have low gradient values.

of the gradients (i.e. distances between two level lines) is very unbalanced as illustrated in Fig. 6. This is a rather good property as many techniques exist to efficiently implement priority queues when priorities are high-quantized but not uniformly distributed. For instance, *bucket queues* are very efficient when the distribution of the priorities is known and can be pre-computed (this is a generalization of the hierarchical queues with non-uniform buckets). *Hybrid queues* enable maintaining a fast-access structure for high-priority items, and a slower structure for low-priority items, and *lazy queues* defer the update of the queue until necessary. Implementation usually combines many of these techniques to achieve a good performance. For instance, the Fast Marching Method has been optimized with approximated bucket queues [15,23]. More recently, in [16], the authors combine bucket queue, lazy queues, and a fast-access cache to compute efficiently the α-tree on extreme dynamic range pixel dissimilarities.

The implementation from [16] could probably be adapted to our problem, but contrary to [16], we need to handle *dynamic* updates of the queue. Indeed, in the α-tree algorithm, once the edges are pushed their priorities are fixed. In our case, the *relative* distances change as soon as the wavefront moves. To come up with this issue, we use a simple hybrid strategy that:

- stores the pixels in a fast `CircularBucketQueue` for nearby distances within a relative distance $d \leq Q_{th}$
- otherwise, fall backs to the `LazyQueue` that is a deferred-priority queue designed to optimize the trade-off between insertion and extraction costs. Rather than maintaining strict heap order with every insertion, elements are simply appended to a list, and the queue defers sorting until a `dequeue` operation is called. Bulk insertions in a heap with Floyd's algorithm have a better amortized complexity than individual insertions.

The *dequeue* operation of our hybrid queue selects the element with the smallest distance from two underlying queues. It compares the top distances from both queues, updates the global distance to the minimum of the two, and pops the corresponding element. If the selected element comes from the lazy queue, it synchronizes the circular queue's internal distance to ensure consistency for future operations. This hybrid design balances performance and flexibility across a wide range of priority distances. Pseudo-code for the two queues is shown in Fig. 7.

Algorithm 4: CircularBucketQueue

```
class CircularBucketQueue:
    queues ← array of 256 lists
    distance ← 0
    count ← 0
    Function push(p, dist):
        queues[dist mod 256].append(p)
        count ← count + 1
    Function top_distance():
        k ← distance mod 256
        d ← 0
        while queues[k] is empty :
            k ← (k + 1) mod 256
            d ← d + 1
        return distance +d
    Function pop():
        // Update the wavefront
           distance
        distance ← top_distance ()
        p ← queues[distance mod 256].pop()
        count ← count −1
        return (p, distance)
```

Algorithm 5: LazyQueue

```
class LazyQueue:
    data ← [ ]
    validItems ← 0
    minDistance ← +∞
    Function push(p, dist):
        if dist < minDistance :
            minDistance ← dist
        data.append((p, dist))
    Function top_distance():
        return minDistance
    Function pop():
        heapify(data[validItems...])
        (p, dist) ← heappop (data)
        validItems ← len (data)
        _, minDistance ← data [0]
        return (p, dist)
```

Fig. 7. Circular bucket queue (left) and lazy queue (right). The circular bucket queue is used for distances within 255 units, while the lazy queue is used for more distant ones.

5 Performance Comparison

In order to evaluate our algorithm, we conducted a performance comparison of the ToS computation based on the proposed distance transform and the one based on the order transform. To this aim, we used three different kinds of data: a set of 50 grayscale aerial pictures encoded on 8 bits whose size range from 2430×3500 to 3289×3500 from the database of the Netherlands Institute of Military History[1], a set of 5 images encoded on 16 bits from the Sentinel-2 database[2] from which the first channel has been extracted, and a set of 50 color images encoded on 8 bits whose values have been used as 24 bits values. The

[1] https://beeldbank.nimh.nl/.

[2] https://data.inpe.br/stac/browser/collections/S2_L2A-1.

CPU of the machine used for the experiments is an Intel Xeon Silver 4000 @ 2.20GHz. As a preliminary step for each computation, a border was added to the images, with its value sets to the median value of the image border. These images were then immersed into a set-valued map. Since these preliminary steps are identical and do not affect the comparison, their performance was not measured.

Table 1. Time distribution and total processing time of the ToS algorithm from [4] depending on the quantization.

Quantization	8 bits	16 bits	24 bits
Immersion	2%	<2%	<2%
Propagation	56%	56%	64%
Max-tree	42%	43%	34%
Proc. Time (MPix/s)	7.47 ± 0.55	1.84 ± 0.13	1.44 ± 0.50

Table 1 shows the performance of the ToS algorithm from [4] depending on the quantization. It can be seen that the *propagation* step is the most time-consuming one, accounting for 56% of the total computation time on 8-bit images, and this part is even more important on high-quantized images when the implementation requires red-black trees to keep the pixels in the front ordered. The (quasi-)linear ToS algorithm is no more linear for high-quantized images as the red-black tree requires logarithmic complexity for insertion and deletion operations. This is a major drawback of the ToS algorithm based on the order map, as it is not suitable for high-dynamic range images.

Table 2. Performance comparison of the ToS algorithm from [4] and the distance transform algorithm. The *hit rate* is the percentage of pixels that were processed using the circular bucket queue when using the hybrid queue.

Quantization	8 bits	16 bits	24 bits
Carlinet et al. [4]	$\mathbf{7.47} \pm 0.55$	1.84 ± 0.13	1.44 ± 0.50
CircularBucketQueue	7.07 ± 0.42	4.44 ± 0.25	N/A
HybridQueue	7.07 ± 0.42	$\mathbf{4.44} \pm 0.25$	$\mathbf{2.26} \pm 0.72$
hit rate	*100%*	*96.9%*	*57.54%*

This urges the need for a more efficient implementation of the propagation step. In table 2, we expose timings of our approach based on a distance transform that relies on a hybrid queue to handle efficiently low edges and postpones the update of the queue for high edges. They show that our implementation takes the best of the two worlds: it is almost as fast as the one from [4] on 8 bits, and it is significantly faster on 16 bits and 24 bits images (two to three times faster). Performance should probably be improved by using a more sophisticated hybrid queue as in [16]. To be short, any strategy that would allow to maximize the hit-rate of the circular bucket queue would be beneficial as it turns most logarithmic operations into constant ones. The hit-rate of the circular bucket queue is shown in table 2 and shows that the hybrid queue is fine on 16-bits images but rather bad on 24-bits images. An analysis of the distribution of the edges could be used to adapt the threshold Q_{th} to the image and would probably improve the hit-rate.

6 Conclusion

We have developed an algorithm based on the distance transform to compute the leveling from which the Tree of Shapes (ToS) is extracted. This method is particularly well-suited for high-dynamic range images, as it enables the use of optimized data structures to maintain pixels sorted according to their distance values, an advantage not feasible with traditional ordering propagation fronts. Our experimental results demonstrate that combining our distance transform computation with a max-tree computation outperforms the ToS computation algorithm based on the order map when computed on high-dynamic range images. However, the current approach still requires two passes over the image to construct the ToS. Future research will focus on exploring the simultaneous computation of the ToS and the distance map using the distance transform, aiming to achieve a more efficient single-pass algorithm.

References

1. Berger, C., et al.: Effective component tree computation with application to pattern recognition in astronomical imaging. In: 2007 IEEE international conference on image processing, vol. 4, pp. IV–41. IEEE (2007)
2. Bosilj, P., Kijak, E., Lefèvre, S.: Beyond MSER: maximally stable regions using tree of shapes. In: Procedings of the British Machine Vision Conference 2015, pp. 169.1–169.13. BMVC 2015, British Machine Vision Association (2015)
3. Bosilj, P., Kijak, E., Lefèvre, S.: Partition and inclusion hierarchies of images: a comprehensive survey. J. Imaging **4**(2), 33 (2018)
4. Carlinet, E., Crozet, S., Géraud, T.: The tree of shapes turned into a max-tree: a simple and efficient linear algorithm. In: 2018 25th IEEE International Conference on Image Processing (ICIP), pp. 1488–1492. IEEE (2018)
5. Carlinet, E., Géraud, T.: A comparative review of component tree computation algorithms. IEEE Trans. Image Process. **23**(9), 3885–3895 (2014)
6. Caselles, V., Monasse, P.: Geometric description of images as topographic maps. Springer, Berlin Heidelberg (2010)
7. Cavallaro, G., Mura, M.D., Benediktsson, J.A., Plaza, A.: Remote sensing image classification using attribute filters defined over the tree of shapes. IEEE Trans. Geosci. Remote Sens. **54**(7), 3899–3911 (2016)
8. Crozet, S., Géraud, T.: A first parallel algorithm to compute the morphological tree of shapes of nd images. In: 2014 IEEE International Conference on Image Processing (ICIP), pp. 2933–2937 (2014)

9. Dial, R.B.: Algorithm 360: shortest-path forest with topological ordering [h]. Commun. ACM **12**(11), 632–633 (1969)
10. Géraud, T., Carlinet, E., Crozet, S., Najman, L.: A Quasi-linear algorithm to compute the tree of shapes of nD images. In: Mathematical Morphology and Its Applications to Signal and Image Processing, pp. 98–110. Springer Berlin Heidelberg (2013)
11. Kimmel, R., Kiryati, N., Bruckstein, A.M.: Sub-pixel distance maps and weighted distance transforms. J. Math. Imaging Vis. **6**, 223–233 (1996)
12. Monasse, P., Guichard, F.: Fast computation of a contrast-invariant image representation. IEEE Trans. Image Process. **9**(5), 860–872 (2000)
13. Moschini, U., Meijster, A., Wilkinson, M.H.: A hybrid shared-memory parallel max-tree algorithm for extreme dynamic-range images. IEEE Trans. Pattern Anal. Mach. Intell. **40**(3), 513–526 (2017)
14. Ôn Vũ Ngọc, M., Boutry, N., Fabrizio, J., Géraud, T.: A minimum barrier distance for multivariate images with applications. Comput. Vis. Image Underst. **197–198**, 102993 (2020)
15. Rasch, C., Satzger, T.: Remarks on the $\mathcal{O}(n)$ implementation of the fast marching method. IMA J. Numer. Anal. **29**(3), 806–813 (2009)
16. Ryu, J., Trager, S.C., Wilkinson, M.H.: A fast alpha-tree algorithm for extreme dynamic range pixel dissimilarities. IEEE Trans. Pattern Anal. Mach. Intell. **46**(5), 3199–3212 (2023)
17. Salembier, P., Oliveras, A., Garrido, L.: Antiextensive connected operators for image and sequence processing. IEEE Trans. Image Process. **7**(4), 555–570 (1998)
18. Sethian, J.A.: A fast marching level set method for monotonically advancing fronts. Proc. Natl. Acad. Sci. **93**(4), 1591–1595 (1996)
19. Song, Y.: A Topdown algorithm for computation of level line trees. IEEE Trans. Image Process. **16**(8), 2107–2116 (2007)
20. Toivanen, P.J.: New geodosic distance transforms for gray-scale images. Pattern Recogn. Lett. **17**(5), 437–450 (1996)
21. Wilkinson, M.H.: A fast component-tree algorithm for high dynamic-range images and second generation connectivity. In: 2011 18th IEEE International Conference on Image Processing, pp. 1021–1024 (2011)
22. Xu, Y., Géraud, T., Najman, L.: Hierarchical image simplification and segmentation based on mumford-shah-salient level line selection. Pattern Recogn. Lett. **83**, 278–286 (2016)
23. Yatziv, L., Bartesaghi, A., Sapiro, G.: O(N) implementation of the fast marching algorithm. J. Comput. Phys. **212**(2), 393–399 (2006)

A Linear Time Algorithm for Local Minimum and Maximum Filters

Arnold Meijster(✉) and Mattia Marziali

Bernoulli Institute for Mathematics, Computer Science, and Artificial Intelligence, University of Groningen, P.O. Box 407, 9700 AK Groningen, The Netherlands
a.meijster@rug.nl

Abstract. We present a linear-time algorithm for computing sliding window maximum and minimum filters over one-dimensional signals. The algorithm is based on a single forward pass using a double-ended queue (deque) and requires only a small amount of memory. It supports in-place and streaming implementations. We provide a self-contained description of the algorithm, including illustrative examples, pseudocode, a practical implementation in C, and an analysis of its runtime behavior. We compare the performance of the algorithm against the well-known Van Herk–Gil–Werman (HGW) algorithm, which performs three scans over the data and offers data-independent performance.

Keywords: Max filter · Min filter · Sliding window · Linear-time algorithm

1 Introduction

In morphological image and signal processing, a common task is to determine local extrema (maximum or minimum values) within a fixed-size window as it slides across a one- or two-dimensional signal. In this paper, we consider the one-dimensional version of the problem: given a signal $x = \{x[0], x[1], \ldots, x[n-1]\}$ and a window of fixed size w, the goal is to compute the sliding window maximum signal $y = \{y[0], y[1], \ldots, y[n-1]\}$, where:

$$y[i] = \mathbf{Max}\,\{x[j] \mid \max(0, i - w + 1) < j \leq i\} \tag{1}$$

Although we focus on the one-dimensional case in this paper, efficient sliding window filters extend naturally to two-dimensional images. This is typically done by applying the one-dimensional operation separately along rows and columns, which corresponds to performing dilation (or erosion) with a rectangular structuring element.

The sliding window minimum signal is defined analogously, by replacing the **Max** operator with **Min**; in the remainder of this paper we restrict our discussion to the maximum case, though all results and algorithms are equally applicable to the minimum case. Note that we consider a *causal* definition of

M. H. F. Wilkinson and J. Kosinka (Eds.): DGMM 2025, LNCS 16296, pp. 315–327, 2026.
https://doi.org/10.1007/978-3-032-09544-2_23

these sliding window filters, where each output $y[i]$ depends only on values $x[j]$ with $j \leq i$. Non-causal definitions are also common in the literature; however, these definitions differ only by a fixed translation of the output and are otherwise equivalent.

Clearly, the naive approach which computes the maximum separately in each translated window, incurs a time complexity of $\mathcal{O}(nw)$, making it impractical for large w. In 1992, van Herk [2] published an $\mathcal{O}(n)$ algorithm for max and min filtering using a block-based decomposition; the method was independently rediscovered by Gil and Werman [4]. In the literature this algorithm is known as the Herk-Gil-Werman (HGW) algorithm. The HGW algorithm, which is widely used for dilation and erosion of images with rectangular structuring elements, is described in detail in Sect. 2. A key strength of the HGW algorithm is that its runtime performance is independent of the input data and window size. It accesses memory in a regular cache-friendly fashion, and avoids branching within loops which is highly beneficial for vectorization. The algorithm needs approximately three comparisons per output sample.

In several publications [1,3,5] improvements to the HGW algorithm have been proposed aimed at reducing the number of comparisons per output sample. However, these refinements are largely of theoretical interest. They often complicate the implementation and introduce branching and irregular memory access patterns, which degrade performance on modern hardware. As a result, despite a slightly higher theoretical comparison count, the original HGW algorithm with its simple, branch-free structure and regular memory access pattern remains the fastest in practice.

In Sect. 3, we introduce the *MM algorithm* (Meijster and Marziali), which uses a *double-ended queue (deque)* to perform a single pass over the data. The algorithm is similar to the approach described by Lemire [6], but represents a practical variation adapted for in-place causal processing. Although deque-based max filtering techniques are known in the algorithmic literature, they appear to have been largely overlooked by the mathematical morphology community.

Like HGW, the MM algorithm has the optimal $\mathcal{O}(n)$ time complexity. However, unlike the HGW algorithm, the practical runtime performance of the MM algorithm does depend on the input and window size. The MM algorithm requires less memory than the HGW algorithm. While the HGW algorithm requires at least two extra arrays of size n to store intermediate results, the MM algorithm only requires memory for the double-ended queue (deque). We will discuss the MM algorithm in Sect. 3 and give a detailed implementation in the C programming language in Sect. 4. We compare its runtime performance performance with the HGW algorithm in Sect. 5.

2 The Van Herk–Gil–Werman (HGW) Algorithm

For the sake of completeness and reproducibility, we present a full description of the HGW algorithm. The algorithm operates in linear time $\mathcal{O}(n)$ with respect to the number of input samples. We follow the original presentation from Van

Algorithm 1. HGW max filter algorithm with window size $w = 2r + 1$

Require: Input array $x[0 \dots n-1]$, radius r
Ensure: Output array $y[0 \dots n-1]$

```
w ← 2r + 1
Allocate arrays P[0...n − 1], S[0...n − 1]
                                          ▷ Prefix maxima (left to right)
for k ← 0 to n − 1 step w do
    P[k] ← x[k]
    for i ← k + 1 to min(k + w − 1, n − 1) do
        P[i] ← max(P[i − 1], x[i])
    end for
end for
                                          ▷ Suffix maxima (right to left)
for k ← n − 1 down to 0 step w do
    S[k] ← x[k]
    for i ← k − 1 down to max(k − w + 1, 0) do
        S[i] ← max(S[i + 1], x[i])
    end for
end for
                          ▷ Output computation with explicit boundaries
for i ← 0 to r − 1 do                                ▷ Left boundary
    y[i] ← max(S[0], P[i + r])
end for
for i ← r to n − r − 1 do                            ▷ Middle part
    y[i] ← max(S[i − r], P[i + r])
end for
for i ← n − r to n − 1 do                            ▷ Right boundary
    y[i] ← max(S[i − r], P[n − 1])
end for
```

Herk [2], which defines the filter using a centered, non-causal window with an odd size $w = 2r + 1$. The sliding maximum is then defined as

$$y[i] = \mathbf{Max}\left\{x[j] \mid i - r \leq j \leq i + r\right\}, \quad \textbf{for } 0 \leq i < n$$

Implicitly, it is assumed that $x[j] = -\infty$ for any $j \notin \{0, 1, \dots, n-1\}$, so that the window is always defined, even near the boundaries. The window is symmetric around the current index i and includes w consecutive elements. Causal variants, such as those defined in Sect. 1, are simple translations of this formulation and do not affect the core algorithmic properties.

The HGW algorithm divides the signal into non-overlapping segments of size w, aligned with multiples of w. It then precomputes local maxima in two passes (a forward and a backward direction pass):

- **Prefix maxima:** for each index i, the prefix maximum is defined as

$$P[i] = \mathbf{Max}\{x[j] \mid s \leq j \leq i\}$$

i	0	1	2	3	4	5	6	7	8	9	10	11	12	13	14
x	3	2	1	0	5	5	2	4	7	2	8	6	3	3	3
P	3	3	3	3	5	5	5	5	7	7	8	8	8	8	8
S	5	5	5	5	5	7	7	7	7	2	8	6	3	3	3
y	3	3	5	5	5	5	7	7	8	8	8	8	8	6	3

Fig. 1. HGW algorithm example with index i, input x, prefix maxima P, suffix maxima S, and output y, using window size $w = 5$. Dashed lines mark segment boundaries.

where s is the start of the segment containing i, i.e., the largest multiple of w less than or equal to i.

- **Suffix maxima:** for each index i, the suffix maximum is defined as

$$S[i] = \mathbf{Max}\{x[j] \mid i \leq j \leq e\}$$

where e is the end of the segment containing i, i.e., one less than the next multiple of w, or $n - 1$ if i is in the final segment.

The prefix and suffix maxima arrays are computed in two passes: one from left to right and one from right to left over the segments. For each index i, the final result is computed in a third pass as

$$y[i] = \max(S[i - r], P[i + r])$$

The values of S and P are assumed to be $-\infty$ for indexes that are out of bound.

Pseudocode of the HGW algorithm is provided in Algorithm 1. The beginning and end segments are handled explicitly so that no bounds checking of array indices is required inside the third pass.

Figure 1 illustrates the algorithm applied to an input sequence with $n = 15$ using a window of size $w = 5$ (hence $r = 2$). The rows show the indices i, the input x, the prefix maxima P, the suffix maxima S, and the final output y. Dashed vertical lines mark the segment boundaries.

To understand how the algorithm produces the correct result, consider index $i = 8$. The centered window of size 5 spans the values $x[6] = 2$, $x[7] = 4$, $x[8] = 7$, $x[9] = 2$, and $x[10] = 8$. The maximum in this window is clearly 8. According to the HGW formula, we compute

$$y[8] = \max(S[8 - 2], P[8 + 2]) = \max(S[6], P[10]) = \max(7, 8) = 8$$

where $S[6] = \mathbf{Max}\{x[6], x[7], x[8], x[9]\} = 7$, and $P[8] = \mathbf{Max}\{x[10]\} = 8$.

Similar reasoning applies at each index. Note that the index ranges of $S[i-r]$ and $P[i+r]$ are either disjoint (as in the case for $i = 8$) or fully overlap. For example, for $i = 7$, we find $P[7+2] = P[9] = \mathbf{Max}\{x[5], x[6], x[7], x[8], x[9]\} = S[5] =$

$S[7-2]$. This means that some of the comparisons performed during the first two passes of the algorithm are redundant due to overlap between the prefix and suffix regions. The variations of the HGW algorithm proposed in [1,3,5] exploit the overlap between the prefix and suffix regions to reduce the total number of comparisons per output location. We do not consider these improvements further, as they tend to complicate the algorithm and introduce conditional branches. In practice, such variants perform worse, since modern processors favor simple, branch-free, and regular control flow.

3 The MM Algorithm Using a Double-Ended Queue

In this section we describe the MM algorithm, which is based on a single forward pass over the data while maintaining a double-ended queue structure, a so-called *deque*, that stores data-index pairs.

Unlike the HGW algorithm, the number of comparisons per output sample depends on the data. The algorithm is a causal and in-place variation of an earlier algorithm proposed by Lemire [6], which appears to have been largely overlooked by the mathematical morphology community. Because Lemire's version stores only indices in the deque, it requires double indirect memory addressing (e.g., accessing values via $x[deque[i]]$); the MM algorithm instead stores value–span pairs directly, avoiding indirect memory accesses and enabling in-place computation.

Unlike the HGW algorithm, the MM algorithm does not require the window size w to be odd (i.e., centered). We revert to the causal version of the filter, as defined in the introduction. That is, for each output index i, we compute the maximum over the past w samples. The definition is:

$$y[i] = \mathbf{Max}\{x[k] \mid i - w + 1 \leq k \leq i\}$$

where values $x[k]$ for $k < 0$ are assumed to be $-\infty$.

The MM algorithm consists of a single loop over the signal data, using an index i that ranges from 0 to $n-1$. The algorithm maintains a double-ended queue (deque) that stores pairs $(x[j], s_j)$, where $x[j]$ is a value from the input signal and $s_j = j + w$ is its *span*. The span indicates the smallest index at which $x[j]$ is no longer included in any window and can be safely discarded.

Initially, the queue is empty. At each iteration i, the algorithm performs the following operations:

- Remove all elements from the back of the queue whose data value $x[j]$ is smaller than or equal to the current value $x[i]$. These values are dominated by $x[i]$ and will never become the maximum in a future window.
- Append the new pair $(x[i], i + w)$ to the back of the queue.
- If the front element has span equal to i, it is removed, as it has just expired and no longer contributes to the current or future outputs.
- The value at the front of the queue is the maximum for the current window and is copied to the output array $y[i]$.

Algorithm 2. MM max filter with window size w using a double-ended queue

```
Require: Input array x[0...n-1], window size w
Ensure: Output array y[0...n-1]
 1: Create empty deque Q
 2: for i ← 0 to n-1 do
 3:     while Q ≠ ∅ and Q.back.value ≤ x[i] do    ▷ Short-circuit boolean evaluation
 4:         Q.pop_back()
 5:     end while
 6:     Append (x[i], i + w) to Q
 7:     if Q.front.span = i then                   ▷ Queue is non-empty here
 8:         Q.pop_front()
 9:     end if
10:     y[i] ← Q.front.value
11: end for
```

The pseudocode is shown in Algorithm 2. The queue is guaranteed to be non-empty before checking for expiration (lines 7–9), since the current value has just been appended. In the expiration check, we test whether the span of the front element equals the current index i; if so, it is removed. At most one element can expire at each step, so there is never more than one element to remove from the front of the queue. Note that the boundary output values $y[0], \ldots, y[w-1]$ are computed using smaller windows. This is not a problem and avoids any additional conditional tests.

The MM algorithm runs in linear time $\mathcal{O}(n)$. This can be seen by analyzing the total number of operations on the queue. Each element $x[i]$ is inserted into the queue exactly once. It can only be removed in one of two ways: either from the front when it expires (i.e., its span ends), or from the back when it is dominated by a new, larger value. Since no element is ever reinserted, and every removal corresponds to a unique earlier insertion, the total number of queue operations is at most $2n$. This guarantees that the entire algorithm completes in linear time, regardless of the input data.

Figure 2 illustrates the MM algorithm applied to a 15-element input with window size $w = 5$. The figure is divided into panels, each summarizing the state after one or more iterations. In the first panel (iterations 0–3), the queue grows by inserting elements from left to right. Since the subsequence $x[0..3]$ is decreasing, no elements expire and they are all stored in decreasing order. The maximum for each prefix is the first inserted value $x[0] = 3$, and this is reflected in the output values $y[0..3] = 3$.

In iteration 4, the value $x[4] = 5$ enters the queue. All previous values are removed from the back because they are smaller, and only the pair $(5, 9)$ remains.

In iterations 5–6, new values are appended while preserving order. No elements have expired yet. At iteration 7, $x[7] = 4$ is added, and no removals are needed since it is less than the current maximum.

In iterations 8–9, a new maximum $x[8] = 7$ is inserted. Previous values are removed from the back because they are smaller.

Iterations 0–3	Iteration 4	Iterations 5–6
$x[0..3] = [3, 2, 1, 0]$	$x[4] = [5]$	$x[5..6] = [5, 2]$
$Q = [(3, 5), (2, 6), (1, 7), (0, 8)]$	$Q = [(5, 9)]$	$Q = [(5, 10), (2, 11)]$
$y[0..3] = [3, 3, 3, 3]$	$y[4] = [5]$	$y[5..6] = [5, 5]$

Iteration 7	Iterations 8–9	Iterations 10–14
$x[7] = [4]$	$x[8..9] = [7, 2]$	$x[10..14] = [8, 6, 3, 3, 3]$
$Q = [(5, 10), (4, 12)]$	$Q = [(7, 13), (2, 14)]$	$Q = [(8, 15), (6, 16), (3, 19)]$
$y[7] = [5]$	$y[8..9] = [7, 7]$	$y[10..14] = [8, 8, 8, 8, 8]$

Fig. 2. Each panel summarizes one or more iterations of the MM algorithm with window size $w = 5$, applied to the input signal $x = [3,\ 2,\ 1,\ 0,\ 5,\ 5,\ 2,\ 4,\ 7,\ 2,\ 8,\ 6,\ 3,\ 3,\ 3]$. The figure shows the state of the double-ended queue and the corresponding output values $y[i]$ after those iterations.

Finally, in the last panel (iterations 10–14), all elements are removed from the queue since they are dominated by $x[10] = 8$, and the new values 6 and 3 form a fresh descending pattern in the queue. The output stabilizes at the maximum 8, which was inserted at iteration 10.

The MM algorithm naturally supports streaming applications. It can be implemented in place (i.e., writing the output to the same array as the input). In-place execution is possible because all values required to compute each output sample $y[i]$ are retained in the queue; the corresponding input location $x[i]$ can thus be safely reused for storing the output.

4 A Concrete Implementation in C

In this section, we provide a concrete implementation of the MM algorithm in the C programming language. The code is intended to be both efficient and portable. It is suitable for both in-place and out-of-place processing.

The double-ended queue is split into two parallel arrays: `deque` for storing the signal values and `span` for storing the corresponding expiration indices. Instead of combining these into an array of pairs (a C struct), we deliberately keep them separate. This avoids unnecessary indirect addressing relative to the start of a structure. Separate arrays allow more efficient structured memory accesses in an array in which consecutive elements have the same data type.

The double-ended queue is managed using two index variables: `front` and `back`. These define a contiguous range of valid entries in the arrays. The element at position `front` is the current maximum candidate, and `back` marks the position where the next element will be inserted at the back end of the queue. Because the actual data type of the signal samples is irrelevant, we introduce a type alias `sampleType`, which can be defined as `int`, `float`, `double`, etc., allowing the same implementation to be reused across different kinds of signals.

```c
typedef int sampleType;

void maxFilter1D(int w, int n, sampleType *x, sampleType *y) {
   sampleType *deque = (sampleType *)malloc(n*sizeof(sampleType));
   int *span = (int *)malloc(n*sizeof(int));
   int front = 0, back = 0;
   for (int i = 0; i < n; i++) {
      // Remove dominated values from the back
      while ((back > front) && (x[i] >= deque[back - 1])) {
         back--;
      }
      // Insert new value and its expiration index
      deque[back] = x[i];
      span[back] = i + w;
      back++;
      // Has maximum expired?
      if (span[front] == i) {
         front++; // remove expired maximum from the front
      }
      // Output the maximum
      y[i] = deque[front];
   }
   free(span);
   free(deque);
}
```

Listing 1.1. C implementation of the MM algorithm

The implementation in C, which is shown in Listing 1.1, is suitable for in-place execution: the user may call the function as `maxFilter1D(w, n, x, x);` or for out-of-place execution using a separate output array: `maxFilter1D(w, n, x, y);` The implementation of the algorithm remains unchanged in both cases; only the call differs.

The `deque` and `span` arrays are allocated with size `n`, the full length of the input signal. However, the number of active elements in the queue never exceeds the window size `w`. To reduce memory usage, these arrays can be implemented as circular buffers. These buffers can have any size greater than `w`. When using circular buffers, queue indices such as `front` and `back` must be wrapped around using the modulus operation. Modulo operations are generally expensive, but when the modulus is a power of two, they can be efficiently implemented using a bitwise AND operation. Therefore, we choose a size that is the smallest power of two greater than `w`. Let `sz` be this size and define `mask = sz - 1`. The modulus operation `index % sz` can then be replaced by a bitwise operation `index & mask`, which is much more efficient. For example, incrementing the front index (`front++`) becomes `front = (front + 1) & mask`. To implement `back--`, we avoid negative values by using the identity $(a-1) \bmod b \equiv (a+b-1) \bmod b$.

This gives `back = (back + mask) & mask`, which is equivalent to decrementing `back` modulo the queue size, without using subtraction that might underflow. These rules ensure that indices wrap around correctly using only fast bitwise operations, without introducing conditional statements and extra branching. In Listing 1.2 an implementation is shown with circular buffers, using bitwise operations to perform moduli efficiently.

```
void maxFilter1D(int w, int n, sampleType *x, sampleType *y) {
  // Determine queue size as smallest power of two > w
  int sz = 1;
  while (sz <= w) {
      sz *= 2;
  }
  int mask = sz - 1;
  // allocate circular buffers
  sampleType *deque = (sampleType *)malloc(sz*sizeof(sampleType));
  int *span = (int *)malloc(sz*sizeof(int));
  int front = 0, back = 0;
  for (int i = 0; i < n; i++) {
    // Remove dominated values from the back
    while ((back != front) && (x[i] >= deque[(back+mask)&mask])) {
       back = (back + mask)&mask;
    }
    // Insert new value and its expiration index
    deque[back] = x[i];
    span[back] = i + w;
    back = (back+1)&mask;
    // Has maximum expired?
    if (span[front] == i) {
      front = (front + 1) & mask; // remove expired maximum
    }
    // Output the maximum
    y[i] = deque[front];
  }
  free(span);
  free(deque);
}
```

Listing 1.2. Max filter with circular buffer implementation.

5 Performance Comparison

In this section we compare the runtime performance of the MM algorithm (with and without circular buffer optimization, denoted MM and MMC, respectively) with the existing algorithms by van Herk and Lemire.

In Sect. 2 we have discussed the HGW algorithm as originally presented by van Herk (in [2]). This version requires three complete passes over the data. As such, it is inherently non-causal and therefore unsuitable for streaming applications. However, it was brought to our attention by one of the anonymous reviewers that the HGW algorithm, since it operates on blocks of size w, can easily be adapted into a streaming variant. To compute the result for an index i, only the block containing i and the subsequent block are needed, which allows processing to be organized on a per-block basis. Each block is processed by performing a forward and a backward pass, yielding a streaming implementation of the HGW algorithm with a latency of $2w$. An additional advantage of this block-based formulation is that it improves cache utilization, greatly improving the runtime of the algorithm. We thank the anonymous reviewer for pointing out this observation.

In this section, we use HGW to denote the original algorithm (as presented in Sect. 2 and in [2]), and HBL to denote the streaming block-based variant suggested by the reviewer. Moreover, we use LM to denote the algorithm by Lemire, and LMC for a circular-queue variant of LM.

When comparing linear-time algorithms, hardware-specific factors such as CPU architecture, cache behavior, and memory access patterns play a major role in determining runtime. These factors influence the constant factors hidden in the $\mathcal{O}(n)$ complexity and may result in significant differences in practical performance. While absolute runtimes may vary across systems, we expect that the observed relative performance trends remain representative.

All experiments were conducted on a first-generation Intel Core i7 processor running at 3 GHz, which is by now rather old; this choice was deliberate, since newer generation Intel processors support AVX extensions that the compiler exploits during optimization, thereby making algorithmic comparison difficult. All experiments were compiled using the `clang` compiler from the LLVM suite with optimization level `-O3`. Each configuration was tested for 500 runs, and average timings were recorded.

We measured the runtime performance of the algorithms on randomly generated one-dimensional signals of type `int` and `double`. The signal lengths were set to $n = 10^5, 10^6, 10^7$, and window sizes $w = 5, 50, 500, 5000$. The results are summarized in Table 1. All experiments presented in this section were performed using out-of-place computation, where the output is written to a separate array. However, we did not observe any significant difference in timings when using in-place computation.

As expected from their theoretical complexity, all algorithms scale (nearly) linearly with the signal length n. In practice, the HGW algorithm is difficult to surpass on `int` data, owing to its highly regular memory access pattern. This effect is amplified in the block-wise formulation (HBL), which achieves superior cache utilization, and overall, HBL consistently outperforms all other algorithms in our comparison. In practice, this implies that HBL is the preferred algorithm for streaming processing, provided that the user is willing to accept a delay of $2w$ samples.

Table 1. Execution time in milliseconds (avg. of 500 runs) for random input.

n	$w = 5$			$w = 50$			$w = 500$			$w = 5000$		
	10^5	10^6	10^7	10^5	10^6	10^7	10^5	10^6	10^7	10^5	10^6	10^7
`int` input												
HGW	0.7	9.2	85.7	0.8	9.6	96.2	0.8	10.5	103.9	0.8	10.4	105.4
HBL	0.5	4.8	48.3	0.4	3.6	36.2	0.4	3.6	37.6	0.3	3.6	37.9
MM	0.8	9.3	97.1	0.8	9.9	107.1	0.8	9.9	105.6	0.8	10.1	106.2
MMC	0.9	9.0	93.6	0.9	9.7	92.9	0.8	8.7	89.5	0.8	8.7	90.3
LM	1.0	10.8	128.3	1.1	11.1	131.3	1.1	10.8	128.7	1.1	10.8	128.1
LMC	1.0	10.4	116.4	1.2	11.9	133.0	1.2	11.8	131.1	1.2	11.7	130.7
`double` input												
HGW	0.8	9.9	175.5	1.0	12.0	194.3	1.2	13.2	208.6	1.2	13.1	210.1
HBL	0.6	5.9	58.5	0.6	6.3	63.6	0.6	6.6	65.7	0.6	6.7	67.7
MM	0.9	9.8	117.2	1.1	11.9	119.5	1.0	11.1	114.6	1.0	11.0	113.0
MMC	0.9	10.2	106.6	1.0	10.4	106.3	0.9	9.2	94.6	0.9	9.3	95.3
LM	1.1	11.4	138.6	1.2	12.4	143.0	1.2	12.1	140.3	1.2	12.1	139.8
LMC	1.4	13.5	144.6	1.5	14.8	156.7	1.5	14.9	158.1	1.5	15.0	157.8

We now turn to the scenario where the user is not willing to accept the delay of $2w$ samples inherent in HBL. In this case, the choice is restricted to the remaining algorithms. Our experiments show that MM and MMC consistently outperform their counterparts LM and LMC. The reason is that the LM algorithms store indexes in the queue rather than the data values themselves. As a consequence, every comparison requires an additional level of indirection: first retrieving an index from the queue, and then using it to access the corresponding data value. We therefore proceed with a detailed performance comparison between HGW and the MM variants.

For window sizes $w \leq 50$, HGW performs slightly better than MM and MMC, but only by a small margin. For larger window sizes, HGW and MM achieve almost identical performance, while in these cases the circular buffer of MMC begins to pay off. This effect can be attributed to improved memory locality: although circular buffer indexing requires a few additional instructions, it confines all elements to a compact, fixed memory region and thereby reduces cache misses compared to the queue traversal in MM. In summary, HGW is preferable for small window sizes, while for larger windows MMC may be better choice.

For floating-point data, HBL again emerges as the absolute winner across all configurations. However, if the delay of $2w$ samples is prohibitive, the choice again lies among the other algorithms. In this setting, the `max` operation on `double` values is not vectorized, and comparisons are executed sequentially. As a result, the performance advantage of HGW disappears and MM-based methods become more competitive. Somewhat surprisingly, the linear scaling behavior of HGW also seems to be lost, and it is not exactly clear why this happens. In contrast, HBL does maintain its linear scaling. Both MM and MMC now consistently outperform HGW, with MMC providing a slight additional gain over MM due to its improved memory locality.

To evaluate performance on real-world data, we repeated the same set of experiments on one-dimensional signals extracted from standard grayscale images. Specifically, we used scanlines from the widely used `cameraman` and `peppers` images. Pixel values were converted to `double` to match the floating-point setup. For longer signals, scanlines were concatenated and repeated as needed to reach the desired lengths. The results of this experiment are nearly identical to the results of the random data setup. We observed that the MM algorithm performed only slightly worse on image data than on random data. This can be explained by the fact that, for random input, roughly half of the samples tend to dominate earlier samples, leading to frequent removals and maintaining a small and localized queue. For image-derived signals, local correlations reduce the number of removals, causing the queue to wander more across memory. This reduces the locality of memory references and can slightly degrade performance. This slight performance degradation was not observed for the MMC algorithm, because this algorithm maintains its queue within a fixed, compact memory region, preserving good locality of memory references.

We also benchmarked the algorithms on a set of real-world int-valued 2D grayscale images of size 6000×4000, using window sizes $w = 5$ and $w = 50$. The 2D operation was performed by first applying the 1D max filter horizontally (row-wise) to the input image, followed by a vertical (column-wise) 1D max filter applied to the intermediate result, effectively implementing a dilation with a rectangular structuring element. The vertical pass was observed to be four to five times slower than the horizontal pass, due to strided memory access causing more cache misses. Otherwise, the timing results were consistent with those obtained on 1D data.

6 Conclusion

In this paper we introduced the MM algorithm for sliding window minimum and maximum filters together with its circular-buffer variant MMC. Both algorithms are variations on an algorithm proposed by Lemire in [6]. We compared these algorithms primarily with the standard HGW algorithm, which is widely used as a reference for efficient sliding-window minimum and maximum filtering. The HGW algorithm in its original form is not streaming, but it can be turned into a streaming variation (HBL) that operates with a delay of $2w$ samples, where w is the window size. By contrast, the other algorithms are streaming by design and incur only a delay of w samples.

From a performance perspective, HBL stands out as the overall winner whenever a delay of $2w$ samples is acceptable. The modification of HGW into HBL makes the algorithm highly cache-friendly and extremely efficient in practice.

Performance measurements further show that the runtime behavior of all algorithms is largely insensitive to the statistical properties of the input data. Results obtained on random signals closely match those obtained on real-world image data.

All algorithms considered in this study run in linear time with respect to the signal length, and their time complexity is independent of the window size w. Furthermore, the MMC algorithm demonstrates that sliding window minimum and maximum filters can be computed using only auxiliary memory proportional to the window size.

An advantage of the MM variants is that they can be used both for in-place processing, where the output overwrites the input, and for out-of-place processing. This flexibility makes them attractive in memory-constrained environments.

The algorithms that are based on a deque all share the property that no padding is required when the signal length is not a multiple of the window size w. Moreover, for these algorithms the parity of w (even or odd) is irrelevant, which simplifies their application in practice.

Finally, although deque-based algorithms cannot surpass the performance of HBL, we believe it is important to draw renewed attention to this technique within the mathematical morphology community. In particular, the paper by Lemire appears to have been largely overlooked. We anticipate that the underlying method may prove useful for other morphological operations involving minima and maxima.

References

1. Gil, J.Y., Kimmel, R.: Efficient dilation, erosion, opening, and closing algorithms. IEEE Trans. Pattern Anal. Mach. Intell. **24**(12), 1606–1617 (2002)
2. van Herk, M.: A fast algorithm for local minimum and maximum filters on rectangular and octagonal kernels. Pattern Recogn. Lett. **13**(7), 517–521 (1992)
3. Yuan, H., Atallah, M.J.: Running max/min filters using 1+o(1) comparisons per sample. IEEE Trans. Pattern Anal. Mach. Intell. **33**(12), 2544–2548 (2011)
4. Gil, J., Werman, M.: Computing 2-D min, median, and max filters. IEEE Trans. Pattern Anal. Mach. Intell. **15**(5), 504–507 (1993)
5. Gevorkian, D.Z., Astola, J.T., Atourian, S.M.: Improving Gil-Werman algorithm for running min and max filters. IEEE Trans. Pattern Anal. Mach. Intell. **19**(5), 526–529 (1997)
6. Lemire, D.: Streaming maximum-minimum filter using no more than three comparisons per element. Nordic J. Comput. **13**(4), 328–339 (2006)

Waterfall-Borvka-Based Algorithm for Binary Partition Tree

Quentin Lebon(✉), Jean Cousty, Thierry Grandpierre, and Benjamin Perret

LIGM, CNRS, Univ Gustave Eiffel, F77454 Marne-la-Vallée, France
quentin.lebon@esiee.fr

Abstract. This article presents an algorithm to construct the binary partition tree by altitude ordering based on Borvka's minimum spanning tree strategy. The binary partition tree by altitude ordering is a data structure to represent an image in the form of a hierarchy of segmentations that is widely used in mathematical morphology processing. Recently, the authors of PANDORA [21] have proposed a novel parallel algorithm for computing single linkage clustering from the minimum spanning tree of a point cloud on a GPU. In this paper, we show that their method can indeed be nicely cast in the framework of watershed cuts and waterfalls for computing the binary partition tree by altitude ordering. More precisely, the method consists in performing a sequence of watershed cuts-basins contractions, seen as a variant of Borvka's algorithm with a complexity of $O(n\log(n))$, followed by a dedicated post-processing. Furthermore, we show that this algorithm can be extended to process any edge-weighted graphs and not only minimum-spanning trees. These results open a new path towards massively parallel algorithms for hierarchical watershed algorithms.

Keywords: Binary partition tree · Watershed-cut · Waterfall Hierarchy · Borvka's algorithm · Minimum spanning tree

1 Introduction

Hierarchies of segmentations are powerful representations that are used in a wide variety of computer vision and data analysis problems. In this context, the binary partition tree by altitude ordering (BPT), obtained by sequentially merging the vertices of a graph by considering its edges in increasing weight order (see 1a), is a fundamental representation that can be used in various morphological (hierarchical) segmentation methods [10] such as quasi-flat zone hierarchies (also known as α-trees) [15,18] and hierarchical watersheds [8,13]. The BPT can also be used to perform hierarchy simplification [19], combination of hierarchies [9,20], and continuous optimization of hierarchical clustering [5].

The BPT is in bijection with the minimum spanning tree (MST) of the underlying graph and can be computed efficiently with a variation of Kruskal's MST algorithm [16].However, Kruskal's algorithm is intrinsically a sequential approach, processing the edge of the graph in increasing order. Each edge depends

M. H. F. Wilkinson and J. Kosinka (Eds.): DGMM 2025, LNCS 16296, pp. 328–340, 2026.
https://doi.org/10.1007/978-3-032-09544-2_24

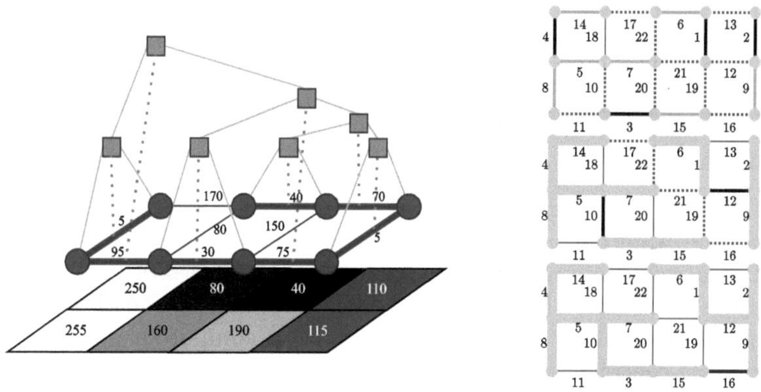

(a) Binary Partition Tree

(b) Waterfall

Fig. 1. (a) Illustration of a BPT constructed over a grayscale image (bottom), its associated adjacency graph (in red), and the BPT (in green). The dotted blue lines depict the bijection between non-leaf nodes and edges of the MST (in bold). (b) Illustration of a three-level waterfall hierarchy. At each level, the diagram displays inner edges (black), border edges (green), and separating edges – solid blue if creating a cycle in the MSF, dotted blue otherwise. The MSF within each watershed basin is shown in bold grey. (Color figure online)

on the order the labels of others. This has shown to be difficult to parallelize efficiently, in particular when the dynamics of the edge weights is large [3, 11].

In [21], the authors recently proposed a novel parallel algorithm to compute single linkage clustering from the MST of a point cloud on GPU. Their approach consists in sequentially contracting the MST and then deducing the hierarchical clustering from the history of contractions. The contractions themselves are performed in a way that guarantees that the subtree in each contracted region has a comb shape (each internal node has at least one leaf child) and that there are in the order of $\log(n)$ contractions, with n the number of vertices.

In this article, we show that the method presented in [21] can be nicely cast in the framework of watershed cuts and waterfalls to compute the binary partition tree by altitude ordering. More precisely, when all edge weights are distinct, we show that each contraction step of their method is indeed a watershed cuts-basins contraction [6] and that the whole sequence of contractions is similar to a waterfall [2, 14]. The whole algorithm can indeed be interpreted as a variant of Borvka's MST algorithm with a runtime complexity of $O(n \times \log(n))$ which leads to our second main contribution: we show that this algorithm can be extended to process any edge weighted graphs and not only minimum-spanning trees.

2 Waterfall, Watershed Cuts and Borvka's Algorithm

In [10], the authors established constructive links between various hierarchies used in mathematical morphology, including BPTs, min-trees, and hierarchical watersheds. In this article, we establish a new constructive link: we show how the BPT of an edge-weighted graph can be obtained by post-processing the waterfall tree of this graph. Such link, given in the form of an algorithm, is presented in Sect. 3.3. In this section, we present the notion of a waterfall and its relation to minimum spanning trees and, in particular, to Borvka's MST algorithm.

Waterfall. From a high-level perspective, given an edge-weighted graph (V, E, w), the waterfall hierarchy (see Fig. 1b), and its tree-based representation called a waterfall tree [2], can be seen as the series of partitions of the vertex space V obtained by iterating the following two steps as long as the initial graph is not contracted to a single vertex: (i) compute a watershed cut of the graph; (ii) contract each catchment basin into a single vertex.

Watershed. Intuitively, a watershed cut [7] of an edge-weighted graph is a cut in the graph such that each connected component induced by that cut, called a *(catchment) basin*, contains exactly one regional minimum of the graph and such that each edge in the cut satisfies the "drop of water principle". An edge in the cut satisfies the drop of water principle if there exist two descending paths (not crossing the cut) that end in two distinct regional minima. It is known from [7] that the watershed cuts of an edge-weighted graph can be equivalently characterized as the cuts induced by the minimum spanning forests rooted in the minima of that graph.

For our purpose, it is important to recall a result stated in [6] that establishes a characterization of watershed cuts from a weight-map transform called a border thinning. This transform relies on a topological classification of the graph edges (see Fig. 1b). In order to introduce this classification, let us call *altitude or weight* of a vertex x, denoted by $w_\bullet(x)$, the minimum weight among all the edges containing the vertex x. Every edge $u = \{x, y\}$ of the graph is either:

- a *border edge* if its weight is greater than the weight of one of its extremities and is equal to the weight of the other, *i.e.*, if we have $w(u) > \min\{w_\bullet(x), w_\bullet(y)\}$ and $w(u) = \max\{w_\bullet(x), w_\bullet(y)\}$; or
- an *inner edge* if its weight is equal to the weight of its two extremities, *i.e.*, if we have $w(u) = w_\bullet(x) = w_\bullet(y)$; or
- a *separating edge* if its weight is greater than the weight of its two extremities, *i.e.*, if we have $w(u) > \max\{w_\bullet(x), w_\bullet(y)\}$.

The *border thinning transform* consists in iteratively lowering the weight of a border edge down to the weight of its lowest extremity, as long as the graph is not free of border edges. The effect of this transform is to progressively expand the regional minima until they form a minimum spanning forest rooted in the minima of the initial graph. As a result of Theorem 17 in [6], the set of edges linking two regional minima of the transformed weight map is a watershed cut of the initial edge-weighted graph whose (catchment) basins are the minima of the transformed map. It can be proven that if any two distinct edges of the graph have distinct weights, then (i) every inner edge one-to-one corresponds to a regional minimum, and (ii) the graph of the minima of the border thinning is given by set comprising every inner edge and every border edge of the graph.

Hence, in this case, the catchment basins can be obtained simply by performing a connected component labeling of the graph of inner and border edges. This strategy will be used to compute watershed cuts in the next section. We furthermore observe that the total number of border and inner edges is equal to $|V| - k$, where k is the number of minima (as the border and inner edges define a forest rooted in the minima). Since every minimum contains at least two vertices, k is at most equal to $|V|/2$ which implies that the number of detected inner and border edges is at least $|V|/2$ and that the number of iterations in a waterfall process is less than $\log_2(|V|)$.

Important Remark. In the remainder of this article, we assume that any two distinct edges of the graph have distinct weights. Thus, the edges of the graph are totally ordered by weight (or altitude).

Borvka's MST Algorithm. Starting from an empty edge set T, this algorithm [4,17] consists in iterating Borvka steps. A Borvka step consists of the following calculations:

1. For each vertex x of the graph, find the edge of minimum weight containing this vertex and add it to a set C;
2. contract the graph along each edge of C;
3. add every edge of C into T.

The algorithm terminates when the contracted graph is made of a single vertex: the set T then defines a MST of the initial edge-weighted graph. It can be seen that the edges selected in first step to form the set C, correspond exactly to the set of all inner edges and of all border edges. From this observation, we deduce the following important propositions:

- Each Borvka step computes a minimum spanning forest rooted in the regional minima of the graph;
- each contracted vertex resulting from Borvka step is a *catchment basin* of its input edge-weighted graph.

As a consequence, the hierarchy given by the successive contractions in Borvka's algorithms is exactly the waterfall hierarchy of the input edge-weighted graph.

3 Binary Partition Tree Construction Algorithm

In this section, we present an algorithm for computing the binary partition tree by altitude ordering based on Borvka's algorithm and the waterfall tree. This algorithm is an adaptation and a generalization of the method proposed in [21].

3.1 Overview of the Method

The workflow of the algorithm is shown in Fig. 2. While following the same pattern as the method presented in [21], it highlights its relation to watershed and waterfall computations, and thus to Borvka's MST algorithm. Additionally,

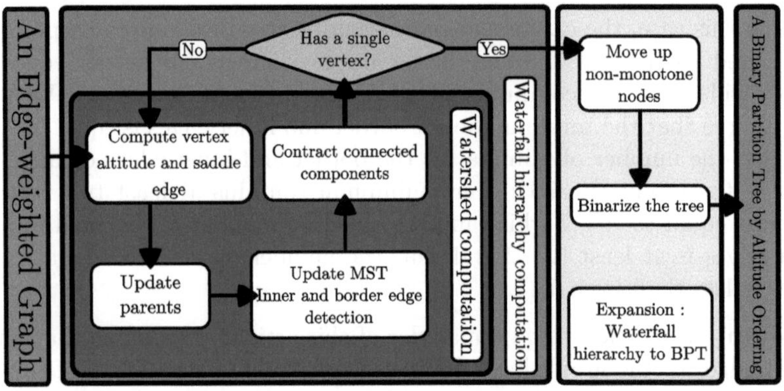

Fig. 2. Workflow of the algorithm.

it extends the approach to process any edge-weighted graph, rather than being limited to a (minimum spanning) tree.

The workflow consists of two main parts: the first one computes the waterfall tree of the input graph while the second one, called the expansion stage, transforms it into a BPT.

As described in the previous section, the waterfall stage iterates watershed computation and basin contraction until reaching a graph made of a single vertex. At each iteration, a tree structure is updated to memorize the hierarchical structure of the waterfall. Each iteration proceeds as follows:

1. Compute the altitude of each vertex/basin in the graph, along with its corresponding saddle edge – *i.e.*, the edge with the minimum weight connecting the vertex/basin to a different one;
2. update the waterfall tree by setting the parents of every node/basin to its saddle edge;
3. build the watershed cut by searching the inner and border edges and adding them to form the new basins, hence expanding the current minimum spanning forest;
4. label the connected components induced by the watershed cut, thus contracting each catchment basin.

When the first stage is completed, we obtain the waterfall tree and the MST of the input graph. Then, in the expansion stage, the waterfall tree is post-processed to produce the BPT with the following two steps:

1. When building the waterfall tree, some nodes may be assigned to a parent with a lower altitude than themselves, we call them *non-monotone nodes.* These nodes correspond to edges that belong to a basin of lower altitude (*e.g.*, a maximum of altitude k included in a basin whose saddle edge has an altitude less than k). The first step of the expansion phase then moves these non-monotone nodes up, such that the altitude strictly increases along any path from a leaf to the root in the tree.

2. So far, the nodes corresponding to several edges appearing in the same basin have been grouped as siblings in the tree, resulting in a non-binary tree (but with a lower depth than a binary tree). The final post-processing step enforces the binary constraint, ensuring that each internal node has exactly two children. This is achieved by reordering and reconnecting the nodes associated with edges in each basin according to altitude, ultimately resulting in a properly structured BPT that is totally ordered by altitude.

3.2 Data Structure

In this section, we describe the input and output of the method, along with the auxiliary data structures required for computing the MST and the BPT.

Input Graph(V, E, w). The vertices (resp. edges) of the graph are identified by integers between 0 and $n = |V|$ (resp. $m = |E|$). The extremities of each edge are stored in an array `E` of length m: `E`$[u] = \{x, y\}$ indicates that the extremities of the edge (of index) u are the vertices (of index) x and y. The weight of each edge is stored in an array `w`: `w`$[u]$ is then the weight of u.

Output Hierarchy$\mathcal{T}$. An important feature of the BPT is that its leaves are in bijection with the vertices of the MST and its internal nodes are in bijection with the edges of the MST. To track this relation, each node of $\mathcal{T}$ is identified with the index of its corresponding vertex or edge in the graph. Since vertices and edges are both indexed from 0, to avoid confusion we represent $\mathcal{T}$ with two arrays: the array `vPar`, of size n, stores the parent of each leaf node (*i.e.*, each vertex) and the array `ePar`, of size m, stores the parent of each internal node (*i.e.*, each edge). Thus, `vPar`$[x]$ gives the parent of the leaf mapped to vertex x, while `ePar`$[u]$ gives the parent of the internal node mapped to edge u.

Output Minimum SpanningTree.`MST`. It is an array of size m, where each entry corresponds to an edge and can take one of the following three values: (i) `in`: the edge is part of the MST; (ii) `undef`: it is not yet known whether the edge is part of the MST or not; and (iii) `out`: the edge is not part of the MST.

In addition, we use four auxiliary data structures.

Component Labels. The array `C` of size n stores vertex labels. The labels are consecutive integers such that any two vertices in the same component have the same label and any two vertices in distinct components have distinct labels

Vertex Altitude. The array `w`$_\bullet$ is of size equal to the number of distinct components in the current waterfall iteration. Each component of label ℓ is a vertex of the contracted graph and the value `w`$_\bullet$ $[\ell]$ is the altitude of this component, *i.e.*, the altitude of the vertex ℓ in the contracted graph.

Paired Saddles. The array `saddles` has the same size as `w`$_\bullet$. For each component label, it stores the index of the edge that determines its altitude, *i.e.*, the edge with the least weight among those linking this component to another.

Basins. The array `B` of size m stores, for each edge, the identifier of the catchment basin in which the edge first appeared during the waterfall iterations. This array is used during the post-processing step.

Algorithm 1: WATERFALL-BORVKA BINARY PARTITION TREE

Data: An edge-weighted graph (V, E, w)
Result: The Binary Partition Tree $\mathcal{T}$ of (V, E, w); and its minimum spanning tree MST

Initialize the parent arrays $\mathcal{T}$.vPar and $\mathcal{T}$.ePar to undef for every element
Initialize two arrays MST and B of size $|E|$ to undef for every element
nB := 0 // set the initial number of catchment basins to 0
Initialize an array C such that C[x] = x for every x in V
w•, saddles := BasinAltitudes-Saddles(MST, C) // Algorithm 2
$\mathcal{T}$.vPar := saddles // set the parents of the tree leaves
while *the number of distinct labels in* C *is greater than* 1 **do**
    MST := InnerBorderDetection(MST, C, w•) // Algorithm 4
    C := ConnectedComponentsLabeling(MST)
    w•, saddles := BasinAltitudes-Saddles(MST, C) // Algorithm 2
    B, $\mathcal{T}$:= UpdateParents(MST, C, B, $\mathcal{T}$) // Algorithm 3
$\mathcal{T}$:= MoveUpNonMonotoneNodes(B, $\mathcal{T}$) // Algorithm 5
$\mathcal{T}$:= BinarizeTree(B, $\mathcal{T}$) // Algorithm 6

3.3 Algorithms

Algorithm 1 gives a pseudocode description of the workflow described in Fig. 2. The subfunctions BasinAltitudes-Saddles, UpdateParents, UpdateMST, MoveUpNonMonotoneNodes and BinarizeTree are presented in Algorithms 2, 3, 4, 5, and 6, respectively.

Algorithm 2: BASIN ALTITUDES AND SADDLES

Data: an array MST tagging the MST edges; a connected component labeling C of the graph induced by the edges already in MST
Result: The vertex/basin altitude map w•; and the array saddles pairing a saddle edge to each vertex/basin

Initialize the array w• such that w•[ℓ] = ∞ for each label appearing in C
foreach *edge* u *from* 0 *to* $|E|$ **do**
    $\{x, y\}$:= E[u]
    c_x := C[x], c_y := C[y]
    if MST[u] = undef *and* $c_x \neq c_y$ **then**
        if w•[c_x] > w[u] **then** w•[c_x] := w[u]; saddles[c_x] := u
        if w•[c_y] > w[u] **then** w•[c_y] := w[u]; saddles[c_y] := u

The implementation of the waterfall iterations follows the principles given in Sect. 3.1, using the datastructures presented in Sect. 3.2. However, for better understanding, two observations need to be highlighted.

First, the graph resulting from the contraction of the basins is never explicitly constructed. Instead, at each waterfall step, we perform a labeling of the

Algorithm 3: UPDATE PARENTS

Data: An array `MST` for tagging MST edges; a connected component labeling `C` of the graph induced by the edges already in `MST`; an array `B` that stores the first basin of each edge; a partial waterfall tree $\mathcal{T}$

Result: An updated version of `B` and of $\mathcal{T}$

foreach *edge* u *from* 0 *to* $|E|-1$ **do**
    $\{x, y\} :=$ `E`$[u]$
    if `MST`$[u]$ = **`in`** *and* `B`$[u]$ = **`undef`** **then**
        $c_x :=$ `C`$[x]$
        `B`$[u] := c_x +$ **`nB`**
        if *the number of distinct labels in* `C` *is greater than* 1 **then**
            $\mathcal{T}$**`.ePar`**$[u] :=$ **`saddles`**$[c_x]$

`nB` $:=$ **`nB`** $+ n$, where n is the number of distinct labels in `C`

Algorithm 4: UPDATE MST - INNER AND BORDER EDGES DETECTION

Data: An array `MST` for tagging MST edges; a connected component labeling `C` of the graph induced by the edges such that `MST`$[u]$ = **`in`**; the altitude map $\mathtt{w}_\bullet$ of the connected components

Result: An updated array `MST`

foreach *edge* u *from* 0 *to* $|E|-1$ **do**
    if `MST`$[u]$ = **`undef`** **then**
        $\{x, y\} := E[u]$; $c_x :=$ `C`$[x]$; $c_y :=$ `C`$[y]$
        if $c_x = c_y$ **then** `MST`$[u] :=$ **`out`** **else if** $\mathtt{w}[u] \leq \mathtt{w}_\bullet[c_x]$ *or* $\mathtt{w}[u] \leq \mathtt{w}_\bullet[c_y]$ **then**
            `// u is either an inner edge or a border edge`
            `MST`$[u] :=$ **`in`**

vertices of the graph into basins, where each basin is a connected component of the MST under construction (*i.e.*, the set comprising each edge u such that `MST`$[u]$= **`in`**). After this labeling, the contracted graph corresponds to the graph in which each basin label becomes a vertex and each edge $\{x, y\}$ in the original graph is replaced by the edge $\{c_x, c_y\}$, where c_x and c_y are the basin labels of x and y, respectively. This correspondence is used in Algorithms 2, 3 and 4. The connected component labeling is computed at line 9 of Algorithm 1. It can be obtained by any standard algorithm, such as those based on graph search or on Tarjan's union-find datastructure. The former achieves a linear-time complexity with respect to the size of the graph per labeling step while the latter offers overall amortized quasi-linear time complexity at the cost of more complex data structures. In the end, the choice between these approaches does not affect the overall complexity of Algorithm 1.

Second, it can also be observed that the first waterfall iteration (lines 4 to 6) is implemented outside the main loop (lines 7 to 11) in Algorithm 1. Indeed, at this point, no graph contraction has been performed yet implying to deal

Algorithm 5: MOVE UP NON-MONOTONE NODES

Data: An array B that stores the first basin of each edge; a partial waterfall tree $\mathcal{T}$

Result: An updated ordered partial waterfall tree $\mathcal{T}$; an updated array B

foreach *edge* u *from* 0 *to* $|E| - 1$ **do**
  node $:= u$
  while $\mathcal{T}$.ePar[*node*] *is not the root and* w[u] > w[$\mathcal{T}$.ePar[*node*]] **do**
    node $:= \mathcal{T}$.ePar[*node*]
  $\mathcal{T}$.ePar[u] $:= \mathcal{T}$.ePar[*node*]
  B[u] := B[*node*]

Algorithm 6: BINARIZE TREE

Data: An array B that stores the first basin of each edge; a partial waterfall tree $\mathcal{T}$

Result: The full BPT $\mathcal{T}$

1 $\mathcal{L} :=$ Sorted edges in lexicographic order according to B first and to w secondly
2 **foreach** i *from* 0 *to* $|V| - 2$ **do**
3   $u := \mathcal{L}[i]$; $u' := \mathcal{L}[i+1]$
4   **if** B[u] = B[u'] **then** $\mathcal{T}$.ePar[u] $:= u'$

directly with the indices of the vertices. Furthermore, the associated tree nodes are leaves and their parents are stored in vPar rather than ePar.

The expansion stage also follows the principles described in Sect. 3.1. Notably, the binarization of the tree (Algorithm 6) implies the construction of a parent-child relationship between the nodes associated to edges appearing in the same basin. Indeed, in the waterfall tree, these nodes are siblings, all pointing to the basin's saddle edge, whereas, in the BPT, they must be ordered by altitude in a binary structure, with the highest one pointing towards the saddle edge. To get this order, we perform a lexicographic sort at line 1 of Algorithm 6. The lexicographic ordering is done first on the index of the basins and then on the edge altitudes. As a result, all edges appearing in the same basin are located consecutively in the sorted array, and, for each basin, they are ordered by altitude. For basins with multiple edges, we set the parent of each edge as the next edge in the sorted array if it belongs to the same basin. Thus, the highest edge of the basin will always keep pointing to the saddle edge of the basin.

Complexity Analysis. Algorithms 2, 3, and 4, which form the core of the waterfall tree construction, browse the edges in E and execute a constant number of constant-time instructions for each edge. Therefore, each of these algorithms runs in linear time with respect to the number of edges in E. Furthermore, the labeling of the connected components, which is also part of the core of the waterfall iterations, can be done in linear time with respect to the size of the graph. Since the number of waterfall iterations is bounded by $\log_2(|V|)$, we deduce that the overall time complexity of the waterfall stage is linearithmic (*i.e.*, $O(n \log n)$) with respect to the size of the graph. Let us now analyze the complexity of the expansion stage. Algorithm 5 consists of climbing up the tree from each internal node. Since the depth of the tree is equal to the number of waterfall iterations, each climb involves visiting at most $\log_2(|V|)$ nodes. Each visit consists of a few constant time instructions. Thus, the overall complexity of Algorithm 5 is linearithmic with respect to the size of the graph. Finally, the complexity of Algorithm 6 is dominated by the complexity of the sorting step, which also results in a linearithmic complexity with respect to the number of edges. Thus, the overall complexity of the Waterfall-Borvka binary partition tree algorithm is linearithmic with respect to the size of the input graph (V, E).

Execution Example. Figure 3 and 4 illustrate the construction of the BPT through a step-by-step execution of Algorithm 1. Starting from the graph of Fig. 3 *(top-left)*, three waterfall iterations are performed, followed by an expansion phase. At each iteration, we present the current state of the graph in Fig. 3 and of the tree structure in Fig. 4, in both graphical and array format. The *initial state* corresponds to the state of the structures and arrays before any computation takes place. *Iteration one* represents the execution of lines 5 to 9 in Algorithm 1, while the following iterations correspond to the loop iterations. After the third one, we present the results of the *move-up* step (Algorithm 5) and of the *binarization* step (Algorithm 6), which is the final BPT.

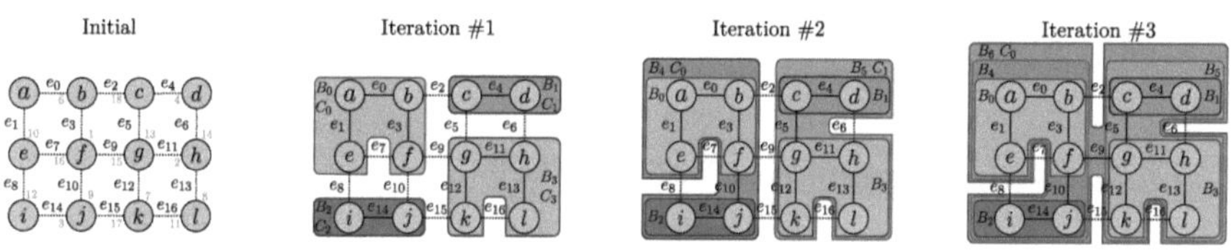

	Initial				Iteration #1				Iteration #2				Iteration t#3			
Vertex	w•	saddles	vPar	C	w•	saddles	vPar	C	w•	saddles	vPar	C	w•	saddles	vPar	C
a	∞	–	–	a	6	e_0	e_0	C_0	–	–	e_0	C_0	–	–	e_0	C_0
b	∞	–	–	b	1	e_3	e_3	C_0	–	–	e_3	C_0	–	–	e_3	C_0
c	∞	–	–	c	4	e_4	e_4	C_1	–	–	e_4	C_1	–	–	e_4	C_0
d	∞	–	–	d	4	e_4	e_4	C_1	–	–	e_4	C_1	–	–	e_4	C_0
e	∞	–	–	e	10	e_1	e_1	C_0	–		e_1	C_0	–	–	e_1	C_0
f	∞	–	–	f	1	e_3	e_3	C_0	–	–	e_3	C_0	–	–	e_3	C_0
g	∞	–	–	g	2	e_{11}	e_{11}	C_3	–	–	e_{11}	C_1	–	–	e_{11}	C_0
h	∞	–	–	h	2	e_{11}	e_{11}	C_3	–	–	e_{11}	C_1	–	–	e_{11}	C_0
i	∞	–	–	i	3	e_{14}	e_{14}	C_2	–	–	e_{14}	C_0	–		e_{14}	C_0
j	∞	–	–	j	3	e_{14}	c_{14}	C_2	–	–	e_{14}	C_0	–	–	e_{14}	C_0
k	∞	–	–	k	7	e_{12}	e_{12}	C_3	–	–	e_{12}	C_1	–	–	e_{12}	C_0
l	∞	–	–	l	8	e_{13}	e_{13}	C_3	–	–	e_{13}	C_1	–	–	e_{13}	C_0
B_0					∞	–	–	–	9	e_{10}	–	–	–	–	–	–
B_1					∞	–	–	–	13	e_5	–	–	–	–	–	–
B_2					∞	–	–	–	9	e_{10}	–	–	–	–	–	–
B_3					∞	–	–	–	13	e_5	–	–	–	–	–	–
B_4									∞	–	–	–	15	e_9	–	–
B_5									∞	–	–	–	15	e_9	–	–
B_6													∞	–	–	–

Fig. 3. Hierarchy and arrays related to vertices across all steps of the method applied to the graph depicted in top left (*Initial*).

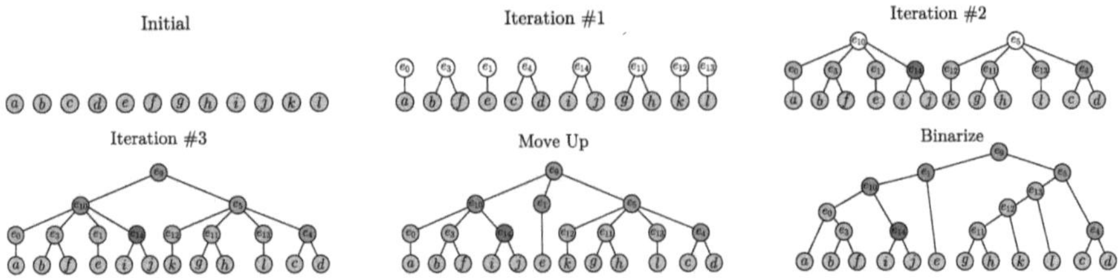

		Initial			Iteration #1			Iteration #2			Iteration #3			Move Up		Binarize
Edge	w	ePar	B	MST	ePar	B	MST	ePar	B	MST	ePar	B	MST	B	ePar	ePar
e_0	6	–	–	–	–	–	in	e_{10}	B_0	in	e_{10}	B_0	in	B_0	e_{10}	e_{10}
e_1	10	–	–	–	–	–	in	e_{10}	B_0	in	e_{10}	B_0	in	B_4	e_9	e_9
e_2	18	–	–	–	–	–	–	–	–	–	–	–	–	–	–	–
e_3	1	–	–	–	–	–	in	e_{10}	B_0	in	e_{10}	B_0	in	B_0	e_{10}	e_0
e_4	4	–	–	–	–	–	in	e_5	B_1	in	e_5	B_1	in	B_1	e_5	e_5
e_5	13	–	–	–	–	–	–	–	–	in	e_9	B_5	in	B_5	e_9	e_9
e_6	14	–	–	–	–	–	–	–	–	–	–	–	out	–	–	–
e_7	16	–	–	–	–	–	–	–	–	out	–	–	out	–	–	–
e_8	12	–	–	–	–	–	–	–	–	–	–	–	out	–	–	–
e_9	15	–	–	–	–	–	–	–	–	–	–	B_6	in	B_6	–	–
e_{10}	9	–	–	–	–	–	–	–	–	in	e_9	B_4	in	B_4	e_9	e_1
e_{11}	2	–	–	–	–	–	in	e_5	B_3	in	e_5	B_3	in	B_3	e_5	e_{12}
e_{12}	7	–	–	–	–	–	in	e_5	B_3	in	e_5	B_3	in	B_3	e_5	e_{13}
e_{13}	8	–	–	–	–	–	in	e_5	B_3	in	e_5	B_3	in	B_3	e_5	e_5
e_{14}	3	–	–	–	–	–	in	e_{10}	B_2	in	e_{10}	B_2	in	B_2	e_{10}	e_{10}
e_{15}	17	–	–	–	–	–	–	–	–	–	–	–	–	–	–	–
e_{16}	11	–	–	–	–	–	–	–	–	out	–	–	out	–	–	–

Fig. 4. Hierarchy and arrays related to edges across all steps of the method applied to the graph depicted in Fig. 3.

4 Discussions and Conclusion

In this work, we have revisited the method proposed in [21] for computing single linkage clustering from a MST, and shown that it admits a natural formulation within the framework of mathematical morphology. By recasting this approach as a sequence of watershed cuts, we construct the waterfall tree through a process similar to Borvka's MST algorithm. This new formulation seamlessly extends to general edge-weighted graphs, broadening its applicability beyond MSTs.Furthermore, we demonstrated how the waterfall tree can be transformed into a binary partition tree by altitude ordering.

The overall complexity of the BPT construction is optimal at $O(n \log n)$, matching that of Kruskal-based methods [16], while being naturally suited for parallel execution [21]. Most steps in the pipeline involve simple array traversals without inter-iteration dependencies, enabling straightforward parallelization. An exception is the required sorting step, for which efficient parallel sorting algorithms–such as radix sort [12], have been extensively studied and can be leveraged in this context. The connected component labeling step can also be addressed using parallel approaches, such as lock-free union-find algorithms [1]. Importantly, the logarithmic bound on contraction steps ($\log_2(n)$) ensures that both the waterfall construction and subsequent processing steps remain efficient even under large edge weight dynamics, which remained challenging in related parallel algorithms [3].

As future work, we aim to develop a fully parallel implementation of this framework on GPU architectures. This will enable scalable computation of hierarchical watersheds and related morphological operators.

References

1. Anderson, R.J., Woll, H.: Wait-free parallel algorithms for the union-find problem. In: ACM symposium on Theory of computing, pp. 370–380 (1991)
2. Beucher, S.: Watershed, hierarchical segmentation and waterfall algorithm. In: ISMM, p. 69–76 (1994)
3. Blin, N., Carlinet, E., Lemaitre, F., Lacassagne, L., Géraud, T.: Max-tree computation on gpus. IEEE TPDS **33**(12), 3520–3531 (2022)
4. Borvka, O.: O jistém problému minimálníÂŋm. Práce Moravské PíÂŋrodovdecké Spolenosti **3**, 37–58 (1926)
5. Chierchia, G., Perret, B.: . NeurIPS **32** (2019)
6. Cousty, J., Bertrand, G., Najman, L., Couprie, M.: Watershed cuts: Thinnings, shortest path forests, and topological watersheds. TPAMI **32**(5), 925–939 (2010)
7. Cousty, J., Bertrand, G., Najman, L., Couprie, M.: Watershed cuts: minimum spanning forests and the drop of water principle. TPAMI **31**(8), 1362–1374 (2008)
8. Cousty, J., Najman, L.: Incremental algorithm for hierarchical minimum spanning forests and saliency of watershed cuts. In: ISMM, pp. 272–283 (2011)
9. Cousty, J., Najman, L., Kenmochi, Y., Guimarães, S.: Hierarchical segmentations with graphs: quasi-flat zones, minimum spanning trees, and saliency maps. JMIV **60**(4), 479–502 (2018)
10. Cousty, J., Najman, L., Perret, B.: Constructive links between some morphological hierarchies on edge-weighted graphs. In: ISMM, pp. 86–97 (2013)
11. Havel, J., Merciol, F., Lefèvre, S.: Efficient tree construction for multiscale image representation and processing. J. Real-Time Image Proc. **16**, 1129–1146 (2019)
12. Merrill, D., Grimshaw, A.: High performance and scalable radix sorting: a case study of implementing dynamic parallelism for GPU computing. Parallel Proc. Lett. **21**(02), 245–272 (2011)
13. Meyer, F.: The dynamics of minima and contours. ISMM pp. 329–336 (1996)
14. Meyer, F.: The waterfall hierarchy on weighted graphs. In: ISMM, pp. 325–336 (2015)
15. Meyer, F., Maragos, P.: Morphological scale-space representation with levelings. In: International Conference on Scale-Space Theories in Computer Vision, pp. 187–198 (1999)
16. Najman, L., Cousty, J., Perret, B.: Playing with kruskal: algorithms for morphological trees in edge-weighted graphs. In: ISMM, pp. 135–146 (2013)
17. Nešetřil, J., Milková, E., Nešetřilová, H.: Otakar borvka on minimum spanning tree problem translation of both the 1926 papers, comments, history. Discret. Math. **233**(1–3), 3–36 (2001)
18. Ouzounis, G.K., et al.: The alpha-tree algorithm. JRC Scientific and Policy Report (2012)
19. Perret, B., Cousty, J., Guimarães, S.J.F., Kenmochi, Y., Najman, L.: Removing non-significant regions in hierarchical clustering and segmentation. PRL **128**, 433–439 (2019)

20. Santana Maia, D., de Albuquerque Araujo, A., Cousty, J., Najman, L., Perret, B., Talbot, H.: Evaluation of combinations of watershed hierarchies. In: ISMM, pp. 133–145 (2017)
21. Sao, P., Prokopenko, A., Lebrun-Grandie, D.: Pandora: a parallel dendrogram construction algorithm for single linkage clustering on GPU. In: ICPP, p. 908–918. ACM (2024)

Learning Based Morphology

Improving Morphological Networks for Learning Image-to-Image Transforms

Antoine Bottenmuller[1(✉)], Guillaume Tochon[2], Romain Hermary[3], Élodie Puybareau[2], and Gustavo Angulo[4]

[1] Center for Mathematical Morphology (CMM), Mines Paris, PSL University, Fontainebleau, France
antoine.bottenmuller@minesparis.psl.eu
[2] EPITA Research Laboratory (LRE), EPITA, Le Kremlin-Bicêtre, France
{guillaume.tochon,elodie.puybareau}@epita.fr
[3] Computer Vision, Imaging and Machine Intelligence Research Group (CVI2), University of Luxembourg, Luxembourg City, Luxembourg
romain.hermary@uni.lu
[4] Center for Applied Mathematics (CMA), Mines Paris, PSL University, Sophia Antipolis, France
jesus.angulo@minesparis.psl.eu

Abstract. Replacing convolution with morphological operations in trainable layers has received significant attention lately. Among the various strategies that have emerged, smooth morphological layers have shown strong potential and flexibility, as a single layer can behave either like a (pseudo-)erosion or a (pseudo-)dilation depending on the sign and value of its trainable control parameter. In this work, we build upon the so-called $\mathcal{S}$Morph layer by introducing a harmonized formulation that addresses previously identified asymptotic limitations when learning grayscale erosion and dilation. We also investigate and compare two strategies (a novel penalty term in the training loss and shared-weight layers) to improve the learning of grayscale opening and closing operations in two-layer networks. Finally, we evaluate the performance of this improved $\mathcal{S}$Morph layer on a salt-and-pepper denoising task in a four-layer network architecture, and compare it with other morphological and convolutional networks.

Keywords: mathematical morphology · morphological network · smooth morphological layer · grayscale transform · salt-and-pepper denoising

1 Introduction

Over the past decade, convolutional neural networks (CNNs) have revolutionized numerous image processing tasks such as object recognition and semantic segmentation. By learning representative multiscale filter banks through

M. H. F. Wilkinson and J. Kosinka (Eds.): DGMM 2025, LNCS 16296, pp. 343–356, 2026.
https://doi.org/10.1007/978-3-032-09544-2_25

trainable convolutional layers, CNNs capture both low-level and high-level features, enabling them to effectively capture complex visual patterns. As universal approximators (provided that the depth of the network is large enough [24]), they offer a powerful framework for approximating a broad class of image operators in a data-driven way. Yet, convolutional layers also suffer from well-known limitations. As a matter of fact, they remain limited in terms of interpretability and they often struggle with small or low-contrast objects. In contrast, morphological operations are particularly well-suited to process such challenging features and offer geometrically meaningful and interpretable operations. This motivates the replacement of standard convolution layers with trainable morphological operations, leading to the development of morphological neural networks (MNNs). These networks thus enable the automatic learning of morphological operations sequences along with their optimal structuring elements. Beside, the Matheron-Maragos-Bannon-Barrera (MMBB) representation theory of non-linear operators [2,12,14] establishes theoretical guarantees regarding the decomposition of any (translation invariant) image operator as a composition of morphological operations. While this property plays a similar role to the universal approximation theory for neural networks, no general construction algorithm exists to derive in practice such decomposition for a given image operator. Thus, MNNs also offer an appealing data-driven approach to uncover such representations.

However, erosion and dilation operations (as well as their compositions) rely on pointwise computations of minimum and maximum over local neighborhoods, which are not differentiable everywhere, unlike standard convolution. A first workaround to this issue explored the use of smooth approximations of these operations. Leveraging the counter-harmonic mean [1], the so-called P-convolution (*PConv*) layer was proposed in [13] and was shown to effectively learn target morphological operations such as erosion, dilation, opening, closing and top-hat. Other smooth approximations of min and max were also explored, including the LogSumExp in [18,19] or the α-softmax which led to the definition of the $\mathcal{S}$Morph layer in [6,8]. This latter was notably shown to outperform the *PConv* layer in learning grayscale morphological operation. A key advantage of these smooth formulations is that their asymptotic behavior is driven by the value of a trainable control parameter. Each layer can act either as an erosion when the parameter tends to $-\infty$, or as a dilation when it tends to $+\infty$. As almost-everywhere differentiable layers such as MaxPooling or ReLU became standard components in neural network architectures, exact morphological layers were proposed as an alternative to smooth approximations by adapting the backpropagation step [5,15,16,20]. This enabled the design of morphological networks performing exact morphological operations, though at the cost of reduced flexibility in learning the optimal sequence of erosions and dilations from data. However, morphological networks, whether based on exact or smooth layers, are still challenging to train and prone to convergence issues [6], which restricts their actual practical applications. Recent studies have therefore investigated how these difficulties relate to the differentiability of the layers and to their initialization strategies [3,4].

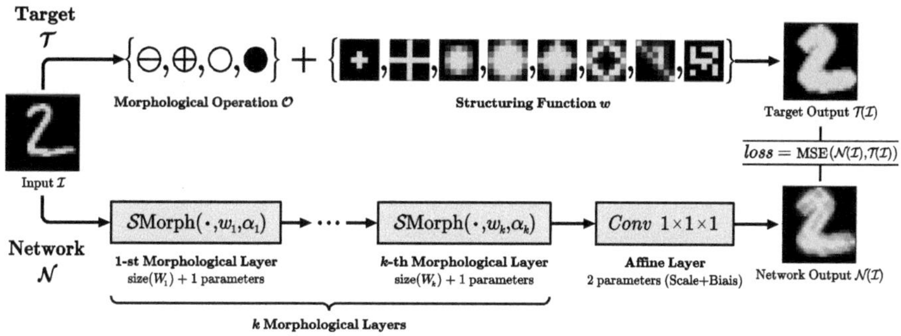

Fig. 1. Architecture of a k-layer morphological network, adapted from [6]. Upper path describes the *target morphological transform* $\mathcal{T} = \mathcal{O}_w$; lower path the *network* $\mathcal{N}$ itself.

In this work, we extend the $\mathcal{S}$Morph layer proposed in [6,8]. Specifically, we introduce a reformulated version that resolves the asymptotic shortcomings of $\mathcal{S}$Morph, and we propose constraint-based strategies that significantly improve convergence when learning opening and closing operations in two-layer networks. Our contributions are organized as follows. In Sect. 2, we recall the definition of the $\mathcal{S}$Morph layer and introduce a novel harmonized formulation, referred to as $\mathcal{HS}$Morph. In Sect. 3, we examine the behavior of $\mathcal{HS}$Morph layers in two-layer networks for learning opening and closing operations. We also propose two architectural improvements, namely a novel penalty term in the training loss and shared-weight layers, to improve network convergence. In Sect. 4, we evaluate the performance of $\mathcal{HS}$Morph in four-layer networks for salt-and-pepper denoising tasks and compare it with other morphological layers. Finally, Sect. 5 concludes the paper and outlines future research directions.

2 Smooth Morphological Layers

Smooth layers are designed as fully differentiable approximations of true morphological operations. These layers rely on smooth approximations of the min and max functions, such as the counter-harmonic mean, LogSumExp, or α-softmax, with each formulation determining the asymptotic behavior of the layer. The behavior of each approximation, and thus the layer itself, depends on a control parameter. Its sign determines which function is approximated (min when negative, with the layer acting like a (pseudo-)erosion, and max when positive, with the layer acting like a (pseudo-)dilation), while its magnitude governs the accuracy of the approximation (the larger in absolute value, the closer the approximation is to the true operation). Smooth layers can then be integrated into neural network architectures in the same way as traditional convolutional layers. Figure 1 illustrates such a morphological network composed of k $\mathcal{S}$Morph layers, used to learn a target morphological operation and its associated structuring element.

2.1 The $\mathcal{S}$Morph Layer

The $\mathcal{S}$Morph layer is based on the α−softmax function [9], denoted by $\mathcal{S}_\alpha$ and defined, for any $\mathbf{x} = (x_1, \ldots, x_n) \in \mathbb{R}^n$ and $\alpha \in \mathbb{R}$, as

$$\mathcal{S}_\alpha(\mathbf{x}) = \frac{\sum_{i=1}^n x_i e^{\alpha x_i}}{\sum_{i=1}^n e^{\alpha x_i}}. \quad (1)$$

The control parameter α governs the behavior of $\mathcal{S}_\alpha(\mathbf{x})$: when $\alpha = 0$, $\mathcal{S}_\alpha(\mathbf{x})$ corresponds to the arithmetic mean of $\mathbf{x}$, while $\lim_{\alpha \to -\infty} \mathcal{S}_\alpha(\mathbf{x}) = \min_i x_i$ and $\lim_{\alpha \to +\infty} \mathcal{S}_\alpha(\mathbf{x}) = \max_i x_i$. Thus, $\mathcal{S}_\alpha$ provides an asymptotic approximation of the min and max functions depending on the value of α.

Let $f\colon V \to \mathbb{R}$ be an image, where $x \in V$ denotes the pixel coordinates on a 2D (or N-D) grid $V \subseteq \mathbb{Z}^2$, and $f(x) \in \mathbb{R}$ the corresponding pixel value, and let $w\colon W \to \mathbb{R}$ be a structuring function (or kernel), with spatial support $W \subseteq \mathbb{Z}^2$. The $\mathcal{S}$Morph operation of image f with kernel w and parameter $\alpha \in \mathbb{R}$ is defined as

$$\mathcal{S}\text{Morph}(f, w, \alpha)(x) = \frac{\sum_{y \in W} \big(f(x+y) + w(y)\big) e^{\alpha(f(x+y)+w(y))}}{\sum_{y \in W} e^{\alpha(f(x+y)+w(y))}}. \quad (2)$$

This formulation directly follows from the definition of the α-softmax function, by substituting the entries x_i in Eq. (1) with $f(x+y)+w(y)$. Thus, the asymptotic behavior of $\mathcal{S}\text{Morph}(f, w, \alpha)$ mirrors that of the α-softmax:

$$\lim_{\alpha \to -\infty} \mathcal{S}\text{Morph}(f, w, \alpha)(x) \& = \min_{y \in W} \{f(x+y) + w(y)\} \& = (f \ominus -w)(x), \quad (3)$$

$$\lim_{\alpha \to +\infty} \mathcal{S}\text{Morph}(f, w, \alpha)(x) \& = \max_{y \in W} \{f(x+y) + w(y)\} \& = (f \oplus \breve{w})(x), \quad (4)$$

with $\breve{w}$ the symmetrized version of the kernel w with respect to the origin 0_n.

Consequently, when $\alpha \to -\infty$, $\mathcal{S}\text{Morph}(f, w, \alpha)$ converges to the erosion of the image f with $-w$. Conversely, when $\alpha \to +\infty$, $\mathcal{S}\text{Morph}(f, w, \alpha)$ converges to the dilation of f with $\breve{w}$. As observed in [6,8], the $\mathcal{S}$Morph layer provides a good approximation of an erosion or a dilation when $|\alpha| > 5$. The performed operation is otherwise only considered to be a pseudo-erosion or pseudo-dilation.

2.2 Proposed $\mathcal{HS}$Morph Layer

The definition of $\mathcal{S}$Morph introduced in [6,8] and recalled in Eq. (2) has two shortcomings regarding its asymptotic behavior: given any kernel w, the layer either approximates the erosion of image f with negated kernel $-w$ when $\alpha < 0$, *i.e.*, $\mathcal{S}\text{Morph}(f, w, \alpha) \simeq f \ominus -w$; or approximates the dilation of f with symmetrized kernel $\breve{w}$ when $\alpha > 0$, *i.e.*, $\mathcal{S}\text{Morph}(f, w, \alpha) \simeq f \oplus \breve{w}$; but never with w itself. While the former behavior was explicitly reported in [6,8], the latter was not discussed, as all filters considered in their experiments were symmetric. We propose here a revised definition of the $\mathcal{S}$Morph layer that ensures consistency,

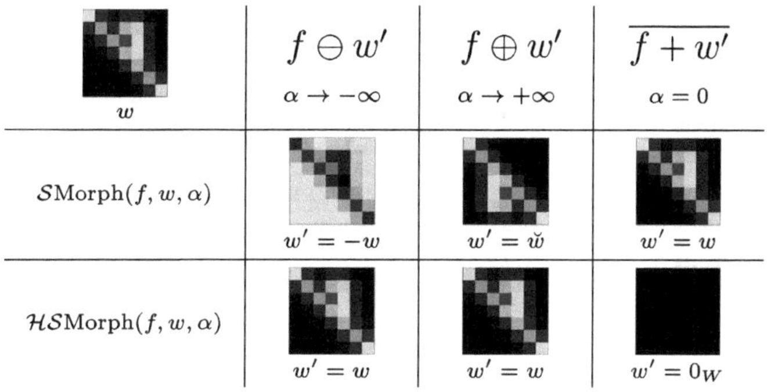

Fig. 2. Asymptotic behavior of $\mathcal{S}$Morph and $\mathcal{HS}$Morph layers. For a given kernel w (top left corner), the $\mathcal{S}/\mathcal{HS}$Morph$(\cdot, w, \alpha)$ layer behaves as a true erosion ($\alpha \to -\infty$) or dilation ($\alpha \to +\infty$) operation of input f with the associated structuring function w'. For $\alpha = 0$, the layer behaves as a local arithmetic mean operation of $f + w'$.

so that it approximates the erosion and dilation of f with the same kernel w. To this end, we replace the filter w in Eq. (2) by the new filter $\tilde{w}_\alpha$ defined as

$$\tilde{w}_\alpha = \tanh(\alpha) \left(\frac{1 + \tanh(\alpha)}{2} \breve{w} + \frac{1 - \tanh(\alpha)}{2} w \right), \tag{5}$$

and define the $\mathcal{HS}$Morph (for Harmonized $\mathcal{S}$Morph) layer as

$$\mathcal{HS}\text{Morph}(f, w, \alpha) = \mathcal{S}\text{Morph}(f, \tilde{w}_\alpha, \alpha). \tag{6}$$

The asymptotic behavior of $\tilde{w}_\alpha$ follows directly from that of the hyperbolic tangent: $\lim_{\alpha \to -\infty} \tilde{w}_\alpha = -w$ and $\lim_{\alpha \to +\infty} \tilde{w}_\alpha = \breve{w}$. Equations (3) and (4) then lead to

$$\lim_{\alpha \to -\infty} \mathcal{HS}\text{Morph}(f, w, \alpha) = \lim_{\alpha \to -\infty} \mathcal{S}\text{Morph}(f, \tilde{w}_\alpha, \alpha) = f \ominus w, \tag{7}$$

$$\lim_{\alpha \to +\infty} \mathcal{HS}\text{Morph}(f, w, \alpha) = \lim_{\alpha \to +\infty} \mathcal{S}\text{Morph}(f, \tilde{w}_\alpha, \alpha) = f \oplus w. \tag{8}$$

Figure 2 illustrates the asymptotic behavior of $\mathcal{S}$Morph and $\mathcal{HS}$Morph layers. Since $\tanh(5) = -\tanh(-5) \approx 0.9999$, the convergence regime of $\mathcal{HS}$Morph is essentially the same as that of $\mathcal{S}$Morph: the performed operation can be regarded as an erosion or dilation when $|\alpha| > 5$, and as a pseudo-operation otherwise.

2.3 Comparison of $\mathcal{S}$Morph and $\mathcal{HS}$Morph Layers

To assess whether the definition of the filter $\tilde{w}_\alpha$ in Eq. (5) affects the ability of the $\mathcal{HS}$Morph layer to accurately learn a target morphological operation and its corresponding structuring function, we conducted a comparative experiment against the standard $\mathcal{S}$Morph layer. More specifically, we independently trained one-layer $\mathcal{S}$Morph and $\mathcal{HS}$Morph networks, following the architecture in Fig. 1,

Two-layer networks that use this strategy contain a single $\mathcal{HS}$Morph*Dual* layer, as it effectively combines the roles of two conventional layers. Architectures composed of such dual layers are denoted $\mathcal{HS}$Morph–S, where "S" indicates "weight sharing" across two consecutive standard layers as described in Eq. (9).

3.2 Constraint on Control Parameters

As originally observed in [6,8], and further illustrated in the top rows of the opening and closing sections in Fig. 4, a consistent pattern in failed training scenarios for learning opening and closing operations is that the control parameters α_1 and α_2 tend to remain close to zero. This results in pseudo-operations, often with an incorrect sign on one of the two control parameters.

To address this issue, we introduce a novel regularization term C (for "constraint") to encourage the control parameters α_1 and α_2 of two consecutive $\mathcal{HS}$Morph layers to move away from zero and from each other during training. This penalty term is defined for any $\alpha_1, \alpha_2 \in \mathbb{R}$ (such that $\alpha_1\alpha_2 \neq 1$) as

$$C(\alpha_1, \alpha_2) = \frac{1}{1 - \alpha_1\alpha_2}. \tag{10}$$

The loss function used to train the network $\mathcal{N}$ with respect to a target morphological transform $\mathcal{T}$ (see Fig. 1) is then given for any input $f\colon V \to \mathbb{R}$ by

$$loss = \mathrm{MSE}(\mathcal{N}(f), \mathcal{T}(f)) + \lambda \cdot C(\alpha_1, \alpha_2), \tag{11}$$

where $\lambda > 0$ is the penalty coefficient, typically set to 0.01 in our experiments.

The function C remains positive as long as $\alpha_1\alpha_2 < 1$, which is always the case in practice since both parameters are initialized at zero. Minimizing $C(\alpha_1, \alpha_2)$ (*i.e.*, $\to 0^+$) is therefore equivalent to minimizing the product $\alpha_1\alpha_2$ toward $-\infty$. This penalty term has appealing gradient properties. Its partial derivatives are

$$\frac{\partial C}{\partial \alpha_1}(\alpha_1, \alpha_2) = \frac{\alpha_2}{(1 - \alpha_1\alpha_2)^2} \quad \text{and} \quad \frac{\partial C}{\partial \alpha_2}(\alpha_1, \alpha_2) = \frac{\alpha_1}{(1 - \alpha_1\alpha_2)^2}. \tag{12}$$

Notably, the partial derivative of C with respect to α_1 depends positively on α_2, and vice versa, with their ratio being exactly α_2/α_1. Thus, the gradient signs are controlled by the opposite parameter, naturally pushing α_1 and α_2 toward opposite signs during training, hence favoring opening or closing. Moreover, the parameter with the smaller magnitude receives a stronger update, encouraging a faster deviation from zero and preventing stagnation. Networks trained with this regularization term, as shown in Eq. (11), are denoted $\mathcal{HS}$Morph–C.

3.3 Comparison of the Proposed Strategies for Two-Layer Networks

To evaluate the effectiveness of the proposed strategies for improving the learning of opening and closing operations in two-layer networks, we conducted a comparative experiment following the same procedure as in Sect. 2.3. More specifically, we compared the performances of the following network configurations:

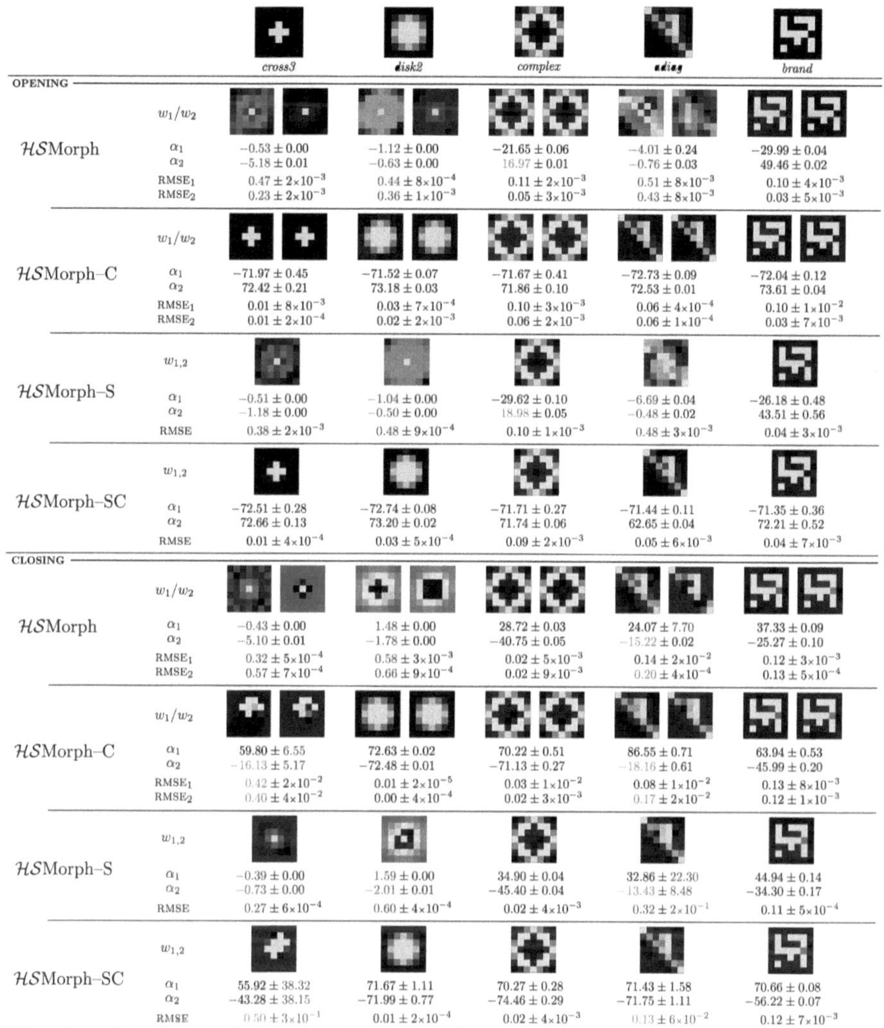

Fig. 4. Learning results for the *opening* (first four rows) and *closing* (last four) operations using four variants of a two-layer $\mathcal{HS}$Morph network on MNIST: learned filters w_1 and w_2, corresponding parameters α_1 and α_2, and RMSE between w_1, w_2 and the target structuring element across five elements (columns). Each metric shows *mean* ± *standard deviation* over five runs. Convergence failures appear in red; medium results in orange.

$\mathcal{HS}$**Morph:** a standard two-layer $\mathcal{HS}$Morph network;

$\mathcal{HS}$**Morph–C:** the same two-layer $\mathcal{HS}$Morph architecture with the penalty term from Eq. (10) added to the training loss, as in Eq. (11);

$\mathcal{HS}$**Morph–S:** architecture with a single $\mathcal{HS}$Morph*Dual* layer from Eq. (9), equivalent to two $\mathcal{HS}$Morph layers with shared weights;

$\mathcal{HS}$**Morph–SC:** with one $\mathcal{HS}$Morph*Dual* layer combined with the penalty term.

Figure 4 presents the obtained results for five target structuring functions: two challenging cases (*cross3* and *disk2*) from [6,8] where standard two-layer $\mathcal{S}$Morph networks failed, one successful case (*complex*) from the same study, and the two asymmetric structuring functions (*adiag* and *brand*) introduced in this work. The three remaining targets (*cross7*, *disk3*, and *diamond3*) from [6,8] were successfully learned by all four network configurations and are omitted from Fig. 4 for brevity. In the three cases where the standard $\mathcal{HS}$Morph network fails (*cross3*, *disk2*, and *adiag*), both $\mathcal{HS}$Morph–C and $\mathcal{HS}$Morph–SC configurations systematically resolve the convergence issue, with the closing of *cross3* as only exception, where the reconstructed cross is incomplete and misaligned, yielding a poor RMSE. In contrast, the $\mathcal{HS}$Morph–S configuration, which relies on a dual layer with shared weights, does not provide any convergence improvements. Thus, the control parameters α_1 and α_2 play a far more decisive role than enforcing a shared kernel w between consecutive layers for learning opening and closing operations. Convergent two-layer networks naturally align their filters w_1 and w_2 to the target filter without requiring weight sharing, while α_1 and α_2 directly determine the nature and quality of the operation performed by each layer. In contrast, enforcing shared weights in the $\mathcal{HS}$Morph*Dual* layer may overly restrict the solution space, making it harder to reach a suitable minimum, which likely explains the lack of improvement with the $\mathcal{HS}$Morph–S configuration. Overall, these results show that promoting divergence of control parameters has a much significant impact than strict weight sharing in ensuring convergence and accurate learning of opening and closing operations.

4 Application to a Denoising Task

We finally evaluate the improvement strategies proposed in the previous Sect. 3 for deeper network architectures. Specifically, we investigate the performance of four-layer $\mathcal{HS}$Morph networks on a salt-and-pepper denoising task, a classical setting for evaluating morphological approaches [5]. Given the nature of this noise, morphological networks are expected to converge toward morphological alternating filters [17], *i.e.*, an opening followed by a closing or vice versa. We compare the same four configurations as in Sect. 3, adapted to a four-layer architecture: a baseline $\mathcal{HS}$Morph network, a $\mathcal{HS}$Morph–C network with constrained control parameters, a $\mathcal{HS}$Morph–S network composed of two dual layers with shared weights, and a $\mathcal{HS}$Morph–SC network combining both the constraint and the dual layers. For the constrained configurations, we replace the penalty term $C(\alpha_1, \alpha_2)$ in Eq. (11) with a new term $C(\alpha_1, \alpha_2, \alpha_3, \alpha_4)$ designed for a four-layer configuration:

$$C(\alpha_1, \alpha_2, \alpha_3, \alpha_4) = \frac{1}{3}\left(C(\alpha_1, \alpha_2) + C(\alpha_2, -\alpha_3) + C(\alpha_3, \alpha_4)\right) \tag{13}$$

Defined as such, this adapted constraint loss encourages the first pair (α_1, α_2) and the last pair (α_3, α_4) of layers to have opposite signs, and the middle pair (α_2, α_3) of layers to have the same sign, thus guiding the four-layer network to a behavior consistent with the expected morphological alternating filtering.

Table 1. Denoising results on the MNIST dataset: PSNR values (dB) between clean images and network outputs on the test set after training for denoising task, for seven four-layer networks (rows) and five salt-and-pepper noise levels (columns). Each cell contains the *mean* $\mu \pm$ *standard deviation* σ (top) and the *max* value (bottom) over five runs per experiment. Best result per column in **bold**, second best <u>underlined</u>.

Noise (%) / Networks	1% 	5% 	10% 	25% 	50% 
$\mathcal{HS}$**Morph**	29.70 ± 0.12 max: 29.81	**28.33** ± 0.24 max: <u>28.61</u>	<u>25.58</u> ± 0.41 max: 26.31	20.52 ± 0.05 max: 20.59	18.44 ± 0.01 max: 18.45
$\mathcal{HS}$**Morph-C**	**31.51** ± 0.02 max: 31.53	<u>27.64</u> ± 0.02 max: 27.66	**25.71** ± 0.26 max: 26.09	**22.38** ± 0.14 max: **22.73**	**20.12** ± 0.03 max: <u>20.16</u>
$\mathcal{HS}$**Morph-S**	28.41 ± 0.22 max: 28.86	24.40 ± 0.01 max: 24.41	22.92 ± 0.01 max: 22.92	20.31 ± 0.07 max: 20.40	18.29 ± 0.01 max: 18.29
$\mathcal{HS}$**Morph-SC**	<u>31.38</u> ± 0.53 max: <u>32.42</u>	24.66 ± 0.47 max: 25.06	22.22 ± 0.01 max: 22.22	19.75 ± 0.01 max: 19.76	19.25 ± 0.32 max: 19.63
MorphoL [21]	24.56 ± 1.22 max: 26.26	23.26 ± 1.72 max: 26.59	24.89 ± 1.51 max: <u>26.61</u>	<u>21.98</u> ± 0.44 max: <u>22.46</u>	19.00 ± 0.37 max: 19.26
PConv [13]	27.88 ± 1.41 max: 30.04	23.75 ± 1.15 max: 25.23	23.54 ± 1.11 max: 25.28	21.05 ± 1.08 max: 22.05	17.78 ± 0.62 max: 18.38
Conv [7]	26.80 ± 8.69 max: **34.70**	18.67 ± 9.90 max: **30.05**	18.92 ± 4.28 max: **26.88**	17.30 ± 2.49 max: 22.29	<u>19.58</u> ± 1.74 max: **20.85**

Table 2. Denoising results on the FashionMNIST dataset: PSNR values (dB) between clean images and network outputs on the test set after training for denoising task, for seven four-layer networks (rows) and five salt-and-pepper noise levels (columns). Each cell contains the *mean* $\mu \pm$ *standard deviation* σ (top) and the *max* value (bottom) over five runs per experiment. Best result per column in **bold**, second best <u>underlined</u>.

Noise (%) / Networks	1% 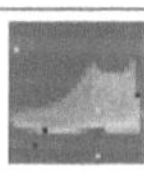	5% 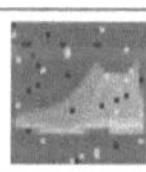	10% 	25% 	50% 
$\mathcal{HS}$**Morph**	26.33 ± 0.01 max: 26.34	24.34 ± 0.01 max: 24.35	<u>23.78</u> ± 0.02 max: 23.80	<u>22.66</u> ± 0.15 max: 22.95	**20.39** ± 0.14 max: **20.51**
$\mathcal{HS}$**Morph-C**	<u>27.81</u> ± 0.41 max: 28.58	**26.37** ± 0.01 max: <u>26.38</u>	**25.42** ± 0.01 max: **25.43**	**23.17** ± 0.03 max: **23.22**	<u>20.27</u> ± 0.02 max: <u>20.29</u>
$\mathcal{HS}$**Morph-S**	**29.99** ± 0.01 max: <u>30.00</u>	<u>25.83</u> ± 0.97 max: 26.32	22.88 ± 0.01 max: 22.89	20.64 ± 0.01 max: 20.65	20.15 ± 0.01 max: 20.16
$\mathcal{HS}$**Morph-SC**	25.30 ± 0.01 max: 25.31	24.75 ± 0.04 max: 24.81	23.75 ± 0.09 max: 23.83	20.28 ± 0.04 max: 20.34	20.17 ± 0.01 max: 20.18
MorphoL [21]	22.79 ± 0.49 max: 23.31	22.29 ± 0.60 max: 23.18	22.83 ± 0.77 max: 24.15	21.33 ± 0.88 max: 22.61	19.51 ± 0.28 max: 19.80
PConv [13]	26.84 ± 1.31 max: 29.33	25.54 ± 0.67 max: 26.18	23.54 ± 0.63 max: 24.70	22.61 ± 0.38 max: <u>23.12</u>	17.61 ± 0.89 max: 18.69
Conv [7]	16.36 ± 9.98 max: **30.95**	15.26 ± 8.46 max: **29.10**	16.26 ± 8.44 max: <u>25.05</u>	16.06 ± 6.58 max: 22.11	15.22 ± 6.09 max: 19.76

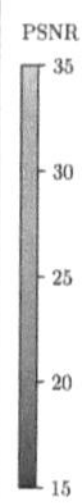

In addition to the four $\mathcal{HS}$Morph configurations, we also evaluate three other four-layer models: an exact morphological network (using the MorphoLayers library [21]), a smooth network using *PConv* layers [13], and a standard convolutional network [7]. Since each exact layer must be explicitly set as either an erosion or a dilation, the MorphoLayers network was configured to perform a closing followed by an opening (which is the configuration toward which $\mathcal{HS}$Morph–C networks naturally converge for most of the runs in these denoising experiments). The *PConv* network includes batch rescaling in [1, 2] before each layer (see [8]). The convolutional baseline uses standard convolution layers, each being followed by the ReLU activation function, with the same basic structure as in [7,11,23], *i.e.*, a network made of four successive convolution-ReLU layers, but without batch normalization layers in order to keep the same architecture as morphological networks (see Fig. 1). All models were trained independently on all 60k images in MNIST and Fashion-MNIST [22] datasets, separately, using salt-and-pepper noisy images as inputs and the corresponding clean images as targets. Performance was then evaluated on their test sets (containing 10k images each).

Tables 1 and 2 summarize the denoising performance of the seven network configurations at various noise levels, for MNIST and Fashion-MNIST, respectively. For each experiment, we report the mean and standard deviation of the peak signal-to-noise ratio (PSNR), computed over five runs and averaged over the entire test set, as well as the maximum PSNR achieved. Across nearly all experiments, the $\mathcal{HS}$Morph–C network consistently achieves the highest mean PSNR, regardless of the dataset or noise level. In the few cases where it does not, it ranks second best. Conversely, $\mathcal{HS}$Morph–S and $\mathcal{HS}$Morph–SC models offer no significant improvement over the baseline $\mathcal{HS}$Morph model, further supporting the conclusion from Sect. 3.3: promoting divergence between the control parameters with the penalty term added to the loss significantly improves the convergence and learning of smooth morphological networks with $\mathcal{HS}$Morph layers.

Moreover, the $\mathcal{HS}$Morph–C model consistently outperforms both the smooth *PConv* network and the exact MorphoLayers approach, suggesting that the flexible nature of smooth morphological layers, when properly regularized, offers a more effective alternative to both fixed morphological operators and earlier smooth formulations. Finally, the convolutional network presents poor performances on average, its high standard deviation indicating frequent convergence issues and unstable training. However, as highlighted by the maximum PSNR over five runs (displayed below the average and standard deviation values in each cell of Tables 1 and 2), successful runs often achieve the highest PSNR among all networks, sometimes significantly. This instability is a known limitation of convolutional architectures without batch normalization layers [7, 11]. While such layers generally improve convergence and denoising performance, they were omitted here to allow a fair comparison with the morphological models which all follow the architecture depicted in Fig. 1. As the convolutional network without batch normalization and the $\mathcal{HS}$Morph network used for these denoising experiments have the same number of learnable parameters (each layer contains the filter weights, plus the biais or the control parameter, associated with the convolutional or smooth morphological layer, respectively), adding new learnable parameters through batch normalization layers only in the convolutional model could significantly advantage its convergence and learning capacities. In future work, new experiments could be conducted that integrate batch normalization layers added before convolutional and morphological layers, in both convolutional and smooth morphological architectures, respectively, to assess the superiority of the former with normalization layers for such denoising tasks.

5 Conclusion

In this work, we introduced a harmonized version of the $\mathcal{S}$Morph layer that overcomes asymptotic limitations while maintaining the performance of one-layer networks when learning erosion and dilation operations. We also proposed a novel penalty term added to the training loss function and shared-weight dual layers to improve the convergence of two-layer networks for learning opening and closing operations, with the former proving to be significantly more effective. Finally, we investigated the proposed $\mathcal{HS}$Morph layer and its improvement strategies in a four-layer network for salt-and-pepper denoising, and compared with other morphological (smooth and exact) and convolutional networks. The obtained results confirmed the superiority of our model with the penalty term, highlighting the importance of control parameter divergence in smooth-layer networks. Future work will focus on integrating these smooth layers into deeper architectures to assess their potential in broader vision tasks, as well as performing a comprehensive benchmarking of morphological networks with classical architectures.

Acknowledgments. This work was conducted during G. Angulo's academic visiting period to NYU, partially funded by the Fondation MINES Paris.

Disclosure of Interests. The authors declare no conflict of interest.

References

1. Angulo, J.: Pseudo-morphological image diffusion using the counter-harmonic paradigm. In: Blanc-Talon, J., Bone, D., Philips, W., Popescu, D., Scheunders, P. (eds.) ACIVS 2010. LNCS, vol. 6474, pp. 426–437. Springer, Heidelberg (2010). https://doi.org/10.1007/978-3-642-17688-3_40
2. Banon, G.J.F., Barrera, J.: Minimal representations for translation-invariant set mappings by mathematical morphology. SIAM J. Appl. Math. **51**(6), 1782–1798 (1991)
3. Blusseau, S.: Training morphological neural networks with gradient descent: some theoretical insights. In: International Conference on Discrete Geometry and Mathematical Morphology, pp. 229–241. Springer, Cham (2024). https://doi.org/10.1007/978-3-031-57793-2_18
4. Dimitrova, M., Blusseau, S., Velasco-Forero, S.: Learning morphological representations of image transformations: Influence of initialization and layer differentiability. In: International Conference on Discrete Geometry and Mathematical Morphology. Springer (2025)

5. Franchi, G., Fehri, A., Yao, A.: Deep morphological networks. Pattern Recogn. **102**, 107246 (2020)
6. Hermary, R., Tochon, G., Puybareau, É., Kirszenberg, A., Angulo, J.: Learning grayscale mathematical morphology with smooth morphological layers. J. Math. Imaging Vis. **64**(7), 736–753 (2022)
7. Ilesanmi, A.E., Ilesanmi, T.O.: Methods for image denoising using convolutional neural network: a review. Complex Intell. Syst0 **7**(5), 2179–2198 (2021). https://doi.org/10.1007/s40747-021-00428-4
8. Kirszenberg, A., Tochon, G., Puybareau, É., Angulo, J.: Going beyond p-convolutions to learn grayscale morphological operators. In: Lindblad, J., Malmberg, F., Sladoje, N. (eds.) DGMM 2021. LNCS, vol. 12708, pp. 470–482. Springer, Cham (2021). https://doi.org/10.1007/978-3-030-76657-3_34
9. Lange, M., Zühlke, D., Holz, O., Villmann, T., Mittweida, S.G.: Applications of lp-norms and their smooth approximations for gradient based learning vector quantization. In: ESANN, pp. 271–276 (2014)
10. LeCun, Y., Bottou, L., Bengio, Y., Haffner, P.: Gradient-based learning applied to document recognition. Proc. IEEE **86**(11), 2278–2324 (1998)
11. Mafi, M., Izquierdo, W., Martin, H., Cabrerizo, M., Adjouadi, M.: Deep convolutional neural network for mixed random impulse and gaussian noise reduction in digital images. IET Image Proc. **14**(15), 3791–3801 (2020)
12. Maragos, P.: A representation theory for morphological image and signal processing. IEEE Trans. Pattern Anal. Mach. Intell. **11**(6), 586–599 (1989)
13. Masci, J., Angulo, J., Schmidhuber, J.: A learning framework for morphological operators using counter–harmonic mean. In: Hendriks, C.L.L., Borgefors, G., Strand, R. (eds.) ISMM 2013. LNCS, vol. 7883, pp. 329–340. Springer, Heidelberg (2013). https://doi.org/10.1007/978-3-642-38294-9_28
14. Matheron, G.: Random Sets and Integral Geometry. Wiley, Hoboken (1975)
15. Mondal, R., Dey, M.S., Chanda, B.: Image restoration by learning morphological opening-closing network. Math. Morphol.Theory Appl. **4**(1), 87–107 (2020)
16. Nogueira, K., Chanussot, J., Dalla Mura, M., Dos Santos, J.A.: An introduction to deep morphological networks. IEEE Access **9**, 114308–114324 (2021)
17. Serra, J.: Image Analysis and Mathematical Morphology, vol. 2. Academic Press (1988)
18. Shen, Y., Shih, F.Y., Zhong, X., Chang, I.C.: Deep morphological neural networks. Int. J. Pattern Recogn. Artif. Intell. **36**(12), 2252023 (2022)
19. Shih, F.Y., Shen, Y., Zhong, X.: Development of deep learning framework for mathematical morphology. Int. J. Pattern Recogn. Artif. Intell. **33**(06), 1954024 (2019)
20. Velasco-Forero, S., Angulo, J.: MorphoActivation: generalizing ReLU activation function by mathematical morphology. In: International Conference on Discrete Geometry and Mathematical Morphology, pp. 449–461. Springer, Cham (2022). https://doi.org/10.1007/978-3-031-19897-7_35
21. Velasco-Forero, S., Pagès, R., Angulo, J.: Learnable empirical mode decomposition based on mathematical morphology. SIAM J. Imag. Sci. **15**(1), 23–44 (2022)
22. Xiao, H., Rasul, K., Vollgraf, R.: Fashion-MNIST: a novel image dataset for benchmarking machine learning algorithms. arXiv preprint arXiv:1708.07747 (2017)
23. Zhang, K., Zuo, W., Zhang, L.: Ffdnet: toward a fast and flexible solution for CNN-based image denoising. IEEE Trans. Image Process. **27**(9), 4608–4622 (2018)
24. Zhou, D.X.: Universality of deep convolutional neural networks. Appl. Comput. Harmon. Anal. **48**(2), 787–794 (2020)

A Mathematical Morphology View of the Universal Representation of Scattering Networks

Gustavo Jesus Angulo(✉)

Mines Paris, PSL University, CMA-Center for Applied Mathematics, Sophia-Antipolis, Paris, France
jesus.angulo_lopez@minesparis.psl.eu

Abstract. Scattering networks involve cascading wavelet transforms, modulus operators, and low-pass filtering. They provide a sound mathematical proxy to model and understand aspects of deep learning such as invariance, stability, hierarchical representation, role of nonlinear activation functions, etc. Linear translation-equivariant filtering based on convolution can be seen as a morphological erosion on a complete inf-semilattice defined by a specific partial ordering in the Fourier domain. In this case, ideal linear filters correspond to morphological openings. Using this framework, we provide a morphological interpretation of the scattering networks from the Fourier-based inf-semilattice. Then, we use the morphological universal representation theory to justify the universal representation of scattering networks of increasing operators on the Fourier-based inf-semilattice. Finally, using our theory we discuss the relevance of learning operators more general than wavelets in the frequency domain.

Keywords: deep learning · mathematical morphology · scattering network · morphological representation theory

1 Introduction

The universal approximation theorem is a foundational result in neural network theory, asserting that feedforward neural networks with at least one hidden layer can approximate any continuous function on compact subsets of $\mathbb{R}^n$, given sufficient width and appropriate activation [6]. Recent research has expanded and refined the original theorem, exploring various network architectures, activation functions, and application domains. See for instance result on ReLU networks using tropical geometry tools [3]. Another key result on the theoretical foundations of deep learning concerns the universal representation theory of linear

This work was partially done during my visiting period to NYU in 2024, funded by the PSL Global Seed Fund 2024. This work was also supported by funding from my AI Chair at PR[AI]RIE-PSAI (ANR-23-IACL-0008, France 2030).

M. H. F. Wilkinson and J. Kosinka (Eds.): DGMM 2025, LNCS 16296, pp. 357–370, 2026.
https://doi.org/10.1007/978-3-032-09544-2_26

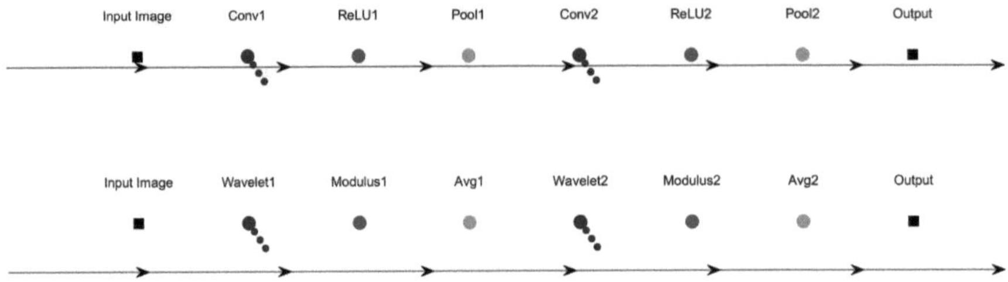

Fig. 1. Comparison of CNN and scattering network architectures. Each layer applies multiple convolutions or wavelet transforms. Colored dots indicate feature maps at different scales and orientations. Red dots are the nonlinear activation layers and in green pooling/averaging. (Color figure online)

operators using convolution. Let $T : L^2(\mathbb{R}^n) \rightarrow L^2(\mathbb{R}^n)$ be a continuous linear operator. Then, under suitable conditions, T can be approximated arbitrarily closely by a convolution operator [18]: $T(f)(x) \approx (k*f)(x) = \int_{\mathbb{R}^n} k(x-y)f(y)\,dy$, where $k \in L^1(\mathbb{R}^n)$ is the convolution kernel. The conditions for approximation are boundedness of the operator and translation-equivariance, i.e., $T(\tau_h f)(x) = \tau_h T(f)(x)$, $\forall h \in \mathbb{R}^d$, where $(\tau_h f)(x) = f(x-h)$ is the translation of f. The rationale behind is based on the fact that T can be represented in terms of its action on a basis of the function space $L^2(\mathbb{R}^n)$, that is related to the link between convolution and the Fourier basis. Under these conditions, the kernel k can be determined through the Fourier transform: $\hat{k}(\omega) = \widehat{T(f)}(\omega)/\hat{f}(\omega)$. Convolutional Neural Networks (CNNs) leverage this principle by using layers of convolutional operations to approximate complex transformations.

Scattering networks provide a method for building informative, invariant and stable representations of signals, particularly images. At its core, a scattering network computes a scattering transform. The scattering transform introduced by Mallat [7] applies a cascade of wavelet transforms followed by modulus and averaging operations. These steps mimic CNNs and offer a non-learned approach to signal representation that shares key architectural similarities, using wavelets instead of learned convolutional filters, see Fig. 1. This approach draws on a strong mathematical background rooted in wavelet theory and harmonic analysis. Their mathematical clarity allows for a deeper understanding of how hierarchical feature extraction, invariance and stability to deformations can be achieved, providing valuable insights into the design and behaviour of their more complex, learned counterparts in deep learning.

Instead of focussing on the convolution-based linear universal theorems to discuss the representation properties of the scattering networks, we propose to use the algebraic perspective from mathematical morphology. The morphological representation of nonlinear operators is based on lattice algebra [11,16]. The main instantiation of the abstract theory has been the representation of increasing translation-equivariant either Boolean operators [10] or numerical operators [9]. We have used this theory to introduce morphological activation functions which can be learnt [17]. Also the morphological representation theory can be stated to the case of equivariant networks and provides a way to model and approximate nonlinear layers (combining for instance group convolution and activation functions) [2]. In this paper we use the morphological representation theory to focus on the role of convolution on CNNs using scattering networks as a proxy [8]. Indeed linear translation-equivariant filtering based on convolution can be seen as a morphological erosion on a complete inf-semilattice defined by a specific partial ordering in the Fourier domain [14]. Using this framework, we provide a morphological interpretation of the scattering networks from the Fourier-based inf-semilattice. Then, we use the morphological universal representation theory to justify the universal representation of scattering networks of operators on the Fourier-based inf-semilattice.

2 From Wavelets to Scattering Networks

A directional wavelet transform extracts the frequency content of a signal within different frequency bands and orientations. Multiscale directional wavelet filters are obtained by scaling and rotating a single band-pass filter (aka mother wavelet), written as $\psi(f)(x)$ when applied to function (or signal) $f(x)$. In 2D, and giving a finite rotation group indexed by $\Theta = \{2k\pi/K : 0 \leq k \leq K\}$, the wavelet transform at a scale 2^j is defined by

$$W(f) = \{(f * \psi_{\theta,\, j})(x)\}_{x\in\mathbb{R}^2,\, \theta\in\Theta,\, j<J}\,, \tag{1}$$

where

$$\psi_{\theta,\, j}(x) = 2^{-2j}\psi(2^{-j}r_{-\theta}x), \tag{2}$$

with r_θ being the rotation matrix for angle θ. It decomposes f along different orientations θ and scales 2^j in the neighborhood of each location x. A typical choice [5] for ψ is the Morlet wavelet (i.e., complex exponential modulated by a Gaussian envelope): $\psi(x) = e^{-\|x\|^2/2\sigma^2}\left(e^{ix\cdot\eta} - C\right)$, where σ of the envelope control the angular sensitivity of ψ, η the frequency of a sine wave and C is adjusted to have $\int \psi(x)dx = 0$. Often only real part is considered. Figure 2 provides examples of Morlet wavelets in spatial and frequency domain and the corresponding image feature extraction by taking the modulus. Their frequency representation highlights the localized band-pass nature of these wavelets. A larger scale σ results in a narrower bandwidth in the frequency domain.

Removing the complex phase in the convolution by taking the modulus $|f * \psi_{\theta,j}|$, like in the Fourier transform, defines a positive envelope which is still equivariant to translation. The pointwise non-linearity of the modulus is important to avoid a linear composition of operators which is not stable with respect to small deformations [7].

In order to introduce a local translation invariance which is stable to small deformations, an averaging filter can be added [7]. The principle is similar to the traditional pooling operator in deep learning architectures. A typical choice of averaging function is the Gaussian filter, i.e., $\phi_{\sigma_G}(x) = (2\pi\sigma_G^2)^{-1}e^{-\|x\|^2/(2\pi\sigma_G^2)}$.

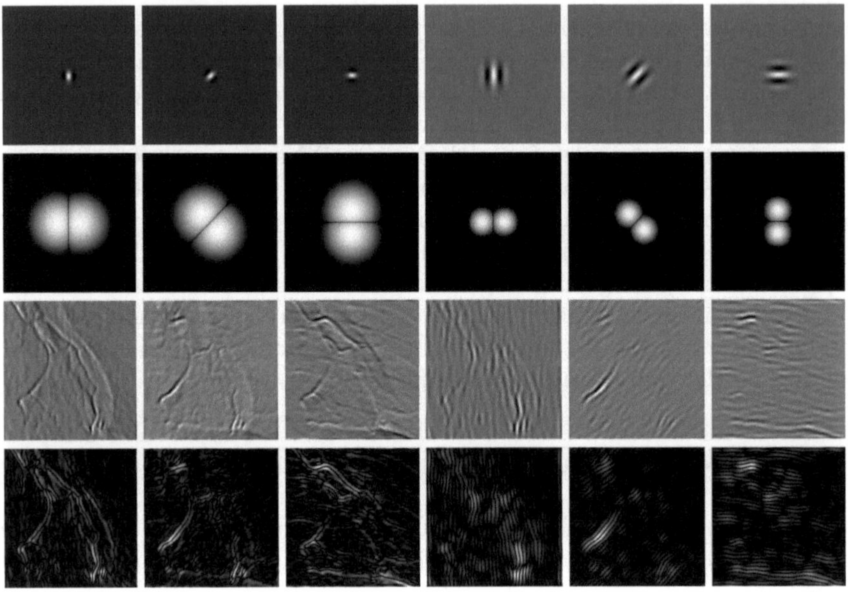

Fig. 2. Examples of Morlet wavelet representation: on the first row, first three columns are $\eta = 0.1$, $\sigma = 3$, with $\theta = 0$, $\theta = \pi/4$ and $\theta = \pi/2$, last three columns $\eta = 0.1$, $\sigma = 7$, with $\theta = 0$, $\theta = \pi/4$ and $\theta = \pi/2$; second row, Fourier transform of wavelets; third row, result of convolution of wavelets on an image; fourth row, modulus of previous one.

This local averaging operator is taking with $\sigma_G = J$ and dilated by 2^J [5]. To introduce the effect of pooling on CNNs, the result can be downsampled of scale 2^J. Using the wavelet transform, modulus and averaging, one gets a zero and first-order representation of the image f:

$$\begin{aligned} U_0(f)(x) &= f; \quad & S_0(f)(x) &= U_0(f) * \phi_J(2^J x); \\ U_1(f)(x,\theta,j) &= |U_0 * \psi_{\theta,\, j}(x)|; \quad & S_1(f)(x,\theta,j) &= U_1(f) * \phi_J(2^J x). \end{aligned}$$

The operator $S_1 f(x,\theta,j)$ provides a local translation invariance but also loses information at multiple scales and frequencies. A scattering successively recovers the information lost by the averaging which computes the invariants. Scattering [7] consists in a cascade of wavelet modulus transforms followed by the non-linearity of the modulus, which can be interpreted as a deep neural network which propagates an image f across multiple layers of the network and which outputs at each layer m scattering invariant coefficients $S_m(f)$. To make notation more compact, let us write $\lambda = (\theta, j)$ and $\Lambda = \{(\theta, j) \,:\, \theta \in [0, 2\pi], j \leq J\}$. The scattering network $\{S_m(f)\}_{\lambda_m \in \Lambda}$ is then defined by induction, for any $m \geq 0$:

$$U_m(f)(x) = |U_{m-1}(x) * \psi_{\lambda_m}(x)| \tag{3}$$
$$= |||f * \psi_{\lambda_1}| * \cdots | * \psi_{\lambda_{m-1}}| * \psi_{\lambda_m}(x)|,$$
$$S_m(f)(x) = U_m(f)(x) * \phi_J(2^J x) \tag{4}$$
$$= |||f * \psi_{\lambda_1}| * \cdots | * \psi_{\lambda_{m-1}}| * \psi_{\lambda_m}| * \phi_J(2^J x).$$

This representation emphasizes the hierarchical nature of the scattering networks similar to a multilayer CNN.

Orthogonal Wavelets. The Morlet wavelet transform produces a sparse set of non-negligible wavelet coefficients [5]. Morlet wavelets are smooth and localized in space and frequency which makes them stable to deformation, however they are not orthogonal. Fourier domain as a representation space is a natural way to construct orthogonal functions with respect to the standard scalar product of functions. It suffices to choose functions with disjoint frequency supports, then using the Parseval–Plancherel theorem, which implies that inner product and norm are preserved under Fourier transform, one gets that for two disjoint frequency support functions, their scalar product is zero. Note that this is not a necessary condition since signals can be orthogonal with the same frequency supports, i.e., sine and cosine functions. The use of kernels with disjoint frequency supports has a particular interpretation in our framework.

3 Morphological Operators on Inf-Semilattices

Morphological operators are mainly formulated in the abstract framework of complete lattices. We adopt here the case of semilattices [12,13] since it will be the appropiate one to relate morphology and linear filtering.

Adjunctions on Complete Semilattices. A partially ordered set $\mathcal{L}$ is an inf-semilattice (resp. sup-semilattice) if every two-element subsets $\{x_1, x_2\}$ in $\mathcal{L}$ has an infimum $x_1 \curlywedge x_2$ (resp. a supremum $x_1 \curlyvee x_2$) in $\mathcal{L}$.

It is called a complete inf-semilattice (or briefly a cisl) when every non-empty subset $\mathcal{K} \subseteq \mathcal{L}$ has an infimum $\curlywedge \mathcal{K}$. The least element (zero element) of a cisl is denoted by $\perp$. In general a cisl does not have a greatest element.

The supremum is defined in an inf-semilattice under the condition that the pair of elements has an upper bound, and that upper bound is the least among all upper bounds for those elements, i.e., for two elements x_1 and x_2, there is an element s such that $x_1 \preccurlyeq s$ and $x_2 \preccurlyeq s$ and that if any other element t satifies $x_1 \preccurlyeq t$ and $x_2 \preccurlyeq t$, then $s \preccurlyeq t$. This is not always guaranteed, as inf-semilattices are only required to have infima (greatest lower bounds).

Example 1. A prominent cisl in the context of self-dual morphology is based on the following partial ordering $\preccurlyeq$ on $\mathbb{R}$, $\forall s, t \in \mathbb{R}$:

$$s \preccurlyeq t \ \text{ if } \ 0 \leq s \leq t \ \text{ or } \ t \leq s \leq 0.$$

Thus $\mathbb{R}$ can be seen as the concatenation of two totally ordered complete lattices (called chains) $(\mathbb{R}_-, \geq)$ and $(\mathbb{R}_+, \leq)$ intersecting at the origin such that 0 is just the least element of the partially ordered set $(\mathbb{R}, \preccurlyeq)$.

Assume that $(\mathcal{L}, \preccurlyeq, \curlywedge)$ is a cisl and $\mathcal{M}$ is some nonempty set. Let $\phi : \mathcal{L} \to \mathcal{M}$ be a bijective mapping. One defines the relation $\preccurlyeq_\phi$ on $\mathcal{M} \times \mathcal{M}$ by

$$y_1 \preccurlyeq_\phi y_2 \iff \phi^{-1}(y_1) \preccurlyeq \phi^{-1}(y_2).$$

Then $(\mathcal{M}, \preccurlyeq_\phi)$ is a cisl with infimum given by [12]:

$$\curlywedge_\phi y_i = \phi\left(\curlywedge \phi^{-1}(y_i)\right). \tag{5}$$

An operator $\varepsilon : \mathcal{L} \to \mathcal{M}$, where both $\mathcal{L}$ and $\mathcal{M}$ are a cisl, is an erosion if for every non empty collection $\{x_i\} \subseteq \mathcal{L}$ we have

$$\varepsilon\left(\curlywedge_{i\in I} x_i\right) = \curlywedge_{i\in I} \varepsilon(x_i).$$

Consider the subset of $\mathcal{M}$ which as an infimum on $\mathcal{L}$ defined as

$$\mathcal{M}[\varepsilon] = \{y \in \mathcal{M} \,:\, \exists x \in \mathcal{L} \text{ such that } y \preccurlyeq_\phi \varepsilon(x)\}.$$

Dilation $\delta : \mathcal{M}[\varepsilon] \to \mathcal{L}$ is defined by adjunction:

$$\delta(y) = \curlywedge \{x \in \mathcal{L} \,:\, y \preccurlyeq_\phi \varepsilon(x)\}, \quad y \in \mathcal{M}[\varepsilon].$$

Inf-Semilattice Erosion and Adjoint Dilation on $\mathbb{C}$. Example 1 of cisl-ordering is naturally extended to the complex plane as follows [12]. The space $\mathbb{C}$ can be seen as an infinite union of chains $\mathbb{C}_\alpha = \left\{re^{i\alpha} \,:\, r \geq 0\right\}$ ordered by the magnitude of the modulus of the complex number. Hence, for two elements $w, z \in \mathbb{C}$, one has

$$w \preccurlyeq z \quad \text{if} \quad \{|w| \leq |z| \quad \text{and} \quad \arg w = \arg z\}. \tag{6}$$

Let us consider now the cisl $(\mathbb{C}, \preccurlyeq)$. The mapping $\varepsilon : \mathbb{C} \mapsto \mathbb{C}$ given by

$$\varepsilon(z) = E\left(|z|\right) e^{iA(\arg z)} \tag{7}$$

defines an erosion if and only if E is an erosion on the cisl $(\mathbb{R}_+, \leq)$ with $E(0) = 0$ and $A : [0, 2\pi) \to [0, 2\pi)$ is an injective mapping. The adjoint dilation is given by

$$\delta(z) = D\left(|z|\right) e^{iA^{-1}(\arg z)}, \tag{8}$$

where D is the dilation on $\mathbb{R}_+[E]$ andjoint to erosion E.

Example 2. The simplest case of cisl complex erosion is given by $E(r) = cr$, $c \geq 0$ and $A(\theta) = \theta + \alpha \mod 2\pi$, which gives the erosion $\varepsilon(z) = ce^{i\alpha}z$ and the adjoint dilation $\delta(w) = c^{-1}e^{-i\alpha}w$ if $c > 0$ and $\delta(w) = 0$ if $c = 0$.

The complex erosion (7) can be generalized to the cisl $(\mathcal{F}(\mathbb{R}^n, \mathbb{C}), \preccurlyeq)$ where $\mathcal{F}(\mathbb{R}^n, \mathbb{C})$ is the set of functions $\mathbb{R}^n \to \mathbb{C}$ and $\preceq$ is (6), as follows: $\forall f \in \mathcal{F}(\mathbb{R}^n, \mathbb{C})$

$$\varepsilon(f)(\omega) = E_\omega\left(|f(\omega)|\right) e^{iA_\omega(\arg f(\omega))}, \tag{9}$$

where for every $\omega \in \mathbb{R}^n$, E_ω and A_ω satisfy the properties of erosion (7).

The classic property of gray-scale shift-equivariance of scalar morphological operators on $\mathcal{F}(\mathbb{R}^n, \mathbb{R})$; e.g., $\varepsilon(f+t)(x) = t + \varepsilon(f)(x)$, $t \in \mathbb{R}$, becomes here the following notion of equivariance to mappings E_ω and A_ω, $\forall \omega \in \mathbb{R}^n$:

$$E_\omega(qr) = qE_\omega(r), \quad q, r > 0$$
$$A_\omega(\theta + \langle a, \omega\rangle) = A_\omega(\theta) + \langle a, \omega\rangle, \quad \theta \in [0, 2\pi)], \; a \in \mathbb{R}^n.$$

Hence, one gets

$$E_\omega(r) = c(\omega) r, \quad A_\omega(\theta) = \theta + \alpha(\omega),$$

where $c : \mathbb{R}^n \to \mathbb{R}_+$ and $\alpha : \mathbb{R}^n \to [0, 2\pi)$. If we write

$$e(\omega) = c(\omega) e^{i\alpha(\omega)} \quad d(\omega) = c^{-1}(\omega) e^{-i\alpha(\omega)}, \tag{10}$$

we obtain

$$\varepsilon(f)(\omega) = e(\omega) f(\omega), \tag{11}$$
$$\delta(f)(\omega) = d(\omega) f(\omega). \tag{12}$$

From these latter expressions of the considered example, we identify a parallel with the convolution on the Fourier domain. From the perspective of equivarince, Fourier transform maps translation equivariance of an operator on the original "spatial" domain onto equivariance under modulations, i.e., $e^{i\langle a, \omega\rangle}$, on the "frequency" domain. That is the fundamental motivation for our morphological perspective of linear filtering using convolution.

4 Fourier Inf-Semilattice of Convolution

In this section we review precisely the morphological view of convolution on the Fourier domain following the ideas of Keshet [14]. Let us first consider the following notation for the convolution of a function $f \in \mathcal{L}^n = \mathrm{Fun}(\mathbb{R}^n, \mathbb{R})$ by a kernel $k \in \mathcal{L}^n$

$$\varepsilon_k^{\mathrm{Conv}}(f)(x) = (f * k)(x),$$

as well as the Fourier transform of the function: $\hat{f}(\omega) = \mathcal{F}(f)(\omega) = |\hat{f}(\omega)| \exp\left(i \angle \hat{f}(\omega)\right)$, with modulus $|\hat{f}(\omega)|$ and phase $\angle \hat{f}(\omega)$, such that using the convolution theorem one has

$$\varepsilon_k^{\mathrm{Conv}}(f)(x) = \mathcal{F}^{-1}\left\{\hat{f}(\omega)\hat{k}(\omega)\right\}(x).$$

We now introduce the spectral partial ordering $\leq_{\mathcal{F}}$ for functions as the application of the complex partial ordering (6) to each frequency of the Fourier domain. For all pairs $f, g \in \mathcal{L}^n$

$$f \leq_{\mathcal{F}} g \iff \forall \omega \in \mathbb{R}^n, \; : \; \hat{f}(\omega) \preccurlyeq \hat{g}(\omega) \iff \begin{cases} |\hat{f}(\omega)| \leq |\hat{g}(\omega)| \text{ and} \\ \angle \hat{f}(\omega) = \angle \hat{g}(\omega). \end{cases}$$

The poset $(\mathcal{L}^n, \leq_\mathcal{F})$ is a complete inf-semilattice: the least element is the null function and the infimum $f \wedge_\mathcal{F} g$ is defined as

$$(f \wedge_\mathcal{F} g)(x) = \begin{cases} \mathcal{F}^{-1}\left\{\min\left(|\hat{f}(\omega)|, |\hat{g}(\omega)|\right) \exp\left(i\angle \hat{f}(\omega)\right)\right\} \text{ if} \\ \qquad \angle \hat{f}(\omega) = \angle \hat{g}(\omega) \\ 0, \qquad\qquad\qquad\qquad \text{otherwise.} \end{cases}$$

Note that inf-semilattice structure of the space domain from the inf-semilattice of the frequency domain is just an example of the isomorphism (5) where the bijective mapping ϕ is the Fourier transform.

We remind now the main result on this framework.

Proposition 1 (Keshet, 2002). *The linear convolution $\varepsilon_k^{Conv}(f)$ of function $f(x)$ by kernel $k(x)$ is an erosion in $(\mathcal{L}^n, \leq_\mathcal{F})$, i.e.,*

- *(increasing) $\forall f, g \in \mathcal{L}^n$, $f \leq_\mathcal{F} g \implies \varepsilon_k^{Conv}(f) \leq_\mathcal{F} \varepsilon_k^{Conv}(g)$,*
- *(commutation with infimum) $\varepsilon_k^{Conv}(f \wedge_\mathcal{F} g) = \varepsilon_k^{Conv}(f) \wedge_\mathcal{F} \varepsilon_k^{Conv}(g)$.*

Proof. The increasing property is direct from the definition of erosion and the fact that

$$|\hat{f}(\omega)| \leq |\hat{g}(\omega)| \implies |\hat{f}(\omega)\hat{k}(\omega)| \leq |\hat{g}(\omega)\hat{k}(\omega)|, \quad \forall \hat{k}(\omega)$$

The second property is obtained by taking the Fourier transform to both sides:

$$\begin{aligned} \mathcal{F}\left\{\varepsilon_k^{\text{Conv}}(f \wedge_\mathcal{F} g)\right\} &= \begin{cases} \min\left(|\hat{f}(\omega)|, |\hat{g}(\omega)|\right) \exp\left(i\angle \hat{f}(\omega)\right)\hat{k}(\omega) \text{ if} \\ \qquad \angle \hat{f}(\omega) = \angle \hat{g}(\omega) \\ 0, \qquad \text{otherwise} \end{cases} \\ &= \begin{cases} \min\left(|\hat{f}(\omega)\hat{k}(\omega)|, |\hat{g}(\omega)\hat{k}(\omega)|\right) \exp\left(i\angle \hat{f}(\omega)\hat{k}(\omega)\right) \text{ if} \\ \qquad \angle \hat{f}(\omega)\hat{k}(\omega) = \angle \hat{g}(\omega)\hat{k}(\omega) \\ 0, \qquad \text{otherwise} \end{cases} \\ &= \mathcal{F}\left\{\varepsilon_k^{\text{Conv}}(f) \wedge_\mathcal{F} \varepsilon_k^{\text{Conv}}(g)\right\}. \end{aligned}$$

Using this Fourier-based semilattice structure, the magnitude activation of the "feature map" associated to the kernel k follows an order structure based on the energy selection by k in the frequency domain. If the spectral erosion in $(\mathcal{L}^n, \leq_\mathcal{F})$ is anti-extensive, i.e., $\forall f \in \mathcal{L}^n$ $\varepsilon_k^{\text{Conv}}(f) \leq_\mathcal{F} f$ the composition is ordered too, i.e., $\varepsilon_{k_2}^{\text{Conv}}\varepsilon_{k_1}^{\text{Conv}}(f) \leq_\mathcal{F} \varepsilon_{k_1}^{\text{Conv}}(f) \leq_\mathcal{F} f$. The effect of successive erosions involves that the signal becomes simpler and simpler by removing frequency information.

The adjoint dilation of a linear convolution filtering $\delta_k^{*,\text{Conv}}$ in $(\mathcal{L}^n, \leq_\mathcal{F})$ is its generalized-inverse filtering operation, i.e.,

$$\delta_k^{*,\text{Conv}}(f)(x) = \mathcal{F}^{-1}\left\{\hat{f}(\omega)\hat{k}^*(\omega)\right\}(x), \quad \text{where}$$

$$\hat{k}^*(\omega) = \begin{cases} \hat{k}(\omega)^{-1}, & \text{if } |\hat{k}(\omega)| > 0 \\ 0, & \text{otherwise.} \end{cases}$$

Openings as Ideal Filters. The opening by adjunction associated to linear convolution filtering, $\gamma_k^{\text{Conv}}(f) = \delta_k^{*,\text{Conv}} \circ \varepsilon_k^{\text{Conv}}(f)$, is an ideal linear filtering operator,

$$\gamma_k^{\text{Conv}}(f)(x) = \mathcal{F}^{-1}\left\{\hat{f}(\omega)\hat{k}^{ind}(\omega)\right\}(x),$$

where $\hat{k}^{ind}(\omega) = 1$ if $|\hat{k}(\omega)| > 0$ and $\hat{k}^{ind}(\omega) = 0$ otherwise. It satisfies naturally the properties of an opening, i.e.,

- (idempotent) $\gamma_k^{\text{Conv}} \circ \gamma_k^{\text{Conv}}(f) = \gamma_k^{\text{Conv}}(f)$,
- (anti-extensive) $\gamma_k^{\text{Conv}}(f) \leq_{\mathcal{L}} f$.

Example 3. Figure 3 provides the effect of a Gaussian-shaped frequency filter $\hat{k}_{\text{Gaussian}}(\omega) = e^{-\|\omega-\omega_0\|^2/(2\sigma^2)}$ of shape σ centered at frequency ω_0. For image $f(x)$, the extracted features are obtained by $\mathcal{F}^{-1}\ \{\hat{f}(\omega)\hat{k}_{\text{Gaussian}}(\omega)\}(x)$. The result is compared to an ideal filter defined as indicator function of a disk of cutoff radius R and center ω_0 on the Fourier domain, i.e., $\hat{k}_{\text{Disk}}(\omega) = 1$ if $\|\omega-\omega_0\|_2 \leq R$ and 0 otherwise.

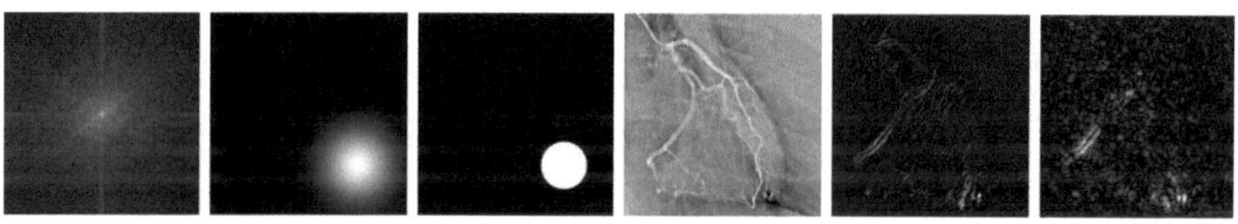

(a) (b) (c) (d) (e) (f)

Fig. 3. Comparing an erosion $\varepsilon_k^{\text{Conv}}(f)$ and an opening $\gamma_k^{\text{Conv}}(f)$ of the Fourier inf-semilattice: (a) Fourier transform of the image, (b) Gaussian-like filter on the Fourier domain, (c) Ideal filter as the indicator function of a disk on the Fourier domain, (d) Original image, (e) Inverse Fourier transform of Gaussian-like filtered image, (d) Inverse Fourier transform of ideal filtered image.

5 Our Contribution: Morphological View of Scattering Network

5.1 Scattering Structure as Morphological Operators

First all, it is obvious that the scattering network for $m = 1$ can be rewritten in terms of the Fourier erosions as follows

$$U_1(f)(x,\theta,j) = |\varepsilon_{\psi_{\theta,j}}^{\text{Conv}}(f)(x)| = |\mathcal{F}^{-1}\{\mathcal{F}(U_0)(\omega)\mathcal{F}(\psi_{\theta,j})(\omega)\}(x)|,$$
$$S_1(f)(x,\theta,j) = \varepsilon_{\phi_J(2^Jx)}^{\text{Conv}}(U_1(f))(x) = \varepsilon_{\phi_J(2^Jx)}^{\text{Conv}} \circ |\varepsilon_{\psi_{\theta,j}}^{\text{Conv}}(f)|(x).$$

However, because the algebraic structure of the Fourier inf-semilattice is formulated on the frequency domain, we consider the scattering network on it too. As we have discussed, the modulus of the wavelet convolution on space introduce a fundamental nonlinearity to the network. Discarding the phase in the Fourier domain provides a nonlinear effect. That nonlinearity is more compatible with our framework. The Fourier modulus can be unstable with respect to deformations on high frequencies. In addition, the Fourier modulus loses much information and may not be able to discriminate different images. The frequency description by the scattering transform is more stable and the redundancy of the representation helps to discriminate. Note that two images having very different modulus can have the same first order scattering coefficients but be separated by the higher orders in [5]. Using (5), if $\varepsilon_k^{\mathrm{Conv}}(f)$ is an erosion then

$$\hat{\varepsilon}_{\hat{k}}(\hat{f})(\omega) = \mathcal{F}\{\varepsilon_k^{\mathrm{Conv}}(f)\}(\omega) = \hat{f}(\omega)\hat{k}(\omega)$$

is an erosion on the Fourier inf-semilattice too, which correspond just to the case (11). We can also see that removing the phase by taking the modulus leads to an erosion too

$$\bar{\hat{\varepsilon}}_{\hat{k}}(\hat{f})(\omega) = |\hat{\varepsilon}_{\hat{k}}(\hat{f})(\omega)| = |\hat{f}(\omega)\hat{k}(\omega)|.$$

We introduce now the scattering (invariant) coefficients on frequency as Fourier erosions:

$$\begin{aligned} U_m^{\mathcal{F}}(\hat{f})(\omega) &= \bar{\hat{\varepsilon}}_{\hat{\psi}_{\lambda_m}}\left(U_{m-1}^{\mathcal{F}}(\omega)\right) \\ &= \bar{\hat{\varepsilon}}_{\hat{\psi}_{\lambda_m}} \circ \bar{\hat{\varepsilon}}_{\hat{\psi}_{\lambda_{m-1}}} \circ \cdots \circ \bar{\hat{\varepsilon}}_{\hat{\psi}_{\lambda_1}}(\hat{f})(\omega), \end{aligned} \tag{13}$$

$$\begin{aligned} S_m^{\mathcal{F}}(\hat{f})(\omega) &= \hat{\varepsilon}_{2^{-J}\hat{\phi}(2^{-J}\omega)}\left(U_{m-1}^{\mathcal{F}}(\omega)\right) \\ &= \hat{\varepsilon}_{2^{-J}\hat{\phi}(2^{-J}\omega)} \circ \bar{\hat{\varepsilon}}_{\hat{\psi}_{\lambda_m}} \circ \bar{\hat{\varepsilon}}_{\hat{\psi}_{\lambda_{m-1}}} \circ \cdots \circ \bar{\hat{\varepsilon}}_{\hat{\psi}_{\lambda_1}}(\hat{f})(\omega). \end{aligned} \tag{14}$$

These coefficients are parameterized by $\lambda_m = (j_m, \theta_m)$, $1 \leq m \leq M$, with $1 \leq j_m \leq J$ (scales) and $\theta_m = 2\pi l/L$ (rotations). Only paths of increasing scale ($j_{m-1} < j_m$) are computed. The order (or depth of the tree) is usually limited to $M = 2$ (higher order scattering has negligible energy for image classification). Note that if the wavelet (2) is centered at a frequency η, its Fourier transform

$$\hat{\psi}_{\theta,\,j}(\omega) = \hat{\psi}(2^j r_{-\theta}\omega),$$

has a support centered at $2^{-j} r_\theta \eta$ with a bandwidth proportional to 2^{-j}. Therefore the hierarchical structure of coefficients $S_m^{\mathcal{F}}(\hat{f})(\omega)$ provides a subdivision of the Fourier space at each level m. In fact, a way to represent the scattering coefficients in the frequency plane is a discretization into piecewise constant functions introduced in [5], where an annulus is partitioned into nested quadrants (divided along the radial axis and the angular variable). The value as each class of the nested partitions corresponds to the norm of the scattering coefficient.

5.2 Universal Representation of Scattering Networks

We now formalize the main theoretical contribution of this work: the universal representation of nonlinear increasing operators on the modulus of the Fourier transform using scattering coefficients. This result is grounded in the morphological representation theory on upper-semicontinuous function lattices.

Let $\widehat{\mathcal{L}} = \text{Fun}(\mathbb{R}^2, \overline{\mathbb{R}}_+)$ denote the set of upper-semicontinuous functions from $\mathbb{R}^2$ to $\mathbb{R}_+$, equipped with the pointwise partial order.

Theorem 1 (Universal Representation via Scattering Coefficients). *Let $\Psi : \widehat{\mathcal{L}} \to \widehat{\mathcal{L}}$ be a nonlinear, increasing operator acting on the modulus of the Fourier transform of an image f. Then, under the assumption that Ψ is increasing and upper-semicontinuous, there exists a (possibly redundant) collection of scattering-based erosions $\{\varepsilon_\psi\}_{\psi \in \text{Ker}(\Psi)}$ such that*

$$\Psi(\widehat{f})(\omega) = \bigvee_{\psi \in \text{Ker}(\Psi)} \varepsilon_\psi(\widehat{f})(\omega),$$

where each ε_ψ is defined as

$$\varepsilon_\psi(\widehat{f})(\omega) = \sup\{v \in \overline{\mathbb{R}}_+ : \psi(\omega)v \leq \widehat{f}\}.$$

Moreover, this representation can be approximated using a finite set of scattering coefficients $\{S_m^{\mathcal{F}}(\hat{f})\}$ as

$$\Psi(\widehat{f})(\omega) \approx \sup_{m, \theta_m, j_m} w_m S_m^{\mathcal{F}}(\hat{f})(\omega),$$

where $w_m \geq 0$ are non-negative weights associated with each scattering path of order m.

Proof The result follows from the general theory of morphological representation of increasing operators on complete lattices of upper-semicontinuous functions, as developed in [4], which extended the original approach for spatially-equivariant morphology [9, 10]. For any such operator Ψ, the kernel $\text{Ker}(\Psi)$ is defined as

$$\text{Ker}(\Psi) = \left\{\psi \, : \, \Psi(\psi(\omega))(\omega) \geq 0, \ \ \forall \omega \in \mathbb{R}^2\right\}. \tag{15}$$

If we define now the erosion of $\widehat{f}$ by ψ, $\widehat{f}, \psi \in \widehat{\mathcal{L}}$, as

$$\varepsilon_\psi(\widehat{f})(w) = \vee\{v \in \overline{\mathbb{R}}_+ \, : \, \psi(\omega)v \leq \widehat{f}\}, \tag{16}$$

Then the supremum over all such erosions yields the operator Ψ:

$$\Psi(\widehat{f})(\omega) = \bigvee_{\psi \in \text{Ker}(\Psi)} \varepsilon_\psi(\widehat{f})(\omega).$$

This representation is very redundant. Indeed, if $\eta \in \text{Ker}(\Psi)$ any function ψ with $\psi \leq \eta$ is also in the kernel of Ψ. The collection of minimal elements of the kernel is defined as the morphological basis of Ψ:

$$\text{Bas}(\Psi) = \{\eta \in \text{Ker}(\Psi) \, : \, [\psi \in \text{Ker}(\Psi) \text{ and } \psi \leq \eta] \implies \psi = \eta\},$$

such that $\Psi(\widehat{f})(\omega) = \bigvee_{\psi \in \mathrm{Bas}(\Psi)} \varepsilon_\psi(\widehat{f})(\omega)$.

To approximate this representation in practice, we restrict the kernel to a finite set of scattering-based erosions. Each scattering coefficient $S_m^{\mathcal{F}}(\hat{f})(\omega)$ corresponds to a structured erosion in the frequency domain, parameterized by scale j_m and orientation θ_m. The approximation then becomes a weighted supremum over these coefficients, where the weights w_m play the role of the scalar multipliers in the erosion definition.

This yields the practical approximation formula:

$$\Psi(\widehat{f})(\omega) \approx \sup_{m,\theta_m,j_m} w_m S_m^{\mathcal{F}}(\hat{f})(\omega),$$

which completes the proof.

5.3 Weight Selection via Minimization and Pratical Approximation Quality

The approximation formula, which expresses a nonlinear increasing operator Ψ acting on the modulus of the Fourier transform as a supremum of erosions by scattering coefficients, is grounded in the morphological representation theory of upper-semicontinuous functions. While this formulation is theoretically sound, its practical effectiveness depends on several factors.

First, the quality of the approximation is influenced by the number of scales, orientations, and the order of scattering paths. Each scattering coefficient captures localized frequency content, and increasing the resolution of this representation (e.g., by increasing the number of wavelets or the depth of the scattering tree) improves the expressiveness of the approximation. However, this also increases redundancy. To mitigate this, sparsity constraints on the weights w_m can be introduced, effectively selecting a minimal subset of scattering paths that best approximate the target operator.

Second, while we do not provide empirical validation in this paper, a practical implementation could involve learning the weights w_m via a shallow neural network or optimization, using a training set of input-output pairs for the operator Ψ.

Thus to approximate the operator Ψ using a supremum over weighted wavelet coefficients, the weights w_λ must be chosen to minimize the discrepancy between the target operator and the approximation. A natural approach is to solve the following optimization problem:

$$\min_{\{w_m \geq 0\}} \left\| \Psi(\widehat{f}) - \sup_{m,\theta_m,j_m} w_m S_m^{\mathcal{F}}(\hat{f}) \right\|_{L^2}.$$

This formulation seeks the optimal set of non-negative weights that best match the action of Ψ on the modulus of the wavelet transform. In practice, this can be implemented using convex optimization techniques or sparse regression methods, particularly when a sparsity constraint is added to promote interpretability and reduce redundancy. For instance, one may include an ℓ_1-penalty on the weights to enforce sparsity:

$$\min_{\{w_m \geq 0\}} \left\| \Psi(\widehat{f}) - \sup_{m,\theta_m,j_m} w_m S_m^{\mathcal{F}}(\hat{f}) \right\|_{L^2} + \lambda \sum_{m \in \Lambda} w_m.$$

This approach aligns with dictionary learning and sparse coding frameworks, where the scattering coefficients serve as a structured dictionary for approximating nonlinear operators. This would allow for quantitative evaluation of the approximation error, potentially measured in L^2 norm or other task-specific metrics.

Finally, this framework suggests a practical utility in hybrid architectures: scattering transforms can serve as fixed, interpretable feature extractors, followed by learnable morphological layers or sparse linear combinations. This could be particularly beneficial in low-data regimes or for interpretable AI applications.

6 Perspectives

The morphological representation theory on the Fourier domain considered in this paper provides a rationale to explore the interest of learning Fourier erosions on that space. Our approximation formula using scattering coefficients is related to hybrid representations in deep learning [15], where the scattering coefficients are used to learn more complex operators. Beyond scattering networks, we think also exploring the interest of learning representations based on openings in the Fourier domain, i.e., ideal filters. Ideal filters have a perfectly sharp cutoff frequency, meaning they can completely eliminate frequencies outside the desired range. This makes them very effective at isolating specific frequency components. However, ideal filters are not realizable in practice in the spatial domain due to their infinite impulse response and the requirement for non-causal operations.

While the current formulation focuses on translation-equivariant operators based on Euclidean convolution, the morphological framework of complex domain is not inherently limited to this group. Recent advances in geometric deep learning have emphasized the importance of equivariance to more general transformation groups, such as rotations, scalings, and non-Euclidean symmetries. Our approach could be extended to these settings where they are naturally defined on the Fourier domain.

In many popular CNN architectures and for efficiency purposes, the convolutions on the spatial domain are just discrete kernels of 3×3, which correspond to 9 weights of the network to be learn. We would like to point out that with 9 parameters we can for instance learnt in the Fourier domain 3 2D frequency locations and 3 radii for the corresponding disks. A small number of parameters can yield high-capacity filters in the Fourier domain but raises questions about inductive bias. While such filters may be expressive, they may also lack the spatial locality or smoothness priors that are implicitly encoded in standard CNN kernels for computer vision tasks. This trade-off between expressiveness and inductive bias is an important consideration for future work as well as the interest of frequency-based biases for other signals such as audio, seismography, and multimodal data.

References

1. Angulo, J.: Some open questions on morphological operators and representations in the deep learning era. In: Lindblad, J., Malmberg, F., Sladoje, N. (eds.) DGMM 2021. LNCS, vol. 12708, pp. 3–19. Springer, Cham (2021). https://doi.org/10.1007/978-3-030-76657-3_1
2. J. Angulo.: Nonlinear Representation Theory of Equivariant CNNs on Homogeneous Spaces Using Group Morphology. In: Proceeding of DGMM 2024 (Discrete Geometry and Mathematical Morphology), LNCS 14605, Springer, 2024
3. Arora, R., Basu, A., Mianjy, P., Mukherjee, A.: Understanding deep neural networks with rectified linear units. arXiv preprint, arXiv:1611.01491, 2016
4. Bouaynaya, N., Schonfeld, D.: Theoretical foundations of spatially-variant mathematical morphology part ii: gray-level images. IEEE Trans. Pattern Anal. Mach. Intell. **30**(5), 837–50 (2008)
5. Bruna, J., Mallat, S.: Invariant scattering convolution networks. IEEE Trans. Pattern Anal. Mach. Intell. **35**(8), 1872–1886 (2013)
6. Cybenko, G.: Approximation by superpositions of a sigmoidal function. Math. Control Signals Systems **2**(4), 303–314 (1989)
7. Mallat, S.: Group invariant scattering. Commun. Pure Appl. Math. **65**(10), 1331–1398 (2012)
8. Mallat, S.: Understanding deep convolutional networks. Phil. Trans. R. Soc. A **374**(2065), 20150203 (2016)
9. Maragos, P.: A representation theory for morphological image and signal processing. IEEE Tran. Pattern Anal. Mach. Intel., **11**(6), 586–599, 1989
10. Matheron, G.: Random sets and integral geometry. John Wiley & Sons, 1974
11. Heijmans, H.J.A.M., Ronse, C.: The algebraic basis of mathematical morphology I. Dilations and erosions. Comput. Vision Graphics Image Process. **50**(3), 245–295, 1990
12. Heijmans, H.J.A.M., Keshet, R.: Inf-semilattice approach to self-dual morphology. J. Math. Imaging Vision **17**, 55–80 (2002)
13. Keshet, R.: Mathematical morphology on complete semilattices and its applications to image processing. Fund. Inform. **41**(1–2), 33–56 (2000)
14. Keshet, R.: A Morphological View on Traditional Signal Processing. In: Mathematical Morphology and its Applications to Image and Signal Processing. Computational Imaging and Vision, vol. 18. Springer, 2002
15. Oyallon, E., et al.: Scattering networks for hybrid representation learning. IEEE Trans. Pattern Anal. Mach. Intell. **41**(9), 2208–2221 (2019)
16. Serra, J., (Ed.).: Image Analysis and Mathematical Morphology: Theoretical Advances. Academic Press, 1988
17. Velasco-Forero, S., Angulo, J.: MorphoActivation: Generalizing ReLU Activation Function by Mathematical Morphology. In: Proceeding of DGMM 2022 (Discrete Geometry and Mathematical Morphology), LNCS 13493, Springer, 2022
18. Weston, J.D.: On the representation of operators by convolution integrals. Pac. J. Math. **10**(4), 1453–1468 (1960)
19. Zarka, J., Thiry, L., Angles, T., Mallat, S.: Deep Network Classification by Scattering and Homotopy Dictionary Learning. ICLR, 2020

Learning Morphological Representations of Image Transformations: Influence of Initialization and Layer Differentiability

Mihaela Dimitrova(✉), Samy Blusseau(✉), and Santiago Velasco-Forero(✉)

Mines Paris - PSL University, Centre for Mathematical Morphology (CMM), Fontainebleau, France
{mihaela.dimitrova,samy.blusseau, santiago.velasco}@minesparis.psl.eu

Abstract. As a combination of two successful paradigms in image processing, namely Mathematical Morphology and Deep Learning, Morphological Networks seem very promising for image analysis. However, their practical applications remain constrained due to possible limitations of expressivity and difficulties related to gradient descent based optimization algorithms. In this paper, we focus on neural architectures inspired by the morphological representation theory, which proves their expressivity. We investigate their optimization difficulties and find out that, rather than the non-differentiability of morphological layers, the sparsity of their gradient (or subgradient) is what limits them the most, and requires an appropriate initialization. Furthermore, we propose a method to reduce the number of model parameters after training, which produces a minimal equivalent operator.

Keywords: Morphological representations · Deep learning · Optimization

1 Introduction

Image transformations are powerful analytical techniques used to extract valuable information from an image, such as contours, key structures, geometric properties, and to reduce noise. However, implementing image transformations in an automated way remains challenging, as efficient representations are generally unknown - making learning them from data using neural networks (NNs) the conventional approach today. Alternative methods rely on nonlinear filtering by mathematical morphology operators. In order to find efficient morphological implementations of image transformations, the morphological operations were integrated as layers in neural networks with learnable parameters. This was achieved through various strategies: subdifferentials to compute the gradient of dilation and erosion operators [5,17], approximation by convolution

M. H. F. Wilkinson and J. Kosinka (Eds.): DGMM 2025, LNCS 16296, pp. 371–383, 2026.
https://doi.org/10.1007/978-3-032-09544-2_27

In turn, the *basis* of Ψ is the set of minimal elements of its kernel, that is

$$\mathcal{B}(\Psi) := \{g \in Ker(\Psi) : [h \in Ker(\Psi) \text{ and } h \leq g] \Rightarrow h = g\}. \tag{8}$$

Then Theorems 3, 4 and 6 of [10] state that any TSEI operator Ψ can be represented by its kernel and, if ψ is upper semicontinuous, it has a non-empty basis and can also be represented by it, as the supremum of erosions

$$\Psi = \sup_{g \in Ker(\Psi)} \varepsilon_g = \sup_{g \in \mathcal{B}(\Psi)} \varepsilon_g. \tag{9}$$

2.3 Two examples: a median filter and an average filter

Despite the existence of the representation 9 for any TSEI operator, there is no general algorithm to construct the basis. Hence, learning such a representation from data is the core idea of the present paper. However, like in any machine learning problem, one is faced with two issues: first, choosing the right parameterized family *a priori* without knowing, for example, the cardinal of the basis, nor the supports of its elements ; second, finding the parameters which achieve the best approximation of the target operator Ψ. In this paper we focus on the latter issue, aiming at understanding the optimization problems that arise with morphological operators. In order to discard the former issue, we focus on the target operators described hereafter, for which the basis is known, yielding an indication of a good parameterized family.

Cross Median Filter. The representation of rank filters by a finite basis was presented in [12]. Here, like in [10], we focus on the median filter in 4-connexity defined by $\text{MEDIAN}_B(f)(x) = \text{RANK}_3(\{f(x+y), y \in B\})$, where $B = \{(i,j) \in \mathbb{Z}^2, |i| + |j| \leq 1\}$ and RANK_3 returns the third largest value of the input set. Note that such a filter is TSEI. Denoting by $A_1, \ldots, A_{10}$ the ten subsets of B with cardinal three, and by $g_k := \log(\mathbb{1}_{A_k})$ for $1 \leq k \leq 10$, then $\mathcal{B}(\text{MEDIAN}_B) = \{g_k, 1 \leq k \leq 10\}$ hence MEDIAN_B is represented by the ten erosions ε_{g_k}, *i.e.* $\text{MEDIAN}_B = \max_{1 \leq k \leq 10} \varepsilon_{g_k}$. Fig. 1(a) shows the ten basis functions.

Two-Pixel Average Filter. In [11] is characterized the basis of any linear TSEI filter with finite number of non zero coefficients in its impulse response. Here we consider the particular case of a moving average filter computing the average of two neighbouring pixels, more precisely: $\text{AVERAGE}_2(f)(x) = 0.5\Big(f(x) + f\big(x + (0,1)\big)\Big)$. For any $a \in \mathbb{R}$, let g_a be the function such that $g_a\big((0,0)\big) = -g_a\big((0,1)\big) = a$ and $g_a(x) = -\infty$ if $x \notin \{(0,0),(0,1)\}$. Then $\mathcal{B}(\text{AVERAGE}_2) = \{g_a, a \in \mathbb{R}\}$ and therefore $\text{AVERAGE}_2 = \sup_{a \in \mathbb{R}} \varepsilon_{g_a}$. Fig. 1(b) shows ten examples of basis functions. Although the basis is infinite, one can show that, to represent AVERAGE_2 on a subset $\mathcal{S} \subseteq \mathcal{F}(\Omega, V)$ of images, the family $\{g_a, a \in \Delta\mathcal{S}\}$ is sufficient, where $\Delta\mathcal{S} := \{[f\big((0,0)\big) - f\big((0,1)\big)]/2, f \in \mathcal{S}\}$. Note that if $\mathcal{S}$ is finite, so is $\Delta\mathcal{S}$ [8].

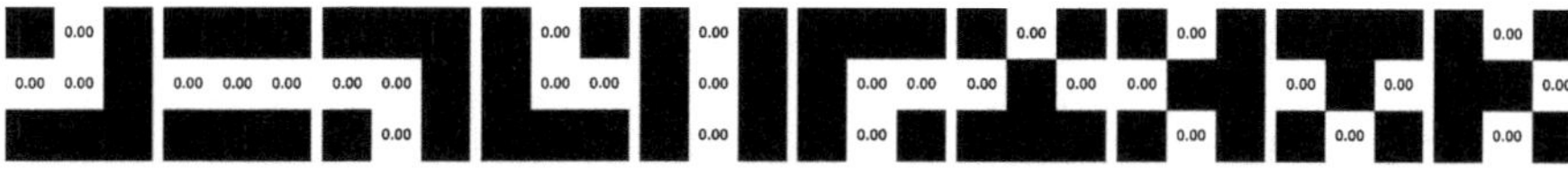

(a) The ten basis functions of the Cross Median Filter.

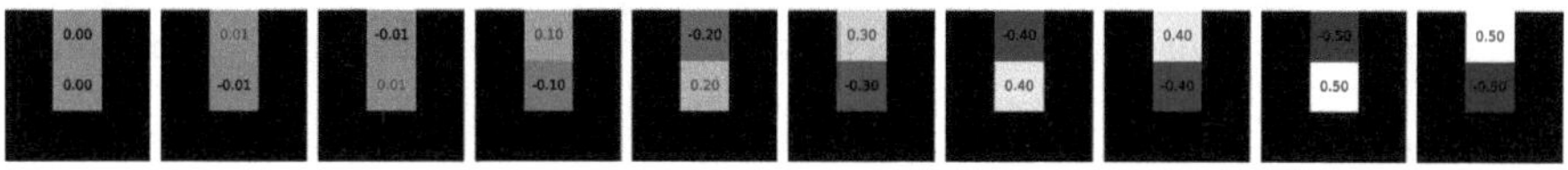

(b) Ten examples of basis functions of the Two-Pixel Average Filter.

Fig. 1. Visualization of basis functions, restricted to the 3×3 centered square $\{-1, 0, 1\}^2$, which contains their supports. Positive values are in blue color, while non-positive values are in red. The value of black pixels is $-\infty$. (Color figure online)

2.4 Minimal representation in the case of finite basis

Removing non-minimal elements from the kernel in the representation (9) is natural since $h \le g \Rightarrow \varepsilon_g \le \varepsilon_h$ and therefore non-minimal elements do not contribute to the supremum over the kernel. It is unclear, however, whether one can reduce further the set of functions to represent Ψ. We now state and prove that this is not possible if $\mathcal{B}(\Psi)$ is finite, a case that is of particular interest in the experimental setting of this paper.

Proposition 1. *Let Ψ a TSEI operator such that $\mathcal{B}(\Psi)$ is finite, and let $G \subseteq \mathcal{F}(\Omega, V)$. Then*

$$G \subseteq \mathcal{B}(\Psi) \text{ and } \Psi = \sup_{g \in G} \varepsilon_g \;\Rightarrow\; G = \mathcal{B}(\Psi). \tag{10}$$

Proof. Let us assume $G \subseteq \mathcal{B}(\Psi)$ and $\Psi = \sup_{g \in G} \varepsilon_g$, and suppose $g_0 \in \mathcal{B}(\Psi) \setminus G$. Then $G \subseteq \mathcal{B}(\Psi) \setminus \{g_0\} \subseteq \mathcal{B}(\Psi)$, therefore $\Psi = \sup_{g \in G} \varepsilon_g \le \sup_{g \in \mathcal{B}(\Psi) \setminus \{g_0\}} \varepsilon_g \le \sup_{g \in \mathcal{B}(\Psi)} \varepsilon_g = \Psi$ which shows that $\sup_{g \in \mathcal{B}(\Psi)} \varepsilon_g = \sup_{g \in \mathcal{B}(\Psi) \setminus \{g_0\}} \varepsilon_g$. Hence in particular $\sup_{g \in \mathcal{B}(\Psi) \setminus \{g_0\}} \varepsilon_g(g_0)(0) = \sup_{g \in \mathcal{B}(\Psi)} \varepsilon_g(g_0)(0) \ge \varepsilon_{g_0}(g_0)(0) = 0$. Since $\mathcal{B}(\Psi) \setminus \{g_0\}$ is finite, this means there exists $g_1 \in \mathcal{B}(\Psi) \setminus \{g_0\}$ such that $\varepsilon_{g_1}(g_0)(0) \ge 0$, which implies $g_1 \le g_0$ by (5). Hence we have found $g_1 \in \mathcal{B}(\Psi)$ such that $g_1 \le g_0$ but $g_1 \ne g_0$, which contradicts that $g_0 \in \mathcal{B}(\Psi)$. □

The practical consequence of Proposition 1 is the following. Suppose $\Psi = \sup_{g \in G} \varepsilon_g$ with G a finite subset of $\mathcal{F}(\Omega, V)$, as will be the morphological networks of Sects. 3 and 4. Then it is straightforward that $Ker(\Psi) = \{h \in \mathcal{F}(\Omega, V), \exists g \in G, h \ge g\}$, and that its basis is the set G' of minimal elements of G. Hence, the representation can be reduced to $\Psi = \sup_{g \in G'} \varepsilon_g$ but, from Proposition 1, no more element of G can be removed.

3 Learning Morphological Representations

As explained earlier, our goal is to learn morphological representations of image transformations from data, within the deep learning paradigm. In order to avoid expressivity issues and focus on optimization ones, we try to learn representations of the two filters introduced in Sect. 2.3, for which we know that suprema of erosions are an appropriate parameterized family of operators.

SupErosions Neural Network. Let $K \in \mathbb{N}^*$ and $W = (w^{(1)}, \ldots, w^{(K)}) \in (\mathbb{R}^{3\times3})^K$ a K-tuple of 3×3 matrices with coefficients indexed by $B = \{-1, 0, 1\}^2$. For $1 \leq k \leq K$ we define the function g_k by $g_k(x) = w_x^{(k)}$ if $x \in B$ and $g_k(x) = -\infty$ otherwise, such that, for $f \in \mathcal{F}(\Omega, V)$ and $x \in \Omega$, $\varepsilon_{g_k}(f)(x) = \min_{y\in B} f(x+y) - w_y^{(k)}$. Then we consider the family of operators parameterized by W, defined by $\Psi_W = \max_{1\leq k\leq K} \varepsilon_{g_k}$, *i.e.* $\forall f \in \mathcal{F}(\Omega, V)$, $\forall x \in \Omega$

$$\Psi_W(f)(x) = \max_{1\leq k\leq K} \min_{y\in B} f(x+y) - w_y^{(k)}. \tag{11}$$

We will call this sequence of operations **SupErosions** block, as it can (and will) be implemented as a feed-forward neural architecture.

Optimization with Gradient Descent. Now given a training set $\{f_1, \ldots, f_N\}$ of N images, we look for the best parameter set W^* minimizing the empirical risk, that is

$$W^* = \underset{W\in(\mathbb{R}^{3\times3})^K}{\arg\min} \sum_{n=1}^{N} \mathcal{L}(\Psi_W(f_n), \Psi^*(f_n)) \tag{12}$$

where Ψ^* is a target operator and $\mathcal{L}$ a loss function penalizing discrepancies between the expected and the actual output. In practice images are functions of a bounded domain $\Omega' \subset \mathbb{Z}^2$, and we choose the mean square error (MSE) loss $\mathcal{L}(\Psi_W(f_n), \Psi^*(f_n)) = \frac{1}{|\Omega'|} \sum_{x\in\Omega'} (\Psi_W(f_n)(x) - \Psi^*(f_n)(x))^2$.

The minimization problem (12) is classically addressed with (variants of) the stochastic gradient descent algorithm. However, in the present case, for a fixed image f, $\Psi_W(f)$ (hence $\mathcal{L}(\Psi_W(f), \Psi^*(f))$ is (*Fréchet*) differentiable with respect to W almost everywhere, but not everywhere. To make this more precise, for any $x \in \Omega'$ let $\kappa(x) := \{k \in \{1, \ldots, K\}, \Psi_W(f)(x) = \varepsilon_{g_k}(x)\}$, and for any $k \in \{1, \ldots, K\}$, let $\mathcal{I}_k(x) := \{y \in B, \varepsilon_{g_k}(x) = f(x+y) - w_y^{(k)}\}$. Conversely, let $\nu_k := \{x \in \Omega', k \in \kappa(x)\}$ and, for $y \in B$, $\mu_y^{(k)} := \{x \in \Omega', y \in \mathcal{I}_k(x)\}$. Then $\Psi_W(f)$ is differentiable with respect to W if and only if for all $x \in \Omega'$, $|\kappa(x)| = 1$ and $|\mathcal{I}_{k_x}(x)| = 1$, where k_x is the only element of $\kappa(x)$. When this condition is fulfilled, then the gradient of the loss with respect to W is

$$\nabla_W \mathcal{L}(\Psi_W(f), \Psi^*(f)) = \left(\frac{2}{|\Omega'|} \sum_{x\in\mu_y^{(k)}\cap\nu_k} [\Psi^*(f)(x) - \Psi_W(f)(x)] \right)_{\substack{y\in B,\\ 1\leq k\leq K}}. \tag{13}$$

Therefore, the update of $w_y^{(k)}$ is proportional to sum of differences $\Psi^*(f) - \Psi_W(f)$ over the pixels for which the kth erosion achieves the maximum in (11), and y achieves the minimum. This expression comes from the gradients of the functions $u \in \mathbb{R}^d \mapsto \varphi(u_1, \ldots, u_d)$, where $d \in \mathbb{N}^*$ and φ is max or min. The gradient exists if and only if the maximum (resp. minimum) is reached only once, and it is equal to $\left(\mathbb{1}_{\varphi(u)}(u_i)\right)_{1 \leq i \leq d}$. Alternatively, when the gradient does not exist, it may be replaced by $\left(\mathbb{1}_{\varphi(u)}(u_i) / \sum_j \mathbb{1}_{\varphi(u)}(u_j)\right)_{1 \leq i \leq d}$, which is an element of the subgradient, as it is done in the Tensorflow framework. Note that the latter vector coincides with the gradient in the differentiable case but, as explained in [3], in general it does not ensure to get a descent direction when included in the backpropagation chain-rule. This alternative leads to the following pseudo-gradient, which we will use in our experiments:

$$\widetilde{\nabla}_W \mathcal{L}(\Psi_W(f), \Psi^*(f)) = \left(\frac{2}{|\Omega'|} \sum_{x \in \mu_y^{(k)} \cap \nu_k} \frac{[\Psi^*(f)(x) - \Psi_W(f)(x)]}{|\mathcal{I}_k(x)| \cdot |\kappa(x)|} \right)_{\substack{y \in B, \\ 1 \leq k \leq K}} . \quad (14)$$

Again, (14) coincides with the gradient (13) in the differentiable case, where $|\kappa(x)| = 1$ and $|\mathcal{I}_k(x)| = 1$ for $x \in \nu_k$; but $-\widetilde{\nabla}_W \mathcal{L}$ is not necessarily a descent direction in the non-differentiable case. Note that in either case, (13) or (14), structuring elements with low values are more likely to be updated because the corresponding erosion is more likely to reach the maximum, as $w^{(k_1)} \leq w^{(k_2)} \Rightarrow \varepsilon_{g_{k_1}} \geq \varepsilon_{g_{k_2}}$.

Minimal Elements Reduction. After training, Ψ_W can be reduced, without any loss, to $\Psi_{W'}$, where W' is the set of minimal elements of W. The fact that $\Psi_W = \Psi_{W'}$ and that we cannot remove any other element from W', is the consequence of Proposition 1, as explained at the end of Sect. 2.4.

4 Morphological Representation Benchmark

The success of the training procedure described in the previous section depends significantly on the choice of various parameters such as the number of structuring elements, their initialization, the optimizer, the learning rate and the batch size. It may also depend on the (non)-differentiability of the **SupErosions** block. Since we noticed a great sensitivity to parameter initialization, in this section we measure and analyze some of its effects on learning accurate representations of the two operators presented in Sect. 2.3, the cross median filter and the two-pixel average filter. We also study the links of these effects with the differentiability of the network.

4.1 Experimental Protocol

Data. We worked with a noisy version of the FashionMNIST dataset [18], obtained by adding zero mean Gaussian noise with standard deviation $\sigma = 40$ to the original images, initially quantized on 8 bits. After adding noise, the images were clipped to $[0, 255]$ and finally normalized to $[0, 1]$. Our training dataset was composed of the first 100 images from the original training set, while our validation dataset consisted of the first 100 images from the original test set. Adding noise allowed to provide more information to the models on the effect of the filters to be learned, without increasing the number of training images.

Architecture and Fixed Hyperparameters. We implemented a **SupErosions** block, as described in Sect. 3, with Tensorflow[1]. The number of structuring elements was set to $K = 100$ for the MEDIAN experiments and $K = 200$ for the AVERAGE experiments, with all elements having a fixed size of 3×3 in both cases. The batch size was set to 50 and 10 images respectively for the MEDIAN and the AVERAGE experiments. We used the mean squared error (`MSE`) as loss function, and Adam optimizer [9] with learning rate $l_r = 0.01$. The maximum number of training epochs was 2000; however, training was stopped early if the validation loss dropped below the threshold $Q = 1/255^2$, corresponding to an average per-pixel error of less than one gray level in 8-bit images.

Varying Initialization Hyperparameters. In every training session, the parameters were randomly initialized according to a uniform distribution over an interval, but we varied the center of that interval (the expected value of the distribution). For each expectation value, twenty repetitions of training were performed and we calculated the mean and the standard deviation of the following two metrics: **(i) The success rate** - the proportion of trainings where the `MSE` of the validation set reached a level below the threshold $Q = 1/255^2$. **(ii) The number of minimal structuring elements** obtained, after training, with the minimal elements reduction described in Sect. 3, as an indication of whether the learned representation is close to the theoretical one.

Additional Experiments and Measurements. In the MEDIAN filter setting, we additionally measured the **proportion of non-differentiable cases** *i.e.* those for which the differentiability condition stated in Sect. 3 was not fulfilled. This measurement was performed for three repetitions with fixed initialization associated with the intervals [–0.35,–0.25] and [0.95,1.05]. For the initialization with the interval [-1.05,-0.95], we performed the measurement for three successful and three unsuccessful repetitions. Furthermore, in order to check the generality of our observations, we also varied the width of the initialization interval, in the MEDIAN filter experiment. Twenty repetitions of training were performed for every pair of initialization hyperparameters.

[1] https://www.tensorflow.org/.

4.2 Results

The results from training with different initializations are shown in Fig. 3 and Fig. 5. Some learned and subsequently reduced representations (which we will further call simply "learned") are illustrated in Fig. 2. The differentiability measurements are reported in Table 1, and we represented the histograms of the weight values at different training stages in Fig. 4.

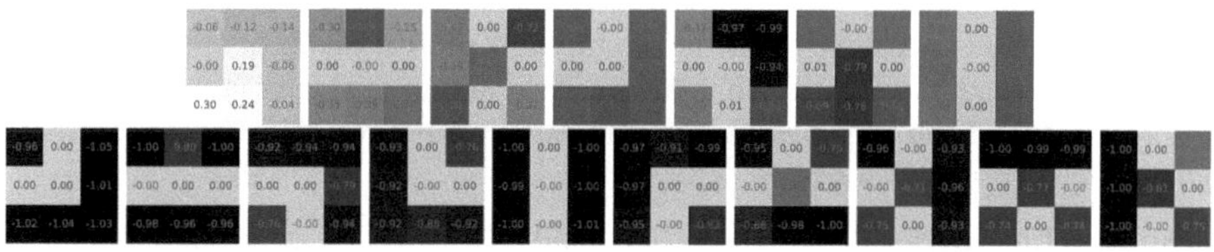

(a) An example of learned representation of the cross median filter. Note the similarity of the bottom row to the theoretical basis illustrated in Fig. 1(a).

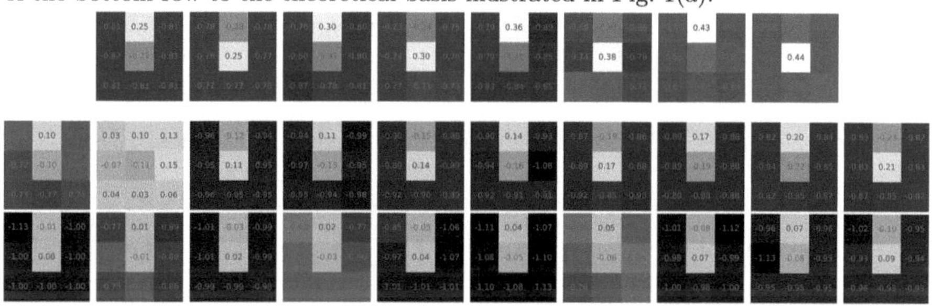

(b) An example of learned representation of the two-pixel average filter. Note it only contains 28 structuring elements, although the basis is theoretically infinite and expected to be a lot larger than 28 for the considered finite training set.

Fig. 2. Representations successfully learned by the **SupErosions** block, after minimal elements reduction.

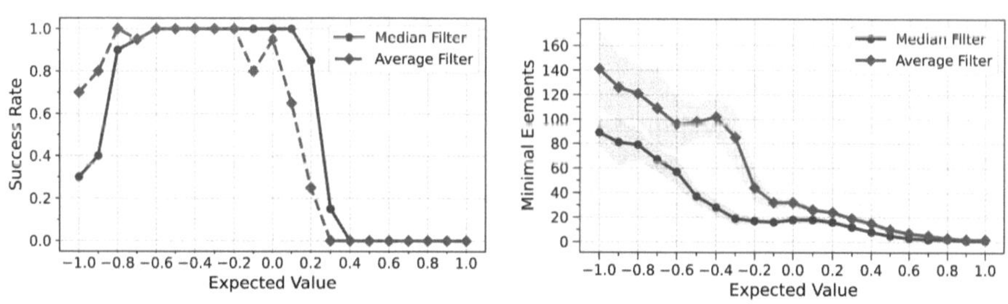

Fig. 3. Success rates (on twenty repetitions) and numbers of minimal structuring elements for different initialization interval centers (expected values) and fixed interval width (0.1 for the MEDIAN experiment and 0.3 for the AVERAGE experiment). The width of the shaded areas is given by twice the standard deviation.

Learned Representations. The high success rate of some initializations and the significant resemblance (notably for the MEDIAN but also for the support pixels of the AVERAGE filter) between the theoretical and the learned representations prove that some image transformations can be successfully approximated by exact morphological neural layers trained with gradient descent. Moreover, the learned representation of the AVERAGE filter is surprisingly lightweight, given its theoretical infinite version, as it can be approximated with as few as 28 filters (see Fig. 2 b). Even considering a finite set $\mathcal{S}$ of 100 noisy Fashion MNIST images, with the reasoning of Sect. 2.3 we would expect $|\Delta\mathcal{S}| \approx 53400$ structuring elements[2]. Furthermore, if we focus on the top and central pixels of the structuring elements in Fig. 2 b, we notice that the positive weight is always smaller than the absolute value of the negative weight, contrary to the theoretical representation, where they are equal (see Sect. 2.3).

Training Limitations. On the other hand, the training outcome is strongly influenced by the initialization parameters. Based on the training performances, we would like to pay closer attention to three groups of initializations with respect to the expected value of the weights distribution.

Table 1. Average proportion of non-differentiability for different initialization intervals (MEDIAN experiment). **(U)**: Unsuccessful and **(S)**: Successful trainings.

Interval	Success rate	Non-differentiability rate
[−1.05,−0.95]	30 %	**(U)** 16.03 ± 8.02 % ; **(S)** 38.87 ± 18.73 %
[−0.35, −0.25]	100 %	5.21 ± 3.40 %
[+0.95,+1.05]	0 %	0.15 ± 0.04 %

Suitable expected value- Initializations with intermediate expected values are associated with high success rate and low uncertainty(i.e., small standard deviation). During training, the process efficiently optimizes a sufficient number of minimal elements, rendering the remaining ones redundant.

Too big expected value- Initializations with expected value ≥ 0.5 result in a 0% success rate. The learned representations have significantly small number of minimal elements which limits their expressivity. If we look at the weight evolution at the top row of Fig. 4, we notice how this inappropriate initialization setting updates a highly restricted number of structuring elements while leaving the rest unchangeably redundant. Recall that, because of Eq. (13) and (14), the structuring elements with low values are more likely to be updated. Consequently here, the minimal elements become smaller, remain minimal and the only ones that are updated. As Table 1 shows, this training limitation is caused by the sparsity of the gradient (only few weights are updated) and not by the non-differentiability of $\Psi_W(f)$. In fact, during these training runs, the function remained differentiable in more than 99.8% of the parameter updates.

[2] This was computed on average over twenty noisy datasets samples.

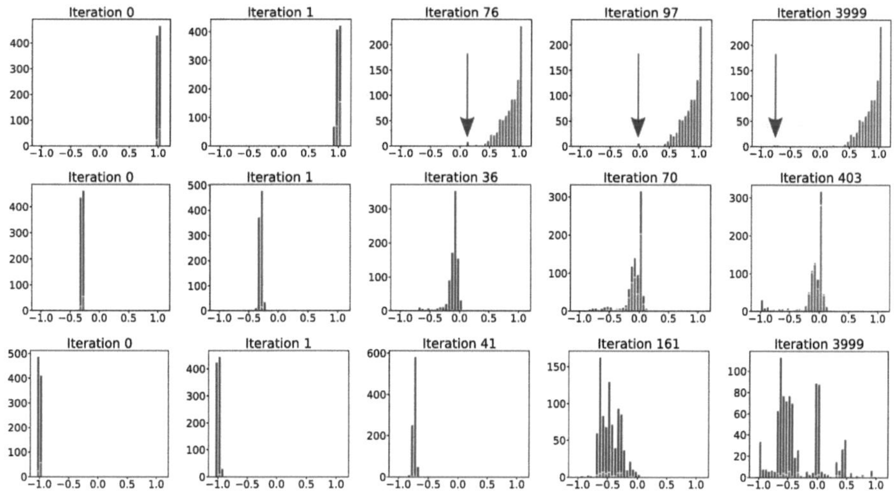

Fig. 4. Histograms showing the evolution of the weight values for different initialization intervals (Median experiment). Top row: [+0.95,+1.05], Middle row:[−0.35,−0.25] and Bottom row:[−1.05,−0.95]. The weights of the minimal structuring elements are in red color, while the weights of others are in blue. Bins with very short bars are indicated by an arrow of the same color. Each iteration corresponds to an update of parameters. (Color figure online)

Too small expected value- As the expected value of the initialization weights decreases, the success rates drops, while the number of minimal elements increases, becoming significantly higher than the expected sufficient number. This can be an indicator that unsuccessful trainings learn too many minimal elements that do not belong to a suitable representation. The bottom row in Fig. 4 illustrates how, at the beginning of the training, minimal elements are forced to become larger, causing others to become minimal. We have a significant portion of minimal elements on each iteration and an unsuitable final distribution. The non-differentiability in this setting appears pronounced. However, its impact on training outcomes remains inconclusive, as successful training runs may still occur despite high non-differentiability, and vice versa.

The gradient descent can fail to optimize consistently an optimal number of minimal elements due to an initialization with inappropriate expected value. Fig. 5 illustrates how the choice of the interval width can additionally strengthen or weaken the expected value influence (for example small intervals can limit the diversity of the weights and make trainings more unsuccessful). The asymmetry of the influence of the initializations with small expected value and big expected value confirms as it leads to different range of inappropriate number of minimal elements.

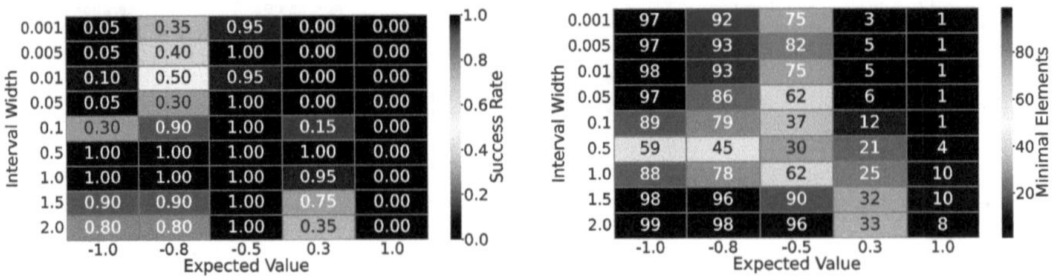

Fig. 5. Success rates and number of minimal structuring elements for different initialization interval centers (expected values) and widths (MEDIAN experiment).

5 Conclusion and Future Work

This paper has demonstrated that **SupErosions** blocks can approximate increasing, translation- and shift-equivariant image transformations with small error by learning structuring elements through gradient descent. Through the theoretical cases of the median and two-pixel average filters, we analyzed the challenges of this learning process. Our findings highlight the critical role of weight initialization in the convergence toward accurate approximations of the desired transformations. Furthermore, we have shown that the primary barrier to effective training is not the non-differentiability of the maximum and minimum operators, as commonly assumed, but rather the sparsity of the gradient during optimization. These insights offer valuable guidance for the design of neural architectures involving morphological operators to represent more general image transformations, and open avenues for improving training stability in such contexts.

Acknowledgments. The authors gratefully acknowledge Gustavo (Jesús) Angulo, who inspired the present work by sharing his knowledge during discussions and presentations at CMM.

References

1. Banon, G.J.F., Barrera, J.: Decomposition of mappings between complete lattices by mathematical morphology, part I. General lattices. Sign. Process. **30**(3), 299–327 (1993)
2. Banon, G.J.F., Barrera, J.: Minimal representations for translation-invariant set mappings by mathematical morphology. SIAM J. Appl. Math. **51**(6), 1782–1798 (1991)
3. Blusseau, S.: Training morphological neural networks with gradient descent: some theoretical insights. In: International Conference on Discrete Geometry and Mathematical Morphology, pp. 229–241. Springer (2024)
4. Blusseau, S., Liu, X., Velasco-Forero, S.: Cell counting with trainable h-maxima and connected component layers. J. of Math. Im. and Vis. **67**(27), 27 (2025)

5. Franchi, G., Fehri, A., Yao, A.: Deep morphological networks. Patt. Recogn. **102**, 107246 (2020)
6. Groenendijk, R., Dorst, L., Gevers, T.: Geometric back-propagation in morphological neural networks. IEEE Trans. Patt. Anal. Mach. Intell. **45**(11), 14045–14051 (2023)
7. Hermary, R., Tochon, G., Puybareau, É., Kirszenberg, A., Angulo, J.: Learning grayscale mathematical morphology with smooth morphological layers. J. Math. Imaging Vision **64**(7), 736–753 (2022)
8. Khosravi, M., Schafer, R.W.: Implementation of linear digital filters based on morphological representation theory. IEEE Trans. Signal Process. **42**(9), 2264–2275 (2002)
9. Kingma, D.P., Ba, J.: Adam: a method for stochastic optimization. In: Bengio, Y., LeCun, Y. (eds.) 3rd International Conference on Learning Representations, ICLR 2015, San Diego, May 7-9, 2015, Conference Track Proceedings (2015)
10. Maragos, P.: A representation theory for morphological image and signal processing. IEEE Trans. Patt. Anal. Mach. Intell. **11**(6), 586–599 (1989)
11. Maragos, P., Schafer, R.: Morphological filters-part I: their set-theoretic analysis and relations to linear shift-invariant filters. IEEE Trans. Acoust. Speech Signal Process. **35**(8), 1153–1169 (1987)
12. Maragos, P., Schafer, R.: Morphological filters-part II: their relations to median, order-statistic, and stack filters. IEEE Trans. Acoust. Speech Signal Process. **35**(8), 1170–1184 (1987)
13. Masci, J., Angulo, J., Schmidhuber, J.: A learning framework for morphological operators using counter–harmonic mean. In: International Symposium on Mathematical Morphology and Its Applications to Signal and Image Processing, pp. 329–340. Springer (2013)
14. Matheron, G.: Random sets and integral geometry. Wiley (1975)
15. Mondal, R., Dey, M.S., Chanda, B.: Image restoration by learning morphological opening-closing network. Math. Morphology-Theory Appl. **4**(1), 87–107 (2020)
16. Nogueira, K., Chanussot, J., Dalla Mura, M., Dos Santos, J.A.: An introduction to deep morphological networks. IEEE Access **9**, 114308–114324 (2021)
17. Velasco-Forero, S., Pages, R., Angulo, J.: Learnable empirical mode decomposition based on mathematical morphology. SIAM J. Imag. Sci. **15**(1), 23–44 (2022)
18. Xiao, H., Rasul, K., Vollgraf, R.: Fashion-mnist: a novel image dataset for benchmarking machine learning algorithms. arXiv preprint arXiv:1708.07747 (2017)

Morphological Perceptron with Competitive Layer: Training Using Convex-Concave Procedure

Iara Cunha[1(✉)] and Marcos Eduardo Valle[2]

[1] Universidade Tecnológica Federal do Paraná (UTFPR), Ponta Grossa, Brazil
iarasilva@utfpr.edu.br
[2] Universidade Estadual de Campinas (UNICAMP), Campinas, Brazil
valle@ime.unicamp.br

Abstract. A morphological perceptron is a multilayer feedforward neural network in which neurons perform elementary operations from mathematical morphology. For multiclass classification tasks, a morphological perceptron with a competitive layer (MPCL) is obtained by integrating a winner-take-all output layer into the standard morphological architecture. The non-differentiability of morphological operators renders gradient-based optimization methods unsuitable for training such networks. Consequently, alternative strategies that do not depend on gradient information are commonly adopted. This paper proposes the use of the convex-concave procedure (CCP) for training MPCL networks. The training problem is formulated as a difference of convex (DC) functions and solved iteratively using CCP, resulting in a sequence of linear programming subproblems. Computational experiments demonstrate the effectiveness of the proposed training method in addressing classification tasks with MPCL networks.

Keywords: Morphological neural network · Convex-concave procedure · Difference of convex optimization

1 Introduction

The perceptron, introduced by Rosenblatt in 1958, is a fundamental model in neural computation and machine learning. It is characterized as a single linear threshold unit, along with a learning rule that effectively trains the perceptron for linearly separable classification tasks. However, due to the limited computational capabilities of Rosenblatt's perceptron, the model was successfully expanded into a multi-layer perceptron (MLP) architecture in the 1980s. The success of MLP networks can be attributed not only to their multi-layer structure but also to the efficacy of the backpropagation algorithm. This algorithm enables the computation of gradients used in first-order optimization techniques to minimize the loss function during the training phase.

M. H. F. Wilkinson and J. Kosinka (Eds.): DGMM 2025, LNCS 16296, pp. 384–398, 2026.
https://doi.org/10.1007/978-3-032-09544-2_28

The morphological perceptron (MP) belongs to the broader category of morphological neural networks, which were developed in the early 1990s by various researchers [6,17,20,22]. Morphological neural networks differ from the traditional perceptron by performing a nonlinear elementary operation based on mathematical morphology before applying the activation function [25]. They continue to attract the attention of the scientific community because of their potential to capture geometrical and topological features [9–11,16,27].

In the MP model, traditional multiplication and addition are substituted with addition and the maximum (or minimum) operations. Using the maximum (or minimum) of sums instead of the sum of products, the computations within the morphological network become non-linear before applying the activation function. This leads to significant geometric and algebraic differences compared to traditional neural network models [3,14]. One drawback of morphological neural networks is that training them using backpropagation can be particularly challenging [3,8]. As a result, many researchers have explored alternative morphological structures or learning rules for these networks [3,9,10,15,26]. Inspired by the insightful work of Charisopoulus and Maragos [3], we propose training a morphological perceptron with a competitive layer (MPCL) network using the convex-concave procedure [13,29]. The motivation for considering this model will be explained in the following paragraphs.

Like the traditional perceptron model, a single morphological perceptron has limited computational capabilities, which can be circumvented by employing a multi-layer structure. The multi-layer morphological perceptron (MLMP) network can theoretically solve any binary classification task [18,24]. Specifically, an MLMP can partition $\mathbb{R}^n$ into the union of hyperboxes. Since any compact set within $\mathbb{R}^n$ can be enclosed by a union of hyperboxes, it follows that a binary classification problem can be resolved by encapsulating all positive training data into hyperboxes [18,19]. This characteristic results in a highly interpretable machine learning model, where a pattern is classified as a positive class if it falls within one of the model's hyperboxes. Moreover, the multi-layer morphological perceptron can be adapted for multi-class classification tasks by incorporating a winner-take-all output layer [25]. Intuitively, the winner-take-all layer enables the hyperboxes to compete in determining the class of an input pattern, resulting in a model referred to as the morphological perceptron with a competitive layer (MPCL). Training an MPCL network is conducted using the learning rule of an MLMP model along with a one-against-all strategy [25].

The first training algorithms for the MLMPs progressively add hyperboxes, expanding the width of the network sequentially until a stopping criterion is met, such as achieving no misclassifications on the training set. Due to the incremental nature of these algorithms, we refer to them as greedy training algorithms for MLMPs. Examples include the algorithms detailed in [18,24,25]. In contrast to the backpropagation-like methods widely used to train traditional neural networks, the greedy algorithms for training an MLMP do not rely on gradients. While these algorithms train quickly, they often lead to large MLMP models with an excessive number of hyperboxes, making them difficult to interpret. In this paper, we apply the convex-concave procedure to train a fixed-architecture MPCL network, enhancing the interpretability of the resulting machine-learning model. Specifically, in our previous conference paper [4], we framed the learning process of a binary MLMP classifier as a difference of convex optimization problem. This was solved using the disciplined convex-concave programming framework, which is available as an extension of `CVXPY` library for `Python` [23]. In this paper, we provide a detailed explanation of the convex-concave procedure for training an MLMP network. Additionally, we introduce a regularization term to limit the size of the hyperboxes and propose an initialization method based on k-means++. Finally, based on [25], we extend the model to address multi-class classification problems by considering a morphological perceptron with a competitive layer.

This paper is organized as follows: Sect. 2 reviews the MPCL network. In Sect. 3, we present an optimization problem for training this model. Section 4 discusses the application of the convex-concave procedure to address the non-convex optimization problem outlined in Sect. 3. We provide the results of computational experiments comparing the performance of the proposed training algorithm with similar algorithms from the literature in Sect. 5. Finally, Sect. 6 offers some concluding remarks.

2 Morphological Perceptron with Competitive Layer

The computations of a morphological perceptron (MP) rely on the algebraic structure $(\bar{\mathbb{R}}, \vee, \wedge, +, +')$ from minimax algebra [5], where $\bar{\mathbb{R}} = \{-\infty, +\infty\}$ and $\vee$ and $\wedge$ denote the supremum and infimum operations, respectively. The operations "+" and "+′" coincide with the real-number addition for finite values, but differ in their treatment of infinities, as specified by $(+\infty) + (-\infty) = (-\infty)+(+\infty) = -\infty$ and $(+\infty)+'(-\infty) = (-\infty)+'(+\infty) = +\infty$. The conjugate of an element $x \in \bar{\mathbb{R}}$ is defined by

$$\texttt{conjugate}(x) = \begin{cases} -x, & x \in \mathbb{R}, \\ -\infty, & x = +\infty, \\ +\infty, & x = -\infty. \end{cases} \tag{1}$$

In this paper, we will focus solely on finite numbers. This allows us to treat $+$ and $+'$ equally (we will only use the symbol $+$) and substitute $\texttt{conjugate}(x)$ with $-x$.

The model considered in this work is a morphological perceptron with a competitive layer (MPCL), as proposed by Sussner and Esmi [25]. Figure 1 illustrates the architecture of this morphological neural network. An MPCL is structured with the following components: a first layer composed of morphological blocks, a hidden layer containing S modules, and a winner-takes-all layer represented by an argmax operation.

The first layer of this neural network consists of blocks, each containing two morphological perceptrons (MPs) that use the identity function as their activation function. One MP performs an erosion, while the other MP implements

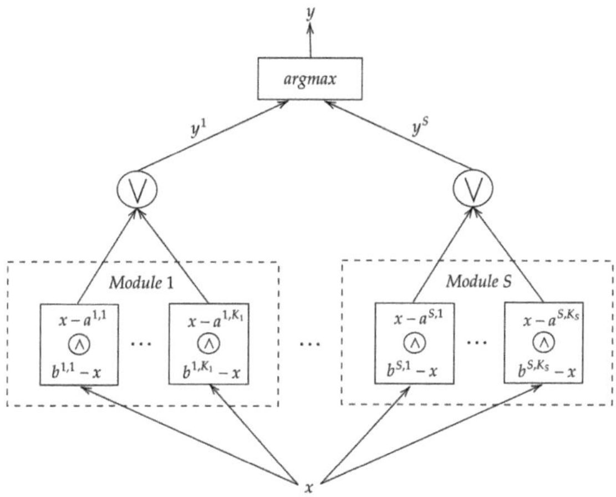

Fig. 1. Architecture of MPCL classifier.

an anti-dilation, both representing elementary operations from mathematical morphology [20,21]. In mathematical terms, let $\boldsymbol{a}^k = (a_1^k, \ldots, a_n^k) \in \mathbb{R}^n$ and $\boldsymbol{b}^k = (b_1^k, \ldots, b_n^k) \in \mathbb{R}^n$ represent the synaptic weights of the two MPs in the k-th block, respectively. The output of the k-th neuron is given by

$$h_k(\boldsymbol{x}) = \left(\bigwedge_{i=1}^{n} (x_i - a_i^k) \right) \wedge \left(\bigwedge_{i=1}^{n} (b_i^k - x_i) \right), \tag{2}$$

where $\boldsymbol{x} = (x_1, x_2, \ldots, x_n) \in \mathbb{R}^n$ is the input vector. Note that the morphological block represented by (2) characterizes a hyperbox in $\mathbb{R}^n$. Specifically, we have that $h_k(\boldsymbol{x}) \geq 0$ if and only if $a_i^k \leq x_i \leq b_i^k$ for all $i = 1, \ldots, n$. Equivalently, we have $h_k(\boldsymbol{x}) \geq 0$ if and only if $\boldsymbol{x} \in [\boldsymbol{a}^k, \boldsymbol{b}^k]$. Geometrically, the hypersurface of the function h_k is a hyper-pyramid with the hyperbox $[\boldsymbol{a}^k, \boldsymbol{b}^k]$ as base.

The hidden layer of an MPCL is composed of S modules. The sth module computes the maximum of K_s morphological blocks. Precisely, the output of the sth module in the hidden layer is y^s given by

$$y^s = \bigvee_{k=1}^{K_s} h_{s,k}(\boldsymbol{x}) = \bigvee_{k=1}^{K_s} \left(\bigwedge_{i=1}^{n} (x_i - a_i^{s,k}) \wedge \bigwedge_{i=1}^{n} (-x_i + b_i^{s,k}) \right), \quad \forall \boldsymbol{x} \in \mathbb{R}^n, \tag{3}$$

where the index pair "s,k" in the morphological block means "$k + \sum_{i=1}^{s-1} K_i$", for $s = 1, \ldots, S$. From a geometrical point of view, the output of the sth module characterizes the union of hyperboxes. Accordingly, we have $y^s \geq 0$ if and only if $\boldsymbol{x} \in [\boldsymbol{a}^{s,k}, \boldsymbol{b}^{s,k}]$ for some $k \in \{1, \ldots, K_s\}$.

Finally, the neurons in the output layer compete, and the MPCL assigns the class label corresponding to the largest output from the modules. Formally, the input vector $\boldsymbol{x}$ is classified as belonging to a class s^* such that $y^{s^*} = \max\{y^1, \ldots, y^S\}$. Since the probability of having two boxes with the same value is nearly zero, we can take $s^* = \arg\max\{y^1, \dots, y^S\}$ as the first index that yields the maximum. Thus, the MPCL assigns the class s associated with the hyperbox $[\mathbf{a}^{s,k}, \mathbf{b}^{s,k}]$ that best encloses the input $\boldsymbol{x}$. Based on this remark, Sussner and Esmi proposed an algorithm that recursively constructs hyperboxes containing only data from one class [25]. Alternative constructive morphological perceptron algorithms are discussed in [18,24]. From an optimization standpoint, these constructive algorithms utilize greedy heuristics to reduce classification error. The following section presents an alternative approach for training an MPCL network.

3 Optimization Problem for Training the MPCL Network

This section presents a constrained convex-concave optimization problem for training the MPCL network to solve multi-class classification tasks. Accordingly, consider a training set

$$\mathcal{T} = \{(\boldsymbol{x}^{(j)}, y^{(j)}) : j = 1, \dots, M\} \subseteq \mathbb{R}^n \times \mathcal{S} \tag{4}$$

where M is the number of samples, $\boldsymbol{x}^{(j)} = (x_1^{(j)}, x_2^{(j)}, \dots, x_n^{(j)}) \in \mathbb{R}^n$ is the jth feature vector, and $y^{(j)} \in \mathcal{S}$ is its label. We will identify the label set with $\mathcal{S} = \{1, \dots, S\}$ for simplicity.

The MPCL training aims to create hyperboxes that efficiently enclose the training data from a single class. The goal is accomplished using a one-against-all strategy. Precisely, let us define the sets of negative and positive samples with respect to a class label $s \in \mathcal{S}$ as follows, respectively:

$$C_0^s = \{\boldsymbol{x}^{(j)} : y^{(j)} \neq s\} \quad \text{and} \quad C_1^s = \{\boldsymbol{x}^{(j)} : y^{(j)} = s\}. \tag{5}$$

The training process aims to find a family of hyperboxes

$$\mathcal{H}_s = \left\{[\boldsymbol{a}^{s,1}, \boldsymbol{b}^{s,1}], [\boldsymbol{a}^{s,2}, \boldsymbol{b}^{s,2}], \dots, [\boldsymbol{a}^{s,K_s}, \boldsymbol{b}^{s,K_s}]\right\}, \tag{6}$$

whose union contains samples from class s while excluding patterns from other classes. In mathematical terms, the hyperboxes are obtained by solving the optimization problem:

$$\begin{aligned}
\min_{\boldsymbol{a}^s, \boldsymbol{b}^s, \boldsymbol{\xi}^s} \; & \sum_{l=1}^{M} \xi_l^s + \gamma \sum_{k=1}^{K_s} \sum_{i=1}^{n} (b_i^{s,k} - a_i^{s,k}), \\
\text{s.t.} \; & \bigvee_{k=1}^{K_s} \left(\bigwedge_{i=1}^{n} (x_i^{(l)} - a_i^{s,k}) \wedge \bigwedge_{i=1}^{n} (-x_i^{(l)} + b_i^{s,k}) \right) \leq \xi_l^s \quad \text{if } \boldsymbol{x}^{(l)} \in C_0^s, \\
& \bigvee_{k=1}^{K_s} \left(\bigwedge_{i=1}^{n} (x_i^{(l)} - a_i^{s,k}) \wedge \bigwedge_{i=1}^{n} (-x_i^{(l)} + b_i^{s,k}) \right) \geq -\xi_l^s \;\; \text{if } \boldsymbol{x}^{(l)} \in C_1^s, \\
& \boldsymbol{a}^s \leq \boldsymbol{b}^s, \quad \mathbf{0} \leq \boldsymbol{\xi}^s,
\end{aligned} \tag{7}$$

where $\boldsymbol{a}^s = [\boldsymbol{a}^1, \ldots, \boldsymbol{a}^{K_s}]$ and $\boldsymbol{b}^s = [\boldsymbol{b}^1, \ldots, \boldsymbol{b}^{K_s}]$ are the lists of lower and upper vertices of the hyperboxes, $\boldsymbol{\xi}^s = (\xi_1^s, \ldots, \xi_M^s) \in \mathbb{R}^M$ is the vector of slack variables, and γ is a regularization parameter.

The optimization problem presented in (7) is based on the approach by Charisopoulos and Maragos [3], who utilize a formulation similar to support vector machines but employ a minimax algebraic structure. The first two constraints address classification errors for negative and positive samples, respectively. Specifically, a sample $\boldsymbol{x}^{(l)}$ that does not belong to class s should not be included in the union of the hyperboxes. Mathematically, this is expressed using (3) as the inequality $\bigvee_{k=1}^{K_s} h_{s,k}(\boldsymbol{x}^{(l)}) < 0$. By incorporating the slack variable ξ_l^s on the right-hand side, we allow for the possibility of violating this inequality, resulting in the first constraint in (7). Conversely, if the sample $\boldsymbol{x}^{(l)}$ belongs to class s, the inequality $\bigvee_{k=1}^{K_s} h_{s,k}(\boldsymbol{x}^{(l)}) > 0$ should hold. Including the slack variable ξ_l^s on the right-hand side of this inequality leads to the second constraint in (7). The final two inequality constraints ensure that $\boldsymbol{a}^s$ and $\boldsymbol{b}^s$ define valid hyperboxes and that the slack variables remain non-negative. It is important to note that $\xi_l^s > 0$ accounts for a misclassified sample $\boldsymbol{x}^{(l)}$. Consequently, the first term in the objective function aims to minimize the slack variables. The second term, which can be interpreted as a regularization term, aims to minimize the size of the hyperboxes.

Although the objective of the optimization problem (7) is linear, the constraints are neither convex nor concave functions, making it a challenging optimization problem. Nevertheless, it is possible to reformulate the constraints of (7) using the difference of convex (DC) functions [4]. Recall that a real-valued function ϕ is called a difference of convex (DC) function if there exist two convex functions f and g such that $\phi = f - g$. A DC optimization problem has the form

$$\begin{aligned} &\min_{\boldsymbol{v}} f_0(\boldsymbol{v}) - g_0(\boldsymbol{v}) \\ &\text{s.t. } f_i(\boldsymbol{v}) - g_i(\boldsymbol{v}) \leq 0, \quad \forall i = 1, \ldots, N_C, \end{aligned} \tag{8}$$

where $\boldsymbol{v}$ is the optimization variable and f_i, g_i are convex functions for all $i \in \{0, 1, \ldots, N_C\}$, with N_C being the number of constraints [13].

The function h_k given by (2), which computes the minimum of linear functions, is concave. Hence, the function $\Psi(\cdot, \boldsymbol{a}^{s,k}, \boldsymbol{b}^{s,k}) = -h_k$ given by the following expression in terms of the input $\boldsymbol{x} \in \mathbb{R}^n$ and the parameters $\boldsymbol{a}^{s,k} \in \mathbb{R}^n$ and $\boldsymbol{b}^{s,k} \in \mathbb{R}^n$ is convex:

$$\Psi(\boldsymbol{x}, \boldsymbol{a}^{s,k}, \boldsymbol{b}^{s,k}) = \left(\bigvee_{i=1}^{n} (a_i^{s,k} - x_i) \right) \vee \left(\bigvee_{i=1}^{n} (x_i - b_i^{s,k}) \right) \equiv \bigvee \begin{bmatrix} \boldsymbol{a}^{s,k} - \boldsymbol{x} \\ \boldsymbol{x} - \boldsymbol{b}^{s,k} \end{bmatrix}. \tag{9}$$

Moreover, we can write the output of a module given by (3) as a function φ, which depends on the input $\boldsymbol{x} \in \mathbb{R}^n$ and the lists $\boldsymbol{a}^s = [\boldsymbol{a}^{s,1}, \ldots, \boldsymbol{a}^{s,K_s}]$ and $\boldsymbol{b}^s = [\boldsymbol{b}^{s,1}, \ldots, \boldsymbol{b}^{s,K_s}]$ of the lower and upper vertices of the hyperboxes, using the equation:

$$\varphi(\boldsymbol{x}, \boldsymbol{a}^s, \boldsymbol{b}^s) = \bigvee_{k=1}^{K_s} \left(\bigwedge_{i=1}^{n} (x_i - a_i^{s,k}) \wedge \bigwedge_{i=1}^{n} (b_i^{s,k} - x_i) \right) = \bigvee_{k=1}^{K_s} \left(-\Psi(\boldsymbol{x}, \boldsymbol{a}^{s,k}, \boldsymbol{b}^{s,k}) \right). \tag{10}$$

From the proof of Proposition 3.1 in [2], the function φ given by (10) can be expressed as a DC function:

$$\varphi(\boldsymbol{x}, \boldsymbol{a}^s, \boldsymbol{b}^s) = f(\boldsymbol{x}, \boldsymbol{a}^s, \boldsymbol{b}^s) - g(\boldsymbol{x}, \boldsymbol{a}^s, \boldsymbol{b}^s) \tag{11}$$

where

$$f(\boldsymbol{x}, \boldsymbol{a}^s, \boldsymbol{b}^s) = \bigvee_{k=1}^{K_s} \left\{ \sum_{j \neq k}^{K_s} \Psi(\boldsymbol{x}, \boldsymbol{a}^{s,j}, \boldsymbol{b}^{s,j}) \right\}, \tag{12}$$

and

$$g(\boldsymbol{x}, \boldsymbol{a}^s, \boldsymbol{b}^s) = \sum_{k=1}^{K_s} \Psi(\boldsymbol{x}, \boldsymbol{a}^{s,k}, \boldsymbol{b}^{s,k}). \tag{13}$$

Hence, the optimization problem (7) can be rewritten as the DC problem [13,23]:

$$\begin{aligned} \min_{\boldsymbol{a}^s, \boldsymbol{b}^s, \boldsymbol{\xi}^s} \quad & \sum_{l=1}^{M} \xi_l^s + \gamma \sum_{k=1}^{K_s} \sum_{i=1}^{n} (b_i^{s,k} - a_i^{s,k}) \\ \text{s.t.} \quad & f(\boldsymbol{x}^{(l)}, \boldsymbol{a}^s, \boldsymbol{b}^s) - g(\boldsymbol{x}^{(l)}, \boldsymbol{a}^s, \boldsymbol{b}^s) \quad -\xi_l^s \leq 0 \;\text{ if } \boldsymbol{x}^{(l)} \in C_0^s \\ & g(\boldsymbol{x}^{(l)}, \boldsymbol{a}^s, \boldsymbol{b}^s) - f(\boldsymbol{x}^{(l)}, \boldsymbol{a}^s, \boldsymbol{b}^s) \quad -\xi_l^s \leq 0 \;\text{ if } \boldsymbol{x}^{(l)} \in C_1^s \\ & \boldsymbol{a}^s - \boldsymbol{b}^s \leq \boldsymbol{0}, \quad \boldsymbol{\xi}^s \geq \boldsymbol{0}. \end{aligned} \tag{14}$$

The optimization problem presented in (14), which is equivalent to problem (7), can be solved using the Disciplined Convex-Concave Programming (DCCP) extension of the `CVXPY` library in `Python` [7], as demonstrated in [4]. In the following section, we eliminate the need for the DCCP framework by detailing the application of the convex-concave procedure (CCP) to solve (14).

4 Convex-Concave Procedure

In this section, we outline the convex-concave procedure (CCP) for solving the optimization problems (14) that incorporate both convex and concave components in the constraints [13,29]. The CCP is a powerful heuristic method used to find local solutions of DC optimization problems in the form (8). The basic idea of the CCP is to replace the convex function g_i (or the concave function $-g_i$) by a linear approximation. Accordingly, recall that $\partial g_i(\boldsymbol{v}_t)$ is a subgradient of g_i at $\boldsymbol{v}_t$ if

$$g_i(\boldsymbol{v}) \geq g_i(\boldsymbol{v}_t) + \langle \partial g_i(\boldsymbol{v}_t), \boldsymbol{v} - \boldsymbol{v}_t \rangle := \mathcal{L}_{g_i, \boldsymbol{v}_t}(\boldsymbol{v}), \tag{15}$$

where $\mathcal{L}_{g_i,\boldsymbol{v}_t}$ denotes a linear approximation of g_i on $\boldsymbol{v}_t$. Given a feasible initial point $\boldsymbol{v}_0$, the CCP computes $\boldsymbol{v}_{t+1}$ by solving iteratively the convex subproblem

$$\begin{aligned} \boldsymbol{v}_{t+1} = \underset{\boldsymbol{v}}{\operatorname{argmin}}\ & f_0(\boldsymbol{v}) - \mathcal{L}_{g_0,\boldsymbol{v}_t}(\boldsymbol{v}) \\ \text{s.t. } & f_i(\boldsymbol{v}) - \mathcal{L}_{g_i,\boldsymbol{v}_t}(\boldsymbol{v}) \leq 0, \quad \forall i = 1, \ldots, N_C, \end{aligned} \tag{16}$$

until a stopping criterion is satisfied. The following presents the convex subproblems derived from applying the CCP to the DC optimization problem (14).

The objective function of the DC optimization problem (14) is linear, so there is no need to linearize it. Similarly, the constraints $\boldsymbol{a}^s - \boldsymbol{b}^s \leq \boldsymbol{0}$ and $\boldsymbol{xi}^s \geq \boldsymbol{0}$ also do not require linearization. Therefore, we only need to focus on simplifying or linearizing the first two inequality constraints.

First of all, recall that $f(\boldsymbol{x}^{(l)}, \boldsymbol{a}^s, \boldsymbol{b}^s) - g(\boldsymbol{x}^{(l)}, \boldsymbol{a}^s, \boldsymbol{b}^s) = \bigvee_{k=1}^{K_s}(-\Psi(\boldsymbol{x}, \boldsymbol{a}^{s,k}, \boldsymbol{b}^{s,k})$. Thus, the first inequality constraint of (14) can be expressed as

$$\bigvee_{k=1}^{K_s} (-\Psi(\boldsymbol{x}, \boldsymbol{a}^{s,k}, \boldsymbol{b}^{s,k}) - \xi_l^s \leq 0, \quad \forall \boldsymbol{x}^{(l)} \in C_0^s. \tag{17}$$

Since the maximum is the greatest element, this inequality holds if and only if

$$-\xi_l^s - \Psi(\boldsymbol{x}^{(l)}, \boldsymbol{a}^{s,k}, \boldsymbol{b}^{s,k}) \leq 0, \quad \forall k = \{1, \ldots, K_s\} \quad \text{and} \quad \forall \boldsymbol{x}^{(l)} \in C_0^s, \tag{18}$$

where Ψ is the convex function defined by (9). Given a training feature vector $\boldsymbol{x}^{(l)} \in \mathbb{R}^n$, we can compute a linear approximation of Ψ on $(\boldsymbol{a}_t^{s,k}, \boldsymbol{b}_t^{s,k})$, where $\boldsymbol{a}_t^{s,k}$ and $\boldsymbol{b}_t^{s,k}$ denote the lower and upper vertexes of the kth hyperbox of module s at iteration t of the CCP algorithm. Using the concept of subgradient, we conclude that a linear approximation of Ψ on $(\boldsymbol{a}_t^{s,k}, \boldsymbol{b}_t^{s,k})$ is given by

$$\mathcal{L}_{\Psi,\boldsymbol{a}_t^{s,k},\boldsymbol{b}_t^{s,k}}(\boldsymbol{x}^{(l)}, \boldsymbol{a}^{s,k}, \boldsymbol{b}^{s,k}) = \begin{cases} a_{i^*}^{s,k} - x_{i^*}, & i^* \leq n, \\ x_{i^*-n} - b_{i^*-n}^{s,k}, & i^* > n, \end{cases} \tag{19}$$

where the index

$$i^* = \arg\max\left\{[\boldsymbol{a}_t^{s,k} - \boldsymbol{x}^{(l)}, \boldsymbol{x}^{(l)} - \boldsymbol{b}_t^{s,k}]\right\} \in \{1, \ldots, 2n\}, \tag{20}$$

yields the maximum of $[\boldsymbol{a}_t^{s,k} - \boldsymbol{x}^{(l)}, \boldsymbol{x}^{(l)} - \boldsymbol{b}_t^{s,k}] \in \mathbb{R}^{2n}$. Note that the linear approximation $\mathcal{L}_{\Psi,\boldsymbol{a}^{s,k},\boldsymbol{b}^{s,k}}$ depends on $\boldsymbol{a}_t^{s,k}$ and $\boldsymbol{b}_t^{s,k}$, as well as on $\boldsymbol{x}^{(l)}$, through the index i^* which depends on s, k, t, and l, that is, $i^* \equiv i^*(s, k, t, l)$. Therefore, the first set of inequalities in (14) leads to the following constraints for the CCP optimization subproblem for all $\boldsymbol{x}^{(l)} \in C_0^s$ and $k = 1, \ldots, K_s$ with i^* given by (20) for each s, k, t, and l:

$$\begin{cases} -a_{i^*}^{s,k} - \xi_l^s \leq -x_{i^*}^{(l)}, & \text{if } i^* \leq n, \\ b_{i^*-n}^{s,k} - \xi_l^s \leq x_{i^*-n}^{(l)}, & \text{if } i^* > n. \end{cases} \tag{21}$$

In a similar fashion, a linear approximation of the convex function f given by (12) on $(\boldsymbol{a}_t^s, \boldsymbol{b}_t^s)$ for a fixed feature vector $\boldsymbol{x}^{(l)}$ can be expressed in terms of the linear approximations of Ψ employing the equation

$$\mathcal{L}_{f,\boldsymbol{a}_t^s,\boldsymbol{b}_t^s}(\boldsymbol{x}^{(l)}, \boldsymbol{a}^s, \boldsymbol{b}^s) = \sum_{j \neq k^*} \mathcal{L}_{\Psi,\boldsymbol{a}_t^{s,j},\boldsymbol{b}_t^{s,j}}(\boldsymbol{x}^{(l)}, \boldsymbol{a}^{s,j}, \boldsymbol{b}^{s,j}), \tag{22}$$

where $k^* \equiv k^*(s,t,l)$ is defined by

$$k^* = \arg\max_{k=1:K_s} \sum_{j \neq k} \Psi(\boldsymbol{x}^{(l)}, \boldsymbol{a}_t^{s,j}, \boldsymbol{b}_t^{s,j}). \tag{23}$$

As a consequence, the second set of inequalities in (14) leads to the following constraints for the CCP optimization subproblem for all $\boldsymbol{x}^{(l)} \in C_1^s$ with k^* given by (23) and for all $i = 1, \ldots, 2n$:

$$\begin{cases} a_i^{s,k^*} - \xi_l^s \leq x_i^{(l)}, & \text{if } i \leq n, \\ -b_{i-n}^{s,k^*} - \xi_l^s \leq -x_{i-n}^{(l)}, & \text{if } i > n. \end{cases} \tag{24}$$

Concluding, the CCP applied to solve the DC optimization problem (14) results in solving iteratively the following sequence of linear programming problems for training the MPCL:

$$\begin{aligned} \boldsymbol{a}_{t+1}^s, \boldsymbol{b}_{t+1}^s = \operatorname*{argmin}_{\boldsymbol{a}^s,\boldsymbol{b}^s,\boldsymbol{\xi}^s} & \sum_{l=1}^{M} \xi_l^s + \gamma \sum_{k=1}^{K_s} \sum_{i=1}^{n} (b_i^{s,k} - a_i^{s,k}) \\ \text{s.t.} & \begin{cases} -a_{i^*}^{s,k} - \xi_l^s \leq -x_{i^*}^{(l)}, & \text{if } i^* \leq n, \quad \forall k = 1, \ldots, K_s, \\ b_{i^*-n}^{s,k} - \xi_l^s \leq x_{i^*-n}^{(l)}, & \text{if } i^* > n, \quad \forall \boldsymbol{x}^{(l)} \in C_0^s, \end{cases} \\ & \begin{cases} a_i^{s,k^*} - \xi_l^s \leq x_i^{(l)}, & \text{if } i \leq n, \quad \forall i = 1, \ldots, 2n, \\ -b_{i-n}^{s,k^*} - \xi_l^s \leq x_{i-n}^{(l)}, & \text{if } i > n, \quad \forall \boldsymbol{x}^{(l)} \in C_1^s, \end{cases} \\ & \boldsymbol{a}^s - \boldsymbol{b}^s \leq \boldsymbol{0}, \quad \boldsymbol{\xi}^s \geq \boldsymbol{0}, \end{aligned} \tag{25}$$

where $k^* \equiv k^*(s,t,l)$ and $i^* \equiv i^*(s,k^*,t,l)$ given respectively by (23) and (20). The linear optimization subproblems are solved iteratively until there is no significant difference in the objective function, or a maximum number of iterations is reached. Also, the sequence of problems is solved for each class label $s = 1, \ldots, S$.

We would like to clarify that we initialized the hyperboxes using the K-means++ algorithm [1]. K-means++ is employed to select initial centroids for K-means clustering. This algorithm ensures that the selected centroids are well-distributed, which helps minimize the risk of poor initialization. By utilizing K-means++, the hyperboxes are initialized with $\boldsymbol{a}^{s,k} = \boldsymbol{b}^{s,k} = \boldsymbol{c}^{s,k}$, where $\boldsymbol{c}^{s,k}$ represents the kth centroid determined by the K-means++ algorithm using the dataset C_1^s, for $s = 1, \ldots, S$.

5 Computational Experiments

In this section, we assess the classification accuracy and computational efficiency of four morphological perceptron-based classifiers: the proposed MPCL-CCP, the MPCL-DCCP method from our previous work [4], the MPCL-Greedy

Table 1. Classification performance on the three-class synthetic dataset.

Method	F1-score		Misclassified (%)	
	Train	Test	Train	Test
MPCL-Greedy	**0.88**	0.76	**11.6**	24.2
MPCL-DCCP	$0.75_{\pm 0.06}$	$0.75_{\pm 0.06}$	$23.7_{\pm 5.5}$	$25.3_{\pm 5.3}$
MPCL-Adam	$0.81_{\pm 0.01}$	$0.81_{\pm 0.02}$	$19.1_{\pm 1.4}$	$19.3_{\pm 1.7}$
MPCL-CCP	$0.82_{\pm 0.01}$	$\mathbf{0.83}_{\pm \mathbf{0.01}}$	$17.2_{\pm 0.6}$	$\mathbf{17.5}_{\pm 1.0}$

a) Boxplot of the F1-score in the test set. b) Diagram of a hypothesis test.

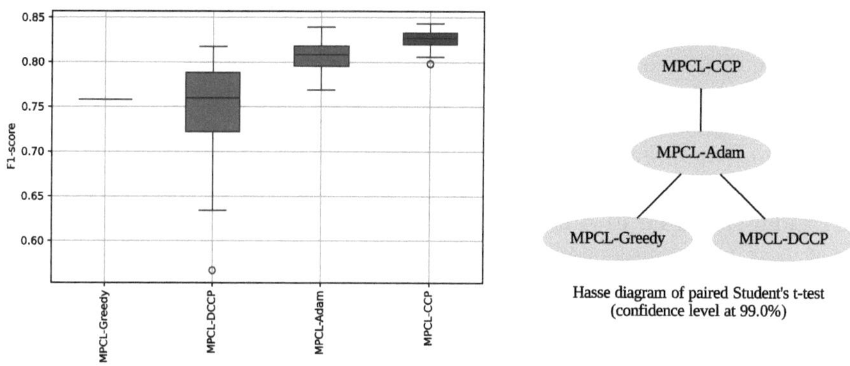

Fig. 2. Test set evaluation: F1-score and Hasse diagram of a hypothesis test.

approach based on Sussner and Esmi [25], and MPCL-Adam, trained using a backpropagation-like technique based on the Adam optimizer [12].

All methods were implemented in Python 3.9, utilizing the CVXPY optimization library with the CLARABEL and MOSEK solvers, as well as TensorFlow for the MPCL-Adam approach. A one-against-all strategy was employed for MPCL-Adam training, as in the other methods. As commonly done in binary classification neural networks, the hyperboxes of the sth module, defined by (3), are determined by minimizing the binary cross-entropy of the logits. The default parameters of Keras, version 3.10, were used for the Adam optimizer, including a learning rate of 0.001, and the number of training epochs was fixed at 100. To reduce sensitivity to initialization [8], the hyperbox parameters were initialized using the K-means++ technique, as described in Sect. 4.

For MPCL-CCP, MPCL-DCCP, and MPCL-Adam, the architecture includes $K_s = 4$ hyperboxes per class. Experiments were conducted on a synthetic dataset with three classes, generated using the `make_blobs` function, containing 1,200 samples with two features, 12 cluster centers, a cluster standard deviation of 1.5, and a fixed random seed of 42. Each method was trained 50 times to account for optimization variability. Performance was measured using the F1-score and misclassification rate. Results are presented in Table 1 and Fig. 2.

Note from Table 1 and Fig. 2 that the MPCL-CCP achieved the best performance, achieving an average F1-score of 0.83 and the lowest error rate of 17.5% on the test set. Moreover, the MPCL-CCP yielded the lowest variability, reinforcing its robustness and stability. The MPCL-Adam obtained similar F1-scores (0.81) but with higher variability, indicating competitive performance but higher sensitivity. In contrast, MPCL-DCCP showed lower performance (average F1-score 0.75) and high variability, as reflected by the standard deviation and wider spread in the boxplot in Fig. 2a). MPCL-Greedy, while achieving a high training F1-score of 0.88, experienced a substantial performance drop on the test set, yielding an F1-score of 0.76 and an error rate of 24.2%. Recall that the MPCL-Greedy method is deterministic and, thus, it does not exhibit any variation in its performance metrics. The Hasse diagram in Fig. 2b) summarizes the outcome of paired Student's t-tests with a 99% confidence level, with pairwise winners indicated above, on the test set [28]. The performance hierarchy highlights MPCL-CCP as the most accurate approach.

Regarding computational efficiency, MPCL-CCP completed training in an average of 1.72 s, approximately 132 times faster than MPCL-DCCP (228.05 s). MPCL-Adam required 40.79 s, over 23 times longer than MPCL-CCP, due in part

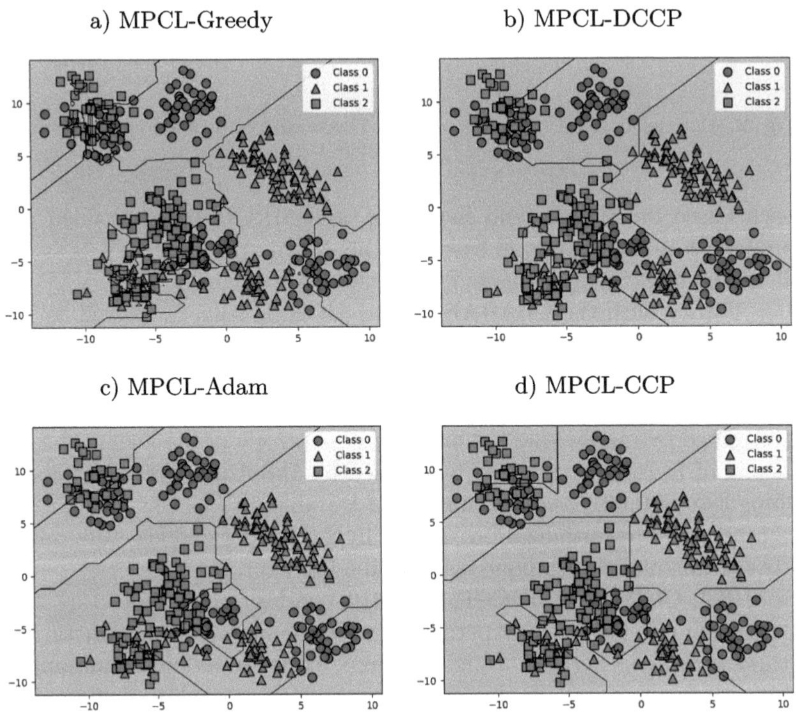

Fig. 3. Decision surfaces generated by the MPCL classifiers on the test set.

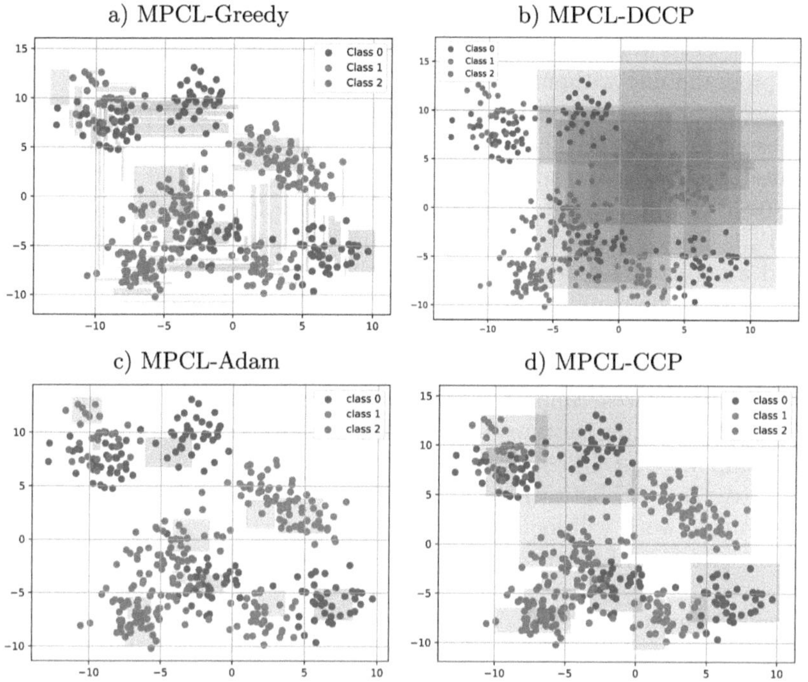

Fig. 4. Hyperboxes generated by the MPCL classifiers on the test set.

to the overhead of managing the TensorFlow computational graph. The MPCL-Greedy was extremely fast, completing training in 0.05 s (≈3.2% of MPCL-CCP's time).

Finally, Figs. 4 and 3 shows the hyperboxes and decision surfaces of the MPCL models. Note from Fig. 4 the significant differences between the hyperboxes produced by the models. In particular, note that the MPCL-CCP features hyperboxes, which enable high interpretability. The MPCL-Adam's also features interpretable hyperboxes, but those produced by the MPCL-CCP better fit the training data, which contributes to its superior generalization capability. Finally, the decision surfaces shown in Fig. 3 confirm that MPCL-CCP provides the most accurate class separation.

6 Concluding Remarks

This paper introduced the MPCL-CCP, a morphological perceptron with a competitive layer trained using the convex-concave procedure (CCP). The training features a difference-of-convex (DC) optimization problem whose goal is to minimize misclassification errors, while incorporating a regularization term that reduces the sizes of hyperboxes. The CCP enables efficient training via iterative linear programming without requiring gradients.

Experimental results on a synthetic dataset demonstrate that MPCL-CCP achieves an excellent predictive performance, with high interpretability and low computational cost. Accordingly, the MPCL-CCP yielded the highest test F1-score (0.83) and the lowest error rate (17.5%) with low variability, outperforming MPCL-Greedy, MPCL-DCCP, and MPCL-Adam. Although MPCL-Adam reached a similar F1-score (0.81), its results were less consistent and computationally more demanding. MPCL-Greedy, while extremely fast and achieving high training scores, showed poor generalization, and MPCL-DCCP exhibited both lower accuracy and higher variability. Qualitative analyses of hyperboxes and decision surfaces confirm that MPCL-CCP produces more interpretable structures that better capture the data distribution than the other approaches.

Overall, MPCL-CCP emerges as a robust and practical method, offering an effective trade-off between accuracy, interpretability, and computational efficiency. Future work will focus on evaluating the method using larger and real-world datasets, and further improving its performance.

Acknowledgments. Marcos Eduardo Valle acknowledges financial support from the National Council for Scientific and Technological Development (CNPq), Brazil, under grant no 315820/2021-7, and the São Paulo Research Foundation (FAPESP), Brazil, under grant no 2023/03368-0.

References

1. Arthur, D., Vassilvitskii, S.: k-means++: the advantages of careful seeding. In: Proceedings of the Eighteenth Annual ACM-SIAM Symposium on Discrete Algorithms, pp. 1027–1035 (2007)
2. Bagirov, A., Taheri, S., Karmitsa, N., Sultanova, N., Asadi, S.: Robust piecewise linear l 1 -regression via nonsmooth dc optimization. Optim. Methods Softw. **37**, 1–21 (2020)
3. Charisopoulos, V., Maragos, P.: Morphological perceptrons: geometry and training algorithms. In: Angulo, J., Velasco-Forero, S., Meyer, F. (eds.) ISMM 2017. LNCS, vol. 10225, pp. 3–15. Springer, Cham (2017). https://doi.org/10.1007/978-3-319-57240-6_1
4. Cunha, I., Valle, M.E.: Training single-layer morphological perceptron using convex-concave programming. In: 2023 IEEE Latin American Conference on Computational Intelligence (LA-CCI), pp. 1–6 (2023)
5. Cuninghame-Green, R.: Minimax Algebra: Lecture Notes in Economics and Mathematical Systems, vol. 166. Springer, New York (1979)
6. Davidson, J.L., Hummer, F.A.: Morphology neural networks: an introduction with applications. Circ. Syst. Signal Process. **12**, 177–210 (1993)
7. Diamond, S., Boyd, S.: CVXPY: a python-embedded modeling language for convex optimization (2016)

8. Dimitrova, M., Blusseau, S., Velasco-Forero, S.: Learning morphological representations of image transformations: influence of initialization and layer differentiability (2025). https://hal.science/hal-05112110, preprint
9. Franchi, G., Fehri, A., Yao, A.: Deep morphological networks. Pattern Recogn. **102** (2020)
10. Groenendijk, R., Dorst, L., Gevers, T.: Geometric back-propagation in morphological neural networks. IEEE Trans. Pattern Anal. Mach. Intell. 1–8 (2023)
11. Hu, Y., Belkhir, N., Angulo, J., Yao, A., Franchi, G.: Learning deep morphological networks with neural architecture search. Pattern Recogn. **131**, 108893 (2022)
12. Kingma, D.P., Ba, J.: Adam: a method for stochastic optimization. In: International Conference on Learning Representations (ICLR) (2015). http://arxiv.org/abs/1412.6980
13. Lipp, T., Boyd, S.: Variations and extension of the convexconcave procedure. Optim. Eng. **17**(2), 263–287 (2016)
14. Maragos, P., Charisopoulos, V., Theodosis, E.: Tropical geometry and machine learning. In: Proceedings of the IEEE (2021)
15. Oliveira, A.L., Valle, M.E.: Linear dilation-erosion perceptron trained using a convex-concave procedure. In: Abraham, A., Ohsawa, Y., Gandhi, N., Jabbar, M.A., Haqiq, A., McLoone, S., Issac, B. (eds.) SoCPaR 2020. AISC, vol. 1383, pp. 245–255. Springer, Cham (2021). https://doi.org/10.1007/978-3-030-73689-7_24
16. Penaud-Polge, V., Velasco-Forero, S., Angulo-Lopez, J.: Group equivariant networks using morphological operators. In: Brunetti, S., Frosini, A., Rinaldi, S. (eds.) DGMM 2024. LNCS, vol. 14605, pp. 165–177. Springer, Cham (2024). https://doi.org/10.1007/978-3-031-57793-2_13
17. Pessoa, L., Maragos, P.: Morphological/rank neural networks and their adaptive optimal design for image processing. In: IEEE International Conference on Acoustics. Speech, and Signal Processing, vol. 6, pp. 3398–3401. IEEE Computer Society, Los Alamitos (1996)
18. Ritter, G., Urcid, G.: Lattice algebra approach to single-neuron computation. IEEE Trans. Neural Netw. **14**(2), 282–295 (2003)
19. Ritter, G.X., G.Urcid: Introduction to Lattice Algebra: With Applications in AI, Pattern Recognition, Image Analysis, and Biomimetic Neural Networks. CRC Press (2021)
20. Ritter, G.X., Sussner, P.: An introduction to morphological neural networks. In: Proceedings of 13th International Conference on Pattern Recognition, vol. 4, pp. 709–717 (1996)
21. Ritter, G.X., Sussner, P., Diaz-De-Leon, J.L.: Morphological associative memories. IEEE Trans. Neural Netw. **9**(2), 281–293 (1998)
22. Salembier, P.: Structuring element adaptation for morphological filters. J. Math.l Imaging Vis. **2**(1), 55–71 (1992)
23. Shen, X., Diamond, S., Gu, Y., Boyd, S.: Disciplined convex-concave programming (2016)
24. Sussner, P.: Morphological perceptron learning. In: Proceedings of the 1998 IEEE International Symposium on Intelligent Control (ISIC) held jointly with IEEE International Symposium on Computational Intelligence in Robotics and Automation (CIRA) Intell, pp. 477–482 (1998)
25. Sussner, P., Esmi, E.L.: Morphological perceptrons with competitive learning: lattice-theoretical framework and constructive learning algorithm. Inf. Sci. **181**(10), 1929–1950 (2011)
26. Valle, M.E.: Reduced dilation-erosion perceptron for binary classification. Mathematics **8**(4), 512 (2020)

27. Velasco-Forero, S., Pages, R., Angulo, J.: Learnable empirical mode decomposition based on mathematical morphology. SIAM J. Imaging Sci. **15**(1), 23–44 (2022)
28. Weise, T., Chiong, R.: An alternative way of presenting statistical test results when evaluating the performance of stochastic approaches. Neurocomputing **147**, 235–238 (2015)
29. Yuille, A.L., Rangarajan, A.: The Concave-Convex Procedure. Neural Comput. **15**(4), 915–936 (2003)

Morphological Filters, PDEs and Multivariate Morphology

dilated/eroded variants of f with disc-shaped structuring elements of radius t. It is worth noticing here that $\pm|\boldsymbol{\nabla} u|$ are the maximum and minimum, respectively, among the directional derivatives of u in all possible spatial directions, see also eqs. (26), (27) of [4]. Therefore (1) are advection PDEs in which the advection direction is determined at each location is that of the greatest or smallest directional derivative.

Several generalisations and modifications of these PDEs have been studied in literature. In this work, we are interested in studying variants in which the spatial gradient is replaced by fractional derivatives.

Fractional-order differentiation goes back to the very origins of calculus itself as already Leibniz has discussed how to do, say, a half-order derivative instead of integer derivative orders, see [15] for a historical overview and [15, 18] for detailed definitions and derivations. One approach to defining a fractional derivative of a (absolutely and square integrable and sufficiently regular) function $f : \mathbb{R} \to \mathbb{R}$ is based on the Fourier transform $\mathcal{F}[f] : \mathbb{R} \to \mathbb{C}$ of f and the identity $\mathcal{F}[\partial_x^k f](\xi) = (\mathrm{i}\xi)^k \mathcal{F}[f](\xi)$ for derivatives of integer order. With the disambiguation $\mathrm{i}^\alpha := \exp(\mathrm{i}\alpha\pi/2)$ (principal value) one can define $\partial_x^\alpha f := \mathcal{F}^{-1}[(\mathrm{i}\xi)^\alpha \mathcal{F}[f]]$ for real $\alpha > 0$. Alternative definitions are the Riemann-Liouville derivative given by an integro-differential operator,

$$\partial_x^\alpha f(x) = \frac{1}{\Gamma(k-\alpha)} \partial^k \int\limits_{-\infty}^{x} (x-y)^{k-\alpha-1} f(y)\,\mathrm{d}y\,, \quad k := \lfloor \alpha \rfloor + 1\,. \tag{2}$$

with Γ denoting the standard gamma function, and $\lfloor \alpha \rfloor$ denotes the greatest integer not greater than α, or the similarly defined Caputo derivative. All three definitions coincide in those cases where all of them are defined. For numerical evaluation on sampled data, discretisations are available, with the Grünwald-Letnikov derivatives (see Sect. 2) as a prominent example.

Unlike ordinary derivatives, fractional derivatives of non-integer order are non-local in that their values at some location x does not depend just on f in a neighbourhood of x but on its values on the entire real half-axis $(-\infty, x]$. In applications, this is a reasonable choice for derivatives w.r.t. the time coordinate which then obey temporal causality. Indeed, time-fractional differential equations have proven useful in applications such as diffusion in porous media [8] or damped oscillators [2, 20], where the non-locality acts to model memory effects or delayed responses. When considering fractional derivatives in space, e.g. for space-fractional diffusion [5], preference for one real half-axis is often unphysical, leading to the consideration of symmetrised fractional derivative concepts [19]. Still, space-fractional PDEs in two or three spatial dimensions are mostly based on combining 1D fractional derivatives w.r.t. coordinate directions [5, 17, 23, 27].

Fractional differential equations have also been considered in the context of image processing [1, 26]. On one hand, time-fractional differential equations can profit from the same benefits of modeling memory or delay by causal time derivatives, see e.g. [13] for time-fractional morphological PDEs. On the other hand, space-fractional derivatives allow to introduce an interesting kind of controlled non-local behaviour into image processing. Space-fractional derivatives have been used e.g. for edge detection [22], denoising and image enhancement [11, 12, 29], regularisation [6, 7, 24], segmentation [21] or image super-resolution [10].

Many of these approaches use symmetric space-fractional derivatives; however, the majority of works we are aware of adhere to the use of 1D fractional derivatives in coordinate directions. Given the importance of spatial symmetries in image processing, we question this approach: Whereas ordinary derivatives in two or three dimensions can be broken down to derivative operators w.r.t. the coordinates, proceeding similarly for non-local fractional derivatives can be expected to introduce strong directional bias towards the coordinate directions.

In an attempt to mitigate these problems, some authors introduce additional directional derivatives in 45° steps [11,21,26]. In [25], the approach is refined by using 22.5° steps by applying a 1D fractional derivative discretisation to a series of intensity values interpolated from grid values.

A rare instance of consequently envisioning rotational invariance is found in [7] where fractional powers of the negative Laplacian are computed via the Fourier domain and used as rotationally symmetric operators for fractional diffusion and regularisation.

Our Contribution. We establish a framework for discrete approximations of space-fractional derivatives for 2D morphology with emphasis on geometric invariance w.r.t. reflections and rotations. In particular, we propose a novel concept of directional fractional derivatives in arbitrary directions from which we construct fractional counterparts of the r.h.s. of Hamilton-Jacobi equations for dilation and erosion. We discuss qualitative properties of the resulting fractional dilation and erosion operations. In particular, the use of space-fractional derivatives extends the classical concept of *morphology* as *shape-based* processing in the sense that due to their non-locality, the resulting filters look to a certain extent *below the surface* of the evolving shapes. We present experiments to confirm and extend these findings, including also composed operators such as opening, closing and morphological shock filters, juxtaposing them with standard morphological operations. As we are interested in the potential of space-fractional morphology, comparisons with time-fractional approaches are not in our scope here.

Structure of the Paper. In Sect. 2, we recall the discretisation of fractional derivatives using the Grünwald-Letnikov formula. We discuss how symmetry can be incorporated when using fractional derivatives for morphological PDEs which is then applied to 1D signals first. Then, the extension to the 2D case is derived based on a truncated general power series and an optimisation over advection directions. In Sect. 3, numerical results for dilation, erosion, opening, closing, and shock filters for a variety of grey-value images with different fractional derivative orders are presented to illustrate their effects. A short conclusion and outlook to future work is given in Sect. 4.

2 Fractional Derivatives

We start by recalling essential facts on 1D fractional derivatives with emphasis on the specifics of morphological image processing. Based on this, we then propose a 2D extension fitting this purpose.

2.1 The 1D Case

Standard (Backward) Grünwald-Letnikov Derivatives. The basic form of a Grünwald-Letnikov discretised derivative of order $\alpha > 0$ for a sampled 1D signal with spatial step size h and values $u_i = u(x_0 + ih)$ is

$$[\partial^{\alpha}_{x-}u]_i = h^{-\alpha}\sum_{k=0}^{\infty}(-1)^k\binom{\alpha}{k}u_{i-k} \tag{3}$$

with the binomial coefficients $\binom{\alpha}{k} := \frac{\alpha(\alpha-1)\cdots(\alpha-k+1)}{k!}$. For a detailed derivation see e.g. [15]. The coefficients of (3) can be read off the power series

$$(1-x)^{\alpha} = \sum_{k=0}^{\infty}(-1)^k\binom{\alpha}{k}x^k\,. \tag{4}$$

In the case of integer order $\alpha > 0$ (where the series is finite) this is a direct consequence of the fact that multiplication of polynomials or power series is equivalent to a convolution of their coefficient sequences and thus to the iterated convolution of the backward finite difference discretisation $[\partial_{x-}u]_i = (u_i - u_{i-1})/h$. For non-integer orders the power series thus offers a natural way to interpolate between the integer order convolutions.

Under suitable regularity assumptions, the Grünwald-Letnikov derivative converges to the continuous-scale (e.g. Riemann-Liouville) derivative as the grid is refined, $h \to 0$. Moreover, for positive integer order α one has a backward-difference approximation for the ordinary derivative $\partial^{\alpha}_x u(x)$.

For non-integer α, the derivative is non-local, and in the given form, it looks backward on the real axis. In applications, this is a reasonable choice for time-fractional differential equations which then are guaranteed to obey temporal causality. When aiming at spatial fractional derivatives for image processing applications, however, it is paramount to ensure compatibility with spatial symmetries and therefore to consider also forward-looking and possibly spatially symmetric derivatives.

Forward and Symmetric Fractional Derivatives. In principle, forward-looking derivatives can be defined by reverting the real axis, i.e. applying the derivatives as before to the function $\tilde{u}(x) := u(-x)$. However, noticing that for ordinary derivatives of integer order α one has $\partial^{\alpha}_x\tilde{u}(-x) = (-1)^{\alpha}\partial^{\alpha}_x u(x)$, the question arises how to treat the sign factor $(-1)^{\alpha}$ for non-integer α. Defining a forward-looking derivative as $\partial^{\alpha}_{x+}u(x) = (-1)^{\alpha}\partial^{\alpha}_{x-}\tilde{u}(-x)$ (where $(-1)^{\alpha}$ requires disambiguation, e.g. as $\exp(\mathrm{i}\pi\alpha)$) makes it complex-valued even for real-valued functions which is cumbersome for applications. Alternatively, one can drop the complex sign factor. One can then either impose even symmetry for all orders α by $\partial^{\alpha}_{x+}u(x) = \partial^{\alpha}_{x-}\tilde{u}(-x)$, or impose odd symmetry for all orders α by $\partial^{\alpha}_{x+}u(x) = -\partial^{\alpha}_{x-}\tilde{u}(-x)$. The first choice, even symmetry, ensures correct embedding of derivatives of even integer order, $\partial^{\alpha}_{x+}u(x) = \partial^{\alpha}_x u(x)$, but leads to a sign-reversal for odd integer orders, $\partial^{\alpha}_{x+}u(x) = -\partial^{\alpha}_x u(x)$. The odd symmetry choice correctly embeds odd integer orders but involves sign reversal for even integer orders. We remark that the even-symmetric variant can be associated with the pseudo-differential operators $-(-\partial_{xx})^{\alpha/2}$, compare [7].

Symmetric derivatives can either be obtained by combining backward and forward derivatives or by directly generalising central finite difference approximations. The difficulties regarding the choice between even and odd symmetry remain and lead to different models proposed in literature [14, 16, 19, 24, 28].

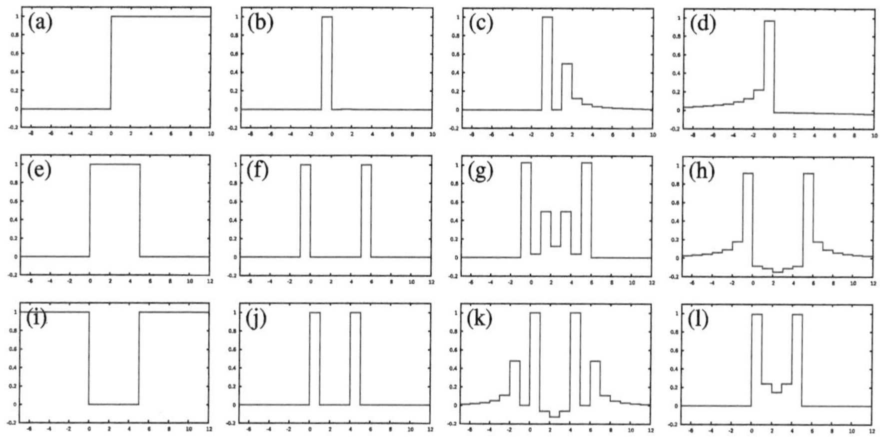

Fig. 1. Dilation PDE r.h.s. of simple discrete 1D signals with spatial step size 1. **Top to bottom:** (a–d) Heaviside signal; (e–h) high impulse of width 5; (i–l) low impulse of width 5. **Left to right:** (a, e, i) signal f; (b, f, j) upwind approximation of $|f_x|$; (c, g, k) fractional derivative $|\partial_x^{1.5} f(x)|_+$; (d, h, l) fractional derivative $|\partial_x^{0.75} f(x)|_+$. Horizontal axis shows space coordinate, vertical axis shows intensity.

Fractional Derivatives for Morphological PDEs. Fortunately, for our goal of generalising PDEs of mathematical morphology we can avoid the aforementioned problems. The quantity we actually want to generalise is $|\partial_x u| = \max\{\partial_x u, -\partial_x u\}$, and it is commonly discretised in upwind schemes by one-sided finite differences, switching between forward and backward differences based on sign. Defining the forward derivative as $\partial^\alpha_{x+} u(x) = -\partial^\alpha_{x-} \tilde{u}(-x)$, we have $|\partial^\alpha_x u(x)|_+ := \max\{-\partial^\alpha_{x-} u(x), \partial^\alpha_{x+} u(x)\}$ as a generalisation of $|\partial_x u|$ that always looks into the half-axis from which higher values are advected, thus suitable for morphological dilation, or $|\partial^\alpha_x u(x)|_- := \max\{\partial^\alpha_{x-} u(x), -\partial^\alpha_{x+} u(x)\}$ which always looks into the half-axis from which lower values are advected, as appropriate for morphological erosion. We can therefore write down the 1D space-fractional morphological dilation PDE $u_t = |\partial^\alpha_x u(x)|_+$ and the 1D space-fractional morphological erosion PDE $u_t = |\partial^\alpha_x u(x)|_-$.

2.2 Behaviour of Fractional Dilation in 1D

We analyse the responses of the r.h.s. of the 1D space-fractional dilation equation $u_t = \max\{\partial^\alpha_{x+} u, -\partial^\alpha_{x-} u\}$ to discrete Heaviside functions ($u_i = 0$ for $i < 0$, $u_i = 1$ for $i \geq 0$), high flip impulses ($u_i = 1$ for $0 \leq i \leq w$, $u_i = 0$ otherwise) and low flip impulses ($u_i = 0$ for $0 \leq i \leq w$, $u_i = 1$ otherwise). Fractional derivative stencils were truncated at distance 100, which is far beyond the length of all involved structures and ensures negligible truncation errors.

We first consider $\alpha = 1$. In this case, the fractional derivative becomes a standard derivative which is approximated by an upwind approximation. This leads to a peak of size 1 shifted one unit to the left for the Heaviside function (see Fig. 1(b)). For the high flip impulse of size 5, we obtain two peaks of size 1. Here, the left peak is shifted one unit to the left whereas the right peak is shifted one unit to the right (Fig. 1(f)). For the low flip impulse of size 5, we also obtain two peaks of size 1. However, the left peak is shifted one unit to the right whereas the left peak is shifted one unit to the left (Fig. 1(j)).

Next, we consider the case $\alpha \in (1, 2)$. As an example, we choose $\alpha = 1.5$. Here, the situation for the Heaviside function leads to the same result as for $\alpha = 1$ except that there is an additional positive decaying tail one unit right at the discontinuity of the Heaviside function (see Fig. 1(c)). This leads to an enhancement of an edge and a violation of the maximum principle. For the high impulse, we obtain the same peaks as for $\alpha = 1$ with the exception that the peaks are slightly larger than 1. Between the two peaks smaller peaks arise (Fig. 1(g)), again entailing a violation of the maximum principle. For the low impulse, we obtain the same two peaks as for $\alpha = 1$ and the same positive decaying tails to the left of the left peak and to the right of the right peak, respectively (Fig. 1(k)). Inbetween the two peaks, the results get negative. Hence, the minimum principle is violated. In total, the expected effects for dilation are enhancement of edges and the enhancement of thin bright line structures. Further, dilation is slowed at thin dark line structures which preserves them longer. Overall, we observe some reduction of numerical dissipation when compared to standard dilation ($\alpha = 1$).

Finally, we consider the case $\alpha \in (0, 1)$. We choose $\alpha = 0.75$ as an example. For the Heaviside function, we obtain a peak left to the discontinuity with a decaying tail on the left of this peak. Interestingly, we also obtain a tail to the right with small negative decaying values (see Fig. 1(d)). Hence, the minimum principle is violated. For the high impulse, we obtain two peaks left and right of the discontinuity, but the height is smaller than one. We observe two decaying tails to the left of the left peak and to the right of the right peak. Inbetween the two peaks, we obtain negative values (Fig. 1(h)). For the low flip impulse, we obtain the same two peaks as for $\alpha = 1$ with the exception that inbetween the peaks mass larger than zero appears (Fig. 1(l)). In total, the expected effects for dilation are contrast reduction as well as increased blurring effects.

2.3 Extension to 2D

As we set out to generalise the PDEs of morphological dilation and erosion in which the gradient norm acts to maximise or minimise, respectively, over directional derivatives, we are led to devise as the first step a fractional counterpart of a directional derivative in direction of a unit vector $\boldsymbol{e}_\varphi := (\cos\varphi, \sin\varphi)^{\mathrm{T}}$, $\varphi \in [0, 2\,\pi]$.

For $\varphi \in [0, \pi/2]$ a spatial discretisation of the first derivative in the direction of $\boldsymbol{e}_\varphi$ with forward finite differences in x and y direction reads as

$$[\partial_{\boldsymbol{e}_\varphi} u]_{i,j} = \cos\varphi\, \frac{u_{i+1,j} - u_{i,j}}{h} + \sin\varphi\, \frac{u_{i,j+1} - u_{i,j}}{h}\,. \tag{5}$$

By a similar reasoning as in the one-dimensional case, a discretised derivative of order α in the same direction is obtained as

$$[\partial^\alpha_{\boldsymbol{e}_\varphi} u]_{i,j} = h^{-\alpha} \sum_{k,l=0}^{\infty} d^\alpha_{\varphi;k,l} u_{i+k,j+l} \tag{6}$$

where the coefficients $d^\alpha_{\varphi;k,l}$ are those from the power series in two variables

$$-\left((\cos\varphi+\sin\varphi)-\cos\varphi\cdot x-\sin\varphi\cdot y\right)^\alpha=\sum_{k,l=0}^{\infty} d^\alpha_{\varphi;k,l}x^k y^l\ , \tag{7}$$

$$d^\alpha_{\varphi;k,l}=(-1)^{k+l+1}\binom{\alpha}{k,l}(\cos\varphi)^k(\sin\varphi)^l(\cos\varphi+\sin\varphi)^{\alpha-k-l}\ , \tag{8}$$

$$k,l=0,1,\ldots,\quad \varphi\in[0,\pi/2] \tag{9}$$

with the trinomial coefficients $\binom{\alpha}{k,l}:=\binom{\alpha}{k}\binom{\alpha-k}{l}$.

This is of course an infinite-size stencil that needs to be cut off at some finite length. Due to the rather slow decay of the coefficients with k and l reasonable stencil sizes involve a significant cut-off error which needs to be compensated to ensure that the numerical approximation of the fractional derivative does not incur an unnatural dependency on absolute intensity levels. We do this by adjusting the coefficient $d^\alpha_{\varphi;0,0}$, i.e. we use $d^\alpha_{\varphi;k,l}$ for $k=0,\ldots,K$, $l=0,\ldots,L$, $(k,l)\neq(0,0)$ and replace $d^\alpha_{\varphi;0,0}$ with $\tilde{d}^\alpha_{\varphi;0,0}=-\sum_{k,l=0;k+l\geq 1}^{K,L} d^\alpha_{\varphi;k,l}$.

For directions $\varphi\in[\pi/2,2\pi]$ the same type of stencil is applied in the other quadrants, i.e. replacing $u_{i+l,j+l}$ in (6) with $u_{i+k,j-l}$, $u_{i-k,j+l}$ or $u_{i-k,j-l}$ as appropriate. On quadrant boundaries, the discretisations from both adjacent quadrants collapse to the same 1D Grünwald-Letnikov discretisation.

In the terms of our discussion of even/odd symmetry of fractional derivatives in the 1D case, we notice that $[\partial_{e_\varphi}u]_{i,j}$ behaves as an odd-symmetric generalisation since one has $[\partial_{-e_\varphi}u]_{i,j}=-[\partial_{e_\varphi}\tilde{u}]_{-i,-j}$ where $\tilde{u}_{i,j}=u_{-i,-j}$ is the point-reflected image.

Altogether this can be used for an approximation of the fractional dilation PDE

$$u_t=|\boldsymbol{\nabla}^\alpha u|_+:=\max_{\varphi\in[0,2\pi)}\partial^\alpha_{e_\varphi}u\ , \tag{10}$$

and similarly $u_t=-|\boldsymbol{\nabla}^\alpha u|_-:=\min_{\varphi\in[0,2\pi)}\partial^\alpha_{e_\varphi}u$ for erosion. As in the 1D case, the maximum makes $|\boldsymbol{\nabla}^\alpha u|_+$ always look into the direction from which the highest values are advected, whereas $-|\boldsymbol{\nabla}^\alpha u|_-$ uses the minimum to detect the direction from which the lowest values are advected.

Unfortunately, the determination of the maximum in (10) requires some effort. Whereas for integer derivative orders, directional derivatives at a given point depend on the orientation angle just by low-order harmonics (such as a single sine wave for first order, and a double sine wave for second order), fractional directional derivatives can have multiple local maxima and minima over the orientation range $[0,2\pi]$. We use therefore a brute-force maximisation over sampled orientations for a full maximisation, which comes at a significant computational expense, and leave efficiency optimisations for future work. As a faster alternative, we consider an approach in which the direction of the standard (first-order) gradient vector is used to guide the direction of the fractional derivative. In effect, this replaces the fractional dilation PDE (10) with

$$u_t=|\partial^\alpha_{e_\varphi}u|_+\ ,\quad e_\varphi=(\cos\varphi,\sin\varphi)=\frac{\boldsymbol{\nabla}u}{|\boldsymbol{\nabla}u|}\ , \tag{11}$$

with $\boldsymbol{\nabla}u$ approximated by central differences, and similarly for erosion. We will compare these two variants of fractional morphological operations in our experiments. To

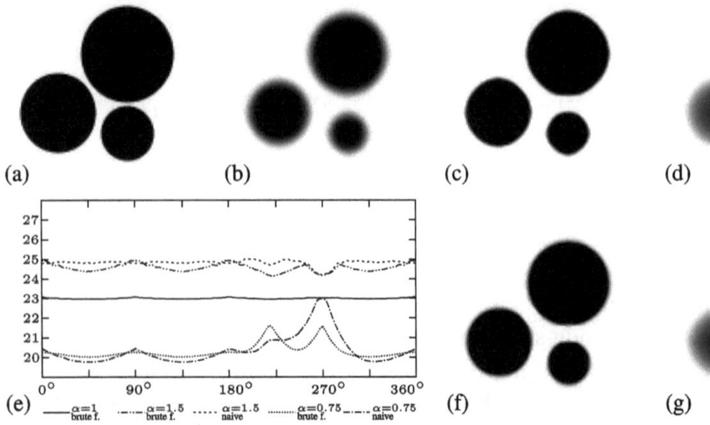

Fig. 2. Standard and fractional PDE-based dilation PDE of a synthetic grey-scale image. All dilations computed by an Euler forward scheme with time step size $\tau = 0.1$ and 50 iterations with brute-force maximisation (top row) or naive method (bottom row). (a) test image, 128×128 pixels; (b) standard dilation ($\alpha = 1$); (c, f) fractional dilation, $\alpha = 1.5$; (d, g) fractional dilation, $\alpha = 0.75$; (e) angular profiles of radii of dilated top-right discs.

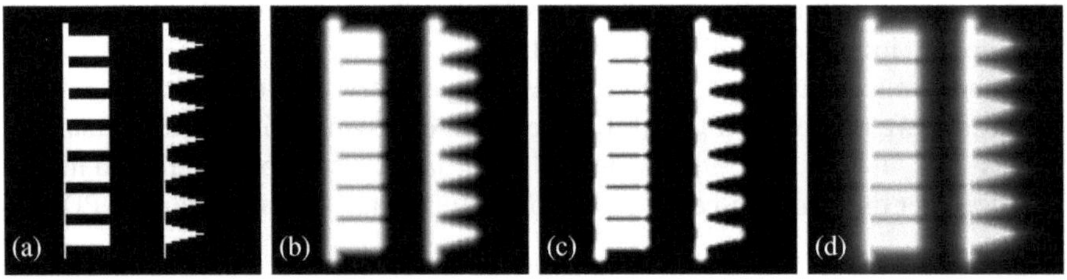

Fig. 3. Standard and fractional PDE-based dilation PDE of a synthetic grey-scale image. All dilations computed by an Euler forward scheme with time step size $\tau = 0.1$ and 30 iterations with brute-force maximisation. **Left to right:** (a) test image, 128×128 pixels; (b) standard dilation ($\alpha = 1$); (c) fractional dilation, $\alpha = 1.5$; (d) fractional dilation, $\alpha = 0.75$.

put the method in context with existing work, we also formulate a *naive* method following the popular approach of combining 1D fractional derivatives in grid directions:

$$u_t = |\cos\varphi| \cdot |\partial^{\alpha}_{(\operatorname{sgn}\cos\varphi,0)} u|_+ + |\sin\varphi| \cdot |\partial^{\alpha}_{(0,\operatorname{sgn}\sin\varphi)} u|_+ \tag{12}$$

with $(\cos\varphi, \sin\varphi)^{\mathrm{T}} = \boldsymbol{\nabla} u / |\boldsymbol{\nabla} u|$, where the axial fractional derivatives collapse to 1D fractional derivatives, and similarly for erosion.

3 Experiments

Synthetic Test Images. In our first series of experiments, see Figs. 2 and 3, we compare standard and fractional dilation on two test images designed to demonstrate theoretically predicted phenomena. The first of these images, Fig. 2(a), shows three black discs against white background that will be shrunk by all variants of PDE-based dilation. We show standard dilation (b), dilation with $\alpha = 1.5$ as an example of $1 < \alpha < 2$ (c), and dilation with $\alpha = 0.75$ as an example of $0 < \alpha < 1$ (d), all computed via brute-force maximisation. Here we truncate the stencils to 30 steps in both x and y direction. The cut-off value is chosen such that both the individual neglected coefficients and their total are small enough to expect no significant influence on the image. For a cut-off of 30 steps, none of the neglected coefficients is greater than 10^{-4}, and their total does not exceed 0.002. Also, experiments with different cut-off values confirm that while there are tiny visual differences in results where the cut-off size is reduced to 10 steps but with cut-offs of 20 and more steps results are visually indistinguishable.

For comparison, we show also results of the naive fractional dilation method with $\alpha = 1.5$ (f), and $\alpha = 0.75$ (g). In Fig. 2(e), we display angular profiles of the radii of the evolved top right disc, obtained by fitting sigmoid functions to the radial intensity profiles where the direction angle is measured from 0° (right) in positive orientation.

As expected, reasonable rotational invariance is observed with the brute-force maximisation for all three values of α, preserving the round shapes during evolution, Fig. 2(b, c, d). The evaluation of radii, Fig. 2(e), confirms the excellent rotational symmetry of the Rouy-Tourin scheme for standard dilation, with an evolved disc radius of 23 in agreement with the theory, given that its radius in the initial image is 28, and 50 iterations with $\tau = 0.1$ amount to a structuring element radius of 5. The radius is shrunk faster with $\alpha < 1$ and slower with $\alpha > 1$. Figure 2 (e) reveals slightly faster shrinkage in diagonal than in axial directions, as visible in the range $0^\circ \ldots 180^\circ$ (top two quadrants) where the disc evolves unperturbed by neighbouring structures, which indicates some further space for improvement in our discretisation of fractional derivatives.

The numerical dissipation due to the finite difference discretisation as evident in the standard dilation result (b) is well-known. As anticipated in Sect. 2.2, $\alpha > 1$ (c) mitigates it whereas $\alpha < 1$ (d) increases the blur. The non-locality of our fractional derivatives implies that they look into the depth of bright structures during fractional dilation. In our example, it is evident that standard dilation (b) evolves the discs roughly independently whereas fractional dilation (c, d) reacts to the initial proximity of the discs by speeding up for $\alpha > 1$ (c) or slowing down for $\alpha < 1$ (d) the inward movement of contours in the regions of proximity, leading to a slight repulsion between the discs in (c) and a slight attraction in (d). In Fig. 2(e) the dents or peaks near directional angles of 218° (bottom left part of the top right disc) and 270° (bottom) reflect this effect.

Regarding the naive dilation results, Fig. 2(f, g), we notice that the radial deviations from rotational symmetry (angular range $0^\circ \ldots 180^\circ$) are smaller for the naive method than for our proposed method in the case $\alpha > 1$ but more pronounced than with our approach in the case $\alpha < 1$. However, the non-locality effects with the naive method are highly direction-dependent: The repulsion in (f) is more pronounced in axial direction but almost absent in the diagonal direction, as confirmed by the different depth of the dents in Fig. 2(e); notice that there is even a slight attraction effect in directions neighbouring to the diagonal 218° direction. The attraction in (g) is highly exaggerated in the axial direction but again much weaker in the diagonal direction; here, one sees also that by the interaction of the 1-D fractional derivatives in x and y direction over the iteration steps an unnatural filling in happens in the roughly triangular region

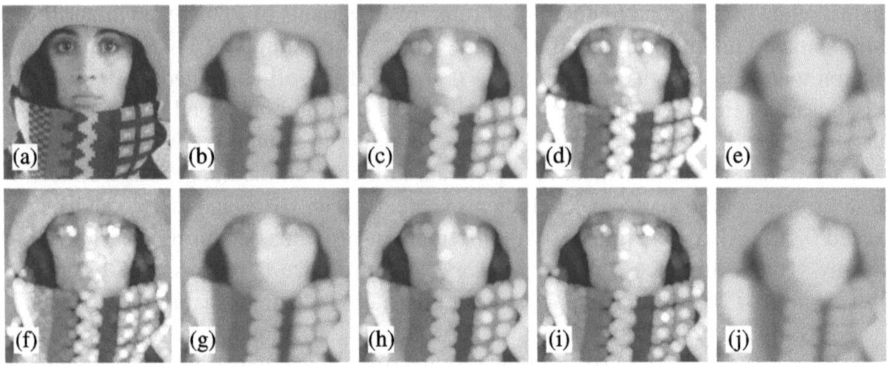

Fig. 4. Standard and fractional PDE-based dilation PDE of a grey-scale image. All dilations computed by an Euler forward scheme with time step size $\tau = 0.1$ and 100 iterations. **Top left to bottom right in rows:** (a) grey-scale image *trui*, 256×256 pixels; (b) standard dilation ($\alpha = 1$), brute-force maximisation; (c) fractional dilation, $\alpha = 1.25$, brute-force maximisation; (d) same but $\alpha = 1.5$; (e) same but $\alpha = 0.75$; (f) fractional dilation, $\alpha = 1.5$, naive composition of x and y derivatives; (g) standard dilation ($\alpha = 1$), gradient-guided computation; (h) fractional dilation, $\alpha = 1.25$, gradient-guided computation; (i) same but $\alpha = 1.5$; (j) same but $\alpha = 0.75$.

between the three discs. Comparing the evidence regarding rotational invariance of our proposed 2-D fractional dilation versus the naive 1-D based fractional dilation, it can be concluded that despite of an apparent slight advantage of the naive method in the case of an isolated contour (top quadrants) and $\alpha > 1$ the 2-D fractional dilation performs clearly better in the overall isotropic treatment of image structures.

The second image, Fig. 3(a), is designed to show the evolution of straight contours (on the left of each of the two white foreground objects) with structures of variable thickness behind, where the "depth-probing" effect of fractional derivatives is expected to appear. We see that standard dilation (b) propagates the straight left object contours at uniform speed. For $\alpha > 1$ (c) the propagation is accelerated where only thin bright layers back up the contour, leading to a progressively bumpy appearance of the propagated contour from top to bottom. For $\alpha < 1$ (d) an opposite effect is observed.

Real-World Test Image. For the following experiments we use a grey-scale image shown in Fig. 4(a). In our first experiment, Fig. 4, we compare different variants of fractional dilation for derivative orders greater and smaller than one with standard dilation. In particular, we compute dilation using brute-force maximisation of directional fractional derivatives and using the directional fractional derivative in the direction given by the standard gradient ("gradient-guided computation"). As expected, the results of both methods coincide for $\alpha = 1$; for other values of α the results are similar such that the gradient-guided method can be considered as an approximation of full maximisation. However, the effects of the fractional derivatives are more pronounced with the full maximisation since it finds the maximum of the fractional derivative with more accuracy. Comparing fractional to standard dilation, we observe the qualitative phenomena suggested by our consideration of the 1D case in Sect. 2.2: for fractional

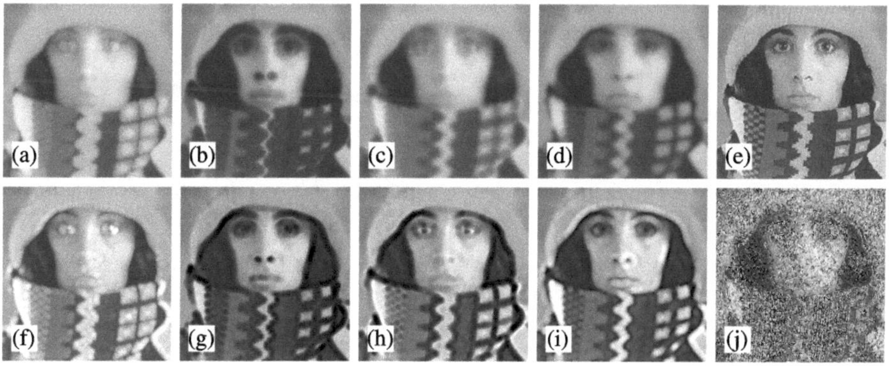

Fig. 5. Standard and fractional morphological filters for the grey-scale image *trui* from Fig. 4(a). All dilations/erosions computed by an Euler forward scheme with time step size $\tau = 0.1$, 50 iterations, and brute-force maximisation. **Top row:** (a–e) standard morphological filters ($\alpha = 1$). **Bottom row:** (f–j) fractional morphological filters, $\alpha = 1.5$. **Left to right:** (a, f) dilation; (b, g) erosion; (c, h) closing; (d, i) opening; (e, j) shock filter.

orders greater than one, an enhancement of edges and thin line structures and a reduction of numerical dissipation, and for a fractional order below one, contrast reduction and blurring. Note that all results reveal a reasonable degree of rotational invariance. To complement our comparison of methods, we give one result for the naive method which confirms its significant directional bias, evident from the blocky appearance of the dilated highlights.

Our next experiment, see Fig. 5, demonstrates the effects of simple and composed morphological operators: dilation, erosion, closing (by feeding the result of the dilation up to the desired evolution time into an erosion evolution that runs for the same time), opening (analogously) and a morphological shock filter. Given the strong dissipation with fractional order below one, we focus here on fractional orders above one as the more interesting case. Moreover, we choose a shorter evolution time than in the previous experiment. The shock filter of fractional order α with full maximisation is defined here as (fractional) dilation for locations where the directional derivative of order $\alpha + 1$ is negative, erosion where it is positive. Note that with our upwind-like discretisation of directional fractional derivatives of order $\neq 1$ the switch can be made simply dependent on the sign of $[|\nabla^\alpha u|]_+ + [|\nabla^\alpha u|]_-$, i.e. just computing the r.h.s. for the dilation and erosion evolution at each point and using the absolutely lesser of the two.

The effects of fractional erosion, closing and opening with order $\alpha = 1.5$ compared to their standard morphological counterparts are along the lines to be expected from our observations on dilation: more pronounced enhancement of contours, and lesser dissipation. In the closing it is interesting to note that due to the contrast enhancement of some minima (dark spots) in the dilation step, the subsequent erosion creates larger dark patches that are darker than in the original image; similarly for opening.

For the shock filter a fractional order $\alpha = 1.5$ is unfavourable in that it leads to the creation of many new extrema that are then enhanced in the process. We compare shock filter results for a broader range of derivative orders in Fig. 6. For $\alpha = 1.25$ the

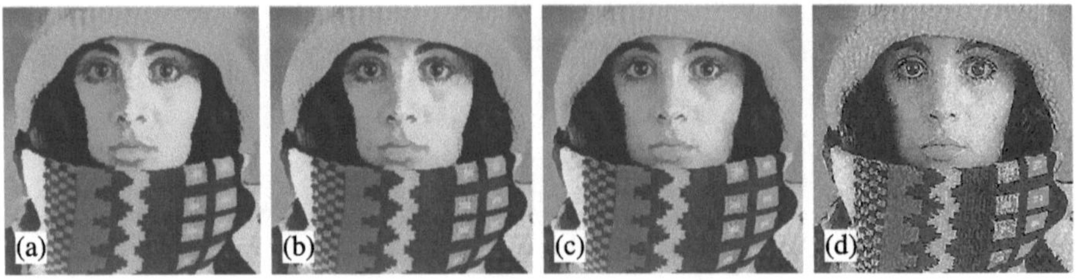

Fig. 6. Standard and fractional morphological shock filtering of the grey-scale image *trui* from Fig. 4(a). All evolutions computed by an Euler forward scheme, $\tau = 0.1$, 50 iterations, and brute-force maximisation. **Left to right:** (a) $\alpha = 0.5$; (b) $\alpha = 0.75$; (c) $\alpha = 1$; (d) $\alpha = 1.25$.

Fig. 7. Standard and fractional morphological filters. All dilations/erosions computed by an Euler forward scheme with time step size $\tau = 0.1$, 100 iterations, and brute-force maximisation. **Top row:** (a) grey-scale image *boats*, 512×512 pixels, (b–e) standard morphological filters ($\alpha = 1$). **Bottom row:** (f–i) fractional morphological filters, $\alpha = 1.5$. **Left to right:** (b, f) dilation; (c, g) erosion; (d, h) closing; (e, i) opening.

creation of additional extrema is already less pronounced than for $\alpha = 1.5$. In contrast, fractional orders below one turn out beneficial here as can be seen from the results for $\alpha = 0.75$ and $\alpha = 0.5$ where the segmentation-like effect of the standard ($\alpha = 1$) filter is even more pronounced, and both segments and their boundaries are visibly smoothed, apparently by the interaction of the blurring tendency of the fractional dilation and erosion for $0 < \alpha < 1$ with the shock-creating effect of the switch.

Finally, we show in Fig. 7 additional numerical results for a well-known test image *boats* comparing standard morphological filters ($\alpha = 1$) such as dilation, erosion, closing, and opening with fractional morphological filters using $\alpha = 1.5$.

4 Conclusion

We have established a framework for space-fractional PDE-based grey-scale morphology by replacing the spatial derivatives in the Hamilton-Jacobi PDEs describing morphological dilation and erosion with a fractional derivative. We have proposed a novel variant of spatial fractional derivatives with spatial discretisations consequently designed to incorporate geometric invariance with respect to reflections and rotations. The discretisation is derived from developing fractional powers of a polynomial associated with an upwind-type first derivative stencil into a power series. As with all fractional derivative concepts, the infinite support requires a truncation. Approximation of fractional dilation and erosion PDEs then involves a one-dimensional optimisation problem for the advection direction. With this at our hand, higher morphological operations such as opening, closing, and shock filter are at our disposal. Numerical results confirm analytically predicted properties of the filters and show that the new method achieves better rotational invariance than conventional approaches.

Future work will be directed on one hand at improvements of the method, in particular more efficient ways to perform the optimisation over advection directions in (10) in order to avoid the brute-force optimisation used in most of our experiments so far, and on the other hand at a more extensive evaluation of the possible benefits of the space-fractional morphological filters for application problems. Moreover the choice of the truncation length for the fractional derivative stencils (which was fixed to 30 in our 2D image experiments) deserves additional investigation analytically and experimentally. Extensions of the approach to colour images and to non-Euclidean norms simulating non-disc-shaped structuring elements are other possible goals. Last but not least, the approach should be embedded into a broader view encompassing space-fractional derivatives for various image processing problems.

References

1. Arora, S., Mathur, T., Agarwal, S., Tiwari, K., Gupta, P.: Applications of fractional calculus in computer vision: a survey. Neurocomputing **489**, 407–428 (2022)
2. Beyer, H., Kempfle, S.: Definition of physically consistent damping laws with fractional derivatives. Zentralblatt für angewandte Mathematik und Mechanik **75**(8), 623–635 (1995)
3. van den Boomgaard, R., Smeulders, A.: The morphological structure of images: the differential equations of morphological scale space. IEEE Trans. Pattern Anal. Mach. Intell. **16**(11), 1101–1113 (1994)
4. Brockett, R.W., Maragos, P.: Evolution equations for continuous-scale morphology. IEEE Trans. Signal Process. **42**(12), 3377–3386 (1994)
5. Chen, S., Liu, F., Turner, I., Anh, V.: A fast numerical method for two-dimensional Riesz space fractional diffusion equations on a convex bounded region. Appl. Numer. Math. **134**, 66–80 (2018)
6. Chowdhury, M.R., Qin, J., Lou, Y.: Non-blind and blind deconvolution under Poisson noise using fractional-order total variation. J. Math. Imaging Vis. **62**, 1238–1255 (2020)
7. Didas, S., Burgeth, B., Imiya, A., Weickert, J.: Regularity and scale-space properties of fractional high order linear filtering. In: Kimmel, R., Sochen, N.A., Weickert, J. (eds.) Scale-Space 2005. LNCS, vol. 3459, pp. 13–25. Springer, Heidelberg (2005). https://doi.org/10.1007/11408031_2
8. Failla, G., Zingales, M.: Advanced materials modelling via fractional calculus: challenges and perspectives. Phil. Trans. R. Soc. A **378**, 20200050 (2020)
9. Guichard, F., Maragos, P., Morel, J.M.: Partial differential equations for morphological operators. In: Bilodeau, M., Meyer, F., Schmitt, M. (eds.) Space, Structure and Randomness: Contributions in Honor of George Matheron in the Field of Geostatistics, Random Sets and Mathematical Morphology, pp. 369–390. Springer, New York (2005)

10. Hao, Z., Sun, Z., Cao, W.: A fourth-order approximation of fractional derivatives with its applications. J. Comput. Phys. **281**, 787–805 (2015)
11. He, N., Wang, J.B., Zhang, L.L., Lu, K.: An improved fractional-order differentiation model for image denoising. Signal Process. **112**, 180–188 (2015)
12. Huang, G., Qin, H., Chen, Q., Shi, Z., Jiang, S., Huang, C.: Research on application of fractional calculus operator in image underlying processing. Fractal and Fractional **8**, 37 (2024)
13. Kleefeld, A., Vorderwülbecke, S., Burgeth, B.: Anomalous diffusion, dilation, and erosion in image processing. Int. J. Comput. Math. **95**, 1375–1393 (2018)
14. Meerschaert, M.M., Tadjeran, C.: Finite difference approximations for two-sided space-fractional partial differential equations. Appl. Numer. Math. **56**, 80–90 (2006)
15. Oldham, K.B., Spanier, J.: The Fractional Calculus, 2nd edn. Dover, New York (2006)
16. Ortigueira, M.D.: Riesz potential operators and inverses via fractional Centred derivatives. Int. J. Math. Math. Sci. **2006**, 48391 (2006)
17. Pan, K., Sun, H.W., Xu, Y., Xu, Y.: An efficient multigrid solver for two-dimensional spatial fractional diffusion equations with variable coefficients. Appl. Math. Comput. **402**, 126091 (2021)
18. Podlubny, I.: Fractional Differential Equations. Academic Press, San Diego (1999)
19. Podlubny, I., Chechkin, A., Skovranek, T., Chen, Y.Q., Vinagre Jara, B.M.: Matrix approach to discrete fractional calculus II: partial fractional differential equations. J. Comput. Phys. **228**, 3137–3153 (2009)
20. Schäfer, I., Kempfle, S.: Impulse responses of fractional damped systems. Nonlinear Dynamica **38**, 61–68 (2004)
21. Shukla, A.K., Pandey, R.K., Pachori, R.B.: A fractional filter based efficient algorithm for retinal blood vessel segmentation. Biomed. Signal Process. Control **59**, 101883 (2020)
22. Sparavigna, A.C.: Fractional differentiation based image processing. Technical report cs.CV:0910.2381v4, arXiv.org (2015)
23. Tuan, N.H., Aghdam, Y.E., Jafari, H., Mesgarani, H.: A novel numerical manner for two-dimensional space fractional diffusion equation arising in transport phenomena. Numerical Methods Partial Differ. Equ. **37**, 1397–1406 (2021)
24. Wu, T., Wan, S., Feng, C., Zhang, H., Zeng, T.: Blind image deconvolution: when patch-wise minimal pixels prior meets fractional-order method. J. Math. Imaging Vis. **67**, 3 (2025)
25. Xu, X., Dai, F., Long, J., Guo, W.: Fractional differentiation-based image feature extraction. Int. J. Signal Process. Image Process. Pattern Recogn. **7**(6), 51–64 (2014)
26. Yang, Q., Chen, D., Zhao, T., Chen, Y.: Fractional calculus in image processing: a review. Fractional Calc. Appl. Anal. **19**(5), 1222–1249 (2016)
27. Yu, Y., Gao, Q., Zhao, X., Ji, Y.: 2-D modeling and analysis of time-domain electromagnetic anomalous diffusion with space-fractional derivative. IEEE Trans. Geosci. Remote Sens. **61**, 2000213 (2023)
28. Zhang, H., Liu, F., Anh, V.: Galerkin finite element approximation of symmetric space-fractional partial differential equations. Appl. Math. Comput. **217**, 2534–2545 (2010)
29. Zhang, X., Li, R., Hong, J., Zhou, X., Xin, N., Li, Q.: Image-enhanced single-pixel imaging using fractional calculus. Opt. Express **30**(1), 81–91 (2022)

On the Representation of Stack Operators by Mathematical Morphology

Diego Marcondes(✉)

Mathematical Sciences Institute and IRL FAMSI, The Australian National University, Canberra, Australia
diego.marcondes@anu.edu.au

Abstract. This paper introduces the class of grey-scale image stack operators as those that (a) map binary-images into binary-images and (b) commute on average with cross-sectioning. Equivalently, stack operators are 1-Lipchitz extensions of set operators which can be represented by applying a characteristic set operator to the cross-sections of the image and adding. In particular, they are a generalisation of stack filters, for which the characteristic set operators are increasing. Our main result is that stack operators inherit lattice properties of the characteristic set operators. We focus on the case of translation-invariant and locally defined stack operators and show the main result by deducing the characteristic function, kernel, and basis representation of stack operators. The results of this paper have implications on the design of image operators, since imply that to solve some grey-scale image processing problems it is enough to design an operator for performing the desired transformation on binary images, and then considering its extension given by a stack operator. We leave many topics for future research regarding the machine learning of stack operators and the characterisation of the image processing problems that can be solved by them.

Keywords: image processing · mathematical morphology · stack filters

1 Introduction

Stack filters, proposed by [17] in the 80 s, is a classical family of filters for signal and image processing. Frameworks for learning stack filters from data have been extensively studied [4,5,7] and many algorithms to design stack filters have been proposed [6,18]. In the context of mathematical morphology, properties of stack filters have been deduced in two important works by Maragos and Schafer [9,10].

It is well-known that an operator ψ acting on functions $f : E \mapsto \{0,\dots,m\}$ (i.e., grey-scale images), for a set E and $m \in \mathbb{Z}_+$, is a stack filter if, and only if, there exists an *increasing* set operator $\tilde{\psi}$, that acts on functions $X : E \mapsto \{0,1\}$ (i.e., binary images), such that

$$\psi(f) = \sum_{t=1}^{m} \tilde{\psi}(T_t[f]) \tag{1}$$

M. H. F. Wilkinson and J. Kosinka (Eds.): DGMM 2025, LNCS 16296, pp. 415–428, 2026.
https://doi.org/10.1007/978-3-032-09544-2_30

for all $f \in \{0,\dots,m\}^E$, in which $T_t[f](x) = \mathbb{1}\{f(x) \geq t\}, x \in E$, is the cross-section of f at level $t \in \{0,\dots,m\}$. We refer to [9,10] for a proof of this fact.

In this paper, we study the class of *stack operators* that are the operators ψ that satisfy (1) for a not necessarily increasing set operator $\tilde{\psi}$. We show that stack operators can be equivalently defined as those that map binary functions into binary functions and satisfy

$$\psi(f) = \frac{1}{m}\sum_{t=1}^{m} \psi(mT_t[f]), \tag{2}$$

that is, $\psi(f)$ is the average of ψ applied to the cross-sections of f rescaled to a function taking values in $\{0,m\}$. We interpret (2) as ψ *commuting on average* with cross-sectioning. In particular, we show that stack operators are an extension of set operators (i.e., binary image operators) to function operators (i.e., grey-scale image operators).

Our main result is that the stack operator ψ inherits many properties of the respective set operator $\tilde{\psi}$, such as being an erosion, a dilation, an anti-erosion, an anti-dilation, increasing, decreasing, extensive, anti-extensive, sup-generating and inf-generating. For simplicity, we focus on the case of translation-invariant and locally defined stack operators and show the main result by deducing the characteristic function, kernel, and basis representation of stack operators.

The results of this paper have implications on the design of image operators, since allow to solve grey-scale image processing tasks by designing an operator for performing the desired transformation on binary images and then considering its extension given by a stack operator via (1). In particular, the results can be leveraged to develop learning methods for grey-scale image processing based on deep neural networks and discrete morphological neural networks [11].

We consider two examples of stack operators designed via (1), one for image denoising and another for boundary recognition, both acting on functions $f : \mathbb{Z}^2 \mapsto \{0,\dots,m\}$ and with associated set operators acting on functions $X : \mathbb{Z}^2 \mapsto \{0,1\}$. The denoising stack operator, denoted by ψ_1, is the extension of the alternating-sequential filter $\tilde{\psi}_1 = \epsilon_W\,\delta_W\,\delta_W\,\epsilon_W$, in which W is the 3×3 square centred at the origin of $\mathbb{Z}^2$, and ϵ_W and δ_W are the set erosion and dilation with structuring element W, respectively. This set operator is obtained by applying the opening $\delta_W\,\epsilon_W$ followed by the closing $\epsilon_W\,\delta_W$, and is known to perform denoising. Since this set operator is increasing, the associated stack operator is actually a stack filter. An example of the action of this set operator, and the associated stack operator, is presented in Fig. 1 (d). Observe that the stack filter ψ_1 inherits the denoising property of $\tilde{\psi}_1$.

The boundary recognition stack operator, denoted by ψ_2, is obtained from a translation-invariant and locally defined set operator[1] satisfying $\tilde{\psi}_2(X)(x) = 0$ if $X(y_1) = X(y_2)\ \forall y_1, y_2 \in x + W$ and equals 1 otherwise, for $x \in \mathbb{Z}^2$, in which $x + W$ is the 3×3 square centred at x. In other words, $\tilde{\psi}_2(X)(x)$ equals

[1] See Sect. 2 for a formal definition of translation-invariant and locally defined set operators.

0 if all pixels of X in a neighbourhood of x have the same value, and equals 1 otherwise. We note that this set operator is not increasing, so the associated stack operator is not a stack filter. Furthermore, as can be seen in the example in Fig. 1 (c), this set operator has the effect of boundary recognition, a property that is inherited by the associated stack operator. In Fig. 1 (e) we present the result obtained by applying $\widetilde{\psi}_2\widetilde{\psi}_1$ and $\psi_2\psi_1$, that is, boundary recognition after denoising. Even though we focus on these examples, the class of stack operators encompasses other operators of interest, such as grey-scale top hat transforms, and morphological gradients, with flat structuring elements.

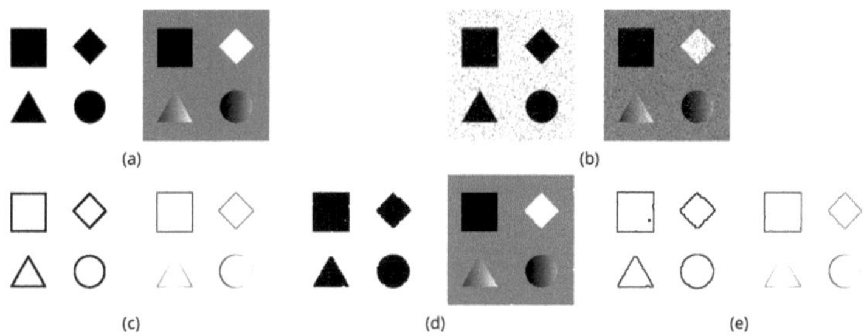

Fig. 1. (a) Binary and grey-scale input images. (b) Binary and grey-scale input images with 2.5% salt and pepper noise. (c) Boundary recognition of (a) by the set operator $\widetilde{\psi}_2$ defined in Sect. 1, and associated stack operator ψ_2. (d) Noise filtering of (b) by the alternating-sequential filter $\widetilde{\psi}_1 = \epsilon_W\, \delta_W\, \delta_W\, \epsilon_W$, and the associated stack filter ψ_1. (e) Result obtained by applying $\widetilde{\psi}_2\widetilde{\psi}_1$ and $\psi_2\psi_1$ to the images in (b). All images are 256×256 pixels and $m = 255$.

In Sect. 2, we define the concepts necessary to introduce the stack operators in Sect. 3. In Sect. 4, we establish the characteristic function, kernel, and basis representation of stack W-operators. In Sect. 5, we state and prove the main result of this paper, and we discuss the implications of our results to the design of image operators. In Sect. 6, we give directions for future research. Proofs for the main results are presented in the Appendix A.

2 Grey-Scale Image and Set Operators

Let E be a countable set, $(E, +)$ be an Abelian group, with zero element $o \in E$, and $M = \{0, \ldots, m\}$ for a positive integer m fixed. A function $f : E \mapsto M$ is a grey-scale image, in which $f(x)$ represents the intensity of the pixel $x \in E$. In particular, a grey-scale binary image is such that $f(x) \in \{0, m\}$ for all $x \in E$. The set M^E of all grey-scale images is equipped with a partial order satisfying $f, g \in M^E, f \leq g \iff f(x) \leq g(x), \forall x \in E$. The *image cross-section, or threshold,* at level t is a mapping $T_t : M^E \mapsto \{0,1\}^E$ given by, for any $f \in M^E, t \in M$ and $x \in E$, $T_t[f](x) = \mathbb{1}\{f(x) \geq t\}$. An image operator is a mapping $\psi : M^E \mapsto M^E$ and the set of all such operators, denoted by Ψ_M, is equipped with a partial order satisfying $\psi_1, \psi_2 \in \Psi_M, \psi_1 \leq \psi_2 \iff \psi_1(f) \leq \psi_2(f), \forall f \in M^E$. The poset $(\Psi_M, \leq)$ is a complete lattice.

A binary image can be represented as a function $X : E \mapsto \{0,1\}$ in which $X(x) = 1$ if, and only if, $x \in E$ is a pixel in the image foreground. The set $\{0,1\}^E$ of all binary images equipped with the partial order satisfying $X, Y \in \{0,1\}^E, X \leq Y \iff X(x) \leq Y(x), \forall x \in E$, is a complete lattice. Mappings from $\{0,1\}^E$ into itself are called set operators, a nomenclature inspired by the fact that $\{0,1\}^E$ is isomorphic to the power-set of E. We denote the space of all set operators by Ψ and equip it with the point-wise partial order $\tilde{\psi}_1, \tilde{\psi}_2 \in \Psi, \tilde{\psi}_1 \leq \tilde{\psi}_2 \iff \tilde{\psi}_1(X) \leq \tilde{\psi}_2(X), \forall X \in \{0,1\}^E$. An image operator is an extension of a set operator if they coincide on binary images. We note that the denoising and boundary recognition operators ψ_1 and ψ_2 introduced in Sect. 1 are extensions of the respective set operator.

Definition 1. *An operator $\psi \in \Psi_M$ is an extension of $\tilde{\psi} \in \Psi$ if $m^{-1}\psi(mX) = \tilde{\psi}(X)$ for all $X \in \{0,1\}^E$. In particular, an image operator that is the extension of a set operator maps grey-scale binary images into grey-scale binary images.*

For each $h \in E$, denote by $\tau_h : E \mapsto E$ the translation by h function that maps $x \mapsto x+h$. An operator $\psi \in \Psi_M$, and a set operator $\tilde{\psi} \in \Psi$, are translation-invariant if they commute with translation, that is, $\psi(f \circ \tau_h) = \psi(f) \circ \tau_h$ and $\tilde{\psi}(X \circ \tau_h) = \tilde{\psi}(X) \circ \tau_h$ for all $f \in M^E$, $X \in \{0,1\}^E$ and $h \in E$, in which $\circ$ means function composition.

For $W \in \{0,1\}^E$, denote by $X|_W$ and $f|_W$ the restriction of $X \in \{0,1\}^E$ and $f \in M^E$ to the set $\chi(W) := \{x \in E : W(x) = 1\}$. An operator $\psi \in \Psi_M$, and a set operator $\tilde{\psi} \in \Psi$, are locally defined within W if[2] $\psi(f)(x) = \psi(f|_{W \circ \tau_{-x}})(x)$ and $\tilde{\psi}(X)(x) = \tilde{\psi}(X|_{W \circ \tau_{-x}})(x)$ for all $f \in M^E$, $X \in \{0,1\}^E$ and $x \in E$. The locally defined condition implies, for example, that the value of the transformed image $\psi(f)$ at point x depends only on the value of f in the neighbourhood $\chi(W \circ \tau_{-x}) = \{h \in E : W(h-x) = 1\}$ of x. Operators that are translation-invariant and locally defined are called W-operators. The collections of image and set W-operators are denoted by $\Psi_M^W \subseteq \Psi_M$ and $\Psi^W \subseteq \Psi$, respectively. We note that the example operators ψ_1 and ψ_2 are W-operators, the latter by construction and the former since erosions and dilations are W-operators, and composition preserves the W-operator properties, but in a greater window[3].

3 Stack Operators

In this paper, we study the collection of stack image operators, defined as follows, and denoted by $\Sigma_S \subseteq \Psi_M$.

[2] The restriction $f|_{W \circ \tau_{-x}}$ should be understood as being equal to $f(y)$ for $y \in \chi(W \circ \tau_{-x})$ and equal to 0 for y not in this set, so it is an element of M^E and $\psi(f|_{W \circ \tau_{-x}})$ is well defined. By an analogous reason, $\tilde{\psi}(X|_{W \circ \tau_{-x}})$ is well defined.

[3] See [3, Corollary 5.1] for a proof of this fact for set operators. That it also holds for stack operators follows from Proposition 2.

Definition 2. *An image operator* $\psi \in \Psi_M$ *is a stack operator if, and only if, there exists a set operator* $\tilde{\psi} \in \Psi$ *such that* (1) *holds for all* $f \in M^E$*, or equivalently, if, and only if, (a) for all* $x \in E$ *and* $X \in \{0,1\}^E$, $\psi(mX)(x) \in \{0, m\}$ *and (b) relation* (2) *holds for all* $f \in M^E$.

Condition (a) of Definition 2 implies that stack operators take grey-scale binary images $mX \in \{0,m\}^E$ into grey-scale binary images. It follows from this condition that $m^{-1}\psi(mT_t[f]) \in \{0,1\}^E$ and therefore the right-hand side of (2) is an element of M^E. Condition (b) then implies that $\psi(f)$ is the average of the transformation by ψ of the cross-sections of f represented as grey-scale binary images. Equivalently, stack operators have an associated set operator $\tilde{\psi} : \{0,1\}^E \mapsto \{0,1\}^E$ satisfying (1), and the collection Ψ of set operators is lattice isomorphic to the collection of stack image operators Σ_S. In particular, $(\Sigma_S, \leq)$ is a complete lattice. We state without proof this result that is a direct consequence of the definitions of $\mathcal{F}$ and $\mathcal{F}^{-1}$ below. This proposition implies that the definitions of stack operators in Definition 2 are indeed equivalent.

Proposition 1. *The poset* $(\Sigma_S, \leq)$ *is lattice isomorphic to* $(\Psi, \leq)$ *with isomorphism* $\mathcal{F} : \Psi \mapsto \Sigma_S$ *given by* $\mathcal{F}(\tilde{\psi})(f) = \sum_{t=1}^{m} \tilde{\psi}(T_t[f])$ *and* $\mathcal{F}^{-1}(\psi)(X) = m^{-1}\psi(mX)$ *for* $\tilde{\psi} \in \Psi$, $\psi \in \Sigma_S$, $f \in M^E$ *and* $X \in \{0,1\}^E$.

We call $\mathcal{F}^{-1}(\psi)$ the *characteristic set operator* of $\psi \in \Sigma_S$. It follows from Proposition 1 that $m^{-1}\psi(mX) = \mathcal{F}^{-1}(\psi)(X)$ and therefore ψ is an extension of its characteristic set operator. Moreover, this extension is 1-Lipschitz considering the L^1-norm in the domain of ψ and the L^∞-norm in its image, that is[4]: $\|\psi(f) - \psi(g)\|_\infty \leq \|f - g\|_1$ for $\psi \in \Sigma_S, f, g \in M^E$. In particular, stack operators are (L^1, L^∞)-continuous. A proof for the next corollary is in Appendix A.

Corollary 1. *A stack operator* $\psi \in \Sigma_S$ *is a 1-Lipschitz extension of its characteristic set operator* $\mathcal{F}^{-1}(\psi)$.

An important class of stack operators is that of the stack W-operators, denoted by $\Sigma_S^W = \Psi_M^W \cap \Sigma_S$. The isomorphism $\mathcal{F}$ preserves the W-operator properties, and hence maps set W-operators to stack W-operators. We state without proof this result that is a direct consequence of the definitions of $\mathcal{F}$ and $\mathcal{F}^{-1}$. Clearly, the example operators ψ_1 and ψ_2 inherit the W-operator properties of their characteristic set operators.

Proposition 2. *An operator* $\psi \in \Sigma_S$ *is a W-operator if, and only if,* $\mathcal{F}^{-1}(\psi)$ *is a W-operator.*

The *stack filters* have the property of commuting with threshold in the following sense:

$$T_t[\psi(f)] = m^{-1}\psi(mT_t[f]) \tag{3}$$

[4] Observe that the Lipschitz inequality is tight since, for example, equality is attained when ψ is the identity operator and $g = f + s\mathbb{1}_{\{x\}}$ for $s \in \{1, \ldots, m\}$ and $x \in E$ fixed.

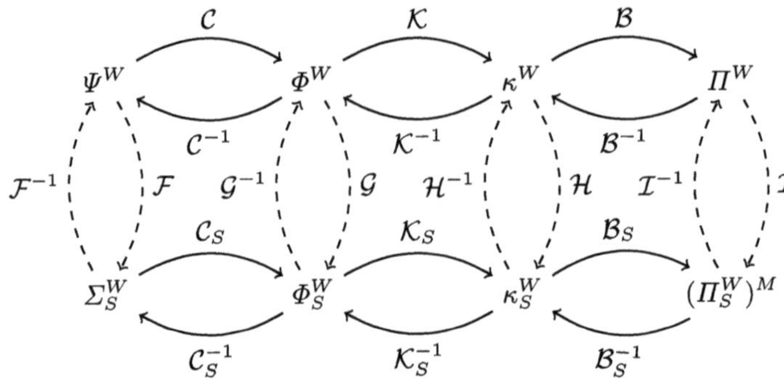

Fig. 2. Lattice isomorphisms between the representations of set (first row) and stack (second row) W-operators. The dashed lines refer to the isomorphisms that associate the representation of set operators with that of stack operators.

for all $t \geq 1$ and $f \in M^E$. Indeed, it follows from the definition of $\mathcal{F}^{-1}$ that a stack operator ψ with characteristic set operator $\tilde{\psi}$ satisfies (3) if, and only if, $T_t[\psi(f)] = \tilde{\psi}(T_t[f])$ for all $t \geq 1$ and $f \in M^E$, a condition that is known to equivalently define stack operators (see [7, Definition 2.1]).

The singularity of stack filters is that they satisfy (3) for all $t \geq 1$ while general stack operators satisfy (3) summing over t (see (2)), and this fact actually implies that the stack filters are the stack operators with increasing characteristic set operator. We state this fact as a proposition whose proof is analogous to [9, Theorem 3].

Proposition 3. *A stack operator $\psi \in \Sigma_S$ is a stack filter if, and only if, $\mathcal{F}^{-1}(\psi)$ is increasing.*

4 Representation of Stack W-Operators

The representation of set and lattice operators by mathematical morphology is a classical research topic, and results about the representation of increasing set operators [9], translation-invariant set operators [1], and general lattice operators [2] have been established. In this section, we deduce the representation of stack W-operators by a characteristic function, a kernel, and a basis, and establish isomorphisms with the respective representation of the associated set W-operator, that are illustrated in Fig. 2.

There are at least two benefits of deducing representations for stack operators. On the one hand, a representation by the combination of simpler operators may yield a more efficient algorithm to compute the operator. This benefit is evident in the representation of the example operator ψ_1 by its characteristic set operator. This representation, that is a consequence of isomorphism $\mathcal{F}$, allows computing ψ_1 by applying erosions and dilations in sequence to the image cross-sections and then adding, what might be more efficient than other representations. Furthermore, if the basis (see Sect. 4.3) of the alternating-sequential set operator $\tilde{\psi}_1$ was computed, then the operator could be implemented by computing $\tilde{\psi}_1$ as the supremum of erosions[5], what is more computationally efficient. This representation is a consequence of the isomorphism $\mathcal{F} \circ \mathcal{C}^{-1} \circ \mathcal{K}^{-1} \circ \mathcal{B}^{-1}$ that maps the basis of the characteristic set operator to the stack operator, as will be proved in the next sections. We also note that certain representations may yield more efficient algorithms for learning operators from data; see for example the algorithm developed in [13,14] to learn the composition of set W-operators from data, relying on the representation via the characteristic function (see Sect. 4.1).

On the other hand, considering the main result of this paper (i.e. Theorem 1), since stack operators inherit properties of the characteristic set operators, in order to design stack operators with desired properties, it is enough to do so on the space of set operators, and it may be easier to enforce desired properties on set operators by considering suitable representations for them. For example, via the basis representation, it is easy to enforce the increasing property by representing the operator as the supremum of erosions. With this representation, it remains to design the structuring elements of these erosions, what can be done manually or by learning from data. Likewise, other restrictions may be considered on the basis of the characteristic set operator to yield desired properties, and the intervals of the basis (see Sect. 4.3) may be designed under these constraints (see Sect. 5 for more details).

4.1 Representation by Characteristic Function

Define $\mathcal{P}(W) := \{0,1\}^{\chi(W)}$ and $\mathcal{P}_M(W) := M^{\chi(W)}$ as the collection of binary and grey-scale patches with support in $\chi(W)$, respectively. For each $\tilde{\psi} \in \Psi^W$ and $\psi \in \Sigma_S^W$, define the characteristic functions $\phi_\psi : \mathcal{P}_M(W) \mapsto M$ and $\tilde{\phi}_{\tilde{\psi}} : \mathcal{P}(W) \mapsto \{0,1\}$ as $\phi_\psi(f) = \psi(f)(o)$ and $\tilde{\phi}_{\tilde{\psi}}(X) = \tilde{\psi}(X)(o)$ for $f \in \mathcal{P}_M(W)$ and $X \in \mathcal{P}(W)$, recalling that o is the zero element of $(E,+)$. The characteristic function determines the value of each pixel $x \in E$ in the transformed image from the values of the pixels of the original image in its neighbourhood $x + W$.

Define $\Phi_S^W := \{\phi_\psi : \psi \in \Sigma_S^W\} \subsetneq M^{\mathcal{P}_M(W)}$ and $\Phi^W := \{0,1\}^{\mathcal{P}(W)}$ as the space of characteristic functions of stack and set W-operators, respectively, and consider in them the point-wise partial orders:

$$\phi_1, \phi_2 \in \Phi_S^W, \phi_1 \preceq \phi_2 \iff \phi_1(f) \preceq \phi_2(f),\ \forall f \in \mathcal{P}_M(W),$$
$$\tilde{\phi}_1, \tilde{\phi}_2 \in \Phi^W, \tilde{\phi}_1 \leq \tilde{\phi}_2 \iff \tilde{\phi}_1(X) \leq \tilde{\phi}_2(X),\ \forall X \in \mathcal{P}(W).$$

The transformations $\mathcal{C} : \Psi^W \mapsto \Phi^W$ and $\mathcal{C}_S : \Sigma_S^W \mapsto \Phi_S^W$, that map stack and set W-operators into their characteristic functions as

$$\mathcal{C}(\tilde{\psi})(X) = \tilde{\psi}(X)(o),\ X \in \mathcal{P}(W), \tilde{\psi} \in \Phi^W,$$
$$\mathcal{C}_S(\psi)(f) = \psi(f)(o),\ f \in \mathcal{P}_M(W), \psi \in \Sigma_S^W,$$

[5] The fact that increasing set operators can be represented as the supremum of erosions has been proved in [8].

respectively, are lattice isomorphisms. The inverse of these mappings are

$$\mathcal{C}^{-1}(\tilde{\phi})(X)(x) = \tilde{\phi}(X|_{W \circ \tau_{-x}} \circ \tau_x) \text{ and } \mathcal{C}_S^{-1}(\phi)(f)(x) = \phi(f|_{W \circ \tau_{-x}} \circ \tau_x).$$

We state this result as a proposition whose proof is a direct consequence of the definition of the isomorphisms and the properties of W-operators.

Proposition 4. *The mappings $\mathcal{C}$ and $\mathcal{C}_S$ are lattice isomorphisms between $(\Psi^W, \leq)$ and $(\Phi^W, \leq)$, and $(\Sigma_S^W, \leq)$ and $(\Phi_S^W, \leq)$, respectively.*

It follows from Propositions 1 and 4 that there exists a lattice isomorphism $\mathcal{G} : \Phi^W \mapsto \Phi_S^W$ given, for example, by $\mathcal{G} = \mathcal{C}_S \circ \mathcal{F} \circ \mathcal{C}^{-1}$. Next lemma gives explicit formulas for $\mathcal{G}$ and $\mathcal{G}^{-1}$ that can be obtained by explicitly computing $\mathcal{C}_S \circ \mathcal{F} \circ \mathcal{C}^{-1}$ and its inverse.

Lemma 1. *For $\tilde{\phi} \in \Phi^W$ and $f \in \mathcal{P}_M(W)$, it holds $\mathcal{G}(\tilde{\phi})(f) = \sum_{t=1}^{m} \tilde{\phi}(T_t[f])$ and, for $\phi \in \Phi_S^W$ and $X \in \mathcal{P}(W)$, it holds $\mathcal{G}^{-1}(\phi)(X) = m^{-1}\phi(mX)$.*

It follows from Lemma 1 that, as stack operators can be expressed by applying a set operator to the cross-sections of an image and adding, their characteristic function can be represented by applying the characteristic function of the associated characteristic set operator to the cross-section of "images" in $\chi(W)$ and adding. Furthermore, lattice isomorphism $\mathcal{G}$ implies an equivalency between the collection of stack W-operators and the functions from $\mathcal{P}(W)$ to $\{0, 1\}$. In particular, it establishes the known (see [7]) equivalency between the translation-invariant and locally defined stack filters and the collection of positive (i.e., increasing) functions from $\mathcal{P}(W)$ to $\{0, 1\}$.

4.2 Kernel Representation

Define as $\kappa^W := \mathcal{P}(\mathcal{P}(W))$ the collection of subsets of $\mathcal{P}(W)$ and $\kappa_M^W := \mathcal{P}(\mathcal{P}_M(W))^M$ as the collection of set-valued functions from M to the collection $\mathcal{P}(\mathcal{P}_M(W))$ of subsets of $\mathcal{P}_M(W)$. These sets are equipped with the partial orders $\mathfrak{X}_1, \mathfrak{X}_2 \in \kappa^W, \mathfrak{X}_1 \leq \mathfrak{X}_2 \iff \mathfrak{X}_1 \subseteq \mathfrak{X}_2$ and $\mathfrak{F}_1, \mathfrak{F}_2 \in \kappa_S^W, \mathfrak{F}_1 \leq \mathfrak{F}_2 \iff \mathfrak{F}_1(t) \subseteq \mathfrak{F}_2(t), \ \forall t \in M$. There exist lattice isomorphisms between Φ^W and κ^W, and Φ_S^W and a subset of κ_S^W, that map each characteristic function to its kernel representation. Formally, these are mappings $\mathcal{K} : \Phi^W \mapsto \kappa^W$ and $\mathcal{K}_S : \Phi_S^W \mapsto \kappa_S^W$ satisfying

$$\mathcal{K}(\tilde{\phi}) = \{X \in \mathcal{P}(W) : \tilde{\phi}(X) = 1\} \quad \text{and} \quad \mathcal{K}_S(\phi)(t) = \{f \in \mathcal{P}_M(W) : t \leq \phi(f)\}$$

for $\tilde{\phi} \in \Phi^W, \phi \in \Phi_S^W$ and $t \in M$, in which $\kappa_S^W := \{\mathcal{K}_S(\phi) : \phi \in \Phi_S^W\} \subsetneq \kappa_M^W$. This definition of kernel is due to [1,2].

The next proposition, that is direct from the definition of $\mathcal{K}$ and $\mathcal{K}_S$, shows that these are indeed lattice isomorphisms. The inverse of these mappings are $\mathcal{K}^{-1}(\mathfrak{X})(X) = \mathbb{1}\{X \in \mathfrak{X}\}$ and $\mathcal{K}_S^{-1}(\mathfrak{F})(f) = \max\{t \in M : f \in \mathfrak{F}(t)\}$.

Proposition 5. *The mappings $\mathcal{K}$ and $\mathcal{K}_S$ are lattice isomorphisms between, respectively, $(\Phi^W, \leq)$ and $(\kappa^W, \leq)$, and $(\Phi_S^W, \leq)$ and $(\kappa_S^W, \leq)$.*

It follows from Lemma 1 and Proposition 5 that there exists a lattice isomorphism $\mathcal{H} : \kappa^W \mapsto \kappa_S^W$, for which we give an explicit expression in the next lemma that is proved in the Appendix A.

Lemma 2. *The mapping* $\mathcal{H} : \kappa^W \mapsto \kappa_S^W$ *given by*

$$\mathcal{H}(\mathcal{X})(t) = \left\{ \sum_{s=1}^{m} X_s : X_s \in \mathcal{P}(W), X_m \leq \cdots \leq X_1, \sum_{s=1}^{m} \mathbb{1}\{X_s \in \mathcal{X}\} \geq t \right\}$$

for $\mathcal{X} \in \kappa^W$ *and* $t \in M$ *is a lattice isomorphism between* $(\kappa^W, \leq)$ *and* $(\kappa_S^W, \leq)$ *with inverse mapping* $\mathcal{H}^{-1}(\mathcal{F}) = \{X \in \mathcal{P}(W) : mX \in \mathcal{F}(m)\}$ *for* $\mathcal{F} \in \kappa_S^W$.

4.3 Basis Representation

For $f_1, f_2 \in \mathcal{P}_M(W)$ and $X_1, X_2 \in \mathcal{P}(W)$, the intervals with limits f_1, f_2 and X_1, X_2 are the sets $[f_1, f_2] = \{f \in \mathcal{P}_M(W) : f_1 \leq f \leq f_2\}$ and $[X_1, X_2] = \{X \in \mathcal{P}(W) : X_1 \leq X \leq X_2\}$, respectively, which are empty if $f_1 \not\leq f_2$ or $X_1 \not\leq X_2$. For collections $\mathcal{X} \in \mathcal{P}(\mathcal{P}(W))$ and $\mathcal{G} \in \mathcal{P}(\mathcal{P}_M(W))$, define the respective maximal collection as formed by the elements in it that are not lesser than any other one: $\mathbf{Max}(\mathcal{X}) = \{X \in \mathcal{X} : Y \in \mathcal{X}, X \leq Y \implies X = Y\}$ and $\mathbf{Max}(\mathcal{G}) = \{f \in \mathcal{G} : g \in \mathcal{G}, f \leq g \implies f = g\}$.

Since set and stack W-operators can be uniquely represented by their kernel, they can be represented by any collection of sets that covers their kernel[6]. A way of covering a subset of a lattice, is by a collection of intervals, and such a cover with the minimum number of intervals is that of maximal intervals. The basis of an operator was proposed by [1,2] based on the idea of covering the kernel (applied at point t in the grey-scale case) with maximal intervals.

Formally, define by $\Pi^W \subseteq \mathcal{P}(\mathcal{P}(W))$ the family of collections of maximal intervals in $\mathcal{P}(W)$ and by $\Pi_S^W \subseteq \mathcal{P}(\mathcal{P}_M(W))$ the family of collections of maximal intervals in $\mathcal{P}_M(W)$. Consider in these collections the partial orders

$$I_1, I_2 \in \Pi^W, I_1 \leq I_2 \iff \forall [X_1, X_2] \in I_1, \exists [Y_1, Y_2] \in I_2 \text{ s.t. } [X_1, X_2] \subseteq [Y_1, Y_2]$$
$$\mathcal{I}_1, \mathcal{I}_2 \in \Pi_S^W, \mathcal{I}_1 \leq \mathcal{I}_2 \iff \forall [f_1, f_2] \subset \mathcal{I}_1, \exists [g_1, g_2] \in \mathcal{I}_2 \text{ s.t. } [f_1, f_2] \subseteq [g_1, g_2].$$

The mappings $\mathcal{B} : \kappa^W \mapsto \Pi^W$ and $\mathcal{B}_S : \kappa_S^W \mapsto (\Pi_S^W)^M$ defined as

$$\mathcal{B}(\mathcal{X}) = \mathbf{Max}(\{[X, Y] : [X, Y] \subseteq \mathcal{X}\}) \quad \mathcal{B}_S(\mathcal{F})(t) = \mathbf{Max}(\{[f, g] : [f, g] \subseteq \mathcal{F}(t)\})$$

for $\mathcal{X} \in \kappa^W, \mathcal{F} \in \kappa_S^W$ and $t \in M$ are lattice isomorphisms, in which the partial order in $(\Pi_S^W)^M$ is $\mathfrak{I}_1, \mathfrak{I}_2 \in (\Pi_S^W)^M, \mathfrak{I}_1 \leq \mathfrak{I}_2 \iff \mathfrak{I}_1(t) \leq \mathfrak{I}_2(t), \forall t \in M$. The inverse of these transformations are $\mathcal{B}^{-1}(I) = \bigcup\{[X, Y] : [X, Y] \in I\}$ and $\mathcal{B}_S^{-1}(\mathfrak{I})(t) = \bigcup\{[f, g] : [f, g] \in \mathfrak{I}(t)\}$ for $I \in \Pi^W, \mathfrak{I} \in (\Pi_S^W)^M$ and $t \in M$. This result is stated as a proposition that follows directly from the definitions of $\mathcal{B}$ and $\mathcal{B}_S$. In particular, $\mathcal{B}^{-1}$ and $\mathcal{B}_S^{-1}$ recover the kernel from its covering by the basis.

[6] In the case of grey-scale image operators, there should actually be a collection of sets for each t, that covers the kernel applied at a point t.

Proposition 6. *The mappings $\mathcal{B}$ and $\mathcal{B}_S$ are lattice isomorphisms.*

We state that there exists a lattice isomorphism $\mathcal{I} : \Pi^W \mapsto (\Pi_S^W)^M$ between the basis representation of set and stack W-operators.

Lemma 3. *The mapping $\mathcal{I} := \mathcal{B}_S \circ \mathcal{H} \circ \mathcal{B}^{-1}$ is a lattice isomorphism between Π^W and $(\Pi_S^W)^M$.*

In particular, the isomorphism $\mathcal{I}$ takes basis of set operators that have one interval to basis of stack operators that have one interval for every t. This simple result is key to proving Theorem 1. This result is proved in Appendix A.

Corollary 2. *If $I = [X, Y]$, then $\mathcal{I}(I)(t) = \{[tX, tY + (m - t)W]\}$ for $t = 1, \dots, m$.*

5 Restricted Classes of Stack Operators and Learning from Data

It follows from Proposition 1 that restricted classes of stack operators can be obtained by constraining Ψ and applying $\mathcal{F}$. Actually, $\mathcal{F}$ preserves many lattice properties of $\tilde{\psi}$, hence restricting Ψ based on these properties will generate analogous restrictions in Σ_S. The next theorem, for which a proof is presented in Appendix A, establishes these properties for the case of stack W-operators.

Theorem 1. *A stack operator $\psi \in \Sigma_S^W$ is (1) increasing, (2) decreasing, (3) extensive, (4) anti-extensive, (5) an erosion, (6) a dilation, (7) an anti-dilation, (8) an anti-erosion, (9) sup-generating, and (10) inf-generating if, and only if, so is $\mathcal{F}^{-1}(\psi)$.*

The isomorphisms depicted in Fig. 2 together with Theorem 1 imply that a stack W-operator can be designed by designing a set operator that performs the desired transformation on sets, what actually reduces to designing a binary function in $\mathcal{P}(W)$. This is what has been done to design the stack operators in Fig. 1.

Prior information about properties that the optimal stack operator satisfies can be used to define a restricted class of set operators that, due to Theorem 1, will generate a restricted class of stack operators with the same properties, from which an operator can be learned from data. For example, applying the isomorphism $\mathcal{F} \circ \mathcal{C}^{-1} \circ \mathcal{K}^{-1} \circ \mathcal{B}^{-1}$ to the basis $I = \{[A_i, B_i] : i = 1, \dots, n\} \in \Pi^W$ of a set operator we get the stack operator given by

$$\psi_I(f) = \sum_{t=1}^{m} \bigvee_{i=1}^{n} \lambda_{[A_i,B_i]}(T_t[f]), \qquad f \in M^E \tag{4}$$

in which $\lambda_{[A_i,B_i]}$ is the set operator with characteristic function $\tilde{\phi}_{[A_i,B_i]}(X) = 1$ if $A_i \leq X \leq B_i$ and equal to zero otherwise for $X \in \mathcal{P}(W)$, that is the operator with basis $\{[A_i, B_i]\}$. This decomposition is based on the sup-generating representation of set operators developed by [1] and $\lambda_{[A_i,B_i]}$ is called a sup-generating operator.

If pairs of input and target images $(f_j, g_j), j = 1, \ldots, N$, are available, then, based on (4), a stack operator may be learned from data by minimising the loss $L_N(I) = \frac{1}{N}\sum_{j=1}^{N}\|\psi_I(f_j) - g_j\|_1$ over I in a subset of Π^W that depends on prior knowledge about the target operator, in which the size n of the basis in (4) controls the *complexity* of the operator. For example, to learn a stack filter, we can consider only basis I with $B_i = W$ for all[7] $i = 1, \ldots, n$. Also, to learn extensive stack operators we can consider only basis which contain the interval $[A, W]$ with $A(o) = 1$ and $A(x) = 0$ for $x \neq o$, what guarantees that $f(x) \leq \psi_I(f)(x)$ for all $x \in E$. Other properties may be enforced on ψ_I by considering an inf-generating representation, under which dilations, decreasing and anti-extensive operators have simplified representations (see [1] for more details).

The minimisation of $L_N(I)$ can be carried out by the lattice descent algorithm proposed for training discrete morphological neural networks (DMNN) in [11] yielding a method that is an extension of DMNN for grey-scale image operators. By dropping the assumption that the intervals in I are maximal, the representation (4) is actually an overparametrisation of the operator, since there are more than one collection of intervals representing the same operator, that are all collections that cover its kernel, but the optimisation problem becomes more efficient (see [12] for more details). If a characteristic function representation for the set operator was considered instead, then, relying on the isomorphism $\mathcal{F} \circ \mathcal{C}^{-1}$, the algorithm proposed in [13,14] could be adapted to learn stack operators without any restriction. Also, a deep neural network might also be considered to represent the characteristic function of the set W-operator and be trained by minimising the error associated with the respective stack operator. We will investigate these learning methods in the future.

6 Future Research

As discussed above, the next step in this research is to develop and implement learning methods for stack operators, but there are also interesting theoretical lines of research. Given the simplicity of the representation of stack W-operators, that are equivalent to Boolean functions in $\mathcal{P}(W)$, they could be the preferred choice when solving a practical image transformation problem. However, it is necessary to characterise what problems they can solve, that is, what image operators they can approximate. Therefore, the main theoretical topic for future research is to characterise the stack operators within the family of 1-Lipschitz extensions of set operators by investigating if they have universal approximating properties, i.e., by characterising spaces in which the stack operators are dense. Stack operators will be suitable to solve a problem when the optimal operator is in such a space.

[7] It follows from the representation deduced in [8] that increasing operators have basis with intervals of form $[A_i, W]$.

Also, it would be interesting to study the representation of stack operators with other kinds of invariance, such as group equivariant operators [15] and scale invariant operators [16]. As a generalisation, other functions that map M into $\{0,1\}$ besides the threshold T_t could be considered in representation (1) yielding more general classes of grey-scale image operators. In general cases, we could obtain an operator that is an extension of a set operator, but with other Lipschitz constant and maybe with better universal approximation properties.

Acknowledgements. This work was partially funded by grant #22/06211-2 São Paulo Research Foundation (FAPESP). I thank the anonymous reviewers for their comments that greatly improved the manuscript, in particular the proof of Corollary 1.

A Proof of Results

Proof (Proof of Corollary 1). Fix $f, g \in M^E$ and $\psi \in \Sigma_S$, and denote $\tilde{\psi} := \mathcal{F}^{-1}(\psi)$. Define $D(f,g) = \{t \in \{1,\dots,m\} : T_t[f] \neq T_t[g]\}$ and observe that $\|\psi(f) - \psi(g)\|_\infty \leq \sum_{t \in D(f,g)} \|(\tilde{\psi}(T_t[f]) - \tilde{\psi}(T_t[g]))\|_\infty \leq |D(f,g)|$. We will show that $|D(f,g)| \leq \|f - g\|_1$. If $x \in E$ is such that $|f(x) - g(x)| = s$, then $\{t_x + 1, \dots, t_x + s\} \subset D(f,g)$ in which $t_x = \min(f(x), g(x))$ so, denoting $\Delta_s = \{x \in E : |f(x) - g(x)| = s\}$, we conclude that $|D(f,g)| \leq \sum_{s=1}^{m} s|\Delta_s|$. The result follows since $\|f - g\|_1 = \sum_{s=1}^{m} s|\Delta_s|$. ■

Proof (Proof of Lemma 2). We will show that $\mathcal{H} = \mathcal{K}_S \circ \mathcal{G} \circ \mathcal{K}^{-1}$ and $\mathcal{H}^{-1} = \mathcal{K} \circ \mathcal{G}^{-1} \circ \mathcal{K}_S^{-1}$ so the result follows from Lemma 1 and Proposition 5. On the one hand, for $\mathfrak{F} \in \kappa_S^W$ and $X \in \mathcal{P}(W)$, it holds $(\mathcal{G}^{-1} \circ \mathcal{K}_S^{-1})(\mathfrak{F})(X) = m^{-1} \max\{t \in M : mX \in \mathfrak{F}(t)\}$, therefore $(\mathcal{K} \circ \mathcal{G}^{-1} \circ \mathcal{K}_S^{-1})(\mathfrak{F}) = \{X \in \mathcal{P}(W) : \max\{t \in M : mX \in \mathfrak{F}(t)\} = m\} = \{X \in \mathcal{P}(W) : mX \in \mathfrak{F}(m)\}$, in which the last equality follows from the fact that $\mathfrak{F}(t)$ is decreasing on t. On the other hand, for $\mathfrak{X} \in \kappa^W$, $f \in \mathcal{P}_M(W)$ and $t \in M$, it holds $(\mathcal{G} \circ \mathcal{K}^{-1})(\mathfrak{X})(f) = \sum_{t=1}^{m} \mathbb{1}\{T_t[f] \in \mathfrak{X}\}$, therefore

$$
\begin{aligned}
(\mathcal{K}_S \circ \mathcal{G} \circ \mathcal{K}^{-1})(\mathfrak{X})(t) &= \left\{ f \in \mathcal{P}_M(W) : t \leq \sum_{s=1}^{m} \mathbb{1}\{T_s[f] \in \mathfrak{X}\} \right\} \\
&= \left\{ \sum_{s=1}^{m} X_s : X_s \in \mathcal{P}(W), X_m \leq \dots \leq X_1, \sum_{s=1}^{m} \mathbb{1}\{X_s \in \mathfrak{X}\} \geq t \right\}.
\end{aligned}
$$

To see that the last equality holds, take $X_s = T_s[f]$ for $f \in \mathcal{P}_M(W)$ and $s \in M$. ■

Proof (Proof of Corollary 2). Fix $I = \{[X,Y]\}$ and observe that, for $t \geq 1$, $(\mathcal{H} \circ \mathcal{B}^{-1})(t) = \left\{ \sum_{s=1}^{t} X_s : X_m \leq \cdots X_1, \sum_{s=1}^{t} \mathbb{1}\{X_s \in [X,Y]\} \geq t \right\}$. Then, $f \in (\mathcal{H} \circ \mathcal{B}^{-1})(t)$ if, and only if, f can be decomposed as $f = \sum_{j=1}^{t} X_{s_j} + g$ with $X_{s_j} \in [X,Y]$ and $g(x) \leq (m-t)W$, assuming that $X_{s_t} \leq \cdots \leq X_{s_1}$. But f can be decomposed as such if, and only if, $f \in [tX, tY + (m-t)W]$. We conclude that $\mathcal{I}([X,Y])(t) = \mathcal{B}_S(\mathcal{H} \circ \mathcal{B}^{-1})(t) = \{[tX, tY + (m-t)W]\}$.

Proof (Proof of Theorem 1). For each $\psi \in \Sigma_S^W$ denote $\tilde{\psi} := \mathcal{F}^{-1}(\psi)$ and $\tilde{\phi} := \mathcal{C}(\tilde{\psi})$, and for each $\tilde{\psi} \in \Psi^W$ denote $\psi := \mathcal{F}(\tilde{\psi})$ and $\phi := \mathcal{C}_S(\psi)$.

(1) *Increasing*: Direct from Proposition 3. (2) *Decreasing*: Analogous to increasing.

(3) *Extensive*: Fix $\tilde{\psi} \in \Psi^W$ extensive and $f \in \mathcal{P}_M(W)$. Then, $\psi(f) = \sum_{t=1}^{m} \tilde{\psi}(T_t[f]) \geq \sum_{t=1}^{m} T_t[f] = f$ in which the inequality follows since $\tilde{\psi}$ is extensive. Fix $\psi \in \Sigma_S^W$ extensive and $X \in \mathcal{P}(W)$. Then, $\tilde{\psi}(X) = \frac{1}{m}\psi(mX) \geq \frac{1}{m}mX = X$. (4) *Anti-extensive*: Analogous to extensive.

(5) *Erosion*: Due to the main result in [2], set and image operators are erosions if, and only if, their kernel is formed by only one interval of form $[X, W]$ and $[f_t, mW]$ for all $t \geq 1$, respectively. On the one hand, it follows from Corollary 2 that if $\tilde{\psi}$ is an erosion with kernel $[X, W]$ then the kernel of ψ equals $[tX, W]$ for all $t \geq 1$ so it is an erosion. On the other hand, if ψ is an erosion then $\mathcal{K}_S(\psi)(m) = [f_m, mW]$ and $\mathcal{K}(\tilde{\psi}) = \{X \in \mathcal{P}(W) : f_m \leq mX\}$. It is clear that $f_m \leq mX$ if, and only if, $T_1[f_m] \leq X$ and therefore $\mathcal{K}(\tilde{\psi}) = [T_1[f_m], W]$, and $\tilde{\psi}$ is an erosion.

(6) *Dilation*: Analogous to erosion by considering the dual partial orders, since dilations according to a partial order are erosions according to the dual partial order.

(7) *Anti-dilation*: Anti-dilations are operators such that $\psi(f \vee g) = \psi(f) \wedge \psi(g)$ and $\tilde{\psi}(X \vee Y) = \tilde{\psi}(X) \wedge \tilde{\psi}(Y)$. Due to the main result in [2], the kernel of an anti-dilation is formed by a unique interval of form $[\emptyset, Y]$, for set W-operators, and $[\emptyset, g_t]$ for all $t \geq 1$, for image operators. A deduction analogous to that of erosions follows by Corollary 2.

(8) *Anti-erosion*: Anti-erosions are operators such that $\psi(f \wedge g) = \psi(f) \vee \psi(g)$ and $\tilde{\psi}(X \wedge Y) = \tilde{\psi}(X) \vee \tilde{\psi}(Y)$. This case is analogous to anti-dilations by considering the dual partial order.

(9) *Sup-generating*: A sup-generating operator is the meet between an erosion and an anti-dilation (ref. [2]). Direct from Corollary 2 since set and image operators are sup-generating if, and only if, their kernel is formed by only one interval.

(10) *Inf-generating*: An inf-generating operator is the join between a dilation and an anti-erosion (ref. [2]). Analogous to sup-generating by considering the dual partial orders, since sup-generating operators according to a partial order are inf-generating according to the dual partial order. ∎

References

1. Banon, G.J.F., Barrera, J.: Minimal representations for translation-invariant set mappings by mathematical morphology. SIAM J. Appl. Math. **51**(6), 1782–1798 (1991)
2. Banon, G.J.F., Barrera, J.: Decomposition of mappings between complete lattices by mathematical morphology, part i. General lattices. Signal Process. **30**(3), 299–327 (1993)

3. Barrera, J., Salas, G.P.: Set operations on closed intervals and their applications to the automatic programming of morphological machines. J. Electron. Imaging **5**(3), 335–352 (1996)
4. Coyle, E.J., Lin, J.H.: Stack filters and the mean absolute error criterion. IEEE Trans. Acoust. Speech Signal Process. **36**(8), 1244–1254 (1988)
5. Coyle, E.J., Lin, J.H., Gabbouj, M.: Optimal stack filtering and the estimation and structural approaches to image processing. IEEE Trans. Acoust. Speech Signal Process. **37**(12), 2037–2066 (1989)
6. Dellamonica, D., Silva, P.J., Humes, C., Hirata, N.S., Barrera, J.: An exact algorithm for optimal MAE stack filter design. IEEE Trans. Image Process. **16**(2), 453–462 (2007)
7. Hirata, N.S., Barrera, J., Dougherty, E.R.: Design of statistically optimal stack filters. In: XII Brazilian Symposium on Computer Graphics and Image Processing (Cat. No. PR00481), pp. 265–274. IEEE (1999)
8. Maragos, P.: A representation theory for morphological image and signal processing. IEEE Trans. Pattern Anal. Mach. Intell. **11**(6), 586–599 (1989)
9. Maragos, P., Schafer, R.: Morphological filters-part i: their set-theoretic analysis and relations to linear shift-invariant filters. IEEE Trans. Acoust. Speech Signal Process. **35**(8), 1153–1169 (1987)
10. Maragos, P., Schafer, R.: Morphological filters-part ii: their relations to median, order-statistic, and stack filters. IEEE Trans. Acoust. Speech Signal Process. **35**(8), 1170–1184 (1987)
11. Marcondes, D., Barrera, J.: Discrete morphological neural networks. SIAM J. Imag. Sci. **17**(3), 1650–1689 (2024)
12. Marcondes, D., Barrera, J.: The lattice overparametrization paradigm for the machine learning of lattice operators. In: International Conference on Discrete Geometry and Mathematical Morphology, pp. 204–216. Springer, Heidelberg (2024). https://doi.org/10.1007/978-3-031-57793-2_16
13. Marcondes, D., Feldman, M., Barrera, J.: An algorithm to train unrestricted sequential discrete morphological neural networks. In: International Conference on Discrete Geometry and Mathematical Morphology, pp. 178–191. Springer, Heidelberg (2024). https://doi.org/10.1007/978-3-031-57793-2_14
14. Marcondes, D., Feldman, M., Barrera, J.: Unrestricted sequential discrete morphological neural networks. J. Math. Imaging Vision **67**(4), 39 (2025)
15. Penaud-Polge, V., Velasco-Forero, S., Angulo-Lopez, J.: Group equivariant networks using morphological operators. In: International Conference on Discrete Geometry and Mathematical Morphology, pp. 165–177. Springer, Heidelberg (2024). https://doi.org/10.1007/978-3-031-57793-2_13
16. Sangalli, M., Blusseau, S., Velasco-Forero, S., Angulo, J.: Scale equivariant neural networks with morphological scale-spaces. In: International Conference on Discrete Geometry and Mathematical Morphology, pp. 483–495. Springer, Heidelberg (2021). https://doi.org/10.1007/978-3-030-76657-3_35
17. Wendt, P., Coyle, E., Gallagher, N.: Stack filters. IEEE Trans. Acoust. Speech Signal Process. **34**(4), 898–911 (1986)
18. Yoo, J., Fong, K.L., Huang, J.J., Coyle, E.J., Adams, G.B.: A fast algorithm for designing stack filters. IEEE Trans. Image Process. **8**(8), 1014–1028 (1999)

Morphological Leveling Based on the Log-Exp-Supremum for Colour Images

Marvin Kahra(✉) and Michael Breuß

Institute for Mathematics, Brandenburg Technical University Cottbus-Senftenberg, 03046 Cottbus, Germany
{marvin.kahra,breuss}@b-tu.de

Abstract. Levelings are morphological filters that are used to simplify and segment images. Depending on the scale, different amounts of detail disappear while the contours of the remaining objects remain sharp.

In this work, we will consider two colour morphological methods to perform the dilation and erosion operations necessary for leveling. The first one computes a direct supremum based on the spectral decomposition of symmetric matrices, which offers some favourable properties, such as transitivity. Based on this, the second method generates a colour ordering of the input colours. We compare the levelings resulting from these two methods experimentally using natural colour images.

Keywords: mathematical morphology · scale space · levelings · partial differential equations

1 Introduction

Mathematical morphology has developed into an extremely powerful toolbox for image processing, see [19,21,23]. The basic idea is to apply a combination of different filters in order to analyse certain areas, shapes, colours or changes in an image. One of these filters are morphological levelings, where an image is sifted more and more finely, i.e. smaller objects gradually disappear. This leads to a series of increasingly coarser representations of the same image. For levelings, the original image and a reference image are required and, depending on whether the reference is larger or smaller than the image with regard to a given order relation, a morphological dilation or erosion is carried out. This method was first introduced by Meyer [17] and further developed together with Maragos [18]. The advantages of leveling over other similar methods are that the contours of the remaining objects remain sharp and localised and the problem of asymmetrical treatment of the foreground and background of the image is avoided.

Due to the structure of the levelings, the order relation plays a central role. While this is still unambiguous in greyscale morphology, the transition to colour morphology provides a greater scope for interpretation and freedom for adaptation. A possible approach in recent years by Breuß, Hoeltgen and Kleefeld

M. H. F. Wilkinson and J. Kosinka (Eds.): DGMM 2025, LNCS 16296, pp. 429–442, 2026.
https://doi.org/10.1007/978-3-032-09541-2_31

[4] uses the Loewner order in combination with a trace order for matrix-valued colour representations. For this purpose, the partial differential equation (PDE) is employed as a basis. The matrix representation has the advantage that all colour channels are considered simultaneously and thus the correlated information between the colour channels is not lost, as is in channel-wise approaches.

In this paper we will build on this, but apply two other methods to calculate the suprema and infima required for dilation and erosion. The first method [11] is an approximation of the maximum by concatenation of summation and logarithmic and exponential functions, going back to Maslov [15,16]. The resulting supremum has some interesting properties, the most notable being the preservation of transitivity from the greyscale morphology and the associativity of the resulting dilation operator. Building on this supremum, in the second method [10] we construct a pre-order that selects the colour that comes closest to the supremum from the available colours. This is extended to a total order by a lexicographic cascade following the example of Angulo [2]. The latter method has the advantage that the "false colour" problem and the colour loss of the first method are avoided while the given direction for determining a 'larger' colour is retained. However, one should bear in mind that the use of a total order in a multidimensional metric space can lead to certain irregularities, see [7].

2 Greyscale Levelings

In this section we start with some concepts of greyscale morphology for the derivation of levelings. We will then turn to the specific levelings and introduce them using a PDE approach, see [5,14] for more details. These fundamentals are used as a basic building block for the later extension to colour images.

The greyscale morphology works on a two-dimensional image domain Ω, which is usually a subset of $\mathbb{R}^2$ or $\mathbb{Z}^2$, and a given single-channel greyscale image $f : \Omega \to [0, 255]$ that assigns a greyscale value to each point in Ω. For the sake of simplicity, we will only focus on flat SEs here. These are described by

$$b(x,y) := \begin{cases} 0 & \text{if } (x,y) \in B, \\ -\infty & \text{otherwise}, \end{cases} \tag{1}$$

where B is our two-dimensional convex SE with compact support. In this paper, we will use a quadratic SE for this purpose. The greyscale operations for dilation δ_B and erosion ε_B are defined as:

$$\begin{aligned} \delta_B[f](x,y) &:= (f \oplus b)(x,y) := \sup_{(p,q)\in\Omega} \{f(x-p, y-q) + b(p,q)\}, \\ \varepsilon_B[f](x,y) &:= (f \ominus b)(x,y) := \inf_{(p,q)\in\Omega} \{f(x+p, y+q) - b(p,q)\}. \end{aligned} \tag{2}$$

The concept of leveling involves the utilisation of a reference function f and a marker function g. In instances where the g is found to be smaller in magnitude than f, the smallest possible dilation is performed until a zone of consistent grey levels is either created or the reference function is reached. Conversely, if the marker function is smaller, the erosion process is employed. This behaviour is illustrated by the following initial boundary value problem:

$$\begin{aligned} \frac{\partial u(x,y,t)}{\partial t} &= \operatorname{sgn}(f(x,y)-u(x,y,t))\|\nabla u(x,y,t)\|, && \forall (x,y)\in\Omega,\ \forall t>0,\\ u(x,y,0) &= (K_\sigma * f)(x,y), && \forall (x,y)\in\Omega,\\ \frac{\partial u(x,y,t)}{\partial \boldsymbol{n}} &= 0, && \forall (x,y)\in\partial\Omega,\ \forall t>0, \end{aligned} \tag{3}$$

where $(K_\sigma * f)$ is a Gaussian convolution with zero mean and given standard deviation σ, which represents our marker function g, and $\boldsymbol{n}$ is the outer normal to the boundary $\partial\Omega$ of Ω. In this context, sgn denotes the signum function. This function thus serves as a switch to toggle between the aforementioned operations, thereby yielding the familiar PDEs for dilation (+) and erosion (–):

$$\frac{\partial u(x,y,t)}{\partial t} = \pm\|\nabla u(x,y,t)\|, \quad \forall (x,y)\in\Omega,\ \forall t>0. \tag{4}$$

In the morphological context, the parameter t can be regarded as the scale for the size of the SE. The norm to be used here depends on the underlying shape of the SE. The use of a convolution with the Gaussian kernel K_σ is also very common in morphological applications, compare [3,4,18]. An alternative to this is the use of morphological smoothing filters, such as opening and closing, see [13]. Finally, we use the Neumann boundary condition for compatibility reasons.

3 The Colour Morphological Setting

The colour morphology is analogous to the greyscale morphology, with the key distinction being the presence of three channels instead of a single one. For a comprehensive overview of various representation formats, we refer to [24].

However, based on the idea of Burgeth and Kleefeld, see [6], we use a modification of the HSL or RGB colour space, which operates with symmetric matrices. This representation has already been successfully implemented in [4] for the construction of levelings using different methods, which include the discretisation of the colour morphology and various approaches for colour-valued PDEs.

3.1 Matrix-Valued Colour Representation

We start by transforming the normalised RGB colour space $[0,1]^3 = [0,1] \times [0,1] \times [0,1]$ into the conventional HCL (hue-chroma-luminance) colour space:

$$H := \begin{cases} \frac{G-B}{6C} \mod 1 & \text{if } M = R,\\ \frac{B-R}{6C} + \frac{1}{3} \mod 1 & \text{if } M = G,\\ \frac{R-G}{6C} + \frac{2}{3} \mod 1 & \text{if } M = B, \end{cases} \quad M := \max\{R,G,B\}, \tag{5}$$

$$C := M - m, \quad m := \min\{R,G,B\} \quad \text{and} \quad L := \frac{1}{2}(M+m).$$

with $\mathcal{X}^p = \{\boldsymbol{X}_1^p, \ldots, \boldsymbol{X}_n^p\}$. By a simple translation of this set, all considerations can also be extended to matrices with negative eigenvalues, see [11].

3.4 LES-Based Colour Ordering

As an extended application of the LES, we now want to specify a total order relation based on it. This was already introduced in work [10], which is why we will only summarise the most important points.

The idea is to introduce a distance term to compare the difference between the LES for a given pixel and the colours available for selection at that pixel, according to the mask. However, since this only leads to a pre-order, we extend this by means of a lexicographic cascade, following the example of Angulo [2].

As distance, we will use the modified HyAB distance

$$\begin{aligned} \Delta E(\boldsymbol{f}_1, \boldsymbol{f}_2) &:= \sqrt{(x_1 - x_2)^2 + (y_1 - y_2)^2} + |z_1 - z_2|\,, \\ \boldsymbol{f}_i &= (x_i, y_i, z_i), \quad i = 1, 2, \end{aligned} \tag{16}$$

with the x, y, z terms according to Subsect. 3.1, because it already showed good results in [10] and [1]. This then leads to the following LES-based colour ordering (LESCO):

$$\boldsymbol{f}_1 <_{\text{LESCO}} \boldsymbol{f}_2 \iff \begin{cases} \Delta E(\boldsymbol{f}_{\text{LES}}, \boldsymbol{f}_1) < \Delta E(\boldsymbol{f}_{\text{LES}}, \boldsymbol{f}_2) \text{ or} \\ \Delta E(\boldsymbol{f}_{\text{LES}}, \boldsymbol{f}_1) = \Delta E(\boldsymbol{f}_{\text{LES}}, \boldsymbol{f}_2) \text{ and} \\ \begin{cases} L_{m_1} < L_{m_2} \text{ or} \\ L_{m_1} = L_{m_2} \text{ and } C_1 < C_2 \text{ or} \\ L_{m_1} = L_{m_2} \text{ and } C_1 = C_2 \text{ and } H_1 \prec_{\text{LES}} H_2, \end{cases} \end{cases} \tag{17}$$

where $\boldsymbol{f}_{\text{LES}}$ is the corresponding colour to the calculated LES $\boldsymbol{S}$ from (14) and the L_{m_i}, C_i, H_i terms are the modified luminance, chroma and hue of the corresponding colours $\boldsymbol{f}_i$. The relation for the hue component is given by

$$\begin{aligned} &H_1 \prec_{\text{LES}} H_2 \iff \begin{cases} -(h_1 \div h_{\text{LES}}) < -(h_2 \div h_{\text{LES}}) \text{ or} \\ -(h_1 \div h_{\text{LES}}) = -(h_2 \div h_{\text{LES}}) \text{ and } (h_1 - h_{\text{LES}}) \bmod 2\pi \leq \pi, \end{cases} \\ &h_i = 2\pi H_i, \quad i = 1, 2, \quad h_{\text{LES}} = 2\pi H_{\text{LES}} \end{aligned} \tag{18}$$

and the angular difference [8]

$$h_1 \div h_2 := \begin{cases} |h_1 - h_2| & \text{if } |h_1 - h_2| \leq \pi, \\ 2\pi - |h_1 - h_2| & \text{otherwise.} \end{cases} \tag{19}$$

However, if the LES is unsaturated, a value for h_{LES} is required. For this, we follow Angulo's proposal [2] by weighting green colours higher than blue or red, based on his previous works on lexicographic ordering, and set it to $h_{\text{LES}} := \frac{2}{3}\pi$.

4 Matrix-Valued Levelings of Colour Images

In order to extend levelings to colour images, we must first convert the underlying greyscale PDE (3) to higher-dimensional colours. To do this, we first formulate it as follows

$$\frac{\partial \boldsymbol{U}(x,y,t)}{\partial t} = \mathrm{Sgn}(\boldsymbol{F}(x,y) - \boldsymbol{U}(x,y,t))\|\nabla \boldsymbol{U}(x,y,t)\|, \quad \forall (x,y) \in \Omega,\ \forall t > 0, \tag{20}$$

where $\boldsymbol{U}, \boldsymbol{F}$ are the matrix-valued representations of the colour vectors corresponding to u, f according to the transformation (6). The interpretation of this equation turns out to be difficult at first, as it is not clear how to interpret the signum function Sgn on the one hand and how to interpret the term $\|\nabla \boldsymbol{U}\|$ on the other. Initially, we only know that the product of the two must result in a symmetric matrix so that the right-hand side matches the left-hand side.

The simplest explanation for Sgn would be to understand it as a scalar according to

$$\mathrm{Sgn}_1(\boldsymbol{X}) := \begin{cases} 1 & \text{if } \boldsymbol{X} \geq_{\mathrm{L}} \boldsymbol{0} \wedge \boldsymbol{X} \neq \boldsymbol{0}, \\ -1 & \text{if } \boldsymbol{X} \leq_{\mathrm{L}} \boldsymbol{0} \wedge \boldsymbol{X} \neq \boldsymbol{0}, \\ 0 & \text{otherwise}, \end{cases} \quad \boldsymbol{X} \in \mathbb{R}^{2\times 2}. \tag{21}$$

In comparison to the greyscale case, this would mean that we perform a matrix-valued dilation or erosion by $\|\nabla \boldsymbol{U}\|$ if the reference is larger or smaller than the marker in the Loewner sense. This means that $\|\nabla \boldsymbol{U}\|$ needs to be a symmetric matrix. The additional possibility compared to the one-dimensional consideration in the greyscale case, which results from the matrix-valued consideration, is the case of an indefinite difference $\boldsymbol{F} - \boldsymbol{U}$. We then interpret this case as an ambiguousness where the function does not change. Since $\boldsymbol{U}$ and as such also $\frac{\partial \boldsymbol{U}}{\partial t}$ can be seen pixel by pixel, i.e. it must represent a specific colour for a given pixel, it is inevitably a symmetric matrix. Therefore, both sides of the equation would represent a symmetric matrix.

An alternative explanation for Sgn is to consider it as a symmetric matrix according to [4]:

$$\mathrm{Sgn}_2(\boldsymbol{X}) := \mathrm{sgn}(\lambda)\boldsymbol{u}\boldsymbol{u}^{\mathrm{T}} + \mathrm{sgn}(\mu)\boldsymbol{v}\boldsymbol{v}^{\mathrm{T}}, \tag{22}$$

where the spectral decomposition of $\boldsymbol{X}$ is given as $\boldsymbol{X} = \lambda \boldsymbol{u}\boldsymbol{u}^{\mathrm{T}} + \mu \boldsymbol{v}\boldsymbol{v}^{\mathrm{T}}$, see (13). Although this has the advantage that we obtain a concrete result in the form of an indefinite symmetrical matrix for an indefinite difference $\boldsymbol{F} - \boldsymbol{U}$, it is still unclear how this is to be interpreted with regard to the PDEs, as neither dilation nor erosion are compatible with it. Furthermore, in this case the product on the right-hand side of equation (20) must also be reinterpreted, for example as the Jordan product [20]

$$\boldsymbol{X} \bullet \boldsymbol{Y} = \frac{1}{2}(\boldsymbol{X}\boldsymbol{Y} + \boldsymbol{Y}\boldsymbol{X}), \quad \boldsymbol{X}, \boldsymbol{Y} \in \mathbb{R}^{2\times 2}, \tag{23}$$

to ensure that the right-hand side yields a symmetrical matrix.

As already mentioned, the norm in equation (20) must result in a symmetric matrix, but otherwise leaves room for interpretation. To ensure the desired form, we apply a scalar-valued norm in the sense of a matrix-valued function to the eigenvalues of the partial derivatives, as in (22). For the norm in question, one could simply use the Euclidean norm, for example, as described in [4]. With regard to our use of quadratic SEs, we consider the norm

$$\|\boldsymbol{x}\| = |x_1| + |x_2|\,, \quad \boldsymbol{x} = (x_1, x_2) \in \mathbb{R}^2, \tag{24}$$

compare [5].

4.1 Discrete Morphological Construction

For this approach, we perform morphological dilation or erosion according to either Subsect. 3.3 or 3.4, depending on the decision of the signum function (21). This is done in the form of the iterative process

$$\boldsymbol{U}_{k+1}(x,y) := \begin{cases} \boldsymbol{\mathcal{D}}_B[\boldsymbol{U}_k](x,y) & \text{if } (\boldsymbol{F} - \boldsymbol{U}_k)(x,y) \geq_{\mathrm{L}} \boldsymbol{0} \wedge (\boldsymbol{F} - \boldsymbol{U}_k)(x,y) \neq \boldsymbol{0}, \\ \boldsymbol{\mathcal{E}}_B[\boldsymbol{U}_k](x,y) & \text{if } (\boldsymbol{F} - \boldsymbol{U}_k)(x,y) \leq_{\mathrm{L}} \boldsymbol{0} \wedge (\boldsymbol{F} - \boldsymbol{U}_k)(x,y) \neq \boldsymbol{0}, \\ \boldsymbol{U}_k(x,y) & \text{otherwise}, \end{cases} \tag{25}$$

where $\boldsymbol{U}_0$ is the original image $\boldsymbol{F}$ after a Gaussian filtering as in (3). Alternatively, if we use the signum function (22), the third case in (25) is replaced by

$$\begin{cases} \boldsymbol{U}_k(x,y) & \text{if } (\boldsymbol{F} - \boldsymbol{U}_k)(x,y) = \boldsymbol{0}, \\ \left(\tilde{\boldsymbol{u}}_k \tilde{\boldsymbol{u}}_k^{\mathrm{T}}\right) \bullet \boldsymbol{\mathcal{D}}_B[\boldsymbol{U}_k](x,y) - \left(\tilde{\boldsymbol{v}}_k \tilde{\boldsymbol{v}}_k^{\mathrm{T}}\right) \bullet \boldsymbol{\mathcal{E}}_B[\boldsymbol{U}_k](x,y) & \text{otherwise}, \end{cases} \tag{26}$$

where we have the spectral decomposition $(\boldsymbol{F} - \boldsymbol{U}_k)(x,y) = \tilde{\lambda}_k \tilde{\boldsymbol{u}}_k \tilde{\boldsymbol{u}}_k^{\mathrm{T}} + \tilde{\mu}_k \tilde{\boldsymbol{v}}_k \tilde{\boldsymbol{v}}_k^{\mathrm{T}}$ with $\tilde{\lambda}_k > 0 > \tilde{\mu}_k$ in the last case. Hence, the final leveling results by $\lim_{k\to\infty} \boldsymbol{U}_k$ for both signum functions.

4.2 PDE-Based Construction

For the numerical treatment of the PDE (20) we follow the approximation using the Rouy-Tourin discretisation [22] as in the paper [4]:

$$\frac{\partial \boldsymbol{U}(x,y,t)}{\partial x} \approx \mathrm{Sup}\left(\mathrm{Sup}(D_x^+ \boldsymbol{U}_{i,j}^k, \boldsymbol{0}), \mathrm{Sup}(-D_x^- \boldsymbol{U}_{i,j}^k, \boldsymbol{0})\right) := [\boldsymbol{U}_x]^\uparrow \tag{27}$$

for dilation, where D_x^-, D_x^+ represent the backward and forward difference operator in x direction and $\boldsymbol{U}(x,y,t) = \boldsymbol{U}_{i,j}^k$ the discretisation with subscript space coordinates and superscript time coordinate. In the case of an erosion, we use for the partial derivative $[\boldsymbol{U}_x]^{\downarrow} = -[-\boldsymbol{U}_x]^{\uparrow}$. The partial derivative to y follows analogously. However, instead of the maximum and minimum given there, we will use the supremum or infimum terms from Subsect. 3.3 or 3.4. The signum functions (21) and (22) can be applied in the discretised PDE as described above and the von Neumann boundary condition is ensured by mirroring the colour values at the boundary of the image. In order to determine the partial time derivative, a straightforward forward difference will be employed.

Fig. 1. Comparison of leveling results after one iteration. **Top: From left to right:** Original image, leveling with the morphologic approach and under usage of the LES (Sgn_1 and Sgn_2) and LESCO (Sgn_1 and Sgn_2). **Bottom: From left to right:** Blurred image, leveling with the PDE approach and under usage of the LES (Sgn_1 and Sgn_2) and LESCO (Sgn_1 and Sgn_2).

5 Experiments

For our experiments, we will use a spatial step size of one in both directions and a temporal step size of 0.25 for the PDE approaches in order to fulfil the stability condition in [18]. One iteration stands for $t = 1$, which means we need four time steps of 0.25 each to perform one iteration. As our test image, we use the pepper image downscaled to 64×64 pixels and a square 3×3 SE. The blurred images were constructed with the *imgaussfilt* function of Matlab with symmetric padding and $\sigma = 2$.

We first analyse how the respective leveling methods behave after one iteration as this is sufficient for a first comparison, see Fig. 1. We can see at first glance that the morphological application with the LES leads to the expected loss of colour, which is particularly evident in the case of the matrix-valued signum. However, the PDE approach with the LES and matrix-valued signum also leads to a strong pixelation with an erratic colour division. Therefore, only the PDE approach with the scalar-valued signum seems to provide a good result using the LES and should be investigated further. If we now look at the LESCO variant, we realise that this leads to more homogeneous images than the LES alone. In general, we can report that the lighter and darker areas of all images were saturated with colour, which is particularly noticeable in the lighter areas.

To examine these five cases in more detail, we look at their behaviour after five iterations of leveling as this yields already reasonable results, see Fig. 2. Here we see signs of increasing fragmentation in the morphological approach, which reacts particularly sensitively to the matrix-valued signum function. With the PDE approach using the LES, we can see that, compared to the other two PDE approaches, it is still most likely to preserve the dark areas in general (top left of the images) and smaller structures within dark areas (bottom right of the images). This results in a weaker scaling effect, in the sense of the coarsening of the structures, than with the other two. In the two LESCO variants, the scaling effect is most pronounced in the variant with the scalar-valued signum. More harmonious colour transitions are also evident in this variant than in the corresponding morphological approach; for instance, around the stem of the long light green pepper. While the matrix-valued signum consistently produced inferior results to the scalar-valued one in previous cases, this is not the case in the PDE LESCO variant. Although we see a weaker scaling effect here, it is still better than in the scalar-valued case with the LES, but has the sharpest contours of all the experiments. In addition, the brighter areas are best preserved here.

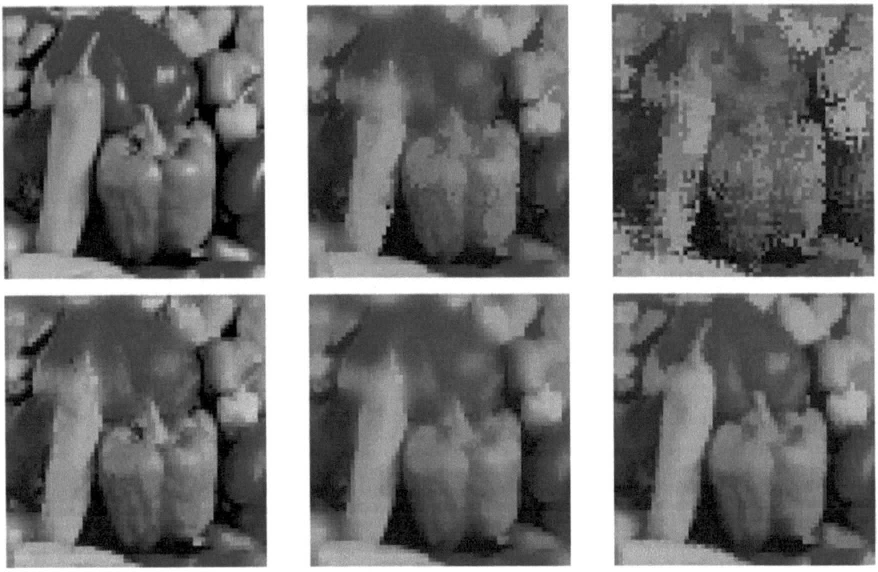

Fig. 2. Comparison of leveling results after five iterations. **Top: From left to right:** Original image, leveling with the morphologic approach and under usage of LESCO (Sgn_1 and Sgn_2). **Bottom: From left to right:** Leveling with the PDE approach and under usage of the LES (Sgn_1) and LESCO (Sgn_1 and Sgn_2).

Finally, we compare the three PDE approaches with regard to their changes after each time step with $\Delta t = 0.25$, see Fig. 3. Here we look at the percentage of pixels that have changed and the extent to which the colours have changed. For the latter, we divide the Frobenius norm of the difference of the image between two time steps and average them over all pixels. In both figures, we see that the differences converge relatively quickly towards zero in the approaches with the scalar-valued signum, whereby it converges somewhat faster with the LES. To be precise, we see that the variant with LESCO always has more pixels that change, but shows smaller differences in colour than the variant with LES for up to two iterations. From the third iteration onwards, the LES variant appears to be superior to the LESCO variant in terms of these properties. However, the matrix-valued signum function with LESCO does not exhibit the desired convergence behaviour. In fact, around 70% of the pixels appear to change at each partial time step for the first five iterations, with the Frobenius colour differences increasing after every iteration. The divergent behaviour for the Frobenius colour difference that is indicated here also continues accordingly after a higher number of iterations, even though the percentage of changing pixels is diminished slowly, i.e. around 36% after 250 iterations. This suggests that, although the matrix signum function can produce visually appealing results, it leads to unstable, divergent behaviour.

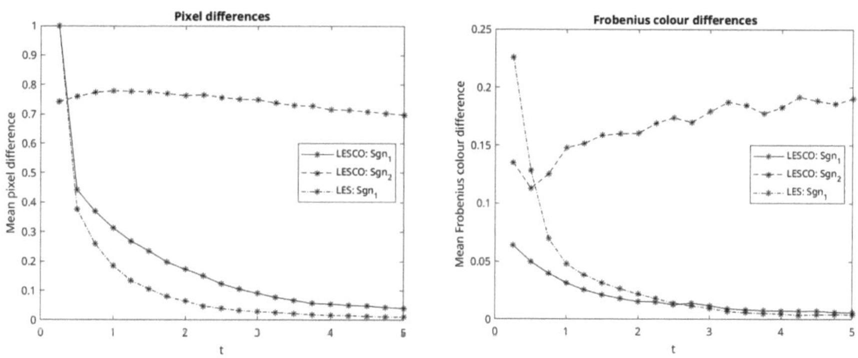

Fig. 3. Change between the first five iterations of the lower three methods of Fig. 2 by performing 20 fractional quarter time steps. **From top to bottom:** Percentage of pixels that have changed and the colour difference using the Frobenius norm averaged over all pixels.

Therefore, we will disregard this approach for now and consider the convergence behaviour of the two remaining scalar-valued sign functions, see Fig. 4. To do this, we perform a total of 20 iterations and look at the results of the last 10 iterations. It is immediately apparent that the LES approach actually converges after 17 iterations, while the LESCO variant, interestingly, also assumes

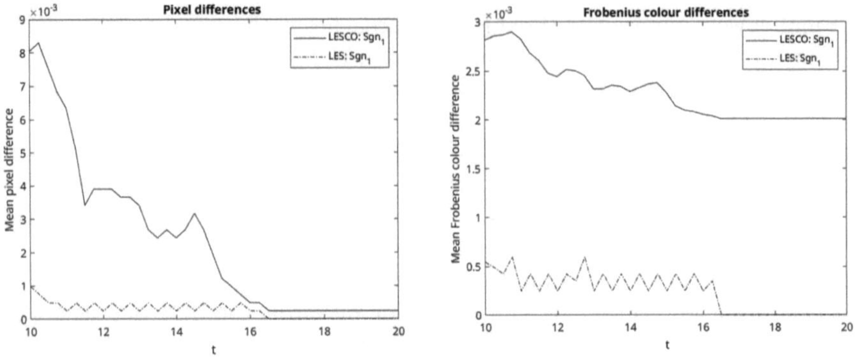

Fig. 4. Change between the 10th and 20th iteration of the two methods with scalar-valued signum functions of Fig. 3 by performing fractional quarter time steps. **From top to bottom:** Percentage of pixels that have changed and the colour difference using the Frobenius norm averaged over all pixels.

stable behaviour after the same number of iterations, albeit not completely convergent. To be precise, we see here that a pixel then permanently changes by two colour values. Since this change is negligible in relation to the overall image, both approaches can be considered convergent in this sense.

6 Conclusion and Future Work

In this work, we have combined the concepts of LES and LESCO with a morphological and a PDE approach to leveling. The PDE approaches proved to be the most promising. For the experiments with the matrix-valued signum function, we saw that this choice of a signum function leads to increasingly unstable and non-convergent behaviour, at least for the methods considered here. However, after appropriate adjustments, this could be suitable for future investigations, as the LESCO variant in particular delivered visually appealing results. Among the scalar-valued PDE approaches, both LES and LESCO showed good convergence behaviour, with the LES variant being the only one that actually converged. However, in our opinion, this is at least compensated for by the greater similarity to morphological levelling for the LESCO variant.

References

1. Abasi, S., Amani Tehran, M., Fairchild, M.D.: Distance metrics for very large color differences. Color. Res. Appl. **45**(2), 208–223 (2020). https://doi.org/10.1002/col.22451
2. Angulo, J.: Morphological colour operators in totally ordered lattices based on distances: application to image filtering, enhancement and analysis. CVIU **107**(1–2), 56–73 (2007). https://doi.org/10.1016/j.cviu.2006.11.008

3. van den Boomgaard, R., Smeulders, A.: The morphological structure of images: the differential equations of morphological scale-space. IEEE PAMI **16**(11), 1101–1113 (1994). https://doi.org/10.1109/34.334389
4. Breuß, M., Hoeltgen, L., Kleefeld, A.: Matrix-valued levelings for colour images. In: Angulo, J., Velasco-Forero, S., Meyer, F. (eds.) ISMM. pp. 296–308. Springer, Cham (2017). https://doi.org/10.1007/978-3-319-57240-6_24
5. Brockett, R.W., Maragos, P.: Evolution equations for continuous-scale morphological filtering. IEEE Trans. Signal Process. **42**(12), 3377–3386 (1994). https://doi.org/10.1109/78.340774
6. Burgeth, B., Kleefeld, A.: An approach to color-morphology based on Einstein addition and Loewner order. Pattern Recogn. Lett. **47**, 29–39 (2014). https://doi.org/10.1016/j.patrec.2014.01.018
7. Chevallier, E., Angulo, J.: The irregularity issue of total orders on metric spaces and its consequences for mathematical morphology. J. Math. Imaging Vision **54**(3), 344–357 (2015). https://doi.org/10.1007/s10851-015-0607-7
8. Hanbury, A.G., Serra, J.: Morphological operators on the unit circle. IEEE Trans. Image Process. **10**(12), 1842–1850 (2001). https://doi.org/10.1109/83.974569
9. Jarre, F.: Convex analysis on symmetric matrices. In: Handbook of Semidefinite Programming: Theory, Algorithms, and Applications, pp. 13–27. Springer (2000). https://doi.org/10.1007/978-1-4615-4381-7_2
10. Kahra, M., Breuß, M.: Colour morphological distance ordering based on the log-exp-supremum. arXiv preprint (2025). https://doi.org/10.48550/arXiv.2503.11329
11. Kahra, M., Breuß, M., Kleefeld, A., Welk, M.: Matrix-valued LogSumExp approximation for colour morphology. arXiv preprint (2024). https://doi.org/10.48550/arXiv.2411.10141
12. Löwner, K.: Über monotone Matrixfunktionen. Math. Z. **38**(1), 177–216 (1934). https://doi.org/10.1007/BF01170633
13. Maragos, P.: Differential morphology and image processing. IEEE Trans. Image Process. **5**(6), 922–937 (1996). https://doi.org/10.1109/83.503909
14. Maragos, P.: Algebraic and pde approaches for lattice scale-spaces with global constraints. IJCV **52**, 121–137 (2003). https://doi.org/10.1023/A:1022999923439
15. Maslov, V.P.: On a new superposition principle for optimization problems. Russ. Math. Surv. **42**(3), 39–48 (1987)
16. Maslov, V.P.: Méthodes opératorielles. Éditions Mir (1987)
17. Meyer, F.: From connected operators to levelings. Comput. Imaging Vision **12**, 191–198 (1998)
18. Meyer, F., Maragos, P.: Nonlinear scale-space representation with morphological levelings. J. Vis. Commun. Image Represent. **11**(2), 245–265 (2000). https://doi.org/10.1006/jvci.1999.0447
19. Najman, J., Talbot, H.: Mathematical Morphology: From Theory to Applications. ISTE-Wiley, London-Hoboken (2010). https://doi.org/10.1002/9781118600788
20. Pizarro, L., Burgeth, B., Breuß, M., Weickert, J.: A directional rouy-tourin scheme for adaptive matrix-valued morphology. In: Wilkinson, M., Roerdink, J. (eds.) ISMM, pp. 250–260. Springer, Heidelberg (2009). https://doi.org/10.1007/978-3-642-03613-2_23
21. Roerdink, J.B.T.M.: Mathematical morphology in computer graphics, scientific visualization and visual exploration. In: Mathematical Morphology and Its Applications to Image and Signal and Image Processing, pp. 367–380. Springer, Heidelberg (2011). https://doi.org/10.1007/978-3-642-21569-8_32
22. Rouy, E., Tourin, A.: A viscosity solutions approach to shape-from-shading. SIAM J. Numer. Anal. **29**(3), 867–884 (1992). https://doi.org/10.1137/0729053

23. Serra, J., Soille, P.: Mathematical Morphology and its Applications to Image Processing, vol. 2. Springer Science & Business Media, Dordrecht (2012)
24. Sharma, G., Bala, R.: Digital Color Imaging Handbook. CRC Press, Boca Raton (2017)
25. Welk, M., Kleefeld, A., Breuß, M.: Quantile filtering of colour images via symmetric matrices. MMTA **1**(1), 136–174 (2016). https://doi.org/10.1515/mathm-2016-0008

Approximating Condorcet Ordering for Vector-Valued Mathematical Morphology

Marcos Eduardo Valle[1](✉), Santiago Velasco-Forero[2], Joao Batista Florindo[1], and Gustavo Angulo[3]

[1] Universidade Estadual de Campinas (UNICAMP), Campinas, Brazil
valle@ime.unicamp.br
[2] Center for Mathematical Morphology (CMM), Mines Paris - PSL University, Fontainebleau, France
[3] Center for Applied Mathematics (CMA), Mines Paris - PSL University, Sophia-Antipolis, France

Abstract. Mathematical morphology provides a nonlinear framework for image and spatial data processing and analysis. Although there have been many successful applications of mathematical morphology to vector-valued images, such as color and hyperspectral images, there is still no consensus on the most suitable vector ordering for constructing morphological operators. This paper addresses this issue by examining a reduced ordering approximating the Condorcet ranking derived from a set of vector orderings. Inspired by voting problems, the Condorcet ordering ranks elements from most to least voted, with voters representing different orderings. In this paper, we develop a machine learning approach that learns a reduced ordering that approximates the Condorcet ordering. Preliminary computational experiments confirm the effectiveness of learning the reduced mapping to define vector-valued morphological operators for color images.

Keywords: Vector-valued mathematical morphology multivariate ordering · Condorcet ordering · neural networks

1 Introduction

Mathematical morphology, a cornerstone of nonlinear image processing, presents unique challenges when applied to vector-valued data, such as color images or hyperspectral remote sensing datasets [1,2,20]. Accordingly, morphological operators can be defined on complete lattices, which are ordered sets with well-defined extrema operations [9,18]. However, there is no natural ordering for vectors, leading to the development of various approaches to vector-valued mathematical morphology. This paper applies concepts from social choice theory to address the challenge of selecting an appropriate ordering for vector-valued mathematical morphology. Social choice theory studies how individual preferences can be

M. H. F. Wilkinson and J. Kosinka (Eds.): DGMM 2025, LNCS 16296, pp. 443–457, 2026.
https://doi.org/10.1007/978-3-032-09544-2_32

aggregated to reach a collective decision, such as electing a candidate through voting [4]. In our context, we aim to establish consensus ordering for vector-valued mathematical morphology.

To our knowledge, integrating different orderings to define vector-valued mathematical morphology was first proposed by Lezoray [13]. Specifically, Lezoray proposed a weighted version of the Borda rule for combining stochastic permutation orderings. Although the Borda rule is computationally straightforward, it may produce an ordering that does not reflect the majority opinion [7]. Instead of using the Borda rule, this paper considers Condorcet's principle, a more robust method for preference aggregation that prioritizes the most preferred candidates [4,12]. The primary limitation of the Condorcet voting method is the computational complexity of the optimization techniques used to reach the consensus ordering. By framing ordering relations as pairwise comparisons between vectors, we propose a learning-based framework to derive simplified representative mappings that approach Condorcet-optimal rankings. This paper is structured as follows: Sect. 2 provides an overview of fundamental concepts in mathematical morphology, Sect. 3 introduces the methodology for learning Condorcet reduced orderings, Sect. 4 presents computational experiments, and Sect. 5 concludes with final remarks and insights.

2 Basic Concepts on Mathematical Morphology

In this paper, we focus on morphological operators for vector-valued images. A vector-valued image, denoted as $\mathbf{I}$, corresponds to a mapping $\mathbf{I} : D \to \mathbb{V}$, where D represents the image domain and $\mathbb{V} \subset \bar{\mathbb{R}}^d$, with $\bar{\mathbb{R}} = \mathbb{R} \cup \{-\infty, +\infty\}$ and $d \geq 2$, denotes a set of vector values. Throughout this paper, we also assume that the domain D is a finite subset of a non-empty discrete additive Abelian group, $(\mathcal{E}, +)$.

Mathematical morphology deals with non-linear operators successfully used for image processing and analysis [9]. Generally speaking, morphological operators explore shapes and geometrical forms present in images. Complete lattices provide suitable frameworks for defining morphological operators [16]. A complete lattice $\mathbb{L}$ is a partially ordered set (poset) where every subset has both a supremum and an infimum [3]. The supremum and infimum of a set $X \subseteq \mathbb{L}$ are denoted by $\bigvee X$ and $\bigwedge X$, respectively. We assume that the vector value set $\mathbb{V}$ is a complete lattice in the following.

Dilations and erosions are two elementary morphological operations [18]. For a given set $S \subset \mathcal{E}$, referred to as a structuring element, the dilation and erosion of the image $\mathbf{I}$ by S are defined by the equations below, respectively:

$$\delta_S(\mathbf{I})(p) = \bigvee_{\substack{s \in S \\ p-s \in D}} \mathbf{I}(p-s) \quad \text{and} \quad \varepsilon_S(\mathbf{I})(p) = \bigwedge_{\substack{s \in S \\ p+s \in D}} \mathbf{I}(p+s), \quad \text{for all } p \in D. \tag{1}$$

Other morphological operators are obtained by combining elementary morphological operators [9]. For instance, the combinations of dilations and erosions yield opening and closing, which exhibit notable topological properties and function as non-linear image filters [18]. Specifically, an opening is characterized by the composition of an erosion succeeded by a dilation, both using the same structuring element S. In contrast, a closing is defined as a dilation followed by an erosion. In mathematical terms, the opening γ_S and the closing ϕ_S of an image $\mathbf{I}$ by a structuring element S are defined, respectively, by

$$\gamma_S(\mathbf{I}) = \delta_S\left(\varepsilon_S(\mathbf{I})\right) \quad \text{and} \quad \phi_S(\mathbf{I}) = \varepsilon_S\left(\delta_S(\mathbf{I})\right). \tag{2}$$

We would like to highlight that dilations and erosions are defined in (1) through supremum and infimum operations. As a result, a partial order with well-defined extremal operations is sufficient for constructing morphological operators. However, unlike the one-dimensional (scalar) case, there is no inherent natural ordering for vectors, and various methods exist for ordering vector values. Examples of partial orderings used in vector-valued mathematical morphology include the marginal or Cartesian product ordering as well as conditional ordering schemes like the RGB-lexicographical ordering [1,2]. Moreover, despite not being partial orders but only pre-orders, reduced orderings have also been successfully used to develop vector-valued morphological operators [8,20].

Several aspects must be considered when selecting a vector ordering for developing morphological operators in image processing tasks, including the *"false color"* problem and the irregularity issue [5,17]. The *"false color"* problem occurs when a morphological operator introduces vector values that do not exist in the original image. This issue can be particularly problematic in applications such as satellite data analysis [17]. One can avoid the *"false color"* problem by ensuring that the partial order is total, that is, either $x \leq y$ or $y \leq x$ holds for any $x, y \in \mathbb{L}$.

With a total ordering, the extreme operators yield elements of the original image, preventing the creation of new vector values. Formally, for any finite set X, we have $\bigvee X \in X$ and $\bigwedge X \in X$ when the supremum and infimum are calculated using a total ordering. However, one drawback of morphological operators based on total orders is that they may introduce irregularities and aliasing in the processed images [5]. While these irregularities can negatively impact specific tasks, such as border detection, this paper focuses on morphological operators defined through total orders. Specifically, our approach mitigates the *"false color"* problem by utilizing a reduced ordering combined with look-up tables (LUTs) [20].

A reduced ordering is defined by a surjective mapping $h : \mathbb{V} \to \mathbb{L}$, which maps the value set $\mathbb{V}$ to a complete lattice $\mathbb{L}$. The partial order $\leq_{\mathbb{L}}$ on $\mathbb{L}$ induces a pre-order on $\mathbb{V}$ as follows:

$$\boldsymbol{x} \leq_h \boldsymbol{y} \iff h(\boldsymbol{x}) \leq_{\mathbb{L}} h(\boldsymbol{y}). \tag{3}$$

In this context, the mapping h is called a reduced mapping or simply an h-mapping. It is important to note that a reduced ordering is not a partial order but rather a pre-order, as it may lack antisymmetry. This means that the inequalities $\boldsymbol{x} \leq_h \boldsymbol{y}$ and $\boldsymbol{y} \leq_h \boldsymbol{x}$ do not necessarily imply that $\boldsymbol{x} = \boldsymbol{y}$. Despite this limitation, reduced orderings can be effectively utilized to develop vector-valued morphological operators. By combining (3) with LUTs, one can define computationally efficient vector-valued morphological operators. Furthermore, alongside a LUT, a reduced ordering yields a total ordering on the image's finite set of vector values. We would like to point out that the use of reduced ordering and LUT for defining morphological operators is detailed in [20], where Algorithm 1 illustrates the implementation of a vector-valued erosion using MATLAB-like notation.

We would like to conclude this section by recalling that the surjective mapping $h : \mathbb{V} \to \mathbb{L}$ can be established using various machine learning techniques, including both supervised and unsupervised methods. In the approach detailed in the following sections, the h-mapping is represented by a neural network trained on an image dataset to generate a consensus ordering.

3 Learning Condorcet Reduced Orderings

In practical scenarios, selecting a reduced mapping for a specific morphological image processing task can be somewhat arbitrary. In this paper, we address the challenge of identifying an appropriate reduced mapping h by examining a family $\mathcal{H} = \{h_1, \ldots, h_m\}$ of possible h-mappings, called a *profile* in social choice literature [24]. Rather than using the common majority voting approach found in ensemble methods [11], our approach seeks to establish an h-mapping that approaches a consensus through the orderings derived from the reduced mappings in $\mathcal{H}$. Such a consensus of orderings is known as the Condorcet ordering in voting methods.

3.1 Condorcet Ordering

The concept of a Condorcet order is inspired by the contributions of *Marie Jean Antonie Nicolas de Caritat* (1743–1794), known as the *Marquis de Condorcet*, on the social choice field during the period of the Enlightenment. A social choice method aims to elect a candidate from a set of voters. Briefly, a candidate is considered a *Condorcet winner* if it is the majority-preferred option in pairwise comparisons [24]. However, we would like to remark that the Condorcet winner is not always well-defined. For instance, the Condorcet voting paradox shows that there is no Condorcet winner when the pairwise majority rule yields a non-transitive relation.

In this paper, we speak of a Condorcet order $\leq^*$ as a total order on $\mathbb{V}$ such that, if a subset $\mathbf{X} \subseteq \mathbb{V}$ has a Condorcet winner, that winner equals $\bigvee^* \mathbf{X}$. Furthermore, the problem of finding a Condorcet ordering is based on the *Kemeny-Young method*, a Condorcet-consistent approach that yields the Condorcet winner when it exists [4,22]. Accordingly, this method selects a total order $\leq^*$ that minimizes the Kemeny score over a family $\mathcal{G} = \{\leq_1, \leq_2, \ldots, \leq_m\}$ of m total orderings [23]. The following reviews a combinatorial linear optimization problem detailed by Marcotorchino and Michaud for finding a Condorcet ordering [14].

First of all, recall that a total order $\leq$ on $\mathbb{V}$ is a binary relation that is reflexive, transitive, antisymmetric, and strongly connected. A binary relation is strongly connected if either $\boldsymbol{x} \leq \boldsymbol{y}$ or $\boldsymbol{y} \leq \boldsymbol{x}$ holds true for all $\boldsymbol{x}, \boldsymbol{y} \in \mathbb{V}$. We can identify the total order on a finite set $\mathbb{V}_n = \{\boldsymbol{x}_1, \ldots, \boldsymbol{x}_n\}$ with a matrix $R \in \{0,1\}^{n \times n}$ such that $r_{ij} = [\boldsymbol{x}_i \leq \boldsymbol{x}_j]$, for all $i, j = 1, \ldots, n$, where

$$[\boldsymbol{x} \leq \boldsymbol{y}] = \begin{cases} 1, & \boldsymbol{x} \leq \boldsymbol{y}, \\ 0, & \text{otherwise}, \end{cases} \tag{4}$$

denotes the indicator function of the binary relation. Furthermore, R is the matrix associated with a total order if, and only if, $r_{ii} = 1$ for all $i = 1, \ldots, n$, and the following holds for distinct indexes i, j, k (i.e., $i \neq j$, $j \neq k$, and $k \neq i$):

$$r_{ij} + r_{jk} - r_{ik} \leq 1 \quad \text{and} \quad r_{ij} + r_{ji} = 1. \tag{5}$$

We would like to note that $r_{ii} = 1$ ensures reflexivity. The first inequality in (5) guarantees the transitivity property. Finally, the second identity in (5) addresses the antisymmetry and strong connectedness properties of a total order [14].

Given a family $\mathcal{G} = \{\leq_1, \leq_2, \ldots, \leq_m\}$ of total orderings on a finite set $\mathbb{V}_n = \{\boldsymbol{x}_1, \ldots, \boldsymbol{x}_n\}$, the average vote margin of $\boldsymbol{x}$ under $\boldsymbol{y}$ is given by

$$\delta(\boldsymbol{x}, \boldsymbol{y}) = \frac{1}{m} \sum_{k=1}^{m} \big([\boldsymbol{x} \leq_k \boldsymbol{y}] - [\boldsymbol{y} \leq_k \boldsymbol{x}] \big), \quad \forall \boldsymbol{x}, \boldsymbol{y} \in \mathbb{V}. \tag{6}$$

Note that $\delta(\boldsymbol{x}, \boldsymbol{y})$ calculates the average difference in the less than or equal count between $\boldsymbol{x}$ and $\boldsymbol{y}$. Therefore, $\delta(\boldsymbol{x}, \boldsymbol{y}) > 0$ if and only if $\boldsymbol{x}$ is less than or equal to $\boldsymbol{y}$ in most orderings in $\mathcal{G}$.

A Condorcet order $\leq^*$ derived from $\mathcal{G}$ corresponds to the total ordering on $\mathbb{V}_n = \{\boldsymbol{x}_1, \ldots, \boldsymbol{x}_n\}$ associated with a matrix $R^* \in \mathbb{R}^{n \times n}$ that solves the following combinatorial optimization problem[1] where $\delta_{ij} \equiv \delta(\boldsymbol{x}_i, \boldsymbol{x}_j)$:

$$\begin{cases} R^* = \underset{R=(r_{ij})}{\text{argmin}} \displaystyle\sum_{i=1}^{n} \sum_{j=1}^{n} \delta_{ij} r_{ij}, \\ \text{subject to } r_{ij} + r_{ij} = 1, \quad \forall i \neq j, \\ \qquad\qquad\quad r_{ij} + r_{jk} - r_{ik} \leq 1, \quad \forall i \neq j, j \neq k, k \neq i, \\ \qquad\qquad\quad r_{ij} \in \{0, 1\}. \end{cases} \tag{7}$$

Unfortunately, the problem described in (7) is NP-hard, making its computational complexity prohibitive for large n. An alternative approach for approximating the solution to (7) is presented in the following subsection, based on [12].

[1] We would like to point out that we obtain a minimization problem by considering a "less than or equal" relation in (6). Dually, a "greater than or equal" relation yields a maximization problem. We considered a minimization problem because deep learning libraries often minimize the loss.

3.2 Soft Condorcet Ordering

In a recent study, Lanctot et al. proposed a ranking method that approximates the Kemeny-Young voting system using score ratings and the logistic function [12]. The following briefly reviews the key steps in their approach to determining the Condorcet ordering.

First, from lattice theory, we know that any finite totally ordered set is isomorphic to $\{1,\ldots,n\}$ [3]. By transitivity, a finite set $\mathbb{V}_n = \{\boldsymbol{x}_1,\ldots,\boldsymbol{x}_n\}$ endowed with a total order can be associated with a corresponding set of scores $\mathcal{S}_n = \{s_1,\ldots,s_n\} \subset \mathbb{R}$ with the natural order $\leq_{\mathbb{R}}$ of real numbers. Formally, we have $\boldsymbol{x}_i \leq \boldsymbol{x}_j$ if and only if $s_i \leq_{\mathbb{R}} s_j$. Thus, determining a total order for $\mathbb{V}_n$ reduces to finding an appropriate score set $\mathcal{S}_n$. In particular, the problem (7) for finding the Condorcet ordering $\leq^*$ using the Kemeny-Young method becomes

$$\mathcal{S}_n^* = \operatorname*{argmin}_{\mathcal{S}_n=\{s_1,\ldots,s_n\}} \sum_{i=1}^{n}\sum_{j=1}^{n} \delta_{ij}[s_i \leq_{\mathbb{R}} s_j], \tag{8}$$

A solution R^* to (7) is derived from $\mathcal{S}_n^*$ by setting $r_{ij}^* = [s_i^* \leq_{\mathbb{R}} s_j^*]$. In this case, the constraints in (7) follow directly from assuming that $\mathcal{S}_n^*$ is equipped with the natural ordering of real numbers.

On the downside, (8) relies on a piecewise constant step function with a discontinuity at $s_i = s_j$, making it extremely difficult to solve. To address this issue, Lanctot et al. replace the step function $[s_i \leq s_j]$ with a smooth approximation using a scaled logistic function:

$$\sigma(s_j - s_i) = \frac{1}{1 + e^{(s_j - s_i)/\tau}}, \tag{9}$$

where $\tau > 0$ is the steepness parameter, controlling the transition speed between the asymptotic values 0 and 1. This substitution transforms the hard optimization problem in (8) into the soft Condorcet optimization (SCO) problem

$$\mathcal{S}_n^* = \operatorname*{argmin}_{\mathcal{S}_n} \sum_{i=1}^{n}\sum_{j=1}^{n} \delta_{ij}\sigma(s_j - s_i), \tag{10}$$

which can be solved via gradient descent or sigmoidal programming [12]. Empirical results indicate that both methods yield similar outcomes, with sigmoidal programming being recommended only for relatively small n due to computational demands. Thus, we adopt a gradient descent-based approach in this paper.

Summarizing, the soft Condorcet optimization problem proposed by Lanctot et al. is applied as follows: Given a family $\mathcal{G} = \{\leq_1, \leq_2, \ldots, \leq_m\}$ of m total orders on a finite set $\mathbb{V}_n = \{\boldsymbol{x}_1,\ldots,\boldsymbol{x}_n\}$, a consensus total order $\leq^*$ is generated by establishing that $\boldsymbol{x}_i \leq^* \boldsymbol{x}_j$ if and only if $s_i^* \leq_{\mathbb{R}} s_j^*$. Here, $\mathcal{S}_n^* = \{s_1^*,\ldots,s_n^*\}$ represents a scoring set that solves the optimization problem (10). It is important to note that the optimal score set $\mathcal{S}_n^*$, and consequently the consensus total order $\leq^*$, depends on both $\mathcal{G}$ and the finite subset $\mathbb{V}_n \subseteq \mathbb{V}$. Specifically, if we were to replace $\mathbb{V}_n$ with a different set $\mathbb{V}'_{n'}$, the soft optimization problem would need to be solved again. Therefore, the consensus total order $\leq^*$ cannot be generalized across different finite subsets of the value set. Using h-orderings, we present a new approach to approximate the Condorcet consensus in the following section. Learning a reduced ordering enables the consideration of subsets (batches) of $\mathbb{V}_n$ as well as the ability to generalize to new values.

3.3 Learning from Reduced Orderings

This paper primarily focuses on identifying a reduced mapping $h^* : \mathbb{V} \to \mathbb{R}$, refered to as a Condorcet h^*-mapping, that captures the consensus of a family $\mathcal{H} = \{h_1, \ldots, h_m\}$ of h-mappings. The Condorcet h^*-mapping is defined by a machine learning model that learns its parameters from a training set denoted as $\mathbb{V}_n = \{\boldsymbol{x}_1, \ldots, \boldsymbol{x}_n\}$. For instance, $\mathbb{V}_n$ can represent the set of colors extracted from a single image or the combined set of colors from multiple images. The finite set $\mathbb{V}_n$ can be considered a finite subset of the larger value set $\mathbb{V}$.

Specifically, let $h : \Theta \to \mathbb{R}^{\mathbb{V}}$ represent a parameterized regressor, where $h(\theta) : \mathbb{V} \to \mathbb{R}$ is an h-mapping for each parameter vector $\theta \in \Theta$, with Θ being the set of all possible parameter vectors. The optimal parameter vector θ^* for the machine learning model is determined by minimizing the soft Kemeny-Young loss function, as introduced by Lanctot et al., across both $\mathbb{V}_n$ and $\mathcal{H}$. Accordingly, based on (6), we define the average voting margin of $\boldsymbol{x}_i$ under $\boldsymbol{x}_j$ by means of the following equation for all $i, j = 1, \ldots, n$:

$$\delta_{ij} = \frac{1}{m} \sum_{k=1}^{m} [h_k(\boldsymbol{x}_i) \leq_{\mathbb{R}} h_k(\boldsymbol{x}_j)] - [h_k(\boldsymbol{x}_j) \leq_{\mathbb{R}} h_k(\boldsymbol{x}_i)]. \tag{11}$$

Note that, instead of considering a family of total orderings, the quantity δ_{ij} is calculated over the family of reduced mappings $\mathcal{H} = \{h_1, \ldots, h_m\}$ using the natural order of real numbers. The soft Kemeny-Young loss function used for training the reduced mapping over $\mathcal{H}$ and $\mathbb{V}_n$ is defined by the following equation as a function of the regressor parameters $\theta \in \Theta$:

$$\mathcal{L}(\mathcal{H}, \mathbb{V}_n)(\theta) = \sum_{i=1}^{n} \sum_{j=1}^{n} \delta_{ij} \sigma\big(h(\theta)(\boldsymbol{x}_j) - h(\theta)(\boldsymbol{x}_i)\big). \tag{12}$$

Concluding, given a family of h mappings $\mathcal{H} = \{h_1, \ldots, h_m\}$ and a finite set of training values $\mathbb{V}_n = \{\boldsymbol{x}_1, \ldots, \boldsymbol{x}_n\} \subset \mathbb{V}$, the Condorcet h^*-mapping is given by $h^* \equiv h(\theta^*) : \mathbb{V} \to \mathbb{R}$, where θ^* solves the optimization problem

$$\theta^* = \operatorname*{argmin}_{\theta \in \Theta} \mathcal{L}(\mathcal{H}, \mathbb{V}_n)(\theta). \tag{13}$$

Finished the training phase, the Condorcet h^*-mapping can be utilized to define morphological operators, which serve as a consensus of the morphological operators derived from the reduced mappings $h_1, \ldots, h_m$. Additionally, due to the model's generalization capability, the Condorcet h^*-mapping can process or analyze new images not included in the training set. The computational experiments reported in the following section illustrate these remarks.

3.4 Borda Rule

The Borda rule, proposed by *Jean-Charles de Borda* (1733-1799) in 1770 during the Enlightenment period, serves as an alternative method to the Condorcet consensus for making social choices [4]. Unlike the Condorcet consensus, which uses pairwise comparisons to establish a binary relation that results in a total order, the Borda rule ranks candidates by aggregating scores that reflect voters' preferences. The following briefly reviews the Borda rule.

Consider a finite set $\mathbb{V}_n = \{\boldsymbol{x}_1, \ldots, \boldsymbol{x}_n\} \subseteq \mathbb{V}$ (set of candidates) and let $\mathcal{G} = \{\leq_1, \leq_2, \ldots, \leq_m\}$ be a family of m total orderings on $\mathbb{V}_n$, reflecting the preferences of voters. The Borda score of a candidate $\boldsymbol{x}_i \in \mathbb{V}_n$ is determined by averaging the ranks of $\boldsymbol{x}_i$ across the ordered lists of all voters' preferences. Using the indicator function defined by (4), the Borda score of $\boldsymbol{x}_i$ is defined by

$$\mathcal{B}(\boldsymbol{x}_i) = \frac{1}{m(n-1)} \sum_{k=1}^{m} \sum_{j \neq i}^{n} [\boldsymbol{x}_j \leq_k \boldsymbol{x}_i], \quad \forall i = 1, \ldots, n. \tag{14}$$

Note that $\mathcal{B}(\boldsymbol{x}_i) \in [0,1]$, where $\mathcal{B}(\boldsymbol{x}_i) = 0$ and $\mathcal{B}(\boldsymbol{x}_i) = 1$ signify that $\boldsymbol{x}_i$ is the least and the most preferred choice for all voters, respectively. The Borda rule establishes a total ordering $\mathbb{V}_n$, denoted as "$\leq_B$", such that $\boldsymbol{x}_i \leq_B \boldsymbol{x}_j$ when $\mathcal{B}(\boldsymbol{x}_i) \leq_{\mathbb{R}} \mathcal{B}(\boldsymbol{x}_j)$. Like reduced orderings, the total order "$\leq_B$" can be computed by combining the Borda scores with a look-up table, as detailed in [20].

The Borda rule is computationally cheaper than our proposed approach, which learns a reduced mapping that approximates the Condorcet consensus. Accordingly, instead of solving the optimization problem (10), the Borda rule requires the computation of the scores by means of (14) and a sorting operation. On the downside, the Borda rule can fail to elect a candidate who is top-ranked by the majority of the voters [4]. For example, consider a set $\mathbb{V}_3 = \{\boldsymbol{x}_1, \boldsymbol{x}_2, \boldsymbol{x}_3\}$ with three candidates and let $\mathcal{G} = \{\leq_1, \ldots, \leq_5\}$ be a family of total orderings representing the preferences of five voters. Suppose the voter's preference are $\boldsymbol{x}_1 \leq_i \boldsymbol{x}_2 \leq_i \boldsymbol{x}_3$ for $i = 1, 2, 3$ and $\boldsymbol{x}_3 \leq_i \boldsymbol{x}_1 \leq_i \boldsymbol{x}_2$ for $i = 4, 5$. Note that $\boldsymbol{x}_3$ is the Condorcet winner because it is the preferred candidate of the majority of the voters. Moreover, following the majority of the voters, the Condorcet consensus yields the total order $\boldsymbol{x}_1 \leq^* \boldsymbol{x}_2 \leq^* \boldsymbol{x}_3$. In contrast, the Borda scores are $\mathcal{B}(\boldsymbol{x}_1) = 2/10$, $\mathcal{B}(\boldsymbol{x}_2) = 7/10$, and $\mathcal{B}(\boldsymbol{x}_1) = 6/10$. As a consequence, the Borda rule yields $\boldsymbol{x}_1 \leq_B \leq \boldsymbol{x}_3 \leq_B \boldsymbol{x}_2$. Note that $\boldsymbol{x}_2$ is the largest element (the elected candidate) despite not being the Condorcet winner (the candidate preferred by the majority).

In conclusion, although the Borda rule is computationally less expensive than the Condorcet method, it may not adequately reflect the majority preference in social choices. In the field of mathematical morphology, Lezoray has successfully applied the Borda rule to define morphological operators that are generated using stochastic permutation orderings [13]. For this reason, we include the Borda rule for comparison purposes in the following section.

4 Computational Experiments

Let us illustrate the approach described in the previous section for learning an h-mapping that closely approximates the Condorcet ordering derived from a family of reduced mappings $\mathcal{H} = \{h_1, \ldots, h_m\}$ and a set of RGB colors $\mathbb{V}_n$. For comparison purposes, we also consider the Borda rule described in Sect. 3.4. The Jupyter Notebook that outlines the computational experiments is available at github.com/mevalle/CondorcetOrdering.

We used the CIFAR dataset to learn the Condorcet h-mapping. The CIFAR dataset includes thousands of RGB color images, each of size 32×32, which are divided into training and testing sets [10]. Although this dataset was originally designed for image classification tasks, the CIFAR images are also suitable for our purposes. Indeed, we selected the first 100 images from the CIFAR-10 training set as our training set, denoted as $\mathcal{T}_{tr} = \{\mathbf{I}_1, \ldots, \mathbf{I}_{100}\}$. Figure 1 shows some illustrative images from the training set $\mathcal{T}_{tr}$. Besides the training set, we also considered a validation set $\mathcal{T}_{val} = \{\mathbf{I}_{101}, \ldots, \mathbf{I}_{200}\}$ with the subsequent hundred images from the CIFAR training set. The training and validation sets $\mathbb{V}_n^{tr}$ and $\mathbb{V}_n^{val}$ consist of all $n = 102,400$ colors from the hundred images in $\mathcal{T}_{tr}$ and $\mathcal{T}_{val}$, respectively.

Besides the set of color images, we considered the family of reduced mappings $\mathcal{H} = \{h_1, h_2, h_3\}$ for learning the Condorcet h^*-mapping and the Borda rule. The following equations give the h-mappings for any $\boldsymbol{x} = [r, g, b] \in \mathbb{V}$, where $\mathbb{V} = \{0, \frac{1}{255}, \ldots, \frac{254}{255}, 1\}^3$ is a discrete set of RGB colors:

$$h_1(\boldsymbol{x}) = 255r + g + \frac{b}{255}, \quad h_2(\boldsymbol{x}) = \frac{r}{255} + 255g + b, \quad \text{and} \quad h_3(\boldsymbol{x}) = r + \frac{g}{255} + 255b. \tag{15}$$

We would like to remark that the h-mappings h_1, h_2, and h_3 yield the lexicographic orderings $R-G-B$, $G-B-R$, and $B-R-G$, respectively. These three h-mappings have been chosen for illustrative purposes in this paper. In practical scenarios, the h-mappings can be defined using supervised or unsupervised approaches, as noted in [13,20].

We adopted a multilayer perceptron (MLP) network with an architecture 3-64-1 for our computational experiments. In other words, the MLP network consists of a single hidden layer containing 64 neurons that utilize the ReLU activation function, along with a single output neuron that has no bias and employs the identity function as its activation function (effectively meaning no activation function is used for the output neuron). The MLP network inputs an RGB color value $\boldsymbol{x} = [r, g, b]$ and produces a score value as the output. The optimization problem outlined in (10), with the loss function defined by (12) with $\tau = 1$, was solved using the Adam optimizer with the default parameters from Keras, version 3.6.0. The training was conducted over a fixed number of 100 epochs, using batches of size 1024. We used the validation set $\mathcal{V}_n^{val}$ to assess the generalization capability of the proposed MLP network. Figure 2 plots the loss values against the number of epochs. Note that the loss levels off after 60 epochs, indicating that the solver has probably reached a minimum of the loss function. Also, the loss curves for the training and validation sets are similar, confirming the proposed network's ability to generalize well. The trained MLP network corresponds to the Condorcet h^*-mapping.

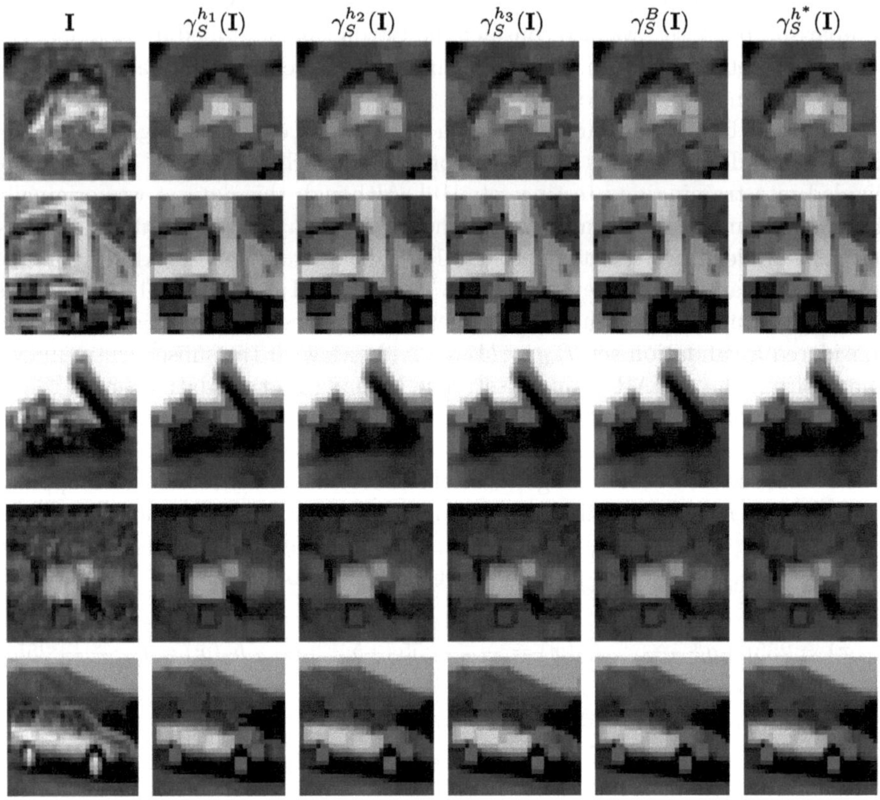

Fig. 1. Illustrative images from the training set $\mathcal{T}$ (first column) and the corresponding opened images obtained using the reduced mappings h_1, h_2, h_3, the Borda rule B, and the Condorcet h^*-mapping.

Figure 1 illustrates the opening operation, denoted as γ_S^h, performed with a 3×3 square structuring element, using the reduced ordering approaches, including the proposed approach, and the Borda rule. Specifically, the first five color images from the training set $\mathcal{T}_{tr}$ were processed with the morphological operation defined by the mappings h_1, h_2, and h_3. The resulting images are displayed in the second, third, and fourth columns of Fig. 1. The last two columns showcase the opening obtained using the Borda rule B and the Condorcet h^*-mapping, respectively. The images produced by Borda rule and the Condorcet h^*-mapping can be interpreted as a consensus generated from the other three morphological approaches.

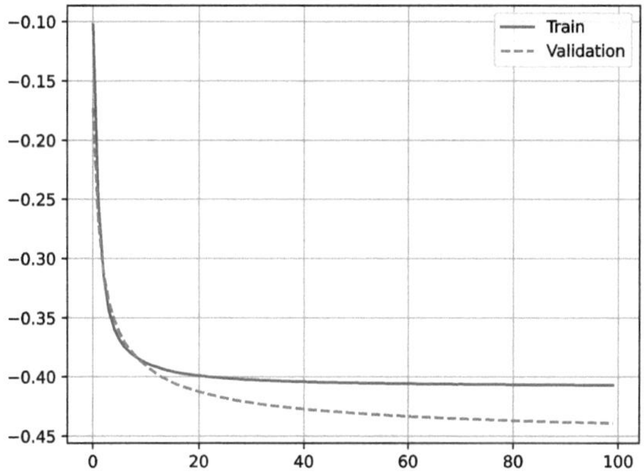

Fig. 2. The loss over epochs during the training of the Condorcet h^*-mapping.

We would like to remark that the morphological operation does not introduce "*false colors*" because we use a reduced ordering combined with a look-up table. Moreover, we observed that the images obtained using the Condorcet h^*-mapping are generally more regular than those obtained using other approaches. Accordingly, Fig. 3a) presents a boxplot illustrating the irregularity index values, $\Phi_1^g(\mathbf{I}_k, \gamma_S^\star(\mathbf{I}_k))$, calculated for all images $\mathbf{I}_k \in \mathcal{T}_{val}$. Here, Φ_1^g denotes the global irregularity index, as detailed in [19]. Notably, the irregularity index values generated by the Condorcet approach display the least variability and are generally lower than those produced by the lexicographic methods and the Borda rule. We confirmed this observation through a Wilcoxon signed-rank test, conducted with a confidence level of 99%. The Hasse diagram shown in Fig. 3b) summarizes the results of the pair-wise Wilcoxon hypothesis test comparing the irregularity indexes of approaches [21]. In the Hasse diagram, the models positioned at the top statistically produced the lowest irregularity index values.

Although the Condorcet h^*-mapping was learned using the CIFAR dataset, it can also be applied to process other color images. For instance, the first row in Fig. 4 displays a color image $\mathbf{J}$ with 321×481 pixels from the BSD dataset [15], along with the closings $\phi_S^B(\mathbf{J})$ and $\phi_S^{h^*}(\mathbf{J})$ obtained using the Borda rule and the Condorcet h^*-mapping. The second row shows the closings obtained using the three lexicographic orderings. All the openings have been computed using a disk of radius equal to ten pixels.

To conclude this section, Fig. 5 shows the rankings of some colors yielded by the reduced mappings h_1, h_2, h_3, the Borda rule, and the Condorcet h^*-mapping. Note that *black* ranks as the least color among the three lexicographical orderings, making it the Condorcet winner (in the sense of "less than or equal"). Therefore, *black* is also the least color in the Condorcet ordering. When *black* is excluded, there is no consensus among the lexicographical orderings for the remaining colors. In this scenario, the h^*-mapping yields *maroon*, the second least element in the lexicographical $G-B-R$ ordering, as the second least color in the Condorcet ordering. Dually, *white* is the Condorcet winner (in the sense

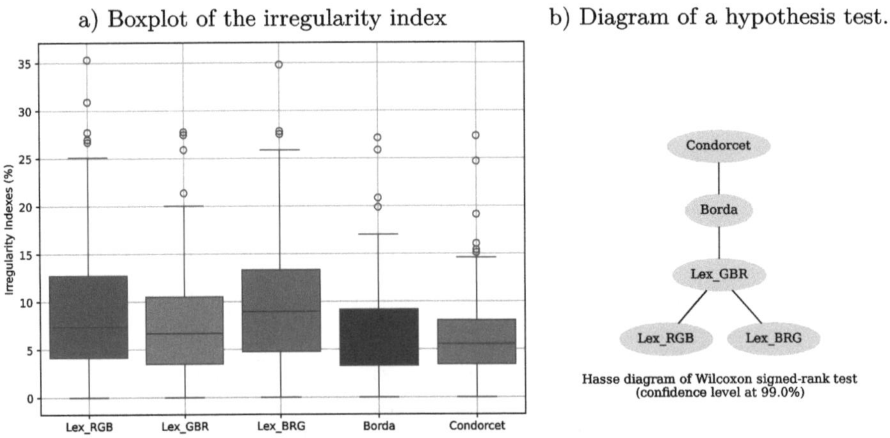

Fig. 3. Boxplot of the global irregularity values and the Hasse diagram of pair-wise Wilcoxon signed-rank test with confidence level at 99%.

of "greater than or equal") since it ranks as the most preferred color in all three lexicographical orderings. Thus, *white* is also the largest color in the Condorcet ordering. Similar to the previous scenario, there is no consensus on the second-largest color among the three lexicographical orderings. The h^*-mapping yields *cyan*, the second largest color in the lexicographical $G-B-R$ ordering, as the second largest color in the Condorcet ordering. In general, the ranking of

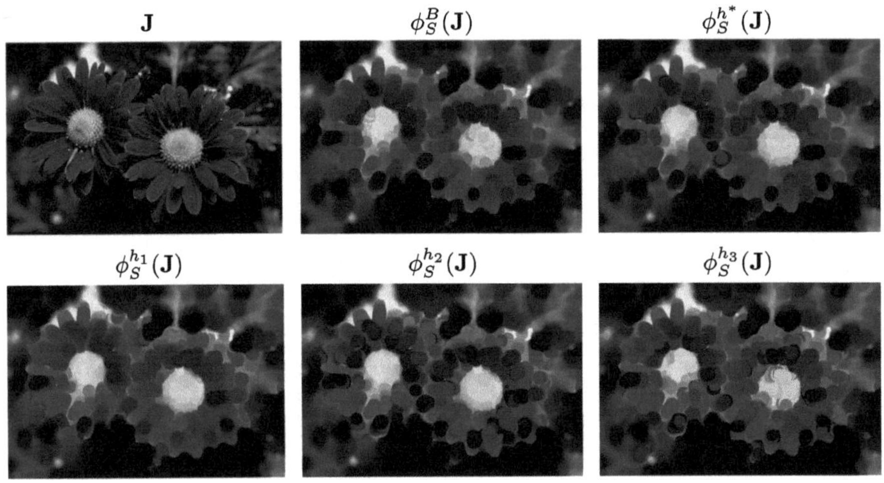

Fig. 4. First row: A color image alongside its closing processed using Borda rule and the Condorcet h^*-mapping. Second row: Closing by Lexicographic orderings h_1 $(R-G-B)$, h_2 $(G-B-R)$, and h_3 $(B-R-G)$.

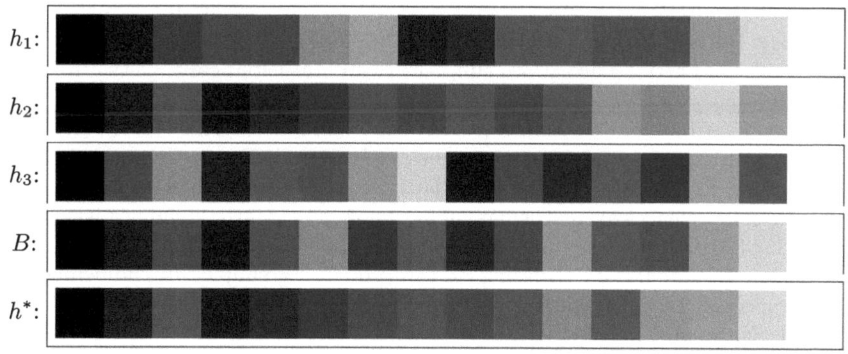

Fig. 5. Illustrative example of the ranking obtained using the lexicographic orderings and the Condorcet h^*-mapping.

colors yielded by h^* agrees with the Condorcet order derived from the three lexicographic orderings, confirming the proposed approach's potential.

5 Conclusions and Perspectives

This paper presents an approach for using gradient descent methods to learn a reduced ordering in the Condorcet sense, where the majority-preferred option is determined through pairwise comparisons. A soft version of the original problem is formulated, enabling the application of learning paradigms from image databases. The experimental results demonstrate that the proposed method is effective on color image datasets, producing reduced orders with a lower irregularity index. Future work will explore extensions of this approach to multichannel images as hyperspectral images in particular represented after source separation into images valued on the simplex of abundances [6], further expanding its applicability and potential impact to remote sensing and medical applications.

Acknowledgments. This research was jointly funded by FAPESP (grant number 2023/03368-0) and ANR (23-CE23-0028-01) in the Deep Ordering for Vector-Valued Mathematical Morphology and Neural Networks – DEEPORDER project.

References

1. Angulo, J.: Morphological colour operators in totally ordered lattices based on distances: application to image filtering, enhancement and analysis. Comput. Vis. Image Underst. **107**(1–2), 56–73 (2007)

2. Aptoula, E., Lefèvre, S.: A comparative study on multivariate mathematical morphology. Pattern Recogn. **40**(11), 2914–2929 (2007)
3. Birkhoff, G.: Lattice Theory, 3rd edn. American Mathematical Society, Providence (1993)
4. Brandt, F., Conitzer, V., Endriss, U., Lang, J., Procaccia, A.D. (eds.): Handbook of Computational Social Choice. Cambridge University Press, Cambridge (2016)
5. Chevallier, E., Angulo, J.: The irregularity issue of total orders on metric spaces and its consequences for mathematical morphology. J. Math. Imaging Vision **54**(3), 344–357 (2016)
6. Franchi, G., Angulo, J.: Ordering on the probability simplex of endmembers for hyperspectral morphological image processing. In: Benediktsson, J.A., Chanussot, J., Najman, L., Talbot, H. (eds.) ISMM 2015. LNCS, vol. 9082, pp. 410–421. Springer, Cham (2015). https://doi.org/10.1007/978-3-319-18720-4_35
7. Gehrlein, W.V., Lepelley, D.: Voting Paradoxes and Group Coherence. Springer, Heidelberg (2011). https://doi.org/10.1007/978-3-642-03107-6
8. Goutsias, J., Heijmans, H.J.A.M., Sivakumar, K.: Morphological operators for image sequences. Comput. Vis. Image Underst. **62**, 326–346 (1995)
9. Heijmans, H.J.A.M.: Mathematical morphology: a modern approach in image processing based on algebra and geometry. SIAM Rev. **37**(1), 1–36 (1995)
10. Krizhevsky, A.: Learning multiple layers of features from tiny images. Technical report, University of Toronto (2009). http://www.cs.toronto.edu/~kriz/cifar.html
11. Kuncheva, L.: Combining Pattern Classifiers: Methods and Algorithms, 2nd edn. John Wiley and Sons, Hoboken (2014)
12. Lanctot, M., et al.: Soft Condorcet Optimization for Ranking of General Agents (2025). https://arxiv.org/abs/2411.00119v3
13. Lézoray, O.: Mathematical morphology based on stochastic permutation orderings. Math. Morphol. Theory Appl. **5**(1), 43–69 (2021)
14. Marcotorchino, F., Michaud, P.: Agrégation de similarités en classification automatique. Revue de statistique appliquée **30**(2), 21–44 (1982)
15. Martin, D., Fowlkes, C., Tal, D., Malik, J.: A database of human segmented natural images and its application to evaluating segmentation algorithms and measuring ecological statistics. In: Proceedings of the 8th International Conference on Computer Vision, vol. 2, pp. 416–423 (2001)
16. Ronse, C.: Why mathematical morphology needs complete lattices. Signal Process. **21**(2), 129–154 (1990)
17. Serra, J.: The "False Colour" problem. In: Wilkinson, M.H.F., Roerdink, J.B.T.M. (eds.) ISMM 2009. LNCS, vol. 5720, pp. 13–23. Springer, Heidelberg (2009). https://doi.org/10.1007/978-3-642-03613-2_2
18. Soille, P.: Morphological Image Analysis. Springer Verlag, Berlin (1999). https://doi.org/10.1007/978-3-662-05088-0
19. Valle, M.E., Francisco, S., Granero, M.A., Velasco-Forero, S.: Irregularity index for vector-valued morphological operators. J. Math. Imaging Vision **64**(7), 754–770 (2022)
20. Velasco-Forero, S., Angulo, J.: Vector ordering and multispectral morphological image processing. In: Celebi, M.E., Smolka, B. (eds.) Advances in Low-Level Color Image Processing. LNCVB, vol. 11, pp. 223–239. Springer, Dordrecht (2014). https://doi.org/10.1007/978-94-007-7584-8_7
21. Weise, T., Chiong, R.: An alternative way of presenting statistical test results when evaluating the performance of stochastic approaches. Neurocomputing **147**, 235–238 (2015)

22. Young, H.P.: Condorcet's theory of voting. Am. Polit. Sci. Rev. **82**(4), 1231–1244 (1988)
23. Young, H.P., Levenglick, A.: A consistent extension of condorcet's election principle. SIAM J. Appl. Math. **35**(2), 285–300 (1978)
24. Zwicker, W.S.: Introduction to the theory of voting. In: Handbook of Computational Social Choice, pp. 23–56. Cambridge University Press (2016)

Mathematical Morphology and Digital Geometry for Applications

Microscopic Image Reconstruction Under Convexity Constraints

Niccolò Di Marco[1] and Andrea Frosini[2](✉)

[1] Department of Legal, Social, and Educational Sciences, Tuscia University, Viterbo, Italy
[2] Department of Mathematics and Informatics, University of Florence, Florence, Italy
andrea.frosini@unifi.it

Abstract. Discrete Tomography fits in the broad framework of inverse problems. One of its main issues concerns the faithful reconstruction of discrete homogeneous objects from internal quantitative information known as *projections*.

In this paper, we consider specific types of projections that count, for each position, the number of points of the discrete objects that lie in its 8 or 4 neighbourhood, including the point itself.

Our final goal is to reconstruct the object from the projections, if possible uniquely, or to determine the minimal number of its points that have to be revealed to do that. This problem is commonly addressed as the *Minimum Surgical Probing* problem (MSP).

Our contribution is to study MSP with the addition of *convexity constraint*. In particular, we focus on a specific class of objects, commonly involved in Discrete Tomography, namely *hv-convex polyominoes*, which are connected sets of points with the further constraint of horizontal and vertical convexity. These constraints allow us to achieve unique reconstruction (except in some corner cases). Additionally, they allow for faster reconstruction algorithms.

Keywords: Image Analysis · Discrete Tomography · Reconstruction Problem · Convexity

1 Introduction

Imaging techniques are widely used to retrieve geometrical information about the interior of unknown discretized objects, usually through the knowledge of aggregate measurements over points in specific areas with the final aim of the full reconstruction of the object itself.

In discrete tomography, a finite object A is represented as a set of discrete points arranged in a 2-dimensional or 3-dimensional space (for example, the lattice $\mathbb{Z}^2$) in which each point is associated with a binary value which is unknown. To uncover these values, the commonly used projections are taken over subsets of the points of A that lie on lines having a discrete slope. Moreover, to drive the reconstruction toward a faithful output, it is commonly

M. H. F. Wilkinson and J. Kosinka (Eds.): DGMM 2025, LNCS 16296, pp. 461–472, 2026.
https://doi.org/10.1007/978-3-032-09544-2_33

assumed that A has specific geometrical constraints, such as connectedness or convexity (see [9] for an overview of the topic).

As an example, a relevant class of objects are hv-convex polyominoes, i.e., 4-connected set of points whose rows and columns are connected. The reconstruction of those polyominoes from horizontal and vertical projections and related problems have been deeply investigated (see [3,5,6,8]).

In this paper, we are concerned with a different notion of projection that considers aggregate measurements of finite sets of positions. This problem is called *Microscopic Image Reconstruction* and has been defined in [7]. In particular, the authors consider projections taken over a rectangular scanning window W.

An extension of this problem, called *Minimum Surgical Probing* (MSP), has been initially introduced in [2]. The authors consider a graph instead of a set of points lying in the $\mathbb{Z}^2$ grid, where each vertex of the $G = (V, E)$ is associated with a real value ℓ_v. The task is to find the minimum number of surgical probes (i.e., the minimum number of known points' values) to uniquely recover ℓ_v for each node $v \in V$ given the neighborhood probes (projections) $P_v = \sum_{u \in N[v]} \ell_u$, with $N[v]$ being the neighbors of v. This question arises from the fact that, in general, there are multiple configurations satisfying the same projections. In particular, the uniqueness of the reconstruction depends on the rank of the associated adjacency matrix of the considered graph G.

As one can easily imagine, MSP can be naturally adapted to binary matrices and polyominoes. In general, different matrices satisfy the same projections, and therefore it becomes relevant to find the minimum number of surgical probes needed to uniquely determine it.

Keeping the framework, our work concerns the MSP problem on polyominoes. More precisely, we investigate the possibility of a fast and faithful reconstruction of hv-convex polyominoes using projections obtained from square or diamond windows, i.e., from the knowledge of the 8 or 4-neighborhood aggregate values, respectively.

1.1 Problem Definition

Consider a finite set of points A in the integer lattice $\mathbb{Z}^2$. If the points of A are 4-connected, then A is said to be a *polyomino*. The dimensions of A are those of its minimal bounding rectangle and therefore, supposing it has dimensions $n \times m$, A can be naturally represented as a binary $n \times m$ matrix where $a_{ij} = 1$ if and only if there is a point in position (i, j). We say that A is h-convex (resp. v-convex) if each row (resp. column) is connected. If A is both h-convex and v-convex, then we say that it is hv-convex. We will indicate with A both the discrete set and the related matrix representation. See Fig. 1 for examples of the previous notions.

0	0	0	1
1	0	0	0
0	1	1	1
0	1	0	0

(*a*)

0	0	0	1
1	1	0	1
0	1	1	1
0	1	0	0

(*b*)

0	0	0	1
1	1	1	1
0	1	1	1
0	1	0	0

(*c*)

Fig. 1. (a) a binary matrix; (b) a polyomino; (c) a $hv-$convex polyomino.

So, provided a binary matrix A of dimension $n \times m$ and a position (i,j) inside, we define $N^8(i,j)$ as the set of the positions in the 8-neighbourhood of (i,j), i.e.,

$$N^8(i,j) = \{(h,k) \mid 1 \leq h \leq n,\ 1 \leq k \leq m,\ \text{for}\ h \in \{i, i \pm 1\},\ k \in \{j, j \pm 1\}\}.$$

Analogously, the set $N^4(i,j)$ is the 4-neighbourhood of (i,j) and contains the vertical neighbors $(i \pm 1, j)$ and the horizontal neighbors $(i, j \pm 1)$ entries, including (i,j) itself. We indicate *probe* $p^8_{i,j}$ (resp. $p^4_{i,j}$) of the matrix A as the sum of the entries of the elements of A in the positions $N^8(i,j)$ (resp. $N^4(i,j)$).

When varying (i,j), we can arrange the probes of the $m \times n$ matrix A in an integer probe matrix of the same dimensions P^8 (resp. P^4) related to N^8 (N^4). The entries of the probe matrix P^8 (resp. P^4) range from 0 to 9 (resp. 0 to 5), i.e. the size of the considered neighbourhood sets. We underline that the sets $N^8(i,j)$ (resp. $N^4(i,j)$) can be regarded as a 3×3 square window (resp. a diamond window) centred in (i,j) in the sense of [7].

In the aim of reconstructing *hv*-convex polyominoes, simply *hv*-polyominoes, using their probe matrices, a main issue is the uniqueness of the obtained binary matrix solution (some examples of *hv*-polyominoes having the same probe are provided in the next sections). To remove this ambiguity, the use of *surgical probes* is proposed in [1]. A surgical probe at position (i,j) returns the value of $a_{i,j}$, i.e. the values of the matrix in that position. Therefore, the following problem arises.

Definition 1 (Minimum Surgical Probing). *Given a probe matrix P, the Minimum Surgical Probing Problem (MSP) asks to find the minimum number of surgical probes needed to uniquely determine a binary matrix consistent with P, if it exists.*

2 The MSP on *hv*-Polyominoes from Square Scans

This section is devoted to studying the MSP problem related to *hv*-polyominoes, say MSP$_{hv}$. We initially consider P^8, while P^4 are considered in another section.

2.1 P^8 Scans to Reconstruct *hv*-Polyominoes

Let A be a *hv*-polyomino of dimension $n \times m$; the general setting of MSP_{hv} aims at finding the minimal number of surgical probes to get a faithful reconstruction of A using the probe matrix P^8 related to the 3×3 window N^8. We define the element $a_{i,j}$ to be *internal* if $i \notin \{1, n\}$ and $j \notin \{1, m\}$; it is a *border element* otherwise.

The following lemma allows determining the entries' values of the internal elements of A from its probe matrix P.

Lemma 1. *Let P^8 be an instance of MSP_{hv} problem. Let (i,j) be an internal position of the related solution, then $a_{i,j} = 0$ if and only if there is an entry $(i',j') \in N^8(i,j)$ such that $p^8_{i',j'} = 0$.*

Proof. ($\Leftarrow$) Assume that there is $(i',j') \in N^8(i,j)$ such that $p_{i',j'} = 0$. It follows that $a_{i,j} = 0$.

($\Rightarrow$) Let us proceed by contradiction assuming $a_{i,j} = 0$, and $p^8_{i',j'} \geq 1$ for all $(i',j') \in N(i,j)$. It follows that from $p^8_{i-1,j-1} \geq 1$, $p^8_{i,j-1} \geq 1$, $p^8_{i+1,j-1} \geq 1$ and *hv*-convexity constraints, we have either $a_{i,j-1} = 1$ or $a_{i,j-2} = 1$.

A similar reasoning applied to $p^8_{i-1,j+1}$, $p^8_{i,j+1}$ and $p^8_{i+1,j+1}$ imply that either $a_{i,j+1} = 1$ or $a_{i,j+2} = 1$. Finally, the convexity implies $a_{i,j} = 1$, a contradiction. □

Figure 2 shows an example of the partial reconstruction of *hv*-polyomino using the previous lemma. The positions in $N^8(i,j)$ of each node lie exactly inside the 3×3 window centered in (i,j). The orange nodes are the positions whose values are not yet determined. On the other hand, yellow nodes represent 1-entries, and grey nodes 0-entries.

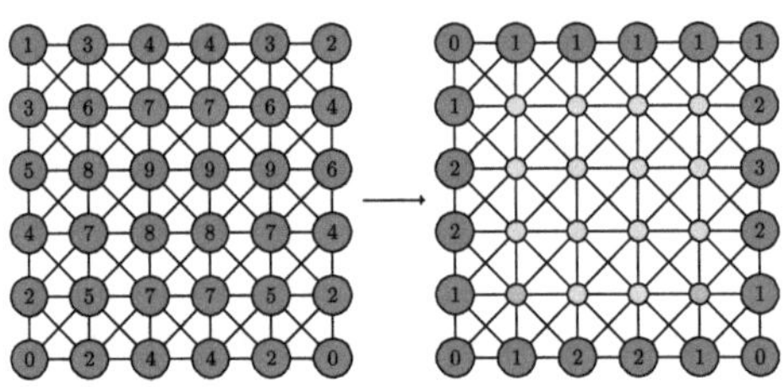

Fig. 2. An example of application of Lemma 1. The 1-elements of the discrete set A in $\mathbb{Z}^2$ are in yellow, the 0-elements are in grey, while in orange are the unknown elements. The updated projections on each of them are computed on the right part. (Color figure online)

After the application of Lemma 1, it remains to retrieve the labels of the border nodes. It is here that some ambiguities may occur, requiring surgical probing. To accomplish the reconstruction task, the idea is to use the updated projections of the border entries after subtracting the values of the already computed internal nodes.

Algorithm 3 is the main routine to detect the positions where a surgical probe has to be performed, if needed. As subroutines, it uses Algorithms 1 and 2 hereafter described. After applying Lemma 1, we choose to start the border reconstruction by setting each border element $a_{i,j} = -1$ meaning that its value is yet unknown. We denote with $d(i,j)$ the number of entries inside the window centered in (i,j) without (i,j) itself and $N(i,j)_k$ is the number of k-nodes, with $k \in \{-1,0,1\}$, in the neighborhood of the node (i,j).

Algorithm 1 performs the border reconstruction in case it has dimension 6×1 where some special cases appear. We treat the six consecutive border elements of A as a 1×6 matrix A' having scan P'.

Algorithm 2 is the core of the reconstruction of A: first, it computes the values of its internal elements, then it moves to those in the border. If n or m have dimension 6, then it runs *Rec-6-line* on them; otherwise, it starts computing the unknown entries of A recursively, based on the updated probes of P^8, until no further changes are possible. If some ambiguities remain, then surgical probes are required.

Figure 3 shows the results of one iteration of Algorithm 2, starting from the probe of Fig. 2, on the left.

Lemma 2. *Algorithm 2,* P^8-Rec, *computes the values of all the entries of a* $n \times m$ *binary matrix* A *from its scan* P^8*, if it exists, except for at most the four corners and their neighbours.*

Proof. At first, Algorithm 2 applies Lemma 1 to retrieve all the internal entries of A. Let us first assume that $n = 6$. Algorithm 1, runs *Rec-6-line* on the border of A of length 6. Maintaining the notation A' and P' for a border and its scans, as in *Rec-6-line*, the following cases hold:

Algorithm 1. Rec-6-line

Require: an unknown 1×6 matrix A' and its probe P'
Ensure: some entries of A' compatible with P'
  if $p'_{1,3} = 1$ **and** $p'_{1,4} = 1$ **then**
    $a'_{1,i} \leftarrow 0, i \in \{1,2,5,6\}$
  else if $p'_{1,3} = 1$ **and** $p'_{1,4} = 2$ **then**
    $a'_{1,i} \leftarrow 0, i \in \{1,2,3\}$
    $a'_{1,i} \leftarrow 1, i \in \{4,5\}$
  else if $p'_{1,3} = 2$ **and** $p'_{1,4} = 1$ **then**
    $a'_{1,i} \leftarrow 0, i \in \{4,5,6\}$
    $a'_{1,i} \leftarrow 1, i \in \{2,3\}$
  else if $p'_{1,3} = 2$ **and** $p'_{1,4} = 2$ **then**
    $a'_{1,i} \leftarrow 0, i \in \{1,2,5,6\}$
    $a'_{1,i} \leftarrow 1, i \in \{3,4\}$
  end if
  Return: A'

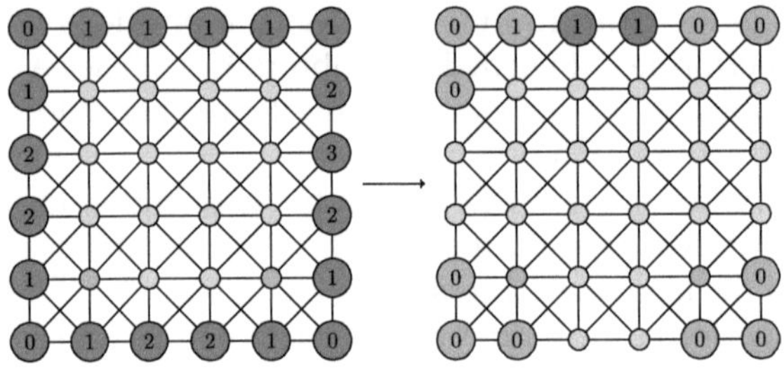

Fig. 3. Starting from the situation in Fig. 2, we show the results after one application of Algorithm 2. The successive iteration of the algorithm returns $a_{1,3} = 1, a_{1,4} = 0$.

- if $p'_{1,3} = p'_{1,4} = 1$, to satisfy the convexity constraints the only 1-entry must lie between $a'_{1,3}$ and $a'_{1,4}$. Therefore all the other entries can be set equal to 0, while $a'_{1,3}$ and $a'_{1,4}$ still have value -1;
- if $p'_{1,3} = 1$ and $p'_{1,4} = 2$, then, by convexity, we know that $a'_{1,3} = 0$, therefore $a'_{1,4} = a'_{1,5} = 1$. The entry $a'_{1,6}$ remains unknown;
- if $p'_{1,3} = 2$ and $p'_{1,4} = 1$ a symmetric result holds;
- if $p'_{1,3} = p'_{1,4} = 2$, by convexity we infer that $a'_{1,3} = a'_{1,4} = 1$ and $a'_{1,j} = 0$ for $j \in \{1,2,5,6\}$.

A simple check confirms that the claim holds in this case.

Algorithm 2. P^8-Rec

Require: an unknown $n \times m$ binary matrix A and its scan matrix P^8
Ensure: some values of the elements of A compatible with P^8
Apply Lemma 1 to a $n \times m$ matrix A and set to -1 its non detected (border) entries
Compute $\overline{P}$ as $\overline{p}_{i,j} \leftarrow p^8_{i,j} - |N(i,j)_1|$, with $1 \leq i \leq n$ and $1 \leq j \leq m$
if $n = 6$ or $m = 6$ **then**
    Apply *Rec-6-line* to the borders of A of length 6 and update $\overline{P}$ accordingly
end if
check = TRUE
while check **do**
    check = FALSE
    Update $\overline{p}_{i,j} \leftarrow \overline{p}_{i,j} - |N(i,j)_1|$, with $1 \leq i \leq n$ and $1 \leq j \leq m$
    for $1 \leq i \leq n$ and $1 \leq j \leq m$ **do**
        if $\overline{p}_{i,j} = |N(i,j)_{-1}|$ **then**
            check = TRUE
            update $a_{k,t} \leftarrow 1$ for each $(k,t) \in N(i,j)_{-1}$
        else if $\overline{p}_{i,j} = 0$ **then**
            check = TRUE
            update $a_{k,t} \leftarrow 0$ for each $(k,t) \in N(i,j)_{-1}$
        else if $\overline{p}_{i,j} = 2$ **and** $i \notin \{1,2,n-1,n\}$ **and** $j \notin \{1,2,m-1,m\}$ **then**
            check = TRUE
            $a_{i,j} \leftarrow 1$
        end if
    end for
end while

Now, let us consider $n > 6$. By the *hv*-convexity of A it holds that along the border of length n it lies a sequence of consecutive values 1. If those values are 3 or more, then there exists an entry of $\overline{P}^8$ on the border having value 3, so P^8-*Rec* inserts the three values 1.

On the other hand, if there are less than three consecutive elements 1, then there are three consecutive values 0, so P^8-*Rec* inserts three entries 0.

In both cases, applying again these rules to the remaining unknown entries with updated projections (as defined before) will eventually find the value of each one.

Finally, the cases with n less than 6 are treated below using exhaustive computation. We will underline that the only unknown values of A lie according to the thesis of the lemma. □

Theorem 1. *If P^8 is a scan whose dimensions are greater than 5, then P^8-*Rec *reconstructs the related hv-polyomino A without surgical probes.*

The proof follows directly from the reconstruction provided in Lemma 2 when the dimensions of the input scan P^8 are greater than five.

Algorithm 3. Probe-Detect

Require: an unknown binary matrix A and its scan P^8
Ensure: the entries' values of A, and the number of the needed surgical probes
check = TRUE
while check **do**
Run P^8-*Rec* on inputs A and $\overline{P}^8$
if for each $1 \leq i \leq n$ and $1 \leq j \leq m$, it holds $a_{i,j} \neq -1$ **then**
check = FALSE
else
Do a surgical probe in (i,j) such that $a_{i,j} = -1$ and $d(i,j) = 5$ (i.e., (i,j) is not a corner element)
end if
end while

The computational complexity of P^8-*rec* can be easily determined after observing that each time the algorithm enters the *while* loop of line 7, the matrix P is updated in $O(m \cdot n)$ and, successively, at least one element of A is determined, so leading to the final complexity of $O(m^2 \cdot n^2)$.

The following example shows the full reconstruction for a 5×5 *hv*-convex matrix that requires one surgical probe. Follow the steps of the reconstruction in Fig. 4 where the 1-entries are colored in yellow, 0 entries are in grey, and -1 entries are in orange.

Example 1. Consider the scan matrix F^8 in Fig. 4, (a). By applying P^8-*Rec* we obtain the matrix in Fig. 4, (b), where the border entries are those of the updated $\overline{P}^8$ matrix.

Note that P^8-*Rec* also detects the values of $a_{1,3} = a_{3,5} = a_{5,3} = a_{3,1} = 1$. Then it performs a surgical probe in one border position with degree equal to 5 (i.e., the red circle in Fig. 4, (b) and it avoids corner nodes). Further runs of P^8-*Rec* lead to the final solution.

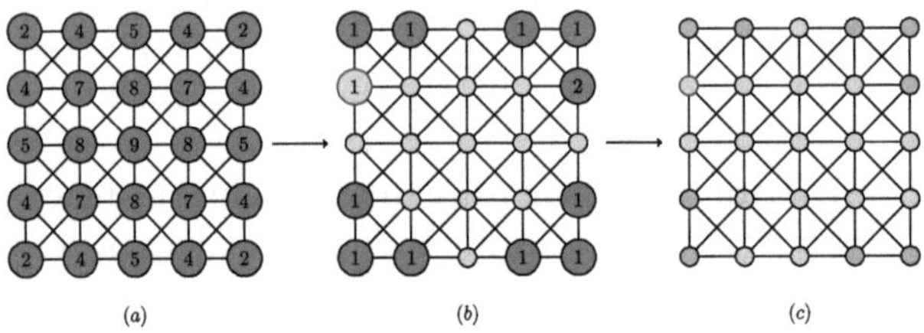

Fig. 4. Left, example of projections of a matrix A. Middle, reconstruction obtained applying Algorithm 2. A surgical probe is highlighted in red. Right, final reconstruction of the labels. (Color figure online)

To complete our study, we finally proceed in the exhaustive computation of small hv-polyominoes sharing the same scan, and the required minimal surgical probes needed to uniquely determine them. We start from dimension 5×5: Fig. 5 shows all the different couples of hv-polyominoes sharing the same scans, up to rotations and symmetries. So, one surgical probe is needed to determine the elements of the couples.

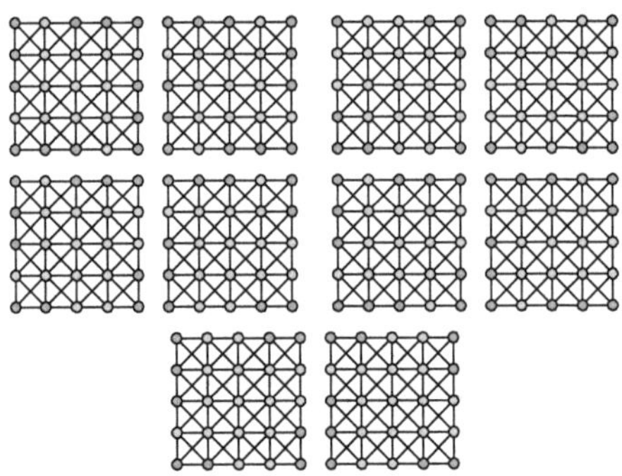

Fig. 5. Each row represents polyominoes sharing the same projections. Note that, to uniquely determine one of them, at most one surgical probe is needed.

Moving to the (still exhaustive) computation of the smaller cases, we get at most four different configurations that share the same scans. Some examples are shown, row-by-row, in Fig. 6 (first and last row for the cases of dimension 2×2 and 2×3, respectively). In those cases, the characterization of each matrix can be performed after two surgical probes at most.

We underline that the border configurations of points whose presence allows multiple matrices with the same projections can also be designed when one of the matrix dimensions exceeds 5, leading to the following result.

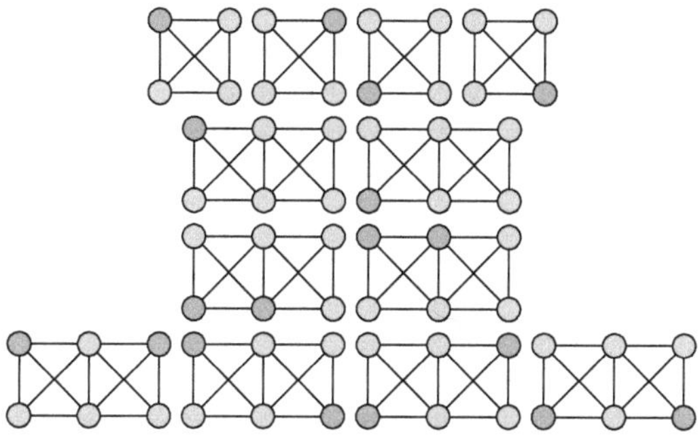

Fig. 6. Each row group matrices share the same projections according to a square window. At most 2 surgical probes are needed to determine each matrix.

Theorem 2. *If a scan P^8 has one dimension lower than 6, then P^8-*Rec *uniquely reconstructs the related hv-polyomino A with at most two surgical probes.*

The examples in Fig. 6 and 5 show cases where at most two surgical probes are required to reconstruct the matrix from the given projection. On these same dimensions, other projections exist so that one only or even no surgical probes are required for the final faithful reconstruction. A trivial example is when the projections are compatible with a polyomino that extends on the whole matrix A.

3 The MSP on *hv*-Polyominoes from Diamond Scans

In this section, we adapt the algorithms for the faithful reconstruction of *hv*-polyominoes from their P^8 scans, together with the needed surgical probe detection, to the case P^4 computed using the four neighbours N^4 of each point.

3.1 P^4 Scans to Reconstruct *hv*-Polyominoes

We highlight that in this section we are using the same notion of internal and border elements. Again, we indicate with A the matrix representation of a *hv*-polyomino. Note that Lemma 1 can be adapted to the case of interest and the proof is similar.

Lemma 3. *Let P^4 be an instance of MSP_{hv} problem. Let (i,j) be an internal position of the related solution, then $a_{i,j} = 0$ if and only if there is an entry $(i',j') \in N^4(i,j)$ such that $p^4_{i',j'} = 0$.*

Lemma 3 allows us to uniquely detect the internal elements of the matrix A. Figure 7 shows an example of the usage of the previous Lemma.

Moving to the border entries, we realize that their detection requires small changes in Algorithm 2 P^8-*Rec* that is updated to Algorithm 4, P^4-*Rec* where the notation introduced in the previous section is maintained. Algorithm 1 *Rec-6-line* and Algorithm 3 *Probe-Detect* are preserved, up to updating P^8-*Rec* with P^4-*Rec*. The main difference between P^8-Rec and P^4-Rec emerges in the border entries whose value is 2. This is due to the fact that the scans of the positions close to a corner do not include each other anymore.

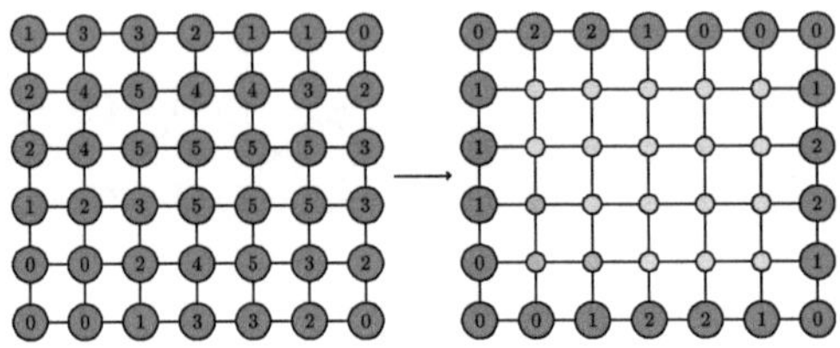

Fig. 7. An example of application of Lemma 3. 1-nodes are highlighted in yellow, 0−nodes in grey and unknown nodes in orange. Note that updated projections are computed on the right.

Algorithm 4. P^4-Rec

Require: an unknown $n \times m$ binary matrix A and its scan matrix P^4
Ensure: some values of the elements of A compatible with P^4
1 : 16 as in Algorithm 2
17: **else if** $\bar{p}_{i,j} = 2$ **and** $i \notin \{1, n\}$ **and** $j \notin \{1, m\}$ **then**
18 : 25 as in Algorithm 2

Theorem 3. *If* P^4 *is a scan whose dimensions are greater than* 4, *then* P^4-rec *reconstructs the related hv-polyomino A without surgical probes.*

Proof. The result easily follows considering that Algorithm 4 works for all the cases with $n, m \geq 6$, since there are no differences with Algorithm 2 in this case. If $n = 5$ or $m = 5$ we can reconstruct the border without surgical probe: w.l.g. let us reconstruct the first row of A. The knowledge of $p_{2,3}$ and all the internal elements of A allows us to get $a_{1,3}$. Updating the scan and using P^4-*Rec*, we have, for $i \in \{2, 4\}$, that $p_{1,i} = 2$ implies $a_{1,i} = 1$, $a_{1,i} = 0$ otherwise. The values $a_{1,i}$ with $i \in \{1, 5\}$ can be obtained by difference. The thesis follows. □

The computational complexity of P^4-*Rec* is inherited by that of P^8-*Rec*.

Example 2. Let us consider the probe P^4 shown in Fig. 8, left. P^4-Rec computes the internal entries of the related matrix A and some elements of each border, according to the proof of Theorem 3. In Fig. 8, center, there are the updated scans. Finally, in Fig. 8, right, the whole matrix is reconstructed.

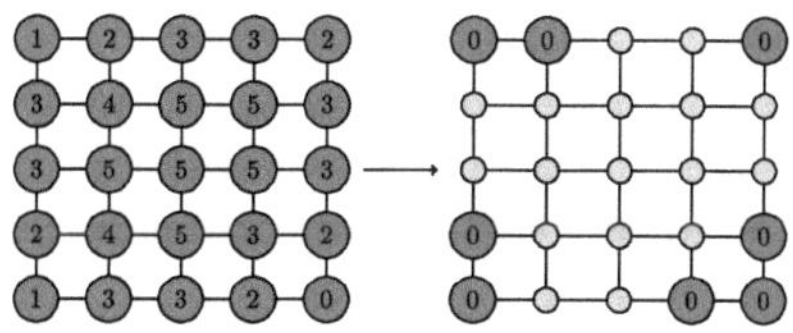

Fig. 8. The steps performed by P^4-*Rec* leading to the reconstruction of a *hv*-polyomino from the P^4 scan on the left. Since the dimensions of P^4 are greater than 4, then no surgical probes are needed.

Again, an analysis of the small cases is required. In particular, the exhaustive computation of the *hv*-polyominoes whose dimensions are smaller than 5 and that share the same P^4 scans reveals that some matrices require surgical probes. As an example, Fig. 9 shows two couples of them.

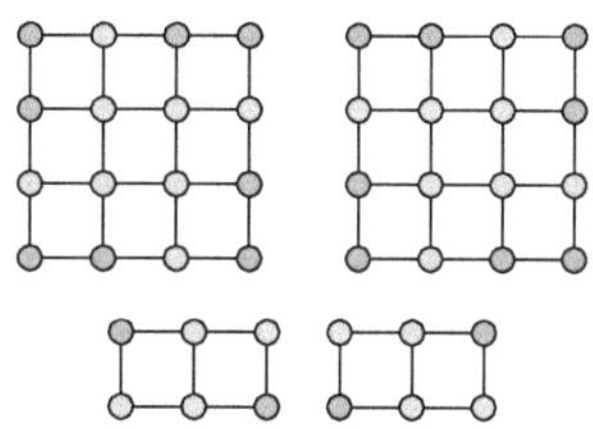

Fig. 9. Two couples of *hv*-polyominoes sharing the same P^4 scan. In both cases, only a single surgical probe is needed to determine them.

Furthermore, no more than two different *hv*-polyominoes share the same P^4 scans, so one surgical probe is required to determine them. The following theorem holds.

Theorem 4. *If a scan* P^4 *has one dimension lower than* 5, *then* P^4-Rec *uniquely reconstructs the related hv-polyomino A with at most one surgical probe.*

4 Final Comments and Future Research

Our research focused on the faithful reconstruction of discrete sets of points from aggregate quantitative measures on the 8 or 4 neighbors of each position, as if a detector or a microscope moving all along the discrete sets, would be impressed by the neighborhood of the scanned position. Unfortunately, as most of the reconstruction problems considered in discrete tomography, the question is basically ill-posed, i.e., there may be different discrete sets sharing the same projections, so, further knowledge about the unknown starting set is usually required to drive the reconstruction process.

This study can be pushed forward by focusing the reconstruction process on different classes of discrete sets defined by relaxing or modifying the geometrical constraints presented here. For example, by considering convexity and connectedness separately, or by including different types of convexity. Each of these scenarios can be further enriched by differentiating the notion of neighborhood scan, excluding the contribution of the central point as in the *exclusive probe model* studied in [1]. In fact, tt turns out that the two inclusive and exclusive models, despite the appearance, have significant differences in some of the most common graph classes.

A final research line concerns the use of scan windows that include neighbors of a position according to different weighted distances, so modifying the shape of the windows themselves, as suggested in [4].

References

1. Bar-Noy, A., Böhnlein, T., Lotker, Z., Peleg, D., Rawitz, D.: Weighted microscopic image reconstruction. In: Bureš, T., et al. (eds.) SOFSEM 2021. LNCS, vol. 12607, pp. 373–386. Springer, Cham (2021). https://doi.org/10.1007/978-3-030-67731-2_27
2. Bar-Noy, A., Böhnlein, T., Lotker, Z., Peleg, D., Rawitz, D.: The generalized microscopic image reconstruction problem. Disc. Appl. Math. **321**, 402–416 (2022)
3. Barcucci, E., Lungo, A.D., Nivat, M., Pinzani, R.: Reconstructing convex polyominoes from horizontal and vertical projections. Theor. Comput. Sci. **155**(2), 321–347 (1996). https://doi.org/10.1016/0304-3975(94)00293-2
4. Battaglino, D., Frosini, A., Rinaldi, S.: A decomposition theorem for homogeneous sets with respect to diamond probes. Comput. Vision Image Understand. **117**(4), 319–325 (2013). https://doi.org/10.1016/j.cviu.2012.09.009. https://www.sciencedirect.com/science/article/pii/S1077314212001804
5. Di Marco, N., Frosini, A.: Properties of sat formulas characterizing convex sets with given projections. In: Baudrier, É., Naegel, B., Krähenbühl, A., Tajine, M. (eds.) Discrete Geometry and Mathematical Morphology, pp. 153–166. Springer, Cham (2022). https://doi.org/10.1007/978-3-031-19897-7_13
6. Dulio, P., Frosini, A., Rinaldi, S., Tarsissi, L., Vuillon, L.: Further steps on the reconstruction of convex polyominoes from orthogonal projections. J. Comb. Optim. **44**(4), 2423–2442 (2021). https://doi.org/10.1007/s10878-021-00751-z
7. Frosini, A., Nivat, M.: Binary matrices under the microscope: a tomographical problem. Theor. Comput. Sci. **370**(1–3), 201–217 (2007)
8. Gerard, Y.: Regular switching components. Theor. Comput. Sci. **777**, 338–355 (2019). https://doi.org/10.1016/j.tcs.2019.01.010
9. Herman, G.T., Kuba, A.: Discrete Tomography: Foundations, Algorithms, and Applications. Springer, Heidelberg (2012)

Efficient Content-Based Time Series Retrieval Using Pattern Spectra

François Merciol[1(✉)], Tom Avellaneda[1], Abdelbadie Belmouhcine[1], and Sébastien Lefèvre[1,2]

[1] IRISA, Université Bretagne Sud, UMR 6074, 56000 Vannes, France
Francois.Merciol@irisa.fr
[2] UiT – The Arctic University of Norway, Tromsø, Norway

Abstract. Pattern spectra are a powerful tool for characterizing image content at both local and global levels. They have been successfully applied to image or patch retrieval, enabling efficient processing of large satellite imagery, for instance. In Earth Observation, multitemporal imaging, or satellite image time series, provides not only spatial but also temporal information. In this context, retrieval aims to identify similar spatio-temporal patterns. In this paper, we address the problem of efficient content-based time series retrieval using pattern spectra. To this end, we revisit the pattern spectra retrieval methodology originally defined in the spatial domain and adapt it to the spatio-temporal case. Preliminary results on the SECOND dataset demonstrate its effectiveness compared to both a deep learning baseline and a uniform guessing baseline, especially when using small patch sizes.

Keywords: Patch Retrieval · Content-Based Spatio-Temporal Pattern Retrieval · Deep Learning · Pattern Spectrum · Content-Based Image Retrieval

1 Introduction

Processing remote sensing data, such as satellite and aerial images, has become an increasingly important research field due to climate change, with applications ranging from ecological assessment and conservation to disaster detection and prevention [8,13]. An important task is remote sensing content-based image retrieval (CBIR), which involves finding images whose content is semantically or structurally similar to a query image [6,11]. More recently, a couple of techniques have been proposed for content-based time series retrieval (CBTSR), which aims to retrieve pixel time series that are similar to a reference [5,17].

In this paper, we propose to address the CBTSR task using pattern spectra (PS), which are histograms describing the distribution of geometric information in an image [7]. PS have been successfully applied to content-based image retrieval [2], but, to the best of our knowledge, no study has investigated their use for retrieving image time series. PS can be efficiently implemented using tree

M. H. F. Wilkinson and J. Kosinka (Eds.): DGMM 2025, LNCS 16296, pp. 473–486, 2026.
https://doi.org/10.1007/978-3-032-09544-2_34

representations of images, also known as morphological hierarchies [3], and can be computed in linear time. Different variants of these constructions have been proposed in the context of satellite image time series [12,15] and validated on various tasks, such as disaster mapping [13] and land-use change detection [14]. However, their relevance in an efficient retrieval scenario remains to be demonstrated. To address this novel task, we introduce a new retrieval model based on corrected pattern spectrum difference and evaluate it on SECOND (SEmantic Change detectiON Dataset) [18] using three well-known attributes: area, Shannon entropy, and depth.

Our experimental results show that CBTSR using PS provides satisfactory performance even when the dataset contains series with no change and when using small patches. For 200 × 200 patches, it outperforms both the uniform guessing and the deep learningâĂŞbased Autoencoder with Early Fusion (AEEF) method [17], which fails to achieve good results and can even perform worse than uniform guessing for some classes.

Our paper is organized as follows. We review related work in Sect .2, then describe how pattern spectra can be used for time series retrieval in Sect. 3, before presenting our experiments in Sect. 4. Finally, we conclude and discuss future perspectives in Sect. 5.

2 Related Work

Content-based image retrieval, and in particular remote sensing retrieval, is typically divided into three stages: (i) computing features (descriptors) for each image, (ii) indexing images using these features, and (iii) finally measuring the similarity between the features of the query image and the indexed features [6].

Recent work has demonstrated competitive results using pattern spectra [2]. Pattern spectra are histogram-like descriptors that summarize the distribution of shapes and geometric attributes within an image. They can be employed directly as feature descriptors for entire images, concatenated with other features, or computed locally on detected regions or using sliding windows [2,4].

It is worth noting that the pattern spectrum of an image can be computed in linear time with respect to the number of pixels. The *sub-pattern spectrum* construction algorithm, introduced in [9], extends this property by allowing the simultaneous computation of the pattern spectra of a set of patches within an image in linear time with minimal overhead. In other words, given a grid of uniformly arranged patches, the pattern spectrum of each patch can be computed in a single pass without increasing the time complexity compared to computing a single pattern spectrum for the entire image.

To our knowledge, only a couple of works have addressed spatio-temporal pattern retrieval in the context of remote sensing data. Bovolo et al. [5] proposed a model that indexes bi-temporal images using clustered change vector analysis (CVA) features. In CVA, the differences between multivariate pixel values of each image are interpreted as directions in a vector space. Despite yielding interesting results, the dataset used for validation is not publicly available, preventing a direct comparison with our method. More recently, Vuran et al. [17]

proposed deep learning models for bi-temporal image retrieval, including deep change vector retrieval (with two variants) and an autoencoder with early fusion (AEEF). Their experimental assessment was performed on SECOND [18], which we also use in this work.

Since pattern spectra provide good descriptors for content-based image retrieval and can be computed in linear time, we study their usefulness for content-based spatio-temporal pattern retrieval and evaluate their performance against a deep learning–based technique.

3 Proposed Method

3.1 Pattern Spectra

Pattern spectra are attractive because they can be efficiently implemented using morphological hierarchies. In this section, we briefly recall the necessary background before specifying the attributes used to compute the pattern spectra.

An image I can be represented as a graph (V, E, w), where V is the set of pixels (nodes), E is the set of edges, and $w : V \to \mathbb{R}^d$ assigns weights to pixels. For grayscale images, we consider intensity values as weights, i.e., $w(x) = I(x)$ for a pixel x. Pixels are connected by an edge when they are adjacent under 4-connectivity. A connected component (CC, also called region or sub-image) is then defined as a subgraph of I in which any two pixels are connected by a path. We define the subset relation $\subseteq$ as an order relation between CCs. A region n satisfying a given criterion is called *maximal* if any other region n' satisfying the same criterion is contained within n.

The constructions introduced above apply to images with real-valued pixels, i.e., grayscale images. The extension of morphological hierarchies to multispectral data has been widely studied [1]. In this work, we adopt a simple approach: projecting multispectral images into a single grayscale band using the *luma* value $Y' = 0.299R + 0.587G + 0.114B$. This method is general and easy to implement, but may lead to reduced performance in tasks that rely on the full spectral information of the image.

Sets of connected components can form hierarchies based on their order relation. These hierarchies are represented by a tree structure, called a component tree. They were first introduced by Salembier et al. [10] with the max-tree. All component trees share several key properties [3]: (i) every node n in a tree $\mathcal{T}$ is a CC of the image, (ii) any two nodes are either disjoint or nested, and (iii) if p is the parent of n, then $n \subseteq p$. Hierarchies can be indexed, meaning that a level λ is assigned to each node such that $n' \subseteq n \Rightarrow \lambda(n') \leq \lambda(n)$. The fundamental relation underlying all types of morphological hierarchies is:

$$n = m_1 \cup \cdots \cup m_l \cup S(n) \quad , \quad S(n) \cap m_i = \emptyset \;\; \forall i \in \{1 \ldots l\} \qquad (1)$$

where m_i are the children of n, with lower levels (or higher, depending on the type of tree), i.e., $\lambda(n) > \lambda(m_i)$ for all $i \in 1 \ldots l$, and $S(n)$ is the set of additional pixels added to the node at level $\lambda(n)$.

Various characteristics, called attributes, can be computed on the CCs of an image. Common attributes include area (number of pixels), mean and variance of grayscale values, moment of inertia, and others. A commonly used attribute to quantify the information contained in a node of a max-tree is the volume, defined as $\mathcal{A}_{vol}(n) = (\lambda(p(n)) - \lambda(n)) \cdot \mathcal{A}_{area}(n)$, where $p(n)$ is the parent of n in the tree. Many attributes can be computed iteratively in linear time.

Pattern spectra are histogram-like descriptors that characterize the distribution of various size and shape attributes of regions in an image. They were first proposed by P. Maragos [7] and were traditionally computed from granulometries, which are parameterized image filters. However, it has been shown that many granulometries and their induced pattern spectra can be calculated more efficiently using component trees [16]. Given an attribute $\mathcal{A}$, the pattern spectrum $PS_{\mathcal{A}}[I]$ maps each possible value h_i of $\mathcal{A}$ to the total amount of information contained in connected components n such that $\mathcal{A}(n) = h_i$. Using the max-tree, $PS_{\mathcal{A}}[I](h_i)$ can be computed iteratively by adding $\mathcal{A}_{vol}(n)$ to $PS_{\mathcal{A}}[I](h_i)$ for each node n satisfying $\mathcal{A}(n) = h_i$. To account for the potential continuum of threshold values h, pattern spectra are typically divided into bins, and the volume of a node is added to the bin $[a, a')$ whenever $a \leq \mathcal{A}(n) < a'$. These bins can be arranged linearly or logarithmically. This computation requires only the attribute $\mathcal{A}$, the area, the volume, and determining the appropriate bin for each node, all of which can be done in a single pass through the tree (for most attributes). Consequently, the complexity of computing pattern spectra depends on the number of nodes, not the number of bins, making the method extremely efficient. Extending this approach to multiple attributes simultaneously is straightforward, resulting in a multi-dimensional histogram.

Recently, pattern spectra have been extended to sub-pattern spectra [9]. For a given window (or patch) S of an image, i.e., a rectangular sub-image, the sub-pattern spectrum $PS_{\mathcal{A}}[I, S]$ is the histogram computed over the set of nodes in the tree of I that are entirely contained within S. The pattern spectrum grid with respect to a *sliding* window (i.e., a window parameterized by a translation x), denoted $(S_x)_x$, is defined as $PS_{\mathcal{A}}^G[I](h, x) = PS_{\mathcal{A}}[I, S_x](h)$, which gives the sub-pattern spectrum for any translation of the window. The pattern spectrum grid is stored in a compact multidimensional array, with two dimensions for spatial binning (corresponding to the values of x) and any number of dimensions for attribute binning (corresponding to the values of h). The computation of the grid $PS_{\mathcal{A}}^G[I]$ can be performed in linear time by visiting each node of the tree once, making pattern spectrum grids extremely efficient and well suited for an interactive, learning-free information retrieval framework.

While pattern spectra were initially defined as shape-size distributions using attributes such as area and compactness, in fact, numerous other attributes can also be considered [12]. In this paper, and for the sake of illustration, we consider the following three attributes, defined for every node n of the tree $\mathcal{T}$ constructed from an image or spatio-temporal cube I:

- *Area*: $\mathcal{A}_{area}(n) = |n|$ (the number of pixels in n);

- *Shannon entropy*: $\mathcal{A}_{entr}(n) = -\sum_{\lambda \in \{I(x), x \in V\}} p_\lambda(n) \log(p_\lambda(n))$, where $p_\lambda(n) = \frac{|\{x \in n, I(x) = \lambda\}|}{\mathcal{A}_{area}(n)}$ is the proportion of pixels in n with level λ;
- *Depth*: $\mathcal{A}_{depth}(n)$ = number of parents of n in the tree.

We apply logarithmic binning for the area attribute $\mathcal{A}_{area}$ and linear binning for the other two attributes.

When computing a pattern spectrum, it is necessary to define the edges of each bin in every dimension, i.e., which attribute values fall into which bins. The bin edges are uniquely determined by the number of bins, as well as the minimum and maximum value of each attribute. In this work, we use a simple implementation in which the attribute bounds are computed dynamically for each scene when building a pattern spectrum or pattern spectrum grid. As a result, histogram bins from different scenes may not represent exactly the same ranges; each bin represents a range of *relative* attribute values. However, we expect this difference to be negligible, as the extrema of the attributes should generally remain similar across scenes.

3.2 Retrieval System

We define the following terms to facilitate understanding of the content-based retrieval task. A *scene* is a large bi-temporal image and serves as the object on which component trees and pattern spectrum grids are built. A *patch* is a single window within a scene, defined by its source time series and spatial support (x, y, x', y'), corresponding to the rectangle with vertices (x, y), (x, y'), (x', y), and (x', y'). A *patch definition* (or patch metadata) specifies the set of patches to be considered across all scenes. It consists of a starting offset (x, y), a patch size (w, h), and a step number n. Any patch of the form $(x + k\frac{w}{n}, y + l\frac{h}{n}, x + (k+1)\frac{w}{n}, y + (l+1)\frac{h}{n})$ conforms to the patch definition, i.e., it lies on a grid starting at (x, y) with steps $\frac{w}{n}$ and $\frac{h}{n}$. Given a set of scenes and a patch definition, the corresponding *patch set* is the set of all patches from the scenes that conform to the definition. A *query patch* is a patch used as the input to a retrieval model. Given a query patch I^q in the *query set* Q and a *candidate set* R, the model outputs a set of *retrieved patches* $R(I^q) \subset R$. A model is parameterized by its sensitivity, which limits the number of retrieved patches. We assume that $R(I^q)$ is ranked (i.e., equipped with a total ordering $\preceq$) and use a rank threshold to represent the model's sensitivity.

We define a retrieval model as a pair (D, s), where D is a *descriptor* function and s is a *distance* function. The descriptor function takes a patch and returns its descriptor. The distance function takes two descriptors and returns their dissimilarity (strictly speaking, it is a dissimilarity function, as s may not satisfy all properties of a metric). The model maps each query I^q to a retrieved set $R(I^q)$, which is ranked according to the distance function ($I_1^r \preceq I_2^r \Leftrightarrow s(I^q, I_1^r) \leq s(I^q, I_2^r)$).

The descriptor function must be *grid-computable*. Specifically, there must exist a *grid descriptor function* D^G that takes a patch definition and a scene and returns an object from which the patch's descriptor can be easily extracted. The grid descriptor computes all possible descriptors for the patch set of the given patch definition. This property reflects the features of the sub-pattern spectrum algorithm introduced in [9], which underlies this work: pattern spectrum grids are efficiently computed with a single aggregation pass through the tree, and sub-pattern spectra can then be trivially extracted.

In contrast to our previous work [9], we aim here to retrieve spatio-temporal patterns rather than purely spatial ones. To this end, we introduce the *naive pattern spectrum difference* model, s_{PSD}, in which the pattern spectra of two consecutive images are subtracted using a vector difference, effectively acting as a *translation* between pattern spectra:

$$s_{PSD}(I, J) = d(PS(I_2) - PS(I_1), PS(J_2) - PS(J_1)) \tag{2}$$

where d represents some distance between pattern spectra, and $I = (I_1, I_2)$ and $J = (J_1, J_2)$ are the two satellite image time series (or spatio-temporal patches) being compared by the model. As noted previously, the pattern spectrum function PS is parameterized by a set of attributes (here, area, Shannon entropy, and depth) and a number of bins for each attribute. The intuition is that regions of the time series that do not change between images will contribute to the same bins in the histograms with similar magnitudes. These bins will therefore have a net contribution close to zero, leaving only the bins corresponding to actual changes as non-zero. In practice, however, this assumption does not always hold, since regions that do not undergo semantic change can still exhibit variations in gray values, boundary shapes, and other characteristics. Consequently, the naive pattern spectrum difference model is expected to fail when a large number of pixel-level changes occur in regions without semantic change, for example, due to differences in off-nadir angle, shadow positions, or illumination variations.

While the problem described above cannot be solved entirely, especially when relying on deterministic, non-learning-based methods, one way to alleviate it is to ensure that pixel-level changes which do not alter object shapes have minimal influence on the difference vector. To this end, we propose the *corrected pattern spectrum difference* (s_{CPSD}). For clarity, let us consider the case of a max-tree and a pattern spectrum PS_γ based only on geometric attributes (i.e., $\gamma = \mathcal{A}_1, \ldots, \mathcal{A}_p$, attributes that do not depend on gray-level values within a node). If two images I and I' can be *matched* (i.e., the transformation from one image to the other does not modify the topology of the tree, and nodes can be matched one-to-one), then the difference between their pattern spectra can be expressed as:

$$PS_\gamma[I'](h) - PS_\gamma[I](h) = PS_\gamma[I' - I](h) = \sum_{n \in \gamma_h(\mathcal{T}')} \mathcal{A}_{area}(n)(\mu(n) - \mu(p(n))) \tag{3}$$

where $\gamma_h(\mathcal{T}')$ denotes the set of nodes n in the max-tree of I' such that $\gamma(n) = h$ (i.e., nodes falling into the hth bin of the histogram), and $\mu(n) = \min\{I'(x)\ ;\ x \in n\} - \min\{I(x)\ ;\ x \in n\}$. We call μ the *transferred contrast* between I and I', and $PS_\gamma[I' - I]$ the *difference pattern spectrum* between I and I'. This expression can

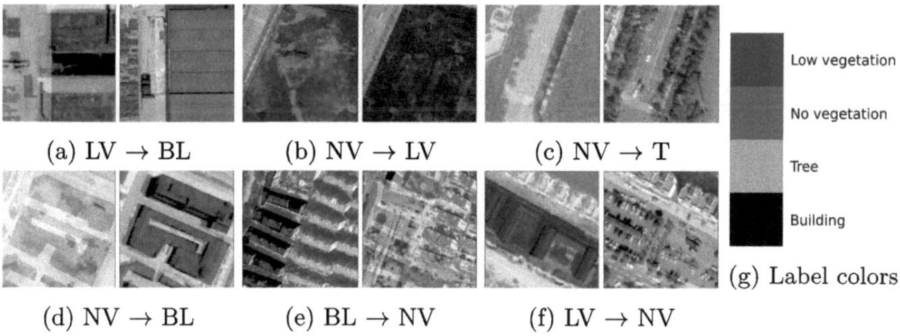

Fig. 1. Examples of 200 × 200 patches from SECOND and their labels. The complete list of labels is provided in Table 1. Pixels that change are overlaid with a color (Fig. 1g) corresponding to their class at each timestamp, while NC pixels have no overlay. Each selected patch contains one or more of the chosen labels.

be computed for any image pair, and subtracted from the raw difference of their pattern spectra, yielding a descriptor that is non-zero only in bins corresponding to topological changes in the tree. In other words, the *corrected* pattern spectrum difference encodes shape-related changes while ignoring intensity variations.

Several important limitations must be emphasized. First, this model is only defined for constructions based on the max-tree (not for tree-of-shapes, α-trees, or ω-trees) and requires the volume attribute. Second, strictly speaking, the formulation of the difference pattern spectrum only holds for purely geometric attributes (i.e., attributes depending solely on the pixel set, not gray levels or tree structure). Finally, the correction only accounts for a narrow class of transformations. In practice, this means that strong assumptions are necessary for s_{CPSD} to hold theoretically, and the volume attribute is essential for it to even be defined. Nevertheless, for illustration, we also evaluate s_{CPSD} with non purely geometric attributes despite the lack of theoretical foundation.

Let us note that the grid-computability property is easily satisfied, since the descriptor is a linear combination of histograms, which can be computed directly on the grids before extracting each patch descriptor. The difference pattern spectrum shares the same structure as standard pattern spectra and can therefore be grid-computed using the same algorithm.

4 Experiments

4.1 Dataset

We assess our method on SECOND[1] (SEmantic Change detectiON Dataset), introduced by Yang et al. [18] to validate change segmentation deep networks. The dataset consists of co-registered pairs of 512 × 512 RGB aerial images captured in various urban areas in China, along with semantic segmentation labels

[1] https://captain-whu.github.io/SCD/.

for each image. The dataset contains 4,662 image pairs, split into 2,968 pairs for training and 1,694 pairs for testing.

There are six pixel classes corresponding to generic urban land-cover categories [18]: non-vegetated ground surface (NV), tree (T), low vegetation (LV), water (W), buildings (BL), and playgrounds (P). Pixels that do not experience a class change between their two images are assigned the no-change (NC) label. Table 1 reports the scene occurrence frequency of each class and class change in our test set. A class change $x \rightarrow y$ is considered to occur in a scene if at least one pixel in that scene changes its class from x to y between images (see below).

Table 1. Pixel change class occurrences in SECOND, with six classes: non-vegetated ground surface (NV), low vegetation (LV), tree (T), water (W), buildings (BL), and playgrounds (P). An entry in row x and column y corresponds to the proportion of scenes that contain at least one pixel changing from class x to class y between the two timestamps. Note that only 28 of the 30 possible transitions are present in the dataset. The last column indicates the proportion of scenes containing the given class at any timestamp.

	LV	NV	T	W	BL	P	Total (scene)
LV	<NA>	0.586	0.228	0.068	0.483	0.012	0.868
NV	0.560	<NA>	0.344	0.077	0.839	0.031	0.996
T	0.141	0.275	<NA>	0.017	0.192	0.003	0.584
W	0.073	0.085	0.023	<NA>	0.046	0.000	0.220
BL	0.290	0.691	0.168	0.022	<NA>	0.006	0.992
P	0.008	0.019	0.003	0.000	0.007	<NA>	0.060

We build our validation set by selecting, for each of the 28 labels, five images that contain at least 1,000 pixels of the corresponding change class, ensuring sufficient examples and some degree of robustness. Figure 1 shows several examples along with their pixel-wise labels. We can qualitatively observe that the co-registration in SECOND is generally good. Note that these examples were selected to illustrate correctly labeled patches; some annotations may be slightly incorrect or debatable (for instance, distinguishing between the LV and T classes, or the precise segmentation of changed regions). Additionally, due to our patch-wise retrieval scenario, some patches are labeled as containing a pattern even though they include only a few pixels of the corresponding class.

4.2 Retrieval Protocol

While SECOND was originally proposed for evaluating change segmentation, it was later used by Vuran et al. [17] to validate bi-temporal remote sensing time series retrieval networks. Since we base our retrieval settings on this previous work, we briefly recall the main steps of their experimental protocol.

- **Label definition** : Each pixel x of an image pair (scene) (I, I') with a support V and a pair of classification maps (l, l') is assigned a unique *change class* $(l(x), l'(x))$. The image pair (I, I') is then associated with the set of change classes of all its pixels, i.e., $\tilde{\Lambda}(I, I') = \{(l(x), l'(x))\ ;\ x \in V\}$.
- **Metrics** : For validation, the authors of [17] used the precision at k ($P@k$) metric, with $k = 5$, under a multi-label evaluation strategy.

The computation of a histogram is carried out in two steps. The first step involves extracting the contributions of the nodes of a tree representing the image, while the second aggregates the contributions of the nodes corresponding to the relevant image region. The first step is described in Algorithm 1 and produces sub-pattern spectra, as illustrated in Fig 2(b, d). The second step differs depending on whether the operation is a query or a retrieval. For a query (Fig. 2a), Algorithm 2 describes how to accumulate the contributions of the nodes within the specified rectangular region of the image. For retrieval, Algorithm 3 first applies a grid partitioning to construct a histogram for each cell (Fig. 2c), and then ranks the cells according to a distance function to identify those closest to the query model.

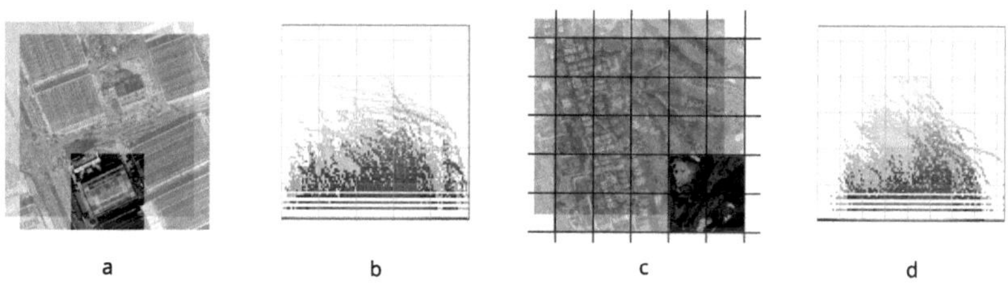

Fig. 2. a) Query patch (in color) within an image (in grayscale). b) Sub-pattern spectrum (in color) of the query patch within the pattern spectrum (in grayscale) of the image. c) Candidate image divided into a grid from which a sliding test window is extracted. d) Sub-pattern spectrum (in color) of the candidate cell within the pattern spectrum (in grayscale) of the candidate image.

Since bi-temporal images in our dataset are associated with multiple change labels, and label frequencies are highly imbalanced, we measured retrieval performance (precision) for each change type separately. Specifically, for a label l, let Q_l denote the set of queries in the test set containing l. For k retrieved images per query, the mean precision $P_l@k$ for a label l is defined as:

$$P_l@k = \frac{1}{Q_l} \sum_{q_l \in Q_l} \frac{1}{k} \sum_{i=1}^{k} s_i^l \tag{4}$$

where $s_i^l = 1$ if the i-th retrieved image has label l, and 0 otherwise. In all experiments, we set $k = 5$.

Algorithm 1. Compute PS items from an image (patch or candidate)

```
procedure CREATEPSITEMFILE(I: image, attrNames: set)
    EDGES: array
    NODES: array
    ATTRS: array
    BINIDX: array
    ITEMVAL: array
                                                        ▷ find neighbours
    for p ∈ getPixels(I) do                             ▷ O(card(pixels))
        EDGES[] ← newEdge(p, p + right)
        EDGES[] ← newEdge(p, p + down)
                                                        ▷ build tree
    for e ∈ EDGES do                                    ▷ O(2 × card(pixels))
        treeConnectNode(e, NODES)
                                                        ▷ compute attributes
    for an ∈ attrNames do                ▷ O(card(NODES) × card(attrNames))
        for n ∈ NODES do
            ATTRS[an][n] ← computeAttrValue(NODES, n)
                                                        ▷ compute PS item
    for n ∈ NODES do                                    ▷ O(card(NODES))
        diff ← abs(ATTRS['weight'][n] − ATTRS['weight'][parent(n)])
        volume ← diff × ATTRS['area'][n]
        BINIDX[n] ← computeId(attrNames, attrs, n)
        ITEMVAL[n] ← volume
    return (BINIDX, ITEMVAL, NODES, ATTRS)
end procedure
```

4.3 Results

We provide in Fig. 3 the experimental results obtained with our method on the SECOND dataset following the retrieval scenario defined by Vuran et al. [17]. To assess the relevance of using 3 attributes in the pattern spectrum, we compare with the more common strategy using 2D pattern spectra, here combining area and Shannon entropy (thus ignoring the information provided by the depth attribute). We also show the scores obtained by a uniform guessing baseline. Moreover, we compared our method with AEEF [17] trained and tested on 200×200 patches. The training and testing patches were obtained by extracting 200×200 patches with a 50% overlap, so that each patch shares half of its area with adjacent patches from images in the training and test sets of the original SECOND dataset. This process yielded 47,488 patches for training and 27,104 patches for testing, as each original image produces 16 patches.

From these preliminary results, several observations can be made. First, the scores obtained with the $P@5$ metric remain far from 100%, underlining the difficulty of the content-based time series retrieval task. A general trend can also be observed: scores follow the distribution of change classes in the dataset, with frequent patterns being better retrieved (precision up to 0.6) than rare

Algorithm 2. CREATE PS FROM QUERY PATCH

```
procedure CREATEPSFROMQUERY(crop: location, BINIDX: array, ITEMVAL: array,
NODES: array, ATTRS: array)
                    ▷ BINIDX, ITEMVAL, NODES, ATTRS come from the query image
    PATCHPS: array
    for n ∈ NODES do                                          ▷ O(card(NODES))
        boundingBox ← ATTRS['boundingbox'][n]
        if boundingboxincrop then
            PATCHPS[BINIDX[n]]+ = ITEMVAL[n]
    return (PATCHPS)
end procedure
```

Algorithm 3. FIND PATCH

```
procedure FINDPATCH(PATCHPS: array, BINIDX: array, ITEMVAL: array, NODES:
array, ATTRS: array)
                                      ▷ PATCHPS comes from the patch image
                 ▷ BINIDX, ITEMVAL, NODES, ATTRS come from the candidate image
    GRIDPS: array
    SELECTION: array
    for n ∈ NODES do    ▷ O(card(NODES)×number of cells covered by the node)
        boundingBox ← ATTRS['boundingbox'][n]
        coveredCells ← convertSizeToGrid(boundingBox)
        for i, j ∈ coveredCells do
            GRIDPS[i][j][BINIDX[n]]+ = ITEMVAL[n]

                                                     ▷ find patch in grid
    for i, j ∈ dim(GRIDPS) do                ▷ O(card(cells in GRIDPS))
        d ← distancePS(PATCHPS, GRIDPS[i][j])
        SELECTION[] ← (d, i, j)
    sort(SELECTION)
return SELECTION
end procedure
```

ones (around 0.2). Interestingly, our model consistently outperforms the uniform guessing baseline by a margin of 0.1–0.3, demonstrating the relevance of the pattern spectrum for efficient time series retrieval. Although AEEF [17] performs well on 512×512 images, as shown in Fig. 4 and in [17], its performance drops considerably on 200×200 patches. This can be explained by the fact that, in the original dataset, all images contain classes of interest and at least one change, whereas considering 200×200 patches led to 3388 pairs without any class of interest and 95 pairs without any change but containing the «*BL*»class. Finally, we also observe that using three complementary attributes enables more effective indexing, as adding the depth attribute leads to significant improvement (up to 0.2 increase of P@5).

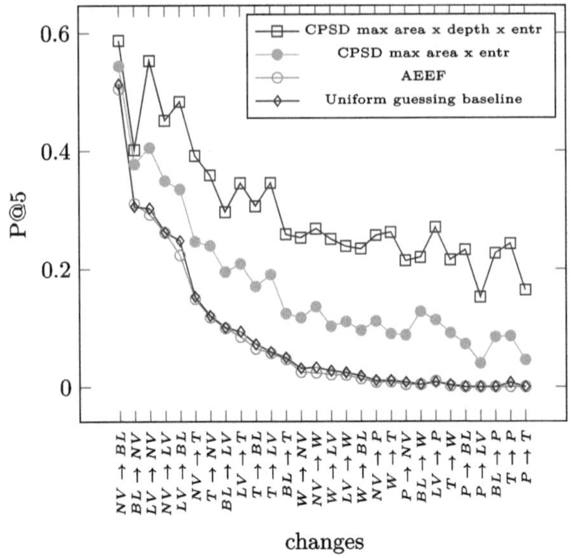

Fig. 3. Quantitative results on the SECOND dataset obtained using 200×200 patches with a 50% overlap, such that each patch shares half of its area with adjacent ones. The proposed model, which relies on pattern spectrum difference and three attributes (area, depth, and Shannon entropy), is compared with (i) a reduced variant limited to two attributes (area and Shannon entropy), (ii) a uniform guessing baseline, and (iii) AEEF trained on 200×200 patches. Performance, measured with the $P@5$ metric, is reported for the 28 change classes listed in Table 1, sorted according to their distribution across all patches.

4.4 Discussion

As pointed out by Vuran et al. [17], AEEF exhibits retrieval performance proportional to the frequency of change types. However, as shown in Fig. 4, it often performs on par with the uniform guessing baseline (and sometimes even worse). Our method, based on pattern spectrum differences over three attributes (area, depth, and Shannon entropy), leads to similar results on large 512×512 patches but performs better on smaller ones.

When applied to 200×200 patches, where local change signals dominate, our approach consistently outperforms the uniform guessing baseline. By contrast, AEEF, trained with 200×200 patches and 50% overlap, remains close to the performance of the uniform guessing baseline and performs even worse for some classes (Fig. 3). Moreover, unlike AEEF, our method retrieves all change types without restriction. In fact, AEEF retrieves only 6 out of the 28 change classes with a precision higher than 0.2.

Being deterministic, our proposed method requires no training and, thanks to the sub-pattern spectrum algorithm [9], it scales efficiently to large satellite image time series archives, with linear complexity.

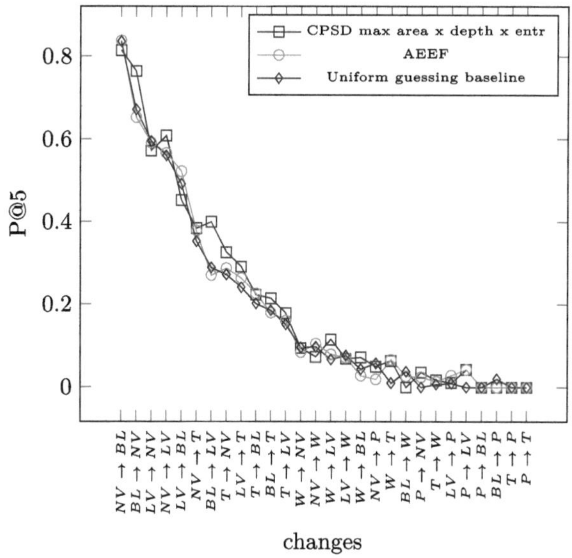

Fig. 4. Comparison between the proposed model, which relies on pattern spectrum differences computed over three attributes (area, depth, and Shannon entropy), AEEF, and a uniform guessing baseline on images of size 512×512. Performance is evaluated using the $P@5$ metric across the 28 change classes listed in Table 1, ordered by their frequency in the dataset.

5 Conclusion

In this study, we tackled the challenging task of CBTSR. We defined a retrieval model based on differences between consecutive pattern spectra and introduced a corrected difference model to account for pixel-level intensity changes. Our approach computes descriptors in linear time using the sub-pattern spectrum algorithm introduced in [9]. Preliminary results on SECOND are promising: when applied to 200×200 patches, our method consistently outperforms the uniform guessing baseline by a 10–30% margin, while the deep learning-based AEEF technique [17] struggles to produce satisfactory results. Moreover, our method is able, to some extent, to retrieve rare change events. Although the performance remains below the level required for operational use, our results are competitive with recent deep learning models [17].

This first attempt to use pattern spectra for CBTSR opens several perspectives. We plan to extend our model with additional attributes beyond the three explored here (area, Shannon entropy, and depth). Furthermore, a promising direction is to investigate how such an efficient morphology-based scheme could be combined with (deep) learning-based systems.

References

1. Aptoula, E., Lefèvre, S.: A comparative study on multivariate mathematical morphology. Pattern Recogn. **40**(11), 2914–2929 (2007)
2. Bosilj, P., Aptoula, E., Lefèvre, S., Kijak, E.: Retrieval of remote sensing images with pattern spectra descriptors. ISPRS Int. J. Geo-Inf. **5**(12), 228 (2016)
3. Bosilj, P., Kijak, E., Lefèvre, S.: Partition and inclusion hierarchies of images: a comprehensive survey. J. Imaging **4**(2), 33 (2018)
4. Bosilj, P., Wilkinson, M.H., Kijak, E., Lefèvre, S.: Local 2D pattern spectra as connected region descriptors. In: International Symposium on Mathematical Morphology (2015)
5. Bovolo, F., Demir, B., Bruzzone, L.: A cluster-based appraoch to content based time series retrieval (CBTSR). In: IEEE International Geoscience and Remote Sensing Symposium, pp. 2793–2796 (2015)
6. Li, Y., Ma, J., Zhang, Y.: Image retrieval from remote sensing big data: a survey. Inf. Fusion **67**, 94–115 (2021)
7. Maragos, P.: Pattern spectrum and multiscale shape representation. IEEE Trans. Pattern Anal. Mach. Intell. **11**(7), 701–716 (1989)
8. Merciol, F., et al.: GEOBIA at the terapixel scale: toward efficient mapping of small woody features from heterogeneous VHR scenes. ISPRS Int. J. Geo Inf. **8**(1), 46 (2019)
9. Mirmahboub, B., Moré, J., Youssefi, D., Giros, A., Merciol, F., Lefèvre, S.: Fast pattern spectra using tree representation of the image for patch retrieval. In: IAPR International Conference on Discrete Geometry and Mathematical Morphology (2021)
10. Salembier, P., Oliveras, A., Garrido, L.: Antiextensive connected operators for image and sequence processing. IEEE Trans. Image Proc. **7**(4), 555–570 (1998)
11. Sumbul, G., Kang, J., Demir, B.: Deep Learning for Image Search and Retrieval in Large Remote Sensing Archives, chap. 11. John Wiley & Sons, Ltd., Hoboken (2021)
12. Tuna, C., Giros, A., Merciol, F., Lefèvre, S.: On Morphological Hierarchies for Image Sequences. In: IAPR International Conference on Pattern Recognition (2021)
13. Tuna, C., Merciol, F., Lefèvre, S.: Analysis of min-trees over sentinel-1 time series for flood detection. In: International Workshop on the Analysis of Multitemporal Remote Sensing Images (2019)
14. Tuna, C., Merciol, F., Lefèvre, S.: Monitoring urban growth with spatial filtering of satellite image time series. In: IEEE/ISPRS Joint Urban Remote Sensing Event (2019)
15. Tuna, C., Mirmahboub, B., Merciol, F., Lefèvre, S.: Component trees for image sequences and streams. Pattern Recogn. Lett. **129**, 255–262 (2020)
16. Urbach, E.R., Roerdink, J.B., Wilkinson, M.H.: Connected shape-size pattern spectra for rotation and scale-invariant classification of gray-scale images. IEEE Trans. Pattern Anal. Mach. Intell. **29**(2), 272–285 (2007)
17. Vuran, O., Akcin, O., Ravanbakhsh, M., Sankur, B., Demir, B.: Deep learning driven content-based image time-series retrieval in remote sensing archives. In: IEEE International Geoscience and Remote Sensing Symposium (2022)
18. Yang, K., et al.: Asymmetric siamese networks for semantic change detection in aerial images. IEEE Trans. Geosci. Remote Sens. **60**, 1–18 (2022)

Morphological Granulometric Analysis of Particle Imagery from Microgravity Experiments

Shima Shabani[1(✉)], Michael Breuß[1], Marvin Kahra[1], Jens Teiser[2], Gretha Swantje Völke[2], and Nico Wenders[2]

[1] Institute for Mathematics, Brandenburg University of Technology, Platz der Deutschen Einheit 1, 03046 Cottbus, Germany
{shima.shabani,breuss,Marvin.Kahra}@b-tu.de

[2] Faculty of Physics, University of Duisburg-Essen, Lotharstr. 1, 47057 Duisburg, Germany
{jens.teiser,gretha.voelke,nico.wenders}@uni-due.de

Abstract. The aim of our work is to analyze size distributions of particles and their agglomerates in imagery from astrophysical microgravity experiments. The data acquired in these experiments are given by sequences consisting of several hundred images. It is desirable to establish an automated routine that helps to assess size distributions of important image structures and their dynamics in a statistical way.

The main technique we adopt to this end is the morphological granulometry. After preprocessing steps that facilitate granulometric analysis, we show how to extract useful information on size of particle agglomerates as well as underlying dynamics. At hand of the discussion of two different microgravity key experiments we demonstrate that the granulometric analysis enables to assess important experimental aspects. We conjecture that our developments are a useful basis for the quantitative assessment of microgravity particle experiments.

Keywords: morphological granulometry · microgravity experiments · pattern formation analysis

1 Introduction

Mathematical morphology is a subfield of image processing in which different filters are constructed to select and change the desired properties of an image, as detailed in [14,16,17]. This involves moving a mask, which is defined by a structuring element (SE) in terms of its shape, size, and centre, over the image. Corresponding morphological operations are performed on the resulting pixels, two of the most fundamental being dilation and erosion. These cause the structure in a binary image to grow or shrink according to the SE.

Image segmentation as a part of mathematical morphology is based on a so-called scale space. Its basic idea is the gradual removal of structures without changing the contours of the objects. However, dilation and erosion are not

M. H. F. Wilkinson and J. Kosinka (Eds.): DGMM 2025, LNCS 16296, pp. 487–500, 2026.
https://doi.org/10.1007/978-3-032-09544-2_35

sufficient in themselves to create such a scale space method, as they enlarge or reduce the contours [10]. For this reason, a combination of these operations is used to partially avoid this undesirable behaviour. The application of an erosion to remove elements, followed by a dilation to partially reconstruct these structures, is referred to as opening. This process selectively removes structures smaller than the SE. By performing openings with increasing SEs, the image is progressively sifted. This process of sifting results in the removal of the next scale of structures at each step, thereby creating a scale space. Because of this sifting, the method is known as granulometry [9,18]. Granulometry is one of the simplest and first morphological scale space approaches that were studied by Matheron [13] and Maragos [12]. Areas of application include medicine [21], biology [6] and geosciences [1].

This paper is about the application of granulometry to imagery from microgravity experiments. In a detailed discussion, we show how to make use of information obtained from the granulometry in order to obtain insights on the underlying physical processes that can be observed. In doing this, we show here first steps in the development of discriminative features that help to assess the large amount of experimental data in an automated way.

2 Microgravity Experiments and Data Sets

Microgravity experiments are a common tool for understanding early phases of planet formation [22]. Planet formation occurs in protoplanetary disks, clouds of gas and dust surrounding a young star, with around 1% of the mass in solids. It starts with dust grains in the micrometer range, which collide and form extremely porous agglomerates in hit-and-stick collisions [7]. With increasing size, the agglomerates decouple from the surrounding gas and the collision velocities increase, leading to the compaction of the particles [2]. From here, it is an open debate on how planetary growth can proceed. With only surface forces acting, dust agglomerates and solid particles in the (sub) millimeter range bounce off each other rather than forming larger agglomerates, leading to the term bouncing barrier in the literature [23].

Once particles have a critical size, hydrodynamic processes in protoplanetary disks are an efficient mechanism to concentrate solid particles. Such processes are referred to as the streaming instability, the baroclinic instability, or vortices in general. All these mechanisms have in common that they require a critical minimum size of the dust agglomerates/solid particles, depending on the model in the decimeter to meter range [4,19].

Here, microgravity experiments are used to investigate this critical growth phase. The key questions to be answered in these experiments are [22]:

- Enable additional forces (such as electrostatic or magnetic forces) growth into the critical size range of centimeters or decimeters?
- Under what conditions can small dust agglomerates grow to larger sizes?
- What are the critical threshold velocities and/or particle sizes for growth, erosion, or catastrophic fragmentation?

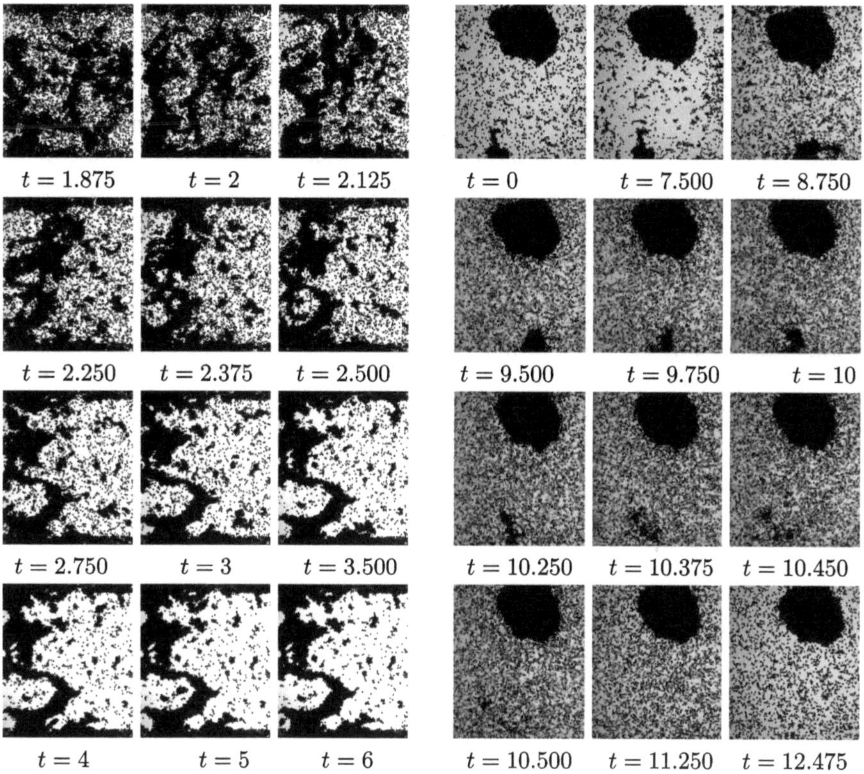

(a) Drop tower experiment (`Ser1`) (b) Sounding rocket experiment (`Ser2`)

Fig. 1. The images in Figs. 1a and 1b are selected frames out of the corresponding image sequences of two different experiments.

- How large can clusters of particles grow and at which growth rate?
- What structures form during these processes, for example in terms of porosity, shape, and fractal dimension?

A series of microgravity experiments have been carried out to address these key questions, including experiments with short duration at the Bremen drop tower (9 s experiment time) [11] and long-term experiments on a sounding rocket (6 min experiment time) [20]. A typical experiment consists of a test cell (volume $50 \times 50 \times 50\,\text{mm}^3$) with a sample of small basalt spheres (sphere diameter 0.5 mm, total amount 8–10 g). The test cell can be agitated in one direction, to induce relative velocities and collisions between the particles.

The particles moving in the test cell are recorded at 40 frames/s. A typical drop-tower experiment has a duration of 9 s, while the experiment on a sounding rocket gives an image sequence of 6 min continuous video data. This makes a detailed analysis challenging because of the amount of data and the difficulty in identifying the physical processes occurring during the experiments. Figure 1 shows a typical example for image data obtained in the drop tower (left series) and in the sounding rocket (right series). We choose these data sets because they refer to two different physical situations, but otherwise are typical examples of the experiments performed.

Figure 1a shows the agglomeration process, which means that over time the particles stick together to form larger-scale structures. More specifically, in the first experiment small (single) particles are absorbed by large-scale aggregates that have formed early in the process. We refer to this setting here as `Ser1`.

Figure 1b illustrates the process of physical erosion of larger-scale aggregates after inducing kinetic energy into the system by shaking the test cell. The scattered particles hit the agglomerates inside the cell and dissolve a larger-scale agglomerate within the observation time. Moreover, as one also might observe, towards the end of the sequence an agglomeration of scattered particles takes place forming small-size agglomerates. We refer to this setting here as `Ser2`.

3 Data Preparation and Elements of Our Granulometry

The purpose of this section is to familiarize the main elements of our granulometric analysis with respect to the data sets we consider. Furthermore, we clarify some design choices in our granulometric pipeline.

Our Basic Setup. While a morphological granulometry can be defined for greyscale imagery, the technique is more easy to interpret when applied with binary images. Therefore, given that the data to be examined are in the form of greyscale images, we opt to perform a binarisation of the input images and work with the binary images for granulometric analysis. The binarisation is achieved by implementing a thresholding process that determines the threshold at which grey values are assigned to the colours black and white.

The approach adopted for this purpose in this paper is a classic method developed by Otsu [15]. It involves the analysis of histograms of grey value images and the search for an optimal threshold based on the maximisation of variance between the classes of grey values that lie above and below this threshold. Otsu's method is mathematically equivalent to a globally optimal k-means clustering [8] applied to the image's intensity histogram.

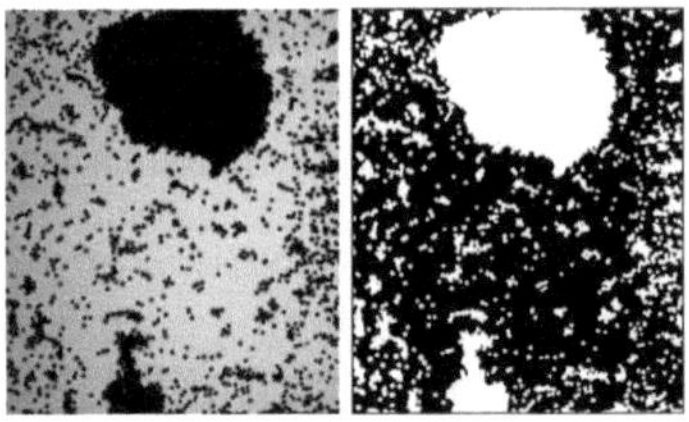

Fig. 2. Input image and its binarised and inverted version

For our granulometric analysis, we then invert the images, so that *white* represents the particles and structures of interest and *black* is the background, compare Fig. 2. In the implementation of the opening that generates our granulometry, we employ the disc-shaped SE defined by the MATLAB command `strel("disk", r)` where r specifies the radius, see Fig. 3 for a visualization.

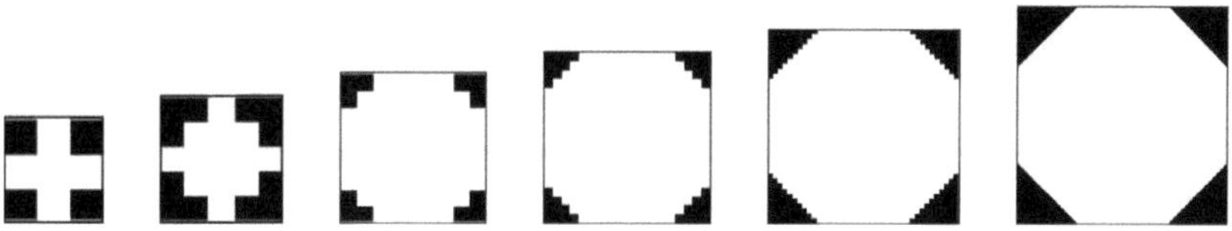

Fig. 3. SE visualization with $r = 1, 2, 5, 10, 20, 140$ in MATLAB.

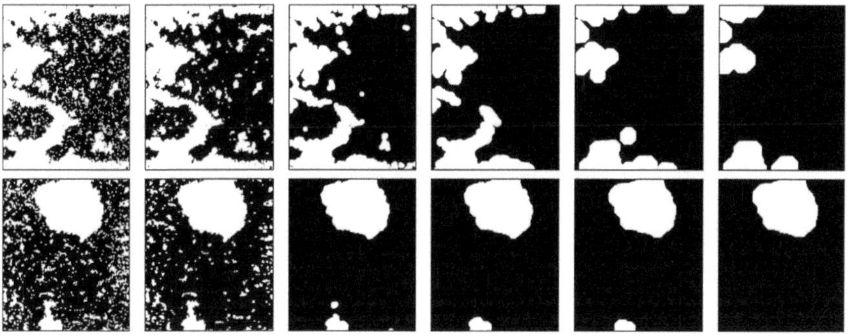

Fig. 4. Opening visualization with SE = 10, 30, 50, 90, 120. Frame 8 of Fig. 1a and frame 2 of Fig. 1b with corresponding opened images in first and second row, respectively.

Figure 4 gives an account of the *opening* of two binarised input images taken from the two image series for our application. Here we would like to remark that the size of structures related to the SE in use is important here. As expected, one observes a gradual simplification of the input image.

By image acquisition, the resolution may vary in practice from image sequence to image sequence, while for one and the same image sequence it is convenient to keep it fixed. For an automated analysis, it is desirable to make comparable quantitative measures obtained from the imagery. Therefore, we normalize the image intensity measure which is a main quantity in granulometry so that it covers the white portion of a given image, meaning the measure is 1 if an image is completely white. See Fig. 5 for an account of two normalised image intensity plots that show how the normalised intensity develops when applying an opening with increasing size of the SE.

Design Choices: Structuring Element Shape and Binarisation. Let us elaborate on our design choice of employing disc-shaped SE. While the disc-shaped SE is surely an intuitive choice in the context of our application, one may consider the question how sensitive the granulometry seems to be in the considered context with respect to that design parameter. There are two such plots displayed in Fig. 5, obtained using a disc-shaped and a diamond-shaped SE, respectively. As observable, the difference between the graphs appears to be negligible, as is also the case when employing box-shaped SE, which we studied but do not display here. Therefore, we opt to restrict the investigation to the use of the disc-shaped SE.

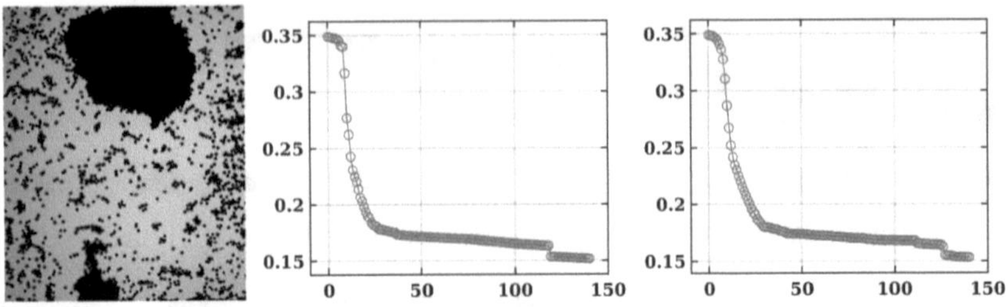

Fig. 5. Input image (left) before subsequent binarisation and inversion. The two graphs to the right show the normalised intensities (**y-axes**) for opening with increasing SE size (0–140, **x-axes**) with disk and diamond, respectively. Let us note that SE $=0$ denotes intensity of input image directly.

Furthermore, as indicated in the introduction, the granulometry may also be realised directly in terms of grey value imagery. To investigate this possibility here in some more detail, we consider in a first step the corresponding normalised intensity distribution at hand of an example, see Fig. 6. Since the background of the input imagery may vary in brightness – compare the difference between `Ser1` and `Ser2` in Fig. 1 – there is depending on the concrete experiment a certain offset in underlying grey values. This is also visible by higher values in normalised intensity distribution, hindering a direct interpretation of corresponding values as a percentage of pixels covering particles and their structure formations. Which may be undesirable for possible subsequent processing or automated classification of image sequences in the physical context. Let us also note that as in the binarised setting, the choice of the SE appears to have no fundamental impact on the form of the granulometry graph.

Continuing upon this example, in order to assess the difference between binarised and grey value data as a basis for the granulometric analysis as we aim to perform it, we study the first-order derivative of corresponding granulometry graphs. Let us note that this is subsequently an important feature in our analysis. Studying this feature in the current context clarifies if the difference between binarised and non-binarised granulometry amounts just to an additive

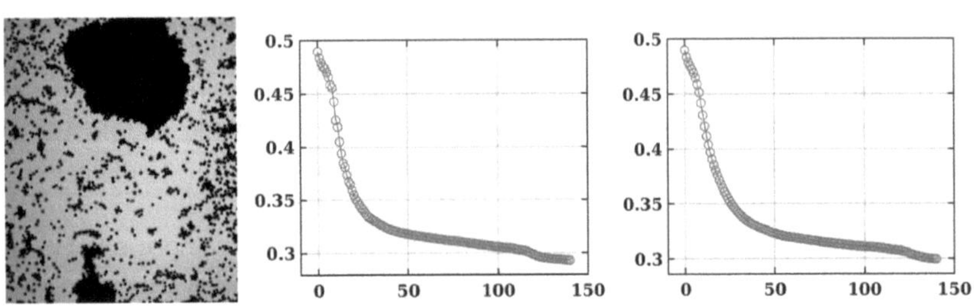

Fig. 6. Input image (left) before subsequent inversion. The two graphs to the right show the normalised intensities (**y-axes**) from greyscale image for opening with increasing SE size (0–140, **x-axes**) with disk and diamond, respectively.

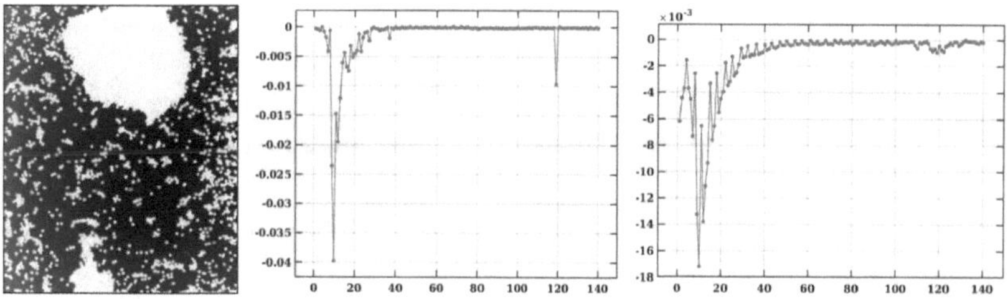

Fig. 7. Inverted input image (left). The two graphs to the right show the first-order derivatives of normalised intensity distributions (**y-axes**) from opening with increasing SE size (1–140, **x-axes**) for inverted binarised and greyscale image, respectively.

offset, or if there are effects that favor a specific design choice in the granulometric pipeline. The first-order derivative is obtained by employing at a given SE size in the normalised intensity distribution a standard finite forward difference operation.

In Fig. 7 we first display the inverted greyscale input image of our example, complementing the information in Fig. 6. Then we display two graphs for the first-order derivatives employing binarised and non-binarised input image as a basis for granulometry, respectively. As a result it is observable that besides a different range of values, the first-order derivative feature based on non-binarised input image appears much more oscillatory. This is supposed to hinder subsequent analysis, especially as we strive to enable in future work an automated approach. In turn, the first-order derivative based on the binarised input image shows specifically a clear first peak, and it appears in general to be a less oscillatory version of its non-binarised counterpart.

Because of the mentioned aspects, meaning more direct interpretability of results and cleaner information in a main feature of our subsequent granulometric analysis, we prefer to work with binarised imagery and rely on only one basic SE shape.

On Scale in Our Granulometric Analysis. Concluding our first discussion of the granulometry in our application, we turn to the filtering and identification of individual particles in an image of our series. The relative amount of such free particles and its temporal evolution is an important feature of the experiment series and allows us to assess the physical processes in an automated way. To this end, it is imperative to know at which scales of the SE the particles are observable. We will extend this kind of investigation to meaningful other scales later on.

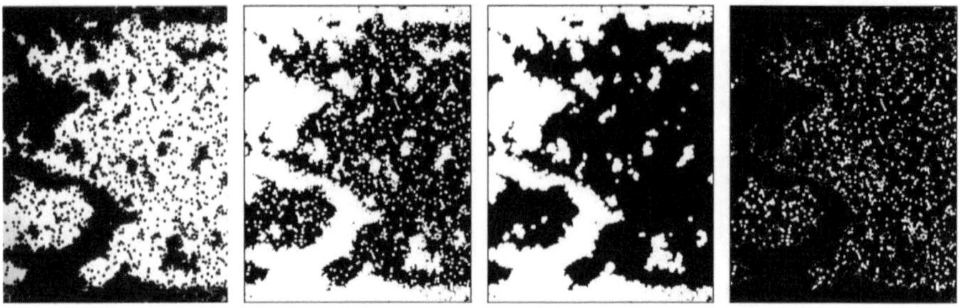

Fig. 8. From left to right: frame 8 of Fig. 1a, opened images after binarisation and inversion with SE $= 6, 12$, respectively, and difference of these opened images. The free particles can be found to a large degree between the depicted SE scales.

In Fig. 8 we give an account of our findings. It is important to note that, in general, the experimental cell and the camera observation point will stay the same for both of our series (and others). So, the setting for identification of the particles can be fixed for both series. As a main result for further use, the free particles can be filtered employing SE scales from six to twelve.

4 Experimental Results

In this section, we present several experiments on images from both series shown in Fig. 1. Furthermore, we draw an individual conclusion after each experiment.

4.1 Experiment 1: Consistency with Physical Theory

Observations in the Solar System (e.g. the asteroid belt) and laboratory experiments give strong empirical evidence that fragmentation processes lead to a characteristic size distribution of particles, typically following a power law [3,5]. The series Ser2 shows a similar fragmentation process, as small (single) particles erode larger agglomerates until they finally vanish. The hypothesis we follow now is that we should find similar particle statistics by 2D imagery, although the empirical evidence in the literature is found in 3D space and a more continuous scale of particle sizes. This is apparently not self-evident since the 2D imagery we evaluate is naturally subject to occlusions that hinder measurements and by the restrictions of image resolution and binarisation. If the size distribution accessible in our imagery still fits well with a power-law assumption, we could be highly confident that our granulometric analysis may convey physically meaningful interpretations. To this end we study if we can fit the power law

$$y = ax^b + c \tag{1}$$

to our normalised image intensity distributions for increasing SE size, an example of which we have already seen in Fig. 5.

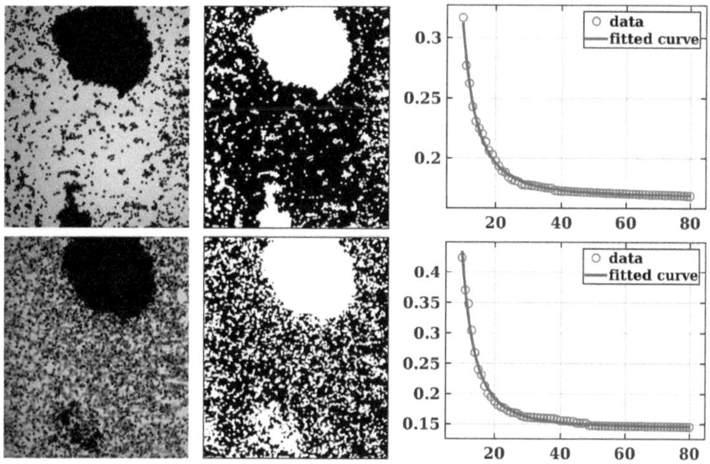

Fig. 9. First column: frames 2, 8 of Ser2. Second column: inverted-binarised versions. Third column: intensities (**y-axes**, red marker) and fitted power law (solid blue line) with opening process for SE = 10–80 (**x-axes**). Fitting parameters are (top) $(a, b, c) = (33.69, -2.368, 0.1681)$ and (bottom) $(a, b, c) = (112.5, -2.588, 0.1447)$. (Color figure online)

Let us comment on the definition and usage of the power law (1), as it is an adoption of the standard power law form $y = ax^b$. As we may already observe in Fig. 5, and also by means of Fig. 8, filtering with low SE sizes (especially of one to five) does not yield information about particle sizes. Therefore, for power-law fitting, the approximate plateau for very small SE sizes bears no information and is skipped. Furthermore, in our finite setting of experimental cell and image resolution, the tail of our distribution inevitably approaches a non-zero value $c > 0$ over interesting ranges of depicted agglomerations.

As we may observe in Fig. 9, the normalised intensities indeed appear to be fitted very well with a power law (1), analogous to empirical evidence in other studies. However, it is important to note that the exponent we find is different from the values found for fragmentation cascades in the literature ($x^{-3.5}$) [3,5]. This is surely due to our adaptations described above and as we are restricted here to 2D imagery.

Conclusion on Experiment 1. The size distribution accessible in our imagery fits qualitatively well with a power-law assumption. Because of this we are confident that our image-based investigation conveys physical meaning.

4.2 Experiment 2: Assessment of Free Particle Movement

In this experiment we come back to the considerations leading to Fig. 8. We have seen that individual particles are mainly observable making use of SE scales six to twelve. In a first step we want to complement the discussion of Fig. 8 by showing

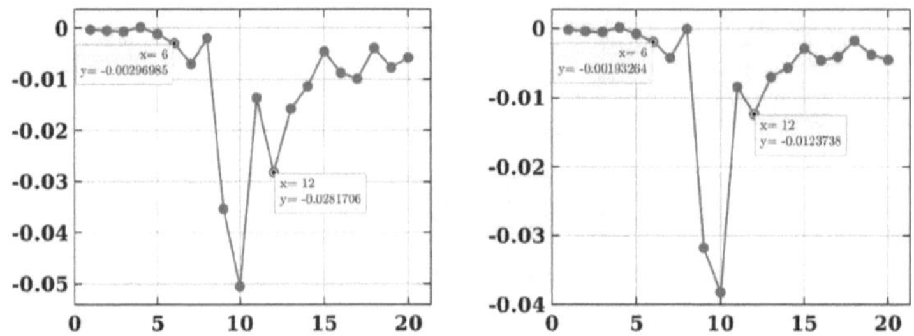

Fig. 10. First-order derivatives of normalised intensities (**y-axes**), based on frame 4 and frame 8 of `Ser1` for `SE` = 1–20 (**x-axes**), respectively.

the first-order derivative (computed by making use of forward differences) of the normalised intensities, compare Fig. 10. The `SE` values contributing in forming the prominent peak exactly identify a source of major loss in intensity. Evidently, the `SE` range from six to twelve describes the peak, and having in mind the information carried by Fig. 8 we have by the peak *sharp* information about the free particles in an image.

Let us also point out that on a more global scale with respect to interesting `SE` sizes, the derivative information as depicted in detail in Fig. 10 contains major information in our application. This remark addresses both the range of the values of first-order derivatives as well as the issue that the largest values of the derivative can easily be assessed (for several possible uses) in an automated way, compare Fig. 11.

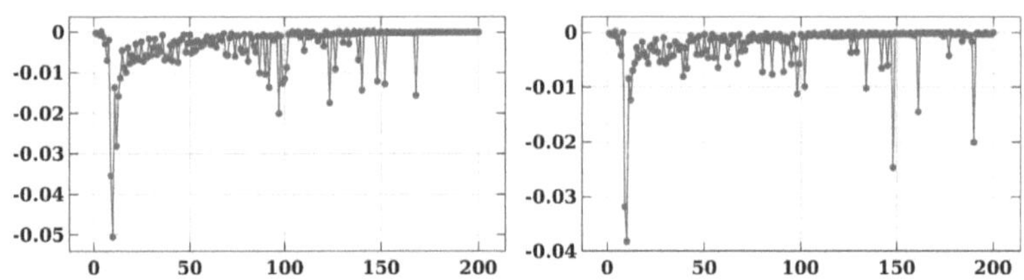

Fig. 11. First-order derivative of normalised intensities (**y-axes**) at a large scale of `SE` sizes (1–200, **x-axes**) based on frame 4 and frame 8 of `Ser1`, respectively. Results for the other frames of `Ser1` and `Ser2` show analogous behaviour. Let us emphasize that especially the prominent peak in the low `SE` regime may easily be identified in several ways based on statistical assessment, e.g. employing quantiles of the range of values in the lower `SE` regime.

Having thus consolidated by use of granulometry that the difference between normalised image intensities after filtering with `SE` = 6, 12, bear significant information about free particles in a given image, we now consider the corresponding *jump size* (i.e. intensity for `SE` = 6 minus intensity for `SE` = 12). More specifically, we are interested in the *temporal evolution* of jump sizes, see Fig. 12, where we observe that evolution for `Ser1` and `Ser2`, respectively. To state it even more clearly, by temporal evolution we mean that we consider the development of the mentioned jump size over the frames of our test image sequences.

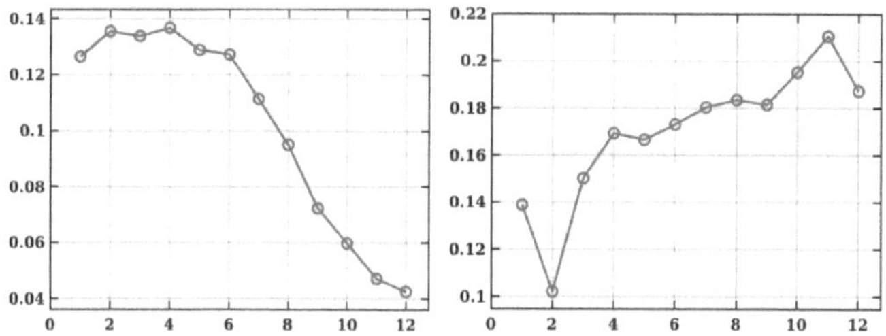

Fig. 12. Jump sizes evolving in time for Ser1 (left, frames 1–12) and Ser2 (right, frames 1–12), obtained by taking difference between normalised intensities for SE = 6, 12 (**y-axes**) in each frame (**x-axes**).

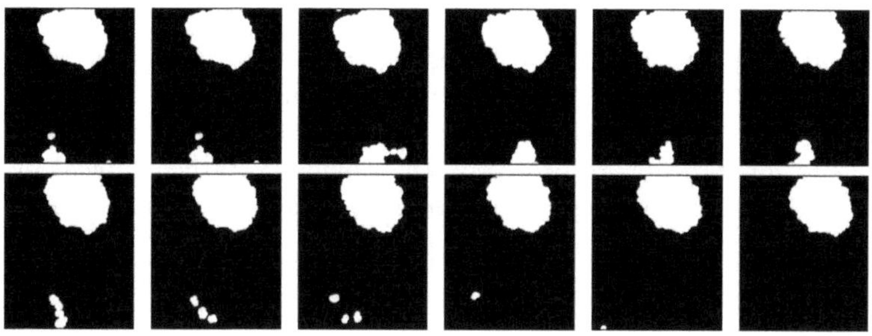

Fig. 13. Opening with SE = 30 to show the way of vanishing mid-sized structure in the process, (first row) frame 1–6 and (second row) frame 7–12 of Ser2.

As one may confirm visually by comparing with Fig. 1, we have obtained here a very characteristic, quantitative account of the significant decrease over time of the free particles in the agglomeration process given by Ser1 as well as of the increase of free particles in the fragmentation process given by Ser2.

Conclusion on Experiment 2. The evolution of *jump size over time*, meaning over considered frames as reported, is apparently an important feature indicative of the type of experiments. In the context of this paper, we consider here fragmentation and agglomeration, which are two fundamental and highly relevant types of physical experiments. This means, that jump size over time may help in an automated classification of corresponding image sequences.

4.3 Experiment 3: Assessing Larger-Scale Structures

We now aim to keep track of larger-scale image structures, such as close to the lower image boundaries in Ser2. In that fragmentation process, the mentioned structure is dissolved over time, which is an important physical process.

In Fig. 13 we give a visual account of the time evolution in Ser2 using a medium-size structuring element SE = 30. Comparing with Ser2 as depicted in Fig. 1, one can clearly observe when the medium-size structure close to the lower image boundaries dissolves. Making use of a sample of SE sizes and studying time evolution of normalised intensities, one can clearly distinguish agglomeration processes as in `Ser1` and fragmentation taking place in certain scales, especially for the large SE sizes 120 and 180 chosen here, see Fig. 14.

Let us note that the left column thereby displays a very simple feature, namely the percentage of white pixels over time, complementing here the large SE sizes. In agglomeration, the number of white pixels declines as free particles are absorbed into larger structures. In fragmentation, there is a significant increase of free particles as observable by this feature, before they again start to agglomerate to small structures, which is expected from the physical point of view and oberservable in some detail in `Ser2`.

Conclusion on Experiment 3. By certain choices of SE sizes, we can identify typical aspects of agglomeration and fragmentation processes, respectively. In turn, having classified an image sequence already as displaying an agglomeration or fragmentation process, respectively, the obviously rather typical evolution characteristics as given in Fig. 14 may help to device an automated routine for identifying important higher regimes of SE size.

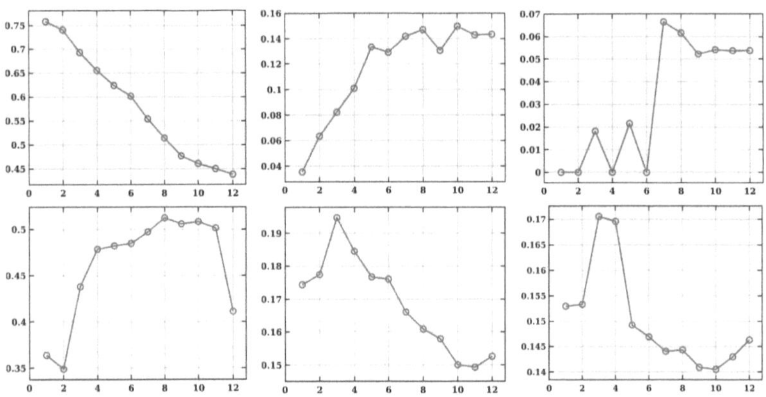

Fig. 14. Normalised intensity (**y-axes**) over time (**x-axes**), for different SE sizes (left to right). First Row: `Ser1` (frames 1–12) with SE $= 0, 120, 180$. We clearly observe agglomeration, meaning that the size of structures at larger scales grow over time. Second row: `Ser2` (frames 1–12) with SE $= 0, 30, 120$. We can clearly observe fragmentation; the slight growth for large SE values in last two frames is due to movement of the big agglomeration in the sequence.

5 Summary and Conclusion

To interpret the date of the image in the astrophysical context, it is crucial to identify the physical processes during the different phases of the experiment. In

Sect. 4.2, we show that granulometry is a powerful tool to distinguish different phases of the experiment from each other, and identify physical growth and erosion phases in the experiment. This is already an important tool for analysis of existing data, but it is absolutely crucial for long-term experiments, as they are currently planned for the International Space Station.

In Sect. 4.1 we show that granulometry is a suitable tool to derive particle statistics from the image data. The conclusions derived in Sects. 4.2 and 4.3 are built upon time evolution characteristics at certain SE sizes identified through granulometric analysis. We have shown that the jump size, that may be easily obtained in an automated way from granulometric analysis, appears to be a powerful feature for classification of an image sequence as showing agglomeration or fragmentation, respectively. Based on our experimental findings, we think that granulometry may help to identify the presence of larger agglomerates in our image sequences in an automated way, since we obtain very characteristic features that could also may help us to give first estimates for the agglomerate sizes. To realize these aspects is the subject of future work.

Acknowledgments. This project is supported by DLR Space Administration with funds provided by the Federal Ministry for Economic Affairs and Climate Action (BMWK) under grant numbers 50WK2270F, 50WK2270C, and 50WM2142. Access to the drop tower ande sounding rocket was provided by the European Space Agency under grant number ESA-HRE-RS-LE-0214.

References

1. Balagurunathan, Y., Dougherty, E.R., Frančišković-Bilinski, S., Bilinski, H., Vdović, N.: Morphological granulometric analysis of sediment images. Image Anal. Stereol. **20**(2), 87–99 (2001). https://doi.org/10.5566/ias.v20.p87-99
2. Blum, J., Wurm, G.: The growth mechanisms of macroscopic bodies in protoplanetary disks. Annual Rev. Astron. Astrophys. **46**(1), 21–56 (2008)
3. Bottke, W.F., Brož, M., O'Brien, D.P., Bagatin, A.C., Morbidelli, A., Marchi, S.: The collisional evolution of the main asteroid belt. In: Michel, P., DeMeo, F.E., Bottke, W.F. (eds.) Asteroids IV, pp. 701–724 (2015)
4. Carrera, D., Simon, J.B.: The streaming instability cannot form planetesimals from millimeter-size grains in pressure bumps. Astrophysical J. Lett. **933**(1), Article no. L10, 11 (2022)
5. Deckers, J., Teiser, J.: Macroscopic dust in protoplanetary disks—from growth to destruction. Astrophysical J. **796**(2), Article no. 99, 11 (2014)
6. Devaux, M.F., Bouchet, B., Legland, D., Guillon, F., Lahaye, M.: Macro-vision and grey level granulometry for quantification of tomato pericarp structure. Postharvest Biol. Technol. **47**(2), 199–209 (2008). https://doi.org/10.1016/j.postharvbio.2007.06.017
7. Drążkowska, J., et al.: Planet formation theory in the era of alma and Kepler: from pebbles to exoplanets. In: Inutsuka, S., Aikawa, Y., Muto, T, Tomida, K., Tamura, M. (eds.) Protostars and Planets VII. ASPC, vol. 534, p. 717. https://doi.org/10.48550/arXiv.2203.09759

8. Gersho, A., Gray, R.M.: Vector Quantization and Signal Compression. Springer, Cham (2012). https://doi.org/10.1007/978-1-4615-3626-0
9. Goutsias, J., Heijmans, H.J.: Mathematical Morphology. IOS Press (2000)
10. Jackway, P.T., Deriche, M.: Scale-space properties of multiscale morphological dilation-erosion. IEEE Trans. Pattern Anal. Mach. Intell. **18**(1), 38–51 (2002). https://doi.org/10.1109/34.476009
11. Jungmann, F., Wurm, G.: Observation of bottom-up formation for charged grain aggregates related to pre-planetary evolution beyond the bouncing barrier. Astronomy Astrophysics **650**, Article no. A77, 7 (2021)
12. Maragos, P.: Pattern spectrum and multiscale shape representation. IEEE Trans. Pattern Anal. Mach. Intell. **11**(7), 701–716 (1989). https://doi.org/10.1109/34.192465
13. Matheron, G.: Random Sets and Integral Geometry. Wiley Series in Probability and Mathematical Statistics, vol. 52. Wiley, New York (1975)
14. Najman, J., Talbot, H.: Mathematical Morphology: From Theory to Applications. ISTE-Wiley, London-Hoboken (2010). https://doi.org/10.1002/9781118600788
15. Otsu, N.: A threshold selection method from gray-level histograms. IEEE Trans. Syst. Mans Cybern. **9**(1), 62–66 (1979)
16. Roerdink, J.B.T.M.: Mathematical morphology in computer graphics, scientific visualization and visual exploration. In: Soille, P., Pesaresi, M., Ouzounis, G.K. (eds.) ISMM 2011. LNCS, vol. 6671, pp. 367–380. Springer, Heidelberg (2011). https://doi.org/10.1007/978-3-642-21569-8_32
17. Serra, J., Soille, P.: Mathematical Morphology and Its Applications to Image Processing 2. Springer, Dordrecht (2012)
18. Soille, P.: Morphological Image Analysis: Principles and Applications, 1st edn. Springer, Heidelberg (1999). https://doi.org/10.1007/978-3-662-03939-7
19. Squire, J., Hopkins, P.F.: Physical models of streaming instabilities in protoplanetary discs. Mon. Not. R. Astron. Soc. **498**(1), 1239–1251 (2020)
20. Teiser, J., et al.: The growth of super-large pre-planetary pebbles to an impact erosion limit. Nat Astron (2025). https://doi.org/10.1038/s41550-024-02470-x
21. Theera-Umpon, N., Dhompongsa, S.: Morphological granulometric features of nucleus in automatic bone marrow white blood cell classification. IEEE Trans. Inf. Technol. Biomed. **11**(3), 353–359 (2007). https://doi.org/10.1109/TITB.2007.892694
22. Wurm, G., Teiser, J.: Understanding planet formation using microgravity experiments. Nat. Rev. Phys. **3**, 405–421 (2021). https://doi.org/10.1038/s42254-021-00312-7
23. Zsom, A., Ormel, C.W., Güttler, C., Blum, J., Dullemond, C.P.: The outcome of protoplanetary dust growth: pebbles, boulders, or planetesimals? II. Introducing the bouncing barrier. Astronomy Astrophysics **513**, Article no. A57, 22 (2010)

Formalizing an Iterated Morphological Erosion for the Discovery of Musical Patterns and Their Variations

Paul Lascabettes[1], Nils Quaetaert[1(✉)], Moreno Andreatta[1(✉)], and Isabelle Bloch[2(✉)]

[1] IRMA, ITI CREAA, Université de Strasbourg, Strasbourg, France
{lascabettes,quaetaert,andreatta}@math.unistra.fr
[2] Sorbonne Université, CNRS, LIP6, Paris, France
isabelle.bloch@sorbonne-universite.fr

Abstract. The discovery of patterns in a point-set representation of music consists in identifying repeating subsets of points. In this task, musical symbolic data is modeled as a discrete set of points, usually in $\mathbb{R}^n$, where each point represents a musical note and its coordinates represent the characteristics of the note, such as its onset or its pitch value. While numerous algorithms have been developed to discover all exact repetitions and extract musically relevant patterns, recent research has turned toward the discovery of patterns that repeat with some variations. Because the morphological erosion of a pattern provides its occurrences, we propose an adaptation of this operation to also obtain its variations with respect to a given approximation. This approach not only reveals certain variations of the pattern, but also enables to associate specific points to the pattern despite the fact that they were not initially present due to the constraints of strict repetition. We demonstrate that the proposed formalism satisfies certain fundamental properties for the musical pattern discovery task, such as the fact that iterating erosion produces cycles of patterns and its translation values. Finally, we apply these operations to the corpus of fugues from Bach's *Well-Tempered Clavier*, highlighting the usefulness of the proposed approach.

Keywords: Mathematical morphology · Pattern discovery · Point-set algorithms · Pattern variations · Erosion · Symbolic music data analysis

1 Introduction

This article proposes a new way of using morphological erosion and iterations of erosions for the discovery of repeating musical patterns and their variations. While many approaches exist for representing music data, we focus here on a multidimensional representation of music [12]. That is, musical data is modeled as a set of points in $\mathbb{R}^n$, where each point represents a musical note and each coordinate corresponds to a musical characteristic of that note, such as its

M. H. F. Wilkinson and J. Kosinka (Eds.): DGMM 2025, LNCS 16296, pp. 501–513, 2026.
https://doi.org/10.1007/978-3-032-09544-2_36

onset or pitch value. Therefore, the musical pattern discovery task in a multidimensional representation of music consists in identifying subsets of points that repeat within the data. Over the past years, a variety of algorithms have been developed for this task: SIA discovers all *Maximal Translatable Patterns* (MTPs) [12], SIATEC finds the occurrences of these MTPs [12], COSIATEC evaluates their salience based on a compactness criterion [9], SIAR restricts the search of MTPs within sliding windows [2], and SIACT identifies compact subpatterns of MTPs [4].

More recently, several algorithms have been proposed to discover not only repeated patterns but also their variations. These include the discovery of *Maximal Transformable Patterns* under musical transformations such as augmentation, diminution, inversion, and retrograde [10,11], as well as repetitions under time-warp invariances [5]. Despite these advances, the considered variations are limited, and certain musical transformations are not considered. Discovering variations of patterns remains a challenging problem.

To address this issue, we propose to use mathematical morphology. This work is based on existing connections between morphological operators and symbolic representations of music [8], particularly in the context of pattern discovery in a multidimensional representation of music [6,7]. We exploit here some preliminary results in which morphological erosion and its iterations are adapted to discover repeated patterns and their variations [13,15].

Our objective is to introduce a new theoretical framework in which key properties of morphological erosion are preserved. In particular, we aim to discover cycles of patterns and their associated translation values, while accounting for approximate repetitions.

The remainder of this article is structured as follows. Section 2 introduces the mathematical properties we aim to generalize and discusses their relevance to the musical pattern discovery task. Section 3 presents our erosion-based operator for discovering repeated patterns and their variations. Section 4 describes the results obtained from this new operation applied to the corpus of fugues from the first book of Bach's *Well-Tempered Clavier* to discover the subjects and their variations. Finally, Sect. 5 concludes this article and proposes future work.

2 Background on the Discovery of Repetitive Patterns in a Discrete Set of Points

In this section, we introduce the mathematical definitions and properties used in this article. Section 2.1 presents mathematical results of morphological erosion that we aim to generalize in the rest of the paper, while Sect. 2.2 demonstrates how these properties are fundamental for the discovery of musical patterns.

2.1 Mathematical Properties of Morphological Erosion

In this paper, we use only the basic operators of binary mathematical morphology and restrict ourselves to the case where the underlying space is $\mathbb{R}^n$. Therefore, the data X is a subset of $\mathbb{R}^n$, whether it represents images or, in our case, musical data. The basic operators of mathematical morphology that we use are defined as follows [14].

Definition 1 (Dilation, Erosion and Opening). *Let $X, S \subset \mathbb{R}^n$. The dilation $\delta_S(X)$, erosion $\varepsilon_S(X)$, and opening $\gamma_S(X)$ of X by S are defined as follows:*

$$\delta_S(X) = \{x + s \mid x \in X, s \in S\}, \tag{1}$$

$$\varepsilon_S(X) = \{t \in \mathbb{R}^n \mid S_t \subseteq X\}, \quad \textit{where } S_t = \{s + t \mid s \in S\}, \tag{2}$$

$$\gamma_S(X) = \delta_S(\varepsilon_S(X)) = \bigcup\{S_t \mid t \in \mathbb{R}^n, S_t \subseteq X\}. \tag{3}$$

In this case, S is called the *structuring element.* These morphological operations have several properties [1,14]. In particular, we often use in this paper the fact that erosion is decreasing with respect to the structuring element: $S \subseteq S' \Rightarrow \varepsilon_{S'}(X) \subseteq \varepsilon_S(X)$.

When applied to music, the interpretation of the basic morphological operators can be summarized as analysis tools for erosion and opening (finding a specific chord or rhythm for instance), and synthesis tools for dilation and closing (adding notes for instance). Examples can be found in [6].

In the context of image processing, the structuring element S is usually very small compared to the image X, and the erosion of X by S can be interpreted as a reduction of the size of the objects or the removal of small details of X. It is then possible to apply again an erosion by S to further reduce object size or remove more details. This is equivalent to applying a single erosion with a structuring element twice as large, if homothetic, i.e., $\varepsilon_S(\varepsilon_S(X)) = \varepsilon_{2S}(X)$. However, we might wonder what happens if, when an erosion is reapplied, the result of the first erosion $\varepsilon_S(X)$ is the structuring element, i.e., $\varepsilon_{\varepsilon_S(X)}(X)$. In this case, the roles of the structuring element S and the set X are reversed during the iteration of morphological erosion. The differences between these two different approaches to iterating morphological erosion are shown in Fig. 1. In particular, Fig. 1a illustrates the iteration of the erosion of the obtained result with S as structuring element, which has the consequence of gradually reducing the size of the set X, assuming that the origin of the space belongs to the structuring element S, whereas Fig. 1b highlights what we are focusing on, by iterating the morphological erosion of X when the structuring element is the previously obtained result.

To better understand this iterative operation, we define here the *iterated erosion* when the result obtained in the previous step is the structuring element of the erosion of X.

Definition 2 (Iterated Erosion). *Let $X, S \subset \mathbb{R}^n$. The iterated erosion $E^k(S, X)$ is defined recursively as follows:*

$$E^k(S, X) = \begin{cases} S, & \textit{if } k = 0, \\ \varepsilon_{E^{k-1}(S,X)}(X), & \textit{if } k \geq 1. \end{cases} \tag{4}$$

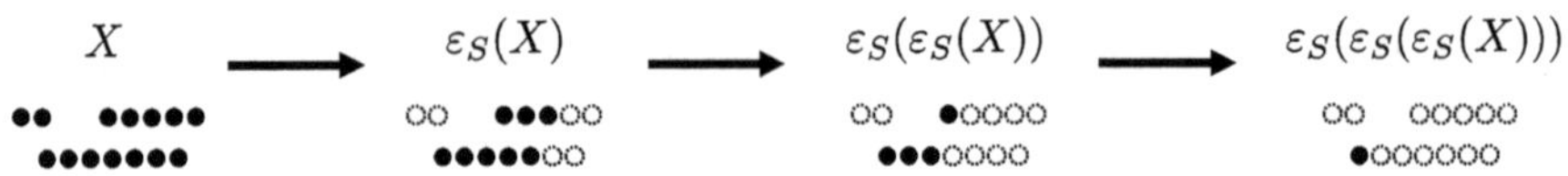

(a) Iteration of a morphological erosion of the obtained result by S.

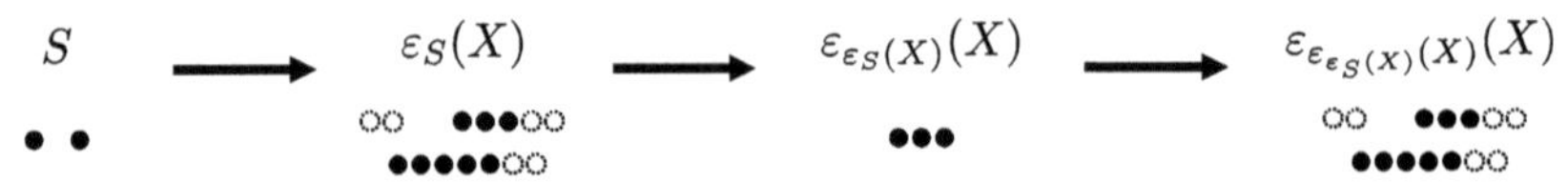

(b) What we propose in this paper: iterate the morphological erosion of X by the obtained result as structuring element.

Fig. 1. Two different approaches to iterate the morphological erosion, illustrated by X and S being two discrete sets of points defined on the left of the two figures.

Remark 1. Let $X, S \subset \mathbb{R}^n$. We have the following equalities for the iterated erosion:

$$\begin{aligned} E^0(S,X) &= S, & E^1(S,X) &= \varepsilon_S(X), \\ E^2(S,X) &= \varepsilon_{\varepsilon_S(X)}(X), & E^3(S,X) &= \varepsilon_{\varepsilon_{\varepsilon_S(X)}(X)}(X). \end{aligned} \tag{5}$$

One might wonder whether there is a relationship between the $E^k(S,X)$ for different values of k. Surprisingly, there is a theorem that provides a very simple answer to this question [6].

Theorem 1. *Let $X, S \subset \mathbb{R}^n$. Then, for all $k \geq 1$, we have:*

$$E^k(S,X) = E^{k+2}(S,X). \tag{6}$$

In addition, we also have the following inclusion for $k = 0$:

$$E^0(S,X) \subseteq E^2(S,X). \tag{7}$$

Remark 2. Let $X, S \subset \mathbb{R}^n$ and $m \in \mathbb{N}$. Theorem 1 leads to the following equalities:

$$\begin{aligned} E^1(S,X) &= E^3(S,X) = E^5(S,X) = \cdots = E^{2m+1}(S,X), \\ E^2(S,X) &= E^4(S,X) = E^6(S,X) = \cdots = E^{2m+2}(S,X). \end{aligned} \tag{8}$$

Moreover, the properties of the iterated erosion from Theorem 1 can be visualized in a cyclic way with the following diagram:

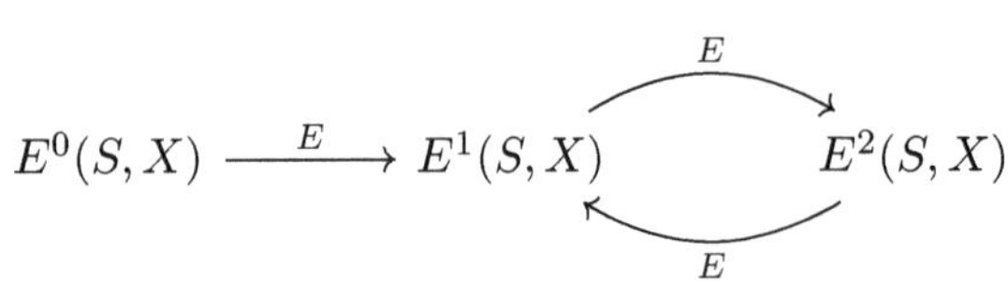

Remark 3. Theorem 1 is equivalent to the following result using traditional erosion notation:

$$\varepsilon_S(X) = \varepsilon_{\varepsilon_{\varepsilon_S(X)}(X)}(X) \quad \text{and} \quad S \subseteq \varepsilon_{\varepsilon_S(X)}(X), \tag{9}$$

which is more complicated to notate because of the increasing morphological erosion index. This is why we have introduced the iterated erosion, to simplify the understanding of this result.

2.2 Musical Interpretations of Morphological Erosion Properties

Theorem 1 is a result directly related to the musical pattern discovery task. Considering data $X \subset \mathbb{R}^n$ and a pattern $P \subset \mathbb{R}^n$, the morphological erosion $O = \varepsilon_P(X)$ of X by P provides the *translation values* of P in X, also called *onsets* of P in X [7]. This can be interpreted as the beginnings of P in the data X if the origin is on the first point of P. Theorem 1 then provides two types of information:

- First, Inclusion 7 reveals that reapplying a morphological erosion, i.e. $P' = \varepsilon_O(X)$, produces a pattern P' that contains P, thus: $P \subseteq P'$.
- Then, Eq. 6 demonstrates that computing the erosion $\varepsilon_{P'}(X)$ is unnecessary because $O = \varepsilon_{P'}(X)$. Therefore, there is no need to reapply a morphological erosion because all the information is already contained in P' and O: we have obtained a pair (P', O) linked by morphological erosion.

This is illustrated in Fig. 2 where O can be interpreted as the beginnings of P or P' in X. Therefore, we can see that $P \subseteq P'$, and if we reapply an erosion, we have $O = \varepsilon_{P'}(X)$.

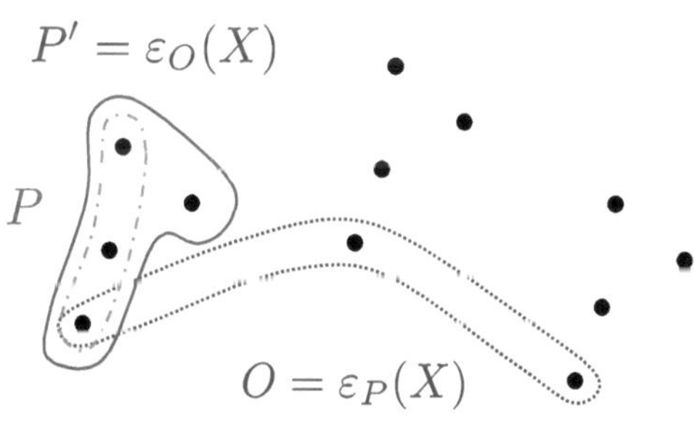

Fig. 2. Illustration of Theorem 1 with the pair (P', O) of pattern P' and its translation values O which are linked by morphological erosion. The data X is the set of points.

In addition, the links between P, O and P' are represented with the following diagram:

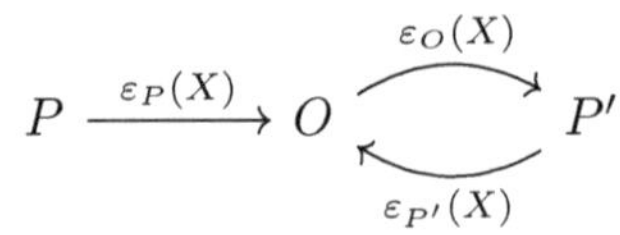

We introduce a category of patterns, called MTEC pattern [3] that allows us to characterize the resulting patterns P' and O illustrated in Fig. 2.

Definition 3 (MTEC Pattern). *Let $P, X \subset \mathbb{R}^n$. P is defined as maximal with respect to its translational equivalence class (MTEC) if:*

$$\forall P' \subset \mathbb{R}^n, P \subset P' \Rightarrow |\varepsilon_{P'}(X)| < |\varepsilon_P(X)|. \tag{10}$$

In other words, when additional points are added to an MTEC pattern, the number of its occurrences necessarily decreases, as $|\varepsilon_P(X)|$ can be interpreted as the number of occurrences of P in X. For instance, in Fig. 3, the pattern P is not an MTEC pattern because $|\varepsilon_P(X)| = 3$, and it is still possible to include the point in the bottom left while maintaining the same number of occurrences in X. Once that point is added, the pattern becomes an MTEC pattern.

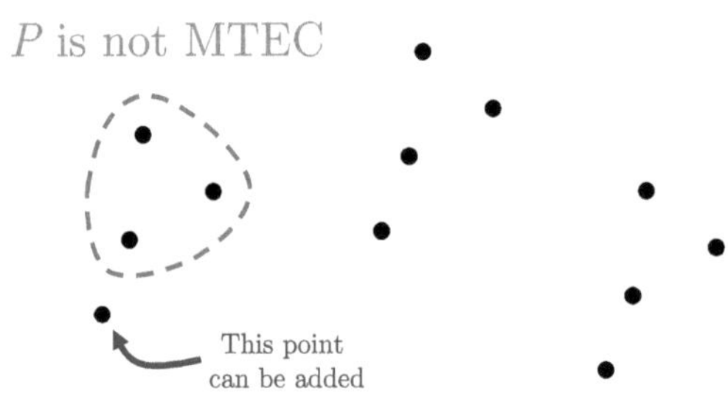

Fig. 3. Illustration of a pattern P that is not an MTEC pattern, since the point in the bottom left can be added without changing the number of its three occurrences in the point set X.

The following theorem provides a link between MTEC patterns and patterns obtained from morphological erosion [6].

Theorem 2. *Let $P, X \subset \mathbb{R}^n$, we have:*

$$P \text{ is MTEC} \iff \exists S \subset \mathbb{R}^n : P = \varepsilon_S(X). \tag{11}$$

Therefore, we characterize the pair (P', O), displayed in Fig. 2, as an MTEC conjugate pair, with the following definition.

Definition 4 (MTEC Conjugate Pair). *Let $P, Q, X \subset \mathbb{R}^n$. The pair (P, Q) is an MTEC conjugate pair if:*

$$P = \varepsilon_Q(X) \quad \text{and} \quad Q = \varepsilon_P(X). \tag{12}$$

Remark 4. Let $P, X \subset \mathbb{R}^n$, from Theorem 2, we have:

$$P \text{ is MTEC} \iff P = \varepsilon_{\varepsilon_P(X)}(X) = E^2(P, X). \tag{13}$$

However, morphological erosion only identifies exact pattern repetitions. Therefore, the previous results are useful for discovering exact repetitions in data, but do not directly support the discovery of patterns with variations. This naturally raises the following question: How can we adapt morphological erosion to also detect pattern variations? Moreover, we constrain this adaptation to preserve the concept of pair presented in Theorem 1. That is, erosion produces a pair (pattern, translation values) where each element is derived from the other with morphological erosion. In the remainder of this article, we propose a method based on morphological erosion that discovers pattern variations, given a certain approximation, and preserves the concept of a pair between pattern variations and translation values obtained with this new operation.

3 Adapting Morphological Erosion to Discover Pattern Variations

Figure 4 illustrates a simple case that we aim to solve. Given a specific pattern P, we observe that the morphological erosion of X by P allows us to discover the occurrences of this pattern. However, if an occurrence of the pattern is slightly altered, it is not discovered by this operation. Our goal, therefore, is to define operations that enable the discovery of such variations.

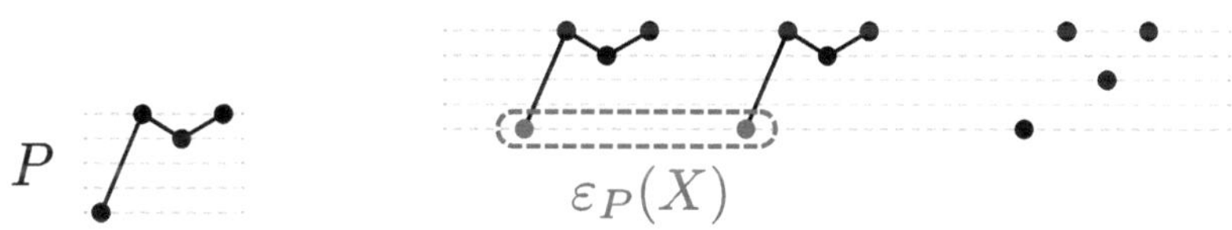

Fig. 4. Limitations of morphological erosion. When applying the erosion of a dataset X by a pattern P (shown on the left, with the origin placed at its bottom-left point), slightly altered occurrences of P are not discovered.

To discover patterns with variation, we then propose to encode the potential variations by a structuring element A, dilate X (or X after removing the exact patterns) by A, and then apply the pattern discovering process. We propose to adapt the iterated morphological erosion by defining the iterated *opening-filtered variational erosion* (or iterated OVE).

Definition 5 (Iterated OVE). *Let $S, A, X \subset \mathbb{R}^n$. The iterated opening-filtered variational erosion $\tilde{E}^k(S, A, X)$ is defined recursively as follows:*

$$\tilde{E}^k(S, A, X) = \begin{cases} S, & \text{if } k = 0, \\ E^k(S, X) \cup \varepsilon_{\tilde{E}^{k-1}(S,A,X)}(X'), & \text{if } k \geq 1, \end{cases} \tag{14}$$

with $X' = \delta_A(X \setminus \gamma_S(X))$.

In this definition, A represents the approximation with which we aim to discover the variations of P. Its role is to extend the domain in which a pattern can be found, according to the admissible variations. For instance a dilation of X by A representing a semi-tone will add to every note of X notes differing by a semi-tone.

Figure 5 illustrates the process. Given a data set X and a pattern P, we first identify the occurrences of P in X through the regular morphological erosion $\varepsilon_P(X)$. We then remove those occurrences, which correspond to the opening $\gamma_P(X)$, and dilate the remaining points with the chosen dilating element A. Finally, we find occurrences of P in this dilated set with $\varepsilon_P(X')$. The result of the operation is then the union of $\varepsilon_P(X)$ and $\varepsilon_P(X')$.

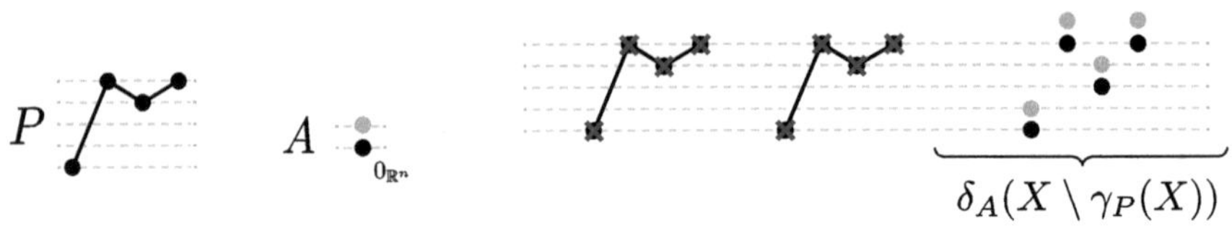

(a) After identifying the occurrences of P in X, we remove $\gamma_P(X)$ from X and dilate the remaining set of points with A to obtain $X' = \delta_A(X \setminus \gamma_P(X))$.

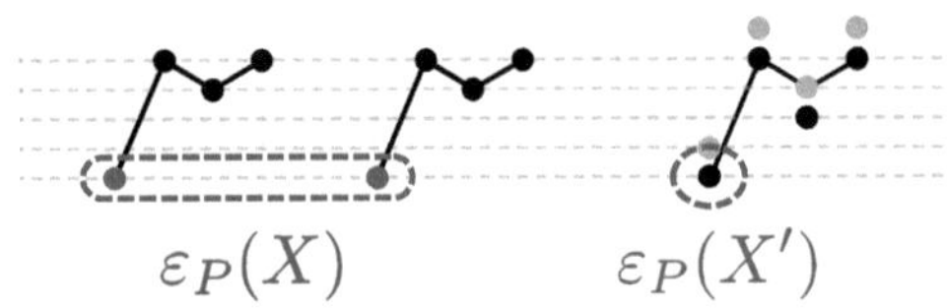

(b) The result of the OVE is the union of $\varepsilon_P(X)$ (origins of exact repetitions of P in X) and $\varepsilon_P(X')$ (origins of variations of P in X): $\tilde{E}^1(P, A, X) = \varepsilon_P(X) \cup \varepsilon_P(X')$.

Fig. 5. Illustration of the OVE applied to the dataset X to identify repetitions of the pattern P and its variations.

Another example is illustrated in Fig. 6, where the variation is now in rhythm.

Note that neither the iterated erosion not the OVE are true algebraic erosions, since they do not commute with the intersection in general.

Starting from an MTEC pattern, the following theorem guarantees the existence of cycles between pattern variations and their associated translation values.

Theorem 3. *Let $S, A, X \subset \mathbb{R}^n$ such that S is an MTEC pattern. Then, for all $k \geq 1$, we have:*

$$\tilde{E}^k(S, A, X) = \tilde{E}^{k+2}(S, A, X). \tag{15}$$

In addition, we also have the following inclusion for $k = 0$:

$$\tilde{E}^0(S, A, X) \subseteq \tilde{E}^2(S, A, X). \tag{16}$$

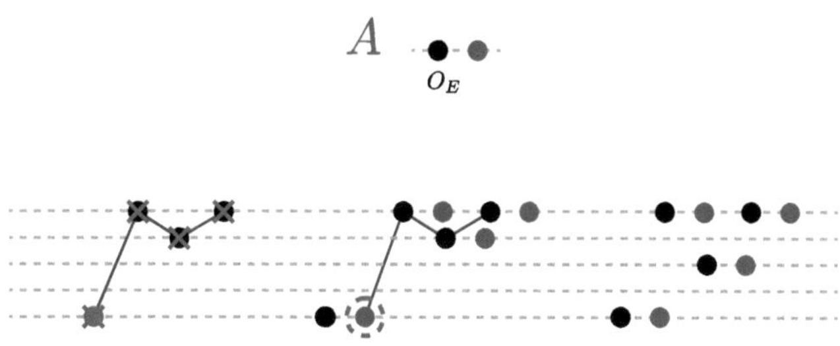

Fig. 6. The pattern on the left is recognized in the middle with a variation in rhythm encoded by A. Variations in both rhythm and pitch would allow detecting three occurrences of the pattern (left: exact, middle: variation in rhythm, right: variation in pitch, as in Fig. 5).

Proof (Proof of Theorem 3). The inclusion in Eq. 16 comes directly from the inclusion of Theorem 1, and $E^0(S,X) = \tilde{E}^0(S,A,X) = S$:

$$\begin{aligned} & E^0(S,X) \subseteq E^2(S,X) && \text{(Theorem 1)} \\ \Rightarrow\ & \tilde{E}^0(S,A,X) \subseteq \tilde{E}^2(S,A,X) && \text{(Definition 5)} \end{aligned}$$

Then, to prove Eq. 15, it is sufficient to prove that it is true for $k = 1$, i.e.:

$$\tilde{E}^1(S,A,X) = \tilde{E}^3(S,A,X). \tag{17}$$

Because we have:

$$\tilde{E}^1(S,A,X) = E^1(S,X) \cup \varepsilon_{\tilde{E}^0(S,A,X)}(X') = E^1(S,X) \cup \varepsilon_S(X'),$$
$$\tilde{E}^3(S,A,X) = E^3(S,X) \cup \varepsilon_{\tilde{E}^2(S,A,X)}(X') = E^1(S,X) \cup \varepsilon_{\tilde{E}^2(S,A,X)}(X'),$$

it is sufficient to prove that:

$$\varepsilon_S(X') = \varepsilon_{\tilde{E}^2(S,A,X)}(X'). \tag{18}$$

Because S is an MTEC pattern, is satisfies $E^2(S,X) = S$ and therefore we have:

$$\tilde{E}^2(S,A,X) = S \cup \varepsilon_{\tilde{E}^1(S,A,X)}(X'). \tag{19}$$

$\boxed{\supseteq}$ The first inclusion in Eq. 18 comes from the decreasingness of erosion with respect to the structuring element:

$$\begin{aligned} & S \subseteq \tilde{E}^2(S,A,X) \\ \Rightarrow\ & \varepsilon_{\tilde{E}^2(S,A,X)}(X') \subseteq \varepsilon_S(X') \end{aligned}$$

$\boxed{\subseteq}$ Let $x \in \varepsilon_S(X')$. We need to prove that $x \in \varepsilon_{\tilde{E}^2(S,A,X)}(X')$, i.e. prove that $\forall s \in E^2(S,A,X), s + x \in X'$.

Let $s \in E^2(S, A, X)$, then $s \in S$ or $s \in \varepsilon_{\tilde{E}^1(S,A,X)}(X')$ (Eq. 19).
If $s \in S$, then $s + x \in X'$ by definition of x because $x \in \varepsilon_S(X')$.
If $s \in \varepsilon_{\tilde{E}^1(S,A,X)}(X')$, then $s + x \in X'$ because $x \in \varepsilon_S(X') \subseteq \tilde{E}^1(S, A, X)$. □

The properties from Theorem 3 can be visualized with a diagram. Given an MTEC conjugate pair (P, O), we can extend either P or O using the iterated OVE, giving us two distinct cycles. In the case where P represents a musical pattern and O its onsets, starting from P allows us to find primarily new occurrences of the pattern, while starting from O allows us to find primarily new points that are part of the pattern.

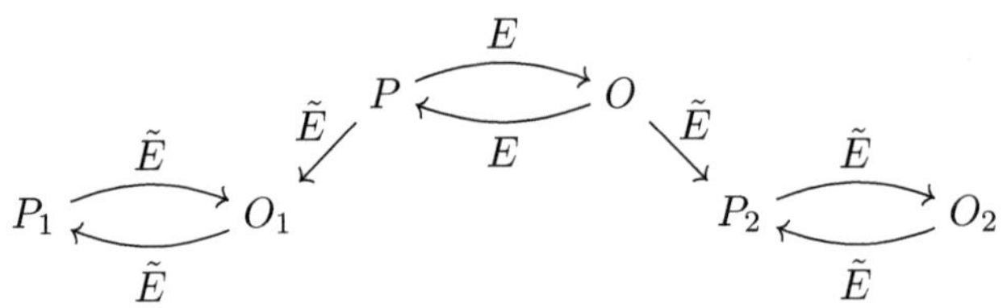

One may wonder if the pairs (P_1, O_1) and (P_2, O_2) from this diagram are related in any way. The following theorem ensures that there is an inclusion relation between P_1 and P_2, and between O_1 and O_2, in particular we have: $P_1 \subseteq P_2$ and $O_2 \subseteq O_1$.

Theorem 4. *Let $S, A, X \subset \mathbb{R}^n$. Then, we have the two following inclusions:*

$$\tilde{E}^2(S, A, X) \subseteq \tilde{E}^1(\varepsilon_S(X), A, X), \tag{20}$$

$$\tilde{E}^2(\varepsilon_S(X), A, X) \subseteq \tilde{E}^1(S, A, X). \tag{21}$$

Proof (Proof of Theorem 4). Let $X' = \delta_A(X \setminus \gamma_{\varepsilon_S(X)}(X))$. First, we prove Inclusion 20. The left side is:

$$\tilde{E}^2(S, A, X) = E^2(S, X) \cup \varepsilon_{\tilde{E}^1(S,A,X)}(X').$$

And the right side can be rewritten as:

$$\begin{aligned}
&\tilde{E}^1(\varepsilon_S(X), A, X) \\
&= E^1(\varepsilon_S(X), X) \cup \varepsilon_{\tilde{E}^0(\varepsilon_S(X),A,X)}(X') && \text{(Definition 5)} \\
&= E^1(\varepsilon_S(X), X) \cup \varepsilon_{\varepsilon_S(X)}(X') && (\tilde{E}^0(\varepsilon_S(X), A, X) = \varepsilon_S(X)) \\
&= E^2(S, X) \cup \varepsilon_{\varepsilon_S(X)}(X') && (E^1(\varepsilon_S(X), X) = E^2(S, X))
\end{aligned}$$

Because $\tilde{E}^1(S, A, X) = \varepsilon_S(X) \cup \varepsilon_{\tilde{E}^0(S,A,X)}(X')$ we can conclude the proof of Inclusion 20:

$$\begin{aligned}
& \varepsilon_S(X) \subseteq \tilde{E}^1(S, A, X) \\
\Rightarrow\ & \varepsilon_{\tilde{E}^1(S,A,X)}(X') \subseteq \varepsilon_{\varepsilon_S(X)}(X') && \text{(Decreasingness of } \varepsilon) \\
\Rightarrow\ & \tilde{E}^2(S, A, X) \subseteq \tilde{E}^1(\varepsilon_S(X), A, X) && \text{(Union with } E^2(S, X))
\end{aligned}$$

A similar reasoning, reversing the role of S and $\varepsilon_S(X)$, proves Inclusion 21. □

4 Application to Symbolic Music Data

In this section, we demonstrate how the mathematical results of the previous section can be applied to a concrete case: the fugues from the *Well-Tempered Clavier* by J.S. Bach. The goal is to discover the subject of a fugue using the iterated OVE with respect to an MTEC pattern. We represent a fugue as a set of notes in $\mathbb{R}^2$, considering only time and pitch dimensions.

Let us illustrate one example by considering the Fugue No. 2 from the first book of the *Well-Tempered Clavier* (BWV 847). The subject of this fugue spans 20 notes and there are 8 full occurrences of the pattern in the piece. Using the iterated erosion of the fugue by its first few notes, one can find an MTEC conjugate pair corresponding to a pattern with 19 notes and 5 occurrences.

Although this result nearly identifies the subject of the fugue, and correctly identifies five of its occurrences, it is nonetheless unsatisfying. By using the iterated OVE on this pair, we can improve the result. Using a structuring element that encodes potential variations of one semitone and a tone, we find a pattern-onset pair with 20 notes and 8 occurrences, which was the expected result. Figure 7 illustrates this result.

This example demonstrates how the iterated OVE can be used to recover the subject of a fugue and its occurrences. This principle can be developed into a systematic approach. An algorithm designed around these ideas was used on all 24 fugues of the first book of the *Well-Tempered Clavier* [13]. Given a fugue, the algorithm returns a list of pattern-onset pairs computed with the iterated OVE with respect to various MTEC pairs. The algorithm is consistently able to find the subject of the fugue and its occurrences, as well as truncated instances of the pattern that appear throughout the piece. Furthermore, our preliminary implementation of the algorithm (using a naive implementation of erosion in linear time) runs very rapidly.

While the point set matching problem could probably be solved in a purely algorithmic way, mathematical morphology provides an elegant way to formalize it, in a general way, taken into account potential variations of the pattern without enumerating all possible cases, and to prove theoretical results that provide guarantees on the properties of the obtained results. Without the use of mathematical morphology, it would have been much more complicated to describe and therefore to guarantee these results.

Fig. 7. Discovering the subject of the Fugue No. 2 of the first book of the *Well-Tempered Clavier* (BWV 847). Using the iterated erosion, we find a pattern with 19 notes and 5 occurrences (illustrated in blue). Then using the iterated OVE, we complete the pattern to 20 notes (the additional note is displayed in purple) and find 3 more occurrences (illustrated in red). (Color figure online)

5 Conclusion

In this article, we have presented how mathematical morphology can be used to discover repeated patterns in a multidimensional representation of music. We first showed that iterated erosion identifies a particular category of patterns, called MTEC patterns, which possess properties that might make them good candidates for representing musically salient patterns. We then introduced the iterated opening-filtered variational erosion (OVE), an extension of the iterated erosion framework that enables the discovery of repeated patterns and their variations by incorporating an approximation parameter via a structuring element. This new operation has been applied to the analysis of fugues from the *Well-Tempered Clavier*, almost successfully identifying subjects and their variations under approximations of one semitone and one tone.

Beyond the methods presented here, the iteration-based formalism opens up possibilities for defining new morphological tools. While this paper focused on the iterated erosion and OVE, other operators such as the pivotal variational erosion (PVE) and the intersectional variational erosion (IVE) [13] can also be expressed and studied within this framework, offering promising directions for future research. On the application side, the method applies also to recorded music, as soon as it can be encoded in a computationally compliant format, such as midi. The fact that our method accounts for potential variations is useful in this context, for instance to accommodate rubato, with slight temporal variations. The proposed formalism applies to any type of composition, as soon as there are patterns that are (approximately) repeated.

Therefore, this approach demonstrates the relevance of developing a mathematical framework, particularly with new operations, for analyzing musical patterns and their variations.

References

1. Bloch, I., Heijmans, H., Ronse, C.: Mathematical Morphology, pp. 857–944. Springer, Cham (2007)
2. Collins, T.: Improved methods for pattern discovery in music, with applications in automated stylistic composition. Ph.D. thesis, Faculty of Mathematics, Computing and Technology, The Open University, UK (2011)
3. Collins, T., Meredith, D.: Maximal translational equivalence classes of musical patterns in point-set representations. In: 4th International Conference on Mathematics and Computation in Music, Montreal, Canada, pp. 88–99 (2013)
4. Collins, T., Thurlow, J., Laney, R., Willis, A., Garthwaite, P.: A comparative evaluation of algorithms for discovering translational patterns in baroque keyboard works. In: 11th International Society for Music Information Retrieval Conference, Utrecht, Netherlands, pp. 3–8 (2010)
5. Laaksonen, A., Lemström, K.: Discovering distorted repeating patterns in polyphonic music through longest increasing subsequences. J. Math. Music **15**(2), 99–111 (2021)
6. Lascabettes, P.: Mathematical models for the discovery of musical patterns, structures and for performances analysis. Ph.D. thesis, Sorbonne Université (2023)
7. Lascabettes, P., Bloch, I.: Discovering repeated patterns from the onsets in a multidimensional representation of music. In: International Conference on Discrete Geometry and Mathematical Morphology. Springer, Florence, Italie (2024)
8. Lascabettes, P., Bloch, I., Agon, C.: Analyse de représentations spatiales de la musique par des opérateurs simples de morphologie mathématique. In: Journées d'Informatique Musicale. Strasbourg, France (2020)
9. Meredith, D.: Point-set algorithms for pattern discovery and pattern matching in music. In: Dagstuhl Seminar Proceedings. Schloss Dagstuhl-Leibniz-Zentrum für Informatik (2006)
10. Meredith, D.: Understanding and compressing music with maximal transformable patterns. In: International Conference on Human-Computer Interaction, pp. 309–325. Springer, Cham (2023)
11. Meredith, D.: A parallel algorithm for finding maximal transformed matches of polyphonic patterns in unvoiced polyphonic music. In: Second International Conference on Computational and Cognitive Musicology (2024)
12. Meredith, D., Lemström, K., Wiggins, G.: Algorithms for discovering repeated patterns in multidimensional representations of polyphonic music. J. New Music Res. **31**(4), 321–345 (2002)
13. Quaetaert, N.: Discovery of musical patterns and their variations with mathematical morphology. Master's thesis, School of Engineering Sciences, Strasbourg, France (2025)
14. Serra, J.: Image Analysis and Mathematical Morphology. Academic Press, London (1982)
15. Tamagna, E.: Découverte de motifs musicaux avec variations en utilisant la morphologie mathématique. Master's thesis, Sorbonne Université, Paris, France (2024)

Digital Plane Detection in a Point Set: Application to the Interactive Extraction of Charcoal Platforms from Airborne LiDAR

Philippe Even(✉) and Phuc Ngo

Université de Lorraine, CNRS, LORIA, Nancy, France
{philippe.even,hoai-diem-phuc.ngo}@loria.fr

Abstract. A digital blurred plane is a finite set of points with controlled thickness. This notion can be used to characterize a real world object surface, considered as planar at a large scale, but actually quite irregular at finer scale. In this paper, we introduce a new framework to detect a blurred plane in a point set, based on the analysis of two orthogonal height profiles using digital geometry tools. It is applied to the interactive extraction of relict charcoal platforms from an airborne LiDAR point cloud. This process was tested on two sets of geolocated platforms obtained by on-site prospecting or visual survey of the LiDAR digital terrain model. Experimental validations show that a large amount of them is successfully extracted and that their estimated extent closely matches the ground truth.

Keywords: Digital plane detection · Point set processing · Airborne LiDAR

1 Introduction

Digital geometry aims at providing efficient processing algorithms to tackle the discrete nature of acquired data by most of today sensors. But many real-world objects are not geometrically perfect. Discrete primitives such as digital straight lines or digital planes are not well adapted to describe their irregular surface. Therefore a notion of blurred digital object, such as blurred line segments [2] or blurred pieces of digital planes [9], has been proposed to incorporate such imperfections, the irregularity being captured by a thickness parameter. In this work, our purpose is to incorporate blurred planes into a new framework to process 3D point clouds, and to test their effectiveness in real context. This could concern a large range of applications, for instance 3D modeling of indoor or outdoor urban scenes using mobile 3D scanners. However, we have chosen to address a quite difficult context involving poorly structured landscapes with hard sensing conditions.

Processing airborne LiDAR data of forest landscapes is such a particularly challenging task. The LiDAR is a survey technique based on the emission of a laser beam swept over the measured scene. In forest scenes, the received signal

M. H. F. Wilkinson and J. Kosinka (Eds.): DGMM 2025, LNCS 16296, pp. 514–527, 2026.
https://doi.org/10.1007/978-3-032-09544-2_37

Fig. 1. A charcoal platform (a) and its extraction (black circle) from the LiDAR ground points superimposed to the hill-shaded DTM (b).

is composed of multiple echoes corresponding to the successive hit obstacles. Lower ground points are interpolated to get a height map called *digital terrain model* (DTM). This provides bare ground visualizations of the landscape (see Fig. 1b). DTM quality depends strongly on the ground point distribution, which can be severely affected in places where the vegetation is dense. This ability to penetrate the canopy is particularly useful for the study of geomorphological structures connected to past human activities. In forest areas, they are often well preserved from water erosion, whereas they have been heavily damaged by agricultural or industrial activities in open landscapes.

In a previous work [4], we proposed a framework to extract linear features such as forest roads, walls or ditches from LiDAR data. The goal addressed in this study is to design a new framework adapted to the processing of planar structures, that are also common: agricultural banks, terraces, pastoral enclosures, and so on. An approach based on the analysis of altimetric profiles obtained from scan lines in different directions can be successful at detecting plane patches [11]. The challenge here is to cope with possibly large terrain irregularities and important artefacts that may alter the surface, such as windthrow remains or effects of forest harvesting activities.

For this purpose, the framework is applied to the extraction of relict charcoal platforms. They were built in many forest areas in connection with industrial development in past centuries. A thorough study of these structures may provide better understanding of the past industrial context as well as the ecological impact of charcoal production. Statistical analysis of their geometric features could help to extrapolate the acquired information on a larger scale. They are characterized by a nearly horizontal circular platform about ten meters wide (see Fig. 1a). On steep slopes, they contrast significantly with the local trend and are often oblong, developed perpendicularly to the slope direction. In flatter areas, they spread more uniformly in all directions, but their discrimination from the natural terrain is more subtle, and sometimes, only a surrounding ditch makes them detectable. This circular ditch may enclose the entire platform or only the upstream part on gentle slopes.

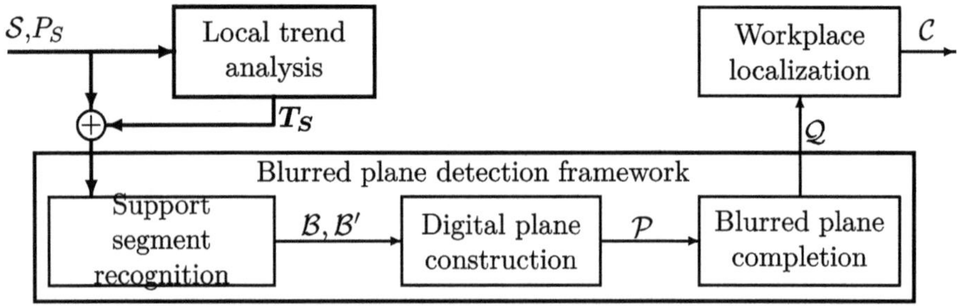

Fig. 2. The charcoal platform extraction process.

Some works have been carried out to automate charcoal platform detection using template matching [10] or convolutional neural networks [8,12,13], but proposed solutions rely on the processing of DTM visualizations, that ignore important features of the original point cloud. They also depend strongly on vegetation cover conditions and LiDAR quality. Beyond this, geometric characterization of these platforms remains an open problem. In our approach, we process the point cloud directly, and the DTM is only used for interaction tasks, such as input seed selection or extracted structure validation.

We have developed and implemented a new blurred plane detection framework. Its workflow is displayed in the lower part of Fig. 2. It takes a point set $\mathcal{S}$, a seed point P_S and a search direction $\boldsymbol{T_S}$ as input. From the recognition of support segments $\mathcal{B}$ and $\mathcal{B}'$ in two orthogonal point profiles, a thick enclosing plane $\mathcal{P}$ is constructed. The output blurred plane $\mathcal{Q}$ is composed of points of $\mathcal{S}$ that lie in $\mathcal{P}$. As depicted in Fig. 2, the input to the overall platform extraction process is a start position P_S. This seed can either be extracted from prior knowledge or manually selected in a DTM view. First, the slope direction $\boldsymbol{T_S}$ is estimated by probing the local terrain around the seed. This provides a favorable direction to find the platform plane $\mathcal{Q}$. Then the blurred plane detection framework is used to determine the possible platform extent. Finally, a circle detector is used to locate the workplace $\mathcal{C}$ where the charcoal mound was built. Charcoal platforms are good candidates to be modeled using the blurred plane concept. As a drawback, it is not possible to perform a separate evaluation of the blurred plane detection part.

The next section details the blurred plane detection framework. Its application to the extraction of charcoal platforms from LiDAR ground points is presented in Sect. 3, with a synthesis of the held experiments. Section 4 gives a conclusion and some perspectives to this work.

2 Blurred Plane Detection

This section introduces the blurred plane detector used to extract charcoal platforms. The inputs are a seed point P_S, a direction $\boldsymbol{D_S}$ orthogonal to the extraction axis $\boldsymbol{Z}$, and a search radius R_S. Two height profiles are extracted from the point set in the planes $(P_S, \boldsymbol{D_S}, \boldsymbol{Z})$ and $(P_S, \boldsymbol{D'_S}, \boldsymbol{Z})$, where $\boldsymbol{D'_S} = \boldsymbol{Z} \times \boldsymbol{D_S}$.

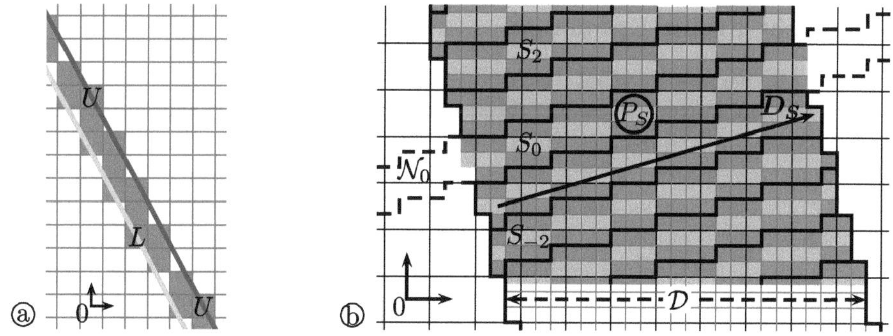

Fig. 3. a) Digital line $\mathcal{L}$: $33 \leq 9x + 5y < 46$ with thickness $\tau(\mathcal{L}) = \frac{4}{3}$, bounding lines in yellow and magenta, lower point L and upper points U. b) Directional scan with subdivision rate $N_G = 3$; the start scan S_0 is defined by point $P_S(5,4)$, direction $\boldsymbol{D_S}(7,2)$, arithmetic width $\mu = 7$ and radius $R_S = 4$ in the original grid $\mathcal{G}$; the scan strip $\mathcal{D}$ is defined by $46 \leq 7x+2y < 215$ in the subdivided grid $\mathcal{G}'$. (Color figure online)

Nearly linear segments are searched in these profiles to detect a candidate plane. The output blurred plane is composed of the points in the set that belong to this plane. In the following, the discrete geometry notions used for the detection framework are introduced. First, we briefly recall the notion of digital hyperplane, and then we explain how height profiles are extracted from the point set using directional scans. Next, we present a procedure for recognizing blurred line segments that fits the assigned maximal thickness to local irregularities of the point set. Finally we describe the blurred plane detection framework.

2.1 Digital Hyperplanes

In the following, we call $\boldsymbol{P}$ the vector from the world reference origin to the point P. Given a non-null, d-dimensional vector $\boldsymbol{N}(n_1, n_2, \ldots, n_d) \in \mathbb{Z}^d$ where $d \geq 2$, a shift value $h \in \mathbb{Z}$ and an arithmetic width value $\mu \in \mathbb{Z}^+$, a **digital hyperplane** [1] is the set of integer points that are bounded by two bounding hyperplanes, which are defined by equations $\boldsymbol{N} \cdot \boldsymbol{P} = h$ and $\boldsymbol{N} \cdot \boldsymbol{P} = h + \mu - 1$:

$$\mathcal{H}_d = \{P(p_1, p_2, \ldots, p_d) \in \mathbb{Z}^d \mid h \leq \boldsymbol{N} \cdot \boldsymbol{P} < h + \mu\} \tag{1}$$

The points of $\mathcal{H}_d$ that lie in $\boldsymbol{N} \cdot \boldsymbol{P} = h$ (resp. $\boldsymbol{N} \cdot \boldsymbol{P} = h + \mu - 1$) are its **lower points**, (resp. **upper points**). The hyperplane's **period** is $p(\mathcal{H}_d) = \frac{max(|n_1|,|n_2|,\ldots,|n_d|)}{gcd(|n_1|,|n_2|,\ldots,|n_d|)}$. When $\mu = p(\mathcal{H}_d)$, the hyperplane is said **naive**. The **thickness** of the digital hyperplane is the l_∞-distance between the bounding hyperplanes:

$$\tau(\mathcal{H}_d) = \frac{\mu - 1}{p(\mathcal{H}_d)} \tag{2}$$

A **digital line** is a digital hyperplane in a two-dimensional space (see Fig. 3a), and a **digital plane** is a digital hyperplane in a three-dimensional space.

2.2 Height Profile Extraction

In this work, we assume that the points of the input set $\mathcal{S}$ are scattered on a non-overlapping surface in height direction $\boldsymbol{Z}$, so that at most one point can be found at each (x, y) position. In the contrary, some preprocessing to split the point cloud (*e.g.*, point classification) is necessary to avoid interruptions in plane detection. Hopefully, there is low probability to find a charcoal platform under a cliff overhang. The points are subdivided into a two-dimensional grid $\mathcal{G}$ in $\boldsymbol{XY}$-plane, which encloses the investigated cloud area. The point set is reorganized to allow fast access from the cell index. The grid $\mathcal{G}$ is used to define the input seed position P_S, the direction vector $\boldsymbol{D_S}$ and the search radius R_S for the plane detection framework. It is also used to build height profiles by sweeping the surface using directional scans.

A **directional scan** [7] is an ordered partition into scans S_i, restricted to the grid domain $\mathcal{G} \subset \mathbb{Z}^2$ of a thick digital line $\mathcal{D}$, called **scan strip**. Each scan S_i is a segment of a digital line $\mathcal{L}_i$, called **scan line**, with arithmetic width μ, and orthogonal to $\mathcal{D}$. Given a start scan S_0 defined by a center point P_S, a direction $\boldsymbol{D_S}$, a radius R_S and an arithmetic width μ, the directional scan is defined as:

$$DS(S_0) = \left\{ S_i = \mathcal{D} \cap \mathcal{L}_i \cap \mathcal{G} \;\middle|\; \begin{array}{l} \boldsymbol{N}(\mathcal{L}_i) \cdot \boldsymbol{N}(\mathcal{D}) = 0 \\ h(\mathcal{L}_i) = h(\mathcal{L}_{i-1}) + \mu \end{array} \right\} \tag{3}$$

In this definition, $\boldsymbol{N}(\mathcal{L}_i) \cdot \boldsymbol{N}(\mathcal{D}) = 0$ expresses the orthogonality between the scan lines $\mathcal{L}_i$ and the scan strip $\mathcal{D}$. The scans S_i can be iteratively parsed from S_0 to both ends. The shift μ between successive scans $\mathcal{L}_{i-1}$ and $\mathcal{L}_i$ guarantees that all points of $\mathcal{D}$ are traversed only once. The thickness of the scan strip $\mathcal{D}$ is defined by the radius value: $\tau(\mathcal{D}) = 2 \cdot R_S$.

For each scan S_i, a **height profile** is obtained by collecting the points of $\mathcal{S}$ stored in each cell and projecting them onto the $(P_S, \boldsymbol{D_S}, \boldsymbol{Z})$ plane. The points P_i are sorted along direction $\boldsymbol{D_S}$, and their 3D embedding $\kappa(P_i) \in \mathcal{S}$ is also registered. Directional scans are used in [7] to scan gray-level images using naive line segments. However, scan line aliasing produces an irregular point distribution in the height profile. In present work, a subdivided grid $\mathcal{G}'$ with N_G-finer resolution is used to get smoother profiles from a directional scan with arithmetic width $\mu' = N_G \cdot \mu$. Figure 3b illustrates this process.

2.3 Blurred Segment Recognition with Controlled Thickness

A **blurred segment** [2] is a finite subset of a digital line $\mathcal{L}$. The covering digital line with minimal thickness is called the **optimal line** of the blurred segment $\mathcal{B}$. Hereafter, we call $\boldsymbol{V}$ its director vector.

Blurred segments can be detected in linear time by a recognition algorithm [2]. The inputs are a set of sorted points along a given axis, a start position P_0 and a maximal thickness value τ_0 for the blurred segment. The points are fed sequentially to the algorithm, from the start position to both ends of the set. At each insertion, the blurred segment and its convex hull are updated if the new thickness is less than τ_0. Otherwise, the recognition process stops. The optimal

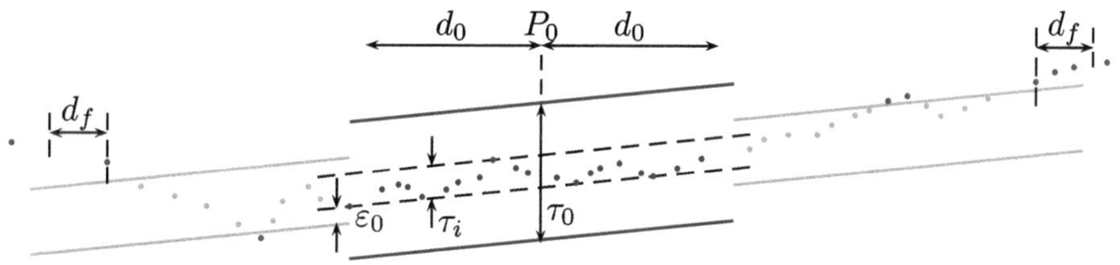

Fig. 4. Blurred segment recognition with controlled thickness: blue dots are first inserted, then green dots after maximal thickness was fit to $\tau_i + \varepsilon_0$; red dots are rejected. (Color figure online)

line is bounded by an edge of the convex hull on one side and by a vertex on the other. The director vector $\boldsymbol{V}$ and the thickness τ can be easily derived from this edge-vertex pair.

In this algorithm, the maximal thickness is set beforehand to account for noise in the point set. This requires a previous analysis of input points. Instead, we propose adapting the initial algorithm to estimate this noise level during the recognition process in order to better fit the segment to the set of points (see Fig. 4). The maximal thickness is initially set to a large value τ_0, and the blurred segment is extended to both sides of a start position P_0. Profile points are submitted to the recognition algorithm from nearest to furthest from P_0. Let d_i be the horizontal distance of the ith point from P_0. Once an observation distance is reached ($d_i > d_0$), the maximal thickness is reduced to the thickness τ_i of the already-detected segment, dilated by a small safety margin ε_0 to incorporate some more terrain roughness. On each side, the detection stops when an d_f-long sequence of successive points is rejected or when the last point is processed. The output is the blurred segment (*i.e.*, the set of accepted points) along with its final thickness τ_S, length λ and director vector $\boldsymbol{V}$.

2.4 Blurred Plane Detection Framework

Similarly to the definition of blurred segments, a blurred plane is a finite subset of a digital plane $\mathcal{P}$. The covering digital plane with minimal thickness is said to be optimal. Since [6] or [9], many algorithms have been proposed to recognize a blurred plane in a point set. However the maximal thickness of the set is a fixed input for this recognition task. For our detection problem, an effective thickness value must be estimated during the recognition itself, so these algorithms are no longer suitable.

The proposed blurred plane detection framework explores the point set in two perpendicular directions using the fast blurred segment recognition. The first direction can be chosen to optimize breaks-in-slope in the height profile. The found segments are enclosed in a digital plane with nearly optimal thickness. This plane is then used to select the points of the set, that compose the output blurred plane. The inputs are a start point P_S, a direction $\boldsymbol{D_S}$ and a search radius R_S in the grid $\mathcal{G}$. The search radius limits the maximal extent of the

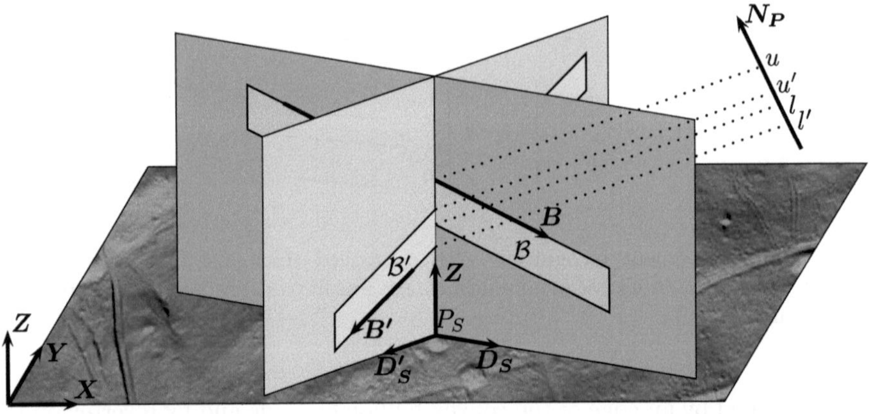

Fig. 5. Construction of the digital plane $\mathcal{P}$ from two orthogonal segments $\mathcal{B}$ and $\mathcal{B}'$.

output plane patch. The parameters are geometric thresholds that are checked in the early stages. Most non-planar inputs are thus rapidly detected.

As depicted in Fig. 5, two orthogonal blurred segments $\mathcal{B}$ and $\mathcal{B}'$ are recognized at position P_S in the height profiles built from the naive scans $S(P_S, \boldsymbol{D_S}, R_S)$ and $S'(P_S, \boldsymbol{D'_S}, R_S)$, where $\boldsymbol{D'_S}$ is orthogonal to $\boldsymbol{D_S}$. Their length is then checked: $\lambda_{min} < \lambda(\mathcal{B}) < \lambda_{max}$ and $\lambda'_{min} < \lambda(\mathcal{B}') < \lambda'_{max}$. If one is too short (resp. too long), the assigned thickness τ_0 and the observation distance d_0 are increased (resp. decreased) by a third, and a new recognition trial is attempted as a possible option (hyper-parameter N_R). If no suitable-length segments are found, the blurred plane detection stops.

Three-dimensional vectors $\boldsymbol{B}$ and $\boldsymbol{B'}$ are then derived from the mapping of two-dimensional vectors $\boldsymbol{V}(\mathcal{B})$ and $\boldsymbol{V}(\mathcal{B}')$ in the world reference system, given the height profile directions. Let $\boldsymbol{N_P} = \boldsymbol{B} \times \boldsymbol{B'}$. The points of each segment are examined to determine the lower and upper shift values along $\boldsymbol{N_P}$ from their 3D embedding:

$$\begin{cases} l = \min_{(P \in \mathcal{B})} \boldsymbol{\kappa(P)} \cdot \boldsymbol{N_P} \\ u = \max_{(P \in \mathcal{B})} \boldsymbol{\kappa(P)} \cdot \boldsymbol{N_P} \end{cases} \quad \text{and} \quad \begin{cases} l' = \min_{(P' \in \mathcal{B}')} \boldsymbol{\kappa(P')} \cdot \boldsymbol{N_P} \\ u' = \max_{(P' \in \mathcal{B}')} \boldsymbol{\kappa(P')} \cdot \boldsymbol{N_P} \end{cases} \tag{4}$$

The digital plane $\mathcal{P}$ is defined by normal vector $\boldsymbol{N_P}$, shift value $h_P = min(l, l')$ and arithmetic width $\mu_P = max(u, u') + 1 - h_P$:

$$\mathcal{P} = \left\{ P \in \mathbb{Z}^3 \;\middle|\; h_P \leq \boldsymbol{P} \cdot \boldsymbol{N_P} < h_P + \mu_P \right\} \tag{5}$$

The **crossing rate** χ_P of $\mathcal{P}$ is defined by the intersection-over-union factor of the blurred segments. The **tilt** θ_P of $\mathcal{P}$ is the deviation of $\boldsymbol{N_P}$ from the vertical direction, given by the unit vector $\boldsymbol{Z}$:

$$\chi_P = \frac{min(u, u') - max(l, l')}{\mu_P - 1}, \qquad \theta_P = \frac{||\boldsymbol{N_P} \times \boldsymbol{Z}||}{||\boldsymbol{N_P}||} \tag{6}$$

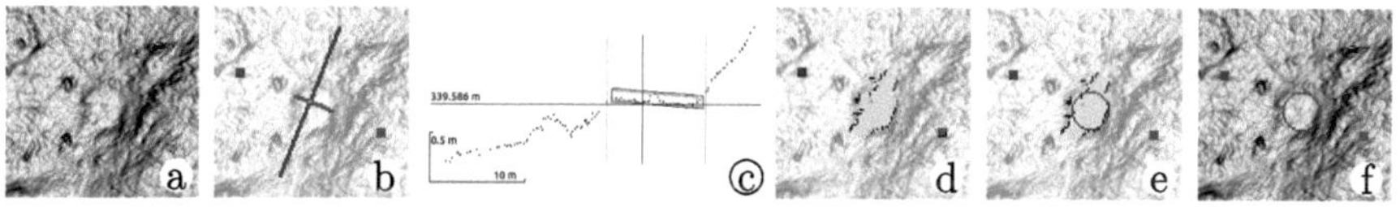

Fig. 6. Extraction steps: a) DTM gradient; b) local trend analysis: extraction seed in green, upper point in red, lower point in blue, main segment in red, orthogonal segment in blue; c) main height profile (Z/X ratio = 10): main segment in green, digital plane in blue; d) extracted platform: extent in yellow, bounds in black; e,f) detected circle (Color figure online)

If $\chi_P < \chi_{min}$ or $\theta_P > \theta_{max}$, where χ_{min} and θ_{max} are threshold parameters, the detection stops.

The output thickness $\tau(\mathcal{P})$ is limited by the maximal assigned thickness τ_0 and the minimal crossing rate χ_{min}. The obtained digital plane $\mathcal{P}$ is not optimal. In the case of a strict intersection between the segments ($[u, l] \subset [u', l']$ or $[u', l'] \subset [u, l]$), a thinner digital plane can be found by rotating it around one of the bounding lines of the largest segment. However, this has low impact on the estimated thickness and tilt of the digital plane when the segments are long enough.

The blurred plane $\mathcal{Q}$ is the local subset of $\mathcal{S}$ belonging to $\mathcal{P}$. To get it, a directional scan is built from S_0 and the scan strip is explored in both directions. In each scan S_i, height profiles are built and inspected from the center to both ends. The 3D embedding of each profile point belonging to the digital plane $\mathcal{P}$ (Eq. 5) is inserted into a plane slice $\mathcal{K}_i$ until a sequence of d'_f outliers or the profile end is met. The obtained slice $\mathcal{K}_i = [\kappa(P_i^j), \ldots, \kappa(P_i^k)]$ is added to the blurred plane. To each slice, we associate a slice length $\lambda(\mathcal{K}_i) = distance(P_i^j, P_i^k)$, a lower bound width $\delta_L(\mathcal{K}_i) = distance(P_i^{j-1}, P_i^j)$ and an upper bound width $\delta_U(\mathcal{K}_i) = distance(P_i^k, P_i^{k+1})$. The plane extension stops when the search radius is reached ($|i| > \frac{R_S}{\tau(S_0)}$) or when the plane slices become narrower ($\lambda(\mathcal{K}_i) < d'_f$).

3 Application to Charcoal Platform Extraction

The blurred plane detection framework was applied to the interactive extraction of charcoal platforms from airborne LiDAR ground points. After the presentation of the extraction process, the experiments held to evaluate achieved performance are introduced.

3.1 The Platform Extraction Process

The process input is a seed position P_S, which is either manually selected in a DTM view or extracted from a database of georeferenced locations of known platforms. Extraction is performed in three steps (see Fig. 6). First the local terrain trend is analyzed to estimate its slope gradient. Next, the platform plane

Table 1. Parameter settings: charcoal platform extraction in upper part, blurred plane detection in lower part

Name	Role	Step	Value or range
Φ	charcoal platform diameter	charcoal platform extraction	[5 m, 20 m]
N_S	maximal number of tested seeds	charcoal platform extraction	$\{1, 9, 17, 29\}$
R_T	local terrain sampling radius	local trend analysis	$2\ \Phi_{max}$
N_T	size of local terrain sampling	local trend analysis	16
Δ_R	accumulator radius cell size	workplace localization	0.1 m
Δ_C	accumulator position cell size	workplace localization	0.2 m
R_A	accumulator radius range	workplace localization	[2.5 m, 10.1 m]
C_A	accumulator position range	workplace localization	[−6.2 m, 6.2 m]
K	number of accumulated slices	workplace localization	$[9, 17]$
R_S	platform search radius	height profile extraction	Φ_{max}
N_G	grid subdivision ratio	height profile extraction	5
N_R	maximal number of trials	blurred segment recognition	$\{1, 2\}$
d_0	observation distance	blurred segment recognition	Φ_{min}
τ_0	maximal assigned thickness	blurred segment recognition	0.15 m
d_f	maximal fail distance	blurred segment recognition	0.3 m
ε_0	safety thickness margin	blurred segment recognition	0.05 m
λ	main support length	blurred plane detection	$[\Phi_{min}, \Phi_{max}]$
λ'	orthogonal support length	blurred plane detection	$[\Phi_{min}, \infty]$
χ	cross ratio between supports	blurred plane detection	[33%, 100%]
θ	blurred plane tilt angle	blurred plane detection	[0, 8%]
d'_f	maximal fail distance	blurred plane detection	0.6 m

is detected from the input seed and the slope direction using the blurred plane detection framework. Finally, a dedicated circle detector is applied to the platform bounds to circumscribe the workplace around the charcoal hearth.

Charcoal platforms have more or less a circular shape with typical diameter ranging from $\Phi_{min} = 5$ m to $\Phi_{max} = 20$ m. Along the local terrain trend, they are often cut by a sharp slope or a ditch. In orthogonal direction, however, they extend along the level line and their bounds are less clear. Therefore it is important to get a rough estimate of the local terrain trend in the seed neighborhood. To that end, the height values of ground points are examined in N_T uniformly spaced DTM cells at distance R_T from point P_S. The low trend position L_T is the cell with the lowest value, and the up trend position U_T the cell with the highest value (see Fig. 6b). The local trend direction $\boldsymbol{T}$ is defined by the vector $L_T U_T$.

The platform plane is detected using the blurred plane detection framework applied to the ground point set from the seed position P_S, trend direction $\boldsymbol{D_S} = \boldsymbol{T}$, and search radius $R_S = \Phi_{max}$. Extraction parameter settings are listed in Table 1. Most parameters depend directly on the application context. The settings of λ_{min}, λ_{max}, λ'_{min}, θ_{max}, τ_0, R_T and R_S are derived from the known geometric features of charcoal platforms. The safety margin ε_0 corresponds to LiDAR vertical accuracy. Then, d'_f is set large enough to cross most obstacles such as a windthrow or a track rut, but not a possible surrounding ditch. d_f is set to $\frac{d'_f}{2}$ because stricter conditions are assumed for blurred segment recognition. Figure 6c displays a typical height profile and Fig. 6d, the extracted platform plane.

In lack of perceptible boundaries perpendicular to the slope, the platform plane may extend quite far along the level line. Therefore a robust circle detector based on parameter space transformation [3] is used to delimit the workplace from well-marked lower and upper bounds (see Fig. 6e–f). The chosen parametrization is based on the shift of the circle center from seed position and on the circle radius. Only the bound pixels of the K central plane slices are accumulated in the parameter space. The accumulator parameters are given in Table 1. The contribution of a bound is distributed over the crossed pixels and set to the slice length λ, so that narrow slices have less impact on the result. The output circle is determined by the cell with the largest sum among those with at least four contributions from each of the lower and upper bounds. This selection excludes the many cells where a sequence of close pixels from only one side of the platform provides a dominant score. In that unpropitious case, wide bounds linked to possible point cloud scattering may give a decisive advantage to large radius values, resulting in overestimation of the diameter.

The result of the blurred segment detection depends strongly on the input start point. Charcoal platform geolocated positions or manual selections are not always accurate. Moreover, some ground perturbation may occur right in the middle, so that even the platform center is not necessarily the best starting point. Therefore, a seed scattering option is proposed, where the extraction process is performed from eight to twenty-eight additional scattered spots around the seed position in a neighborhood radius of five pixels, until a valid solution is found (hyper-parameter N_S).

3.2 Experiments

The charcoal platform extraction tool was tested on two LiDAR data sets mapping forested landscapes in which charcoal production had been developed on an industrial scale in the last centuries. Both data sets feature areas with a large amount of charcoal hearth platforms, typically around 150 per square kilometer in very dense sectors. They show different topographic contexts.

Bitche forest is located in northeastern Lorraine. It is a sandstone area in the northern part of Vosges massif, featuring deep valleys enclosed by steep slopes with flat plateaux on the top. Most of the platforms are on the slopes, where they form a striking contrast to the local trend. They are easily detected on site. Many of them are oblong in shape and extend perpendicular to the slope. Quite often, the expected planarity is not fully realized across the entire platform, possibly due to upper bank erosion. On the plateaux, they are large and circular. Their detection is much more difficult, sometimes aided by the presence of a ditch 30–40 cm deep all around.

Spincourt forest is located in the Woëvre plain (northwestern Lorraine), on clayey terrain with gentle slopes. Platforms are difficult to find during field

Table 2. Experimental site features and platform diameter comparisons (mean and standard deviation); A: site area, d: ground point mean density, N_M: number of on-site measured platforms, N_A: number of annotated platforms, Φ_M: on-site measured diameter, Φ_A: annotated diameter, Φ_E: extracted diameter, t: extraction time

	$A(km^2)$	$d(m^{-2})$	N_M	N_A	$\Phi_A - \Phi_M$ (m)	$\Phi_E - \Phi_M$ (m)	t (ms)
Bitche	144	6.7	164	11754	0.64 ± 1.20	-0.12 ± 2.09	10.69 ± 2.17
Spincourt	49	16.3	43	1378	0.13 ± 0.65	1.26 ± 3.06	9.53 ± 0.49

work. Many of them feature a subtle embankment upstream, while their downstream edge is often lost in low vegetation. Their detection is much easier in DTM gradient views.

The Bitche LiDAR acquisition, designed for large-scale ecological studies, features a standard ground point density. The Spincourt one, focused on a smaller area, is more detailed to meet the requirement of archaeological prospection. Forest area and corresponding ground point density are given in Table 2. Both were completed with a database of platform locations obtained by DTM survey and field verification. In complement, a large set of platform dimensions was collected by manual annotation and a smaller set by on-site surveys. Both were mostly based on the available locations (see Table 2). On-site measurements may be very inaccurate in flat areas where platform boundaries are often hard to discriminate. On the other hand, annotation involves delineating the platform contour in DTM gradient views. Visualization parameters have a strong impact on the result. A sharp slope gradient was used for the Spincourt set to better discriminate the platforms from the background. Oppositely, for the Bitche set, a relaxed slope gradient with larger cut threshold was preferable to outline the entire platform. Despite of the great care put into this task, errors are still significant. Two annotation sets of the same 6 km^2 area created by two different people showed an average diameter difference of 2.5 m.

To assess the reliability of the obtained diameters, we compared the annotated and extracted values to on-site surveys. The extraction seed used for this test was the estimated position of the platform based on the available geolocations. The results of Table 2 for Bitche set show reliable mean values, but a larger standard deviation for extracted diameters compared to annotated ones. An overestimation of one meter appears in the small set surveyed in Spincourt forest. In this smooth relief area, the platform boundary may not be identifiable and the blurred segment recognition may stop at a further break-in-slope. This error source is not compensated by fail cases caused by too short blurred segments, which are much fewer as they are more easily diagnosed in the extraction process.

The execution time for a complete extraction was also recorded, and was found to be quite compatible with interactive applications. A mean extraction time of 10 ms was obtained on a standard portable computer with an Intel Core i7 processor.

Table 3. Comparison of annotated and extracted diameters on the whole set of annotated platforms; N_R: maximal number of blurred segment recognition trials, N_S: maximal number of tested seeds, S: successful extraction rate, $\Phi_E - \Phi_A$: difference between extracted and annotated diameters, R: recall, P: precision

N_R	N_S	Bitche				Spincourt			
		$S(\%)$	$\Phi_E - \Phi_A(m)$	$R(\%)$	$P(\%)$	$S(\%)$	$\Phi_E - \Phi_A(m)$	$R(\%)$	$P(\%)$
1	1	62.2	0.79 ± 1.81	85.5	71.8	62.7	1.70 ± 2.65	88.1	69.1
1	9	84.2	0.83 ± 1.93	84.4	70.3	91.6	2.20 ± 3.10	87.4	64.9
1	17	86.4	0.84 ± 1.95	84.2	70.0	95.6	2.28 ± 3.17	87.3	64.3
1	29	87.7	0.85 ± 1.99	84.1	69.8	97.8	2.35 ± 3.23	87.2	63.8
2	1	66.5	0.81 ± 1.83	85.5	71.4	65.1	1.75 ± 2.74	88.1	68.8
2	9	86.0	0.88 ± 1.96	84.7	69.9	94.2	2.19 ± 3.11	87.2	64.8
2	17	87.9	0.89 ± 2.00	84.6	69.6	97.8	2.26 ± 3.21	87.0	64.2
2	29	88.9	0.90 ± 2.02	84.5	69.4	99.1	2.28 ± 3.22	86.9	64.1

We then performed extraction tests on the complete sets (see Table 3). The extraction seed used was the center of the annotated platform. Different values were tested for each parameter of the blurred plane detection process to analyze their influence on the result. A large part of the platforms were successfully extracted. For most parameters (θ_{max}, χ_{min}, d_f, ε_0), small variations did not significantly alter the output values. Seemingly, no careful setting is required. On the contrary, the seed scattering option has a strong effect. The method depends a lot on the selected seed. Only two thirds of the platforms were extracted when testing with only the input seed. Much more platforms were found when the seed scattering option was used with N_S additional near positions. This highlights the sensitivity of the approach to the local context around the input seed. The other main factor is the second trial performed when a blurred segment recognition produces a segment that is too long or too short. The results show that the settings of underlying parameters (*i.e.*, assigned maximal thickness τ_0 and observation distance d_0) depend heavily on the local terrain context.

We also compared the diameters and areas of the extracted and annotated platforms. Diameter values show similar results to those obtained for the smaller set of platforms surveyed on site. Circle areas were compared based on recall and precision values. For each platform, the area A_I of the intersection between annotated and extracted circles is computed and compared to the area A_D of the annotated circle, and to the area A_E of the extracted circle. Recall is defined as $R = \frac{A_I}{A_D}$ and precision as $P = \frac{A_I}{A_E}$. Mean values are displayed in Table 3. Performance decreases when more platforms are found using additional neighbor seeds or a second blurred segment recognition trial. These options lead to the integration of lower quality platforms, which contributes to lower the coincidence rate between extracted and annotated platforms. In the Spincourt set, recall is much higher than precision. This observation could be linked to the previously observed overestimation of platform extent. In many cases, the annotated platform is probably covered by a larger extracted one.

On the whole, the extraction results fit rather well to those of a manual annotation. The extraction process could be used advantageously to speed up this task in semi-supervised mode. Moreover, it delivers additional information on the platforms, such as the deviation from the horizontal plane or the surface roughness.

4 Conclusion

In this paper, we propose a new framework to detect a digital blurred plane in a set of points. The blurred plane is constructed as the union of two blurred line segments in orthogonal directions, and then completed with points collected in directional scans. This framework has been successfully applied to the extraction of charcoal platforms from LiDAR ground points. The local trend is analyzed beforehand to get a favorable direction to facilitate the recognition of the main blurred segment. After platform plane detection in the point set, a dedicated circle detector is used to determine the workplace. Experimental work showed that platforms are quickly and accurately localized, so that the produced tool can be used to annotate them quickly and reliably in DTM views. Moreover, in the scope of a large-scale study, the produced geometric information can be valuable when collated with other factors to reveal associated geometric patterns.

The blurred plane detection framework may be used to extract other structures with plane surfaces and applied to a wide range of applications dealing with modeling or processing of more structured scenes using 3D sensors. As it does not optimally enclose the blurred segments, restrictions should be applied to cases where the plane normal vector deviates significantly from the depth direction.

Future work will involve more advanced geometric analyses to provide a finer estimation of the platform extent based on an elliptic approximation. It will also seek to better exploit the local topographical context in order to adjust the extraction parameters. Then, similarly to earlier work on forest roads [5], large-scale automatic detection of charcoal platforms will be studied. To this end, detecting circular spots in DTM gradient views could provide input seeds for the extraction framework. In the longer term, a combination of the approach with deep-learning-based solutions will be studied for training set preparation as well as output validation.

Acknowledgments. The authors would like to thank all the organisms and persons who contributed to this work by making available all the necessary data:

– SRA-Lorraine, ONF and *Association des Amis de Senon et de Spincourt*, for the LiDAR acquisition of Spincourt forest,

– LIEC, *Université de Lorraine*, for the LiDAR acquisition of Bitche area,

– Simon Ritz, INRAP / ArTeHiS, *Université de Bourgogne*, for the database of geolocated charcoal platforms in Spincourt forest,

– Vincent Robin, LIEC, *Université de Lorraine*, for the database of geolocated charcoal platforms in Bitche area,

– Anne Gebhardt, INRAP/LIEC, *Université de Lorraine*, for her lead participation to the platform survey in both sites and to the annotation work.

Disclosure of Interests. The authors declare no conflicts of interest.

References

1. Andres, E., Acharya, R., Sibata, C.: Discrete analytical hyperplanes. Graph. Models Image Process. **59**(5), 302–309 (1997). https://doi.org/10.1006/gmip.1997.0427
2. Debled-Rennesson, I., Feschet, F., Rouyer-Degli, J.: Optimal blurred segments decomposition of noisy shapes in linear time. Comput. Graph. **30**(1), 30–36 (2006). https://doi.org/10.1016/j.cag.2005.10.007
3. Duda, R.O., Hart, P.E.: Use of the Hough transform to detect lines and curves in pictures. Commun. ACM **15**(1), 11–15 (1972). https://doi.org/10.1145/361237.361242
4. Even, P., Ngo, P.: Live extraction of curvilinear structures from LiDAR raw data. ISPRS Ann. Photogrammetry Remote Sens. Spatial Inf. Sci. 211–219 (2020). https://doi.org/10.5194/isprs-annals-V-2-2020-211-2020
5. Even, P., Ngo, P.: Automatic forest road extraction from lidar data of mountainous areas. In: Lindblad, J., Malmberg, F., Sladoje, N. (eds.) DGMM 2021. LNCS, vol. 12708, pp. 93–106. Springer, Cham (2021). https://doi.org/10.1007/978-3-030-76657-3_6
6. Gerard, Y., Debled-Rennesson, I., Zimmermann, P.: An elementary digital plane recognition algorithm. Discret. Appl. Math. **151**, 169–183 (2005). https://doi.org/10.1016/j.dam.2005.02.026
7. Kerautret, B., Even, P.: Blurred segments in gray level images for interactive line extraction. In: Wiederhold, P., Barneva, R.P. (eds.) IWCIA 2009. LNCS, vol. 5852, pp. 176–186. Springer, Heidelberg (2009). https://doi.org/10.1007/978-3-642-10210-3_14
8. Oliveira, C., Aravecchia, S., Pradalier, C., Robin, V., Devin, S.: The use of remote sensing tools for accurate charcoal kilns' inventory and distribution analysis: comparative assessment and prospective. Int. J. Appl. Earth Obs. Geoinf. **105**, 102641 (2021). https://doi.org/10.1016/j.jag.2021.102641
9. Provot, L., Buzer, L., Debled-Rennesson, I.: Recognition of blurred pieces of discrete planes. In: Kuba, A., Nyúl, L.G., Palágyi, K. (eds.) DGCI 2006. LNCS, vol. 4245, pp. 65–76. Springer, Heidelberg (2006). https://doi.org/10.1007/11907350_6
10. Schneider, A., Takla, M., Nicolay, A., Raab, A., Raab, T.: A template-matching approach combining morphometric variables for automated mapping of charcoal kiln sites. Archaeol. Prospect. **22**(1), 45–62 (2015). https://doi.org/10.1002/arp.1497
11. Sithole, G., Vosselman, G.: Automatic structure detection in a point-cloud of an urban landscape. In: 2nd GRSS/ISPRS Joint Workshop on Data Fusion and Remote Sensing over Urban Areas, pp. 67–71 (2003). https://doi.org/10.1109/DFUA.2003.1219959
12. Trier, Ø.D., Reksten, J.H., Løseth, K.: Automated mapping of cultural heritage in Norway from airborne lidar data using faster R-CNN. Int. J. Appl. Earth Obs. Geoinf. **95**, 102241 (2021). https://doi.org/10.1016/j.jag.2020.102241
13. Verschoof-van der Vaart, W., Bonhage, A., Schneider, A., Ouimet, W., Raab, T.: Automated large-scale mapping and analysis of relict charcoal hearths in Connecticut (USA) using a deep learning YOLOv4 framework. Archaeol. Prospection **30**, 251–266 (2023). https://doi.org/10.1002/arp.1889

Author Index

M. H. F. Wilkinson and J. Kosinka (Eds.): DGMM 2025, LNCS 16296, pp. 529–530, 2026.
https://doi.org/10.1007/978-3-032-09544-2

MIX
Papier aus verantwortungsvollen Quellen
Paper from responsible sources
FSC® C105338
FSC
www.fsc.org

If you have any concerns about our products,
you can contact us on
ProductSafety@springernature.com

In case Publisher is established outside the EU,
the EU authorized representative is:
Springer Nature Customer Service Center GmbH
Europaplatz 3, 69115 Heidelberg, Germany

Printed by Libri Plureos GmbH
in Hamburg, Germany